# ELECTRONICS AND NUCLEONICS DICTIONARY

BOOKS by NELSON M. COOKE

*ELECTRONICS AND NUCLEONICS DICTIONARY*

*MATHEMATICS ESSENTIAL TO ELECTRICITY AND RADIO*

*MATHEMATICS FOR ELECTRICIANS AND RADIOMEN*

BOOKS by JOHN MARKUS

*ELECTRONICS AND NUCLEONICS DICTIONARY*

*ELECTRONICS FOR ENGINEERS*

*ELECTRONICS MANUAL FOR RADIO ENGINEERS*

*ELECTRONICS FOR COMMUNICATION ENGINEERS*

*HANDBOOK OF INDUSTRIAL ELECTRONIC CIRCUITS*

*HANDBOOK OF INDUSTRIAL ELECTRONIC CONTROL CIRCUITS*

*HANDBOOK OF ELECTRONIC CONTROL CIRCUITS*

*HOW TO GET AHEAD IN THE TELEVISION AND RADIO SERVICING BUSINESS*

*TELEVISION AND RADIO REPAIRING*

# ELECTRONICS AND NUCLEONICS DICTIONARY

*An illustrated dictionary giving up-to-date definitions, abbreviations, and synonyms for over 13,000 terms used in television, radio, medical electronics, industrial electronics, space electronics, military electronics, avionics, radar, nuclear science, and nuclear engineering*


**NELSON M. COOKE**

*Lieutenant Commander, United States Navy (Retired)*
*President, Cooke Engineering Company*
*Senior Member, Institute of Radio Engineers*

**JOHN MARKUS**

*Technical Director, Dictionary Department, McGraw-Hill Book Company, Inc.*
*Senior Member, Institute of Radio Engineers*



McGRAW-HILL BOOK COMPANY, INC.

New York Toronto London

1960

ELECTRONICS AND NUCLEONICS DICTIONARY

 *Library of Congress Catalog Card Number: 60–10605*

V

12542

# PREFACE

Concise and accurate definitions of over 13,000 electronic and nucleonic terms, nade even clearer by 450 carefully chosen illustrations, are presented here for every-lay use by engineers, technical writers, advertising copy writers, technicians, tudents, and stenographers working in these fields. Synonyms are identified and ross-referenced to the term that has become generic through common usage or hrough standardization by IRE, AIEE, ASA, and other engineering organizations.

Although this book is actually the second edition of *Electronics Dictionary,* very ew of the 6,400 definitions in that volume could be used verbatim. As the electronics ndustry matured and expanded in the past 15 years, familiar old terms took on new neanings. Hundreds of new terms evolved in this same period, including asroc, stron, cifax, cozi, delrac, dovap, iraser, laddic, lodar, miran, ordir, padar, pyrotron, ıbal, snivet, spheredop, tonlar, tridop, triductor, tunnel diode, and twistor.

The terminology of nucleonics is now interlocked with that of electronics, making impossible to achieve a logical separation of terms. The new title, *Electronics and 'ucleonics Dictionary,* reflects the interrelation of these two fields, which are defined s follows:

**lectronics** The science that deals with electron devices, including electron tubes, magnetic amplifiers, transistors, and other devices that do the work of tubes.

**ucleonics** The science that deals with the release and utilization of energy from the nuclei of atoms.

A consistent policy has been followed throughout on spelling and hyphenation f controversial words, compound terms, and abbreviations, to reflect current usage nd trends. This dictionary can therefore be adopted as a style manual with the ssurance that it embodies the rules followed by the majority of writers and publishers n these fields today. When used in this manner on the desk of each writer, editor, nd stenographer, the book pays for itself over and over again in time saved through limination of style arguments and reduction of editing and correction costs.

The first *Electronics Dictionary* was adopted as the style guide by Electronics nagazine and by many government, military, and commercial organizations. It also erved as the guide for reports at the first Bikini atom bomb tests.

A compound term starts out as two words, takes on a hyphen next, then becomes ne word, as in push button, push-button, and pushbutton. Usage governs the ransition from one form to another. Much time was devoted to determination of he form that best represents current usage for each such term, using as guides the atest Government Printing Office Style Manual, McGraw-Hill Book Company styles, ecent issues of technical publications, and other up-to-date references.

Acronyms such as conelrad, loran, sarah, and radar have been made entirely lower-case for consistency, as also have practically all abbreviations. For compound terms like direct current, the same abbreviation is used for both noun and adjective forms.

A definition is given only once, to keep down the size of the *Dictionary*. Synonyms are listed in their own alphabetical order, followed by the generic or more commor term, in italics, after which the definition will be found. Compound terms ar alphabetized as if all the words were run together, ignoring hyphens and spaces.

An accurate compilation of the language of electronics and nucleonics would b impossible without the help of a great many individuals and organizations. Th collections of definitions prepared by engineering societies and government organiz tions proved invaluable as references for phrasing many of the definitions. Particul thanks are extended to The Institute of Radio Engineers for permission to u definitions and illustrations, to National Radio Institute for lending original drawin for illustrations, and to The American Society of Mechanical Engineers for permissi to extract and condense terms from their publication, *Glossary of Terms in Nucle Science and Technology*. Credit for illustrations also goes to Aerovox Corporatio Allied Radio Corporation, Bell Telephone Laboratories, Central Scientific Compan General Electric Company, National Bureau of Standards, Philco Corporation, RC Review, United States Navy, Westinghouse Electric Corporation, Weston Instr ments, and to the authors of McGraw-Hill books from which illustrations were take Finally, to Marjorie Appert goes appreciation for three long years of typing, retypin editing, and cross-checking definitions—a process that eventually consumed a tot of 65,000 four by six cards and several dozen typewriter ribbons.

A dictionary is a growing thing, never quite complete no matter how much tin is spent in its compilation. The authors would therefore appreciate receiving cc rections and new definitions as they come to the attention of the users of this boo If enough new terms are received between editions, it is hoped to make the available as published addendas. The authors can be addressed in care of McGra Hill Book Company.

*NELSON M. COOKE*
*JOHN MARKUS*

# ELECTRONICS AND NUCLEONICS DICTIONARY

## A

Abbreviation for *angstrom*.

- [A minus] The negative terminal of an battery or other source of filament volt- ɡe for an electron tube, or the tube fila- ıent terminal to which this source should ə connected. Also called F−.

- [A plus] The positive terminal of an battery or other source of filament volt- ɡe for an electron tube, or the tube fila- ıent terminal to which this source should e connected. Also called F+.

**and R display** An A display, any portion f which may be expanded. Also called . and R scan.

**and R scan** *A and R display.*

**and R scope** A radar indicator that pro- ides an A and R display on the screen of cathode-ray tube.

**-** A prefix used to identify cgs electro- ıagnetic units, as in abampere, abcoulomb, bfarad, abhenry, abmho, abohm, and ab- olt.

**ac** *Nomograph.*

**ampere** The cgs electromagnetic unit of urrent. One abampere is equal to 10 am- eres.

**battery** The battery that supplies power or filaments or heaters of electron tubes in attery-operated equipment.

**c** Abbreviation for *automatic brightness ontrol.*

**coulomb** The cgs electromagnetic unit f electricity. An abcoulomb is the quan- ity of electricity that passes a point in an lectric circuit in 1 second when the cur- ent is 1 abampere. One abcoulomb is qual to 10 coulombs.

**erration** An image defect that occurs vhen an optical lens or mirror does not ring all light rays to the same focus, or vhen an electron lens does not bring the lectron beam to the same sharp focus at ll points on the screen of a cathode-ray ube.

**abfarad** The cgs electromagnetic unit of capacitance. A capacitor has a capacitance of 1 abfarad when a charge of 1 abcoulomb produces a voltage of 1 abvolt between the terminals. One abfarad is equal to $10^9$ farads.

**abhenry** The cgs electromagnetic unit of inductance. A coil has an inductance of 1 abhenry when 1 abvolt is induced across its terminals by a current changing at the rate of 1 abampere per second. One ab- henry is equal to $10^{-9}$ henry.

**abmho** The cgs electromagnetic unit of conductance. A conductor has a conduc- tance of 1 abmho when 1 abvolt between its terminals will cause a current of 1 ab- ampere to flow through the conductor. One abmho is equal to $10^9$ mhos.

**abnormal glow discharge** A glow dis- charge characterized by the fact that the voltage drop increases as the current in- creases. It occurs when the current is in- creased beyond the point at which the cathode of the gas tube is completely covered with glow.

**abnormal reflection** A sharply defined re- flection of radio waves from an ionized layer of the ionosphere, occurring at fre- quencies higher than the critical or pene- tration frequency of the layer. Also called sporadic reflection.

**abohm** The cgs electromagnetic unit of resistance. A conductor has a resistance of 1 abohm when 1 abvolt across its termi- nals will cause a current of 1 abampere to flow. One abohm is equal to $10^{-9}$ ohm.

**A-bomb** *Atomic bomb.*

**abort** A failure of a missile to achieve its objective, or a missile that so fails.

**AB power pack** A packaged power source that provides the required A (filament) and B (anode) supply voltages for an electron-tube circuit. It may be a combination of individual batteries or a rectifier unit operating from an a-c power line.

**abscissa** The horizontal distance from a point on a graph to the zero reference line. The units of this distance are indicated on a scale at the bottom or top of the graph.

**absolute address** An address assigned to a particular storage location in a computer. Also called specific address.

**absolute altimeter** An altimeter that registers the absolute altitude of an aircraft above the earth or sea over which the aircraft is flying. The frequency-modulated altimeter and the radar altimeter are the commonest examples in current use.

**absolute altitude** The height or altitude of an aircraft above the surface or terrain over which it is flying.

**absolute cutoff frequency** The lowest frequency at which a waveguide will propagate energy without attenuation.

**absolute delay** The predetermined time interval between the transmission of two synchronized radio, radar, or loran signals from the same station or from different stations.

**absolute drift** The amount of inherent unbalance in a magnetic amplifier, measured in terms of the watts, amperes, or ampere-turns of input signal required for rebalancing.

**absolute efficiency** The ratio of the output of a transducer under specified conditions to the output of a corresponding ideal transducer.

**absolute humidity** The mass of water vapor per unit volume in the atmosphere at a given temperature.

**absolute pressure pickup** An instrument that compares an unknown source of pressure with zero pressure (a vacuum) and translates this information into an electrical quantity such as a change in inductance, resistance, or voltage.

**absolute temperature scale** A temperature scale in which zero is the absolute zero of temperature, −273.16°C or −459.69°F. The most commonly used scale is the Kelvin scale, which uses centigrade degrees; here absolute zero is 0°K, water freezes at 273.16°K and boils at 373.16°K. The less-used Rankine scale is based on Fahrenheit degrees; here water freezes at 491.69°R and boils at 671.69°R.

**absolute unit** A unit defined in terms of fundamental units of mass, length, time, and charge, such as the centimeter-gram-second electromagnetic and electrostatic units and the meter-kilogram-second-ampere electromagnetic units.

**absolute value** The numerical value of a number without regard to sign. Vertical lines on each side of a symbol specify that its absolute value is intended. Thus, the absolute value of Z is written $|Z|$.

**absolute zero** The lowest temperature that can exist, corresponding to a complete absence of molecular motion. Absolute zerc is approximately −273.16°C or −459.69°F

**absorbed dose** *Dose.*

**absorbed dose rate** The dose per unit o time, measured in rads per unit time.

**absorber** 1. A material or device that take up and dissipates radiated energy. It ma be used to shield an object from that er ergy, prevent reflection of the energy, de termine the nature of the radiation, c selectively transmit one or more compc nents of the radiation. Examples are acous tic absorbers and microwave absorbers. 2 In a nuclear reactor, a material that absorb neutrons without reproducing them.

**absorber control** Control of a nuclea reactor by a material that absorbs neutron such as a movable cadmium or boron roc

**absorptance** The ratio of the radiant en ergy absorbed in a body of material to th incident radiant energy.

**absorptiometer** An instrument for deter mining the concentration of substances b their absorption of nearly monochromati radiation at a wavelength selected by filter or by a simple radiation-dispersing system

**absorption** The dissipation of energy b radiation passing through a medium. Thus some electromagnetic energy is lost whe radio waves travel through the atmosphere Acoustic energy is lost when sound wave pass through an object. The kinetic energ of a nuclear particle is reduced when passes through a body of matter. In anothe nuclear example of absorption, a particle i absorbed by a nucleus in the medium with a different type of particle sometime being emitted as a result.

**absorption band** A region of the absorp tion spectrum of a material in which th amount of absorption passes through maximum.

**absorption circuit** A series resonant cir cuit used to absorb power at an unwante signal frequency. The circuit provides low impedance to ground at this frequency

**absorption coefficient** The fraction of th intensity of a radiation that is absorbed b a unit thickness of a particular substance

**absorption control** Control of a nuclea

reactor by a neutron absorber such as cadmium or boron steel.

**absorption cross section** The sum of the cross sections for all neutron reactions with an atom except elastic and inelastic collisions.

**absorption current** The component of dielectric current that is proportional to the rate of accumulation of electric charges within the dielectric.

**absorption curve** A graph in which intensity of transmitted radiation is plotted as a function of material thickness. Also called transmission curve.

**absorption discontinuity** A discontinuity in the absorption coefficient of a substance for a particular type of radiation. Also called absorption edge and absorption limit.

**absorption edge** *Absorption discontinuity.*

**absorption frequency meter** *Absorption wavemeter.*

**absorption limit** *Absorption discontinuity.*

**absorption loss** 1. That part of the transmission loss which is converted into heat when radiated energy is transmitted or reflected by a material. 2. Power loss in a transmission circuit caused by coupling to an adjacent circuit.

**absorption mesh** A filter used in a waveguide to absorb electromagnetic energy at undesired frequencies.

**absorption modulation** A system of amplitude modulation in which a variable-impedance device is inserted in or coupled to the output circuit of the transmitter, to absorb carrier power in accordance with the intelligence to be transmitted. In one system the modulator tubes control the absorption of the transmission line directly by means of stub connections, to achieve the same result. Also called loss modulation.

**absorption peak** Abnormally high attenuation at a particular frequency as a result of absorption loss.

**absorption spectrum** The spectrum obtained when continuous radiation is passed through an absorbing medium before entering a spectroscope. The resulting recorded spectrum shows dark lines at wavelengths corresponding to maximum absorption.

**absorption trap** A parallel-tuned circuit used to absorb and thereby attenuate interfering signals.

**absorption wavemeter** A wavemeter consisting of a calibrated tuned circuit and a resonance indicator. When lightly coupled to a signal source and tuned to resonance, maximum energy is absorbed from the source. The unknown wavelength or frequency is then read on the calibrated tuning dial. With waveguides, a cavity-type resonant circuit is used. Also called absorption frequency meter. When a vacuum-tube oscillator is a part of the resonance indicator, the instrument is usually called a grid-dip meter.

**absorptive attenuator** A waveguide section containing dissipative material that gives a desired transmission loss.

**AB test** A method of comparing two sound systems by switching inputs so the same recording is heard in rapid succession over one system and then the other.

**abundance ratio** The ratio of the number of atoms of various isotopes in a mixture.

**abvolt** The cgs electromagnetic unit of voltage. A voltage of 1 abvolt exists between two points when 1 erg of work is required to transfer 1 abcoulomb of positive electricity from the point of lower potential to the point of higher potential. One abvolt is equal to $10^{-8}$ volt.

**a-c** Abbreviation for *alternating current.*

**accelerating anode** *Accelerating electrode.*

**accelerating chamber** An evacuated glass, metal, or ceramic envelope in which charged particles are accelerated.

**accelerating electrode** An electrode used in cathode-ray tubes and other electron tubes to increase the velocity of the electrons that constitute the space current or form a beam. Such an electrode is operated at a high positive potential with respect to the cathode. Also called accelerating anode and accelerator.

**accelerating tube** A tubular accelerating chamber. It may be toroidal as in a betatron, or in the form of a long cylinder as in a linear accelerator.

**acceleration** The rate at which the velocity of a body changes.

**acceleration effect** The difference between the output value of a device without acceleration and the output value measured during a specified steady-state acceleration in a specified axis at any one input value.

**acceleration space** The region just outside of the output aperture of the electron gun in an electron tube, in which electrons are accelerated to a desired higher velocity.

**accelerator** 1. A device that accelerates charged particles to high velocities so they have high kinetic energy. It can be used for electrons, protons, deuterons, and helium ions. Also called particle accelerator. Examples include the betatron, cyclotron, linear accelerator, synchrocyclotron, synchrotron, and Van de Graaff electrostatic accelerator. 2. *Accelerating electrode.*

**accelerometer** A device for measuring the acceleration of a moving body and translating it into a corresponding electrical quantity.

**accentuation** *Preemphasis.*

**acceptance angle** The solid angle within which all received light reaches the cathode of a phototube in its housing.

**acceptor** An impurity element that increases the number of holes in a semiconductor crystal such as germanium and silicon. Current flow is then due essentially to transfer of holes. Since these holes are equivalent to positive charges, the resulting alloy is called a p-type semiconductor. Aluminum, gallium, and indium are examples of acceptors. Also called acceptor impurity.

**acceptor circuit** A series resonant circuit that has a low impedance at the frequency to which it is tuned and a higher impedance at all other frequencies. Used in series with a signal path to pass the desired frequency.

**acceptor impurity** *Acceptor.*

**acceptor level** An intermediate level close to the normal band in the energy-level diagram of an extrinsic semiconductor. It is empty at absolute zero. At other temperatures some electrons corresponding to the normal band can acquire energies corresponding to this intermediate level.

**accessory** A part, subassembly, or assembly that contributes to the effectiveness of a piece of equipment without changing its basic function. An accessory may be used for testing, adjusting, calibrating, recording, or other purposes.

**access time** The read time or the write time in a computer.

**accidental coincidence** Coincidence due to the chance occurrence of unrelated counts in separate radiation detectors. Also called chance coincidence and random coincidence.

**accidental coincidence correction** The correction made in coincidence counting to offset the chance occurrence of unrelated signals within the resolving time of the apparatus.

**accidental jamming** Jamming due to transmission of radio or radar signals by friendly equipment.

**a-c coupling** A coupling arrangement that will not pass direct current or a d-c component of a signal.

**accumulating stimulus** A current that is increased gradually, so it is less effective than if suddenly increased to final intensity. Used in electrobiology.

**accumulator** 1. A computer device that stores a number and, on receipt of another number, adds it to the number already stored and stores the sum. In another version, stored integers can be increased by unity or by an arbitrary integer. An accumulator can be reset either to zero or to an arbitrary integer. Also called counter. 2. British term for *storage battery.*

**accuracy** 1. The quality of being free from errors. 2. The extent to which the indications of an instrument approach the true values of the quantities measured.

**accurate range marker** An adjustable range calibrator used on some radar ppi displays. By turning a control, a bright range circle of variable diameter is made to run through an observed target. A counter geared to the control then indicates directly the range to the target.

**a-c/d-c receiver** A radio receiver designed to operate from either an a-c or d-c power line. Also called universal receiver.

**a-c dump** The removal of all a-c power from a computer intentionally, accidentally, or conditionally. It usually results in the removal of all power.

**a-c erase** Use of alternating current to energize an erasing head.

**a-c erasing head** A magnetic head that uses alternating current to produce the gradually decreasing magnetic field necessary for erasing recorded signals.

**acetate** *Cellulose acetate.*

**acetate base** A transparent backing film for magnetic recording tape and motion-picture film, made from cellulose acetate. Also called safety base.

**acetate disk** A mechanical recording disk, either solid or laminated, made of various acetate and cellulose nitrate compounds.

**a-c generator** A rotating electric machine that converts mechanical power into a-c electric power.

**achromatic** 1. Without color. 2. Capable of transmitting light without breaking it up into constituent colors.

**achromatic antenna** An antenna whose characteristics are uniform in a specified frequency band.

**achromatic color** A shade of gray.

**achromatic lens** A lens combination that

gives correction for chromatic aberration. It usually consists of a convex lens of crown glass and a concave lens of flint glass, designed so one lens corrects for the errors of the other. The combination brings all colors of light rays nearer to the same focus point.

**achromatic locus** An area on a chromaticity diagram that contains all points representing acceptable reference white standards. Also called achromatic region.

**achromatic point** A point on a chromaticity diagram that represents an acceptable reference white standard.

**achromatic region** *Achromatic locus.*

**achromatic stimulus** A visual stimulus that gives the sensation of white light, having no hue.

**a-c interruption** Intermittent operation of a metallic rectifier in which the a-c input circuit of the rectifier is opened.

**aclinic line** *Isoclinic line.*

**a-c magnetic biasing** Magnetic biasing with alternating current, usually well above the signal frequency range, in magnetic recording.

**acorn tube** A uhf vacuum tube resembling an acorn in shape and size. Leads come out directly through the sides of the tube. Small electrodes give low interelectrode capacitances, and close electrode spacings give low electron transit time.

**acoustic** Containing, producing, arising from, actuated by, related to, or associated with sound. The adjective acoustic is used (rather than acoustical) when the term being qualified designates something that has the properties, dimensions, or physical characteristics associated with sound waves.

**acoustic absorption coefficient** *Sound absorption coefficient.*

**acoustic absorption loss** Energy lost by conversion into heat or other forms when sound passes through or is reflected by a medium.

**acoustic absorptivity** *Sound absorption coefficient.*

**acoustical** Containing, producing, arising from, actuated by, related to, or associated with sound. The adjective acoustical is used (rather than acoustic) when the term being qualified does not explicitly designate something that has the properties, dimensions, or physical characteristics associated with sound waves.

**acoustical attenuation constant** The real part of the acoustical propagation constant. The commonly used unit is the neper per section or per unit distance.

**acoustical ohm** A unit of acoustic resistance, acoustic reactance, or acoustic impedance. The magnitude is 1 acoustical ohm when a sound pressure of 1 dyne per sq cm (1 microbar) produces a volume velocity of 1 cu cm per second.

**acoustical phase constant** The imaginary part of the acoustical propagation constant. The commonly used unit is the radian per section or per unit distance.

**acoustical propagation constant** A rating for a sound medium. It is the natural logarithm of the complex ratio of particle velocities, volume velocities, or pressures at two points in the path of a sound wave. The ratio is determined by dividing the value at the point nearer the sound source by the value at the more remote point. The real part of this constant is the acoustical attenuation constant, and the imaginary part is the acoustical phase constant.

**acoustical reciprocity theorem** A theorem applying to an acoustic system. The theorem states that a simple sound source at point A in a region will produce the same sound pressure at another point B as would have been produced at A had the source been located at B.

**acoustic burglar alarm** A burglar alarm that is responsive to sounds produced by an intruder. Microphones concealed in the rooms to be protected are connected to audio amplifiers that trip an alarm when sounds exceed a predetermined normal level. Also called acoustic intrusion detector.

**acoustic clarifier** A system of cones loosely attached to the baffle of a loudspeaker, designed to vibrate and absorb energy during sudden loud sounds in order to suppress these sounds.

**acoustic compensator** A device for matching acoustical path lengths in binaural or stereophonic audio equipment.

**acoustic compliance** The reciprocal of acoustic stiffness.

**acoustic delay line** A device capable of transmitting and delaying sound pulses by recirculating them in a liquid or solid medium. For computers the pulses are usually in binary form. Also called sonic delay line.

**acoustic depth finder** *Fathometer.*

**acoustic dispersion** The separation of a complex sound wave into its frequency components. It is usually caused by variation of the wave velocity of the medium with frequency. The rate of change of the velocity with frequency is a measure of the dispersion.

**acoustic dissipation element** An element used to dissipate some or all of the acoustic energy reaching it.

**acoustic feedback** The feedback of sound waves from a loudspeaker to a preceding part of an audio system, such as to the microphone, in such a manner as to aid or reinforce the input. When feedback is excessive, a howling sound is heard from the loudspeaker. Also called acoustic regeneration.

**acoustic filter** A sound-absorbing device that selectively suppresses certain audio frequencies.

**acoustic generator** A transducer that converts electrical, mechanical, or other forms of energy into sound. Buzzers, headphones, and loudspeakers are examples.

**acoustic homing** Homing on sound sources. Used in torpedoes to home on sounds made by the propellers of an enemy ship or submarine.

**acoustic horn** *Horn.*

**acoustic impedance** The sound pressure on a unit area of surface divided by the sound flux through that surface, expressed in acoustical ohms. The real component of acoustic impedance is acoustic resistance, and the imaginary component is acoustic reactance. The two types of acoustic reactance are acoustic compliance and acoustic inertance.

**acoustic inertance** *Acoustic mass.*

**acoustic interferometer** An instrument for measuring the velocity of sound waves in a liquid or gas. Variations of sound pressure are observed in the medium between a sound source and a reflector, as the reflector is moved or the frequency is varied. Interference between direct and reflected waves produces standing waves that are related to the velocity of sound in the medium.

**acoustic intrusion detector** *Acoustic burglar alarm.*

**acoustic labyrinth** A loudspeaker baffle consisting of a long absorbent-walled duct folded into the volume of a cabinet, with a loudspeaker mounted at one end. The other end is open to the air in front of or underneath the cabinet. Used to reinforce bass response and prevent cavity resonance.

**acoustic lens** An array of obstacles that refracts sound waves in the same way that an optical lens refracts light waves. The dimensions of the obstacles are small compared to the wavelengths of the sounds being focused.

**acoustic mass** The quantity that, when multiplied by $2\pi$ times the frequency, gives the acoustic reactance associated with the kinetic energy of a medium. The unit is the gram per centimeter to the fourth power. Also called acoustic inertance.

**acoustic memory** A computer memory that uses an acoustic delay line in which a train of pulses travels through a medium such as mercury or quartz.

**acoustic mine** An underwater mine that is detonated by sound waves, such as from a ship's propeller or engines. Also called sonic mine.

**acoustic mirage** The distortion of a sound wavefront by a large temperature gradient in air or water, creating the illusion of two sound sources.

**acoustic pickup** A pickup that transforms phonograph-record groove modulations directly into sound, as in early phonographs. The phonograph needle is mechanically linked to a flexible diaphragm. Also called sound box and mechanical reproducer.

**acoustic radiator** A vibrating surface that produces sound waves, such as a loudspeaker cone and a headphone diaphragm.

**acoustic radiometer** An instrument for measuring sound intensity by determining the unidirectional steady-state pressure caused by the reflection or absorption of a sound wave at a boundary.

**acoustic reactance** The imaginary component of acoustic impedance. The unit is the acoustical ohm.

**acoustic reflection coefficient** *Sound reflection coefficient.*

**acoustic reflectivity** *Sound reflection coefficient.*

**acoustic refraction** A bending of sound waves when passing obliquely from one medium to another in which the velocity of sound is different, as from warm water to cool water in the ocean or from warm air to cool air.

**acoustic regeneration** *Acoustic feedback.*

**acoustic resistance** The real component of acoustic impedance. The unit is the acoustical ohm.

**acoustic resonator** A resonator in the form of an enclosure that exhibits resonance at a particular frequency of acoustic energy.

**acoustics** 1. The science that deals with the production, transmission, and effects of sound, including its absorption, reflection, refraction, diffraction, and interference. 2. The properties of a room or location that control reflections of sound waves and therefore determine the character of sounds heard in that location.

**acoustic scattering** The irregular and diffuse reflection, refraction, or diffraction of sound in many directions.

**acoustic shock** Dizziness, physical pain, and sometimes also nausea produced by a sudden loud sound.

**acoustic stiffness** The quantity that, when divided by $2\pi$ times the frequency, gives the acoustic reactance associated with the potential energy of a sound medium. The unit is the dyne per centimeter to the fifth power. The reciprocal of acoustic stiffness is acoustic compliance.

**acoustic transmission coefficient** *Sound transmission coefficient.*

**acoustic transmission system** An assembly of elements adapted for the transmission of sound.

**acoustic transmittivity** *Sound transmission coefficient.*

**acoustic treatment** The use of sound-absorbing materials to give a room a desired degree of freedom from echo and reverberation.

**acoustic velocity** *Velocity of sound.*

**a-c power supply** A power supply that provides one or more a-c output voltages, such as an a-c generator, dynamotor, inverter, or transformer.

**acquisition and tracking radar** A radar set that locks onto a strong signal, tracks the object emitting the signal, and feeds position data directly and continuously to gun or missile control systems.

**acquisition radar** A radar set that detects an approaching target and feeds approximate position data to a fire-control or missile-guidance radar, which takes over the function of tracking the target.

**a-c receiver** A radio receiver designed to operate only from an a-c power line.

**a-c resistance** *High-frequency resistance.*

**acrylic resin** A glasslike thermoplastic resin made by polymerizing esters of acrylic or methacrylic acid. Widely used for transparent parts. Trademark names include Lucite and Plexiglas.

**a-c tacho-generator** An a-c generator whose output voltage and output frequency are proportional to rotational speed.

**actinides** A name proposed for the series of elements having atomic numbers 89 through 102.

**actinium** [symbol Ac] A radioactive element. Atomic number is 89.

**actinium series** The series of nuclides resulting from the decay of $U^{235}$, including actinium A, B, C, C′, C″, D, K, and X. Mass numbers of all members are given by $4n + 3$, where $n$ is an integer. The sequence is also known as the $4n + 3$ series or the actinouranium series.

**actinometer** An instrument that measures the intensity of radiation by determining the amount of fluorescence produced by that radiation.

**actinon** [symbol An] The common name for 3.92s $Em^{219}$, a member of the actinium series. Actinon is an isotope of radon.

**actinouranium** [symbol AcU] A common name for uranium isotope $U^{235}$, the natural parent of the actinium series.

**actinouranium series** *Actinium series.*

**action current** A brief and very small electric current flowing in a nerve during a nervous impulse.

**action potential** The instantaneous value of the voltage between excited and resting portions of an excitable living structure.

**action spike** The greatest in magnitude and briefest in duration of the characteristic negative waves in an action potential.

**activated water** Water having ions, atoms, radicals, or molecules that are temporarily in a chemically reactive state because of exposure to ionizing radiation.

**activation** 1. The process of treating the cathode or target of an electron tube to create or increase its emission. Also called sensitization. 2. The process of inducing radioactivity by bombardment with neutrons or other types of radiation.

**activation analysis** A method of chemical analysis in which the material being analyzed is bombarded with nuclear particles and the resulting characteristic radionuclides are detected.

**activation cross section** The cross section for the formation of a specified radionuclide, generally by a neutron-induced reaction.

**activation energy** The excess energy required for a particular nuclear process. An example is the energy needed by an electron to reach the conduction band in a semiconductor.

**activator** 1. An impurity atom that increases the luminescence of a solid material, such as copper in zinc sulfide and thallium in potassium chloride. 2. An impurity atom used to activate the target of a camera tube. Also called sensitizer.

**active** 1. *Fissionable.* 2. *Radioactive.*

**active air defense** Air defense concerned with combatant action taken to prevent or interfere with a hostile attack by aircraft or guided missiles. It includes electronic countermeasures, air-to-air guided missiles, and surface-to-air guided missiles.

**active area** The portion of the rectifying junction of a metallic rectifier that carries forward current.

**active deposit** A radioactive decay product deposited on a surface.

**active electric network** An electric network or circuit containing one or more sources of energy.

**active electronic countermeasures** Electronic countermeasures involving actions of such nature that their use in jamming or otherwise disrupting enemy radio, radar, or sonar transmissions is detectable by the enemy.

**active filter** A filter used for smoothing data. The time delay and/or phase lag introduced by such a low-pass filter is canceled by use of an identical reciprocal filter in the feedback circuit of the associated amplifier.

**active homing** Homing in which the missile contains both the source of energy for illuminating the target and the receiver for energy reflected from the target.

**active jamming** Intentional radiation or reradiation of electromagnetic energy in such a way as to impair use of a specific band of frequencies.

**active line** A horizontal line that carries picture information in television, as opposed to retrace lines that are blanked out during horizontal and vertical retrace.

**active material** 1. A fluorescent material used in screens for cathode-ray tubes. Examples include calcium tungstate, zinc phosphate, and zinc silicate. 2. The lead oxide or other energy-storing material used in the plates of a storage battery.

**active network** A network whose output is dependent on a source of power other than that associated with the input signal.

**active product** A radioactive decay product of a radionuclide.

**active sonar** Underwater sonar equipment that generates bursts of ultrasonic sound and picks up echoes reflected from submarines, fish, and other objects within range.

**active transducer** A transducer containing one or more sources of power that contribute to its output.

**active water homing** Homing on the trail of radioactive sea water left by a nuclear-powered submarine.

**activity** 1. The intensity of a radioactive source. It can be expressed as the number of atoms disintegrating in unit time or as the number of scintillations or other effects observed per unit time. One unit of activity is the curie, equal to $3.7 \times 10^{10}$ disintegrations per second. 2. A measure of the amplitude of vibration of a crystal unit, generally expressed as the rectified grid current of the oscillator circuit in which the crystal is used. 3. Short form of *radioactivity.*

**activity dip** A decrease in the value of crystal activity, other than a band-break, that occurs over a small temperature interval. It is usually the result of loose coupling to other modes of vibration or the result of variations in the mounting system.

**a-c transmission** A mode of television transmission in which a fixed setting of the controls makes any instantaneous value of signal correspond to the same value of brightness for only a short time.

**actual frequency** The measured frequency of a crystal-controlled oscillator, as distinguished from the nominal frequency value that is marked on the crystal unit.

**actuating signal** The reference input minus the primary feedback in a control system.

**actuating transfer function** The transfer function that relates a feedback control loop actuating signal to the corresponding loop input signal.

**AcU** Symbol for *actinouranium.*

**acute angle** An angle numerically smaller than a right angle, hence less than 90°.

**acute exposure** Exposure to occasional large doses of nuclear radiation.

**a-c voltage** *Alternating voltage.*

**a-c welder** A welding machine utilizing alternating current for welding purposes.

**acyclic** Following no regularly repeated cycles of variations.

**acyclic machine** *Homopolar generator.*

**adapter** A device used to make electric or mechanical connections between items not originally intended for use together.

**Adcock antenna** A directional antenna consisting of two vertical wires spaced one half-wavelength apart or less, connected in phase opposition to give a figure-of-eight radiation pattern.

**Adcock direction finder** A radio direction

finder utilizing one or more pairs of Adcock antennas.

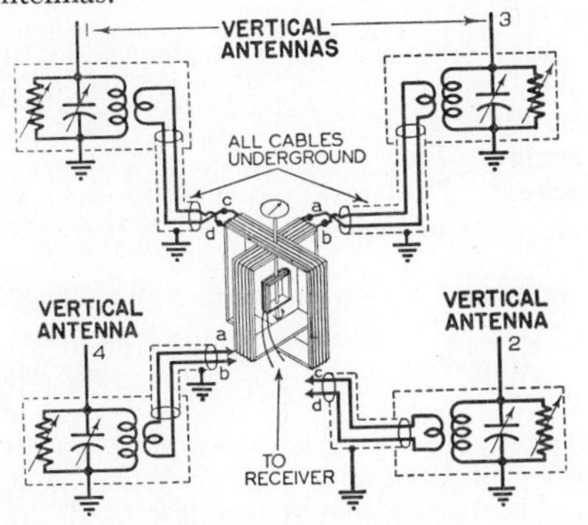

Adcock direction finder.

**Adcock radio range** An A-N radio range that uses Adcock antennas arranged at the four corners of a square on the ground. The vertical antennas at one set of opposite corners transmit the letter A in international Morse code and the other two antennas transmit the letter N.

**adder** 1. A circuit in which two or more signals are combined to give an output signal amplitude that is proportional to the sum of the input signal amplitudes. In a color television receiver, the adder combines the chrominance and luminance signals. 2. A computer device that can form the sum of two or more numbers or quantities.

**additive color system** A system that adds two colors to form a third.

**additive primaries** Sources of color or light which, by additive mixture in varying proportions, can be made to match a large range of colors. The three additive primaries used are red, green, and blue.

**additron** A form of beam-switching tube used as a binary adder in digital computers.

**address** A numerical or other expression that designates a particular location in a storage device or other source or destination of information in a computer.

**adf** Abbreviation for *automatic direction finder*.

**adhesion** *Bond strength.*

**adiabatic** Occurring without change in heat content.

**adiactinic** Not transmitting photochemically active rays.

**adion** An ion that has been adsorbed on a surface and cannot move out of it.

**A display** A radarscope display in which targets appear as vertical deflections from a line representing a time base. Target distance is proportional to the horizontal position of the deflection from one end of the time base, and target echo signal intensity is proportional to the amplitude of the vertical deflection. On some scopes the display is rotated 90°, so the time base is vertical and the signal pips increase from left to right horizontally. Also called A scan.

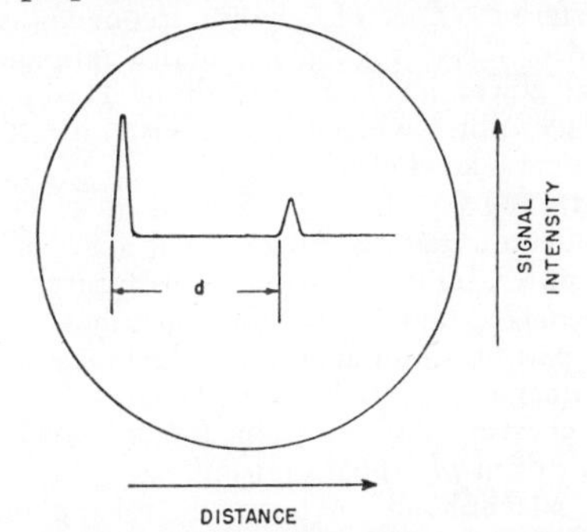

A display, with transmitted pulse at left. Distance d to echo pulse gives range of target.

**adjacent channel** The channel immediately above or below the channel under consideration.

**adjacent-channel attenuation** *Selectance.*

**adjacent-channel interference** Interference caused by a transmitter operating in an adjacent channel. It is recognized as a peculiar garbled sound heard along with the desired program when the sidebands of the adjacent-channel transmitter beat with the carrier signal of the desired station. Also called monkey chatter, sideband interference, and sideband splash.

**adjacent-channel selectivity** The ability of a receiver to reject signals on channels adjacent to that of the desired station.

**adjacent sound carrier** The r-f carrier that carries the sound modulation for the television channel immediately below that to which the receiver is tuned.

**adjacent video carrier** The r-f carrier that carries the picture modulation for the television channel immediately above the channel to which the receiver is tuned.

**adjustable resistor** A wirewound resistor having a sliding contact whose positions can be changed by loosening a locking screw. Extra sliders can be added if desired. Used chiefly as a voltage divider.

**adjustable transformer** *Variable transformer.*

**adjustable voltage divider** A wirewound resistor having one or more movable terminals that can be slid along the length of the exposed resistance wire until the desired voltage values are obtained.

**adjusted decibel** [abbreviated dba] A unit used to show the relationship between the interfering effect of a noise frequency, or band of noise frequencies, and a reference noise power level of −85 dbm. This unit replaces dbrn, which was based on a reference noise level of −90 dbm.

**admittance** [symbol $Y$] A measure of how readily alternating current will flow in a circuit. Admittance is the reciprocal of impedance, and is expressed in mhos. The real part of admittance is conductance, and the imaginary part is susceptance.

**adp crystal** Abbreviation for *ammonium dihydrogen phosphate crystal.*

**adp microphone** A crystal microphone using an ammonium dihydrogen phosphate crystal having piezoelectric properties.

**Advance** Trademark of Driver-Harris Co. for an alloy of copper and nickel, used in the construction of electric instruments.

**advance ball** A rounded support, often sapphire, that rides ahead of or alongside the cutting stylus when making a mechanical recording such as a phonograph record. The ball maintains a uniform depth of cut regardless of irregularities in the surface of the disk.

**advantage factor** The ratio of radiation dosage received at two points in a nuclear reactor during a specified time interval. The first point is one at which an enhanced effect is obtained, so the ratio is greater than 1.

**AEC** Abbreviation for *Atomic Energy Commission.*

**aeolight** A glow discharge lamp employing a cold cathode and a mixture of inert gases. The intensity of illumination varies with the applied signal voltage. Used to produce a modulated light for motion-picture sound recording.

**aerial** British term for *antenna.*

**aerial torpedo** A missile guided through the air to its target by remote control, as a glide bomb.

**Aerobee** A high-altitude research rocket.

**aerograph** A meteorograph carried aloft by a balloon, kite, or airplane. Also called aerometeorograph.

**aerometeorograph** *Aerograph.*

**aeronautical mile** A unit of length equal to 6,080 feet or 1.15155 miles, the same length as a nautical mile. Also called air mile.

**aerophare** *Radio beacon.*

**aew radar** Abbreviation for *airborne early-warning radar.*

**a-f** Abbreviation for *audio frequency.*

**afc** Abbreviation for *automatic frequency control.*

**a-f noise** Any electric disturbance, in the audio-frequency range, that is introduced from a source extraneous to the signal.

**afterglow** *Phosphorescence.*

**afterheat** Heat resulting from residual activity after a nuclear reactor has been shut down.

**after pulse** A spurious pulse induced in a multiplier phototube by a previous pulse.

**agc** Abbreviation for *automatic gain control.*

**aggregate recoil** The ejection, from the surface of a sample, of a cluster of atoms attached to one that is recoiling as the result of alpha-particle emission.

**aging** 1. Allowing a permanent magnet, capacitor, meter, or other device to remain in storage for a period of time, sometimes with voltage applied, until the characteristics of the device become essentially constant. 2. Changes in the characteristics of a device during use.

**aided tracking** A radar antenna control system in which a constant rate of motion of the tracking mechanism is maintained by a d-c motor and selsyn system so an equivalent constant rate of movement of a target in bearing, elevation, distance, or any combination of these variables can be followed. An operator adjusts the rate of motion from time to time with a potentiometer in the d-c motor circuit, as required to compensate for target speed and course changes.

**AIEE** Abbreviation for *American Institute of Electrical Engineers.*

**a-i radar** Abbreviation for *airborne intercept radar.*

**airborne beacon** *Radar safety beacon.*

**airborne early-warning radar** [abbreviated aew radar] An early-warning radar carried by aircraft. The radar signals are relayed from the aircraft to surface stations, or their significance is reported by radio.

**airborne intercept radar** [abbreviated a-i radar] Airborne radar used for detecting and tracking other aircraft at night or in clouds. It may also include computers that provide fire-control data. Also called aircraft intercept radar.

**airborne moving-target indicator** A moving-target indicator system for airborne radar operating close to the ground, where moving targets are obscured by ground clutter and both the ground and the target are moving with respect to the radar in the airplane.

**airborne radar** A self-contained radar installed in aircraft. It may provide information about ground landmarks, ships at sea, shoreline contours, other aircraft, storm clouds, or weather fronts.

**air capacitor** A capacitor having only air as the dielectric material between its plates.

**air cell** A cell in which depolarization at the positive electrode is accomplished chemically by reduction of the oxygen in the air.

**air column** The air space within a horn or acoustic chamber for a loudspeaker.

**aircom** [AIR COMmunication] An Air Force system for furnishing communication requirements of future space and missile programs.

**air conduction** The process by which sound is conducted to the inner ear through the air in the outer ear canal.

**air-cooled tube** An electron tube in which the generated heat is dissipated to the surrounding air directly, through metal heat-radiating fins, or with the aid of channels or chimneys that increase air flow.

**air-core coil** A coil wound on a fiber, plastic, or other nonmagnetic form, with no iron in its vicinity.

**air-core transformer** A transformer having two or more coils wound on a fiber or other nonmagnetic form, and having no iron in its magnetic circuits. Usually designed for use as an r-f transformer, i-f transformer, antenna coil, or oscillator coil.

**aircraft** A vehicle designed to travel through the air when given lift by its own buoyancy or by dynamic reaction of air particles with its surfaces.

**aircraft controller** A person who controls the movements of aircraft, including guided missiles, by means of radio communication or electronic devices.

**aircraft db rating** A rating in decibels assigned to each type of aircraft to indicate its approximate radar cross-section or echo area. Used primarily with a radar coverage indicator.

**aircraft flutter** Sudden, flickering changes in the contrast of a television picture, caused by reflection of the television signal from an aircraft flying somewhere over the direct path between transmitter and receiver. The reflected signal alternately reinforces and cancels the normal signal at the receiving antenna. Also called airplane flutter.

**aircraft instrument** A mechanical, electric, or electronic device used aboard an aircraft for indicating engine performance, aircraft performance, or navigation data.

**aircraft intercept radar** *Airborne intercept radar.*

**air defense controller** An aircraft controller responsible for controlling and vectoring friendly aircraft during air defense and coordinating the operations of antiaircraft artillery.

**air dose** The x-ray dose in roentgens at a point in free air, including only the radiation of the primary beam and that scattered from surrounding air.

**air equivalent** A thickness of material having the same stopping power as air for nuclear particles. Applied chiefly to materials used for walls and electrodes of ionization chambers.

**air-equivalent ionization chamber** *Air-wall ionization chamber.*

**airframe** The complete structure of an aircraft or guided missile, including the framework and skin but not the engines.

**air gap** 1. A short gap or equivalent filler of nonmagnetic material across the core of a choke, transformer, or other magnetic device. The gap prevents the core from being saturated by direct current or permits required mechanical movement of coils or an armature. 2. A spark gap consisting of two conducting electrodes separated by air.

**air-gap crystal unit** A crystal unit in which the electrodes are separate metallic plates rigidly spaced apart by an amount slightly greater than the thickness of the quartz plate.

**air intercept radar** An airborne radar that searches for, acquires, and tracks a target to provide data needed for control of an air-to-air guided missile.

**air log** A distance-measuring device used in certain guided missiles to control range.

**air mile** *Aeronautical mile.*

**air monitor** A device for detecting and measuring airborne radioactivity for warning and control purposes.

**air navigation** The science or action of plotting and directing from within an aircraft its movement through the air from one place to another. Also called avigation.

**air navigation aid** A radar beacon, radio range, or other system, instrument, or device used in air navigation.

**airplane** A heavier-than-air craft supported by the dynamic reaction of air flowing over fixed or rotating plane surfaces, including piston-driven and jet airplanes, gliders, helicopters, gyroplanes, and winged guided missiles.

**airplane dial** A round radio receiver dial having a pointer rotating over a scale to indicate the frequency of the station to which the receiver is tuned.

**airplane flutter** *Aircraft flutter.*

**airplane insulator** A streamlined insulator once used for radio antennas on aircraft.

**airport** A defined area on land or water, including any buildings and installations, normally used for the takeoff and landing of aircraft.

**air-portable** Readily carried in aircraft with only minor or no dismantling and reassembly.

**airport surface detection equipment** A short-range ground radar used to show the positions of all aircraft and vehicles on the surface of an airport. Runways, taxiways, and ramps also show clearly on the radar screen. Also called taxi radar.

**airport surveillance radar** A radar located on or near an airport to provide an indication of the bearing and distance of each aircraft within the terminal area. It is used by itself for air traffic control, and is used with precision approach radar to form a ground-controlled approach system. Also called surveillance radar element.

**air-position indicator** An airborne computing system that presents a continuous indication of aircraft position on the basis of aircraft heading, airspeed, and elapsed time. Position is indicated in latitude and longitude values or other coordinates. True heading and air mileage flown are also shown.

**air sounding** Measuring atmospheric pressure, humidity, and other characteristics of the atmosphere with instruments carried aloft.

**airspeed** The speed of an aircraft, measured along its longitudinal axis relative to the air through which it moves.

**airspeed computer** A computer used to determine true airspeed from indicated airspeed, temperature, and pressure data.

**airspeed indicator** A flight instrument that shows the approximate speed of an aircraft relative to the air through which it flies.

**air-to-air guided missile** A guided missile designed to be fired at an airborne target from an airborne aircraft. Examples include Falcon, Sidewinder, and Sparrow.

**air-to-surface guided missile** A guided missile designed to be fired at a surface vessel from an airborne aircraft. Examples include Bullpup, Dove, and Rascal.

**air traffic control** A service operated by appropriate authority to promote the safe, orderly, and expeditious flow of air traffic.

**air-wall ionization chamber** An ionization chamber in which the materials of the wall and electrodes are selected to produce ionization essentially equivalent to that in a free-air ionization chamber. Also called air-equivalent ionization chamber.

**air warning system** A system for warning of hostile aircraft approaching a defended area. An air warning system may include radar and communication facilities.

**ajax** A frequency-dispersal type of radar.

**alabamine** Former name for *astatine.*

**Alamogordo bomb** The first atomic bomb, detonated July 16, 1945, at Alamogordo, New Mexico. Also called Trinity bomb.

**alarm signal** The international radiotelegraph alarm signal, transmitted on 500 kc as twelve 4-second dashes 1 second apart, to actuate automatic devices that sound an alarm indicating that a distress message is about to be broadcast.

**albedo** The reflection factor of a surface for neutrons.

**Alexanderson alternator** A high-speed a-c generator used in the early days of radio to generate r-f energy for transmission. Improved versions today generate a-c power for low-frequency induction heating.

**Alford loop** A multielement antenna having approximately equal in-phase currents uniformly distributed along each of its peripheral elements. The radiation pattern is very nearly circular in the plane of polarization.

**Alford slotted tubular antenna** A horizontally polarized antenna consisting of a metal cylinder having a full-length slot. Currents flow in horizontal circles, simulating the operation of a vertical stack of in-phase loop antennas. Originally developed for f-m broadcasting.

**alice** [ALaska Integrated Communications Exchange] A network of radio stations, generally using scatter propagation equipment, used to link early-warning radar stations. Also called White Alice.

**align** To adjust two or more sections of a circuit or system so their functions are properly synchronized. Trimmers, padders, or variable inductances in tuned circuits are adjusted to give a desired response for fixed-tuned equipment or to provide tracking for tunable equipment. Search radar antenna orientation is aligned to coincide with ppi sweep.

**aligned-grid tube** A multigrid vacuum tube in which at least two of the grids are aligned one behind the other to give such effects as beam formation or noise suppression.

**aligning plug** The plug in the center of the base of an octal, loktal, or other tube, having a single vertical projecting rib that prevents the tube from being inserted incorrectly in its socket.

**aligning tool** A small screwdriver, socket wrench, or special tool constructed partly or entirely of nonmagnetic materials, used to align tuned circuits.

**alignment** The process of aligning.

**alignment chart** *Nomograph.*

**alive** *Energized.*

**alkali metal** An alkali-producing metal such as lithium, cesium, or sodium, having photoelectric characteristics. Used in phototubes and camera tubes.

**alkaline storage battery** A storage battery in which the electrolyte consists of an alkaline solution, usually potassium hydroxide.

**Allen screw** A screw having a hexagonal hole in the head.

**Allen wrench** A wrench made from a straight or bent hexagonal rod, used to turn an Allen screw.

**alligator clip** A long, narrow spring clip with meshing jaws, used with test leads to make temporary connections quickly.

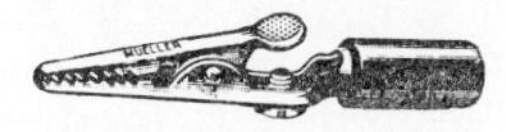

Alligator clip for test lead.

**allocate** To assign storage locations to the main routines and subroutines in a computer, thereby fixing the absolute values of any symbolic addresses.

**allochromatic** Having photoelectric properties due to microscopic particles occurring naturally in a crystal or resulting from exposure to certain forms of radiation.

**allowed band** A band containing a group of energy levels that electrons may occupy in a given material, such as a conduction band and a valence band.

**allowed transition** The most probable type of transition between two states of a quantum-mechanical system.

**alloy** A material having metallic properties and consisting of two or more elements, of which at least one is a metal.

**alloy-diffused transistor** A transistor in which diffusion and alloy techniques are combined in a different manner than for a diffused-alloy transistor.

**alloy junction** A junction produced by alloying one or more impurity metals to a semiconductor. A small button of impurity metal is placed at each desired location on the semiconductor wafer, heated to its melting point, and cooled rapidly. The impurity metal alloys with the semiconductor material to form a p or n region, depending on the impurity used.

**alloy-junction photocell** A photodiode in which an alloy junction is produced by alloying an indium disk into a thin wafer of n-type germanium.

**alloy-junction transistor** A transistor having an alloy junction. Also called fused-junction transistor.

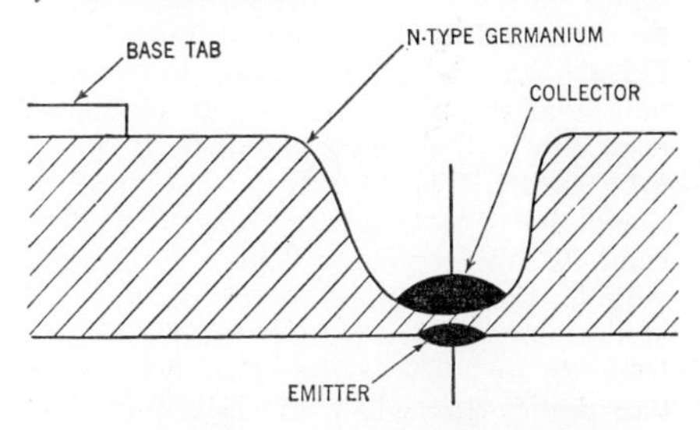


Alloy-junction transistor construction.

**all-pass network** A network designed to introduce phase shift or delay without introducing appreciable attenuation at any frequency.

**all-wave antenna** A radio receiving antenna so constructed that it responds reasonably well to a wide range of frequencies, including the short-wave bands as well as the broadcast band.

**all-wave receiver** A radio receiver capable of being tuned from about 535 kc to at least 20 mc. Some all-wave receivers go above 100 mc and thus cover the f-m band also.

**alnico** [ALuminum NIckel CObalt] An alloy consisting chiefly of iron, aluminum, nickel, and cobalt, having high retentivity. Used to make permanent magnets, as required in loudspeakers, magnetrons, and other devices requiring strong magnetic fields. Not a trademark. Usually used with an alloy number; thus, alnico V gives stronger permanent magnets than earlier alnico alloys.

**alpha** [Greek letter $\alpha$] 1. Symbol for the current amplification factor of a transistor in a grounded-base circuit. It is the ratio of an incremental change in collector current to an incremental change in emitter current when collector voltage is held constant. 2. The ratio of radiative capture to fission cross section for a fissionable element. 3. Symbol for *attenuation constant.*

**alphabetic coding** A system of abbreviations used in preparing information for input to a computer, so information may be handled in letters and words as well as in numbers.

**alpha chamber** *Alpha counter tube.*

**alpha counter** An instrument for counting alpha particles, comprising an alpha counter tube, amplifier, pulse height discriminator, scaler, and recorder.

**alpha counter tube** An electron tube consisting chiefly of a chamber for detecting alpha particles. It is usually operated in the nonmultiplying or proportional region. Pulse height selection is used to discriminate against pulses due to beta or gamma rays. Also called alpha chamber.

**alpha cutoff** The high frequency at which the alpha of a transistor drops 3 decibels from its low-frequency value. The current amplification at alpha cutoff is thus 0.7 of the alpha rating for the transistor.

**alpha decay** The radioactive transformation that occurs when an alpha particle is emitted by a nuclide. The decay product is a new nuclide having a mass number four units smaller and an atomic number two units smaller than the original nuclide.

**alpha disintegration energy** The disintegration energy of an alpha disintegration process, equal to the sum of the kinetic energy of the alpha particle and the kinetic energy of recoil of the product atom.

**alpha emitter** A radionuclide that undergoes transformation by alpha-particle emission.

**alphanumeric coding** Coding having provisions for 26 alphabetic characters in addition to 10 numeric characters, usually with additional codes for special punctuation and machine-control characters.

**alpha particle** A positively charged particle having two protons and two neutrons. It is emitted from certain radioactive elements or isotopes, has high ionizing power but little penetrating ability, and can damage living tissue. An alpha particle is identical in all measured properties with the nucleus of a helium atom.

**alpha-particle binding energy** The energy required to remove an alpha particle from a nucleus.

**alpha-particle spectrum** A line chart showing the distribution, with respect to energy or momentum, of the alpha particles emitted by a radionuclide.

**alpha radiation** Alpha particles emerging from radioactive atoms.

**alpha ray** A stream of alpha particles. It is only slightly deflected by a magnetic field.

**alpha-ray spectrometer** An instrument used to determine the energy distribution of alpha particles emitted by a radionuclide.

**alpha-ray vacuum gage** An ionization gage in which the ionization is produced by alpha particles emitted by a radioactive source, instead of by electrons emitted from a hot filament. Used chiefly for pressures above $10^{-3}$ mm Hg, where filament life of the conventional ionization gage is seriously shortened by positive-ion bombardment and by chemical reaction with the residual gas.

**alpha uranium** An allotropic modification of uranium metal that is stable below approximately 660°C.

**alpha wave** A brain wave having a frequency of 9 to 14 cps.

**AlSb** Symbol for *aluminum antimonide.*

**alternating current** [abbreviated a-c] An electric current that is continually varying in value and is reversing its direction of flow at regular intervals, usually in a sinusoidal manner. Each repetition, from zero to a maximum in one direction and then to a maximum in the other direction and back to zero, is called a cycle. The number of cycles occurring in one second is called the frequency. The average value of an alternating current is zero.

**alternating quantity** A periodic quantity whose average value is zero over a complete cycle.

**alternating voltage** The voltage generated by an alternator or developed across a resistance or impedance through which alternating current is flowing. This voltage is continually varying in value and reversing its direction at regular intervals. Also called a-c voltage.

**alternation** Half of an a-c cycle, consisting of a complete rise and fall of voltage or current in one direction. There are 120 alternations per second in 60-cycle a-c power.

**alternator** A machine that generates an alternating voltage when its armature or field is rotated by a motor, an engine, or other means. The output frequency is directly proportional to the speed at which the generator is driven. Also called synchronous generator.

**alternator transmitter** A radio transmitter that utilizes power generated by an r-f alternator.

**altigraph** A recording altimeter.

**altimeter** [pronounced al-tim'e-ter] An instrument used in air navigation to indicate altitude above sea level, above ground level at the point of measurement, or above ground level at some other point for which the altimeter was calibrated. An ordinary pressure altimeter uses an aneroid barometer to measure changes in barometric pressure with altitude. An absolute altimeter determines altitude above ground or water by measuring the time it takes for a radio or radar wave to travel straight down and be reflected back to the aircraft.

**altimetric flareout** An aircraft descent path in which the rate of descent is reduced as the touchdown point is approached. Also called exponential flareout.

**altitude** The height of an aircraft or other object above a given level, as above the sea or ground. Also called elevation.

**altitude hole** The small circle in the center of an airborne ppi radar presentation of ground terrain, corresponding to the time required for the radar signal to travel from the aircraft to the ground and back.

**altitude signal** The radar signal returned to an airborne radar set by the ground or water surface directly beneath the aircraft.

**alumina** A ceramic used for insulators in electron tubes because it is easily degassed. It can withstand temperatures up to 1,400°C continuously, and has low dielectric loss over a wide frequency range.

**aluminized screen** A television picture tube screen that has a thin coating of aluminum on the back of its phosphor layer. Electrons in the beam readily penetrate the coating and activate the phosphors to produce an image. The aluminum serves to reflect outward the light that would otherwise go back inside the tube, thereby improving the brilliance and contrast of the image. Also called metal-backed screen, metallized screen, and mirror-backed screen.

**aluminum** [symbol Al] A light-weight silvery white metal having atomic number 13. Widely used in electronic equipment for capacitor foil, tuning capacitor plates, shields, and equipment housings. Used as structural material in low-temperature nuclear reactors because it has a fairly low neutron-capture cross section and is easily worked.

**aluminum antimonide** [symbol AlSb] A semiconductor having a forbidden band gap of 2.2 electron-volts and a maximum operating temperature of 500°C when used in a transistor.

**aluminum electrolytic capacitor** An electrolytic capacitor that uses plain or etched aluminum foil for both electrodes.

**a-m** Abbreviation for *amplitude modulation.*

**amateur** A person holding a license, issued by the Federal Communications Commission or by corresponding authorities in other countries, authorizing that person to operate a licensed amateur radio station for pleasure and service only, not for profit. Also called ham (slang).

**amateur band** A band of frequencies assigned exclusively to amateur operators.

**ambient condition** The condition of the surrounding medium.

**ambient fuze** A proximity fuze that is activated by some characteristic of the environment in which the target is normally found, rather than by the presence of the target.

**ambient light** Normal room light, such as that reaching a television picture tube screen from the outside.

**ambient-light filter** A filter used in front of a television picture tube screen to reduce the amount of ambient light reaching the screen and to minimize reflections of light from the glass face of the tube. The filter is generally a dull color. It can be incorporated in the glass faceplate of the tube or in the safety-glass window, or can be a separate sheet of plastic.

**ambient noise** Noise associated with a given environment, made up of more or less continuous sounds from near and far sources. It is generally undesirable.

**ambient temperature** The temperature of the immediately surrounding medium. If the device is giving off heat, the air in its vicinity will be heated and the ambient temperature will then be higher than room temperature.

**ambiguity** 1. The condition in which navigation coordinates define more than one point, direction, line of position, or surface of position. 2. The condition in which a synchro or servosystem seeks more than one null position.

**American Institute of Electrical Engineers** [abbreviated AIEE] A nonprofit professional organization of engineers and scientists established for the advancement of the theory and practice of electrical engineering.

**American Morse code** A dot-dash code used in wire telegraphy. It has a different spacing method from the international Morse code used in radio and has entirely different letter codes.

**American Standards Association** [abbreviated ASA] A nonprofit organization supported by industry for the purpose of establishing uniform standards.

**American wire gage** [abbreviated AWG] A gage used chiefly for specifying nonferrous wire and sheet metal. Sizes range from 0000 for the largest (0.46 inch) to 0 (0.325 inch) and 1 (0.289 inch) to 50 (0.001 inch). Also called Brown and Sharpe gage.

**americium** [symbol Am] An unstable transuranic radioactive element produced artificially by bombarding plutonium with helium ions. Atomic number is 95.

**ammeter** An instrument for measuring current flow. Its scale may be calibrated in amperes or smaller units. An ammeter that indicates milliampere values is often called a milliammeter. An ammeter that indicates microampere values is often called a microammeter.

**ammonia maser clock** A gas maser that utilizes the transition of high-energy ammonia molecules to generate a stable microwave output signal for use as a time standard.

**ammonium dihydrogen phosphate crystal** [abbreviated adp crystal] A piezoelectric crystal used in sonar transducers.

**amp** Abbreviation for *ampere.*

**amperage** The amount of current in amperes.

**ampere** [abbreviated amp] The practical unit of electric current. A voltage of 1 volt will send a current of 1 ampere through a resistance of 1 ohm.

**ampere-hour** A unit of quantity of electricity. Multiplying current in amperes by time of flow in hours gives ampere-hours. Used chiefly to indicate the amount of energy that a storage battery can deliver before it needs recharging, or that a primary battery can deliver before it needs replacing.

**Ampere's law** The magnetic intensity at any point near a current-carrying conductor can be computed on the assumption that each infinitesimal length of the conductor contributes an amount which is directly proportional to that length and to the current it carries, inversely proportional to distance, and directly proportional to the sine of the angle.

**Ampere's rule** The magnetic field surrounding a conductor will have a counterclockwise direction when the electron flow is away from the observer.

**ampere-turn amplification** The ratio of the change in output ampere-turns of a magnetic amplifier to the change in control ampere-turns.

**ampere-turns** [abbreviated a-t] A measure of magnetomotive force, equal to turns multiplied by coil current in amperes.

**amplidyne** [AMPLIfier DYNE] A rotating magnetic amplifier consisting of a combination d-c motor and generator having special windings and brush connections to give power amplification, so small changes in power input to the field coils produce large changes in power output. Widely used as a power amplifier in servosystems.

**amplification** The process of increasing the strength (current, voltage, or power) of a signal.

**amplification factor** [symbol $\mu$] The ratio of the change in anode voltage of an electron tube to a change in control-electrode voltage, when other tube voltages and currents are held constant.

**amplified agc** An automatic gain-control circuit in which the control voltage is amplified before being applied to the tube whose gain is to be controlled in accordance with the strength of the incoming signal.

**amplifier** A device using an electron tube, transistor, magnetic unit, or other amplification-producing component to increase the strength of a signal without appreciably altering its characteristic waveform. An amplifier transfers power to the signal from an external source, whereas a transformer changes signal voltage or current without adding power.

**amplify** To increase in magnitude.

**Amplitron** Trademark of Raytheon Mfg. Co. for their line of platinotrons.

**amplitude** The value of a varying quantity at a specified instant.

**amplitude balance control** *Sensitivity-time control.*

**amplitude discriminator** A circuit whose output is a function of the relative magnitudes of two signals.

**amplitude distortion** *Frequency distortion.*

**amplitude-frequency distortion** *Frequency distortion.*

**amplitude-frequency response** A graph

showing how the gain or loss of a device or system varies with frequency. Also called frequency characteristic, frequency response, response, response characteristic, and sine-wave response.

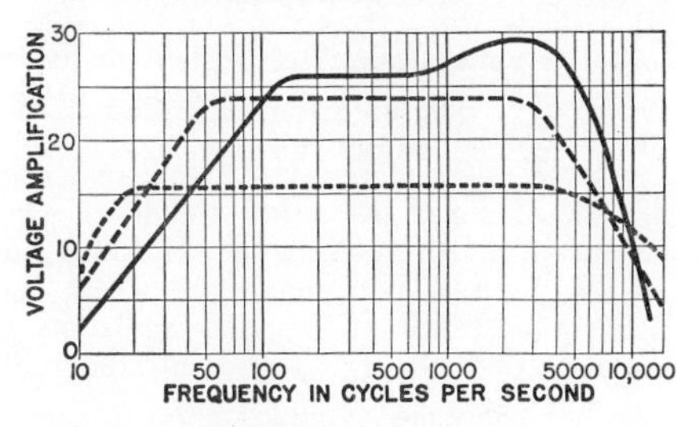


Amplitude-frequency response curves for three different a-f amplifiers.

**amplitude-modulated transmitter** A transmitter that transmits an amplitude-modulated wave.

**amplitude-modulated wave** A sinusoidal wave whose envelope contains a component similar to the waveform of the signal to be transmitted.

**amplitude modulation** [abbreviated a-m] Modulation in which the amplitude of the carrier-frequency current is varied above and below its normal value in accordance with the audio, picture, or other intelligence signal to be transmitted.

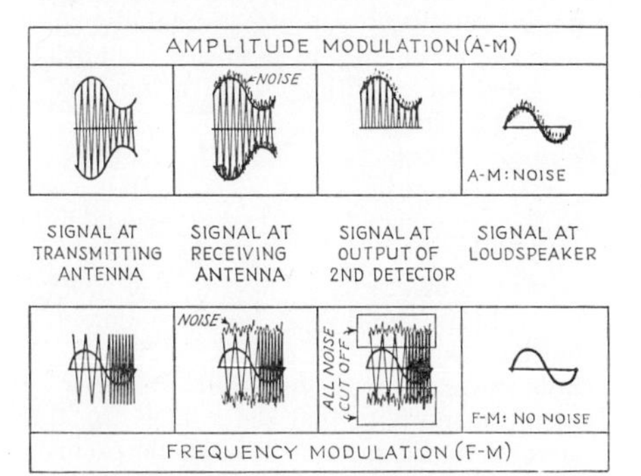


Amplitude modulation compared with frequency modulation in noise-suppressing action.

**amplitude-modulation noise level** The noise level produced by undesired amplitude variations of an r-f signal in the absence of any intended modulation.

**amplitude range** The ratio between the upper and lower limits of audio signal amplitudes that contain all significant energy contributions.

**amplitude selection** The selection of the portion of a waveform that lies above or below a given value, or between two given values.

**amplitude separator** A circuit used to isolate the portion of a waveform that is above or below a given value, or is between two given values.

**amu** Abbreviation for *atomic mass unit.*

**An** Symbol for *actinon.*

**analog** The representation of numerical quantities by physical variables such as translation, rotation, voltage, and resistance.

**analog computer** A computer that solves problems by setting up equivalent electric circuits and making measurements as the variables are changed in accordance with the corresponding physical phenomena. An analog computer gives approximate solutions, whereas a digital computer gives exact solutions.

**analyzer** 1. A Nicol prism or other device for detecting and testing polarized light. 2. *Set analyzer.*

**ancillary equipment** Auxiliary equipment.

**and-circuit** *And-gate.*

**Anderson bridge** A six-branch modification of the Maxwell-Wien bridge, used to measure self-inductance in terms of capacitance and resistance. Bridge balance is independent of frequency.

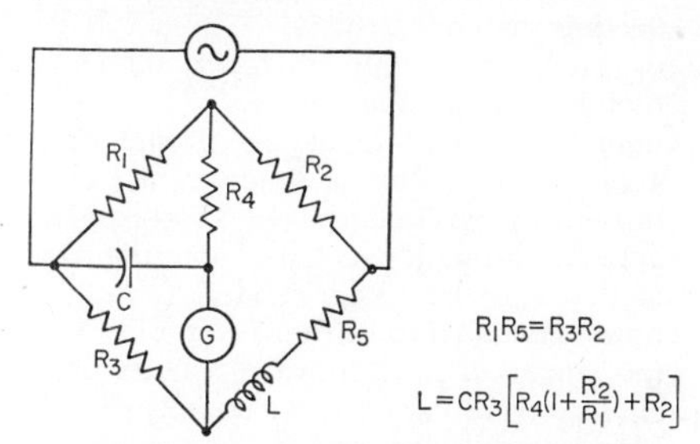


Anderson bridge circuit and equations.

**and-gate** A multiple-input gate circuit whose output is energized only when every input is energized simultaneously in a specified manner. Used in digital computers. Also called and-circuit.

**anechoic room** A room in which sound reflections from the walls, ceiling, and floor have been reduced to a minimum by covering them with sound-absorbing material. Also called dead room and free-field room.

**anelectrotonus** Reduced excitability of a nerve or muscle that is near the anode during passage of direct current through living tissue.

**anemograph** An instrument used for recording wind velocity.

**anemometer** An instrument used for measuring wind velocity.

**aneroid** An evacuated chamber having flexible metal walls, used to convert changes in atmospheric pressure into corresponding mechanical movements. When atmospheric pressure increases, the chamber becomes smaller.

**aneroid altimeter** An aircraft altimeter in which an aneroid is used to sense changes in atmospheric pressure and actuate an indicator calibrated in feet of altitude above sea level or above some other reference level. Atmospheric pressure decreases with altitude.

**aneroid barometer** A barometer that uses an aneroid to drive a pointer that indicates atmospheric pressure.

**A neutron** Obsolete term for a neutron of such energy that it is strongly absorbable by silver.

**angel** 1. A radar echo coming from an invisible and sometimes unknown origin. High-flying birds and swarms of insects are some causes. Unusual atmospheric conditions can also cause radar signal reflections that result in target indications on a radar screen when the sky is perfectly clear. 2. A corner reflector or other metallic reflecting material suspended from a balloon or kite to simulate a radar target and thereby confuse the enemy.

**angle** The measure of the progression of a sine wave in time or space from a chosen instant or position, or the corresponding amount through which the rotating vector of the wave has progressed.

**angle modulation** Modulation in which the angle of a sine-wave carrier is the characteristic varied from its normal value. Phase modulation and frequency modulation are particular forms of angle modulation.

**angle of elevation** The angle between the horizontal plane and the line ascending to the object.

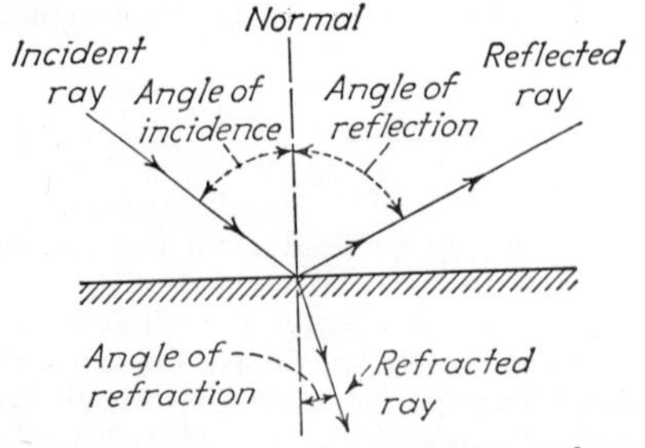


Angle of incidence, angle of reflection, and angle of refraction for a light ray or electromagnetic wave.

**angle of incidence** The angle between a wave or beam arriving at a surface and the perpendicular to the surface at the point of arrival.

**angle of reflection** The angle between a wave or beam leaving a surface and the perpendicular to the surface.

**angstrom** [abbreviated A] A unit of wavelength of light and other radiation. One angstrom is $10^{-8}$ centimeter. The range of visible light is between about 4,000 and 7,500 angstroms. Ultraviolet radiation is below 4,000 angstroms, and infrared radiation is above 7,500 angstroms.

**angular acceleration** The rate of change of angular velocity about a rotational axis.

**angular deviation loss** The ratio of the response of a microphone or loudspeaker on its principal axis to the response at a specified angle from the principal axis, expressed in decibels.

**angular deviation sensitivity** The ratio of the change in a course indicator reading to the actual angular change in the course of an aircraft or ship.

**angular distance** Distance expressed in radians or degrees. It is equal to the distance in wavelengths multiplied by $2\pi$ radians or by 360°.

**angular distribution** The distribution in angle, relative to an experimentally specified direction, of the intensity of particles or photons resulting from a nuclear or extranuclear process.

**angular frequency** The frequency expressed in radians per second. It is equal to the frequency in cycles per second multiplied by $2\pi$.

**angular momentum** The angular velocity of a body multiplied by its moment of inertia.

**angular-momentum quantum number** A quantum number that determines the total angular momentum of a molecule exclusive of nuclear spin.

**angular-position pickup** An instrument that translates variations in angular position into an electrical quantity, such as a change in inductance, resistance, or voltage.

**angular resolution** The ability of a radar to distinguish between two targets solely by the measurement of angles. It is generally expressed as the minimum angle by which targets must be spaced to be separately distinguishable.

**angular velocity** [symbol $\omega$] 1. The speed

of a rotating object measured in radians per second. It is equal to revolutions per second multiplied by $2\pi$. 2. The rate of change of phase of an alternating quantity. It is equal to the frequency in cycles per second multiplied by $2\pi$.

**angular width** *Course width.*

**anion** A negative ion.

**annihilation** 1. A process in which a pair of antiparticles meet and convert spontaneously into one or more photons; it is the inverse of pair production. The commonest example is the annihilation of an electron and a positron, the rest masses of which are converted into photons. 2. The conversion of rest mass into electromagnetic radiation.

**annihilation radiation** Electromagnetic radiation produced by the union, and consequent annihilation, of a positron and an electron. Each such annihilation usually produces two photons, which have the same properties as gamma rays.

**AN nomenclature system** A joint Army-Navy-Air Force code system for designating communication and electronic equipment. The prefix AN indicates that the designation was assigned under this system.

**annunciator** An electric remote signaling device, such as a buzzer or signal lamp.

**anode** 1. [symbol P] The positive electrode through which a principal stream of electrons leaves the interelectrode space in an electron tube. Also called plate. 2. The positive electrode of a battery or other electrochemical device.

**anode bend** The curved portion at the bottom of the anode-current/grid-voltage characteristic for an electron tube.

**anode-bend detector** A detector that uses a triode vacuum tube operating over the lower curved portion of its anode-current/grid-voltage characteristic. Positive swings of the carrier signal are then amplified much more than negative swings, giving the effect of rectification as required for detection.

**anode breakdown voltage** The anode voltage required to cause conduction across the main gap of a gas tube when the starter gap is not conducting and all other tube elements are at cathode potential.

**anode bypass capacitor** A capacitor connected between anode and ground in an electron-tube circuit to bypass high-frequency currents and keep them out of the load. Also called plate bypass capacitor.

**anode characteristic** A graph plotted to show how the anode current of an electron tube is affected by changes in anode voltage.

**anode circuit** A circuit including the anode voltage source and all other parts connected between the cathode and anode of an electron tube. Also called plate circuit.

**anode current** The electron current flowing through an electron tube from the cathode to the anode. Also called plate current.

**anode dark space** A narrow dark zone next to the surface of the anode in a gas tube.

**anode detection** Detection in which rectification of r-f signals takes place in the anode circuit of an electron tube. The grid bias is made sufficiently negative to bring anode current nearly to cutoff for no signal, so average anode current follows changes in signal amplitude. Also called plate detection.

**anode dissipation** Power dissipated as heat in the anode of an electron tube because of bombardment by electrons and ions.

**anode efficiency** The ratio of the a-c load circuit power to the d-c anode power input for an electron tube. Also called plate efficiency.

**anode fall** The voltage between the positive column and the anode of a gas tube. It may be either positive, zero, or negative.

**anode firing** A method of initiating conduction of an ignitron, in which the ignitor is connected through a rectifying element to the anode of the ignitron to obtain power for the firing current pulse.

**anode follower** A tube circuit with heavy feedback from anode to grid, such that the output voltage is nearly equal and opposite to the input voltage. The input impedance is then very high.

**anode glow** A narrow bright zone on the anode side of the positive column in a gas tube.

**anode input power** The product of the direct anode voltage applied to the tubes in the last radio stage of a transmitter and the total direct current flowing to the anodes of these tubes, measured without modulation. Also called plate input power.

**anode keying** Keying of a radiotelegraph transmitter by interrupting the anode supply circuit. Also called plate keying.

**anode load impedance** The total impedance between the anode and cathode of an electron tube, exclusive of the electron stream. Also called plate load impedance.

**anode modulation** Modulation produced

by introducing the modulating signal into the anode circuit of any tube in which the carrier is present. Also called plate modulation.

**anode neutralization** Neutralization in which a portion of the anode-cathode a-c voltage is shifted 180° and applied to the grid-cathode circuit through a neutralizing capacitor. Also called plate neutralization.

**anode power input** The d-c power delivered to the anode of an electron tube by the source of supply. It is the product of average anode voltage and average anode current. Also called plate power input.

**anode pulse modulation** Modulation produced in an amplifier or oscillator by application of externally generated pulses to the anode circuit. Also called plate pulse modulation.

**anode pulsing** An r-f oscillator circuit arrangement in which the anode voltage is normally reduced to such a low value that no anode current flows and no oscillations occur. A pulse equal to the full anode voltage is then introduced in series with the anode. Oscillations begin and last for the duration of the pulse. This circuit requires a modulator capable of supplying full anode power.

**anode rays** Positive ions coming from the anode of an electron tube. They are generally due to impurities in the metal of the anode.

**anode region** The positive column, anode glow, and anode dark space in a gas tube.

**anode resistance** The resistance value obtained when a small change in the anode voltage of an electron tube is divided by the resulting small change in anode current.

**anode saturation** The condition in which the anode current of an electron tube cannot be further increased by increasing the anode voltage. The electrons are then being drawn to the anode at the same rate as they are emitted from the cathode. Also called current saturation, plate saturation, saturation, and voltage saturation.

**anode sheath** A layer of electrons that surrounds the anode of a gas tube when the anode current is high.

**anode shield** A shield that partially surrounds the anode in a mercury-arc rectifier. It protects the anode from excessive ionization or radiation.

**anode sputtering** The emission of fine particles from the anode of an electron tube as a result of electron bombardment.

**anode strap** A metallic connector used between selected anode segments of a multicavity magnetron, principally for mode separation.

**anode supply** The direct voltage source used in an electron tube circuit to place the anode at a high positive potential with respect to the cathode. Also called plate supply.

**anode voltage** The direct voltage existing between anode and cathode of an electron tube. Also called plate voltage and high tension (British).

**anode voltage drop** The voltage existing between anode and cathode in a cold-cathode gas tube after conduction has been established in the main gap.

**anodizing** An electrolytic process for producing a protective or decorative film on certain metals, chiefly aluminum and magnesium.

**anomalous propagation** Freak propagation of vhf radio waves beyond the horizon, apparently due to temperature inversion in the lower atmosphere.

**A-N radio range** A radio range that provides four radial lines of position for aircraft guidance, each identified aurally as a continuous tone resulting from the interlocking of equal-amplitude A and N International Morse code letters. The sense of deviation from one of these lines is indicated by deterioration of the steady tone into audible A or N code signals. The two types of A-N radio ranges are the Adcock radio range and the loop-type radio range.

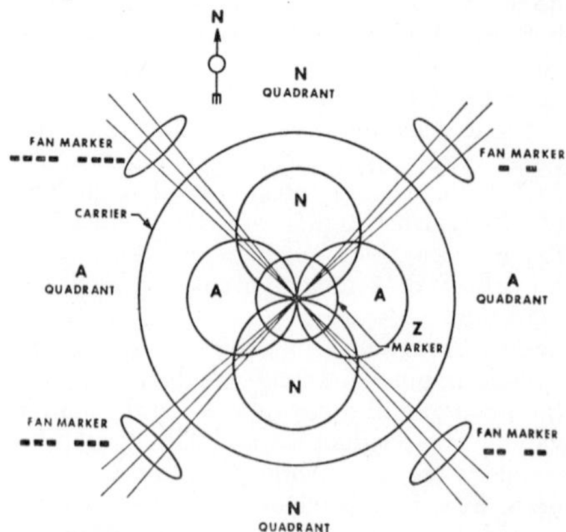


A-N radio range courses, radiation patterns, and fan marker codes.

**antenna** A device used for radiating or receiving radio waves. British term is aerial.

**antenna array** An arrangement of two or

more antennas spaced apart and coupled together in such a way as to give desired directional characteristics. Also called array and beam antenna.

**antenna beam width** The angle in degrees between two opposite half-power points of an antenna beam.

**antenna coil** The first coil in a receiver, through which antenna current flows. When this coil is inductively coupled to a secondary coil, the combination of two coils should be called an antenna transformer or r-f transformer.

**antenna counterpoise** *Counterpoise.*

**antenna coupler** An r-f transformer, tuned line, or other device used to transfer energy efficiently from a transmitter to a transmission line or from a transmission line to a receiver.

**antenna cross-section** A microwave receiving-antenna rating, expressed as the area, perpendicular to incident radiation, that intercepts an amount of energy equal to that delivered to a receiver by the antenna.

**antenna crosstalk** A measure of undesired power transfer through space from one antenna to another. It is the ratio of the power received by one antenna to the power transmitted by the other, usually expressed in decibels.

**antenna current** The r-f current that flows in a transmitting antenna. It is generally measured when there is no modulation.

**antenna detector** An antenna and associated receiving equipment used in an aircraft to warn crew members that they are being observed by an enemy radar set. The equipment is usually mounted in the nose or tail of the aircraft and connected to one or more warning lamps.

**antenna drive** A motor or other device that rotates or positions an antenna, such as for tracking a radar target.

**antenna effect** 1. An undesired output signal resulting from a directional array acting as a nondirectional antenna in an electronic navigation system. Also called height effect. 2. A spurious effect caused by the capacitance of a loop antenna to ground.

**antenna eliminator** A device that plugs into a wall electric outlet and provides connections through capacitors to the power-line wires, to serve as an antenna connection for a radio receiver. Widely used before the development of efficient built-in loop antennas.

**antenna feed** The device used for supplying energy to a transmitting antenna. It may be a transmission line for direct feed, or a dipole or horn for indirect feed to a reflector.

**antenna field gain** A Federal Communications Commission rating for a transmitting antenna. It is the effective free-space field intensity in millivolts per meter that is produced in the horizontal plane at a distance of 1 mile by an antenna input power of 1 kilowatt, divided by 137.6 millivolts per meter (the value for a half-wave dipole).

**antenna gain** The effectiveness of a directional antenna as compared to a standard nondirectional antenna. It is usually expressed as the ratio in decibels of standard antenna input power to directional antenna input power that will produce the same field strength in the desired direction. For a receiving antenna, the ratio of signal power values produced at the receiver input terminals is used. The more directional an antenna is, the higher is its gain. Also called gain.

**antenna height above average terrain** A Federal Communications Commission rating for transmitting antennas. It is the average of the antenna heights above the terrain from 2 to 10 miles from the antenna for the eight directions, spaced evenly for each 45° of azimuth starting with true north. The averages for each direction are averaged to get the final value.

**antenna lens** An arrangement of shaped metal vanes or dielectric material used in front of a microwave antenna to concentrate the beam of transmitted or received radio waves.

**antenna loading** Use of lumped reactances to tune an antenna.

**antenna power** A transmitter rating equal to the square of the antenna current multiplied by the antenna resistance at the point where the current is measured.

**antenna power gain** A transmitting antenna rating equal to the square of the antenna gain, expressed in decibels.

**antenna relay** A relay used in radio communication stations to switch an antenna from a receiver to a transmitter and vice versa.

**antenna resistance** A transmitting antenna rating expressing the total resistance of the antenna system at the operating frequency. The antenna resistance in ohms is equal to the power in watts supplied to the entire antenna circuit, divided by the square of the effective antenna current in amperes measured at the point where power is supplied to the antenna. Components of antenna resistance include radiation resistance, ground resistance, r-f resistance of conductors in the antenna circuit, and the equivalent resistance due to corona, eddy currents, insulator leakage, and dielectric power loss.

**antenna resonant frequency** The frequency or frequencies at which an antenna appears to be a pure resistance.

**antenna series capacitor** A capacitor used in series with an antenna to shorten the electrical length of the antenna.

**antiaircraft missile** A guided missile launched from the surface against an airborne target.

**antibonding orbital** An orbital electron whose energy increases monotonically as the two atoms to which it belongs move closer, so it does not lead to closer binding of the molecule.

**anticapacitance switch** A switch designed to have low capacitance between its terminals when open.

**anticathode** The anode or target of an x-ray tube.

**anticipation mode** A mode of storing binary digits in a cathode-ray memory tube, wherein one type of digit is represented by a continuous line of excitation on the screen and the other type by gaps in the line.

**anticlutter circuit** A radar circuit that attenuates undesired reflections so as to permit detection of targets otherwise obscured by such reflections.

**anticlutter gain control** *Sensitivity-time control.*

**anticoincidence** Occurrence of a count in a specified detector unaccompanied, simultaneously or within an assignable time interval, by a count in one or more other specified detectors.

**anticoincidence circuit** A circuit that produces a specified output pulse when one of two inputs receives a pulse and the other receives no pulse within an assigned time interval.

**anticoincidence counter** An arrangement of counters and associated circuits that will record a count only if an ionizing particle passes through certain of the counters but not through the others.

**anticoincidence counting** The recording, from one or more counters of a coincidence circuit, of all counts except coincidences.

**anticollision radar** A radar set designed to give warning of possible collisions during movements of ships or aircraft.

**anticyclotron** A type of traveling-wave tube.

**antifading antenna** An antenna designed to confine radiation mainly to small angles of elevation, so as to minimize radiation of the sky waves that cause fading.

**antiferromagnetic material** A material in which spontaneous magnetic polarization occurs in equivalent sublattices. The polarization in one sublattice is aligned antiparallel to the other.

**antihunt circuit** A stabilizing circuit used in a closed-loop feedback system to prevent self-oscillations.

**antihunt device** An electric or mechanical device used in positioning systems to prevent hunting or oscillation of the load around an ordered position. It usually involves some form of feedback.

**antihunt transformer** A transformer used as a stabilizing network in a d-c feedback system. Its primary winding is in series with the load. Its secondary winding provides a voltage, proportional to the derivative of the primary current, that is fed back into some other part of the loop to prevent self-oscillation.

**antijamming** A device, method, system, or technique that reduces or eliminates jamming.

**antilambda** A particle of antimatter corresponding to a lambda particle.

**antilog** Abbreviation for *antilogarithm.*

**antilogarithm** [abbreviated antilog] The number corresponding to a given logarithm. Example: If the logarithm of 563.2 is 2.75066, then 563.2 is the antilogarithm of 2.75066.

**antimatter** A form of matter in which protons, electrons, and other particles have charges opposite those with which they are normally associated.

**antimicrophonic** Not affected by vibrations or by sound waves from a loudspeaker to the extent that feedback and howling occur. Also called nonmicrophonic.

**antimissile missile** An explosive missile launched to intercept and destroy another missile in flight.

**antimony** [symbol Sb] A metallic element. Atomic number is 51.

**antineutrino** A hypothetical particle presumed to be emitted during radioactive decay. It can be differentiated from a neutrino only by its momentum characteristics.

**antineutron** A hypothetical particle having the same mass and the same zero charge as a neutron, and capable of combining with a neutron to give complete conversion of both particles into mesons.

**antinode** A point, line, or surface in a standing-wave system at which some characteristic of the wave has maximum amplitude. Also called loop.

**antinoise microphone** A microphone having characteristics that discriminate against undesired noise, such as a close-talking microphone, lip microphone, or throat microphone.

**antiplugging relay** A relay used in some control systems to prevent plugging.

**antiproton** An elementary particle that differs from a proton only in that its charge is negative. Also called negative proton.

**antiresonance** *Parallel resonance.*

**antiresonant circuit** *Parallel resonant circuit.*

**antistickoff voltage** A small voltage applied to the rotor winding of the coarse synchro control transformer in a two-speed control system to eliminate the possibility of ambiguous behavior.

**antitank guided missile** A surface-to-surface guided missile whose flight path to an enemy tank is controlled by command signals produced by an automatic computer from data obtained by optical sighting, with the control signals being fed to the missile over multiple wire command links. The missile may also contain an infrared homing device for final range correction.

**anti-transmit-receive tube** *Atr tube.*

**anti-tr tube** *Atr tube.*

**aperiodic** Not responsive to any particular frequency.

**aperiodic antenna** An antenna having essentially constant impedance over a wide range of frequencies. Examples are terminated rhombic antennas and terminated wave antennas. Also called untuned antenna.

**aperiodic circuit** A circuit in which it is not possible to produce free oscillation.

**aperiodic damping** Damping so great that a disturbed system or instrument comes to a position of rest without passing through this position. The point of change between aperiodic and periodic damping is called critical damping.

**aperture** An opening through which electrons, light, radio waves, or other radiation can pass. The aperture in the electron gun of a cathode-ray tube determines the size of the electron beam. The aperture in a television camera is the effective diameter of the lens that controls the amount of light entering the camera tube. The dimensions of the horn mouth or parabolic reflector determine the aperture of a microwave antenna.

**aperture distortion** Attenuation of the high-frequency components of a television picture signal due to the finite cross-sectional area of the scanning beam in the camera. The beam then covers several mosaic globules in the camera simultaneously, causing loss of picture detail.

**aperture illumination** The strength distribution of an electromagnetic wave in an aperture.

**aperture mask** *Shadow mask.*

**apex step** A flexible step or corrugation used at the center of a loudspeaker cone to modify high-frequency response.

**apochromatic lens** A lens that has been corrected for chromatic aberration for three colors.

**apogee** 1. The point in an elliptical orbit of an earth satellite farthest from the earth. 2. The point in the trajectory of a ballistic missile farthest from the earth.

**apostilb** A unit of luminance equal to one ten-thousandth of a lambert.

**apparent power** The power value obtained in an a-c circuit by multiplying the effective values of voltage and current. The result is expressed in volt-amperes, and must be multiplied by the power factor to secure the average or true power in watts.

**apparent precession** The relative angular movement of the spinning axis of a gyroscope in relation to a line on the earth, resulting from the rotation of the earth.

**appearance potential** The minimum energy that the electron beam in the ion source of a mass spectrometer must have in order to produce ions of a particular species when a molecule is ionized.

**Applegate diagram** A diagram illustrating the behavior of the electrons in a velocity-modulation tube. The distances of electrons from the buncher in the drift space are plotted as vertical coordinates against time on the horizontal axis. Close spacing of these vertical lines indicates bunching of electrons.

**Appleton layer** *F layer.*

**appliance** A piece of equipment that consumes current and produces a desired work-saving or other result, such as an electric heater, radio, or electronic range.

**applicator** An appropriately shaped metal electrode used to establish an alternating electric field in material to be heated by dielectric heating.

**applied research** Research directed toward practical use of knowledge gained by basic research.

**applied shock** Excitation that produces shock motion within a system.

**applique circuit** A circuit that can be added to existing equipment to increase or change the possible applications. For example, some carrier telephone equipment designed for dial signaling can be converted to ringdown signaling through the use of an applique circuit.

**approach control radar** Any radar set or system used in a ground-controlled approach system, such as an airport surveillance radar, a precision approach radar, or a combination of both.

**approach path** The portion of the flight path, in the immediate vicinity of a landing area, that terminates at the touchdown point.

**Aquadag** Trademark of Acheson Colloids Co. for their brand of colloidal graphite in water. Widely used to produce a conductive coating on the inside surface of the glass envelope for cathode-ray picture tubes, where it collects secondary electrons emitted by the fluorescent screen. Also used on the outside of some picture tubes, where it serves as the final capacitor of the high-voltage filter circuit.

**A quadrant** One of the two quadrants in which the A signal of an A-N radio range is heard.

**aquarium reactor** A swimming-pool reactor arranged so the core can be seen during operation.

**aqueous homogeneous reactor** A nuclear reactor in which the fuel is in a solution, such as uranyl sulfate dissolved in water.

**arbitrary constant** A constant to which various values may be assigned by decision alone, with these values being unaffected by any of the variables in an equation.

**arbitrary function generator** A function generator in which a mask having the desired waveform outline is inserted in a photoelectric scanning system, as in the photoformer.

**arc** *Electric arc.*

**arcback** The flow of a principal electron stream in the reverse direction in a mercury-vapor rectifier tube because of formation of a cathode spot on an anode. This results in failure of the rectifying action. Also called backfire. The action corresponds to reverse emission in electron tubes.

**arc baffle** A baffle used in mercury-pool tubes to prevent mercury from splashing the anode and causing an electric arc. Also called splash baffle.

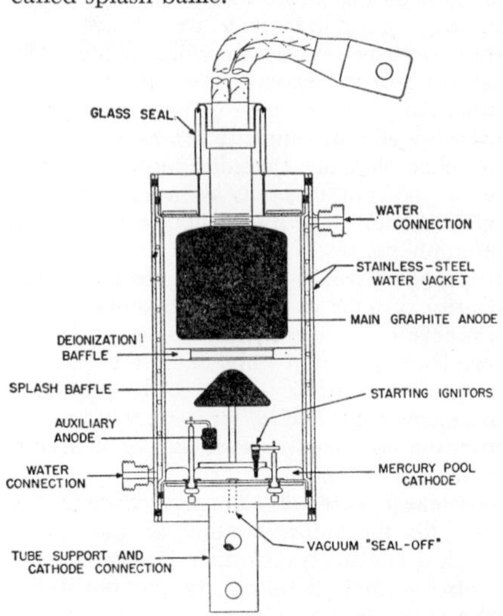


Arc baffle is mushroom-shaped cap, also called splash baffle, over mercury-pool cathode of ignitron.

**arc cathode** A cathode in which electron emission is self-sustaining at a low voltage drop, approximately equal to the ionization potential of the gas.

**arc converter** A form of oscillator utilizing an electric arc as the generator of alternating or pulsating current. Used in some induction heating generators.

**arc discharge** A discharge between electrodes in a gas or vapor, characterized by a relatively low voltage drop and a high current density.

**arc-discharge tube** A discharge tube in which a high-current arc discharge passes through the gas between the electrodes, generally for the purpose of producing an intense flash of light as in a strobotron.

**arc drop** The voltage drop between the anode and cathode of a gas rectifier tube during conduction.

**arc-drop loss** The product of the instantaneous values of tube voltage drop and current averaged over a complete cycle of operation of a gas tube.

**arcing** The production of an arc, as at the brushes of a motor or the contacts of a switch.

**arc lamp** An electric lamp in which the light is produced by an arc made when current flows through ionized gas between two electrodes.

**arc resistance** The time required for a

given electric current to render the surface of a material conductive because of carbonization by the arc flame.

**arc spectrum** The spectrum of light produced by vaporizing an element in an electric arc.

**arcthrough** In multielectrode gas tubes, the loss of control resulting from the flow of a principal electron stream in the normal direction during a scheduled nonconducting period.

**arc transmitter** A radio transmitter that uses an arc to generate r-f carrier signals.

**arc welding** A fusion welding process in which welding heat is obtained from an electric arc struck between an electrode and the metal being welded, or between two separate electrodes as in atomic hydrogen welding.

**area control radar** A radar set or system used for air traffic control over a relatively large area, to provide a smooth flow of air traffic to the approach control radar.

**area monitor** A device used for detecting and/or measuring radiation levels at a given location for warning or control purposes.

**area monitoring** Routine monitoring of levels of ionizing radiation in an area in which radiation hazards are present or suspected.

**argon** [symbol A] An inert gas element that gives a purple glow when ionized. Atomic number is 18. Sometimes used in electron tubes, electric lamps, and neon signs.

**argon glow lamp** A glow lamp containing argon gas. It produces a pale blue-violet light. Wattages of standard sizes range from ¼ watt to 2 watts.

**arithmetical element** *Arithmetical unit.*

**arithmetical operation** A digital computer operation in which numerical quantities are added, subtracted, multiplied, divided, or compared.

**arithmetical shift** A computer operation in which a quantity is multiplied or divided by a power of the number base. Thus, binary 1011 represents decimal 11, and therefore two arithmetical shifts to the left is binary 101100, which represents decimal 44.

**arithmetical unit** The section of a computer that performs adding, subtracting, multiplying, dividing, and comparing operations. Also called arithmetical element.

**arithmetic mean** *Mean.*

**arm** *Branch.*

**armature** The moving portion of a magnetic circuit, such as the rotating section of a generator or motor, the movable iron part of a relay, or the spring-mounted iron part of a vibrator or buzzer.

**armature chatter** Undesired vibration of the armature of a relay in an a-c circuit.

**armature contact** *Movable contact.*

**armature reaction** Interaction between the magnetic flux produced by armature current and that of the main magnetic field in an electric motor or generator. The resulting distortion of the main magnetic field affects the speed of motors and the voltage regulation of generators.

**armature relay** The commonest type of relay, having a pivoted iron armature that is attracted by a coil and thereby completes the magnetic circuit that passes through the coil. The relay contacts are mounted on the armature or actuated by it through mechanical linkage.

**armature travel** The total distance traveled during relay operation by a point on the armature that is nearest the pole-face center.

**armature voltage control** Control of electric motor speed by changing the voltage applied to its armature windings.

**arming** The act of changing an explosive fuze from a safe condition to a state of readiness for initiation. A fuze may be armed manually, by acceleration, by rotation, by a clock mechanism, or by air travel.

**armored cable** A cable provided with a sheath of metal primarily for mechanical protection.

**Armstrong frequency-modulation system** A phase-shift modulation system developed by E. H. Armstrong. A low-frequency carrier is modulated at a low level, and the signal is passed through several frequency-multiplier amplifier stages to obtain the desired high carrier frequency and high level. The low-level modulation is produced by a balanced modulator that has a quadrature carrier reinserted in its output.

**Armstrong oscillator** A tuned-grid, tuned-anode oscillator circuit developed by E. H. Armstrong. A parallel resonant circuit serves as the required inductive anode load when tuned slightly above the resonant fre-

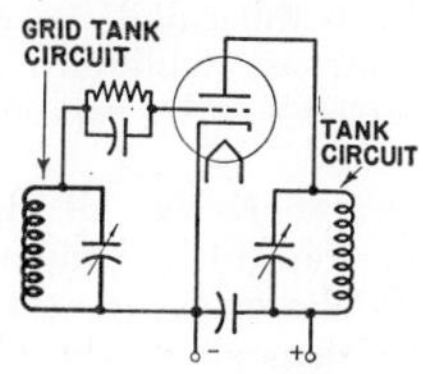


Armstrong oscillator circuit.

quency of the grid tank circuit or crystal. The interelectrode capacitance of the oscillator tube serves as the feedback path.

**array** *Antenna array.*

**arrester** *Lightning arrester.*

**ARRL** American Radio Relay League, an organization of licensed amateur radio operators.

**arsenic** [symbol As] An element. Atomic number is 33.

**articulation** The percentage of speech units understood correctly by a listener in a communication system. Also called intelligibility. Articulation generally applies to unrelated words, as in code messages. Intelligibility is customarily used for regular messages where the context aids the listener's perception.

**artificial antenna** *Dummy antenna.*

**artificial delay line** *Delay line.*

**artificial dielectric** A three-dimensional arrangement of metallic conductors, which are usually small compared to a wavelength. The resulting medium acts as a dielectric to electromagnetic waves.

**artificial ear** A device that presents an acoustic impedance to an earphone equivalent to the impedance presented by the average human ear. It is equipped with a microphone for measuring the sound pressures developed by the earphone.

**artificial earth satellite** A manmade earth satellite, as distinguished from the moon.

**artificial echo** A radar echo signal generated by artificial means for test purposes, as by an echo box.

**artificial horizon** A device that indicates the attitude of an aircraft with respect to the true horizon. Also called flight indicator and gyro horizon.

**artificial line** A network that simulates the electrical characteristic of a transmission line.

**artificial load** A dissipative but essentially nonradiating device having the impedance characteristics of an antenna, transmission line, or other practical load.

**artificial radioactive element** A radioactive element produced from another element, or from an isotope of the same element, by bombarding the element with protons, neutrons, deuterons, gamma rays, or other particles or by exposing it to radiation.

**artificial radioactivity** Radioactivity induced in an element by bombarding it with nuclear particles or exposing it to radiation. Radioactive elements produced in this way usually have a short life, and emit electrons, positrons, other particles, and gamma rays as they decompose. Also called induced radioactivity.

**artificial satellite** A manmade body placed in orbit around the earth, moon, or other solar body. It usually carries recording and transmitting instruments. Also called satellite.

**artificial voice** A small loudspeaker mounted in a baffle that simulates the acoustical constants of the human head. Used for calibrating and testing close-talking microphones.

**ASA** Abbreviation for *American Standards Association.*

**asbestos** A noninflammable fibrous mineral used in electronics for heat-insulating and fireproofing purposes, as in a line-cord resistor.

**A scan** *A display.*

**A scope** 1. A radarscope that produces an A display. 2. *Waveform monitor.*

**asdic** [Anti-Submarine Detection Investigation Committee] British term for sonar and underwater listening devices.

**A signal** A dot-dash signal heard either in a bisignal zone or in an A quadrant of a radio range.

**aspect ratio** The ratio of the frame width to the frame height in television. It is 4 to 3 in the United States and Great Britain.

**asroc** [Anti-Submarine ROCket] A Navy underwater-to-air-to-underwater guided missile system in which an acoustic homing torpedo is launched upward from under water, covers most of the distance to the target through the air under rocket power, then reenters the water to home on an enemy submarine.

**assembly** A number of parts or subassemblies joined together to perform a specific function.

**assembly machine** A machine used for inserting components automatically in printed wiring boards, such as an in-line assembly machine or a single-station assembly machine.

**associated corpuscular emission** The full complement of secondary charged particles associated with an x-ray or gamma-ray beam in its passage through air.

**astable circuit** A circuit that alternates automatically and continuously between two unstable states at a frequency dependent on circuit constants. It can readily be synchronized at the frequency of any repetitive input signal. Blocking oscillators and certain multivibrators are examples.

**astable multivibrator** A multivibrator in

which each tube alternately conducts and is cut off for intervals of time determined by circuit constants, without use of external triggers. Also called free-running multivibrator.

**astatic** Without orientation or directional characteristics; having no tendency to change position.

**astatic galvanometer** A sensitive galvanometer consisting of two small magnetized needles mounted parallel to each other inside the galvanometer coil. The needles reduce errors caused by the earth's magnetic field.

**astatic microphone** *Omnidirectional microphone.*

**astatic wattmeter** An electrodynamic wattmeter designed to be insensitive to uniform external magnetic fields.

**astatine** [symbol At] A radioactive element. Atomic number is 85. Called alabamine before 1940.

**A station** The loran station whose signal is always transmitted more than half a repetition period before the signal from the slave or B station of the pair. Also called master station.

**astigmatism** 1. A type of spherical aberration in which light rays from a single point of an object do not converge at the corresponding point in the image. 2. An electron-beam tube defect in which electrons in the beam come to a focus in different axial planes as the beam is deflected so that the spot on the screen is distorted in shape and the image is blurred.

**ASTM index** A card index published by the American Society for Testing Materials, in which the x-ray powder diffraction lines obtained from a large number of substances are recorded and classified in terms of the three most intense lines. Other information that might assist in the x-ray identification of an unknown substance is also included.

**Aston dark space** The dark space next to the cathode of a glow discharge tube, in which the emitted electrons do not have enough velocity to excite the gas.

**astor** [Anti-Submarine Torpedo Ordnance Rocket] A Navy rocket-powered torpedo having a nuclear warhead that gives a large kill radius against submarines.

**astrionics** The science of electronics as applied to space travel.

**astrodome** A transparent plastic dome projecting from the upper surface of an aircraft, used for celestial navigation.

**astron** A system involving use of a cylindrical sheath of rotating electrons, moving at speeds approaching that of light, to produce plasma pinch. An electron gun having several million electron-volts of energy shoots the plasma beam into a vacuum chamber containing deuterium, to produce controlled thermonuclear reactions.

**astronautics** The science of designing, building, and operating space vehicles.

**astronomical unit** A unit of distance equal to the mean distance of the earth from the sun. One astronomical unit is 92,897,000 miles.

**A supply** The A battery, transformer filament winding, or other voltage source that supplies power for heating the cathode of an electron tube.

**asymmetrical** Not symmetrical.

**asymmetrical-sideband transmission** *Vestigial-sideband transmission.*

**asynchronous** Not synchronous.

**asynchronous computer** A computer in which the performance of any operation starts as a result of a signal that the previous operation has been completed, rather than on signal from a master clock.

**asynchronous machine** An a-c machine whose speed is not proportional to the frequency of the power line.

**asynchronous multiplex** A multiplex transmission system in which two or more transmitters occupy a common channel without provision for preventing simultaneous operation.

**a-t** Abbreviation for *ampere-turns.*

**AT-cut crystal** A quartz crystal slab cut at a 35° angle to the $z$ axis of the mother crystal. Temperature variations have little effect on its natural vibration frequency.

**athermanous** Opaque to infrared radiation.

**Atlas** An Air Force surface-to-surface intercontinental ballistic missile having a range of about 6,000 miles and capable of carrying nuclear warheads. Speed is well above 10,000 mph.

**atm** Abbreviation for *atmosphere.*

**atmosphere** [abbreviated atm] 1. The mixture of gases, chiefly oxygen and nitrogen, that surrounds the earth for a distance of about 100 miles from the surface. Atmospheric pressure at sea level is approximately 14.7 pounds per square inch; this value is often used as a unit of pressure, called 1 atmosphere.

**atmospheric absorption** The attenuating effect of gases and moisture in the earth's atmosphere on the propagation of microwaves.

**atmospheric duct** An atmospheric layer that conducts radio waves in the same manner as a waveguide under certain conditions of temperature and humidity, to give signal transmission far outside the usual reception area. The effect is most noticeable above 3,000 mc.

**atmospheric interference** Static caused by natural electric disturbances such as lightning and northern lights. Heard as crackling and hissing noises in radios. Also called atmospherics and sferics.

**atmospheric noise** Noise heard during radio reception due to atmospheric interference.

**atmospheric radio wave** A radio wave that reaches its destination after reflection from the upper ionized layers of the atmosphere.

**atmospherics** *Atmospheric interference.*

**atom** The smallest particle into which an element can be divided and still retain the chemical properties of that element. There are only 102 known kinds of atoms, each having a different arrangement of electrons and protons.

**atom bomb** *Atomic bomb.*

**atomic** Pertaining to atoms.

**atomic absorption coefficient** The fractional decrease in intensity of a beam of photons or particles, per number of atoms per unit area of an element. It is equal to the linear absorption coefficient divided by the number of atoms per unit volume, or to the mass absorption coefficient divided by the number of atoms per unit mass. For the absorption of photons it is equal to the atomic cross section.

**atomic age** The period characterized by the use of nuclear energy. It may be dated from December 2, 1942, when a uranium pile was first operated successfully in Chicago; from July 16, 1945, when the first experimental atomic bomb was exploded at Alamogordo, New Mexico; or from August 6, 1945, when the first atomic bomb was used in warfare, at Hiroshima, Japan.

**atomic air burst** The explosion of an atomic weapon in air at a height greater than the maximum radius of the fireball.

**atomic battery** *Nuclear battery.*

**atomic-beam frequency standard** A frequency standard that provides one or more precise frequencies derived from an element such as cesium. Frequency determination is accomplished by measuring the nuclear magnetic moment of the element by resonance absorption techniques. When the element is acted on by an applied frequency at its atomic spectrum line frequency, the element absorbs a quantum of energy, indicating coincidence between the two frequencies. The applied frequency thus becomes a standardized frequency, and may be utilized at its fundamental or harmonics for calibration of other equipment.

**atomic bomb** A nuclear bomb in which the explosive consists of a nuclear-fissionable radioactive material such as uranium-235 or plutonium-239. Also called A-bomb, atom bomb, and fission bomb. The other type of nuclear bomb is the hydrogen bomb.

**atomic charge** The electric charge of an ion. It is equal to the number of electrons the atom has gained or lost in its ionization multiplied by the charge on one electron.

**atomic clock** A highly accurate source of frequency or time that depends on the unchanging nuclear resonance of atoms of elements such as cesium when subjected to an r-f electromagnetic field. Accuracy of one part in 10 million is achievable. Another type of atomic clock utilizes the ammonia molecule in a gas maser.

**atomic cross section** The cross section of an atom for a particular process. For the absorption of photons it is equal to the atomic absorption coefficient.

**atomic device** An explosive device that uses active nuclear material to produce a chain reaction when detonated.

**atomic disintegration** Conversion of the nucleus of an atom of one element into that of some other element.

**atomic energy** *Nuclear energy.*

**Atomic Energy Commission** [abbreviated AEC] A civilian government agency established in 1946 to supervise and control the production and use of nuclear-fissionable radioactive materials in the United States.

**atomic energy level** The energy corresponding to one of the stationary states of an isolated atom.

**atomic fission** *Nuclear fission.*

**atomic frequency** The natural vibration frequency of an atom.

**atomic fusion** *Nuclear fusion.*

**atomic hydrogen welding** An a-c arc-welding process in which welding heat is obtained from an arc between two suitable electrodes in an atmosphere of hydrogen. The arc changes atomic hydrogen into molecular hydrogen in a reaction that produces the intense heat required for fusing metals having high melting temperatures.

**atomic mass** The mass of a neutral atom

of a nuclide, usually expressed in atomic mass units. Also called nuclidic mass. Formerly called isotopic mass.

**atomic mass conversion factor** The experimentally determined ratio of the atomic weight unit to the atomic mass unit. Also called mass conversion factor.

**atomic mass unit** [abbreviated amu] A unit of mass equal to one-sixteenth the mass of $O^{16}$, the lightest naturally occurring isotope of oxygen. One amu is $1.657 \times 10^{-24}$ gram. Also called mass unit and physical mass unit.

**atomic migration** The transfer of the valence bond of one atom to another atom in the same molecule.

**atomic nucleus** *Nucleus.*

**atomic number** [symbol Z] The number of elementary positive charges in the nucleus of an atom. It is a different number for each element, and ranges from 1 for hydrogen to 102 for the heaviest known element. For a neutral atom, the atomic number is also the number of electrons outside the nucleus of the atom. In the symbol of a nuclide, the atomic number is given as a subscript preceding the element symbol; thus, in ${}_{26}Fe_{33}{}^{59}$, the atomic number for iron is 26, the neutron number is 33, and the mass number is 59.

**atomic orbital** A wave function assumed to be concentrated about independent atoms.

**atomic photoelectric effect** Ejection of a bound electron from an inner orbit of a gas atom by a photon of light or ultraviolet radiation. Also called photoionization.

**atomic pile** *Nuclear reactor.*

**atomic power** *Nuclear energy.*

**atomic reactor** *Nuclear reactor.*

**atomic refraction** The specific refraction of an element multiplied by its atomic weight.

**atomic scattering factor** A factor representing the efficiency with which x-rays are scattered by a given atom, as compared to the scattering by a single electron under the same conditions.

**atomic spectrum** The spectrum of radiation emitted by an excited atom because of changes within the atom. An atomic spectrum is characterized by more or less sharply defined lines at certain frequencies, representing quanta of energy.

**atomic stopping power** The average energy loss suffered by a particle per atom of the material through which it passes. Also called stopping cross section.

**atomic structure** The internal structure of the atom, including its nucleus and the electrons that surround the nucleus.

**atomic surface burst** The explosion of an atomic weapon at an elevation such that the fireball touches the ground.

**atomic theory** The concept that an atom consists of a central positively charged nucleus having considerable mass but minute dimensions surrounded by a number of electrons moving in orbits at a relatively great distance from the nucleus.

**atomic underground burst** The explosion of an atomic weapon with its center beneath the surface of the ground.

**atomic underwater burst** The explosion of an atomic weapon with its center beneath the surface of the water.

**atomic weapon** A bomb, shell, guided missile, or the like in which the warhead consists of nuclear-fissionable radioactive material such as uranium-235 or plutonium-239. Also called nuclear weapon.

**atomic weight** The relative weight of a neutral atom of an element, based on an atomic weight of 16 for the neutral oxygen atom. On this basis, hydrogen has an atomic weight of 1.0078.

**atomic weight unit** [abbreviated awu] A unit of weight equal to exactly one-sixteenth of the average weight of $O^{16}$, $O^{17}$, $O^{18}$, the three kinds of naturally occurring oxygen atoms. One atomic weight unit is equal to $1.660 \times 10^{-24}$ gram or 1.000272 atomic mass units.

**atom smasher** Any of the accelerators used in nuclear physics, such as a cyclotron or a Van de Graaff generator.

**atran** [Automatic Terrain Recognition And Navigation] A map-matching system for missile navigation. One version uses a tape or film recording of radar or other data for terrain covered in the flight, previously obtained by reconnaissance. A continuous comparison of desired position with actual position is made to provide appropriate corrections.

**atr box** *Atr tube.*

**atr switch** *Atr tube.*

**atr tube** [Anti-Transmit-Receive tube] A gas-filled r-f switching tube used in radar to isolate the transmitter from the antenna during the interval for pulse reception. Also called atr box, atr switch, and anti-tr tube. Deprecated terms are rt box and rt switch.

**attack plotter** An instrument used with navigation and sonar to show, on the screen of a cathode-ray tube, a plot of the course of the ship making an antisubmarine attack, the path of each sonar search beam from the ship, the position of the underwater target when each sound contact is made, the course of the target as successive target positions appear, and the range and bearing of the target.

**attack time** The interval required, after a sudden increase in input signal amplitude to a system or component, to attain a specified percentage (usually 63%) of the ultimate change in amplification or attenuation due to this increase.

**attenuation** A reduction in energy. Attenuation occurs naturally during wave travel through lines, waveguides, spaces, or a medium such as water. It also occurs naturally when nuclear radiation passes through a medium, but may be produced intentionally by inserting an attenuator in a circuit or placing an absorbing device in the path of the radiation. The amount of attenuation is generally expressed in decibels or decibels per unit length.

**attenuation band** The rejection band of a uniconductor waveguide.

**attenuation coefficient** The fraction of a beam of light that is removed by scatter or absorption in passing through unit thickness of a material.

**attenuation constant** [symbol $\alpha$] A rating for a line or medium through which a plane wave of a given frequency is being transmitted. It is the real part of the propagation constant, and is the relative rate of decrease of amplitude of a field component (or of voltage or current) in the direction of propagation, in nepers per unit length. The imaginary part of the propagation constant is the phase constant.

**attenuation distortion** A deviation from uniform amplification or attenuation over a required frequency range.

**attenuation equalizer** An equalizer used to make the total transmission loss of a line or circuit essentially the same for all frequencies in the range being transmitted.

**attenuation factor** 1. The ratio of incident intensity to transmitted intensity for radiation passing through a layer of material. 2. The ratio of input current to output current for a transmission line or network.

**attenuation ratio** The magnitude of the propagation ratio.

**attenuator** An arrangement of fixed and variable resistive elements used to reduce the strength of an r-f or a-f signal a desired adjustable amount without introducing appreciable distortion. It is designed so the impedance of the attenuator will match that of the circuit in which it is connected, regardless of the amount of attenuation introduced. The corresponding nonadjustable device is usually called a pad.

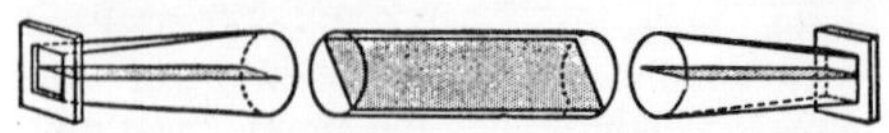

Attenuator for insertion in rectangular waveguide, consisting of rotatable resistive film mounted between two fixed resistive films. When all three films are in same plane, attenuation is zero.

**attenuator tube** A gas-filled r-f switching tube in which a gas discharge, initiated and regulated independently of r-f power, is used to control this power by reflection or absorption.

**attitude** The position of an aircraft as determined by the inclination of its axes to some frame of reference, such as the earth.

**attitude control** A control system or mechanism, as an automatic pilot, that puts or keeps an aircraft or missile in a desired attitude.

**attitude gyro** A gyro used in attitude control of aircraft.

**attitude indicator** An indicator that shows the roll and pitch angles of an aircraft in relation to the earth.

**audibility** A measure of the strength of a specified sound as compared with the strength of a sound that can just be heard. Usually expressed in decibels.

**audibility limit** A threshold of hearing. The lower limit is the minimum effective sound pressure that can be heard at a specified frequency. The upper limit is the minimum effective sound pressure that causes pain in the ear.

**audibility threshold** The lower limit of audibility. For a specified signal, it is the minimum effective sound pressure capable of producing an auditory sensation in a specified fraction of the trials. The characteristics of the signal, the manner in which it is presented to the listener, the point at which the sound pressure is measured, and the ambient noise all affect the value. Usually expressed in decibels above 0.0002 microbar or above 1 microbar.

**audible** Capable of being heard by the average human ear. The approximate range of human hearing is between 15 and 20,000 cps, but actual limits vary greatly with different individuals.

**audible doppler enhancer** A doppler circuit that serves as a third detector for the received pulses. It separates the envelope from the pulses, thus giving an audio signal that can be amplified and fed to a loudspeaker.

**audio** [Latin for "I hear"] 1. Pertaining to signals, equipment, or phenomena involving frequencies in the range of human hearing. 2. Slang for *sound.*

**audio amplifier** *Audio-frequency amplifier.*

**audio frequency** [abbreviated a-f] A frequency that can be detected as a sound by the human ear. The range of audio frequencies extends approximately from 15 to 20,000 cps. Also called sonic frequency and sound frequency.

**audio-frequency amplifier** An amplifier having one or more electron-tube or transistor amplifier stages designed for amplifying an a-f signal. In a superheterodyne receiver it follows the second detector, and serves to amplify the a-f signal after demodulation. Also used separately to amplify the a-f output of a microphone, phonograph, magnetic tape recorder, or other a-f signal source. Also called audio amplifier.

**audio-frequency harmonic distortion** Distortion in which integral multiples of a single a-f input signal are generated by the amplifier.

**audio-frequency oscillator** An oscillator circuit using an electron tube, transistor, or other nonrotating device to produce an a-f alternating current. Also called audio oscillator.

**audio-frequency peak limiter** A circuit used in an a-f system to cut off signal peaks that exceed a predetermined value. Also called audio peak limiter.

**audio-frequency shift modulation** A facsimile system in which picture tones are represented by audio frequencies. In one example a 1,500-cps tone represents black, a 2,300-cps tone represents white, and frequencies in between represent shades of gray.

**audio-frequency signal generator** A signal generator that can be set to generate a sinusoidal a-f signal voltage at any desired frequency in the audio spectrum. Also called audio signal generator.

**audio-frequency transformer** An iron-core transformer used for coupling between a-f circuits. Also called audio transformer.

**audiogram** A graph showing hearing loss, per cent hearing loss, or per cent hearing as a function of frequency.

**audio level meter** An instrument that measures a-f power with reference to a predetermined level. Its scale is usually calibrated in decibels.

**audio masking** *Masking.*

**audiometer** An instrument used to measure hearing ability. In one form it consists of an audio oscillator having variable calibrated output and capable of generating a wide range of audio tone frequencies. Recorded speech sounds may also be used.

**audiometry** The study of hearing ability by means of audiometers.

**audion** The original three-element vacuum tube invented by Dr. Lee de Forest.

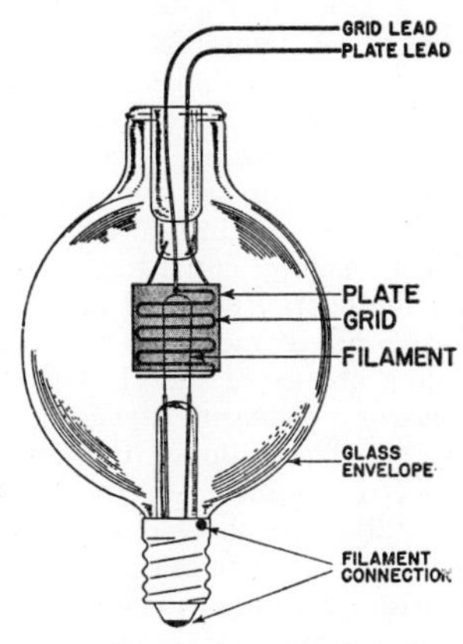


Audion tube construction.

**audio oscillator** *Audio-frequency oscillator.*

**audio peak limiter** *Audio-frequency peak limiter.*

**audiophile** A person interested in listening to broadcasts and recordings that are reproduced with high fidelity.

**audio signal** An electric signal having an audio frequency.

**audio signal generator** *Audio-frequency signal generator.*

**audio spectrum** The continuous range of frequencies extending from the lowest to the highest audio frequency (from about 15 to 20,000 cps).

**audio subcarrier** A subcarrier whose frequency lies within the audio range.

**audio system** *Sound-reproducing system.*

**audio taper** A taper that gives a semilogarithmic change in resistance with rotation. Used in tone controls, to compensate for the fact that the frequency range of the human ear is less at low volume levels.

**audio transformer** *Audio-frequency transformer.*

**audition** A preliminary studio test of a performer, act, or complete program for a television or radio show.

**auditory perspective** Three-dimensional

realism of sound, as produced by an actual orchestra or by a stereophonic sound system.

**auditory sensation area** The region enclosed by curves defining the threshold of feeling and the threshold of audibility as functions of frequency.

**Auger coefficient** The ratio of Auger yield to fluorescence yield, or the ratio of the number of Auger electrons to the number of x-ray photons emitted from a large number of similarly excited atoms.

**Auger effect** A nonradiative transition of an atom from an excited energy state to a lower energy state, accompanied by the emission of an electron.

**Auger electron** An electron ejected from an atom by a photon in the Auger effect. It has a kinetic energy equal to the difference between the energy of the x-ray photon of the corresponding radiative transition and the binding energy of the ejected electron.

**Auger shower** *Extensive shower.*

**Auger yield** The ratio of the number of Auger electrons emitted to the number of events resulting in an electron vacancy in the inner shell of an atom.

**augmentation distance** The distance between the extrapolated boundary and the true boundary of a nuclear reactor.

**aural** Pertaining to the ear or the sense of hearing.

**aural center frequency** The average frequency of a carrier modulated by an a-f signal or the carrier frequency without modulation.

**aural harmonic** A harmonic generated in the human ear.

**aural masking** *Masking.*

**aural null** The condition of weakest sound when tuning or otherwise adjusting a circuit having an audio output.

**aural-null direction finder** A radio direction finder consisting of a radio receiver and rotatable loop antenna. When the loop is rotated to give an aural null, the plane of the loop is at right angles to the direction of the transmitted signal.

**aural radio range** A radio range station providing lines of position by virtue of aural identification or comparison of signals at the output of a receiver, as in the A-N radio range.

**aural signal** 1. A signal that can be heard. 2. The sound portion of a television signal; the picture portion is called the visual signal.

**aural transmitter** The radio equipment used to transmit the aural portion of a television program. The aural and visual transmitters together are called a television transmitter.

**auroral reflection** Reflection of radio waves back to earth by a rapidly fluctuating ionized layer in the upper atmosphere of the polar regions, usually during magnetic storms accompanied by auroral displays.

**auroral storm** *Ionospheric storm.*

**autoalarm** *Automatic alarm receiver.*

**autocorrelation function** A mathematical quantity defined as the time average of the product of a function of time and a delayed version of that function of time.

**autocorrelator** A correlator in which the input signal is delayed, then multiplied by the undelayed signal. The product is then smoothed in a low-pass filter to give an approximate computation of the autocorrelation function. It can be used to detect a weak periodic signal hidden in noise, if the chosen time delay is equal to the period of the signal. The autocorrelator may also be applied to the detection of nonperiodic signals.

**autocovariance function** The covariance between $X(t)$ and $X(t+r)$ as a function of the lag $r$. If averages of these two terms are zero, the covariance is equal to the average value of their product.

**autodyne circuit** A circuit in which the same tube elements serve as oscillator and detector simultaneously. The output frequency is equal to the difference between the frequencies of the received signal and the oscillator signal, just as in a conventional superheterodyne circuit.

**autoelectric emission** The emission of electrons from a cold cathode under the influence of an intense electric field.

**automatic** Having a self-acting mechanism that performs a required act at a predetermined time or in response to certain conditions.

**automatic aiming** *Automatic tracking.*

**automatic alarm receiver** A complete receiving, selecting, and warning device capable of being actuated automatically by intercepted r-f waves forming the international automatic alarm signal. Also called autoalarm.

**automatic-alarm-signal keying device** A device capable of keying the radiotelegraph transmitter on board a vessel automatically to transmit the international automatic alarm signal.

**automatic back bias** Use of one or more automatic gain control loops in a radar receiver to prevent overloading by strong radar echoes or jamming signals.

**automatic background control** *Automatic brightness control.*

**automatic bass compensation** A circuit used in some radio receivers and audio amplifiers to make bass notes sound more

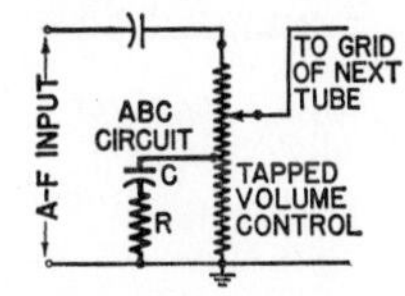


Automatic bass compensation.

natural at low volume-control settings. The circuit usually consists of a resistor and capacitor in series, connected between ground and a tap on the volume control. This circuit automatically compensates for the poor response of the human ear to weak low-frequency sounds.

**automatic bias** *Self-bias.*

**automatic brightness control** [abbreviated abc] A circuit used in a television receiver to keep the average brightness of the reproduced image essentially constant. Its action is like that of an automatic volume control circuit in a sound receiver. Also called automatic background control.

**automatic carrier landing system** A combination radio and radar system used in landing aircraft on carriers during adverse weather conditions. Radar is used to locate the aircraft and determine its exact position at each instant with respect to the landing deck. A computer converts the radar data into course correction data that is fed to the approaching aircraft by a radio transmitter. If the approach is incorrect, the system automatically waves off the aircraft.

**automatic C bias** *Self-bias.*

**automatic chart-line follower** A device that automatically derives error signals proportional to the deviation of the track of a vehicle from a predetermined course line drawn on a chart. Also called automatic track follower.

**automatic check** A check performed by equipment built into a computer specifically for that purpose, and automatically accomplished each time the pertinent operation is performed.

**automatic chrominance control** A color television receiver circuit that automatically controls the amplitude of the chrominance signal.

**automatic computer** A computer that performs long sequences of operations in accordance with a predesigned program, without human intervention.

**automatic contrast control** A circuit that varies the gain of the r-f and video i-f amplifiers in such a way that the contrast of the television picture is maintained at a constant average level. Control is achieved by varying the bias on one or more variable-mu tubes. The manual contrast control determines the average level and the automatic contrast control maintains this average, despite variations in signal strength as different stations are tuned in.

**automatic control** Control in which regulating and switching operations are performed automatically in response to predetermined conditions.

**automatic controller** An instrument that continuously measures the value of a variable quantity or condition, then automatically acts on the controlled equipment to correct any deviation from a desired preset value. Also called controller.

**automatic control system** A control system having one or more automatic controllers connected in closed loops with one or more processes. Also called regulating system.

**automatic cutout** A device, usually operated by centrifugal force or by an electromagnet, that automatically shorts part of a circuit at a particular time. Used on some induction motors to cut out the starting winding when operating speed is reached.

**automatic degaussing control system** A system that automatically changes the degaussing current for a ship to compensate for changes in the ship's heading. A synchro transmitter installed in the gyrocompass equipment of the ship provides a signal that is rectified and amplified for use in controlling the degaussing current source.

**automatic direction finder** [abbreviated adf] A direction finder that automatically and continuously indicates the direction of arrival of a radio signal. The directional antenna is automatically aimed in the direction of minimum signal from the radio station being used for navigation guidance, by means of electronic circuits used alone or in conjunction with a motor drive for the antenna. Modern versions are generally used with an additional fixed antenna to provide a unidirectional bearing indication telling whether the aircraft or ship is traveling toward or away from the radio station. Also called automatic loop radio compass, automatic radio compass, automatic radio direction finder, and radio compass.

**automatic flight control** A control system that includes an automatic pilot and the additional equipment needed to maintain automatically a desired aircraft altitude and heading.

**automatic focusing** Electrostatic focusing in which the focusing anode of a television picture tube is internally connected through a resistor to the cathode so that no external focusing voltage is required.

**automatic frequency control** [abbreviated afc] 1. A circuit used to maintain the frequency of an oscillator within specified limits, as in a transmitter. 2. A circuit used to keep a superheterodyne receiver tuned accurately to a given frequency by controlling its local oscillator, as in an f-m receiver. 3. A circuit used in radar superheterodyne receivers to vary the local oscillator frequency so as to compensate for changes in the frequency of the received echo signal. 4. A circuit used in television receivers to make the frequency of a sweep oscillator correspond to the frequency of the synchronizing pulses in the received signal.

**automatic gain control** [abbreviated agc] A control circuit that automatically changes the gain (amplification) of a receiver or other piece of equipment so the desired output signal remains essentially constant despite variations in input signal strength.

**automatic grid bias** *Self-bias.*

**automatic loop radio compass** *Automatic direction finder.*

**automatic machine equipment** Equipment that provides automatic control for any type of rotating machine or rectifier.

**automatic modulation control** A transmitter circuit that reduces the gain for excessively strong audio input signals without affecting the strength of normal signals. This permits higher average modulation without overmodulation, equivalent to an increase in carrier-frequency power output.

**automatic peak limiter** *Limiter.*

**automatic phase control** A circuit used in color television receivers to keep both frequency and phase of the 3.58-mc color oscillator synchronized with the burst signal.

**automatic pilot** *Autopilot.*

**automatic programing** Any technique whereby a digital computer itself transforms programing from a form that is easy for a human being to produce into a form that is efficient for the computer to carry out.

**automatic radio compass** *Automatic direction finder.*

**automatic radio direction finder** *Automatic direction finder.*

**automatic record changer** An electric phonograph that automatically plays a number of records one after another.

**automatic-scanning receiver** A receiver that automatically sweeps back and forth through a preselected frequency range. It may be set to plot signal occupancy in the range or stop when a signal is found.

**automatic-search jammer** A search receiver and jamming transmitter used together to search for and automatically jam enemy signals having specified radiation characteristics.

**automatic selectivity control** A circuit that makes a receiver less selective when the received signal is strong and more selective when the signal is weak. The reduction in the anode-cathode resistance of a tube, when handling a strong signal, is used to damp a tuned circuit and thereby make the receiver less selective.

**automatic sensitivity control** A circuit used for maintaining receiver sensitivity at a predetermined level.

**automatic sequencing** Ability of a computer to perform successive operations without human intervention.

**automatic shutoff** A switch incorporated in some tape recorders to stop the machine when the tape breaks or runs out.

**automatic track follower** *Automatic chart-line follower.*

**automatic tracking** Tracking in which a servomechanism keeps the radar beam trained on the target. The servomechanism is actuated by circuits that respond to some characteristic of the echo signal from the target. Also called automatic aiming and autotrack.

**automatic tuning system** An electric, mechanical, or electromechanical system that tunes a radio receiver or transmitter automatically to a predetermined frequency when a button or lever is pressed, a knob

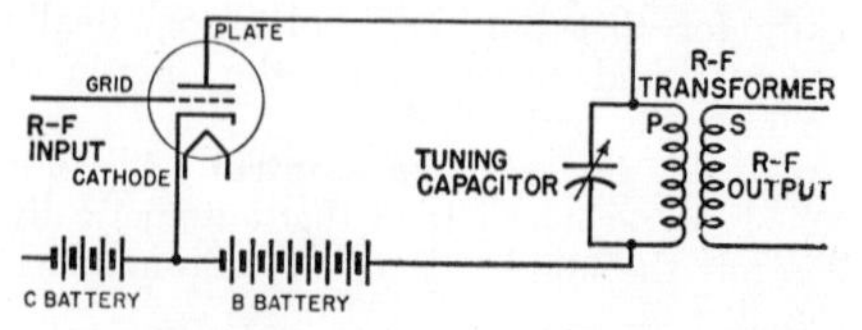


Automatic tuning system using latching switches controlled by pushbuttons.

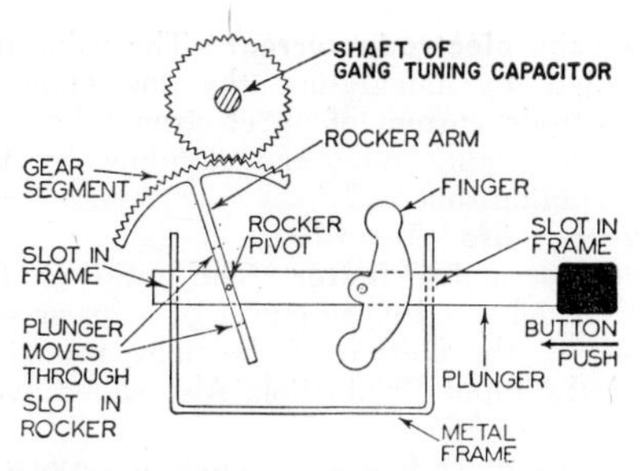

Automatic tuning system using mechanical gear and finger arrangement. Widely used in auto radios.

is turned, or a telephone-type dial is operated.

**automatic voltage regulator** *Voltage regulator.*

**automatic volume compressor** *Volume compressor.*

**automatic volume control** [abbreviated avc] An automatic gain control that keeps the output volume of a radio receiver essentially constant despite variations in input signal strength during fading or when

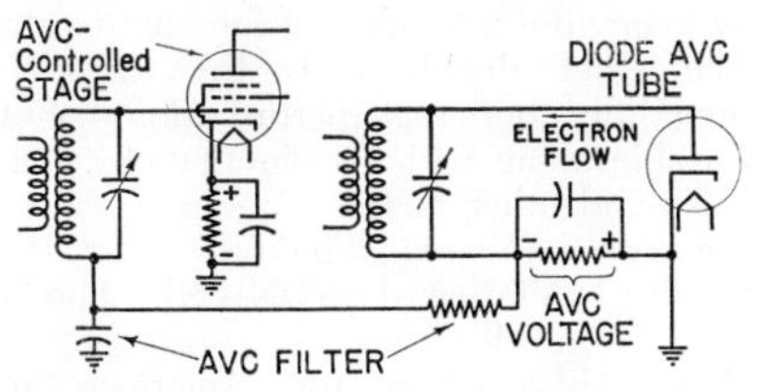

Automatic volume control circuit for radio receiver, in which the diode detector provides the avc voltage.

tuning from station to station. A d-c voltage proportional to audio output signal strength is obtained from the second detector and used to change the bias of one or more preceding r-f and i-f amplifier stages.

**automatic volume expander** *Volume expander.*

**automation** Continuous automatic operation in which control functions are performed by mechanisms instead of men.

**autopilot** An arrangement of gyroscopes combined with amplifiers and servomotors to detect deviations in the flight of an aircraft and apply the required corrections directly to the controls. Also called automatic pilot.

**autopilot coupler** A coupling system used to link the output of the navigation system receiver to the automatic pilot in an aircraft.

**autoradar plot** *Chart comparison unit.*

**auto radio** A radio receiver designed to be installed in an automobile and operated from the automobile's storage battery.

**autoradiograph** A photographic record of radiation from radioactive material in an object, made by placing the object against photographic film. Also called radioautograph (deprecated).

**autoregulation induction heater** An induction heater in which a desired control is effected by the change in characteristics of a magnetic charge as it is heated at or near its Curie point.

**Autosyn** [AUTOmatically SYNchronous] Trademark of Bendix Aviation Corp. for their line of synchros.

**autotrack** *Automatic tracking.*

**autotransductor** British term for a magnetic amplifier in which the same windings serve both as control windings and as power windings.

**autotransformer** A power transformer having one continuous winding that is tapped. Part of the winding serves as the primary and all of it serves as the secondary, or vice versa.

**auxiliary anode** An anode located adjacent to the pool cathode in an ignitron. It helps the main anode to maintain a cathode spot.

**auxiliary relay** A relay that assists another relay or device in the performance of a function.

**auxiliary transmitter** A transmitter maintained only for transmitting the regular programs of a station in case of failure of the main transmitter.

**available conversion gain** The ratio of the available output-frequency power of a conversion transducer to the available input-frequency power from the driving generator.

**available line** The portion of the length of a facsimile scanning line that can be used specifically for picture signals. Also called useful line.

**available power** The maximum power that can be delivered to a load. At a specified frequency, maximum power transfer takes place when the impedance of the load is equal to that of the source and the sign of the load reactance is opposite that of the source reactance.

**available power efficiency** The ratio of the electric power delivered by a microphone to the acoustic power reaching the microphone.

**available power gain** The ratio of the

signal power available from the output of a transducer to the signal power available from the input source. This gain is a maximum at a specified frequency when the transducer is conjugately matched to source and load.

**available power response** The ratio of the mean-square sound pressure 1 meter from a loudspeaker or other electroacoustic transducer to the electric power available from the source. Usually expressed in decibels.

**avalanche** 1. The cumulative process in which an electron or other charged particle accelerated by a strong electric field collides with and ionizes gas molecules, thereby releasing new electrons which in turn have more collisions. The discharge is thus self-maintained. Also called avalanche effect, cascade, cumulative ionization, Townsend avalanche, and Townsend ionization. 2. Cumulative multiplication of carriers in a semiconductor as a result of avalanche breakdown. Also called avalanche effect.

**avalanche breakdown** Nondestructive breakdown in a semiconductor diode when the electric field across the barrier region is strong enough so current carriers collide with valence electrons to produce ionization and cumulative multiplication of carriers. Often confused with zener breakdown, in which the electric field across the barrier region becomes high enough to produce a form of field emission that suddenly increases the number of carriers in this region.

**avalanche diode** A silicon diode that has a high ratio of reverse to forward resistance until avalanche breakdown occurs. After breakdown the voltage drop across the diode is essentially constant and independent of current. Used for voltage regulating and voltage limiting. Originally called zener diode, before it was found that the Zener effect had no significant role in the operation of diodes of this type. Also called breakdown diode.

**avalanche effect** *Avalanche.*

**avalanche noise** Noise produced when a p-n junction diode is operated at the onset of avalanche breakdown.

**avc** Abbreviation for *automatic volume control.*

**average calculating operation** A computer calculating operation that is longer than an addition and shorter than a multiplication. Often taken as the mean of nine additions and one multiplication.

**average electrode current** The value obtained by integrating the instantaneous electrode current of an electron tube over an averaging time and dividing by the averaging time.

**average life** *Mean life.*

**average noise factor** The ratio of the total delivered noise power of a linear system to the portion of the noise produced by the input termination. Also called average noise figure.

**average noise figure** *Average noise factor.*

**average pulse amplitude** The average of the instantaneous amplitudes taken over the duration of a pulse.

**average speech power** The average of instantaneous speech power values during a given time interval.

**average value** The average of many instantaneous amplitude values taken at equal intervals of time during half a cycle of alternating current. For a sine wave, the average value is 0.637 times the peak value.

**averaging** A means for improving the precision of measurement of a given quantity by averaging a number of measured values. The chief drawback is the danger of smoothing out real perturbations of the variable along with the fluctuations due to noise and other factors.

**avigation** *Air navigation.*

**avionics** [AVIation electrONICS] The field of airborne electronics.

**AWG** Abbreviation for *American wire gage.*

**A wind** Magnetic tape wound on the reel with the dull oxide-coated side of the tape toward the inside. This wind is now almost universally used. Recorder design determines whether A or B wind tape is required.

**awu** Abbreviation for *atomic weight unit.*

**axial lead** A wire lead coming out from the end along the axis of a resistor, capacitor, or other component.

**axial ratio** The ratio of the major axis to the minor axis of the polarization ellipse of a waveguide. Also called ellipticity (deprecated).

**axiotron** An axially controlled magnetron, in which the magnetic field produced by the filament current serves to control the anode current.

**Ayrton shunt** A shunt used to increase the range of a galvanometer without changing the damping. Also called universal shunt.

**azel indicator** *Expanded plan position indicator.*

**azel scope** A cathode-ray tube on which both azimuth and elevation of a target are presented at the same time. Used in precision approach systems such as gca.

**azimuth** *Bearing.*

**azimuth marker** A radar receiver circuit used to produce a bright radial line on a ppi display, at an angle that can be adjusted by means of a control dial so the line passes through a target indication on the screen.

**azimuth rate** Rate of change of true bearing.

**azimuth rate computer** A computer that calculates the rate of change of horizontal angular measurements from a base line.

**azimuth-stabilized ppi** A plan position indicator that is stabilized by a gyrocompass so either true or magnetic north is always at the top of the scope regardless of equipment orientation. Also called north-stabilized ppi.

**azon** [AZimuth ONly] A bomb having movable control surfaces in the tail that can be adjusted by radio to control the bomb in azimuth.

**azusa** A guidance or range instrumentation system in which directive antennas and phase comparison (coherent carrier) techniques provide angle determination, while multichannel subcarriers provide range measurements by means of time delay. The equipment includes an elaborate ground antenna array, a transmitting and receiving station, and a missile-born transponder.

# B

**B** Symbol for *base*. Used on transistor circuit diagrams.
***B*** Symbol for *magnetic induction*.
**B−** [B minus] The negative terminal of a B battery or other anode voltage source for an electron tube, or the anode circuit terminal to which this negative source terminal should be connected.
**B+** [B plus] The positive terminal of a B battery or other anode voltage source for an electron tube, or the anode circuit terminal to which this positive source terminal should be connected.
**babble** The aggregate crosstalk from a large number of channels.
**babs** [Blind Approach Beacon System] A pulse-type ground-based navigation beacon used for runway approach at airports. When interrogated by an aircraft making an instrument approach, the beacon sends out signals that produce range and runway position information on the L-scan cathode-ray indicator in the aircraft. Also called beam approach beacon system (British).
**back conductance** The inverse of the back resistance of a contact rectifier.
**back contact** A normally closed stationary contact on a relay. Its circuit is opened when the relay is energized. Also called break contact.
**back echo** An echo signal produced on a radar screen by one of the minor back lobes of a search radar beam.
**backed stamper** A thin metal stamper attached to a solid backing material for use in molding phonograph records.
**back electromotive force** *Counterelectromotive force*.
**back emission** *Reverse emission*.
**backfire** *Arcback*.
**background** Ever-present effects in physical apparatus above which a phenomenon must manifest itself in order to be measured or observed.
**background count** A count caused by ionizing radiation coming from sources other than that being measured.
**background music** Music used as an accompaniment to a performance.
**background noise** Undesired noise heard along with a desired program.
**background response** The response caused by ionizing radiation coming from sources other than that to be measured by a radiation detector.
**background return** *Clutter*.
**backheating** Heating of a magnetron cathode by the high-velocity electrons that return to the cathode surface. Once the tube is in operation, backheating is often sufficient to keep the cathode at emitting temperature without heater current.
**back lobe** A radiation pattern lobe that is directed away from the intended direction.
**backplate** The electrode to which the stored charge image of a camera tube is capacitively coupled.
**back porch** The portion of a composite picture signal that follows the horizontal sync pulse and extends to the trailing edge of the corresponding blanking pulse. The color burst, if present, is not considered part of the back porch.
**back-porch effect** The continuation of collector current in a transistor for a short time after the input signal has dropped to zero. The effect is due to storage of minority carriers in the base region. The effect also occurs in junction diodes.
**back resistance** The contact resistance that opposes the inverse current of a contact rectifier.
**backscattering** 1. Deprecated term for radar echoes, which may include both clutter and desired echoes from a target. 2. Deflection of radiation or nuclear particles by scattering processes through angles greater than 90° with respect to the original direction of travel. 3. Undesired radiation of energy to the rear by a directional antenna.
**backscattering coefficient** The ratio of reflected power to incident power for a plane wave. Also called echoing area.
**back-shunt keying** A method of keying a transmitter in which the r-f energy is fed to the antenna when the telegraph key is closed and to an artificial load when the key is open.

**backstop** The part of a relay that limits the movement of the armature away from the core.

**back-to-back connection** A method of connecting a pair of tubes so that each operates on one half of an a-c cycle, thereby permitting control of alternating current. The anode of one tube is connected to the cathode of the other, and vice versa. Used chiefly for thyratrons and ignitrons.

**back-to-back repeater** A repeater in which the output of the receiver is connected directly to the input of the transmitter.

**back-wall photovoltaic cell** A photovoltaic cell constructed so light must pass through a transparent metal electrode and a semiconductor layer before reaching the barrier layer.

**backward wave** A wave traveling opposite to the normal direction, such as a wave whose group velocity is opposite to the direction of electron-stream motion in a traveling-wave tube, or the reflected wave in a mismatched transmission line.

**backward-wave magnetron** A magnetron oscillator in which the electron beam travels in a direction opposite to the flow of r-f energy. Characterized by high power output, and can be voltage-tuned over a wide band.

**backward-wave oscillator** An oscillator using a special vacuum tube in which electrons are bunched by an r-f magnetic field as they flow from cathode to anode.

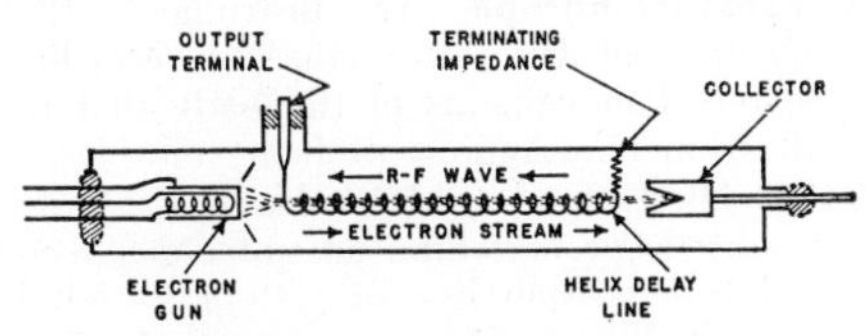


Backward-wave oscillator construction.

This bunching action is such as to produce a backward wave that becomes larger as it progresses toward the electron-gun end of the tube. The magnetic field is produced by a folded-line structure centered on the axis of the electron beam. The output signal is taken from the gun end of this folded line.

**backward-wave tube** A traveling-wave tube in which the electrons travel in a direction opposite to that in which the wave is propagated.

**back wave** *Spacing wave.*

**baffle** 1. A cabinet or partition used with a loudspeaker to increase the effective length of the air path from the front to the rear of the moving diaphragm. By reducing interaction between sound waves produced simultaneously by the two surfaces of the diaphragm, a baffle improves the fidelity of reproduction. 2. An auxiliary member placed in the arc path of a gas tube and having no external connection. It may be used for controlling the flow of mercury particles or deionizing the mercury vapor following conduction.

**baffle plate** A metal plate inserted in a waveguide to reduce the cross-sectional area for wave conversion purposes.

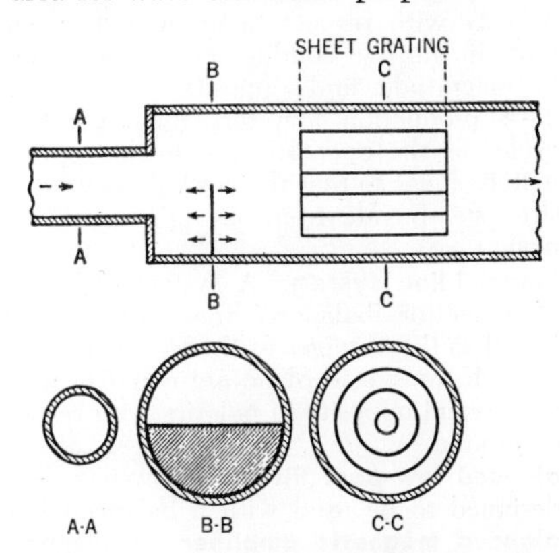


Baffle plate at B-B, used with sheet grating at C-C consisting of concentric cylinders to convert transverse magnetic wave entering from left to transverse electric wave leaving at right.

**Bakelite** Trademark of Bakelite Corp. for their plastics and resins. Originally it applied only to their phenolic compound, widely used in radio parts because of its excellent insulating qualities.

**balance control** A control used in a stereo sound system to vary the volume of one loudspeaker system relative to the other while maintaining their combined volume essentially constant.

**balanced amplifier** An amplifier circuit in which there are two identical signal branches connected to operate in phase opposition, with input and output connections each balanced to ground. A push-pull amplifier is an example.

**balanced circuit** A circuit whose two sides are electrically alike and symmetrical with respect to a common reference point, usually ground.

**balanced converter** *Balun.*

**balanced currents** Currents flowing in the two conductors of a balanced line which, at every point along the line, are equal in magnitude and opposite in direction. Also called push-pull currents.

**balanced detector** A detector used in f-m receivers. In one form the audio output is the rectified difference between the voltages produced across two resonant circuits, one circuit being tuned slightly above the carrier frequency and the other slightly below.

**balanced line** 1. A transmission line consisting of two conductors capable of being operated in such a way that the voltages of the two conductors at any transverse plane are equal in magnitude and opposite in polarity with respect to ground. The currents in the two conductors are then equal in magnitude and opposite in direction. 2. A production line for which the time cycles of the operators are made approximately equal so that the work flows at a desired steady rate from one operator to the next.

**balanced-line system** A system consisting of generator, balanced line, and load adjusted so the voltages of the two conductors at each transverse plane are equal in magnitude and opposite in polarity with respect to ground.

**balanced low-pass filter** A low-pass filter designed to be used with a balanced line.

**balanced magnetic amplifier** A magnetic amplifier formed by mixing the outputs of two identical single-ended magnetic amplifiers in such a way that the output polarity can be reversed. The bias for each single-ended amplifier can be chosen to give class A, class B, or class C operation for the balanced amplifier.

**balanced mixer** A hybrid junction with crystal detectors in one pair of uncoupled arms, the arms of the remaining pair being fed from a signal source and a local oscillator. The resulting i-f signals from the crystals are added in such a manner that the effect of local-oscillator noise is minimized.

**balanced modulator** A modulator in which the carrier and modulating signal are introduced in such a way that the output contains the two sidebands without the carrier. Used in color television transmitters to apply the I and Q signals to the subcarriers, as well as in suppressed-carrier communication transmitters.

**balanced oscillator** An oscillator in which the impedance centers of the tank circuits are at ground potential and the voltages between either end and their centers are equal in magnitude and opposite in phase. A push-pull oscillator is an example.

**balanced termination** A two-terminal load in which both terminals present the same impedance to ground.

**balanced transmission line** *Twin-line.*

**balanced voltages** Voltages that are equal in magnitude and opposite in polarity with respect to ground. Also called push-pull voltages.

**balance method** *Null method.*

**ballast resistor** A resistor that increases in resistance when current increases, thereby maintaining essentially constant current despite variations in line voltage. Also called barretter (British).

**ballast tube** A ballast resistor mounted in an evacuated glass or metal envelope like that of a vacuum tube, to reduce radiation of heat from the resistance element and thereby improve the voltage-regulating action.

**ballistic element** The element in a fire-control system that provides the missile trajectory data required by other elements of the system.

**ballistic galvanometer** An instrument that measures the total quantity of electricity in a transient current such as the discharge current of a capacitor.

**ballistic missile** A missile that is guided during powered flight in the upward part of its trajectory but becomes a free-falling body in the latter stages of its flight toward its target. The German V-2 was an example.

**ballistocardiograph** An instrument that measures cardiac performance by recording the recoil movements of the body that result from contractions of heart muscles.

**balun** [BALanced to UNbalanced] A device used for matching an unbalanced coaxial transmission line or system to a balanced two-wire line or system. It is a quarter-wavelength cylindrical sleeve placed over the end of a coaxial cable feed to an antenna, and serves to isolate the outer conductor of the cable from ground. Also called balanced converter, bazooka (slang), and line-balance converter.

**banana jack** A jack that accepts a banana plug. Generally designed for panel mounting.

**banana plug** A plug having a spring-metal tip somewhat resembling a banana, used on test leads or as terminals for plug-in components.

**band** 1. A range of frequencies between two definite limits. By international agreement, the eight bands that cover the radio spectrum are numbered as follows:

| No. | Band Name and Range |
|---|---|
| 4 | vlf (very low frequency): 10–30 kc |
| 5 | l-f (low frequency): 30–300 kc |
| 6 | m-f (medium frequency): 300–3,000 kc |
| 7 | h-f (high frequency): 3–30 mc |
| 8 | vhf (very high frequency): 30–300 mc |
| 9 | uhf (ultrahigh frequency): 300–3,000 mc |
| 10 | shf (superhigh frequency): 3,000–30,000 mc |
| 11 | ehf (extremely high frequency): 30,000–300,000 mc |

2. A group of tracks on a magnetic drum in a computer. 3. The set of closely spaced spectral lines produced by molecules of one kind when there is a transition between two electronic-vibrational states possessing rotational fine structure.

**band-break** A discontinuity in the crystal unit characteristic correlating frequency and/or activity with temperature. It is usually the result of close coupling to other modes of vibration or of resonances in the mounting system.

**band-edge energy** The energy of the edge of the conduction band or valence band in a solid. It is the minimum energy needed by an electron to move freely in a semiconductor or the maximum energy it may have as a valence electron.

**band-elimination filter** A filter that attenuates alternating currents whose frequencies are between given upper and lower cutoff values while transmitting frequencies above and below this band. It is the opposite of a bandpass filter. The band rejected is generally much wider than that suppressed by a trap. Also called band-rejection filter, bandstop filter, and rejector circuit.

**band-ignitor tube** A glow-discharge tube in which conduction is initiated by applying a high voltage between the cathode and a metal band wrapped around the envelope.

**band-limited function** A function whose Fourier transform is very small or vanishes outside some finite interval.

**bandpass amplifier** An amplifier designed to pass a definite band of frequencies with essentially uniform response.

**bandpass filter** A filter that transmits alternating currents whose frequencies are between given upper and lower cutoff values, while substantially attenuating all frequencies outside this band.

**bandpass response** A response characteristic in which a definite band of frequencies is transmitted with essentially uniform response. In i-f transformers, this response is obtained by tuning the primary and secondary resonant circuits to slightly different frequencies. The response curve then usually has two humps. Also called double-hump response and flat-top response.

**band pressure level** The effective sound pressure level of the sound energy contained within a specified frequency band.

**band-rejection filter** *Band-elimination filter.*

**band selector** A switch used to select any one of the bands in which a receiver, signal generator, or transmitter is designed to operate. It usually has two or more sections, to make the required changes in all tuning circuits simultaneously. Also called band switch.

**B and S gage** Abbreviation for *Brown and Sharpe gage.*

**band spectrum** A spectrum giving the appearance of bands rather than separate lines. Usually applied to the spectra of molecules, even though the bands of most molecules have now been resolved into their separate lines. In nuclear physics, band spectra are useful in determining nuclear spin, nuclear statistics, and isotopic abundances.

**bandspread tuning control** A separate tuning control provided on some receivers to spread stations in a single band of frequencies over an entire tuning dial. It controls small variable capacitors that are connected in parallel with each main tuning capacitor.

**bandstop filter** *Band-elimination filter.*

**band switch** *Band selector.*

**bandwidth** 1. The width of a band of frequencies used for a particular purpose. Thus, the bandwidth of a television station is 6 mc. 2. The range of frequencies within which a performance characteristic of a device is above specified limits. For filters, attenuators, and amplifiers these limits are generally taken to be 3 decibels below the average level. Half-power points are also used as limits.

**bang-bang control** A missile or pilotless airplane control system in which corrective action is always applied to the full extent of servomotion.

**bank** 1. A number of similar devices connected together for use as a single device.

A bank of resistors is an example. 2. A planned accumulation of subassemblies stored at a point on a production line to permit reasonable fluctuations in line speed before and after the bank.

**banked winding** A method of winding r-f coils in which single turns are wound one over the other in a flat outward spiral. The entire coil consists of many such spirals side by side, giving a multilayer coil without going back to the starting point. This construction reduces the distributed capacitance of the coil.

**bar** 1. A metric unit of pressure equal to 1,000,000 dynes per sq cm, or slightly less than 1 atmosphere. In acoustics the bar was formerly 1 dyne per sq cm; to avoid confusion, the unit most commonly used for sound pressure today is the microbar, which is 1 dyne per sq cm. 2. A subdivision of a crystal slab. 3. A black vertical or horizontal line used as a test pattern in television.

**bare homogeneous thermal reactor** A nuclear reactor in which the fissions are induced by thermal neutrons. The fuel is homogeneously distributed throughout the moderator, and there is no reflector.

**bar generator** A signal generator that delivers pulses uniformly spaced in time and synchronized to produce a stationary bar pattern on a television screen. A color-bar generator produces these bars in different colors on the screen of a color television set.

**barium** [symbol Ba] An element used in cathode coatings of electron tubes. Atomic number is 56.

**barium titanate** A ceramic having piezoelectric properties and capable of withstanding much higher temperatures than Rochelle salt crystals. Used in crystal pickups and sonar transducers.

**barium titanate microphone** A crystal microphone that uses a barium titanate ceramic bar having piezoelectric properties.

**Barkhausen effect** The succession of abrupt changes in magnetization occurring when the magnetizing force acting on a piece of iron or other magnetic material is varied.

**Barkhausen-Kurz oscillator** An oscillator of the retarding-field type in which the frequency of oscillation depends solely on the electron transit time within the tube. Electrons oscillate about a highly positive grid before reaching the less positive anode. Also called Barkhausen oscillator.

**Barkhausen oscillation** An undesired oscillation in the horizontal output tube of a television receiver, causing one or more ragged dark vertical lines on the left side of the picture.

**Barkhausen oscillator** *Barkhausen-Kurz oscillator.*

**bar magnet** A bar of hard steel that has been strongly magnetized and holds its magnetism, thereby serving as a permanent magnet.

**barn** A unit of nuclear cross section, equal to $10^{-24}$ sq cm.

**barograph** A barometer that produces a continuous record of atmospheric pressure on a graph.

**barometer** An instrument for measuring the pressure of the atmosphere. Two common types are the aneroid barometer and the mercury barometer.

**barometric altimeter** An altimeter that uses a barometer for measuring height.

**barometric pressure** Atmospheric pressure as measured by a barometer.

**barometric switch** A switch activated by an aneroid reacting to barometric pressure.

**baroresistor** A device in which electric resistance varies as a function of atmospheric pressure. It generally consists of a barometric element mechanically linked to a resistance element.

**baroswitch** A device that performs electric switching functions by mechanical actuation resulting from changes in atmospheric pressure. In one form an aneroid diaphragm is mechanically linked to a contact arm that slides over a commutator whose separate segments are connected to the signal circuits used to modulate a radiosonde.

**barrage jamming** The simultaneous jamming of a number of radio or radar channels.

**barrel distortion** Distortion in which all four sides of a received television picture bulge outward like a barrel.

**barrel-stave reflector** A parabolic radar antenna reflector having the horizontal top third and bottom third cut away, to give a high vertical beam so that roll of the ship will not cause the beam to miss the target.

**barretter** 1. A bolometer that consists of a fine wire or metal film having a positive temperature coefficient of resistivity, so that resistance increases with temperature. Used for making power measurements in microwave devices. It is the opposite of a thermistor, in which resistance decreases with temperature. 2. British term for *ballast resistor.*

**barretter mount** A waveguide mount in which a barretter can be inserted to measure electromagnetic power.

**barrier** *Potential barrier.*

**barrier layer** Deprecated term for *depletion layer.*

**barrier-layer capacitance** *Depletion-layer capacitance.*

**barrier-layer cell** Deprecated term for *photovoltaic cell.*

**base** 1. [symbol B] The transistor electrode that corresponds to the grid of a tube. 2. The part of an electron tube that has the pins, leads, or other terminals to which external connections are made either directly or through a socket. 3. The plastic, ceramic, or other insulating board used to support a printed wiring pattern. 4. A plastic film used as a support for the magnetic powder of magnetic tape or for the emulsion of photographic film. 5. An integer to which all digits are related in a positional notation system for computers. The successive digits are coefficients of successive powers of the base. The commonest base values are 2, 8, and 10. Also called radix. 6. The number on which a system of logarithms is based, such as 10 or 2.718.

**baseband** The frequency band occupied by all of the transmitted signals that are used to modulate a particular carrier. The band used for transmitting picture and synchronizing signals in television is one example. Another is the band containing all the modulated subcarriers in a carrier system.

**base bias** The direct voltage that is applied to the base electrode of a transistor.

**base electrode** An ohmic or majority carrier contact to the base region of a transistor.

**basegroup** A group of carrier channels in a particular frequency range that forms a basic unit for further modulation to a final frequency band in a carrier communication system.

**base insulator** A heavy-duty insulator used to support the weight of an antenna mast and insulate the mast from the ground or some other surface.

**baseline** 1. A line joining the two stations between which electric phase or time is compared in determining navigation coordinates, such as a line joining a master and a slave station in a loran system. 2. The line produced on the screen in the absence of an echo in certain types of radar displays.

**base-loaded antenna** A vertical antenna whose electrical height is increased by adding inductance in series at the base.

**base pin** *Pin.*

**base point** *Point.*

**base region** The interelectrode region of a transistor into which minority carriers are injected.

**basic frequency** The frequency of the sinusoidal component considered to be the most important, such as the fundamental frequency of an oscillator or the driving frequency of an acoustic transducer.

**basic repetition rate** The lowest pulse repetition rate of each of the several sets of closely spaced repetition rates in a loran system.

**basket winding** A crisscross coil winding in which adjacent turns are far apart except at points of crossing, giving low distributed capacitance.

**bass** Sounds corresponding to frequencies at the lower end of the audio range, below about 250 cps.

**bass boost** A circuit that emphasizes the lower audio frequencies, generally by attenuating higher audio frequencies.

**bass compensation** A circuit that offsets the lowered sensitivity of the human ear to weak low frequencies. This is done by making the bass frequencies relatively stronger than the high audio frequencies as volume is lowered.

**bass control** A manual tone control that has the effect of changing the level of bass frequencies in an audio amplifier.

**bass reflex baffle** A loudspeaker baffle having an opening below the loudspeaker of such size that bass frequencies from the rear emerge to reinforce those radiated directly forward.

**bass response** The extent to which an a-f amplifier, loudspeaker, or other device handles low audio frequencies.

**bassy** Pertaining to overemphasis of bass notes in sound reproduction.

**bat-handle switch** A toggle switch having an actuating lever shaped like a baseball bat.

**bathtub capacitor** A paper capacitor enclosed in a metal housing having broadly rounded corners like those on a bathtub.

**bathythermograph** A sensitive recording thermometer lowered into the water to determine temperatures at different levels, as required in predicting sound conditions for sonar. The data may be either recorded or transmitted up the support cable as modulation on a carrier signal.

**battery** A d-c voltage source made up of one or more cells that convert chemical, thermal, nuclear, or solar energy into electric energy.

**battery charger** A rectifier unit used to

change a-c power to d-c power for charging a storage battery. Also called charger.

**battery clip** A terminal having spring jaws that can be quickly snapped on a battery terminal or other point to which a temporary wire connection is desired.

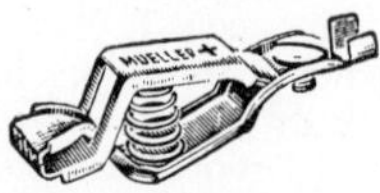

Battery clip with screw terminal for connecting lead.

**battery receiver** A radio receiver that obtains its d-c operating voltages from one or more batteries.

**batwing antenna** *Superturnstile antenna.*

**baud** A unit of telegraph signaling speed. The speed in bauds is the number of code elements (pulses and spaces) per second. A teleprinter handling 22.5 pulses per second is operating at 45 bauds.

**Baudot code** A teleprinter code that uses combinations of five or six marking and spacing intervals of equal duration. The five-unit code gives 32 possible characters, and the six-unit version gives 64. Used in radio and wire teleprinter operation.

**bay** 1. One segment of an antenna array. 2. A housing for equipment. 3. A space formed by structural partitions on an aircraft.

**bayonet base** A tube or lamp base having two projecting pins on opposite sides of a smooth cylindrical base, to engage in corresponding slots in a bayonet socket and hold the base firmly in the socket.

**bayonet socket** A socket for bayonet-base tubes or lamps, having J-shaped slots on opposite sides and one or more contact buttons at the bottom.

**bazooka** Slang term for *balun.*

**B battery** The battery that supplies the d-c voltages required by the anode and grid electrodes of electron tubes in battery-operated equipment.

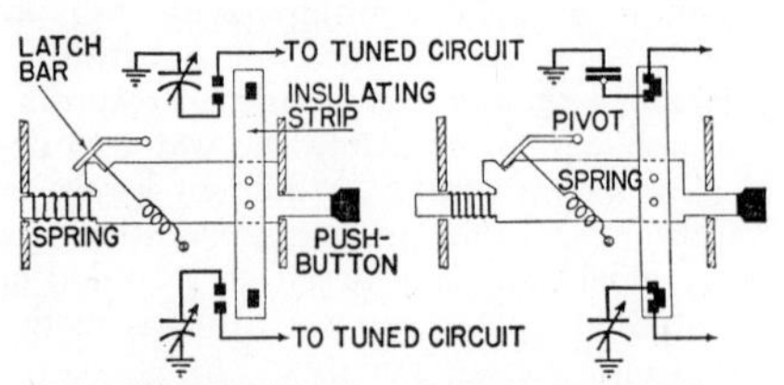


B battery connections in r-f amplifier stage.

**bdi** Abbreviation for *bearing deviation indicator.*

**B display** A rectangular radarscope display in which targets appear as bright spots, with target bearing indicated by the horizontal coordinate and target distance by the vertical coordinate. Also called B scan and range-bearing display.

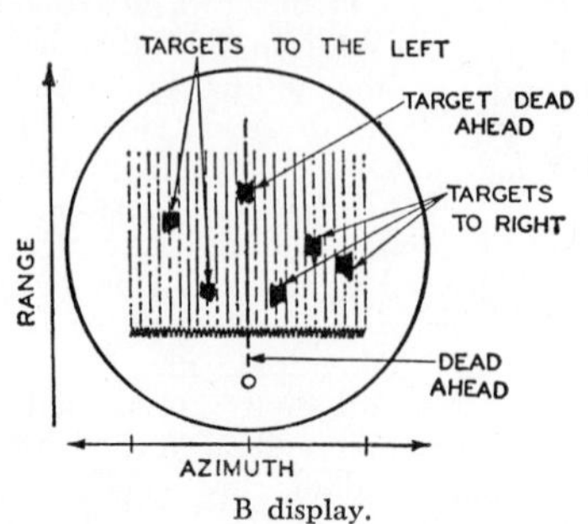


B display.

**beacon** 1. *Radio beacon.* 2. *Radar beacon.*

**beacon presentation** The radarscope presentation resulting from r-f waves sent out by a radar beacon.

**bead** A glass, ceramic, or plastic insulator with a hole through its center, used for supporting the inner conductor of a coaxial line in the exact center of the line.

**bead thermistor** A thermistor consisting of a small bead of semiconducting material such as germanium, placed between two wire leads. Used for microwave power measurement, temperature measurement, and as a protective device.

**beam** 1. A concentrated unidirectional stream of particles, such as electrons or protons. 2. A concentrated unidirectional flow of electromagnetic waves, as from a radar antenna, a microwave relay antenna, or an A-N radio range antenna array. The beam here is a major lobe of the antenna radiation pattern and is restricted to a small solid angle in space. 3. A concentrated unidirectional flow of acoustic waves. 4. A parallel arrangement of light rays. Also called ray.

**beam alignment** An adjustment of the electron beam in a camera tube, performed on tubes employing low-velocity scanning, to cause the beam to be perpendicular to the target at the target surface.

**beam antenna** *Antenna array.*

**beam approach beacon system** British term for *babs.*

**beam bender** *Ion-trap magnet.*

**beam bending** Deflection of the scanning beam by the electrostatic field of the charges stored on the target of a camera tube.

**beam capture** Entry of a missile into the beam of a radar beam-rider guidance system so that it can receive coded guidance signals.

**beam convergence** The adjustment that makes the three electron beams of a three-gun color picture tube meet or cross at a shadow-mask hole.

**beam coupling** The production of an alternating current in a circuit connected between two electrodes that are close to, or in the path of, a density-modulated electron beam.

**beam-coupling coefficient.** The ratio of alternating driving current produced in a resonator to the alternating component of the initiating beam current.

**beam current** The electric current determined by the number and velocity of electrons in an electron beam.

**beam-deflection tube** An electron-beam tube in which current to an output electrode is controlled by the transverse movement of an electron beam.

**beam hole** A hole extending through the shield and usually through the reflector of a nuclear reactor to permit the escape of a beam of fast neutrons or other radiation for experimental purposes.

**beam-indexing color tube** A color picture tube in which a signal, generated by an electron beam after deflection, is fed back to a control device or element in such a way as to provide an image in color.

**beam jitter** A small oscillatory, angular movement of a radar antenna array and hence of the radar beam, required to develop accurate error signals for automatic tracking.

**beam magnet** *Convergence magnet.*

**beam pattern** *Directivity pattern.*

**beam-power tube** An electron-beam tube in which use is made of directed electron beams to contribute substantially to its power-handling capability. The control grid and screen grid are essentially aligned, and special deflecting electrodes are generally used to concentrate the electrons into beams. Also called beam tetrode.

**beam rider** A missile that is directed to its target by beam-rider guidance.

**beam-rider guidance** A missile guidance system in which a radar or other type of fixed or moving beam is directed into space to form a path along which it is desired to guide a missile to a target. Equipment in the missile detects deviations from the center of the beam and makes appropriate corrections to missile control settings.

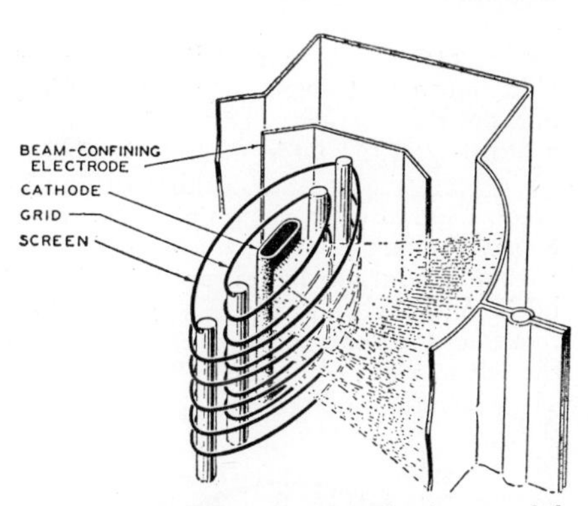

Beam-power tube construction, showing one of the pair of beam-confining electrodes that face each other on opposite sides of the grid structure.

**beam signal** A signal in a radio beam, or the beam itself, used as a navigation aid.

**beam switching** *Lobe switching.*

**beam-switching tube** A vacuum tube having 10 identical arrays of electrodes around a central cathode. These electrodes act with a ring-shaped permanent magnet surrounding the glass envelope to provide crossed electric and magnetic fields that switch the electron beam sequentially or in any desired manner from one electrode array to another. Used as a ten-position electronic switch. Also called counter tube, counting tube, decade counter tube, magnetron beam-switching tube, and trochotron.

**beam tetrode** *Beam-power tube.*

**beam transadmittance** The ratio of the fundamental component of beam current through the output gap of a velocity-modulated tube to the fundamental component of the voltage applied to the electron beam at the input gap.

**beam width** The angular width of a radar, radio, or other beam, measured in azimuth between points of half-power intensity.

**beam-width error** A radar error that occurs because the width of the scanning beam makes the target appear wider than in actuality. By covering two targets in a single sweep, beam width can also make two targets appear as one on a radarscope.

**bearing** Angular position in a horizontal plane, expressed as the angle in degrees from true north in a clockwise direction. Also called azimuth. In navigation, azimuth and bearing have the same meaning; however, bearing is preferred for terrestrial navigation and azimuth for celestial navigation.

**bearing cursor** A bearing line on a rotatable transparent screen mounted in front of a ppi radar screen. The line is placed on a target, and the target bearing is read on an outer circular scale calibrated in degrees. Used also for determining the course in navigation.

**bearing deviation indicator** [abbreviated bdi] A sonar indicator used with a split transducer to show whether the target is to the left or right of the transducer heading. The magnitude of the aiming error is also shown.

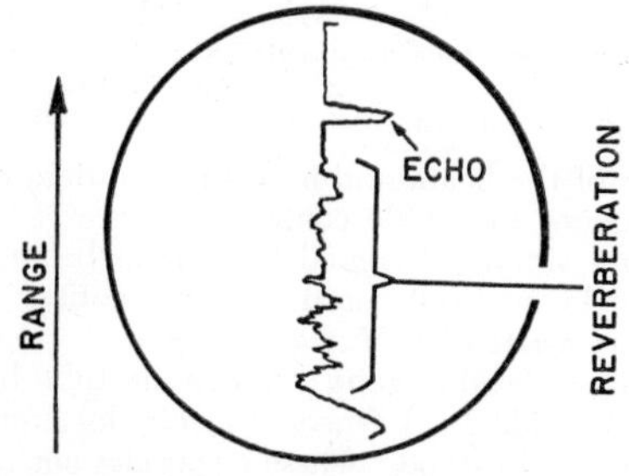


Bearing deviation indicator display. Echo pulse deflects to right when target is at right of sonar beam, and to left when target is at left of beam. Operator adjusts sonar projector until spot brightens at echo location without deflecting to right or left.

**bearing resolution** The smallest angular difference in bearing at which a given radar is able to distinguish between targets close together having the same range.

**beat frequency** The difference frequency obtained when signals of two different frequencies are combined in a nonlinear circuit.

**beat-frequency oscillator** [abbreviated bfo] An oscillator in which a desired audio difference frequency is produced by combining two different r-f signals. Used in audio signal generators for test purposes. Used in communication receivers to produce an audible signal when tuned to continuous-wave signals.

**beating** A phenomenon in which two or more periodic quantities of different frequencies produce a resultant having pulsations of amplitude.

**beat note** The difference frequency created when two sinusoidal waves of different frequencies are fed to a nonlinear device.

**beat-note detector** A detector that incorporates an oscillator or is fed by an external oscillator having a frequency sufficiently close to the unmodulated incoming carrier frequency so an audible signal frequency is produced.

**beat reception** *Heterodyne reception.*

**beats** Beat notes, generally at a sufficiently low audio frequency to be counted.

**beavertail** A fan-shaped radar beam, wide in the horizontal plane and narrow in the vertical plane. The beavertail is swept up and down for height-finding.

**bedspring array** *Billboard array.*

**beeper** 1. A simple remote control system used to control target drones and other vehicles, in which the carrier is modulated with a different audio frequency for each desired on-off control function. 2. One who directs a pilotless aircraft or missile by remote control.

**bel** [after Alexander Graham Bell] A unit of sound level, equal to the logarithm to the base 10 of the ratio of two amounts of power. One power value is a reference value. The decibel, a smaller unit equal to one-tenth bel, is more commonly used.

**B eliminator** A separate power supply that changes the a-c power-line voltage to the d-c anode voltages required by electron tubes. It eliminated the need for B batteries in early radios.

**bell wire** Cotton-covered or plastic-covered copper wire, usually No. 18, used chiefly for doorbell and thermostat connections.

**bend** 1. A smooth change in the direction of the longitudinal axis of a waveguide. 2. Departure of a defined navigation course from a straight line.

**benito** A continuous-wave navigation system in which the distance to an aircraft is determined on the ground by a phase-difference measurement of an audio signal transmitted from the ground and retransmitted by the aircraft. Bearing information is obtained from a direction finder at the ground station.

**bent-gun ion trap** An ion trap that consists of a bend in the electron gun of a cathode-ray tube. The electrons are successfully bent by a small external permanent magnet so they pass through the gun. The ions, being heavier and hence less sensitive to the magnetic field, strike the sides of the gun harmlessly and are trapped.

**berkelium** [symbol Bk] A transuranic radioactive element that was initially synthesized by bombarding americium-241 with helium ions. Atomic number is 97.

**beryllium** [symbol Be] A metallic element of importance as a moderator and reflector in nuclear reactors because of its low neutron-absorption cross section, high slowing-down power, and high thermal conductivity. Atomic number is 4.

**beryllium oxide** [symbol BeO] A ceramic used as a moderator or reflector in nuclear reactors. Its low neutron-capture cross section makes it useful also for other reactor parts.

**beta** [Greek $\beta$] The current gain of a transistor when connected as a grounded-emitter amplifier. It is the ratio of a small change in collector current to the resulting change in base current, collector voltage being constant.

**beta-absorption gage** *Beta gage.*

**beta activity** Radioactivity in which beta particles are emitted.

**beta cutoff frequency** The frequency at which the beta of a transistor is down 3 decibels from its low-frequency value.

**beta decay** Radioactive transformation of a nuclide in which the atomic number increases or decreases by 1 and the mass number remains unchanged. The atomic number increases when a negative beta particle (negatron) is emitted and decreases when a positive beta particle (positron) is emitted or an electron is captured.

**beta disintegration** The disintegration energy of a beta-decay process. For negatron emission it is equal to the sum of the kinetic energies of the beta particle, the neutrino, and the recoil atom. For positron emission, the energy equivalence of two electron rest masses must be added. For electron capture, the disintegration energy is equal to the sum of the kinetic energy of the neutrino and the electronic excitation energy of the product atom.

**beta emitter** A radionuclide that disintegrates by beta-particle emission.

**beta gage** An instrument that measures the thickness or density of a sample by measuring the absorption of beta rays in the sample. Also called beta-absorption gage.

**beta particle** A negative electron (negatron) or a positive electron (positron) emitted from a nucleus during beta decay.

**beta ray** A stream of beta particles.

**beta-ray spectrometer** An instrument used to determine the energy distribution of beta particles and secondary electrons.

**beta-ray spectrum** The distribution in energy or momentum of the beta particles emitted in a beta-decay process.

**betatron** An accelerator consisting of a horizontal doughnut-shaped vacuum enclosure, an electron gun serving as a source of electrons inside, and an external a-c electromagnet that produces magnetic lines of force passing vertically through the enclosure. As emitted electrons travel in a circular path around the doughnut, they are continuously accelerated by the rapidly changing magnetic field. The resulting high-energy electron beam is then deflected out through the window for direct use as beta rays or directed against a target to produce high-energy x-rays. Also called induction accelerator. Formerly called rheotron.

**beta uranium** An allotropic modification of uranium that is stable between approximately 660°C and 770°C.

**beta wave** A brain wave having a frequency above 14 cps.

**Bethe-hole directional coupler** A directional coupler in which the amounts of electric and magnetic coupling are balanced by rotating one waveguide with respect to the other about a coupling hole that is common to a broad face of each waveguide.

**bev** Abbreviation for *billion electron-volt.*

**bevatron** *Proton synchrotron.*

**Beverage antenna** *Wave antenna.*

**beyond-the-horizon communication** *Scatter propagation.*

**bfo** Abbreviation for *beat-frequency oscillator.*

***B-H* curve** A characteristic curve showing the relation between magnetic induction $B$ and magnetizing force $H$ for a magnetic material. It shows the manner in which the permeability of a material varies with flux density. Also called magnetization curve.

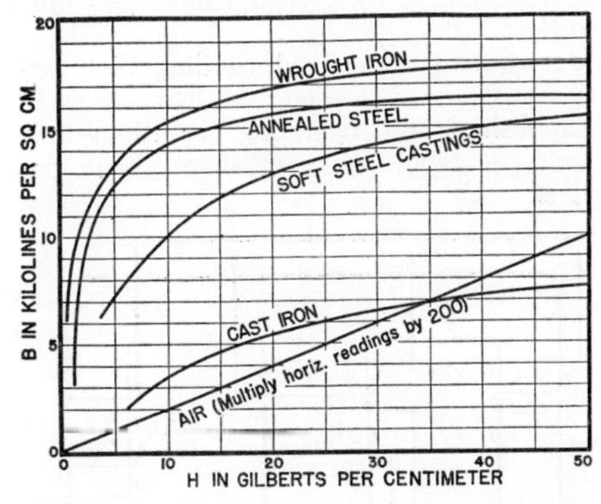

*B-H* curves for four different ferrous materials.

**bias** 1. A d-c voltage applied to a transistor control electrode to establish the desired operating point. 2. *Grid bias.*

**bias cell** A small dry cell used singly or in series to provide the required negative bias voltage for the grid circuit of a vacuum tube. A bias cell will last indefinitely if no current is drawn from it. Also called grid-bias cell.

**bias current** An alternating electric current above about 40,000 cps, added to the audio current being recorded on magnetic tape to reduce distortion.

**bias distortion** *Bias telegraph distortion.*

**biased automatic gain control** *Delayed automatic gain control.*

**bias lighting** A method of illuminating the mosaic of an iconoscope from the rear to increase its sensitivity.

**bias modulation** Amplitude modulation in which the modulating voltage is superimposed on the bias voltage of an r-f amplifier tube.

**bias oscillator** An oscillator used in a magnetic recorder to generate an a-c signal having a frequency in the range of 40 to 80 kc, as required for magnetic biasing to give a linear recording characteristic. The bias oscillator usually also serves as the erase oscillator.

**bias resistor** A resistor used in the cathode or grid circuit of an electron tube to provide a voltage drop that serves as the bias voltage.

**bias telegraph distortion** A distortion that causes telegraph mark-and-space pulses to be lengthened or shortened. Often caused by changes in the amplitude of incoming pulses. Also called bias distortion.

**bias winding** A control winding that carries a steady direct current that serves to establish desired operating conditions in a magnetic amplifier or other magnetic device.

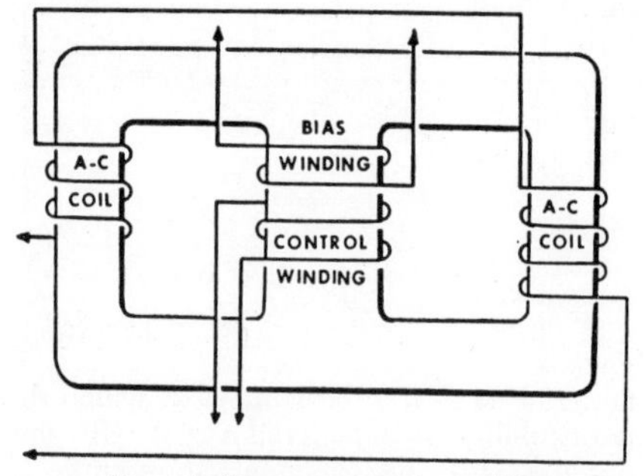


Bias winding on magnetic amplifier.

**biconical antenna** An antenna consisting of two metal cones having a common axis, with their vertices coinciding or adjacent, and with coaxial cable or waveguide feed to the vertices. The radiation pattern is circular in a plane perpendicular to the axis.

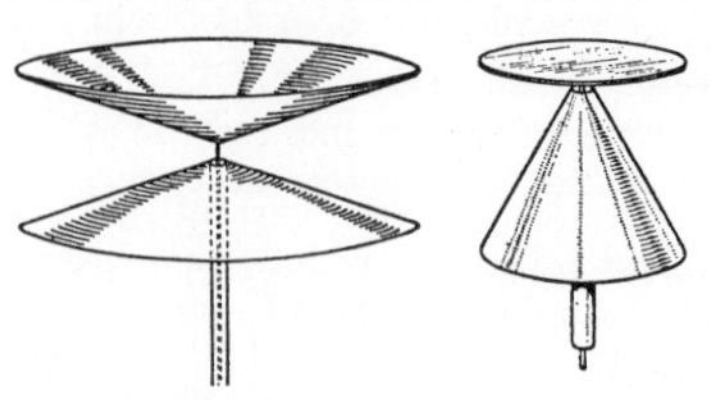

Biconical antenna and discone antenna, both fed by coaxial cable.

Also called biconical horn antenna. When the vertex angle of one of the cones is 180°, the antenna is called a discone antenna.

**biconical horn antenna** *Biconical antenna.*

**biconjugate network** A linear network having four resistances, conjugate in pairs so that a voltage in series with one resistance causes no current to flow in the other of a pair. Examples include directional couplers, hybrid junctions, and hybrid rings.

**bicotar** A trajectory-measuring system that uses a cotar at each of two separated ground stations. The intersection of the pointing vectors generated by the cotars gives target location.

**bidirectional** Responsive in two opposite directions.

**bidirectional antenna** An antenna that radiates or receives most of its energy in only two directions.

**bidirectional coupler** A directional coupler that has terminals available for sampling both directions of transmission.

**bidirectional microphone** A microphone that responds equally well to sounds coming from its front and rear, corresponding to sound incidences of 0 and 180°.

**bidirectional pulse** A pulse in which the variation from the normally constant value occurs in both directions.

**bidirectional pulse train** A pulse train, some pulses of which rise in one direction and the remainder in the other direction.

**bifilar resistor** A resistor wound with a wire doubled back on itself to reduce the inductance.

**bifilar winding** A winding consisting of two insulated wires side by side, with currents traveling through them in opposite directions. Used to get maximum coupling between two circuits, to get minimum inductance in a resistor, or to get minimum pulse voltage between the two wires when they carry heater current to a tube whose cathode is driven by a pulse transformer.

**bilateral antenna** An antenna having maximum response in exactly opposite directions, 180° apart, such as a loop.

**bilateral-area track** A photographic sound track having the two edges of the central area modulated according to the signal.

**bilateral network** A network in which a given current flow in either direction causes the same voltage drop.

**bilateral transducer** A transducer capable of transmission simultaneously in both directions between at least two terminations.

**billboard array** A broadside antenna array consisting of stacked dipoles spaced ¼ to ¾

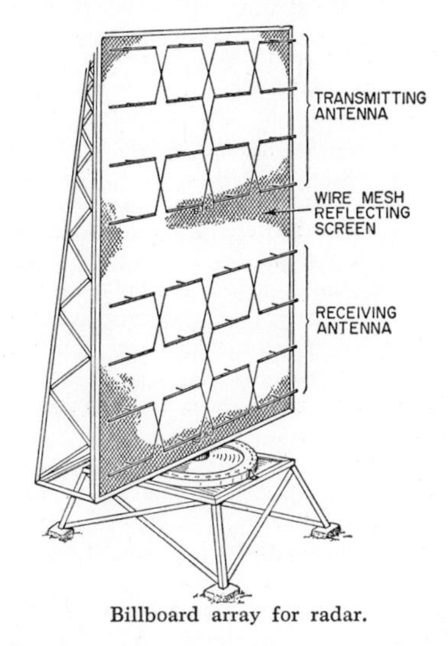


Billboard array for radar.

wavelength apart in front of a large sheet-metal reflector. Also called bedspring array and mattress array.

**billicycle** *Kilomegacycle.*

**billion electron-volt** [abbreviated bev] A unit of energy equal to $10^9$ electron-volts in the United States and France. Outside of France in Europe, $10^9$ electron-volts is called 1 giga-electron-volt, while $10^{12}$ electron-volts is called 1 billion electron-volts; to avoid confusion, it is proposed that $10^{12}$ electron-volts be called 1 tetra-electron-volt.

**bimetallic strip** A strip formed of two dissimilar metals welded together. The metals have different temperature coefficients of expansion, causing the strip to bend or curl when the temperature changes. Used in thermostats and thermal time-delay switches.

**bimorph cell** Two piezoelectric plates cemented together in such a way that an applied voltage causes one to expand and the other to contract, so that the cell bends in proportion to the applied voltage. Conversely, applied pressure will generate double the voltage of a single cell. Used in pickups and microphones.

**binary** Having only two possible alternatives.

**binary cell** An elementary unit of computer storage that can have one or the other of two stable states and can thus store one bit of information.

**binary code** *Binary number system.*

**binary-coded decimal system** A system of number representation in which each decimal digit is represented by a group of binary digits. Thus, in the 8-4-2-1 coded decimal notation the number 17 is 0001 0111 for 1 and 7 respectively.

**binary counter** *Binary scaler.*

**binary digit** *Bit.*

**binary number** A numerical value expressed in binary digits as a sequence of 0's and 1's representing 1, 2, 4, 8, 16, 32, 64, 128, and other powers of 2 according to position from right to left in a group. These positional values are added to get the equivalent decimal number. Thus, 010 is 2; 101 is 5; 1010 is 10; 01000 is 16; 11110 is 30.

**binary number system** A system of positional notation in which the successive digits are interpreted as coefficients of the successive powers of the base 2. Also called binary code.

**binary point** The character, or the location of an implied symbol, that separates the integral part of a numerical expression from its fractional part in binary notation.

**binary scaler** A scaler that produces one output pulse for every two input pulses. Two binary scaler stages in sequence give an output pulse for every four input pulses; three in sequence give one for eight; four in sequence give one output pulse for every sixteen input pulses. Also called binary counter and scale-of-two circuit.

**binary-to-decimal conversion** The mathematical process of converting a number written in binary notation to the equivalent number written in ordinary decimal notation.

**binaural** Pertaining to hearing with both ears.

**binaural effect** The ability to determine the direction from which a sound is coming by sensing the difference in arrival times of a sound wave at each ear.

**binaural sound system** A sound-reproducing system in which two microphones mounted on a dummy simulating a human head feed two independent channels, each of which terminates in one earphone of a pair worn by the listener, so as to give a binaural effect. Seldom used.

**binder** A resin or other cementlike material used to hold particles together and provide mechanical strength. Used in phonograph records, carbon resistors, and fluorescent screens.

**binding energy** The energy required to remove a neutron, proton, alpha particle, or electron from an atom. Also called separation energy.

**binding post** A manually turned screw terminal used for making electric connections.

**binomial antenna array** A broadside array having major lobes in opposite directions and no side lobes. This is achieved by spacing the antennas of the array at half-wavelength intervals and feeding them all in phase, with the relative current amplitudes of the various elements being proportional to the coefficients of successive terms in a binomial series.

**biological hole** A cavity in a nuclear reactor, used for animals or plants in research on the effects of radiation or neutron bombardment.

**biological shield** A radiation-absorbing shield used to protect personnel from the effects of nuclear particles or radiation in the vicinity of a nuclear reactor.

**Biot-Savart's law** A law giving the intensity of the magnetic field produced by a current-carrying conductor.

**bipolar** Having two poles.

**bipolar transistor** A transistor that uses both positive and negative charge carriers.

**biquinary notation** A mixed-base notation system in which the first of each pair of digits counts 0 or 1 unit of five, and the second counts 0, 1, 2, 3, or 4 units. Thus, decimal number 7 is biquinary 12; 43 is 04 03; 901 is 14 00 01; 4719 is 04 12 01 14. This is the code of the Chinese abacus, and is used in binary digit form in some computers.

**bird** A missile, target drone, earth satellite, or other inanimate object that flies.

**birdie** A high-pitched whistle sometimes heard while tuning a radio receiver. It is due to beating between two carrier frequencies differing by about 10,000 cps.

**biscuit** Deprecated term for *preform.*

**bisignal zone** *Equisignal zone.*

**bismuth** [symbol Bi] An element. Atomic number is 83. It is a brittle, heavy metal with low melting point and low capture cross section for neutrons. High absorption for gamma rays makes it useful as a filter or window to reduce gamma rays but transmit neutrons.

**bismuth telluride** [symbol $Bi_2Te_3$] An intermetallic compound having high thermoelectric power and showing promise for thermoelectric refrigeration and power generation.

**bistable** Having two stable states.

**bistable multivibrator** *Flip-flop circuit.*

**bistable unit** A physical element that can be made to assume either of two stable states. A binary cell is an example.

**bit** [Binary digIT] A single character of a language employing exactly two distinct kinds of characters. They correspond to on and off conditions in a digital computer, and are usually designated as 0 and 1. Computer word size and computer storage capacity can be expressed in bits. Thus, one computer has a storage capacity of 2,048 36-bit words. Also called binary digit.

**$Bi_2Te_3$** Symbol for *bismuth telluride.*

**black after white** A television receiver defect in which an unnatural black line follows the right-hand contour of any white object on the picture screen. The same defect also causes a white line to follow a sudden change from black to a lighter background. It is caused by receiver misalignment.

**black-and-white television** *Monochrome television.*

**blackbody** A perfect absorber of all incident radiant energy. It radiates energy solely as a function of its temperature.

**black box** A unit of electronic equipment, such as a receiver or amplifier, that can be put into or removed from an aircraft or other location as a single package. So named because the housings of such units are usually painted black.

**black compression** A reduction in television picture-signal gain at levels corresponding to dark areas in a picture. The effect is to reduce contrast in the dark areas of the picture as seen on monitors and receivers. Also called black saturation.

**blacker-than-black region** The portion of the standard television signal in which the electron beam of the picture tube is cut off and synchronizing signals are transmitted. These synchronizing signals have greater peak power than those for the blackest portions of the picture.

**black level** The level of the television picture signal corresponding to the maximum limit of black peaks. This level is generally set at 75% of the maximum signal amplitude of the synchronizing pulses.

**black light** Deprecated term for invisible *ultraviolet radiation* or *infrared radiation.*

**black negative** A television picture signal in which the voltage corresponding to black is negative with respect to the voltage corresponding to the white areas of the picture.

**blackout** *Radio fadeout.*

**blackout effect** A temporary loss of sensitivity of an electron tube after handling a strong, short pulse.

**black peak** A peak excursion of the television picture signal in the black direction.

**black positive** A television picture signal in which the voltage corresponding to black is positive with respect to the voltage corresponding to the white areas of the picture.

**black recording** Facsimile recording in which the maximum received power corresponds to the maximum density of the record medium for amplitude modulation or to the lowest received frequency for frequency modulation.

**black saturation** *Black compression.*

**black signal** The signal at any point in a facsimile system produced by the scanning of a maximum-density area of the subject copy.

**black transmission** Facsimile transmission using black recording.

**blade** A flat moving conductor in a switch.

**blank** 1. The result of the final cutting operation on a natural crystal. 2. To cut off the electron beam of a cathode-ray tube.

**blanked picture signal** The signal resulting from blanking a television picture signal. Adding the sync signal to the blanked picture signal gives the composite picture signal.

**blanket** A layer of fertile material placed outside the core of a nuclear reactor.

**blanket area** The area in the immediate vicinity of a broadcast station, where the signal of that station is so strong (above 1 volt per meter) that it interferes with reception of other stations.

**blanketing** Interference due to a nearby transmitter whose signals are so strong that they override other signals over a wide band of frequencies.

**blank groove** *Unmodulated groove.*

**blanking** The process of cutting off the electron beam of a television picture tube, camera tube, or cathode-ray oscilloscope tube during retrace by applying a rectangular pulse voltage to the grid or cathode during each retrace interval. The opposite action is called gating.

**blanking level** The level that separates picture information from synchronizing information in a composite television picture signal. It coincides with the level of the base of the synchronizing pulses. Also called pedestal and pedestal level.

**blanking pulse** One of the pulses that make up the blanking signal in television

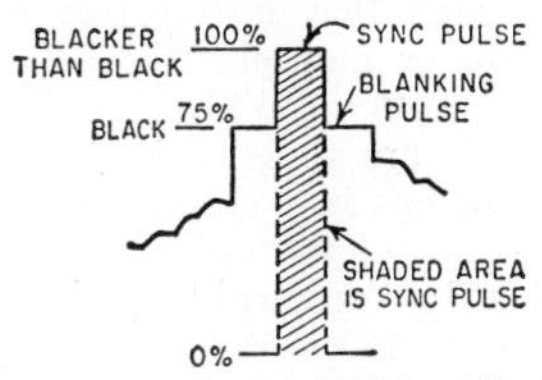


Blanking pulse at end of line, with horizontal sync pulse in its center.

**blanking signal** A wave of recurrent pulses, related in time to the scanning process, used to effect blanking in television. The pulses occur at both the line and field frequencies and serve to cut off the electron beam during retrace at both transmitter and receiver.

**blasting** Distortion due to overloading of any part of a radio transmitter, receiver, or audio amplifier.

**bleeder current** Current drawn continuously from a voltage source to lessen the effect of load changes or to provide a voltage drop across a resistor.

**bleeder resistor** A resistor connected across a power pack or other voltage source to improve voltage regulation by drawing a fixed current value continuously. Also used to dissipate the charge remaining in filter capacitors when equipment is turned off.

**blemish** An imperfection of the storage surface that produces a spurious output in charge-storage tubes.

**blind approach** An aircraft approach for a landing when visibility is poor, usually made with the aid of instruments and radio.

**blind bombing** *Bombing through overcast.*

**blind flying** *Instrument flying.*

**blind landing** An aircraft landing made with the aid of ground-controlled approach, an instrument landing system, or some other guidance system designed for use when visibility is poor.

**blind navigation** Navigation of an aircraft with the aid of instruments, radio, or electronic equipment when visibility is poor.

**blind sector** A shadow on a radarscope screen due to a fixed obstruction near the radar antenna, such as the funnels on a ship.

**blind speed** A radar target speed at which a moving target cannot be distinguished from a stationary target. With a radar moving-target indicator, blind speeds are those at which the radial velocity of the target is such that it traverses one half-wavelength, or multiples thereof, between successive pulses. With doppler radar, blind speeds are those at which the doppler shift in cps is a multiple of the pulse repetition rate of the radar.

**blind takeoff** An airplane takeoff under conditions of no visibility, usually made with the aid of instruments or radio.

**blinking** A method of providing information in pulse systems by modifying the signal at its source so the signal presentation on the display alternately appears and disappears. In loran this indicates that a station is malfunctioning.

**blip** 1. The target echo indication on a radarscope. It may be a bright spot of light as on a ppi display, or a sharply peaked pulse as on an A display. Also called pip. The Navy uses pip, while the Air Force, IRE Standards, and British texts generally use blip. 2. [Background-Limited Infrared Photoconductor] An ideal infrared radiation detector that detects with unit quantum efficiency all of the radiation in the signal for which the detector was designed, and responds only to the background radiation noise that comes from the field of view of the detector. Response to all other noise should be zero.

**blip-scan ratio** The number of scans necessary to produce one recognizable blip on a radarscope. The blip-scan ratio of a radar set varies with range, antenna tilt, ability of operator, set performance, target aspect, wind, and other factors.

**blister** 1. A radome that protrudes beyond the normal skin contours of an aircraft. 2. A protruding housing for a sonar transducer on a ship or submarine.

**Bloch band** *Energy band.*

**block** A group of words considered or transported as a unit in a computer.

**block diagram** A diagram in which the principal divisions of an electronic system are indicated by rectangles or other geometric figures and the signal paths are represented by lines.

**blocked-grid keying** A method of keying a radiotelegraph transmitter in which sufficient grid bias is provided to block one or more tubes when the transmitting key is open. Closing the key removes this bias, applying full transmitter power to the antenna.

**blocked impedance** The impedance at the input of a transducer when the impedance of the output system is made infinite, as by blocking or clamping the mechanical system.

**blocked resistance** The real part of blocked impedance.

**blocking** 1. Applying a high negative bias to the grid of an electron tube to block its anode current. 2. Overloading of a receiver by an unwanted signal so the automatic gain control reduces the response to a desired signal.

**blocking capacitor** *Coupling capacitor.*

**blocking layer** Deprecated term for *depletion layer.*

**blocking oscillator** An oscillator in which the negative grid bias increases gradually during oscillation as a capacitor is charged, until a point is reached where anode current is cut off and oscillations stop. The capacitor then discharges until the grid is unblocked and oscillation is resumed. This process produces a sawtooth voltage waveform that may be used as the sweep voltage for a cathode-ray tube. Also called squegging oscillator.

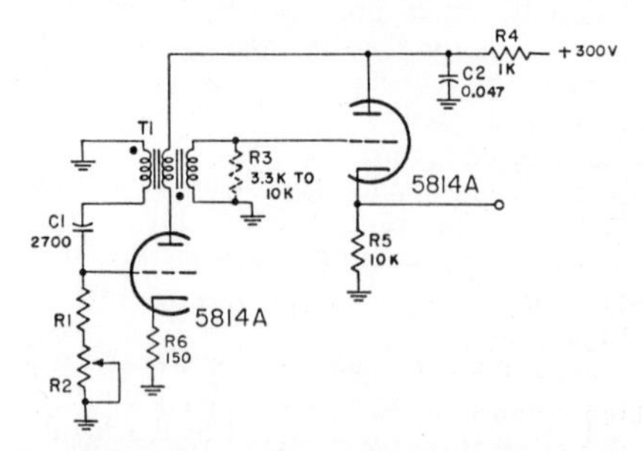

Blocking oscillator using preferred circuit design by National Bureau of Standards. Pulse repetition frequency is in range from 200 to 2,000 pulses per second, depending on values of C1, R1, and R2, which are best determined experimentally.

**blocking-oscillator driver** A blocking oscillator that develops and shapes an essentially square pulse for driving radar modulator tubes.

**blocking period** The portion of the idle period in the cycle of a mercury-arc rectifier in which the anode is positive but the start of commutation is blocked by phase control.

**blooming** 1. Defocusing of television picture areas where brightness is excessive, owing to enlargement of spot size and halation of the fluorescent screen. 2. An increase in radarscope spot size due to an increase in signal intensity. Blooming may be used to convey information in navigation systems having intensity modulation.

**blower** An electric fan used to supply air for cooling purposes.

**blown-fuse indicator** A neon warning light connected across a fuse so that it lights when the fuse is blown.

**blue-beam magnet** A small permanent magnet used as a convergence adjustment to change the direction of the electron beam for blue phosphor dots in a three-gun color television picture tube.

**blue gain control** A variable resistor used in the matrix of a three-gun color television receiver to adjust the intensity of the blue primary signal.

**blue glow** A glow normally seen in electron tubes containing mercury vapor, due to ionization of the molecules of mercury vapor. A blue glow near the electrodes of a vacuum tube means that the tube is gassy and hence defective. A soft blue fluorescent glow is normal on the glass envelopes of some vacuum tubes.

**blue gun** The electron gun whose beam strikes phosphor dots emitting the blue primary color in a three-gun color television picture tube.

**blue restorer** The d-c restorer for the blue channel of a three-gun color television picture tube circuit.

**blue video voltage** The signal voltage output from the blue section of a color television camera, or the signal voltage between the receiver matrix and the blue gun grid of a three-gun color television picture tube.

**bobbin** An insulated spool serving as a support for a coil.

**bobbin core** A tape-wound core in which the ferromagnetic tape has been wrapped on a form or bobbin that provides mechanical support for the tape.

**Bode diagram** A diagram in which the phase shift or the gain of an amplifier, servomechanism, or other device is plotted against frequency to show frequency response.

**body capacitance** Capacitance existing between the hand or body and a circuit.

**body-section radiography** *Laminography.*

**bogie** An indication of an enemy or unidentified aircraft on a radar screen.

**Bohr atom** An atom model conceived by Bohr and Rutherford, consisting of a positive nucleus about which circulate a number of orbital electrons.

**Bohr magneton** A unit of magnetic moment used in specifying the magnetic moment of an atomic particle or system of particles. The electronic Bohr magneton is $9.27 \times 10^{-21}$ erg-gauss$^{-1}$. For the nuclear Bohr magneton or nuclear magneton, this value is divided by 1,836. Also called magneton.

**Bohr radius** The radius of the lowest-energy electron orbit in the Bohr model of the hydrogen atom.

**boiling-water reactor** [abbreviated bwr] A pressurized-water nuclear reactor in which the cooling water is allowed to boil in the reactor core, producing steam that can be used to generate electricity.

**Bold Orion** An air-to-surface guided missile having a range of about 1,000 miles, to be launched from bombers.

**bolometer** A device used to measure microwave and infrared energy. It contains a resistance element that changes in resistance when heated by the radiant energy. Used in infrared search and homing equipment for missiles. Used also in waveguides to measure microwave power and standing-wave ratio. Two types of bolometers are the barretter and the thermistor. Also called thermal detector.

**bolometer mount** A waveguide termination in which a bolometer can be inserted to measure electromagnetic power.

**Boltzmann's equation** The equation for particle conservation, based on the description of individual collisions.

**Bomarc** An Air Force surface-to-air guided missile having supersonic speed, launched from the ground and designed to seek out and destroy enemy aircraft. Slant range is about 300 miles.

**bombard** To direct a stream of high-energy particles or photons against a target.

**bombardment** 1. The process of directing high-speed electrons at an object, causing secondary emission of electrons, heating, fluorescence, disintegration, or production of x-rays. 2. The process of directing electrons or other high-speed particles at atoms or smaller particles. 3. Use of induction heating for heating electrodes of tubes to drive out gases during evacuation.

**bombing through overcast** [abbreviated bto] Blind bombing through clouds, using radar, infrared equipment, or other electronic equipment for guidance. Also called blind bombing.

**bonded-barrier transistor** A transistor made by alloying the base with material that is on the end of a wire.

**bonding** Connecting together the metal parts of an aircraft, vehicle, structure, or housing to prevent interference-producing static or r-f voltage buildup between adjacent metal parts.

**bond strength** A measure of the force required to separate a layer of material from an adjoining surface, such as printed wiring from its base. Also called adhesion.

**bone conduction** The process by which sound is conducted to the inner ear through the cranial bones.

**bone seeker** A compound or ion that migrates preferentially into living bone. Used in radiobiology.

**Boolean algebra** Algebra that deals with classes, propositions, on-off circuit elements, and other nonnumerical elements associated by such operators as "and," "or," "not," "except," and "if . . . then." It permits computation and demonstration, as in any mathematical system, by making use of symbols efficient in calculation.

**Boolean calculus** Boolean algebra that has been modified to include time, to permit handling such elements as: (a) states and events; (b) operators such as "after," "while," "happen," "delay," and "before"; (c) classes whose members change with time; (d) circuit elements whose on-off state changes from time to time, such as delay lines, flip-flops, and sequential circuits; (e) step-functions and their combinations.

**Boolean function** A mathematical function in Boolean algebra.

**boom** A movable mechanical support used to suspend a microphone within range of actors but above the field of view of television cameras.

**boost** To increase or amplify.

**booster** 1. A separate r-f amplifier connected between an antenna and a television receiver to amplify weak signals. 2. An r-f amplifier that amplifies and rebroadcasts a received television or communication radio signal at higher power without change in carrier frequency, for reception by the general public.

**booster amplifier** An audio amplifier used between mixer controls and the master volume control of a studio audio console. It serves to compensate for mixing-circuit losses.

Boom for microphone in television studio.

**booster voltage** The additional voltage supplied by the damper tube to the horizontal output, horizontal oscillator, and vertical output tubes of a television receiver to give greater sawtooth sweep output.

**bootstrap circuit** A single-stage amplifier in which the output load is connected between the negative end of the anode supply and the cathode, while signal voltage is applied between grid and cathode. A change in grid voltage changes the input signal voltage with respect to ground by an amount equal to the output signal voltage.

**bootstrap driver** A vacuum-tube circuit used to produce a square pulse that drives a radar modulator tube. The duration of the square pulse is determined by a pulse-forming line. The circuit is called a bootstrap driver because voltages on both sides of the pulse-forming line are raised simultaneously with voltages in the output pulse, but their relative difference (on both sides of the pulse-forming line) is not affected by the considerable voltage rise in the output pulse.

**bootstrapped sawtooth generator** A circuit capable of generating a highly linear positive sawtooth waveform through the use of bootstrapping.

**bootstrapping** A technique for lifting a generator circuit above ground by a voltage value derived from its own output signal.

**boresighting** Initial alignment of a directional microwave or radar antenna system, using an optical procedure or a fixed target at a known location.

**boron** [symbol B] An element used in instruments for detecting and measuring neutrons because the isotope $B^{10}$, on absorbing a neutron, breaks into two charged particles that can be detected easily. Atomic number is 5.

**boron chamber** An ionization chamber that is lined with boron or boron compounds and/or filled with a gaseous boron compound.

**boron counter tube** A counter tube filled with boron fluoride and/or having electrodes coated with boron or boron compounds. Used for detecting slow neutrons.

**borrow** A computer carry signal that is produced when the difference between digits being subtracted is less than zero.

**bottoming** A limiting action that occurs on positive grid-voltage peaks in a beam-power tube or pentode because of the formation of a virtual cathode under certain operating conditions. The virtual cathode limits anode current and hence flattens the corresponding anode voltage peak.

**boundary marker beacon** A fan marker beacon located near the approach end of the runway in an instrument landing system. Used at military airports but generally omitted at civil airports. Also called inner marker beacon.

**bound electron** An electron bound to the nucleus of an atom by electrostatic attraction.

**boxcar detector** A circuit used to sample and store the instantaneous value of a varying voltage. Also called sampler.

**boxcar function** A function that is zero except over a finite interval, during which it takes a constant value, often +1.

**boxcar lengthener** A pulse-lengthening circuit that lengthens a series of pulses without changing their height.

**boxcars** Long pulses separated by very short intervals.

**Bragg angle** The glancing angle for x-rays at the reflecting planes of a crystal. Used in x-ray orientation of quartz crystals for radio use.

**Bragg curve** 1. A curve showing the average number of ions per unit distance along a beam of initially monoenergetic ionizing particles, usually alpha particles, passing through a gas. 2. A curve showing the average specific ionization of an ionizing particle of a particular kind as a function of its kinetic energy, velocity, or residual range.

**Bragg scattering** Scattering of x-rays and neutrons by the regularly spaced atoms in a crystal, for which constructive interference occurs only at definite angles called Bragg angles.

**Bragg's law** A statement of the conditions under which a crystal will reflect a beam of x-rays with maximum intensity.

**Bragg spectrometer** An instrument for x-ray analysis of crystal structure, in which a homogeneous beam of x-rays is directed on the known face of a crystal and the reflected beam is detected in a suitably placed ionization chamber. As the crystal is rotated, the angles at which Bragg's law is satisfied are identified as sharp peaks in the ionization current. Also called crystal spectrometer and ionization spectrometer.

**Bragg's rule** An empirical rule, according to which the mass stopping power of an element for alpha particles is inversely proportional to the square root of the atomic weight. The atomic stopping power is then directly proportional to the square root of the atomic weight.

**brain wave** A rhythmic fluctuation of voltage between parts of the brain, ranging from about 1 to 60 cps and 10 to 100 microvolts. It is called a delta wave when the frequency is below 9 cps, an alpha wave when the frequency is 9 to 14 cps, and a beta wave when the frequency is above 14 cps.

**branch** 1. A portion of a network consisting of one or more two-terminal elements in series. Also called arm. 2. A product resulting from one mode of decay of a radioactive nuclide that has two or more modes of decay. 3. *Conditional jump.*

**branching** The occurrence of two or more modes by which a radionuclide can undergo radioactive decay. Also called multiple decay and multiple disintegration.

**branching fraction** That fraction of the total number of atoms involved which follows a particular branch of the disintegration scheme. Usually expressed as a percentage.

**branching ratio** The ratio of the number of parent atoms decaying by one mode to the number decaying by another mode, or the ratio of two specified branching fractions.

**branch point** A terminal common to two or more branches of a network, or a terminal on a branch of a network. Also called junction point.

**Braun tube** *Cathode-ray tube.*

**breadboard model** An experimental arrangement of a circuit on a board or other flat surface without regard for final locations of components, to prove the feasibility of the circuit and to facilitate changes when necessary.

**break** 1. Interruption of a radio transmission, as for sending in the opposite direction. 2. A fault in a circuit. 3. British term for a reflected radar pulse appearing on a radarscope as a line perpendicular to the baseline.

**break-before-make contacts** Contacts that interrupt one circuit before establishing another.

**break contact** *Back contact.*

**breakdown** 1. A disruptive discharge through insulation, involving a sudden and large increase in current through the insulation due to complete failure under electrostatic stress. Also called puncture. 2. Initiation of a desired discharge between two electrodes in a gas, occurring at a voltage dependent on gas density, electrode shape, electrode spacing, and polarity. 3. An undesired runaway increase in an electrode current in a gas tube.

**breakdown diode** *Avalanche diode.*

**breakdown transfer characteristic** The relation between the breakdown voltage of an electrode and the current to another electrode in a gas tube.

**breakdown voltage** The voltage at which breakdown occurs in a dielectric or in a gas tube.

**break-in keying** A method of operating a radiotelegraph communication system in which the receiver is capable of receiving signals during transmission spacing intervals.

**break-in operation** A method of radio communication involving break-in keying, allowing the receiving operator to interrupt the transmission.

**break-in relay** A relay used for break-in operation.

**break point** A place in a computer routine at which a special instruction is inserted to stop a digital computer for a visual check of progress, if desired.

**breeder reactor** A nuclear reactor that produces at least as much fuel as it consumes, or that produces more fissionable atoms than it consumes.

**breeding gain** The excess of fissionable atoms produced per fissionable atom consumed in a breeder reactor. The breeding gain is equal to the breeding ratio minus one.

**breeding rate** The net gain in fissionable material per unit time in a breeder reactor.

**breeding ratio** The ratio of the fissionable atoms produced from fertile material to fissionable atoms destroyed in a breeder reactor.

**Breit-Wigner formula** An equation relating the cross section of a particular nuclear reaction to the energy of the incident particle and the energy of a resonance level of the compound nucleus.

**bremsstrahlung** Electromagnetic radiation produced when a fast charged particle (usually an electron) is deflected by another charged particle (usually a nucleus). The radiation corresponds to a continuous spectrum of x-rays.

**brevium** Uranium $X_2$, one of the decay products in the uranium series.

**Brewster angle** The angle of incidence for which a wave polarized parallel to the plane of incidence is wholly transmitted, with no reflection.

**bridge** An instrument or circuit having four or more arms, by means of which one or more of the electrical constants of an unknown component may be measured.

**bridge circuit** A circuit consisting basically of four sections connected in series to form a diamond. An a-c voltage source is connected between one pair of opposite junctions, and an indicating instrument or output circuit is connected between the other pair of junctions. When the bridge is balanced, the output is zero.

**bridged-T network** A T network with a fourth branch connected across the two series arms of the T, between an input terminal and an output terminal.

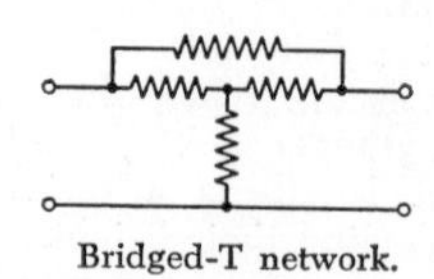

Bridged-T network.

**bridge hybrid** *Hybrid junction.*

**bridge magnetic amplifier** A magnetic amplifier in which each of the gate windings is connected in series with an arm of a bridge rectifier. The rectifiers provide self-saturation and d-c output.

**bridge rectifier** A full-wave rectifier with four elements connected in series as in a bridge circuit. Alternating voltage is applied to one pair of opposite junctions, and direct voltage is obtained from the other pair of junctions.

**bridging** 1. Connecting one electric circuit in parallel with another. 2. Selector switch action in which the movable contact is wide enough to touch two adjacent contacts so that the circuit is not broken during contact transfer.

**bridging amplifier** An amplifier with an input impedance sufficiently high that its input may be bridged across a circuit without substantially affecting the signal level.

**bridging gain** The ratio of the power a transducer delivers to a specified load impedance under specified operating conditions to the power dissipated in the reference impedance across which the input of the transducer is bridged. Usually expressed in decibels.

**bridging loss** The reciprocal of the bridging gain ratio. Usually expressed in decibels.

**brightening pulse** A pulse applied either to the grid or cathode of a radar cathode-ray tube at the beginning of the sweep to intensify the beam during the sweep.

**brightness** 1. The characteristic of light that gives a visual sensation of more or less light. 2. Former name for *luminance.*

**brightness control** A control that varies the brightness of the fluorescent screen of a cathode-ray tube by changing the grid bias of the tube, thereby changing the beam current. Used in television receivers, radar receivers, and cathode-ray oscilloscopes. Also called brilliance control and intensity control.

**brilliance** The degree to which higher audio frequencies are present when a sound recording is played back.

**brilliance control** *Brightness control.*

**broadband amplifier** An amplifier having essentially flat response over a wide range of frequencies.

**broadband antenna** An antenna that will function satisfactorily over a wide range of frequencies, such as for all 12 vhf television channels.

**broadband klystron** A klystron having three or more resonant cavities that are externally loaded and stagger-tuned to broaden the bandwidth.

**broadband tube** A gas-filled fixed-tuned tr or pre-tr tube having a built-in bandpass filter suitable for r-f switching.

**broadcast** A television or radio transmission intended for public reception.

**broadcast band** The band of frequencies extending from 535 to 1,605 kc, corresponding to assigned carrier frequencies that increase in multiples of 10 kc between 540 and 1,600 kc for the United States. Also called standard broadcast band.

**broadcasting** Transmission of television and radio programs by means of radio waves, for reception by the public.

**broadcast station** A television or radio station used for transmitting programs to the general public. Also called station.

**broadcast transmitter** A transmitter designed for use in a commercial a-m, f-m, or television broadcast channel.

**broad dimension** The wider of the two cross-section dimensions of a rectangular waveguide. It determines the critical frequency. Also called critical dimension.

**broadside** Perpendicular to an axis or plane.

**broadside array** An antenna array whose direction of maximum radiation is perpendicular to the line or plane of the array.

**broad tuning** Poor selectivity in a radio receiver, causing reception of two or more stations at a single setting of the tuning dial.

**bromine** [symbol Br] A nonmetallic liquid element. Atomic number is 35.

**Brown and Sharpe gage** [abbreviated B and S gage] *American wire gage.*

**brush** A conductive metal or carbon block used to make sliding electric contact with a moving part.

**brush discharge** A luminuous electric discharge that starts from a conductor when its potential exceeds a certain value but remains too low for the formation of an actual spark. Generally accompanied by a hissing sound.

**BS** Symbol for base shield connection. Used on tube-base diagrams.

**B scan** *B display.*

**B scope** A radarscope that produces a B display.

**B station** The loran station whose signal is always transmitted more than half a repetition period after the signal from the master or A station of the pair. Also called slave station.

**B supply** A power source used to provide a positive voltage for the anode and other electrodes of an electron tube.

**BT-cut crystal** A crystal plate cut from a plane that is rotated about an $x$ axis so the angle made with the $z$ axis is approximately $-49°$. This cut has an essentially zero temperature coefficient.

**bto** Abbreviation for *bombing through overcast.*

**bubble chamber** A chamber in which the paths of nuclear particles appear as tracks of ionized bubbles in liquid hydrogen.

**bucking coil** A coil connected and positioned in such a way that its magnetic field opposes the magnetic field of another coil. The hum-bucking coil of an excited-field loudspeaker is an example.

**bucking voltage** A voltage having exactly opposite polarity to that of another voltage against which it acts.

**Buckley gage** A type of ionization gage used for measuring very low gas pressures.

**buckling** A measure of the curvature of neutron density distribution in a reactor.

**Bucky diaphragm** *Potter-Bucky grid.*

**buffer** 1. An isolating circuit used in a digital computer to avoid any reaction of a driven circuit on the corresponding driving circuit. 2. A computer circuit that has an output and a multiplicity of inputs so designed that the output is energized when one or more inputs are energized. The circuit function of a buffer is thus equivalent to the logical "or." 3. A storage device used to compensate for differences in rates of data flow when transmitting information from one computer device to another. 4. *Buffer amplifier.*

**buffer amplifier** An amplifier used after an oscillator or other critical stage to isolate it from the effects of load impedance variations in subsequent stages. Also called buffer and buffer stage.

**buffer capacitor** A capacitor connected across the secondary of a vibrator transformer or between the anode and cathode of a cold-cathode rectifier tube to suppress voltage surges that might otherwise damage other parts in the circuit.

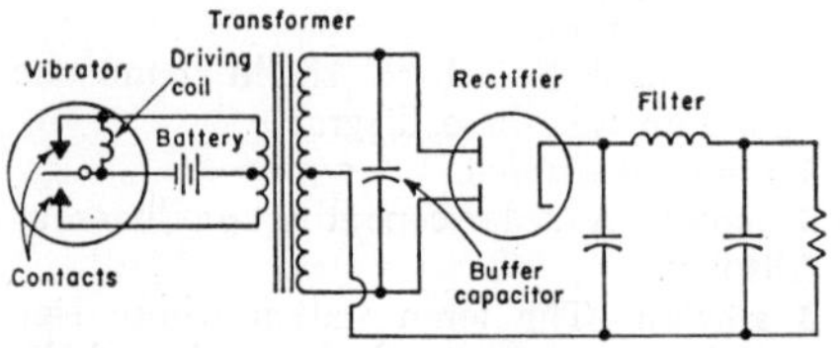

Buffer capacitor used across secondary of vibrator transformer.

**buffer stage** *Buffer amplifier.*

**buffer storage** A synchronizing element used between two different forms of storage in a computer. Computation continues while transfers take place between buffer storage and the secondary or internal storage.

**bug** 1. A semiautomatic code-sending key in which movement of a lever to one side produces a series of correctly spaced dots and movement to the other side produces a single dash. 2. Slang term for trouble in a piece of equipment.

**building-out section** A short section of transmission line, either open or short-circuited at the far end, shunted across another transmission line for tuning or matching purposes.

**built-in antenna** An antenna located inside the cabinet of a radio or television receiver.

**built-up mica** Large laminated plates of mica made by bonding thin splittings of natural mica with shellac, glyptol, or some other suitable adhesive.

**bulb** *Envelope.*

**bulk eraser** A device used to erase an entire reel of magnetic tape at once without running it through a recorder. The reel of tape is placed over a strong 60-cps a-c magnetic field, rotated a few times, then slowly withdrawn at least 2 feet before turning off the power.

**Bulldog** A Navy air-to-surface guided missile that is essentially a larger version of Bullpup.

**Bull Goose** A surface-to-air decoy missile containing electronic countermeasure equipment, used to confuse enemy radar systems.

**bull horn** A large, high-power multicellular horn used primarily for speech reproduction.

**Bullpup** A Navy air-to-surface guided missile having a range of about 3 miles, intended for use against small targets such as tanks and bridges. Speed is about mach 1.8.

**Bumblebee** A surface-to-air guided missile identical to Terrier except for its guidance system.

**buncher resonator** The first or input cavity resonator in a velocity-modulated tube, next to the cathode. Here the faster electrons catch up with the slower ones to produce bunches of electrons. Also called input resonator.

**bunching** The flow of electrons from cathode to anode of a velocity-modulated tube as a succession of electron groups rather than as a continuous stream. It is a direct result of the differences of electron transit time produced by the velocity modulation.

**bunching angle** The average transit angle between the processes of velocity modulation and energy extraction at the same or different gaps in a given electron stream drift space.

**bunching parameter** One-half the product of the bunching angle (in the absence of velocity modulation) and the depth of velocity modulation. In a reflex klystron tube the effective bunching angle is used.

**burial ground** A place for burying unwanted radioactive objects to prevent escape of their radiations. The earth then acts as a shield. Such objects must be placed in watertight, noncorrosive containers so that radioactive material cannot be leached out and get into an underground water supply.

**burned-in image** An image that persists in a fixed position in the output signal of a television camera tube after the camera has been turned to a different scene.

**burnishing surface** The portion of the cutting stylus, directly behind the cutting edge, that smooths the groove in mechanical recording.

**burnout** Failure of a device due to excessive heat produced by excessive current.

**burnup** Destruction of atoms by neutrons in reactor operation.

**burst** 1. A sudden increase in the strength of a signal being received from beyond line-of-sight range. It is believed due to meteors passing through the upper atmosphere and momentarily affecting the ionized layers that reflect radio waves back to earth. 2. An exceptionally large pulse observed in an ionization chamber, signifying the arrival of several ionizing particles simultaneously. It may be caused by a cosmic-ray shower. 3. *Color burst.*

**burst gate** The stage in a color television receiver that is gated to pass burst signals to color oscillator control circuits but is cut off during other portions of the composite color signal.

**burst pedestal** *Color-burst pedestal.*

**burst separator** The circuit in a color television receiver that separates the color burst from the composite video signal.

**bus** 1. One or more conductors used as a path for transmitting information from any of several sources to any of several destinations in an electronic computer. 2. *Busbar.*

**busbar** A heavy, rigid metallic conductor, usually uninsulated, used to carry a large current or to make a common connection between several circuits. Also called bus.

**butterfield** A confusion reflector made of wire mesh.

**butterfly capacitor** A variable capacitor having stator and rotor plates shaped like butterfly wings. Each stator plate has an outer ring that forms an inductance which varies with rotor position. Both inductance and capacitance are a minimum when the stator and rotor plates form the four quarters of a circle, and increase simultaneously to a maximum when the plates are rotated to the fully meshed position. This greatly increases the tuning range when the capacitor is used as a tuned circuit in vhf and uhf circuits.

**butt joint** 1. A connection giving physical contact between the ends of two waveguides to maintain electric continuity. 2. A connection formed by placing the ends of two conductors together and joining them by welding, brazing, or soldering.

**button** 1. A small round piece of metal that is alloyed to the base wafer of an alloy-junction transistor. Also called dot. 2. The container that holds the carbon granules of a carbon microphone. Also called carbon button.

**buzz** *Dither.*

**buzzer** An electromagnetic device having an armature that vibrates rapidly, producing a buzzing sound.

**B wind** Magnetic tape wound with the oxide surface facing outward. Seldom used today.

**bwr** Abbreviation for *boiling-water reactor.*

**BX cable** Insulated wires in flexible metal tubing, used for bringing electric power to electronic equipment.

**bypass capacitor** A capacitor connected to provide a low-impedance path for r-f or a-f currents around a circuit element.

**bypassed mixed highs** The mixed-highs signal, containing frequencies between 2 and 4 mc, that is shunted around the chrominance-subcarrier modulator or demodulator in a color television system.

**bypassed monochrome** Deprecated term for *shunted monochrome.*

**B — Y signal** A blue-minus-luminance color-difference signal used in color television. It is combined with the luminance signal in a receiver to give the blue color-primary signal.

# C

**C** 1. Symbol for *capacitor.* 2. Symbol for the grid-bias voltage source for an electron tube. 3. Abbreviation for *centigrade* or *Celsius.* 4. Symbol for *collector.* Used on transistor circuit diagrams.

**C** Symbol for *capacitance.*

**c** 1. Abbreviation for *candle.* 2. Abbreviation for *curie.* 3. Deprecated abbreviation for *cycle per second.*

**C−** [C minus] The negative terminal of a C battery or other grid-bias voltage source for an electron tube, or the grid-circuit terminal to which this source terminal should be connected.

**C+** [C plus] The positive terminal of a C battery or other grid-bias voltage source for an electron tube, or the grid-circuit terminal to which this source terminal should be connected.

**$C^{14}$** The artificially produced radioisotope of carbon, used as a tracer in research.

**cabinet** The housing for a radio receiver, television receiver, or other piece of electronic equipment.

**cable** A transmission line, group of transmission lines, or group of insulated conductors mechanically assembled in compact flexible form.

**cactus needle** A phonograph needle made from the thorn of a cactus plant.

**cadmium** [symbol Cd] A metallic element of importance as a control absorber and shield in nuclear reactors because of its very high capture cross section for neutrons with energies of less than 0.5 electron-volt. Widely used also as a plated coating on steel hardware for electronic equipment because it improves solderability, improves surface conductivity, and prevents corrosion. Atomic number is 48.

**cadmium cell** A standard cell used as a voltage reference. At 20°C its voltage is 1.0186 volts.

**cadmium ratio** The ratio of the neutron-induced saturated activity in an unshielded foil to the saturated activity of the same foil when it is covered with cadmium.

**cadmium sulfide** [symbol CdS] A semiconductor having a forbidden band gap of 2.4 electron-volts and a maximum operating temperature of 870°C in a transistor.

**cage** To lock the gyroscope of a gyro-controlled instrument in a fixed position with reference to its case.

**cage antenna** An antenna consisting of long parallel wires or rods arranged in the form of a cylinder.

**cake wax** A thick disk of wax on which an original mechanical disk recording may be inscribed.

**calcium** [symbol Ca] A silver-white soft metallic element used in cathode coatings for some types of phototubes. Atomic number is 20.

**calibrate** 1. To determine, by measurement or comparison with a standard, the correct value of each scale reading on a meter or other device. 2. To determine the settings of a control that correspond to particular values of voltage, current, frequency, or some other characteristic.

**calibration curve** A plot of calibration data, giving the correct value for each indicated reading of a meter or control dial.

**calibration marker** A marker line or circle used to divide a radar screen into accurately known intervals for determination of range, bearing, height, or time.

**californium** [symbol Cf] A transuranic radioactive element produced by bombardment of curium isotope $Cm^{242}$ with helium atoms. Atomic number is 98.

**call** A radio transmission made to identify the transmitting station and designate the station for whom the transmission is intended.

**call in** To transfer control of a digital computer temporarily from a main routine to a subroutine that is inserted in the sequence of calculating operations temporarily to fulfill a subsidiary purpose.

**call letters** Identifying letters, sometimes including numerals, assigned to radio and television stations by the Federal Communications Commission and other regulatory authorities throughout the world.

**call number** A set of characters identifying a computer subroutine and containing information concerning parameters to be inserted in the subroutine, information to be used in generating the subroutine, or information related to the operands.

**call word** A call number that exactly fills one word in a computer routine.

**calorescence** The production of visible light by infrared radiation. The transformation is indirect, the light being produced by heat and not by any direct change of wavelength.

**calorie** The unit of quantity of heat in the metric system. It is approximately the amount of heat required to raise the temperature of 1 gram of water 1°C.

**calorimeter** A device that measures quantity of heat. Used to measure microwave power in terms of its heating effect.

**calutron** [CALifornia University + TRON] An electromagnetic apparatus for separating isotopes of uranium and other elements according to their masses, using the principle of the mass spectrograph.

**camera** *Television camera.*

**camera chain** A television camera, associated amplifiers, a monitor, and the cable needed to bring the camera output signal to the control room.

**camera signal** The video output signal of a television camera.

**camera spectral characteristic** The sensitivity of each of the camera color-separation channels with respect to wavelength.

**camera tube** An electron-beam tube used in a television camera to convert an optical image into a corresponding charge-density electric image and scan the resulting electric image in a predetermined sequence to provide an equivalent electric signal. Examples are the iconoscope, image dissector, image orthicon, and vidicon. Also called pickup tube and television camera tube.

**Campbell bridge** A bridge specifically designed for comparison of mutual inductances.

**Campbell-Colpitts bridge** An a-c bridge designed to measure capacitance by the substitution method.

**can** *Jacket.*

**canal** A water-filled trench or conduit associated with a nuclear reactor, used for removing and sometimes storing radioactive objects taken from the reactor. The water acts as a shield against radiation.

**canal ray** A ray of positive ions passing through a hole in the cathode of a discharge tube.

**canceled video** The radar video output remaining after moving-target-indicator cancellation.

**cancellation ratio** The ratio of a fixed radar target signal voltage after moving-target-indicator cancellation to the same target voltage without such cancellation.

**candela** *Candle.*

**candle** [abbreviated c] The unit of luminous intensity. One candle is defined by international agreement (1948) as the luminous intensity of 1/60th sq cm of a blackbody radiator operating at the temperature of solidification of platinum. Also called candela and new candle.

**candlepower** Luminous intensity expressed in candles.

**Candohm** Trademark of Muter Co. for their line of metal-encased fixed and tapped wirewound resistors.

**canned music** Music on phonograph records or magnetic tape.

**cannibalize** To remove serviceable parts from one piece of equipment for use in repairing another piece of equipment.

**canning** Placing a sheath around a slug of uranium before inserting the slug in a nuclear reactor.

**capacitance** [symbol *C*] 1. The electric size of a capacitor. The basic unit is the farad, but the smaller microfarad and micromicrofarad units are commonly used. Also called capacity (deprecated) and permittance (obsolete). 2. The property that exists whenever two conductors are separated by an insulating material, permitting the storage of electricity.

**capacitance altimeter** An absolute altimeter in which altitude is determined by the effect of the ground on the capacitance between two widely separated electrodes on a plane. Developed in commercial form in Great Britain but largely replaced by radio altimeters.

**capacitance bridge** A bridge for comparing two capacitances, such as a Schering bridge.

**capacitance-loop directional coupler** A directional coupler in which a coupling link much shorter than a quarter-wavelength is placed lengthwise in a waveguide.

**capacitance meter** An instrument used to measure capacitance values of capacitors or of circuits containing capacitance.

**capacitance-operated intrusion detector** A boundary alarm system in which the approach of an intruder to an antenna wire encircling the protected area a few feet above ground changes the antenna-ground capacitance and sets off the alarm.

**capacitance relay** An electronic relay that responds to a small change in capacitance, such as that created by bringing a hand near a pickup wire or plate.

**capacitance standard** *Standard capacitor.*

**capacitive coupling** Use of a capacitor to transfer energy from one circuit to another.

**capacitive diaphragm** A resonant window used in a waveguide to provide the equivalent of capacitive reactance at the frequency being transmitted.

**capacitive load** A load in which the capacitive reactance exceeds the inductive reactance. Such a load draws a leading current.

**capacitive post** A metal post or screw extending across a waveguide at right angles to the E field, to provide capacitive susceptance in parallel with the waveguide for tuning or matching purposes.

**capacitive reactance** [symbol $X_C$] Reactance due to the capacitance of a capacitor or circuit. Capacitive reactance is measured in ohms and is equal to 1 divided by 6.28 $fC$, where $f$ is in cps and $C$ is in farads.

**capacitive tuning** Tuning involving use of a variable capacitor.

**capacitive window** A conducting diaphragm extending into a waveguide from one or both sidewalls, producing the effect of a capacitive susceptance in parallel with the waveguide.

**capacitor** [symbol C] A device consisting essentially of two conducting surfaces separated by a dielectric material such as air, paper, mica, ceramic, glass, or Mylar. A capacitor stores electric energy, blocks the flow of direct current, and permits the flow of alternating current to a degree dependent on its capacitance and the frequency. Also called condenser (deprecated).

**capacitor bank** A number of capacitors connected together in series or in parallel.

**capacitor color code** A method of marking the value on a capacitor by means of dots or bands of colors as specified in the EIA color code.

**capacitor hydrophone** A capacitor microphone that responds to waterborne sound waves.

**capacitor-input filter** A power-supply filter in which a shunt capacitor is the first element after the rectifier.

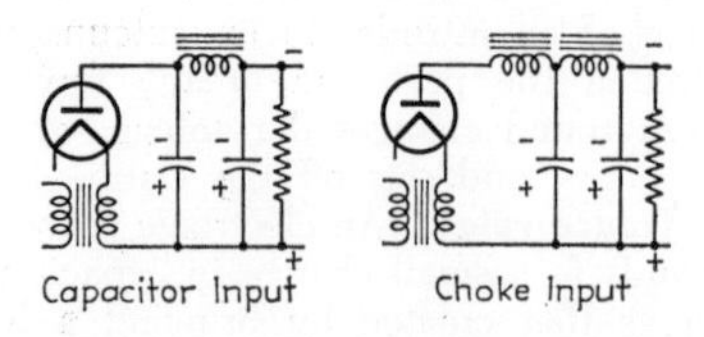


Capacitor-input filter and choke-input filter.

**capacitor integrator** An integrator circuit in which a current proportional to the function to be integrated is fed into a capacitor and the capacitor voltage is read after the desired time interval of integration.

**capacitor ionization chamber** *Capacitor r-meter.*

**capacitor loudspeaker** *Electrostatic loudspeaker.*

**capacitor microphone** A microphone consisting essentially of a flexible metal diaphragm and a rigid metal plate that together form a two-plate air capacitor. Sound waves set the diaphragm in vibration, producing capacitance variations that are converted into a-f signals by a suitable amplifier circuit. Also called condenser microphone (deprecated) and electrostatic microphone.

**capacitor motor** A single-phase induction motor having a main winding connected directly to a source of a-c power and an auxiliary winding connected in series with a capacitor to the source of a-c power.

**capacitor pickup** A phonograph pickup in which movements of the stylus in a record groove cause variations in the capacitance of the pickup.

**capacitor-resistor unit** *Rescap.*

**capacitor r-meter** An r-meter consisting of an electrometer and a detachable ionization chamber embodying a capacitor. The chamber is attached to the electrometer and charged, then detached and partly discharged by exposure to the radiation being measured. The chamber is then returned to the electrometer for reading the decrease in charge. The instrument can be calibrated to read directly in roentgens. Also called capacitor ionization chamber, condenser ionization chamber (deprecated), and condenser r-meter (deprecated).

**capacitor-start motor** A capacitor motor in which the capacitor is in the circuit only during the starting period. The capacitor and its auxiliary winding are disconnected automatically by a centrifugal switch or other device when the motor reaches a predetermined speed, after which the motor runs as an induction motor.

**capacitron** An externally fired pool tube in which each conducting cycle is started by applying a high-voltage pulse between the mercury-pool cathode and a metal band on the outside of the glass envelope just above the pool level. Current control is achieved by varying the time at which each cycle is started.

**capacity** 1. The rated or maximum load or capability of a machine or device, such as the upper and lower limits of the numbers that may be processed or stored in a digital computer. 2. Deprecated term for *capacitance.*

**capristor** *Rescap.*

**capstan** The shaft that rotates against the tape in a magnetic-tape recorder, pulling the tape past the head at a constant speed during recording and playback.

**capstan idler** A rubber-tired roller that holds the magnetic tape against the capstan by means of spring pressure.

**capture** 1. A process in which an atomic or nuclear system acquires an additional particle, such as the capture of electrons by positive ions, of electrons by nuclei, or of neutrons by nuclei. The gamma rays emitted as a result of capture are called capture gamma rays. 2. The act of placing a missile under control of a guidance system after flight speed has been reached.

**capture cross section** The cross section that is effective for radiative capture.

**capture effect** The effect wherein a strong f-m signal in an f-m receiver completely suppresses a weaker signal on the same or nearly the same frequency.

**capture efficiency** The time available during a frequency-modulation cycle of a synchrocyclotron for starting particles into stable orbits, divided by the total time for repetition of the cycle.

**capture gamma rays** Gamma rays emitted during capture.

**capture spot** The area of a point-contact transistor near the point where the collector makes contact with the germanium. Also called trapping spot.

**carbon** [symbol C] An element widely used in the construction of resistors and dry cells. Atomic number is 6. Its low neutron-capture cross section and low atomic weight make it useful as a moderator in a nuclear reactor.

**carbon button** *Button.*

**carbon cycle** A series of thermonuclear reactions, with release of energy, that presumably occurs in the sun and other stars. The net accomplishment is the synthesis of four hydrogen atoms into a helium atom with attendant release of nuclear energy.

**carbon hydrophone** A carbon microphone that responds to waterborne sound waves.

**carbonized filament** A thoriated tungsten filament treated with carbon to form a coating of tungsten carbide. This permits higher filament temperatures and correspondingly greater electron emission without excessive evaporation of thorium.

**carbon microphone** A microphone in which a flexible diaphragm moves in response to sound waves and applies a varying pressure to a container filled with carbon granules, causing the resistance of the microphone to vary correspondingly.

**carbon-pressure recording** Electromechanical recording in which a stylus or other pressure device acts on carbon paper placed over the record sheet.

**carbon resistor** A resistor consisting of carbon particles mixed with a binder, molded into a cylindrical shape, and baked.

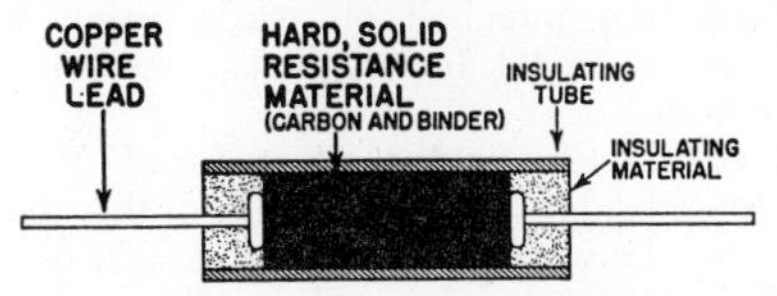


Carbon resistor construction.

Terminal leads are attached to opposite ends. The resistance of a carbon resistor decreases as temperature increases. Also called composition resistor.

**carcinotron** A voltage-tuned backward-wave oscillator tube used to generate frequencies ranging from uhf values up to over 100,000 mc. The desired output wave travels in the opposite direction to the electron beam, so the output termination is at the electron-gun end of the tube.

**card** *Punched card.*

**card column** A punched-card column into which information is entered by punches. Also called column.

**card field** A set of punched-card columns fixed as to number and position, into which the same item of information is regularly entered.

**cardiogram** *Electrocardiogram.*

**cardioid diagram** A polar diagram in the shape of a heart, such as the radiation pattern of a dipole antenna with reflector.

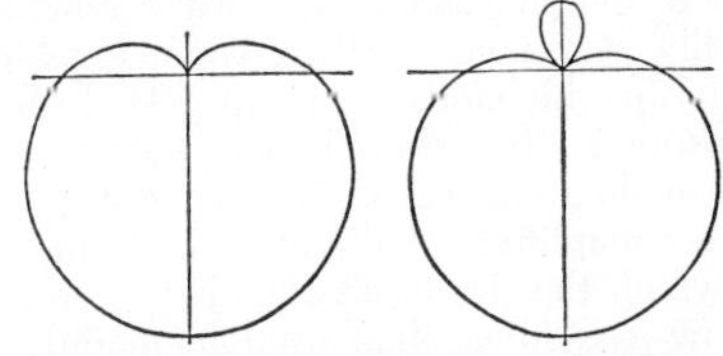

Cardioid diagram representing pickup pattern of directional microphone, and modified cardioid diagram (right) having still greater suppression of sounds arriving from sides and rear. Microphone is at intersection of horizontal and vertical lines.

**cardioid microphone** A microphone having a heart-shaped or cardioid response pattern, so it has nearly uniform response for a range of about 180° in one direction and minimum response in the opposite direction. In one form it is a combination of dynamic-microphone and ribbon-microphone elements.

**cardiotachometer** An electronic amplifier that times and records pulse rates of the heart.

**card punch** A mechanism or machine that punches information into cards in code form.

**card reader** A mechanism that senses information punched on cards, using wire brushes, metal feelers, or a photoelectric system.

**Carey-Foster bridge** A type of Wheatstone bridge used to measure the difference between nearly equal resistances. Coils connected by a slide wire serve for two ratio arms.

**carnauba wax** A natural wax used for insulating purposes, having a melting point of 85°C.

**carnotite** An ore of uranium, consisting of hydrous vanadate of uranium, potassium, and other elements.

**carpet** A noise-modulated airborne jamming transmitter or other electronic device used for radar jamming.

**carriage** A device that moves in a predetermined path and carries some other part, such as a recorder head.

**carrier** 1. The radio wave produced by a transmitter when there is no modulating signal, or any other wave, recurring series of pulses, or direct current capable of being modulated. Also called carrier wave. 2. A mobile conduction electron or hole in a semiconductor. 3. A substance that, when associated with a radioactive trace of another substance, will carry the trace with it through a chemical or physical process. An isotope is often used for this purpose. Also called isotopic carrier. 4. A wave generated locally at a receiver that, when combined with the sidebands of a suppressed-carrier transmission in a suitable detector, produces the modulating wave. 5. *Carrier system.*

**carrier amplifier** A direct-current amplifier in which the d-c input signal is filtered by a low-pass filter, then used to modulate a carrier so it can be amplified conventionally as an a-c signal. The amplified d-c output is obtained by rectifying and filtering the rectified carrier signal. A more common version of the carrier amplifier is the chopper amplifier, which uses either one or two choppers to convert the d-c input signal into a square-wave a-c signal and synchronously rectify the amplified square-wave signal.

**carrier-amplitude regulation** The change in amplitude of the carrier wave in an amplitude-modulated transmitter when modulation is applied symmetrically. Also called carrier shift (deprecated).

**carrier beat** An undesirable heterodyne of facsimile signals, each synchronous with a different stable reference oscillator, causing a pattern in received copy. When one or more of the oscillators is controlled by a tuning fork, the effect is also called fork beat.

**carrier channel** The equipment and lines that make up a complete carrier-current circuit between two or more points.

**carrier chrominance signal** *Chrominance signal.*

**carrier-controlled approach system** An aircraft-carrier radar system providing information by which aircraft approaches may be directed by radio.

**carrier current** A higher-frequency alternating current superimposed on ordinary telephone, telegraph, and power-line frequencies for communication and control purposes. The carrier current is modulated with voice signals to provide telephone communication between points on the power system or is tone-modulated to actuate switching relays or convey data.

**carrier-free isotope** A minute quantity of a radioisotope of an element, essentially undiluted by a carrier.

**carrier frequency** The frequency generated by an unmodulated radio, radar, carrier communication, or other transmitter, or the average frequency of the emitted wave when modulated by a symmetrical signal. Also called center frequency and resting frequency.

**carrier-frequency pulse** A carrier that is amplitude-modulated by a pulse. The amplitude of the modulated carrier is zero before and after the pulse.

**carrier-frequency stereo disk** A stereo phonograph record in which one channel is cut laterally in the usual manner. The second channel is used to frequency-modulate a supersonic carrier frequency that is also cut laterally. The playback cartridge delivers the signal for one channel plus the carrier containing the other channel. The carrier must then be demodulated to furnish the second channel.

**carrier leak** The carrier frequency remaining in a suppressed-carrier system.

**carrier level** The strength or level of an unmodulated carrier signal at a particular point in a radio system, expressed in decibels in relation to some reference level.

**carrier line** Any transmission line used for multiple-channel carrier communication.

**carrier mobility** The average drift velocity of carriers per unit electric field in a homogeneous semiconductor. The mobility of electrons is usually different from that of holes.

**carrier modulation** The process of varying some characteristic of a carrier in accordance with a modulating wave.

**carrier noise level** The noise level produced by undesired variations of an r-f signal in the absence of any intended modulation. Also called residual modulation.

**carrier primary flow** The current flow that is responsible for the major properties of a semiconductor device.

**carrier repeater** A one-way or two-way repeater used in a carrier channel.

**carrier shift** Deprecated term for *carrier-amplitude regulation.*

**carrier signaling** Use of tone signals for ringing and other signaling functions in a carrier communication system.

**carrier suppression** 1. Suppression of the carrier frequency after conventional modulation at the transmitter, with reinsertion of the carrier at the receiving end before demodulation. 2. Suppression of the carrier when there is no modulation signal to be transmitted. Used on ships to reduce interference between transmitters.

**carrier swing** The total deviation of a frequency-modulated or phase-modulated wave from the lowest instantaneous frequency to the highest instantaneous frequency.

**carrier system** A system permitting a number of independent communications over the same circuit. Also called carrier.

**carrier telegraphy** Telegraphy in which a single-frequency carrier wave is modulated by the transmitting apparatus for transmission over wire lines.

**carrier telephony** Telephony in which a single-frequency carrier wave is modulated by a voice-frequency signal, for transmission over wire lines.

**carrier-to-noise ratio** The ratio of the magnitude of the carrier to that of the noise after selection and before any nonlinear process such as amplitude limiting and detection.

**carrier transmission** Transmission in which a single-frequency carrier wave is modulated by the signal to be transmitted.

**carrier wave** *Carrier.*

**carry** A signal produced in a computer when the sum of two digits in the same column equals or exceeds the base of the number system in use or when the difference between two digits is less than zero.

**Cartesian control** A guided-missile control that is dependent on two sets of control surfaces, each producing movement perpendicular to that of the other.

**cartridge** 1. *Phonograph pickup.* 2. *Tape cartridge.*

**cascade** *Avalanche.*

**cascade amplifier** A vacuum-tube amplifier containing two or more stages arranged in the conventional series manner.

**cascade-amplifier klystron** A klystron having three resonant cavities to provide increased power amplification and output. The extra resonator, located between the input and output resonators, is excited by the bunched beam emerging from the first resonator gap and produces further bunching of the beam.

**cascade connection** A series connection of amplifier stages, networks, or tuning circuits in which the output of one feeds the input of the next. Also called tandem connection.

**cascade control** An automatic control system in which various control units are linked in sequence, each control unit regulating the operation of the next control unit in line.

**cascaded carry** A carry that uses the normal adding circuit in a computer.

**cascade limiter** A limiter circuit that uses two vacuum tubes in series to give improved limiter operation for both weak and strong signals in an f-m receiver. Also called double limiter.

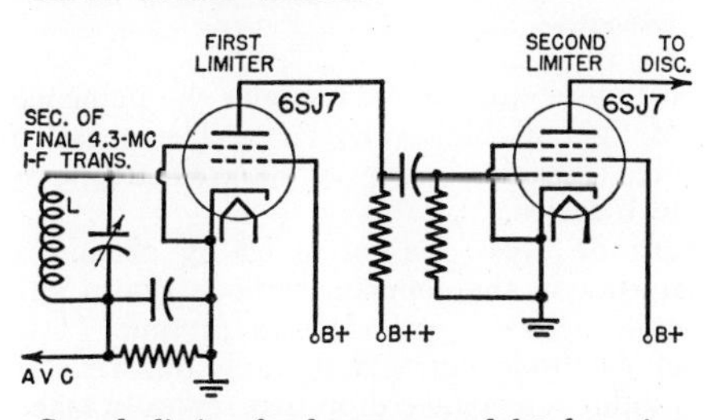


Cascade limiter for frequency-modulated receiver.

**cascade shower** A cosmic-ray shower that is initiated when a high-energy electron, in its passage through matter, produces one or more photons having energies comparable with its own. These photons convert into electrons and positrons by the process of pair production. The secondary electrons in turn produce the same effects as the primary. The cascade shower of electrons and positrons builds up until the level of energy is so low that photon emission and pair production can no longer occur.

**cascode amplifier** An amplifier consisting of a grounded-cathode input stage that drives a grounded-grid output stage. Advantages include high gain and low noise. Widely used in television tuners.

**casket** *Coffin.*

**cataphoresis** *Electrophoresis.*

**catcher** *Output resonator.*

**catcher resonator** *Output resonator.*

**catcher space** The space between the output resonator grids in a velocity-modulated tube, where the density-modulated electron beam excites oscillations in the output resonator.

**catching diode** A diode connected to act as a short-circuit when its anode becomes positive. The diode then prevents the voltage of a circuit terminal from rising above the diode cathode voltage.

**cathode** 1. [symbol K] The primary source of electrons in an electron tube. In directly heated tubes the filament is the cathode. In indirectly heated tubes a coated metal cathode surrounds a heater. Other types of

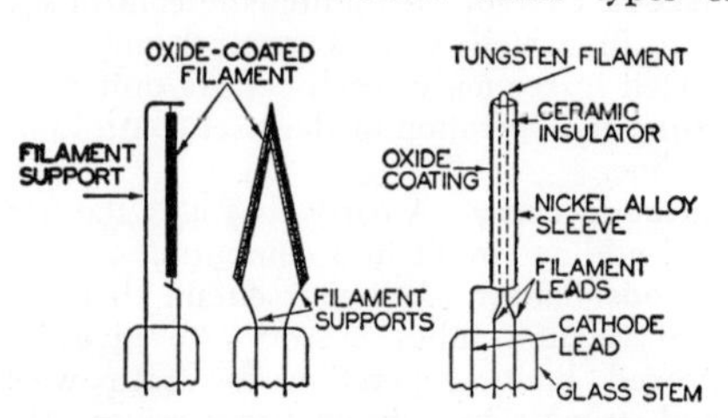


Cathodes for electron tubes. At left are two directly heated types, and at right is heater-type cathode.

cathodes emit electrons under the influence of light or high voltage. 2. The negative electrode of a battery or other electrochemical device.

**cathode bias** Bias obtained by placing a resistor in the common cathode return circuit, between cathode and ground. Flow of electrode currents through this resistor produces a voltage drop that serves to make the control grid negative with respect to the cathode.

**cathode-coupled amplifier** A cascade amplifier in which the coupling between two stages is provided by a common cathode resistor.

**cathode dark space** The relatively nonluminous region between the cathode glow and the negative glow in a glow-discharge cold-cathode tube. Also called Crookes dark space.

**cathode disintegration** The destruction of the active area of a cathode by positive-ion bombardment.

**cathode follower** A vacuum-tube circuit in which the input signal is applied between the control grid and ground, and the load is connected between the cathode

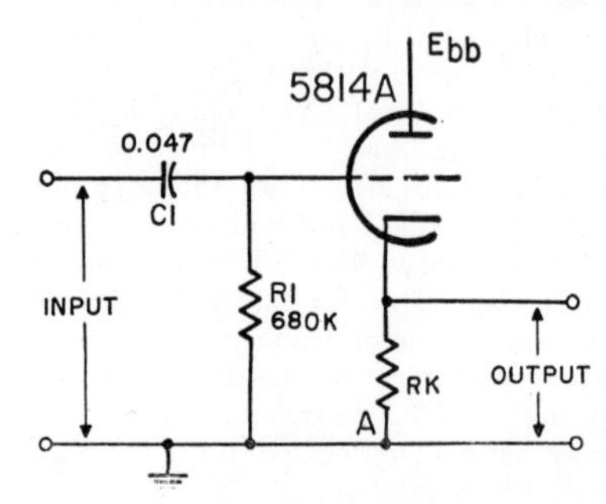


Cathode follower for pulses, using preferred circuit design by National Bureau of Standards. Value of RK is 10,000 ohms when anode voltage is 150 volts.

and ground. A cathode follower has a low output impedance, high input impedance, and a gain of less than unity. The anode is at ground potential at the operating frequency. Also called grounded-anode amplifier.

**cathode glow** The luminous glow that covers all or part of the cathode in a glow-discharge cold-cathode tube.

**cathode interface impedance** The impedance between the cathode base and coating in an electron tube, due to a high-resistivity layer or a poor mechanical bond.

**cathode keying** Transmitter keying by means of a key in the cathode lead of the keyed vacuum-tube stage, opening the d-c circuits for the grid and anode simultaneously.

**cathode luminous sensitivity** The photoelectric emission current divided by the luminous flux on a photocathode under specified conditions of illumination.

**cathode modulation** Amplitude modulation accomplished by applying the modulating voltage to the cathode circuit of an electron tube in which the carrier is present.

**cathode poisoning** The chemical effect of

residual gases on the emissivity of the cathode of an electron tube.

**cathode preheating time** The minimum period of time during which heater voltage should be applied before electrode voltages are applied in an electron tube.

**cathode pulse modulation** Modulation produced in an amplifier or oscillator by applying externally generated pulses to the cathode circuit.

**cathode radiant sensitivity** The photoelectric emission current divided by the radiant flux on a photocathode at a given wavelength, under specified conditions of irradiation.

**cathode ray** A stream of electrons, such as that emitted by a heated filament in a tube, or that emitted by the cathode of a gas-discharge tube when the cathode is bombarded by positive ions.

**cathode-ray oscillograph** A cathode-ray oscilloscope in which a photographic or other permanent record is produced by the electron beam of the cathode-ray tube. Also called oscillograph.

**cathode-ray oscilloscope** [abbreviated cro] A test instrument that uses a cathode-ray tube to make visible on a fluorescent screen the instantaneous values and waveforms of electrical quantities that are rapidly varying as a function of time or another quantity. Also called oscilloscope and scope.

**cathode-ray tube** [abbreviated crt] An electron-beam tube in which the electrons emitted by a hot cathode are formed by an electron gun into a narrow beam that can

FLUORESCENT SCREEN
+ ++
F K G A1 A2 V H
ELECTRON BEAM
NECK
SPOT
ELECTRON GUN
DEFLECTION PLATES

F--FILAMENT--Pins 1 & 11
K--CATHODE--Pin 11
G--CONTROL GRID--Pin 10
A1--FIRST ANODE--Pin 4
A2--SECOND ANODE--Pin 7
V--VERT. DEFL. PLATES--Pins 6 & 9
H--HORIZ. DEFL. PLATES--Pins 3 & 8

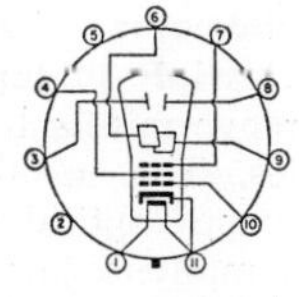

Cathode-ray tube having electrostatic deflection, with symbol at lower right.

be focused to a small cross-section on a fluorescent screen. The beam can be varied in position and intensity by internal electrostatic deflection plates or external electromagnetic deflection coils to produce a visible trace, pattern, or picture on the screen. Originally called Braun tube. Also called kinescope and picture tube when used in television receivers.

**cathode-ray tube display** The presentation of a received signal on the screen of a cathode-ray tube.

**cathode-ray tuning indicator** A small cathode-ray tube having a fluorescent pattern whose size varies with the voltage

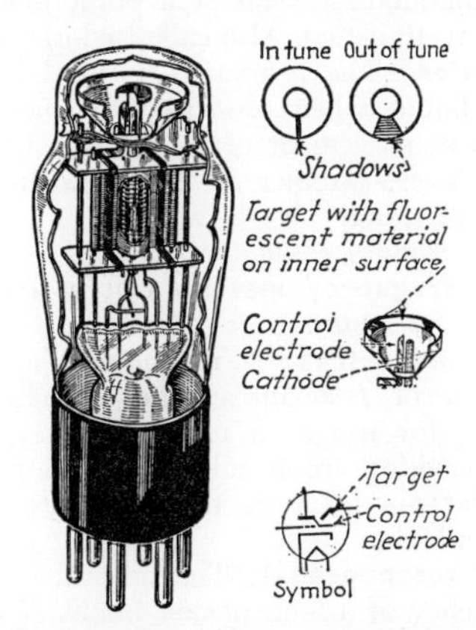


Cathode-ray tuning indicator construction, operation, and symbol.

applied to the grid. Used in radio receivers to indicate accuracy of tuning. Also used as a modulation indicator in some tape recorders, and used in place of a meter in some instruments. Also called electron-ray indicator.

**cathode resistor** A resistor used in the cathode circuit of a vacuum tube, having a resistance value such that the voltage drop across it due to tube current provides the correct negative grid bias for the tube.

**cathode spot** The small cathode area from which an arc appears to originate in a discharge tube.

**cathode sputtering** *Sputtering.*

**cathodoluminescence** Luminescence produced by high-velocity electrons. When these electrons bombard a metal in a vacuum, small amounts of the metal are vaporized in an excited state and emit radiation characteristic of the metal.

**cathodophosphorescence** Phosphorescence produced when high-velocity electrons bombard a metal in a vacuum.

**cation** A positive ion.

**catwhisker** A sharply pointed flexible wire used to make contact with the surface of a

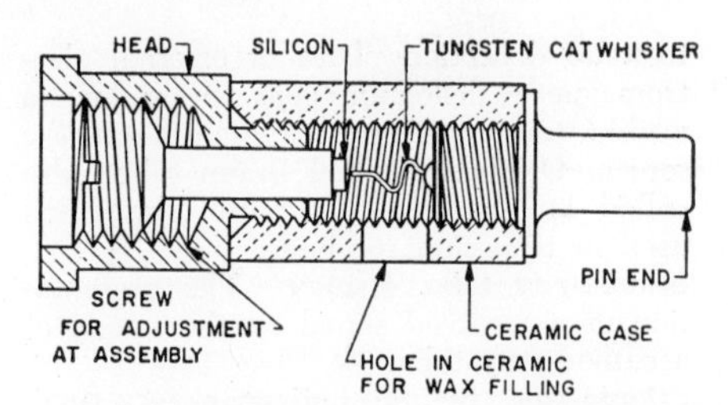


Catwhisker in microwave crystal diode.

semiconductor crystal at a point that provides rectification. Also called whisker.

**cavitation** The formation of local cavities in a liquid when pressure is reduced because of movement of a body through the fluid. These bubbles are beneficial in ultrasonic cleaning.

**cavity** *Cavity resonator.*

**cavity frequency meter** *Cavity-resonator frequency meter.*

**cavity magnetron** A magnetron having a number of resonant cavities forming the anode, for use as a microwave oscillator. The cavities are usually radial slots in a circular anode, with the openings facing the cathode.

**cavity resonance** 1. The natural resonant frequency of a loudspeaker baffle. If in the audio range, it is evident as unpleasant emphasis of sounds at that frequency. 2. The resonant frequency of a cavity resonator.

**cavity resonator** A space totally enclosed by a metallic conductor and excited in such a way that it becomes a source of electromagnetic oscillations. The size and shape of the enclosure determine the resonant frequency. Used with klystrons, magnetrons, and other microwave devices. Also called cavity, resonant cavity, resonant chamber, resonant element, rhumbatron, tuned cavity, and waveguide resonator.

**cavity-resonator frequency meter** A variable cavity resonator used to determine the frequency of an electromagnetic wave. Also called cavity frequency meter and cavity-resonator wavemeter.

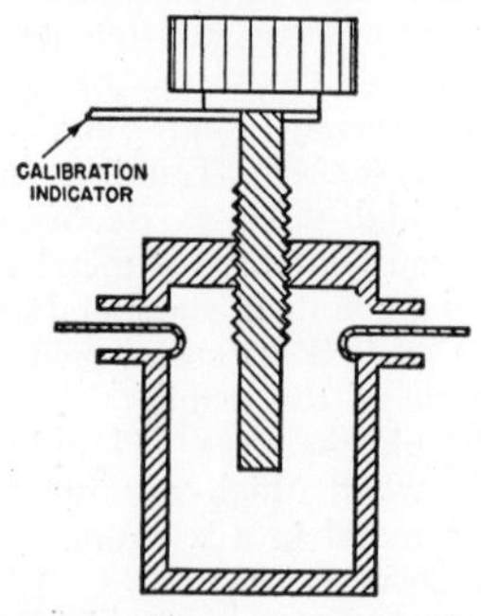


Cavity-resonator frequency meter for use with coaxial cable.

**cavity-resonator wavemeter** *Cavity-resonator frequency meter.*

**cavity-type diode amplifier** *Diode amplifier.*

**C band** The r-f band from 3.9 to 612 kilomegacycles, equivalent to wavelengths from 11.8 to 7.3 cm.

**C battery** The battery that supplies the steady bias voltage required by the control-grid electrodes of electron tubes in battery-operated equipment.

**C bias** *Grid bias.*

**C core** A spirally wound magnetic core that is formed to a desired rectangular shape before being cut into two C-shaped pieces and placed around a transformer or magnetic amplifier coil.

**C display** A rectangular radarscope display in which targets appear as bright spots,

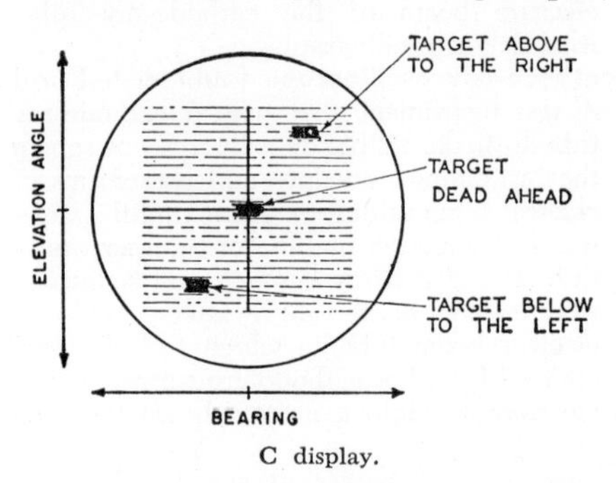


C display.

with target bearing indicated by the horizontal coordinate and target angle of elevation by the vertical coordinate. Also called C scan.

**ceiling-height indicator** A photoelectric instrument for measuring the height of a cloud ceiling with the aid of a vertical beam of light. Also called ceilometer and cloud-height detector.

**ceilometer** *Ceiling-height indicator.*

**celestial guidance** Guidance of a long-range missile by reference to celestial bodies. The missile is equipped with gyroscopes, optical or radio star trackers, servos, computers, and other devices that together sight stars, calculate positions, and direct the missile. The Snark is an example. Also called stellar guidance.

**celestial-inertial guidance** An inertial guidance system into which supplementary position and/or velocity information is fed by celestial navigation equipment that is also built into the missile.

**celestial navigation** Navigation by means of observations of celestial bodies.

**celestial radio tracking** A navigation technique wherein the microwave emanations of the sun, moon, or certain stars are used to determine the position of a missile or other object in space.

**cell** 1. A single unit of a primary or secondary battery that converts chemical energy into electric energy. Also called electric cell. 2. A single unit of a device that converts radiant energy into electric energy, such as a nuclear cell, solar cell, or photovoltaic cell. 3. A single unit of a device whose resistance varies with radiant energy, such as a selenium cell. 4. An elementary unit of storage in a computer, such as a binary cell or decimal cell.

**cell-type tube** A gas-filled tr, atr, or pre-tr tube that operates in an external resonant circuit. A tuning mechanism may be incorporated in the external resonant circuit or in the tube.

**cellular horn** *Multicellular horn.*

**cellulose acetate** A thermoplastic material that is widely used as the base for magnetic tape and movie film. It can be made transparent or opaque in various colors. It is tough, flexible, slow-burning, and long-lasting. Used for molding receiver cabinets. Also called acetate.

**cellulose nitrate disk** *Lacquer disk.*

**Celsius** [abbreviated C; after Anders Celsius] *Centigrade.* Actually, the two scales differ slightly because the Celsius scale uses the triple point of water, at 0.01° centigrade, in place of the ice point as a reference because of its greater convenience in theoretical computations. Although the term centigrade was officially changed to Celsius at an international conference in 1948, public acceptance has been slow in Europe and still slower in the United States.

**cent** 1. The interval between two sounds whose basic frequency ratio is the twelve-hundredth root of two. The interval, in cents, between any two frequencies is 1,200 times the logarithm to the base 2 of the frequency ratio. Thus, 1,200 cents = 12 equally tempered semitones = 1 octave. 2. A unit of nuclear reactivity equal to one-hundredth of a dollar.

**center-coupled loop** A coupling loop in the center of one of the resonant cavities of a multicavity magnetron.

**center expansion** *Expanded-center ppi display.*

**center frequency** *Carrier frequency.*

**center-frequency stability** The ability of a transmitter to maintain an assigned center frequency in the absence of modulation.

**centering** The process of adjusting the position of the trace or image on a cathode-ray tube screen so it is centered on the screen.

**centering control** One of the two controls used for positioning the image on the screen of a cathode-ray tube. The horizontal centering control moves the image horizontally, while the vertical centering control moves the image vertically. Centering is achieved by adjusting the d-c voltage applied to deflection plates or by adjusting the direct current flowing through deflection coils.

**centering diode** A clamping circuit used in some types of ppi indicators.

**center line** The locus of points equidistant from two reference points or lines. In loran the center line is the perpendicular bisector of the baseline connecting the master and slave stations.

**center tap** [symbol CT] A terminal at the electrical midpoint of a resistor, coil, or other device.

**center-tap keying** A method of keying a radiotelegraph transmitter by interrupting the anode return lead going to the filament transformer center-tap. Used when modulating a filament-type tube.

**centi-** A prefix representing $10^{-2}$.

**centigrade** [abbreviated C] The metric temperature scale, in which the interval between the freezing and boiling points of water is divided into 100 equal parts or degrees, with 0°C as the freezing point and 100°C as the boiling point. Absolute zero is −273.16°C. Also called Celsius, although the Celsius scale uses 0.01°C as the cold reference. To change degrees centigrade to degrees Fahrenheit, multiply by 9/5 and add 32 to the result. Fahrenheit and centigrade scales agree at one point: −40°F = −40°C.

**centimeter** [abbreviated cm] A unit of length in the metric system, equal to 0.01 meter or 0.394 inch.

**centimeter-gram-second electromagnetic unit** [abbreviated cgs emu] A cgs unit based on the assignment of unity to the strength of each of two like magnetic poles that repel each other with a force of 1 dyne at a distance of 1 centimenter in a vacuum.

**centimeter-gram-second electrostatic unit** [abbreviated cgs esu] A cgs unit based on assignment of unity to the constants in an equation for the electrostatic force between unit charges in a vacuum.

**centimeter-gram-second unit** [abbreviated cgs unit] An absolute unit based on the centimeter, gram, and second as fundamental units.

**centimeter waves** Microwaves having wavelengths between 1 and 10 centimeters, corresponding to frequencies between 3 and 30 mc.

**central force** A nuclear force that is a simple attraction or repulsion directed along the line joining a pair of nucleons.

**central potential** A nuclear potential that is spherically symmetric, so the potential energy of a particle is the same in all directions and is a function only of the distance from the center of the field.

**centrifugal separation** Separation of isotopes by spinning a mixture in gas or vapor form at high speed. Heavier isotopes then concentrate near the outside of the container, while lighter isotopes concentrate near the axis of the centrifuge.

**centrifugal switch** A switch that is opened or closed by centrifugal force. Used on some induction motors to open the starting winding when the motor has almost reached synchronous speed.

**ceramet** *Metal-ceramic.*

**ceramic capacitor** A capacitor whose dielectric is a ceramic material such as steatite or barium titanate, the composition of which can be varied to give a wide range of temperature coefficients. The electrodes are usually silver coatings fired on opposite sides of the ceramic disk or slab, or fired on the inside and outside of a ceramic tube. After connecting leads are soldered to the electrodes, the unit is usually given a protective insulating coating.

Ceramic capacitor types, and methods of applying color codes to indicate capacitance values in micromicrofarads and other significant characteristics.

**ceramic cartridge** A device containing a piezoelectric ceramic element, used in phonograph pickups and microphones. Ceramic cartridges deliver somewhat lower output voltage than crystal cartridges but are less affected by heat and humidity.

**ceramic microphone** A microphone using a ceramic cartridge.

**ceramic pickup** A phonograph pickup using a ceramic cartridge.

**ceramic reactor** A nuclear reactor in which the fuel and moderator assemblies are made from high-temperature-resistant ceramic materials such as metal oxides, carbides, or nitrides.

**ceramic tube** An electron tube having a ceramic envelope capable of withstanding operating temperatures of over 500°C, as required to withstand reentry temperatures of guided missiles.

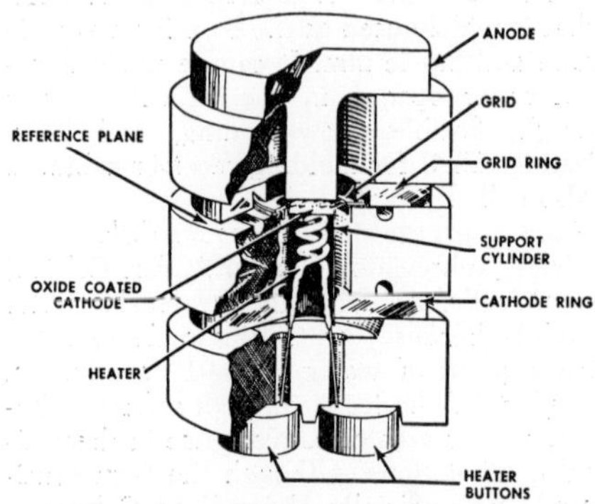


Ceramic tube construction.

**ceramoplastic** A high-temperature insulating material made by bonding synthetic mica with glass.

**Cerenkov radiation** Visible light produced when charged particles traverse a transparent medium with a velocity exceeding the velocity of light in the medium. The index of refraction of the medium must be considerably greater than unity.

**cerium** [symbol Ce] A rare-earth metallic element. Atomic number is 58.

**cesium** [symbol Cs] An alkali metallic element used in forming the cathodes of certain types of phototubes. Atomic number is 55.

**cesium-ion engine** An ion engine that uses

a stream of cesium ions to produce a thrust for space travel.

**cesium phototube** A phototube having a cesium-coated cathode. It has maximum sensitivity in the infrared portion of the spectrum.

**cesium-vapor lamp** A lamp in which light is produced by the passage of current between two electrodes in ionized cesium vapor.

**cesium-vapor rectifier** A gas tube in which cesium vapor serves as the conducting gas and a condensed monatomic layer of cesium serves as the cathode coating. The tube is heated to about 180°C to give the desired vapor pressure.

**cgs emu** Abbreviation for *centimeter-gram-second electromagnetic unit.*

**cgs esu** Abbreviation for *centimeter-gram-second electrostatic unit.*

**cgs unit** Abbreviation for *centimeter-gram-second unit.*

**chadless tape** Paper tape in which partial perforations are used to represent code characters so that corresponding letters and numerals can be printed on the tape.

**chad tape** Paper tape in which perforations for code characters are completely punched.

**chaff** A confusion reflector consisting of narrow metallic or metal-coated paper strips cut to various lengths that are resonant at expected enemy radar frequencies. When dropped in clusters, chaff gives strong echo signals on enemy radarscopes. Chaff is one type of window.

**chain** A network of radio, television, radar, navigation, or other similar stations connected by special telephone lines, coaxial cables, or radio relay links so all can operate as a group for broadcast purposes, communication purposes, or determination of position.

**chain decay** *Series disintegration.*

**chain disintegration** *Series disintegration.*

**chain fission yield** The sum of the independent fission yields for all isobars of a particular mass number.

**chain home** An early-warning and aircraft height-finding radar system used in the British Isles. The word home refers to the British Isles.

**chain reaction** A reaction in which one of the agents necessary to the reaction is itself produced by the reaction, thereby causing additional reactions. In the neutron-fission chain reaction, a neutron plus a fissionable atom cause a fission, resulting in a number of neutrons that in turn cause other fissions.

**challenge** *Interrogation.*

**challenger** *Interrogator.*

**challenging signal** *Interrogation.*

**chance coincidence** *Accidental coincidence.*

**chance failure** Failure within the operational time period of a piece of equipment after all efforts have been made to eliminate design defects and unsound components, but before any foreseen wearout phenomena have time to appear.

**channel** 1. A band of radio frequencies allocated for a particular purpose. A standard broadcasting channel is 10 kc wide, whereas a television channel is 6 mc wide. 2. A path for a signal. An audio amplifier may have several input channels. A stereo amplifier has at least two complete channels. 3. A path for carrier-current signals. 4. A passage for fuel slugs or heat-transfer fluid in a reactor. 5. A path along which digital or other information may flow in a computer. 6. The section of a storage medium that is accessible to a given reading station in a computer, such as a path parallel to the edge of a magnetic tape or drum or a path in a delay-line memory.

**channel effect** A leakage current flowing over a surface path between the collector and emitter in some types of transistors.

**channeling** 1. A type of multiplex transmission in which the separation between communication channels is accomplished by the use of carriers or subcarriers. 2. The extra transmission of particles through a medium in a nuclear reactor due to the presence of the voids in the medium.

**channeling-effect factor** Attenuation per unit length of equivalent homogeneous material divided by attenuation per unit length of the material containing voids.

**channel selector** A switch or other control used to tune in the desired channel in a television receiver.

**channel shifter** A radiotelephone carrier circuit that shifts one or two voice-frequency channels from normal channels to higher voice-frequency channels to reduce crosstalk between channels. The channels are shifted back by a similar circuit at the receiving end.

**character** A letter, digit, elementary mark, or event that may be used in various combinations to express information that a computer can read, store, or write. A group of characters in one context may become a single character in another, as in the binary-coded decimal system.

**characteristic curve** A curve plotted on graph paper to show the relation between two changing values.

**characteristic impedance** The impedance that, when connected to the output terminals of a transmission line of any length, makes the line appear to be infinitely long. There are then no standing waves on the line, and the ratio of voltage to current is the same for each point on the line. For a waveguide, the characteristic impedance is the ratio of rms voltage to total rms longitudinal current at specified points on a diameter when the guide is match-terminated. For an acoustic device it is the ratio of the effective sound pressure at a point to the effective particle velocity at that point. Also called surge impedance.

**characteristic radiation** Radiation originating in an atom following removal of an electron. The wavelength of the emitted radiation depends only on the element concerned and the energy levels involved.

**characteristic telegraph distortion** Distortion that does not affect all signal pulses alike. The effect on each transition depends on the signal previously sent, owing to remnants of previous transitions or transients that persist for one or more pulse lengths.

**characteristic wave impedance** The ratio of the transverse electric vector to the transverse magnetic vector at a point traversed by a traveling electromagnetic wave.

**characteristic x-rays** Electromagnetic radiation emitted as a result of rearrangements of the electrons in the inner shells of atoms. The spectrum of the radiation consists of lines that are characteristic of the element in which the x-rays are produced. The target of an x-ray tube will in general emit both continuous x-rays and characteristic x-rays.

**character reader** A photoelectric device that scans printed alphanumeric characters and delivers corresponding machine-readable code characters that can be fed to a computer, tape punch, line printer, electric typewriter, or other machine. Also called photoelectric character reader.

**character-writing storage tube** A character-writing tube that retains its display as long as the necessary operating voltages are supplied, but can be erased by lowering one electrode voltage momentarily.

**character-writing tube** A cathode-ray tube that forms alphanumeric and symbolic characters on its screen for viewing or recording purposes. The Charactron, Scriptron, and Typotron are examples. Also called display tube.

**Charactron** Trademark of Stromberg-Carlson for their cathode-ray tube that produces a display in the form of letters or numbers.

**charge** 1. The quantity of electric energy stored in a capacitor, battery, or insulated object. Also called electric charge. 2. The material or part to be heated by induction or dielectric heating. 3. The fissionable material or fuel placed in a reactor to produce a chain reaction.

**charge density** The charge per unit area on a surface or per unit volume in space.

**charge-exchange phenomenon** The phenomenon in which a positive ion possessing sufficient kinetic energy is neutralized by colliding with a molecule and capturing an electron from it. The molecule is transformed into a positive ion.

**charge-mass ratio** The ratio of the electric charge of a particle to its mass.

**charger** *Battery charger.*

**charger-reader** An auxiliary device used to charge and read small portable ionization chambers.

**charge storage tube** A storage tube in which information is retained on a surface in the form of electric charges.

**charge-transfer spectrum** The spectrum caused by transition of an electron from a bonding orbital to an antibonding orbital.

**charging current** The current that flows into a capacitor when a voltage is first applied.

**chart** The paper or other material on which a graphic record is made by a recording instrument.

**chart comparison unit** A device that permits simultaneous viewing of a radar ppi display and a navigation chart in such a manner that one appears superimposed on the other. Also called autoradar plot.

**chassis** The metal frame on which circuit components are mounted. Plural is same as singular.

**chassis punch** A hand tool used to make round or square holes in sheet metal.

**chatter** 1. Prolonged undesirable opening and closing of electric contacts, as on a relay. Also called contact chatter. 2. Vibration of a disk recorder cutting stylus in a direction other than that in which it is driven.

**check** Partial or complete testing of the accuracy of a computer operation.

**check digit** A redundant digit in a self-checking code for a computer, used to indicate a malfunction. Thus, a given check digit may be zero if the sum of other digits in the word is odd, or one if the sum of other digits in the word is even. When a single error occurs in a word, the sum of the word digits no longer agrees with the check digit.

**check point** *Way point.*

**check problem** *Check routine.*

**check routine** A routine or problem designed primarily to indicate whether a fault exists in a computer, without giving detailed information on the location of the fault. Also called check problem, test program, and test routine.

**cheese antenna** An antenna having a parabolic reflector between two metal plates, dimensioned to permit propagation of more

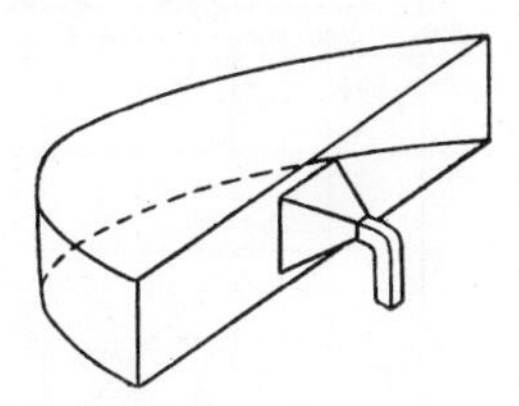

Cheese antenna fed by horn at end of rectangular waveguide.

than one mode in the desired direction of polarization. It is fed by either a dipole or flared-out waveguide facing into the reflector cavity.

**chemical binding effect** The dependence of the neutron cross sections of a material on the chemical binding of the atoms composing the material.

**chemically deposited printed circuit** A printed circuit formed on a base by the reaction of chemicals alone. Dielectric, magnetic, and conductive circuits can be applied.

**chemical tracer** A tracer having chemical properties similar to those of the substance with which it is mixed.

**chemosphere** The portion of the earth's atmosphere between altitudes of about 20 and 50 miles, in which photochemical activity is present. Here nitric oxide forms an ionized cloud that reflects radio signals.

**Child's law** An equation stating that the current in a thermionic diode varies directly with the three-halves power of anode voltage and inversely with the square of the distance between the electrodes, provided operating conditions are such that the current is limited only by the space charge.

**chip** The material removed from the recording medium by the recording stylus while cutting the groove in mechanical recording. Also called thread.

**chlorine** [symbol Cl] A gaseous element. Atomic number is 17.

**choke** 1. An inductance used in a circuit to present a high impedance to frequencies above a specified frequency range without

Chokes used in electronic circuits.

appreciably limiting the flow of direct current. Also called choke coil. 2. A groove or other discontinuity in a waveguide surface so shaped and dimensioned as to impede the passage of guided waves within a limited frequency range.

**choke coil** *Choke.*

**choke coupling** Coupling between two parts of a waveguide system that are not in direct mechanical contact with each other.

**choke flange** A waveguide flange having in its mating surface a slot so shaped and dimensioned as to restrict leakage of microwave energy within a limited frequency range.

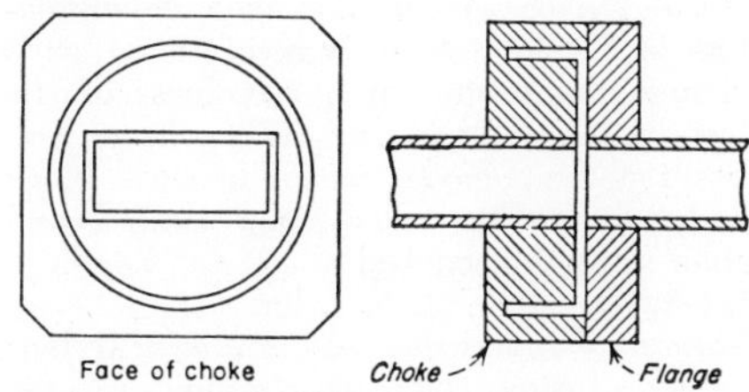


Choke flange construction.

**choke-input filter** A power-supply filter in which the first filter element is a series choke.

**choke joint** A connection between two waveguides that uses two mating choke flanges to provide effective electric continuity without metallic continuity at the inner walls of the waveguide.

**choke piston** A piston in which there is no metallic contact with the walls of the waveguide at the edges of the reflecting surface. The short-circuit for high-frequency currents is achieved by a choke system. Also called noncontacting piston and noncontacting plunger.

**chopper** A device for interrupting a current, beam of light, or beam of infrared radiation at regular intervals, to permit amplification of the associated electrical quantity or signal by an a-c amplifier.

**chopper amplifier** A carrier amplifier in which the d-c input is filtered by a low-pass filter, then converted into a square-wave a-c signal by either one or two choppers. After amplification, the square-wave signal is synchronously rectified by chopper action to obtain the desired amplified d-c output.

**chopper-stabilized amplifier** A direct-current amplifier in which a direct-coupled amplifier is in parallel with a chopper amplifier. The chopper amplifier provides increased stability against drift in the direct-coupled amplifier, particularly when negative feedback is used. Conversely, the direct-coupled amplifier extends the frequency range beyond that of the chopper amplifier.

**Christmas-tree pattern** *Optical pattern.*

**chroma** The dimension of the Munsell system of color that corresponds most closely to saturation, which is the degree of vividness of a hue. Chroma is frequently used, particularly in English works, as the equivalent of saturation. Also called Munsell chroma.

**chroma control** The control that adjusts the amplitude of the carrier chrominance signal fed to the chrominance demodulators in a color television receiver, so as to change the saturation or vividness of the hues in the color picture. When in its zero position, the received picture becomes black and white. Also called color control and color-saturation control.

**chromatic** Relating to color.

**chromatic aberration** 1. An optical lens defect causing color fringes, because the lens material brings different colors of light to a focus at different points. An achromatic lens corrects for this error. 2. An electron-gun defect causing enlargement and blurring of the spot on the screen of a cathode-ray tube, because electrons leave the cathode with different initial velocities and are hence deflected differently by the electron lenses and deflection coils.

**chromaticity** The color quality of light that can be defined by its chromaticity coordinates. Chromaticity depends only on hue and saturation of a color, and not on its luminance (brightness). Chromaticity applies to all colors, including shades of gray, whereas chrominance applies only to colors other than grays.

**chromaticity coordinate** One of the two coordinates ($x$ or $y$) that precisely specify the exact identity or chromaticity of a color on the CIE chromaticity diagram. Also called color coordinate and trichromatic coefficient.

**chromaticity diagram** A diagram in which one of the three chromaticity coordinates is plotted against another. The most common version is the CIE chromaticity diagram used in color television.

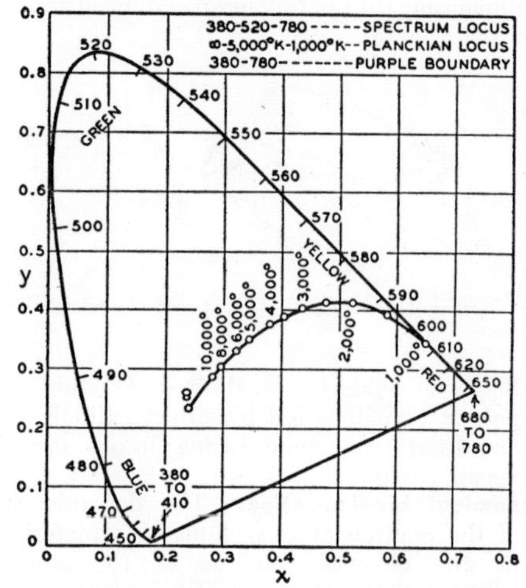


Chromaticity diagram as prepared by CIE.

**chromaticity flicker** Flicker in a color television receiver due to fluctuation of chromaticity only.

**chromatron** A single-gun color picture tube having color phosphors deposited on the screen in strips instead of dots. The red, green, and blue color signals are applied in sequence to the single grid of the tube as the beam is deflected to the correct color strip by horizontal grid wires adjacent to the screen. Also called Lawrence tube.

**chrominance** The difference between any color and a specified reference color of equal brightness. In color television, this reference color is white having coordinates $x = 0.310$ and $y = 0.316$ on the chromaticity diagram.

**chrominance bandwidth** *Chrominance-channel bandwidth.*

**chrominance carrier** *Chrominance subcarrier.*

**chrominance-carrier reference** A continuous signal having the same frequency as the chrominance subcarrier in a color television system and having fixed phase with respect to the color burst. This signal is the reference with which the phase of a chrominance signal is compared for the purpose of modulation or demodulation. In a color receiver it is generated by a crystal-controlled oscillator. Also called chrominance-subcarrier reference, color-carrier reference, and color-subcarrier reference.

**chrominance channel** Any path that is intended to carry the chrominance signal in a color television system.

**chrominance-channel bandwidth** The bandwidth of the path intended to carry the chrominance signal in a color television system. Also called chrominance bandwidth.

**chrominance demodulator** A demodulator used in a color television receiver for deriving the I and Q video-frequency chrominance components from the chrominance signal and the chrominance-subcarrier frequency. Also called chrominance-subcarrier demodulator.

**chrominance modulator** A modulator used in a color television transmitter to generate the I or Q chrominance signal from the video-frequency chrominance components and the chrominance subcarrier. Also called chrominance-subcarrier modulator.

**chrominance primary** The nonphysical color represented by either the I or Q chrominance signal in a color television system. These chrominance signals are chosen to be electrically convenient components remaining after removal of the luminance signal from a full color signal at the transmitter.

**chrominance signal** One of the two components, called the I signal and Q signal, that add together to produce the total chrominance signal in a color television system. Also called carrier chrominance signal.

**chrominance subcarrier** The 3.579545-mc carrier whose modulation sidebands are added to the monochrome signal to convey color information in a color television receiver. The chrominance subcarrier is transmitted unmodulated in the form of color bursts that are used for synchronizing purposes in the receiver. Also called chrominance carrier, color carrier (deprecated).

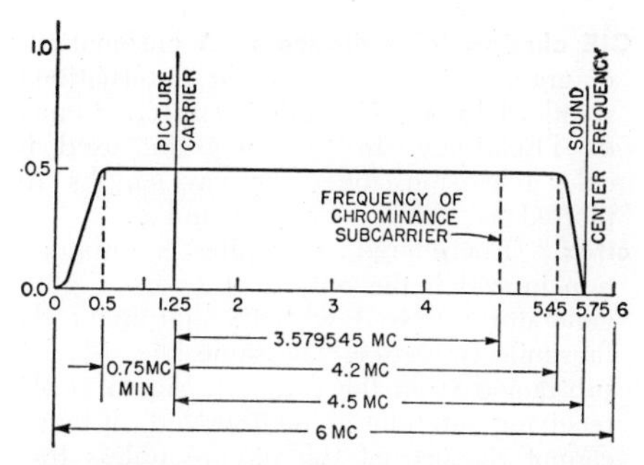


Chrominance subcarrier in standard 6-mc television channel. Vertical scale gives relative maximum radiated field strength with respect to picture carrier.

color subcarrier (deprecated), and subcarrier.

**chrominance-subcarrier demodulator** *Chrominance demodulator.*

**chrominance-subcarrier modulator** *Chrominance modulator.*

**chrominance-subcarrier oscillator** The oscillator, usually crystal-controlled, that generates the subcarrier signal in a color television receiver. Also called color-subcarrier oscillator.

**chrominance-subcarrier reference** *Chrominance-carrier reference.*

**chromium** [symbol Cr] A metallic element. Atomic number is 24.

**chronic exposure** Continued exposure to small doses of nuclear radiation.

**chronistor** A subminiature elapsed-time indicator that uses electroplating principles to totalize operating time of equipment up to several thousand hours.

**chronograph** 1. An instrument that records intervals of time with a high degree of accuracy. 2. An instrument that measures and records projectile velocity by measuring the time required for the projectile to travel a known distance.

**chronometer** A highly accurate clock or watch used in air and surface navigation.

**chronoscope** An electronic instrument used for measuring extremely short intervals of time, such as the time of passage of a rifle bullet between two points.

**chronotron** A device that measures millimicrosecond time intervals between pulses. In one type, interval-determining pulses are fed to a transmission line and the spacing between the pulses on the line is measured.

**chuck** The clamping device that holds the stylus or needle of a phonograph pickup.

**CIE** Abbreviation for *Commission Internationale de l'Eclairage.*

**CIE chromaticity diagram** A chromaticity diagram established as an international standard by the Commission Internationale de l'Eclairage. In this diagram, used in color television, the color wavelengths are plotted as coordinates of $x$ and $y$.

**cifax** Enciphered facsimile communication in which the output of a keyed pulse generator is mixed with the output of the facsimile converter. The same key signal is subtracted from the facsimile signal at the receiving station. Unauthorized listeners cannot reconstruct the picture unless they have an identical key generator and the daily key setting.

**cipher** A transposition or substitution code for transmitting messages secretly.

**ciphony equipment** [CIPHer + telephONY] Any equipment attached to a radio transmitter, radio receiver, or telephone for scrambling or unscrambling voice messages. In one form, speech is converted into a series of on-off pulses that are mixed with pulses obtained from a key generator. To recover the speech at the receiving terminal, an identical key generator must be set to the daily key setting and its output subtracted from the received signal. The on-off pulses can then be converted to the original speech patterns.

**circle-dot mode** A mode of cathode-ray-tube storage of binary digits in which one kind of digit is represented by a small circle of excitation of the screen, and the other kind by a similar circle with a concentric dot.

**circuit** 1. An arrangement of one or more complete paths for electron flow. 2. A complete wire, radio, or carrier communications channel.

**circuit analyzer** *Volt-ohm-milliammeter.*

**circuit breaker** An electromagnetic device that opens a circuit automatically when the current exceeds a predetermined value. It can be reset by operating a lever or by other means.

**circuit diagram** *Schematic circuit diagram.*

**circuit gap admittance** The admittance of the circuit at a gap in the absence of an electron stream.

**circuit-noise meter** An instrument for measuring the electric noise level in a communication circuit, in decibels above reference noise. Also called noise-measuring set.

**circuitry** The complete combination of circuits used in a piece of electronic equipment.

**circular antenna** A folded dipole that is bent into a circle, so the transmission line and the abutting folded ends are at opposite ends of a diameter. When mounted horizontally, it radiates uniformly in all directions and has very little vertical radiation.

**circular electric wave** A transverse electric wave for which the lines of electric force form concentric circles.

**circularly polarized wave** An electromagnetic wave for which the electric and/or magnetic field vectors at a point describe a circle. This term is usually applied to transverse waves.

**circular magnetic wave** A transverse magnetic wave for which the lines of magnetic force form concentric circles.

**circular mil** A unit of area equal to the area of a circle whose diameter is 1 mil (0.001 inch). Used chiefly in specifying cross-sectional areas of round conductors.

**circular polarization** Polarization in which the vector representing the wave has a constant magnitude and rotates continuously about a point.

**circular probable error** The radius of a circle within which half of a given number of guided missiles can be expected to fall.

**circular scanning** Scanning in which the radar antenna rotates in a complete circle, so that the beam generates a plane or a cone whose vertex angle is close to 180°.

**circular system** A system for controlling guided missiles in which the missile automatically sends out signals to two radio beacon stations and times their echoes to keep itself on course.

**circular trace** A time base produced by applying sine waves of the same frequency and amplitude, but 90° out of phase, to the horizontal and vertical deflection plates of a cathode-ray tube. The trace is then a circle, and signals give inward or outward radial deflections from the circle.

**circular waveguide** A waveguide whose cross-sectional area is circular.

**circulating memory** A digital computer device that uses a delay line to store information in the form of a pattern of pulses in a train. The output pulses are detected electrically, amplified, reshaped, and reinserted in the delay line at the beginning. Also called circulating register and circulating storage.

**circulating reactor** A nuclear reactor in which the fissionable material circulates through the core in fluid form or as small particles suspended in a fluid.

**circulating register** *Circulating memory.*

**circulating storage** *Circulating memory.*

**circulator** A waveguide component having a number of terminals so arranged that energy entering one terminal is transmitted

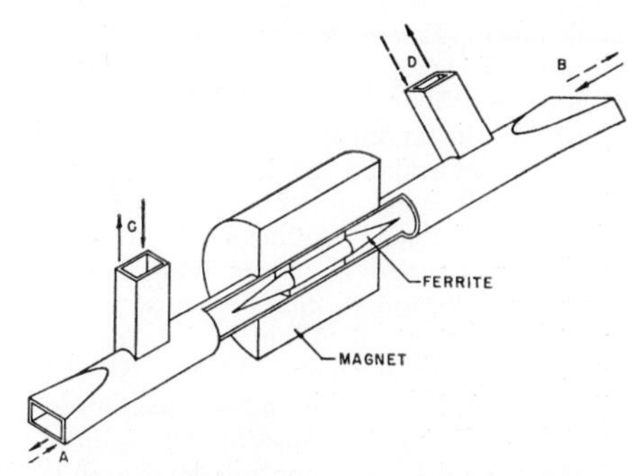


Circulator using ferrite rod positioned in longitudinal magnetic field in circular section of waveguide to rotate plane of polarization in traveling plane wave.

to the next adjacent terminal in a particular direction. Also called microwave circulator.

**circumlunar rocket** A rocket designed and fired to orbit around the moon.

**cislunar space** The space between the earth and the moon.

**citizens' radio service** A radio communication service intended for private or personal radio communication, including radio signaling, control of objects by radio, and other purposes. The frequency bands are 460–470 mc for general use and 27.23–27.28 mc for on-off unmodulated or tone-modulated carrier for remote control.

**cladding** A process of covering one metal with another, usually by bringing the two metals together and rolling, extruding, drawing, or swaging until a bond is produced.

**clamper** *D-c restorer.*

**clamping** The introduction of a reference level that has some desired relation to a pulsed waveform, as at the negative peaks or at the positive peaks. Also called d-c reinsertion and d-c restoration.

**clamping circuit** A circuit that reestablishes the d-c level of a waveform. Used in the d-c restorer stage of a television receiver to restore the d-c component to the video signal after its loss in capacitance-coupled a-c amplifiers, to reestablish the average light value of the reproduced image.

**clamping diode** A diode used to clamp a voltage at some point in a circuit.

**clamp-on ammeter** *Snap-on ammeter.*

**clamp-tube modulation** Amplitude modulation in which the a-f signal is applied to the screen grid of a tetrode or pentode operated as a modulator.

**Clapp oscillator** A series-tuned Colpitts oscillator, having low drift.

**Clark cell** An early form of standard cell, having a voltage of 1.433 volts at 15°C, now largely replaced by the Weston standard cell as a voltage standard.

**class A amplifier** An amplifier in which the grid bias and alternating grid voltages are such that anode current in a specific tube flows at all times. The suffix 1 is added to the letter or letters of the class

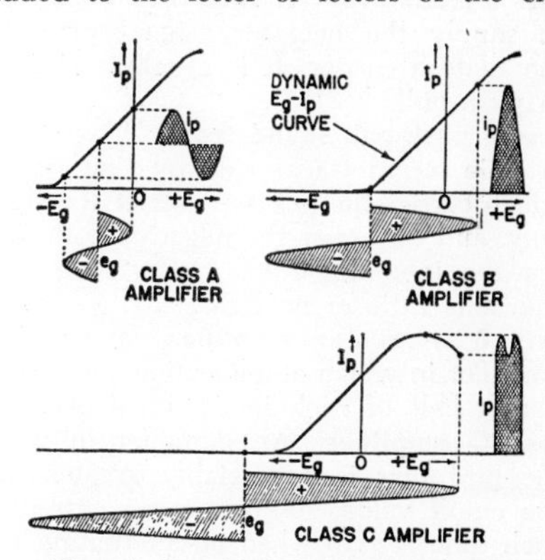


Class A, B, and C amplifier operation.

identification to denote that grid current does not flow during any part of the input cycle. The suffix 2 denotes that grid current flows during some part of the cycle.

**class AB amplifier** An amplifier in which the grid bias and alternating grid voltages are such that anode current in a specific tube flows for appreciably more than half but less than the entire electric cycle. The suffix 1 denotes that grid current does not flow. The suffix 2 denotes that grid current flows during some part of the cycle.

**class AB transistor amplifier** A transistor amplifier in which collector current or voltage is zero for less than half but not all of each input signal cycle.

**class A modulator** A class A amplifier used to supply the necessary signal power to modulate a carrier.

**class A push-pull sound track** Two single photographic sound tracks side by side, the transmission of one being 180° out of phase with the transmission of the other. Both positive and negative halves of the sound wave are linearly recorded on each of the two tracks.

**class A transistor amplifier** A transistor amplifier in which operation is in the linear region of the collector characteristic at all times.

**class B amplifier** An amplifier in which the grid bias is approximately equal to the cutoff value, so that anode current is approximately zero when no exciting grid voltage is applied, and flows for approximately half of each cycle when an alternating grid voltage is applied. The suffix 1 denotes that grid current does not flow. The suffix 2 denotes that grid current flows during some part of the cycle.

**class B modulator** A class B amplifier used to supply the necessary signal power to modulate a carrier. It is usually connected in push-pull.

**class B push-pull sound track** Two photographic sound tracks side by side, one of which carries the positive half of the signal only, and the other the negative half. During the inoperative half-cycle, each track transmits little or no light.

**class B transistor amplifier** A transistor amplifier in which amplification occurs only during half of each input signal cycle.

**class C amplifier** An amplifier in which the grid bias is appreciably greater than the cutoff value, so that anode current in each tube is zero when no alternating grid voltage is applied, and flows for appreciably less than half of each cycle when an alternating grid voltage is applied. The suffix 1 denotes that grid current does not flow. The suffix 2 denotes that grid current flows during some part of the cycle.

**class C transistor amplifier** A transistor amplifier in which collector current or voltage is zero for more than half of each input signal cycle.

**classical scattering cross section** *Scattering cross section.*

**classical system** *Nonquantized system.*

**classified** Containing information whose disclosure to a prospective enemy is not in the best interests of the nation.

**cleanup** Gradual disappearance of gases from an electron tube during operation, due to absorption by getter material or the tube structure.

**clear** To restore a storage device, memory device, or binary stage to a prescribed state, usually that denoting zero. Also called reset.

**clear channel** A standard broadcast channel in which the dominant station or stations render service over wide areas. Stations on a clear channel are cleared of objectionable interference within their primary service areas and over all or a substantial portion of their secondary service areas.

**click filter** *Key-click filter.*

**clipper** *Limiter.*

**clipper amplifier** An amplifier designed to limit the instantaneous value of its output to a predetermined maximum.

**clipper-limiter** A device whose output is a function of the instantaneous input amplitude for a range of values lying between two predetermined limits but is approximately constant, at another level, for input values above the range.

**clipping** 1. Perceptible mutilation of speech syllables during transmission. 2. *Limiting.*

**clipping level** The amplitude level at which a waveform is clipped.

**clipping time** The time constant of a limiter.

**clock** A source of accurately timed pulses for a digital computer, used to form signals corresponding to data that are to be processed.

**clock frequency** The master frequency of the periodic pulses that schedule the operation of a digital computer.

**clockwise capacitor** A variable capacitor whose capacitance increases with clockwise rotation of its rotor, as viewed from the end of the control shaft.

**clockwise polarized wave** *Right-hand polarized wave.*

**close control** The control exercised by an air controller, using radio and radar, over friendly aircraft in a tactical air situation.

**close-control radar** Ground radar used with radio to position an aircraft over a target that is normally difficult to locate or is invisible to the pilot.

**close coupling** The coupling obtained when the primary and secondary windings of an r-f or i-f transformer are close together.

**closed circuit** A complete path for current.

**closed-circuit television** Any application of television that does not involve broadcasting for public viewing. Theater television and industrial television are examples. The programs can be seen only on specified receivers connected to the television camera by circuits that may or may not include microwave relays and coaxial cables. Used also to permit large groups of medical students or doctors to watch surgical operations.

**closed-circuit voltage** The voltage at the terminals of a source when a specified load current is being drawn.

**closed-cycle control system** A control system in which changes in the quantity being controlled are utilized to actuate the controller directly.

**closed-loop control system** *Feedback control system.*

**closed magnetic circuit** A complete circulating path for magnetic flux around a core of ferromagnetic material.

**closed shell** A shell containing the maximum number of nucleons in the independent particle model of the nucleus.

**closed subroutine** A computer subroutine that is stored away from the routine that refers to it. Such a subroutine is entered by a jump, and provision is made to jump back to the proper point in the main routine at the end of the closed subroutine.

**close-talking microphone** A microphone designed for use close to the mouth, so noise from more distant points is suppressed. Also called noise-canceling microphone.

**cloud and collision warning system** An aircraft radar that gives a cathode-ray display of storm clouds and high ground in the intended course, at ranges sufficient to permit course changes when necessary.

**cloud chamber** An enclosure containing air supersaturated with water vapor by sudden expansion, in which moving ionizing particles are revealed by streaks of droplets called cloud tracks. Also called expansion chamber and fog chamber.

**cloud-height detector** *Ceiling-height indicator.*

**cloud pulse** The output resulting from space charge effects produced by turning the electron beam on or off in a charge-storage tube.

**clover-leaf antenna** A nondirectional vhf transmitting antenna consisting of a number of horizontal four-element radiators stacked a half-wave apart vertically. Each horizontal unit has four loops arranged much like a four-leaf clover. The units are energized in such a way as to give maximum radiation in the horizontal plane.

**clutter** Unwanted echoes on a radar screen, such as those caused by ground clutter, sea clutter, rain clutter, stationary objects, chaff, enemy jamming transmissions, and grass. Also called background return and radar clutter.

**cm** Abbreviation for *centimeter.*

**C network** A network composed of three impedance branches in series, the free ends of the network being connected to one pair of terminals, and the junction points being connected to another pair of terminals.

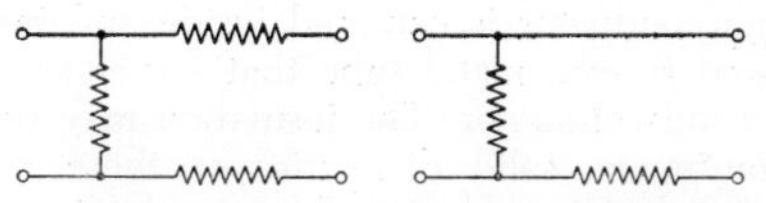

C network has three impedance branches, whereas L network at right has only two branches.

**C neutron** A neutron of such energy, up to about 0.3 electron-volt, that it is strongly absorbable in cadmium.

**coarse chrominance primary** The less important of the two chrominance primaries in a color television signal, called the Q signal. Because its bandwidth is limited to 0.5 mc, this signal affects only the larger, coarser variations in the color picture. The other primary is the fine chrominance primary or I signal, going up to 1.5 mc.

**coarse control** A control used to make a rough adjustment of some characteristic or quantity.

**coastal refraction** The bending of the path of a direct radio wave as it crosses the coast at or near the ground, due to changes in electrostatic conditions between earth and water.

**coated cathode** A cathode that has been coated with compounds to increase electron emission. An oxide-coated cathode is an example.

**coated filament** A vacuum-tube filament that has been coated with metal oxides to provide increased electron emission.

**coated tape** *Magnetic powder-coated tape.*

**coax** *Coaxial cable.*

**coaxial antenna** An antenna that consists of a quarter-wave extension of the inner conductor of a coaxial line and a radiating sleeve that is in effect formed by folding back the outer conductor of the coaxial line for a length of approximately one quarter-wavelength.

**coaxial attenuator** An attenuator that has a coaxial construction and terminations suitable for use with coaxial cable.

**coaxial cable** A transmission line in which

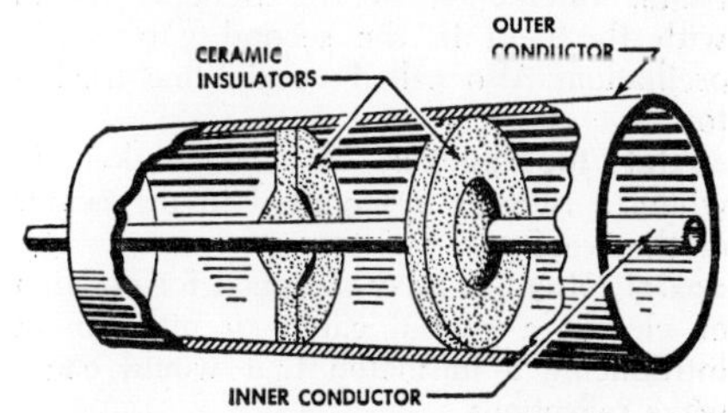


Coaxial cable using dielectric spacers and essentially rigid conductors.

one conductor is centered inside and insulated from a metal tube that serves as the second conductor. The insulation may be a continuous solid dielectric, or there may be dielectric spacers and an insulating gas. Also called coax, coaxial line, coaxial transmission line, concentric line, and concentric transmission line.

**coaxial cavity** A cylindrical resonating cavity having a central conductor in contact with its pistons or other reflecting devices. The conductor serves to pick up a desired wave.

**coaxial-cylinder magnetron** A magnetron in which the cathode and anode consist of coaxial cylinders.

**coaxial-dipole antenna** A dipole antenna having lengths of metal tubing as the radiating elements. The twin-lead transmission line connects conventionally to the inner ends of the tubing. The outer ends of the tubing are connected by a metal rod centered in the radiating elements.

**coaxial dry load** A sand load for a coaxial cable.

**coaxial filter** A section of coaxial line having reentrant elements that provide the inductance and capacitance of a filter section.

**coaxial line** *Coaxial cable.*

**coaxial-line frequency meter** A shorted section of coaxial line that acts as a resonant circuit and is calibrated in terms of frequency or wavelength. Also called coaxial wavemeter.

**coaxial-line oscillator** *Coaxial-line tube.*

**coaxial-line resonator** A resonator consisting of a length of coaxial line short-circuited at one or both ends.

**coaxial-line tube** A tube used for r-f generation in the range of 4,000 to 6,000 mc. It consists of a quarter-wave coaxial line, part of which is cut away to permit an electron beam to be fired through it. The beam thus crosses the gap between outer and inner conductors twice. The field in the first gap causes velocity modulation of the beam. Interaction of the bunched beam with the field in the second gap causes oscillation. Also called coaxial-line oscillator.

**coaxial loudspeaker** A loudspeaker in which a tweeter is mounted in the center of the woofer.

**coaxial relay** A relay designed for opening or closing a coaxial cable circuit without introducing a mismatch that would cause wave reflections.

**coaxial sheet grating** A sheet grating consisting of concentric metal cylinders each about 1 wavelength long, centered in a coaxial waveguide to suppress undesired modes of propagation.

**coaxial stub** A length of nondissipative cylindrical waveguide or coaxial cable branched from the side of a waveguide to produce some desired change in its characteristics.

**coaxial transistor** A point-contact transistor in which the emitter and collector are point electrodes making pressure contact at the centers of opposite sides of a thin disk of semiconductor material serving as base.

**coaxial transmission line** *Coaxial cable.*

**coaxial wavemeter** *Coaxial-line frequency meter.*

**cobalt** [symbol Co] A metallic element having weak magnetic characteristics, sometimes combined with iron and steel to make special alloys used in permanent magnets. Atomic number is 27. Cobalt-60 is a radioisotope having a half-life of 5.3 years, widely used in place of x-ray equipment.

**cobalt-beam therapy** Therapy involving the use of gamma radiation from a cobalt-60 source mounted in a cobalt bomb.

**cobalt bomb** 1. An atomic bomb or hydrogen bomb that is encased in cobalt. This bomb is considered too dangerous to use, because the deadly cobalt-60 radioactive dust that it releases would act on friend, foe, and neutrals alike. 2. A quantity of cobalt-60 mounted in a housing with walls having up to 8 inches of lead for protection, and with means for removing a lead plug to release a beam of gamma rays for use in cobalt-beam therapy.

**cobs** Bell-shaped deflections produced on an oscilloscope by frequency-modulated continuous-wave jamming.

**co-channel interference** Interference between two signals of the same type in the same radio channel.

**Cockcroft-Walton accelerator** An early accelerator in which rectified alternating voltage is used to charge a series of capacitors to a high voltage for use in accelerating protons.

**codan** [Carrier-Operated Device AntiNoise] A device that silences a receiver except when a modulated carrier signal is being received.

**code** A system of symbols and rules for expressing information, such as the Morse code, EIA color code, and the binary and other machine languages used in digital computers.

**coded decimal digit** A decimal digit that is expressed by a pattern of four ones and zeros.

**coded decimal notation** A form of notation in which each decimal digit is converted separately into a pattern of binary ones and zeros. For example, in the 8-4-2-1 coded decimal notation, the number 13 is represented as 0001 0011, whereas in pure binary notation it is represented as 1101.

**code delay** An arbitrary interval of time introduced, in addition to other time intervals, between pulsed signals sent by master and slave transmitters in loran and similar systems. The code delay is used as a security measure or to allow a navigator to resolve ambiguity in plotting a line of position. Also called coding delay (deprecated) and suppressed time delay.

**coded interrogator** An interrogator whose output signal forms the code required to trigger a specific radio or radar beacon. The iff interrogator is an example.

**code distinguishability** The quality of a coded radio beacon that permits it to be distinguished from all other emissions from the beacon, such as those giving distance.

**coded passive reflector** A radar reflector whose reflecting properties can be varied according to a predetermined code, to produce a recognizable indication on a radarscope.

**coded program** A program that has been expressed in the required code for a computer.

**coded stereo** A stereophonic sound system used in theaters, in which a single audio channel is accompanied by a subsonic code signal that controls the volume of sound fed to loudspeakers at the left, center, and right. Thus, if drums are to seem to be at the right of the listener, the coded subsonic signal causes the system to supply relatively more power to the loudspeakers on the right during loud drum passages.

**code element** One of the separate elements or events constituting a code message, such as the presence or absence of a pulse, dot, dash, or space.

**code group** A combination of letters or numerals, or both, assigned to represent one or more words of plain text in a coded message.

**code name** A generic code name assigned to each guided missile to permit convenient reference to it in unclassified correspondence and oral discussions, such as Nike, Talos, Atlas, and Bomarc.

**code practice oscillator** An oscillator used with a key and either headphones or a loudspeaker to practice sending and receiving Morse code.

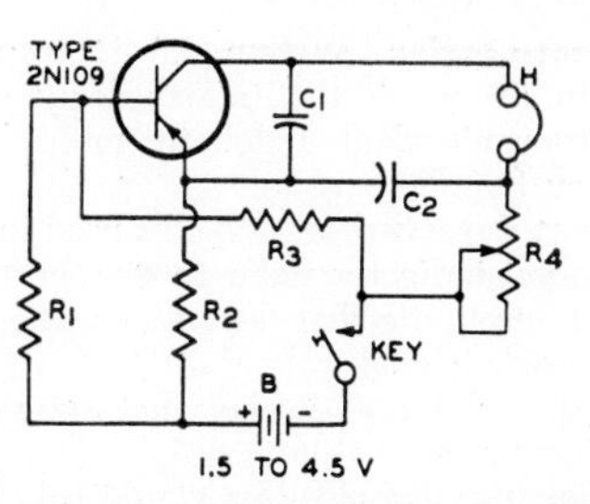


Code practice oscillator circuit using magnetic headphones. C1 and C2–0.01 microfarad; R1–2,200 ohms; R2–27,000 ohms; R3–3,000 ohms; R4–50,000-ohm potentiometer.

**coder** 1. A device that generates a code by generating pulses having varying lengths and/or spacings, as required for radio beacons and interrogators. Also called moder, pulse coder, and pulse-duration coder. 2. A person who translates a sequence of computer instructions into codes that are acceptable to the machine.

**code recorder** An instrument that makes a permanent record of code messages received by radio or wire, as by punching holes in a tape or by making dot-and-dash marks on a tape.

**coding delay** Deprecated term for *code delay.*

**coefficient of coupling** A numerical rating between 0 and 1 that specifies the degree of coupling between two circuits, or the corresponding percentage value. Maximum coupling is 1, and no coupling is 0. Also called coupling coefficient.

**coercive force** The magnetizing force required to bring the flux density to zero in a magnetic material that has been magnetized alternately by equal and opposite magnetizing forces. It is the reverse magnetizing force needed to remove the residual magnetism.

**coercivity** The property of a magnetic material that is measured by the coercive force corresponding to the saturation induction for the material.

**coffin** A box of heavy shielding material, usually lead, used for transporting radioactive objects and having walls thick enough to attenuate radiation from the objects within it to an allowable level. Also called casket.

**cohered video** The video detector output signal in a coherent moving-target-indicator radar system.

**coherent-carrier system** A transponder system in which the interrogating carrier is retransmitted at a definite multiple frequency for comparison.

**coherent detector** A detector used in moving-target-indicator radar to give an output signal amplitude that depends on the phase of the echo signal instead of on its strength, as required for a display that shows only moving targets.

**coherent oscillator** [abbreviated coho] An oscillator used in moving-target-indicator radar to serve as a reference by which changes in the r-f phase of successively received pulses may be recognized. The coho is locked to the radar transmitter frequency, so beating of this signal with the echo signal will reveal phase changes.

**coherent-pulse radar** A radar in which the r-f oscillations of recurrent pulses bear a constant phase relation to those of a continuous oscillation.

**coherent radiation** Radiation in which there are definite phase relationships between different points in a cross-section of the beam. In noncoherent radiation these relationships are random. Interference bands are observed only between coherent beams.

**coherent scattering** Scattering in which there is a definite phase relationship between incoming and scattered particles or photons.

**coherent system** A navigation system in which the signal output is obtained by demodulating the received signal after mixing with a local signal having a fixed phase relation to that of the transmitted signal, to permit use of the information carried by the phase of the received signal.

**coherer** A cell containing a granular conductor between two electrodes. The cell becomes highly conducting when subjected to an electric field, and conduction can then be stopped only by jarring the granules. Formerly used as a detector in wireless telegraphy.

**coho** Abbreviation for *coherent oscillator.*

**coil** [symbol L] A number of turns of wire used to introduce inductance into an electric circuit, to produce magnetic flux, or to react mechanically to a changing magnetic flux. In high-frequency circuits a coil may be only a fraction of a turn. The electrical size of a coil is called inductance and is expressed in henrys. The opposition that a coil offers to alternating current is called impedance and is expressed in ohms. The impedance of a coil increases with frequency. Also called inductance and inductor.

**coil form** The tubing or spool of insulating material on which a coil is wound.

**coil loading** The insertion of loading coils at regular intervals in a transmission line to improve its transmission characteristics over the required frequency band.

**coil neutralization** *Inductive neutralization.*

**coil winder** A manual or motor-driven mechanism for winding coils individually or in groups.

**coincidence** The occurrence of counts in two or more nuclear-particle detectors simultaneously or within an assignable time interval.

**coincidence circuit** A circuit that produces a specified output pulse when and only when a specified number or combination of two or more input terminals receive pulses within an assigned time interval. Also called coincidence counter and coincidence gate.

**coincidence correction** *Dead-time correction.*

**coincidence counter** *Coincidence circuit.*

**coincidence counting** A method of distinguishing particular types of events from background events by means of coincidence circuits.

**coincidence gate** *Coincidence circuit.*

**coincidence loss** Loss of counts due to the occurrence of ionizing events at intervals less than the resolution time of the counting system.

**coincident-current selection** The selection of a particular magnetic cell, for reading or writing in computer storage, by simultaneously applying two or more currents.

**cold area** An area in a plant or laboratory in which there is little or no contact with radioactive chemicals or radiations.

**cold cathode** A cathode whose operation does not depend on its temperature being above the ambient temperature.

**cold-cathode counter tube** A counter tube having one anode and three sets of 10 cathodes. Two sets of cathodes serve as guides that direct the glow discharge to each of the 10 output cathodes in correct sequence in response to driving pulses. Used for data storage, preset counting, tuning, and gating. Also called counter tube, counting tube, decade counter tube, and decade glow counting tube. Trademark of one type is Dekatron.

**cold-cathode rectifier** A cold-cathode gas tube in which the electrodes differ greatly in size so electron flow is much greater in one direction than in the other. An example is the 0Z4 full-wave gas rectifier tube. Also called gas-filled rectifier.

**cold-cathode tube** An electron tube containing a cold cathode, such as a cold-cathode rectifier, mercury-pool rectifier, neon tube, phototube, and voltage regulator.

**cold clean reactor** A reactor in which a chain reaction of any appreciable power has never been established. It has no induced radioactivity and no poisons other than those present when the reactor was constructed.

**cold cutting stylus** A stylus having its cutting edge burnished at a plane substantially different from the cutting face, so as to cut and polish the groove in an acetate disk at normal room temperature.

**cold junction** The junction of thermocouple wires with conductors leading to the measuring instrument. This junction is normally at room temperature.

**cold test** A microwave system test with the microwave tube in place but not in operation, for measurement of such quantities as resonant frequency, loaded $Q$, unloaded $Q$, and driving-point admittance.

**collate** To combine two or more similarly ordered sets of values into one set that may or may not have the same order as the original sets. Merging is collating without changing the nature of the order.

**collateral series** A radioactive decay series, initiated by transmutation, that eventually joins into one of the four radioactive decay series.

**collator** A punched-card machine that can be used for merging two similarly sequenced sets of cards into one sequence or for matching detail cards with master cards.

**collector** 1. [symbol C] An electrode at which a primary flow of carriers leaves the interelectrode region of a transistor. It corresponds to the anode of a tube. 2. An electrode that collects electrons or ions which have completed their functions within an electron tube. A collector receives electrons after they have done useful work, whereas an anode receives electrons whose useful work is to be done outside the tube. Also called electron collector.

**collector capacitance** The depletion-layer capacitance associated with the collector junction of a transistor.

**collector characteristic curves** A set of characteristic curves of collector voltage versus collector current, for a fixed value of transistor base current.

**collector-current runaway** The continuing increase in collector current as collector junction temperature is increased by collector current flow.

**collector efficiency** The ratio of useful power output to d-c power input for a transistor, usually expressed as a percentage.

**collector junction** A semiconductor junction located between the base and collector electrodes of a transistor. It is normally biased in the high-resistance direction, and the current through it is controlled by introducing minority carriers.

**collimator** A device for confining the elements of a beam within an assigned solid angle.

**collinear array** *Linear array.*

**collinear-electrode plasma accelerator** A plasma engine for space travel, in which the plasma is produced by electrode erosion.

**collision** A close approach of two or more particles, photons, atomic systems, or nuclear systems, during which there is an interchange of quantities such as energy, momentum, and charge.

**collision-course homing** Homing in which an offset antenna is used on the missile in conjunction with built-in computers that anticipate the motion of the target and direct the missile ahead of the present position of the target on a converging course that gives a collision in minimum missile travel time.

**collision density** The number of neutron collisions with matter per unit volume per unit time.

**collision excitation** The excitation of a gas by collisions of moving charged particles.

**collision ionization** The ionization of atoms or molecules of a gas or vapor by collision with other particles.

**colloidal graphite** Extremely fine flakes of graphite suspended in water, petroleum oil, castor oil, glycerine, or other liquids. Used to provide conductive shields on the inside or outside surfaces of electron tubes.

**color** A characteristic of light that can be specified in terms of luminance, dominant wavelength, and purity. Luminance is the magnitude of brilliance. Wavelength determines hue, and ranges from about 4,000 angstroms for violet to 7,000 angstroms for red. Purity corresponds to chroma or saturation, and specifies vividness of a hue.

**color balance** Adjustment of the circuits feeding the three electron guns of a color picture tube to compensate for differences in light-emitting efficiencies of the three color phosphors on the screen of the tube.

**color-bar generator** A signal generator that delivers to the input of a color television receiver the signal needed to produce a color-bar test pattern on one or more channels.

**color-bar test pattern** A test pattern made up of different colors of vertical bars, used to check the performance of a color television receiver.

**color breakup** Momentary separation of a color picture into its primary components as a result of a sudden change in the condition of viewing, such as fast movement of the head or blinking of the eyes.

**color burst** A short series of oscillations at the chrominance subcarrier frequency of 3.579545 mc, following each transmitted horizontal sync pulse in a color television system. This burst serves as a frequency reference for generating a continuous wave that is at the burst frequency and locked to it in phase. Also called burst.

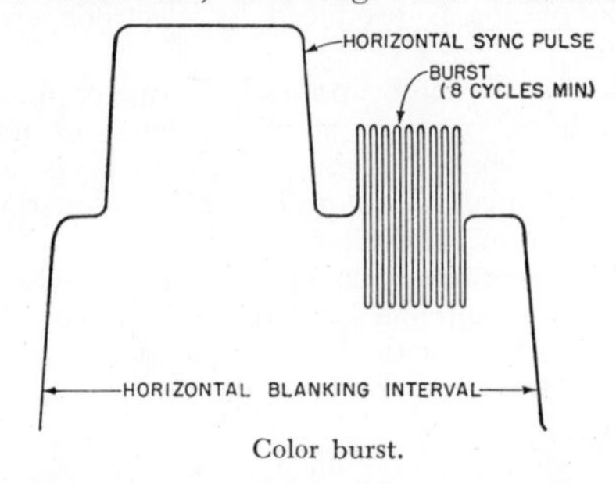


Color burst.

**color-burst pedestal** The rectangular pulse-like component that is part of the color burst when the axis of the color-burst oscillations does not coincide with the back porch. Also called burst pedestal.

**color carrier** Deprecated term for *chrominance subcarrier.*

**color-carrier reference** *Chrominance-carrier reference.*

**colorcast** A television broadcast in color.

**color cell** The smallest area that contains a complete set of primary phosphors on the screen of a color television picture tube.

**color center** A point or region through which an electron beam must pass in order to strike the phosphor array of one primary color in a color television picture tube.

**color code** A system of colors used to indicate the electrical value of a component or to identify terminals and leads. The EIA color codes are the standards for the electronics industry.

**color coder** *Matrix.*

**color comparator** A photoelectric instrument that compares an unknown color with that of a standard color sample for matching purposes. The sample and unknown can be placed alternately in the measuring position, or two identical measuring systems can be used to feed a null indicator. Also called photoelectric color comparator.

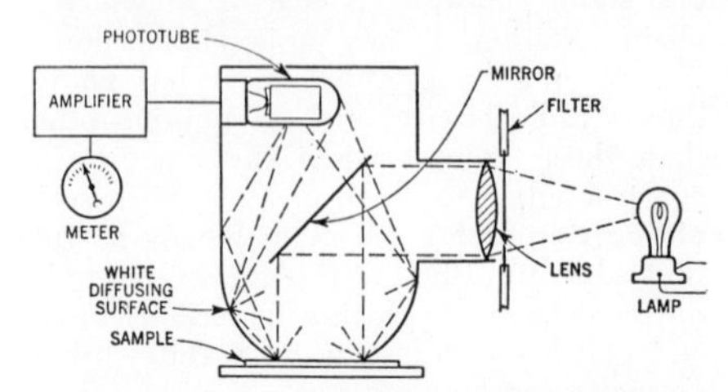


Color comparator in which sample and unknown are alternately placed in measuring position.

**color control** *Chroma control.*

**color coordinate** *Chromaticity coordinate.*

**color-coordinate transformation** Changing the chromaticity coordinates of a color from a system using one set of primaries to a system using another set of primaries. The transformation is performed electrically in a color television system.

**color decoder** *Matrix.*

**color-difference signal** A signal that is added to the monochrome signal in a color television receiver to obtain a signal representative of one of the three tristimulus values needed by the color picture tube. There are three color-difference signals; the $G-Y$ signal feeds the green channel, the $R-Y$ signal feeds the red channel, and the $B-Y$ signal feeds the blue channel.

**color disk** A rotating circular disk having red, green, and blue filter sections to produce the individual red, green, and blue pictures in a field-sequential color television system. The disk in front of the receiving screen must rotate in synchronism with a corresponding disk in front of the television camera.

**color edging** Spurious color at the boundaries of differently colored areas in a color television picture. Color edging includes color fringing and misregistration.

**color encoder** *Matrix.*

**color fidelity** The degree to which a color television system is capable of reproducing faithfully the colors in an original scene.

**color field corrector** A device used outside a color picture tube to produce an electric or magnetic field that acts on the electron beam after deflection to produce more uniform color fields.

**color filter** A sheet of material that absorbs certain wavelengths of light while transmitting others.

**color flicker** Flicker due to fluctuations of both chromaticity and luminance in a color television receiver.

**color fringing** Spurious chromaticity at boundaries of objects in a color television picture. Small objects may appear separated into different colors. It can be caused by a change in position of the televised object from field to field or by misregistration.

**color gate** A circuit used in a single-gun sequential color television receiver to pass a signal for one primary color when the electron beam is striking the phosphor for that color.

**colorimeter** An instrument that measures color by determining the intensities of the three primary colors that will give that color.

**colorimetry** The science of color measurement.

**color killer circuit** The circuit in a color television receiver that biases chrominance amplifier tubes to cutoff during reception of monochrome programs.

**color-mixture data** Deprecated term for *tristimulus values.*

**color phase** The difference in phase between a chrominance signal (I or Q) and the chrominance-carrier reference in a color television receiver.

**color phase alternation** [abbreviated cpa] The periodic changing of the color phase of one or more components of the chrominance subcarrier between two sets of assigned values after every field in a color television system.

**color picture signal** The electric signal that represents complete color picture information, excluding all synchronizing signals. In one form it consists of a monochrome component plus a subcarrier modulated with chrominance information.

**color picture tube** A cathode-ray tube having three different colors of phosphors. When these are appropriately scanned and excited in a color television receiver, a color picture is obtained. Also called tricolor picture tube and color television picture tube.

**color plane** A surface approximating a plane containing the color centers in a multibeam color picture tube.

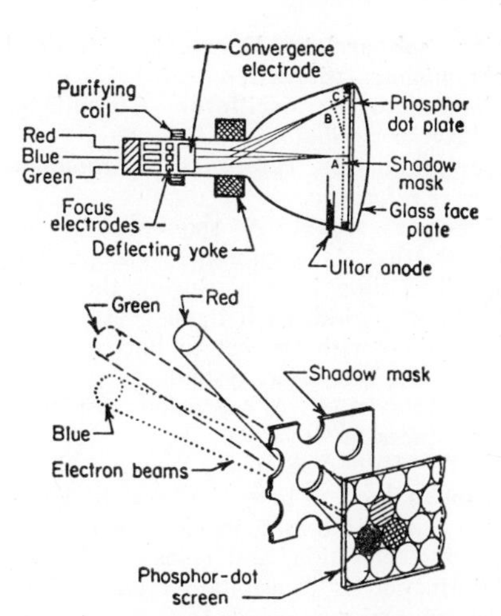


Color picture tube having three electron guns and shadow mask.

**colorplexer** The section of a color television transmitter in which red, green, and blue signals are combined by matrixing and multiplexing to produce a single compatible color television signal.

**color primaries** The red, green, and blue primary colors that are mixed in various proportions to form all the other colors on the screen of a color television receiver.

**color-purity magnet** A magnet used on the neck of a color picture tube to improve color purity by changing the path of the electron beam.

**color registration** The accurate superimposing of the red, green, and blue images used to form a complete color picture in a color television receiver.

**color response** The sensitivity of a device to different wavelengths of light.

**color sampling rate** The number of times per second that each primary color is sampled in a color television system.

**color saturation** The degree to which a color is mixed with white. High saturation means little or no white, as in a deep red color. Low saturation means much white, as in light pink. The amplitudes of the I and Q chrominance signals determine color saturation in a color television receiver. Also called saturation.

**color-saturation control** *Chroma control.*

**color signal** Any signal that controls the chromaticity values of a color television picture, such as the color picture signal and the chrominance signal.

**color subcarrier** Deprecated term for *chrominance subcarrier.*

**color-subcarrier oscillator** *Chrominance-subcarrier oscillator.*

**color-subcarrier reference** *Chrominance-carrier reference.*

**color-sync signal** A sequence of color bursts that is continuous except for a specified time interval during the vertical blanking period, each burst occurring at a fixed time with respect to horizontal sync in a color television system.

**color television** A television system that reproduces an image in its original colors. In the United States a compatible dot-sequential color television system is used. The bandwidth is 6 mc just as for monochrome television, and color synchronizing information is transmitted on a 3.579545-mc subcarrier.

**color television picture tube** *Color picture tube.*

**color television signal** The entire signal used to transmit a full-color picture. It consists of the color picture signal and all the synchronizing signals.

**color temperature** The temperature of a blackbody radiator that produces the same chromaticity as the light under consideration.

**color transmission** The transmission of a signal wave for controlling both the luminance and chromaticity values in a picture.

**color triad** A color cell of a three-color phosphor-dot screen in a color picture tube.

**color triangle** A triangle drawn on a chromaticity diagram, representing the entire range of chromaticities obtainable as additive mixtures of three prescribed primaries.

**Colpitts oscillator** An electron-tube oscillator in which a parallel-tuned tank circuit

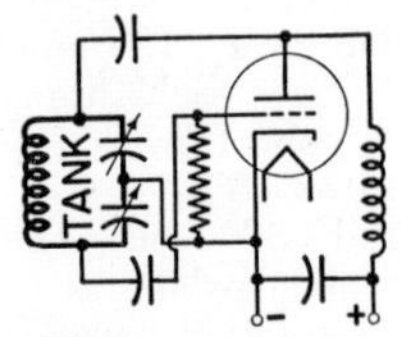

Colpitts oscillator circuit.

is connected between grid and anode. The tank capacitance consists of two voltage-dividing capacitors in series, with their common connection at cathode potential.

**columbium** Former name for *niobium.*

**column** 1. *Card column.* 2. *Place.*

**columnar ionization** Regions of such dense ionization that even a strong external electric field cannot prevent some recombination. The term refers usually to ionization produced by alpha particles.

**columnar recombination** Recombination that takes place before ions have left the track. The ionization then takes place along a column, as in the case of the dense ionization produced along the track of an alpha particle.

**coma** A cathode-ray tube image defect that makes the spot on the screen appear comet-shaped when away from the center of the screen.

**coma lobe** A side lobe that occurs in the radiation pattern of a microwave antenna when the reflector alone is tilted back and forth to sweep the beam through space. The coma lobe is produced under these conditions because the feed is no longer always at the center of the reflector. Used to eliminate the need for a rotary joint in the feed waveguide.

**combat information center** A shipboard location at which tactical information from radar, sonar, and other equipment is received, displayed for rapid analysis, and evaluated. The results are distributed over wire lines or other communication links to appropriate points for action.

**comb filter** A wave filter whose frequency spectrum consists of a number of equi-spaced elements resembling the teeth of a comb.

**combination microphone** A microphone consisting of two or more dissimilar microphone elements.

**combiner circuit** The circuit that combines the luminance and chrominance signals with the synchronizing signals in a color television camera chain.

**command** 1. A signal that initiates a predetermined type of computer operation that is defined by an instruction. 2. The control of an airborne guided missile or other pilotless aircraft by electronic signals. 3. An independent signal in a feedback control system, from which the dependent signals are controlled in a predetermined manner.

**command-destruct signal** A radio signal used to detonate the destruction device of a guided missile in the event of a malfunction.

**command guidance** A missile guidance system in which flight direction information is transmitted to the missile from a point external to the missile. The information needed about missile performance may be derived by telemetering, ground-based radar, optical tracking, or other means. The commands may be transmitted to the missile by radio, radar, or other means.

**command resolution** The maximum change in command that can be made in a feedback control system without causing a change in the ultimately controlled variable.

**command set** A radio set used to receive or give commands, as between one aircraft and another or between an aircraft and the ground.

**commentator** One who edits and broadcasts news at a radio or television station, often interspersed with personal comments.

**commercial** An advertising message that is broadcast by television or radio.

**Commission Internationale de l'Eclairage** [abbreviated CIE] An international group that has set most of the basic standards of light and color now used in color television. Also abbreviated ICI (deprecated).

**common-base connection** *Grounded-base connection.*

**common cathode** A cathode that serves for two or more sections of an electrode tube, as in a duodiode pentode.

**common-collector connection** *Grounded-collector connection.*

**common-emitter connection** *Grounded-emitter connection.*

**common-mode rejection** The ability of an amplifier to cancel a common-mode signal while responding to an out-of-phase signal. Also called in-phase rejection.

**common-mode signal** A signal applied equally to both ungrounded inputs of a balanced amplifier stage or other differential device. Also called in-phase signal.

**common return** A return conductor that serves two or more circuits.

**common system** A system of air navigation and traffic control intended for common use by all civil and military aviation.

**communication** The transmission of intelligence between two or more points over wires or by radio. The terms telecommunication and communication are often used interchangeably, but telecommunication is usually the preferred term when long distances are involved.

**communication band** The band of frequencies effectively occupied by a radio transmitter for the type of transmission and the speed of signaling used.

**communication channel** The wire or radio channel that serves to convey intelligence between two or more terminals.

**communication countermeasure** Any electronic countermeasure against communications.

**communication jackal** A frequency-modulated airborne barrage jammer covering the band from 27 to 57 mc, with power up to 150 watts. It operates unattended during flight and serves to jam enemy amplitude-modulated signals in its band.

**communication receiver** A receiver designed especially for reception of voice or code messages transmitted by radio communication systems.

**communications deception** Use of devices or techniques that confuse or mislead the user of a communication link or radio navigation system.

**communications engineer** An engineer who specializes in the design, construction, and operation of all types of equipment used for radio, wire, or other types of communication.

**community television system** A television receiving system that distributes signals over coaxial cables to homes in an entire community. Used chiefly in areas where a high and elaborate receiving antenna is required to obtain signals of sufficient strength for adequate reception, or in valleys where the antenna installation must be on a nearby mountain top.

**commutating capacitor** A capacitor used in gas-tube rectifier circuits to prevent the anode from going highly negative immediately after extinction.

**commutating reactance** An inductive reactance placed in the cathode lead of a three-phase mercury-arc rectifier to insure that tube current holds over during transfer of conduction from one anode to the next. Without this reactance the arc would go out, and the tube would have to be restarted by auxiliary means.

**commutation** 1. The transfer of current from one anode to another in a gas tube. 2. The sampling of various quantities in a repetitive manner, for transmission over a single channel in telemetering.

**commutation factor** The product of the rate of current decay and the rate of the inverse voltage rise immediately following such current decay in a gas tube. The rates are commonly stated in amperes per microsecond and volts per microsecond.

**commutator** A circular assembly of conducting members, individually insulated in a supporting structure, with an exposed surface for contact with current-collecting brushes.

**commutator switch** A switch, usually rotary and mechanically driven, that performs a set of switching operations in repeated sequential order such as is required for telemetering many quantities.

**companding** A process in which compression is followed by expansion. Often used for noise reduction, in which case compression is applied before noise exposure and expansion after exposure.

**compandor** A system for improving the signal-to-noise ratio by compressing the volume range of the signal at a transmitter or recorder by means of a compressor, and restoring the normal range at the receiving or reproducing apparatus with an expandor.

**comparative lifetime** The product of the half-life of a beta disintegrator and a function that expresses the probability per unit time that a beta transition will take place in a given nucleus.

**comparator** 1. A device that compares two transcriptions of the same information to verify the accuracy of transcription, storage, arithmetical operation, or some other process in a computer, and delivers an output signal of some form to indicate whether or not the two sources are equal or in agreement. 2. An electronic instrument that measures a quantity and compares it with a precision standard.

**comparison** A computer operation in which two numbers are compared as to identity, relative magnitude, or sign.

**compass bearing** A bearing measured relative to compass north.

**compatibility** The ability of a new system to serve users of an old system.

**compatible color television system** A color television system that permits the substantially normal monochrome reception of the transmitted color picture signal on a typical unaltered monochrome receiver. This is accomplished in the U. S. color television system by dividing the color video information into a luminance signal and two chrominance signals. The luminance signal is the equivalent of a monochrome television picture signal, and is utilized alone by a monochrome receiver.

**compatible single-sideband system** A single-sideband system that can be received by an ordinary amplitude-modulation radio receiver without distortion.

**compatible stereo system** A stereo system that gives satisfactory single-channel sound for those having single-channel phonographs, tape equipment, or radio receivers.

**compensated amplifier** A broadband amplifier in which the frequency range is extended by choice of tubes and circuit constants.

**compensated ionization chamber** An arrangement of two ionization chambers in parallel, with potentials reversed, used as a radiation null indicator. One chamber has an adjustable source of ionizing radiation such as uranium. This is adjusted until both chambers have the same ionization. The instrument then serves as a null indicator that shows both increases and decreases from normal ionization in the main chamber.

**compensated semiconductor** A semiconductor in which one type of impurity or imperfection partially cancels the effects of the other type of impurity or imperfection. Thus, a donor impurity may offset an acceptor impurity.

**compensated volume control** *Loudness control.*

**compensating element** The element in a fire-control system that corrects for variation of the mechanical reference frame from the basic reference plane of the system.

**compensation** Modification of the amplitude-frequency response of an amplifier to broaden the bandwidth or to make the response more nearly uniform over the existing bandwidth. Also called frequency compensation.

**compensation signal** A signal recorded on magnetic tape, along with computer data, for use during playback to correct for the effects of tape speed errors.

**compensator** A component that offsets an error or other undesired effect.

**compiling routine** A digital computer routine by means of which a computer can construct its own program for solving a problem by assembling, fitting together, and copying other programs stored in its library of routines.

**complement** A number whose representation for a computer is derived from the finite positional notation of another by subtracting each digit from one less than the base, then adding 1 to the least significant digit and executing all carries required. This is the true complement. Thus, the two's complement of binary 10010 is 01110; the ten's complement of decimal 2,546 is 7,454. In many machines, a negative number is represented as a complement of the corresponding positive number.

**complementary color** A color that, when

added to another given color in proper proportion, produces white.

**complementary symmetry** A circuit using both p-n-p and n-p-n transistors in a symmetrical arrangement that permits push-pull operation without an input transformer or other form of phase inverter.

**complementary wavelength** The wavelength of light that, when combined with a sample color in suitable proportions, matches a reference standard light. The purples that have no dominant wavelengths, including nonspectral violet, purple, magenta, and nonspectral red colors, are specified by use of their complementary wavelengths.

**complete carry** The condition wherein a carry resulting from the addition of carries is allowed to propagate in a computer.

**complex operator** The letter *j*, used to designate the reactive component of a complex impedance.

**complex reflector** A structure or group of structures having many radar-reflecting surfaces facing in different directions.

**complex target** A radar target composed of a number of reflecting surfaces that, in the aggregate, are smaller in all dimensions than the resolution capabilities of the radar.

**complex tone** A sound wave produced by the combination of simple sinusoidal components of different frequencies.

**compliance** The acoustical and mechanical equivalent of capacitance. It is the opposite of stiffness. In a phonograph, compliance is the force needed to move the stylus a given distance.

**component** Any electric device, such as a coil, resistor, capacitor, generator, line, or electron tube, having distinct electrical characteristics and having terminals at which it may be connected to other components to form a circuit. Also called element.

**composite color signal** The color picture signal plus all blanking and synchronizing signals. The composite color signal thus includes the luminance signal, the two chrominance signals, vertical and horizontal sync pulses, vertical and horizontal blanking pulses, and the color-burst signal.

**composite color sync** The signal comprising all the sync signals necessary for proper operation of a color receiver. This includes the horizontal and vertical sync and blanking pulses and the color-burst signal.

**composite controlling voltage** The voltage of the anode of an equivalent diode combining the effects of all individual electrode voltages in establishing the space-charge-limited current.

**composited circuit** A circuit used simultaneously for voice communication and telegraphy, with frequency-discriminating networks serving to separate the two types of signals.

**composite modulation voltage** The combined output voltage of the subcarrier oscillators in a telemetering system, applied as modulation to the transmitter.

**composite picture signal** The signal obtained by combining a monochrome television picture signal with the horizontal and vertical blanking and synchronizing signals. Also called composite signal.

**composite pulse** A pulse composed of a series of overlapping pulses received from the same source over several paths in a pulse navigation system.

**composite signal** *Composite picture signal.*

**composite wave filter** A combination of two or more low-pass, high-pass, bandpass, or band-elimination filters.

**composition resistor** *Carbon resistor.*

**compound-connected transistors** An arrangement of two transistors in which the base of one is connected to the emitter of the other and the two collectors are connected together. The combination may be considered as a single transistor having a high current amplification factor.

**compound horn** An electromagnetic horn of rectangular cross-section, the four sides of which diverge in such a way as to coincide with or approach four planes, with the provision that the line of intersection of two opposite planes does not intersect the line of intersection of the remaining planes.

**compound modulation** Modulation in which one or more signals are used to modulate their respective subcarriers and these subcarriers are used to modulate the carrier.

**compound nucleus** An excited nucleus formed as an intermediate stage in an induced nuclear reaction. It is characterized by a long lifetime compared with the normal transit times of nuclear particles across the nucleus.

**compound target** A radar target composed of a number of randomly disposed reflecting surfaces the aggregate extent of which exceeds any of the dimensions of the pulse packet.

**compressed-air loudspeaker** A loudspeaker having an electrically actuated valve that modulates a stream of compressed air.

**compression** Reduction of the volume range of an a-f signal. Weak signal components are made stronger so that they will not be lost in background noise, whereas loud passages are reduced in strength so that they will not overload any part of the system. This action is achieved by making the effective gain vary automatically as a function of signal magnitude. The same action is used in disk recording to prevent the cutting stylus from going into adjacent grooves at high volume levels, and in facsimile to reduce the white-to-black amplitude range or frequency swing.

**compressional wave** A wave in an elastic medium that causes an element of the medium to change its volume without undergoing rotation. A compressional plane wave is a longitudinal wave.

**compression molding** Molding a record or transcription by compressing a preform of plastic.

**compression ratio** The ratio of the amplification at a reference signal level to that at a higher stated signal level.

**compressor** The part of a compandor that is used to compress the intensity range of signals at the transmitting or recording end of a circuit. The compressor amplifies weak signals and attenuates strong signals, so as to produce a smaller amplitude range.

**compromise net** A network used with a hybrid junction to balance a connected communication circuit such as a subscriber's loop, other lines, or equipment. It is designed to give a compromise between the extremes of impedance balance.

**Compton absorption** The absorption of an x-ray or gamma-ray photon in the Compton effect. The energy of the electromagnetic radiation is not completely absorbed since another photon of lower energy is simultaneously created.

**Compton effect** The elastic scattering of photons by electrons. Since the total energy and total momentum are conserved in the collisions, the wavelength of the scattered radiation is changed by an amount that depends on the angle of scattering. Also called Compton scattering.

**Compton recoil electron** An electron that has been set in motion through interaction with a photon in the Compton effect.

**Compton recoil particle** Any particle that has acquired its momentum in a scattering process similar to that in the Compton effect.

**Compton scattering** *Compton effect.*

**Compton shift** The change in wavelength of scattered radiation due to the Compton effect.

**Compton wavelength** A wavelength characteristic of a particle of given mass. Its value for the electron is $2.43 \times 10^{-10}$ cm, and for the proton it is $1.32 \times 10^{-13}$ cm.

**computer** A machine for carrying out arithmetical calculations and logical operations. Analog computers and digital computers are examples.

**computer code** The code representing the operations built into the hardware of a particular computer.

**computer operation** The electronic action required in a computer to give a desired computation.

**computing linkage** An assembly of rigid links, pivots, and sliding members so arranged that the motion of a selected output link is a predetermined function of the motions of all the input links.

**comraz** [COMmunication Range AZimuth] A system for determining the ground-to-air or air-to-air range between any two stations equipped with radio communication and range/azimuth equipment.

**concentric groove** *Locked groove.*

**concentric line** *Coaxial cable.*

**concentric transmission line** *Coaxial cable.*

**condensed-mercury temperature** The temperature measured on the outside of a glass tube envelope in the region where mercury is condensing, or measured at a designated point on the outside of a metal tube.

**condenser** 1. A system of lenses designed to concentrate or focus light rays to a point. Also called condensing lens. 2. Deprecated term for *capacitor.*

**condenser ionization chamber** Deprecated term for *capacitor r-meter.*

**condenser microphone** Deprecated term for *capacitor microphone.*

**condenser r-meter** Deprecated term for *capacitor r-meter.*

**condensing lens** *Condenser.*

**conditional** Subject to the result of a comparison made during computation in a computer, or subject to human intervention.

**conditional breakpoint instruction** A conditional jump instruction that, if some specified switch is set, will cause a computer to stop. The routine may then be continued as coded or a jump may be forced.

**conditional jump** A computer instruction

that will cause the proper one of two or more addresses to be used in obtaining the next instruction, depending on some property of a numerical expression that may be the result of some previous instruction. Also called branch, conditional transfer, and discrimination.

**conditional stability** A property of a system that causes it to be stable for certain values of input signal and gain, and unstable for other values. Also called limited stability.

**conditional transfer** *Conditional jump.*

**condor** A continuous-wave navigation system, similar to benito, that automatically measures bearing and distance from a single ground station. The distance is determined by phase comparison and the bearing by automatic direction finding. Distance and bearing are displayed on a cathode-ray indicator.

**conductance** [symbol $G$] A measure of the ability of a material to conduct electric current. It is the reciprocal of the resistance of the material, and is expressed in mhos. Conductance is the resistive or real part of admittance.

**conduction** The transmission of energy by means of a medium without movement of the medium itself.

**conduction band** An energy band in which electrons can move freely in a solid. The conduction band is normally empty or partly filled with electrons.

**conduction current** A current due to a flow of conduction electrons through a body.

**conduction electron** An electron in the conduction band of a solid, where it is free to move under the influence of an electric field. Also called outer-shell electron and valence electron.

**conductive pattern** A design formed of any conductive or resistive material.

**conductivity** The ability of a material to conduct electric current, as measured by the current per unit of applied voltage. It is the reciprocal of resistivity.

**conductivity modulation** Variation of the conductivity of a semiconductor by variation of the charge carrier density.

**conductivity-modulation transistor** A transistor in which the active properties are derived from minority carrier modulation of the bulk resistivity of a semiconductor.

**conductor** A wire, cable, or other body or medium that is suitable for carrying electric current.

**conductor pattern** A conductive pattern having low electric resistance.

**conduit** Solid or flexible metal or other tubing through which insulated electric wires are run.

**cone** The cone-shaped paper or fiber diaphragm of a loudspeaker.

**cone antenna** *Conical antenna.*

**cone loudspeaker** A loudspeaker employing a magnetic driving unit that is mechanically coupled to a paper or fiber cone.

**conelrad** [CONtrol of ELectromagnetic RADiation] A system for providing official civil defense information and instructions by radio in an emergency without providing radio homing guidance for the enemy. During an emergency, all commercial broadcast stations operate on either 640 or 1,240 kc in such a way that enemy aircraft cannot take a bearing on any one station.

**cone of nulls** A conical surface formed by directions of negligible radiation from an antenna.

**cone of silence** A cone-shaped region, directly over the antenna of a radio-beacon transmitter, in which no signal is heard by the pilot of an aircraft.

**cone resonance** The frequency at which the diaphragm or cone of a loudspeaker vibrates most easily. Cone resonance must be minimized by careful design and by placing the loudspeaker in a proper enclosure, to prevent abnormally high acoustic output at the resonant frequency.

**confidential** Having such security status that unauthorized disclosure could be prejudicial to the defense interests of the nation.

**configuration control** Control of the reaction rate of a nuclear reactor by changing the configuration of its core.

**confusion reflector** An electromagnetic-wave reflector that is dropped from aircraft to create false signals on enemy radarscopes. It usually consists of strips of aluminum foil or metallized paper such as chaff or window, cut to lengths that are approximately resonant to the expected enemy radar frequency. The strips give strong echo signals when dropped in clusters.

**conical antenna** A wideband antenna in which the driven element is conical in shape. Also called cone antenna.

**conical horn** A horn having a circular cross-section and straight sides.

**conical horn antenna** A horn antenna

having a circular cross-section and straight sides, as in a cone. It is energized by a circular waveguide that feeds the smaller end of the horn.

**conical scanning** Scanning in which the radar beam describes a cone whose axis coincides with the axis of the reflector.

**conjugate bridge** A bridge in which the detector circuit and the supply circuits are interchanged as compared with a normal bridge of the given type.

**conjugate impedances** Impedances having resistance components that are equal and reactance components that are equal in magnitude but opposite in sign.

**connected network** A network in which there exists at least one path, composed of branches of the network, between every pair of nodes of the network.

**connection** A direct wire path for current between two points in a circuit.

**connector** A complete electric connecting device, consisting of a mating plug and receptacle for cables or of mechanically mating flanges for waveguides.

**conoscope** An optical instrument used to locate the optical or $z$ axis of a quartz crystal.

**consol** *Sonne.*

**consolan** A long-range directional navigation system that transmits a slowly rotating keyed radio field pattern, from which a line of position can be obtained that is accurate within 20 miles at 1,500 miles range. Only a standard radio receiver is required in the aircraft. Consolan is an American version of the German sonne and British consol, using two radiators instead of three to minimize night-effect errors.

**console** 1. A large cabinet for a radio or television receiver, standing on the floor rather than on a table. 2. A main control desk for electronic equipment, as at a radar station, radio or television station, or airport control tower. Also called control desk.

**console receiver** A television or radio receiver in a console.

**constant** A value that does not change during a particular process.

**constant-amplitude recording** A sound-recording method in which all frequencies having the same intensity are recorded at the same amplitude. The resulting recorded amplitude is independent of frequency.

**constantan** An alloy containing 60% copper and 40% nickel, used in making precision wirewound resistors because of its low temperature coefficient of resistance. It is often used with iron or copper in thermocouples.

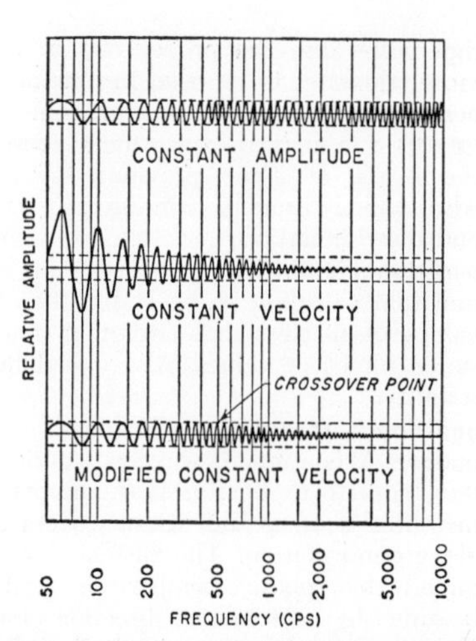


Constant-amplitude, constant-velocity, and modified constant-velocity recording having constant amplitude below crossover frequency of 500 cps.

**constant-current characteristic** The relation between the voltages of two electrodes in an electron tube when the current to one of them is maintained constant and all other electrode voltages are constant.

**constant-current generator** A vacuum-tube circuit, generally containing a pentode, in which the a-c anode resistance is so high that anode current remains essentially constant despite variations in load resistance.

**constant-current modulation** A system of amplitude modulation in which the output circuits of the signal amplifier and the carrier-wave generator or amplifier are connected through a common coil to a constant-current source. Changes in anode current of the signal amplifier thus produce equal and opposite changes in anode current of the r-f carrier stage, thereby giving the desired modulation of the carrier. Also called Heising modulation.

**constant-current transformer** A transformer that automatically maintains a constant current in its secondary circuit under varying loads, when supplied from a constant-voltage source.

**constant-delay discriminator** *Pulse demoder.*

**constant-*k* filter** A filter in which the product of the series and shunt impedances is a constant that is independent of frequency.

**constant-*k* network** A ladder network in which the product of the series and shunt impedances is independent of frequency within the operating frequency range.

**constant-luminance transmission** That type of color television transmission in which the transmission primaries are a luminance primary, controlled only by the monochrome signal, and two chrominance primaries. This system is currently being used in the United States.

**constant-potential accelerator** An accelerator in which constant d-c voltage is applied to an accelerating tube to produce high-energy ions or electrons.

**constant-resistance structure** A network whose iterative impedance, in at least one direction, is a resistance and is independent of the frequency.

**constant-velocity recorder** A disk recorder having a turntable that rotates in such a manner that constant velocity is effected at the recording stylus irrespective of diameter.

**constant-velocity recording** A sound-recording method in which, for input signals of a given amplitude, the resulting recorded amplitude is inversely proportional to the frequency. The velocity of the cutting stylus is then constant for all input frequencies having that given amplitude.

**constraint** The condition wherein a particle or group of particles has less than $3N$ degrees of freedom, where $N$ is the number of particles in the group.

**contact** 1. A conducting part of a relay or switch that coacts with another such part to make or break a circuit. 2. Initial detection of an enemy aircraft, ship, submarine, or other object on a radarscope or other detecting equipment.

**contact bounce** The uncontrolled making and breaking of contact one or more times, but not continuously, when relay contacts are moved to the closed position.

**contact chatter** *Chatter.*

**contact electromotive force** *Contact potential.*

**contact follow** The distance two contacts travel together after just touching. Also called contact overtravel.

**contact gettering** The absorption of gas by contact with a dispersed getter film in an electron tube.

**contact-making meter** *Instrument-type relay.*

**contact microphone** A microphone designed to pick up mechanical vibrations directly and convert them into corresponding electric currents or voltages. When used with wind, string, and percussion musical instruments, it is attached to the housing of the instrument. When used in vibration analysis of machinery, it is held against various parts of the machinery. When used as a throat microphone, it is strapped against the throat of the speaker. When used as a lip microphone, it is held against the lip of the speaker.

**contact-modulated amplifier** An amplifier having a chopper at its input, to change d-c and very-low-frequency a-c signals to a higher frequency such as 60 cps or 400 cps. The resulting modulated wave is amplified in an a-c amplifier to a suitable level, then demodulated, sometimes by the same contact system used to accomplish the original modulation.

**contact noise** The fluctuating electric resistance observed at the junction of two metals or at the junction of a metal and a semiconductor.

**contactor** A heavy-duty relay used to control electric power circuits.

**contact overtravel** *Contact follow.*

**contact piston** A waveguide piston that makes contact with the walls of the waveguide. Also called contact plunger.

**contact plunger** *Contact piston.*

**contact potential** 1. The voltage due to contact between two different metals, between bodies in different physical states, or between materials having different chemical compositions. Also called contact electromotive force and Volta effect. 2. The voltage existing between the control grid and cathode of an electron tube when there is no external grid bias, due to the difference in work functions of the electrode surfaces.

**contact-potential barrier** The potential hill at the contact surfaces of two bodies, due to formation of a barrier layer.

**contact-potential difference** The difference between the work functions of two materials in contact, divided by the electronic charge.

**contact rectifier** *Metallic rectifier.*

**contact resistance** The resistance in ohms between the contacts of a relay, switch, or other device when the contacts are touching each other. The value is generally a small fraction of an ohm.

**contaminated** Made radioactive by the addition of radioactive material.

**contamination** The deposit of radioactive materials, such as fission fragments or radiological warfare agents, on any objective or surface.

**continuity** The presence of a complete path for current flow.

**continuity test** An electrical test used to determine the presence and location of a broken connection.

**continuous control** Automatic control in which the controlled quantity is measured continuously and corrections are a continuous function of the deviation.

**continuous-duty rating** The rating that defines the load which can be carried for an indefinite time without exceeding a specified temperature rise.

**continuous film scanner** A television film scanner in which the motion-picture film moves continuously while being scanned by a flying-spot kinescope.

**continuous linear antenna array** An antenna array consisting of an infinite number of infinitesimally spaced sources, as in some dielectric antennas.

**continuous loading** Loading in which the added inductance is distributed uniformly along a line by wrapping magnetic material around each conductor.

**continuously adjustable transformer** *Variable transformer.*

**continuous power spectrum** A power spectrum representable by the indefinite integral of a suitable spectral density function. All power spectra of physical systems are continuous.

**continuous recorder** A recorder whose record sheet is a continuous strip or web rather than individual sheets.

**continuous spectrum** The spectrum of a wave whose components are continuously distributed over a frequency region.

**continuous tuner** *Spiral tuner.*

**continuous wave** [abbreviated c-w] A radio or radar wave that maintains a constant amplitude and a constant frequency.

**continuous-wave doppler radar** *Continuous-wave radar.*

**continuous-wave radar** A radar system in which a transmitter sends out a continuous flow of radio energy. The target reradiates a small fraction of this energy to a separate receiving antenna so located and oriented that only a small fraction of the transmitted power leaks directly into the receiver. The reflected wave is distinguished from the transmitted signal by a slight change in radio frequency called the doppler shift. Continuous-wave radar can distinguish moving targets against a stationary reflecting background, and is more conservative of bandwidth than pulse radar. Also called continuous-wave doppler radar.

**continuous x-rays** The electromagnetic radiation, having a continuous spectral distribution, that is produced when high-velocity electrons strike a target. For a given x-ray tube current, the intensity associated with each wavelength is dependent on the material and thickness of the target and on the voltage applied to the tube.

**contrast** The degree of difference in tone between the lightest and darkest areas in a television or facsimile picture. Contrast is measured in terms of gamma, a numerical indication of the degree of contrast. Pictures with high contrast have deep blacks and brilliant whites, while pictures with low contrast have an over-all gray appearance.

**contrast control** A manual control that adjusts the range of brightness between highlights and shadows on the reproduced image in a television receiver. Usually, the contrast control varies the gain of a video amplifier tube. In a color television receiver a dual control may be used, with one section controlling the luminance signal and the other section controlling the chrominance signals; this permits adjustment of contrast without changing color.

**contrast range** The ratio of the brightness of the whitest portion of a picture to that of the blackest portion.

**control** 1. A component that is used to start, stop, or adjust a piece of equipment. 2. The section of a digital computer that carries out instructions in proper sequence, interprets each coded instruction, and applies the proper signals to the arithmetic unit and other parts in accordance with this interpretation. 3. A mathematical check used in some computer operations. 4. A test made to determine the extent of error in experimental observations or measurements.

**control accuracy** The degree of correspondence between the ultimately controlled variable and the ideal value in a feedback control system.

**control characteristic** 1. The relation, usually shown by a graph, between critical grid voltage and anode voltage of a gas tube. 2. The relation between control ampere-turns and output current of a magnetic amplifier.

**control circuit** 1. The circuit that feeds the control winding of a magnetic amplifier. 2. One of the circuits that responds to the instructions in the program for a digital computer. 3. A circuit that controls some function of a machine, device, or piece of equipment.

**control desk** *Console.*

**control electrode** An electrode used to initiate or vary the current between two or more electrodes in an electron tube.

**control element** The portion of a feedback control system that acts on the process or machine being controlled. In a nuclear reactor the control element is commonly a control rod.

**control grid** A grid, ordinarily placed between the cathode and an anode, that serves to control the anode current of an electron tube.

**controlled-carrier modulation** A method of modulation in which the percentage modulation is held constant at all times by varying the amplitude of the carrier wave automatically to offset the variations produced by conventional amplitude modulation of the carrier wave. Also called floating-carrier modulation and variable-carrier modulation.

**controlled-devices countermeasure** Any electronic countermeasure against guided missiles, pilotless aircraft, proximity fuzes, or similar controlled devices.

**controlled mercury-arc rectifier** A mercury-arc rectifier in which one or more electrodes control the start of the discharge in each cycle and thereby control output current.

**controlled rectifier** A rectifier having means for controlling its output current.

**controlled variable** The quantity or condition that is measured and controlled.

**controller** *Automatic controller.*

**control point** The value of controlled variable that is maintained by an automatic control system.

**control ratio** The ratio of the change in anode voltage to the corresponding change in critical grid voltage in a gas tube.

**control register** *Program register.*

**control rod** Any rod used to control the reactivity of a nuclear reactor. It may be a fuel rod or a part of the moderator; in a thermal reactor it is commonly a neutron absorber. The control rod changes the effective multiplication constant and hence the time derivative of reactivity. Examples include power control rod, regulating rod, safety rod, and shim rod.

**control room** A room from which engineers and production men control and direct a television or radio program. It is adjacent to the main studios and separated from them by large soundproof double glass windows.

**control signal** The signal applied to the device that makes corrective changes in a controlled process or machine.

**control synchro** *Control transformer.*

**control system** 1. An arrangement of a sensing element, amplifier, and control device acting together to control some condition of a process or machine. 2. A system used in a ballistic or guided missile to maintain altitude stability during powered flight and to correct deflections caused by gusts or other disturbances. A control system uses jet vanes and other devices in common with the guidance system.

**control track** A supplementary sound track that is usually placed on the same motion-picture film with the sound track carrying program material. It usually contains tone signals that control the reproduction of the sound track, such as by changing feed levels to loudspeakers in a theater to achieve stereophonic effects.

**control transformer** A synchro in which the electric output of the rotor is dependent on both the shaft position and the electric input to the stator. Also called control synchro.

**control winding** A winding used on a magnetic amplifier or saturable reactor to apply control magnetomotive forces to the core.

**convection current** The time rate at which the electric charges of an electron stream are transported through a given surface.

**convection-current modulation** The time variation in the magnitude of the convection current passing through a surface, or the process of directly producing such a variation in a microwave tube.

**convective discharge** The movement of a visible or invisible stream of charged particles away from a body that has been charged to a sufficiently high voltage. Also called electric wind and static breeze.

**convenience receptacle** *Outlet.*

**convergence** A condition in which the electron beams of a multibeam cathode ray tube intersect at a specified point, such as at an opening in the shadow mask of a three-gun color television picture tube. Both static and dynamic convergence are required.

**convergence coil** One of the coils used to obtain convergence of electron beams in a three-gun color television picture tube.

**convergence control** A control used in a color television receiver to adjust the potential on the convergence electrode of the three-gun color picture tube to achieve convergence.

**convergence electrode** An electrode whose electric field converges two or more electron beams.

**convergence magnet** A magnet assembly whose magnetic field converges two or more electron beams. Used in three-gun color picture tubes. Also called beam magnet.

**convergence plane** A plane containing the points at which the electron beams of a multibeam cathode-ray tube appear to experience a deflection applied for the purpose of obtaining convergence.

**convergence surface** The surface generated by the point of intersection of two or more electron beams in a multibeam cathode-ray tube during the scanning process.

**conversion coefficient** *Conversion fraction.*

**conversion efficiency** 1. The ratio of a-c output power to the d-c power input to the electrodes of an electron tube. 2. The ratio of the output voltage of a converter at one frequency to the input voltage at some other frequency.

**conversion electron** An electron emitted by internal conversion during deexcitation of a nucleus.

**conversion fraction** The ratio of the number of internal conversion electrons to the total number of quanta plus the number of conversion electrons emitted in a given mode of deexcitation of a nucleus. Also called conversion coefficient, conversion ratio, and internal conversion coefficient.

**conversion gain** The conversion ratio minus one in a nuclear reactor.

**conversion gain ratio** The ratio of signal power output to signal power input for a frequency converter or mixer.

**conversion ratio** 1. The number of fissionable atoms produced per fissionable atom destroyed in a converter type of nuclear reactor. 2. *Conversion fraction.*

**conversion transconductance** The output current of a heterodyne converter at the desired intermediate frequency divided by the r-f input signal voltage, the output impedance being negligibly small for all frequencies that may affect the result.

**conversion transducer** An electric transducer in which the input and output frequencies are different. An example is the converter used in superheterodyne receivers.

**conversion voltage gain** The output voltage of a conversion transducer divided by the input voltage.

**converter** 1. The section of a superheterodyne radio receiver that converts the desired incoming r-f signal to the i-f value. The converter section includes the oscillator and the mixer-first detector. Also called heterodyne conversion transducer and oscillator-mixer-first detector. 2. An auxiliary unit used with a television or radio receiver to permit reception of channels or frequencies for which the receiver was not originally designed. 3. In facsimile, a device that changes the type of modulation delivered by the scanner. 4. A computer unit that changes numerical information from one form to another, as from decimal to binary or vice versa, from fixed point to floating-point representation, or from punched cards to magnetic tape. 5. A nuclear reactor that converts fertile atoms into fuel by neutron capture, using one kind of fuel and producing another. 6. *Remodulator.* 7. *Dynamotor.* 8. *Synchronous converter.*

**converter tube** An electron tube that combines the mixer and local-oscillator functions of a heterodyne conversion transducer.

**coolant** A substance, ordinarily fluid, used for cooling any part of a reactor in which heat is generated.

**Coolidge tube** An x-ray tube in which the needed electrons are produced by a hot cathode.

**cooling** Setting aside a highly radioactive material until the radioactivity has diminished to a desired level.

**cooperative system** A missile guidance system that requires transmission of information from a remote ground station to a missile in flight, processing of the information by the missile-borne equipment, and retransmission of the processed data to the originating and/or other remote ground stations, as in azusa and dovap.

**coordinate** Any one of two or more magnitudes that determine position relative to the reference axes of a coordinate system.

**coordinate data receiver** A receiver specifically designed to accept the signal of a coordinate data transmitter and reconvert this signal into a form suitable for input to associated equipment such as a plotting board, computer, or a radar set.

**coordinate data transmitter** A transmitter that accepts two or more coordinates, such as those representing a target position, and converts them into a form suitable for transmission.

**coordinate system** A system for specifying the location of a point, using two coordinates if on a surface and three coordinates if in space.

**coplanar electrodes** Electrodes mounted in the same plane.

**copper** [symbol Cu] A metallic element having good conductivity. Atomic number is 29.

**copper loss** Power loss in a winding due to current flow through the resistance of the copper conductors. Also called $I^2R$ loss.

**copper-oxide photovoltaic cell** A photovoltaic cell in which light acting on the surface of contact between layers of copper and cuprous oxide causes a voltage to be produced.

**copper-oxide rectifier** A metallic rectifier in which the rectifying barrier is the junction between metallic copper and cuprous oxide. A disk of copper is coated with

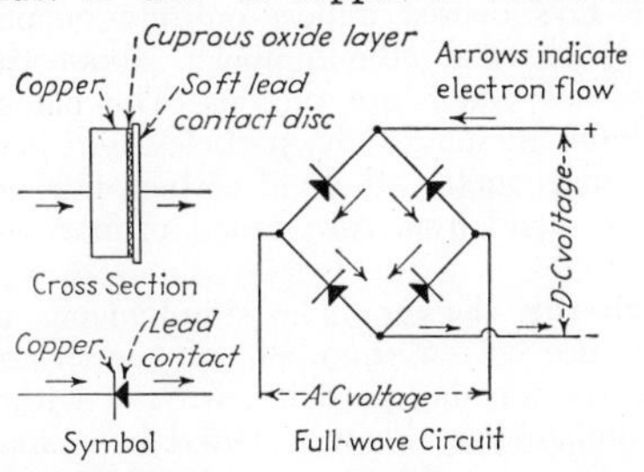


Copper-oxide rectifier construction, symbol, and typical circuit.

cuprous oxide on one side, and a soft lead washer is used to make contact with the oxide layer.

**copper sulfide rectifier** A semiconductor rectifier in which the rectifying barrier is the junction between magnesium and copper sulfide.

**copy** 1. To transfer information to a new location in a computer. 2. *Subject copy.*

**cord** A small, very flexible insulated cable.

**core** 1. The active portion of a nuclear reactor, containing the fissionable material. 2. *Magnetic core.*

**core iron** A grade of soft iron suitable for cores of chokes, transformers, and relays.

**coreless-type induction furnace** *High-frequency furnace.*

**coreless-type induction heater** A device in which a charge is heated directly by induction, with no magnetic core material linking the charge to the induction coil.

**core loss** The power loss in an iron-core transformer or inductor due to eddy currents and hysteresis effects in the iron core. Also called iron loss.

**core memory** *Magnetic-core storage.*

**core-type induction heater** A device in which a charge is heated by induction, with a magnetic core being used to link the induction coil to the charge.

**corner** A sharp bend in a waveguide. Also called elbow.

**corner antenna** *Corner-reflector antenna.*

**corner reflector** A radar reflector consisting of three conducting surfaces mutually intersecting at right angles. This reflector returns electromagnetic radiation to its source. Used to make a position more conspicuous for radar observations.

**corner-reflector antenna** An antenna consisting of two conducting surfaces intersecting at an angle that is usually 90°, with a dipole or other antenna located on the bisector of the angle. The surfaces are often made of wire mesh to reduce wind resistance. Maximum pickup is along the bisector of the reflector angle. Used as a uhf television and radio receiving antenna. Also called corner antenna.

**corona** A discharge of electricity appearing as a bluish-purple glow on the surface of and adjacent to a conductor when the voltage gradient exceeds a certain critical value. It is due to ionization of the surrounding air by the high voltage.

**corona shield** A shield placed about a point of high potential to redistribute electrostatic lines of force.

**corona tube** A gas-discharge voltage-reference tube employing a corona discharge.

**corona voltmeter** A voltmeter in which the crest value of a voltage is indicated by the inception of corona at a known electrode spacing.

**Corporal** An Army surface-to-surface guided missile having a range of about 100 miles and a speed several times that of sound. It can carry either conventional or nuclear warheads. Used to engage tactical targets beyond the range of artillery.

**corrected compass course** *Magnetic course.*

**corrected compass heading** *Magnetic heading.*

**correction** A quantity added to a calculated or observed value to obtain the true value.

**correction element** The element in a fire-control system that introduces corrections based on variations of conditions from standard firing table conditions, along with arbitrary corrections based on observations of previous firings.

**correction time** The time required for the

controlled variable to reach and stay within a predetermined band about the control point following any change of the independent variable or operating condition in a control system. Also called settling time.

**corrective network** An electric network inserted in a circuit to improve its transmission properties, its impedance properties, or both. Also called shaping network.

**correlation-type receiver** *Correlator.*

**correlator** A device that detects weak signals in noise by performing an electronic operation approximating the computation of a correlation function. Examples include the autocorrelator and crosscorrelator. Also called correlation-type receiver.

**corrugated-surface antenna** A microwave antenna consisting of a waveguide feed to a mode transformer or horn launcher, and a transversely corrugated metal surface that acts as a guide for surface waves.

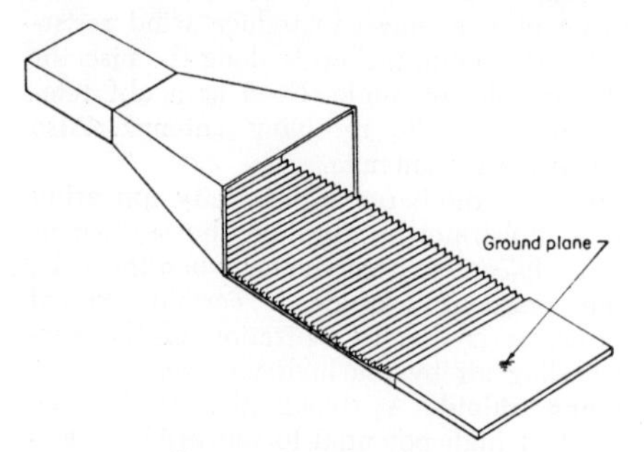

Corrugated-surface antenna.

**cortical stimulator** An electronic instrument used in nerve and mental therapy to deliver an electric shock of prescribed strength by means of a pulsating current. A low-frequency relaxation oscillator is sometimes used to produce the pulses.

**Corvus** A Navy air-to-surface guided missile intended for use against surface ships and for attacking heavily defended areas.

**cosecant antenna** An antenna that gives a beam whose amplitude varies as the cosecant of the angle of depression below the horizontal. Used in navigation radar. It may use a cheese antenna with a line source or a distorted parabolic reflector with a point source.

**cosecant-squared antenna** An antenna having a cosecant-squared pattern.

**cosecant-squared pattern** A ground radar antenna radiation pattern that sends less power to nearby objects than to those farther away in the same sector. The field intensity varies as the square of the cosecant of the elevation angle. The pattern is achieved either by bending the top portion of the parabolic reflector forward or by using a spoiler on the reflector. With this pattern, approximately equal echo signals are received from objects at the same altitude but at varying distances. Also used in airborne antennas to produce a uniform electric field along a line on the earth's surface.

**cosine winding** A winding used in the deflection yoke of a cathode-ray tube to prevent changes in focus as the beam is deflected over the entire area of the screen.

**cosmic noise** Radio static caused by a phenomenon outside the earth's atmosphere, such as sunspots.

**cosmic rays** High-energy radiation originating outside the atmosphere of the earth, capable of producing ionizing events in passing through the air or other matter. The rays consist almost entirely of positively charged atomic nuclei, about two-thirds of which are protons. The balance includes mesons, alpha particles, and heavier nuclei such as those of carbon, nitrogen, oxygen, and iron. Also called primary cosmic rays.

**cosmic-ray shower** The simultaneous appearance of a number of downward-directed ionizing particles, with or without accompanying photons, caused by a single cosmic ray. Cosmic-ray showers reveal themselves by simultaneous actuation of separated counters. They can be roughly classified according to their properties as cascade shower, extensive shower, and penetrating shower. Also called shower.

**cosmonautics** The science of travel beyond the solar system.

**cosmotron** *Proton synchrotron.*

**cotar** [Correlated Orientation Tracking And Range] A passive system used for tracking a vehicle in space by determining the line of direction between a remote ground-based receiving antenna and a telemetering transmitter in the missile, using phase comparison techniques.

**Coulmer antenna array** A high-gain planar antenna array consisting of nonresonant elements stacked vertically and horizontally to produce both vertically and horizontally polarized waves.

**coulomb** A unit of electric charge, defined as the amount of electric charge that crosses a surface in one second when a steady current of one absolute ampere is flowing across the surface. This is the absolute coulomb, and has been the legal standard of quantity of electricity since 1950. The previous standard was the international coulomb, equal to 0.999835 absolute coulomb.

**coulomb barrier radius** The nuclear radius deduced from the rate of alpha disintegration or from cross sections of nuclear reactions involving charged particles. Also called Gamow barrier radius.

**coulomb force** The electrostatic force of attraction or repulsion exerted by one charged particle on another.

**coulomb potential** A scalar point function equal to the work per unit charge done against or by the coulomb force in transferring a particle bearing an infinitesimal positive charge from infinity to the field of a charged particle in a vacuum.

**Coulomb's law** The attraction or repulsion between two electric charges is proportional to the product of their magnitudes and is inversely proportional to the square of the distance between them. Also called law of electrostatic attraction.

**count** 1. A single response of the counting system in a radiation counter. 2. The total number of events indicated by a counter.

**count-down** 1. The ratio of the number of interrogation pulses not answered by a transponder to the total number received. 2. The step-by-step process of preparing a missile for launching.

**count-down circuits** Circuits that are connected to a guided missile through its umbilical cord to actuate and check the firing controls during the final audible counting of seconds before firing.

**counter** 1. A complete instrument for detecting, totalizing, and indicating a sequence of events. 2. *Accumulator.* 3. *Radiation counter.* 4. *Scaler.*

**counter circuit** A circuit that receives uniform pulses representing units to be counted and produces a voltage proportional to the total count.

**counterclockwise capacitor** A variable capacitor whose capacitance increases with counterclockwise rotation of its rotor, as viewed from the end of the control shaft.

**counterclockwise polarized wave** *Left-hand polarized wave.*

**counter-controlled cloud chamber** A cloud chamber whose expansion is triggered by a counter, for studying particular events.

**counter dead time** The time interval between the start of a counted event and the earliest instant at which a new event can be counted by a radiation counter.

**counterelectromotive force** The voltage developed in an inductive circuit by a changing current. The polarity of the voltage is at each instant opposite that of the generated or applied voltage. Also called back electromotive force.

**countermeasures** The use of devices and/or techniques intended to impair the operational effectiveness of enemy activity.

**counter overvoltage** The amount by which the applied voltage of a radiation counter exceeds the Geiger-Mueller threshold.

**counter plateau** The region of a radiation-counter characteristic curve in which the counting rate is substantially independent of voltage. The counter is normally operated in this region.

**counterpoise** A system of wires or other conductors that is elevated above and insulated from the ground to form a lower system of conductors for an antenna. Used as a substitute for a ground connection. Also called antenna counterpoise.

**counter range** Low reaction rate values in a nuclear reactor, occurring during start-up, that must be measured by counters because the neutron flux is too weak to be measured by the permanently installed control instruments.

**counter tube** 1. An electron tube that converts an incident particle or burst of incident radiation into a discrete electric pulse, generally by utilizing the current flow through a gas that is ionized by the radiation. Used in radiation counters. Also called radiation-counter tube. 2. An electron tube having one signal input electrode and 10 or more output electrodes, with each input pulse serving to transfer conduction sequentially to the next output electrode. Beam-switching tubes and cold-cathode counter tubes are examples. Used for preset counting, data storage, timing, and gating. Also called counting tube and decade counter tube. 3. *Beam-switching tube.* 4. *Cold-cathode counter tube.*

**counting circuit** A circuit that counts pulses by frequency-dividing techniques, by charging a capacitor in such a way as to produce a voltage proportional to the pulse count, or by other means.

**counting efficiency** The ratio of the average number of photons or ionizing particles that produce counts to the average number incident on the sensitive area of a radiation counter.

**counting ionization chamber** *Pulse ionization chamber.*

**counting loss** The counting error due to events occurring within the dead time of a radiation detector.

**counting rate** The average rate of occurrence of events as observed by means of a counting system.

**counting-rate curve** A curve showing how counting rate varies with applied voltage in a radiation counter. It generally starts with a sharp rise at the threshold voltage, has a flat region known as the plateau, then ends with a sudden sharp rise. The counter is usually operated in the plateau region, where the counting rate is not appreciably affected by changes in applied voltage.

**counting-rate meter** An instrument that indicates the time rate of occurrence of input pulses to a radiation counter, averaged over a time interval. Also called rate meter.

**counting-rate—voltage characteristic** *Plateau characteristic.*

**counting tube** 1. *Beam-switching tube.* 2. *Cold-cathode counter tube.* 3. *Counter tube.*

**Couplate** Trademark of Centralab for printed-circuit units containing two or more printed elements of resistance, capacitance, and sometimes also inductance.

**couple** 1. Two metals placed in contact, as in a thermocouple. 2. To connect two circuits so signals are transferred from one to the other.

**coupler** 1. The portion of a navigation system that receives signals of one type from a sensor and transmits signals of a different type to an actuator. 2. A component used to transfer energy from one circuit to another.

**coupling** 1. A mutual relation between two circuits that permits energy transfer from one to the other. Coupling may be direct

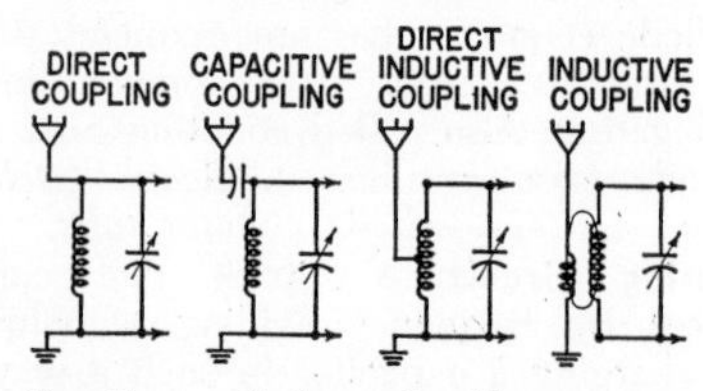


Coupling methods used with an antenna.

through a wire, resistive through a resistor, inductive through a transformer or choke, or capacitive through a capacitor. 2. A flexible or rigid device used to fasten together two shafts end to end.

**coupling aperture** An aperture in the wall of a waveguide or cavity resonator, designed to transfer energy to or from an external circuit. Also called coupling hole and coupling slot.

**coupling capacitor** A capacitor used to block the flow of direct current while allowing alternating or signal currents to pass. Widely used for joining two circuits or stages. Also called blocking capacitor and stopping capacitor.

**coupling coefficient** 1. The ratio of the maximum change in energy of an electron traversing an interaction space to the product of the peak alternating gap voltage and the electronic charge. 2. *Coefficient of coupling.*

**coupling hole** *Coupling aperture.*

**coupling loop** A conducting loop projecting into a waveguide or cavity resonator, designed to transfer energy to or from an external circuit.

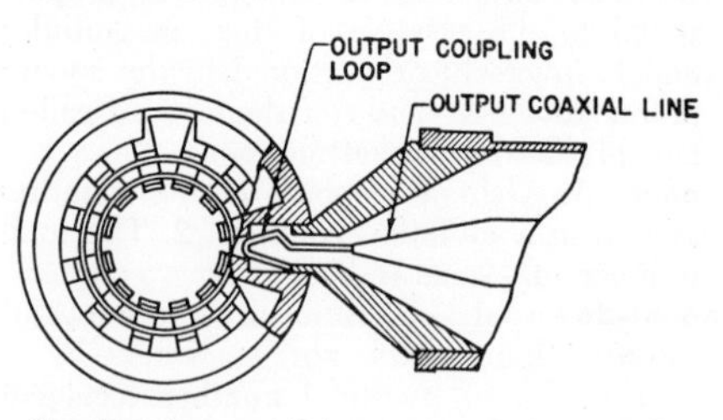


Coupling loop used between magnetron and coaxial line.

**coupling probe** A probe projecting into a waveguide or cavity resonator, designed to transfer energy to or from an external circuit.

**coupling slot** *Coupling aperture.*

**course** 1. The intended direction of travel, expressed as an angle in the horizontal plane between a reference line and the course line, usually measured clockwise from the reference line. Also called course line and desired track. 2. A radio range beam.

**course error** Deprecated term for *drift angle.*

**course line** *Course.*

**course-line computer** An airborne computer that continually computes an aircraft's position, in terms of its departure from course and its distance from destination or an intermediate point, by utilizing omnirange and distance-measuring-equipment transmission. Also called offset-course computer.

**course-line deviation** The angular or linear difference between the actual track of a vehicle and the intended course line.

**course-line deviation indicator** An instrument that indicates deviation from a desired course line.

**course made good** The resultant direction of actual travel of a vehicle, equivalent to its bearing from the point of departure.

**course pull** *Course push.*

**course push** An erroneous deflection of the indicator of a navigation aid, due to a change in the attitude of the receiving antenna. Also called course pull.

**course sensitivity** The amount of displacement of a vehicle from the course line that produces a given change of course indication.

**course softening** An intentional decrease in course sensitivity upon approaching a navigation aid, such that the ratio of indicator deflection to linear displacement from the course line tends to remain constant.

**course width** The arithmetic sum of the plus and minus lateral deviations from the course line within which the course-defining parameters do not vary by a detectable amount. Also called angular width.

**coverage** *Service area.*

**coverage diagram** A diagram depicting the service area of a navigation aid.

**cozi** [COmmunications Zone Indicator] A system that indicates how well Voice of America and other long-distance high-frequency broadcasts are reaching their destinations. The broadcast is interrupted at unnoticeably short intervals to send out a radar beam from the station's transmitting antenna and pick up the returned echo as the radar pulse travels out and back along the same path taken by the radio waves. Travel time and intensity of the received echo give the skip distance of the wave and its probable strength at destination, and sometimes also reveal jamming.

**cpa** Abbreviation for *color phase alternation.*

**cps** Abbreviation for *cycle per second.*

**crab angle** *Drift correction angle.*

**crater lamp** A glow-discharge tube used as a point source of light whose brightness is proportional to the signal current sent through the tube. Used for photographic recording of facsimile signals. The glow discharge is concentrated in a crater-shaped depression at one end of the cathode.

**crest factor** The ratio of the peak value to the effective value of any periodic quantity such as a sinusoidal alternating current.

**crest value** *Peak value.*

**crest voltmeter** A voltmeter reading the peak value of the voltage applied to its terminals.

**crit** The mass of fissionable material that is critical under a given set of conditions. Sometimes applied to the mass of an untamped critical sphere of fissionable material.

**critical** Capable of sustaining a chain reaction at a constant level.

**critical absorption wavelength** The wavelength, characteristic of a given electron energy level in an atom of a specified element, at which an absorption discontinuity occurs.

**critical angle** The smallest angle away from the vertical at which a radiated radio wave of a given frequency will still be reflected by the ionosphere. At smaller angles the radio waves penetrate the ionosphere and are not returned to earth.

**critical anode voltage** The anode voltage at which breakdown occurs in a gas tube.

**critical coupling** The degree of coupling that provides maximum transfer of signal energy from one r-f resonant circuit to another when both are tuned to the same frequency. Also called optimum coupling.

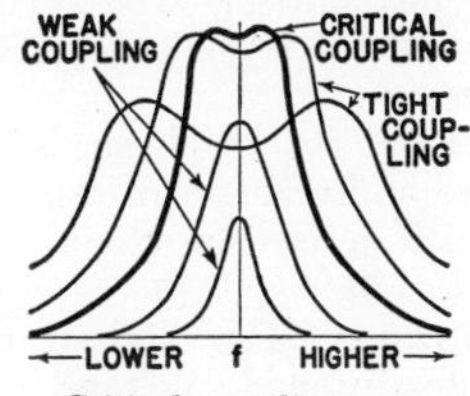


Critical coupling curve.

**critical damping** The degree of damping required to give the most rapid transient response without overshooting or oscillation. Thus, a critically damped meter moves its pointer to a new value without going past it.

**critical dimension** *Broad dimension.*

**critical equation** Any equation relating parameters of a reactor that must be satisfied for the reactor to be critical.

**critical experiment** An experiment in which fissionable material is assembled gradually until the arrangement will support a self-sustaining chain reaction. It is carried out at substantially zero power so forced cooling is unnecessary and fission-product activity is negligible. Used in testing proposed new designs on an experimental basis.

**critical field** The smallest theoretical value of steady magnetic flux density that would prevent an electron emitted from the cathode of a magnetron at zero velocity from reaching the anode. Also called cutoff field.

**critical frequency** 1. The limiting frequency below which a radio wave will be reflected by an ionospheric layer at vertical incidence at a given time. Higher frequencies will penetrate the layer. Also called penetration frequency. 2. *Cutoff frequency.*

**critical grid current** The current in a gas tube at the instant when anode current starts to flow.

**critical grid voltage** The grid voltage at which anode current starts to flow in a gas tube.

**critical high-power level** The r-f power level in an attenuator tube at which ionization is produced in the absence of a control-electrode discharge.

**critical inductance** The minimum input choke inductance required to prevent the input choke current from going to zero during any part of the cycle in a choke input filter for a full-wave rectifier.

**critical mass** The mass of fissionable material of a particular shape that is just sufficient to sustain a nuclear chain reaction.

**critical reactor** A nuclear reactor in which the ratio of moderator to fuel is either subcritical or just critical. Used to study the properties of the system and determine critical size.

**critical size** A set of physical dimensions for the core and reflector of a nuclear reactor at which a critical chain reaction is maintained.

**critical voltage** The highest theoretical value of steady anode voltage, at a given steady magnetic flux density, at which electrons emitted from the cathode of a magnetron at zero velocity would fail to reach the anode. Also called cutoff voltage.

**critical wavelength** The free-space wavelength corresponding to the critical frequency.

**cro** Abbreviation for *cathode-ray oscilloscope.*

**Crookes dark space** *Cathode dark space.*

**Crookes radiometer** A radiometer used to demonstrate that radiant energy from the sun can produce motion. A miniature four-vane windmill is mounted in a glass-envelope vacuum tube. Each vane is polished on one side and black on the other side. Absorption of radiant energy by the black sides warms these sides and makes adjacent residual molecules of gas rebound more rapidly than from the polished sides. The black sides then rotate away from the source of radiation.

**Crookes tube** An early form of discharge tube.

**crossband** Two-way communication in which one radio frequency is used in one direction and a frequency having different propagation characteristics is used in the opposite direction.

**crossbanding** Use of one interrogation frequency with several reply frequencies or one reply frequency with several interrogation frequencies.

**crossband transponder** A transponder that replies in a different frequency band from that of the received interrogation.

**crossbar switch** A switch having a three-dimensional arrangement of contacts and a magnet system that selects individual contacts according to their coordinates in the matrix.

**cross-bombardment** A method for assigning the mass of a radioactive species based on its production by several nuclear bombardments using different projectiles and/or different target materials.

**cross-channel communication** Two-way communication in which one radio frequency is used in one direction and a different frequency having similar propagation characteristics is used in the opposite direction.

**cross-color interference** Interference produced in the chrominance channel of a color television receiver by crosstalk from the monochrome signal.

**cross-control circuit** A compandor circuit in which input signals to the compressor also control the operation of the expandor at the same end of the circuit.

**crosscorrelation function** A mathematical quantity defined as the product of two functions of time.

**crosscorrelator** A correlator in which a locally generated reference signal is multiplied by the incoming signal and the result is smoothed in a low-pass filter to give an approximate computation of the crosscorrelation function. It can be used for detecting weak signals in noise in cases where the important signal characteristics are known prior to detection. Also called synchronous detector.

**cross-coupling** A measure of the undesired power transferred from one channel to another in a transmission medium.

**crossed-field accelerator** A plasma engine for space travel in which plasma serves as a conductor to carry current across a magnetic field, so that a resultant force is exerted on the plasma.

**crossed-pointer indicator** A two-pointer indicator used with an instrument-landing system to indicate the position of an airplane with respect to the glide path.

**cross-modulation** A type of interference in which the carrier of a desired signal becomes modulated by the program of an undesired signal on a different carrier frequency. The program of the undesired station is then heard in the background of the desired program. Cross-modulation occurs because the first tube circuit in the receiver is nonlinear and hence acts as a detector for the strong undesired signal.

**cross-neutralization** A method of neutralization used in push-pull amplifiers, in which a portion of the a-c anode-cathode voltage of each tube is applied to the grid-cathode circuit of the other tube through a neutralizing capacitor.

**crossover frequency** 1. The frequency at which a dividing network delivers equal power to the upper and lower frequency channels when both are terminated in specified loads. 2. *Transition frequency.*

**crossover network** A selective network used to divide the audio-frequency output of an amplifier into two or more bands of frequencies. The band below the crossover frequency is fed to the woofer loudspeaker, while the high-frequency band is fed to the tweeter. Also called dividing network and loudspeaker dividing network.

**crossover region** A zone in space, close to the localizer on-course line or guide slope of an instrument approach system, in which the pointer of the indicator is in a position between the full-scale indications.

**crossover spiral** *Leadover groove.*

**cross-polarization** The component of the electric field vector normal to the desired polarization component.

**cross-section** 1. A section at right angles to an axis. 2. Deprecated term for *echo area.*

**cross section** The probability, per unit flux and per unit time, that a given ionization or capture reaction will occur in a nuclear reactor. The cross section of an atom or nucleus has the dimensions of an area and is actually the effective target area presented to an incident particle or photon for a particular reaction. If the reaction cannot take place, the cross section is zero. A commonly used unit of area for cross sections is the barn, equal to $10^{-24}$ sq cm.

**cross section per atom** The microscopic cross section for a given nuclear reaction referred to the natural element, even though the reaction involves only one of the natural isotopes.

**crosstalk** 1. The sound heard in a receiver along with a desired program because of cross-modulation or other undesired coupling to another communication channel. 2. Interaction of audio and video signals in a television system, causing video modulation of the audio carrier or audio modulation of the video signal at some point. 3. Interaction of chrominance and luminance signals in a color television receiver. 4. Deprecated term for *magnetic printing.*

**crosstalk unit** [abbreviated cu] A measure of the coupling between two circuits. The number of crosstalk units is 1,000,000 times the ratio of the current or voltage at the observing point to the current or voltage at the origin of the disturbing signal, the impedances at these points being equal.

**crt** Abbreviation for *cathode-ray tube.*

**cruciform core** A transformer core in which all windings are on one center leg and four additional legs arranged in the

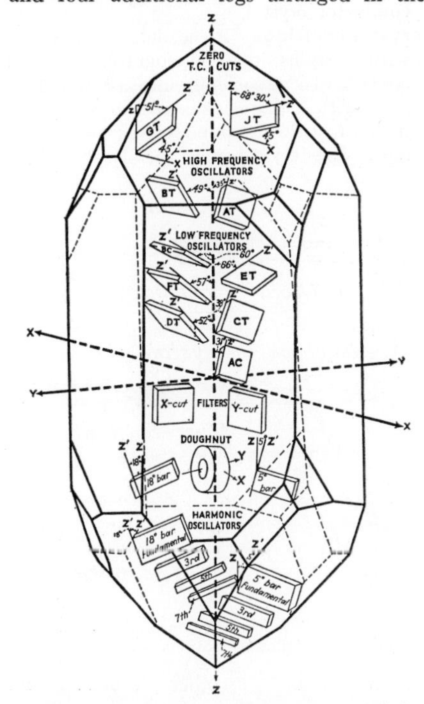


Crystal blank positions for different cuts in natural quartz. The angle of the cut with respect to the natural faces of the crystal determines the electrical characteristics of the finished crystal blank.

form of a cross serve as return paths for magnetic flux.

**cryogenic gyroscope** A gyroscope employing spinning electrons near absolute zero in temperature instead of a conventional spinning flywheel.

**cryogenics** The science of physical phenomena at very low temperatures, approaching absolute zero.

**cryosar** A computer component whose operation is based on impact ionization of impurities in germanium.

**cryotron** A switch that operates in liquid helium, consisting of a short wire on which is wound a control wire. When current is sent through the control wire to produce a magnetic field, the short main wire changes from a superconductive zero-resistance state to its normal resistive state.

**crystal** A natural or synthetic piezoelectric or semiconductor material.

**crystal activity** A measure of the amplitude of vibration of a piezoelectric crystal plate under specified conditions.

**crystal blank** The result of the final cutting operation on a piezoelectric or semiconductor crystal.

**crystal cartridge** A piezoelectric unit used with a stylus in a phonograph pickup to convert disk recordings into a-f signals, or used with a diaphragm in a crystal microphone to convert sound waves into a-f signals. The crystal may be Rochelle salt, barium titanate, or other piezoelectric material.

Crystal cartridge construction, and twisting action when stylus follows groove of record.

**crystal control** Control of the frequency of an oscillator by means of a quartz crystal unit.

**crystal-controlled oscillator** An oscillator whose frequency of operation is controlled by a crystal unit.

**crystal-controlled transmitter** A transmitter whose carrier frequency is directly controlled by the electromechanical characteristics of a quartz crystal unit.

**crystal counter** A counter utilizing one of several known crystals that are rendered momentarily conducting by ionizing events. One example of such a crystal is a diamond.

**crystal current** The actual alternating current flowing through a crystal unit.

**crystal cutter** A cutter in which the mechanical displacements of the recording stylus are derived from the deformations of a crystal having piezoelectric properties.

**crystal detector** A crystal diode or equivalent earlier crystal-catwhisker combination used to rectify a modulated r-f signal to obtain the audio or video signal directly.

Crystal detector and crystal set.

Crystal diodes are also used in microwave receivers to combine an incoming r-f signal with a local oscillator signal to produce an i-f signal.

**crystal diode** A two-electrode semiconductor device that utilizes the rectifying properties of a junction between p-type and n-type material in a semiconductor, as in a junction diode, or the rectifying properties of a sharp wire point in contact with a semiconductor material, as in a point-contact diode. Also called crystal rectifier, diode, and semiconductor diode.

**crystal filter** A highly selective tuned circuit employing one or more quartz crystals. Sometimes used in i-f amplifiers of communication receivers to improve the selectivity.

**crystal headphones** Headphones using Rochelle-salt or other crystal elements to convert a-f signals into sound waves.

**crystal holder** A housing designed to provide proper support, mechanical protection, and connections for a quartz crystal plate. When the crystal plate is installed, the combination is called a crystal unit.

**crystal hydrophone** A crystal microphone that responds to waterborne sound waves.

**crystal loudspeaker** A loudspeaker in which movements of the diaphragm are produced by a piezoelectric crystal unit that twists or bends under the influence of

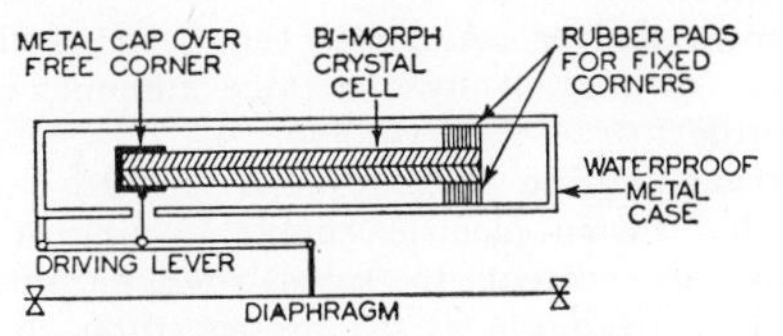

Crystal loudspeaker construction.

the applied a-f signal voltage. Also called piezoelectric loudspeaker.

**crystal microphone** A microphone in which deformation of a piezoelectric bar by the action of sound waves or mechanical vibrations generates the output voltage between the faces of the bar. Rochelle-salt crystals or barium-titanate ceramic bars are most often used. Also called piezoelectric microphone.

**crystal mixer** A mixer that uses the nonlinear characteristic of a crystal diode to mix two frequencies. Widely used in radar receivers to convert the received radar signal to a lower i-f value by mixing it with a local oscillator signal.

**crystal oscillator** An oscillator in which the frequency of the a-c output is deter-

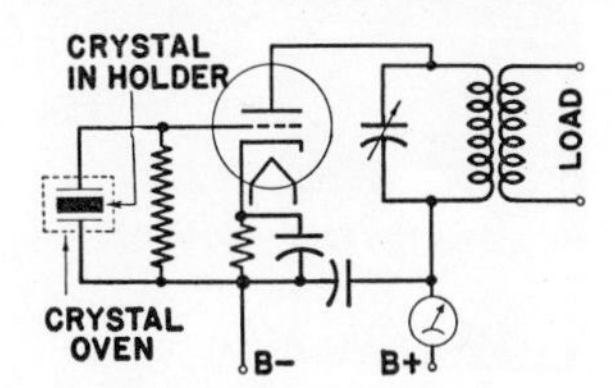

Crystal oscillator circuit.

mined by the mechanical properties of a piezoelectric crystal.

**crystal oven** A temperature-controlled oven in which a crystal unit is operated to stabilize its temperature and thereby minimize frequency drift.

**crystal pickup** A phonograph pickup in which movements of the needle in the record groove cause deformation of a piezoelectric crystal, thereby generating an a-f output voltage between opposite faces of the crystal. The piezoelectric material used is generally Rochelle salt or barium titanate. Also called piezoelectric pickup.

**crystal plate** A precisely cut slab of quartz crystal that has been lapped to final dimensions, etched to improve stability and efficiency, and coated with metal on its major surfaces for connecting purposes. Also called quartz plate.

**crystal pulling** A method of crystal growing in which the developing crystal is gradually withdrawn from a melt.

**crystal rectifier** *Crystal diode.*

**crystal set** A radio receiver having a crystal detector stage for demodulation of the received signals, but no amplifier stages.

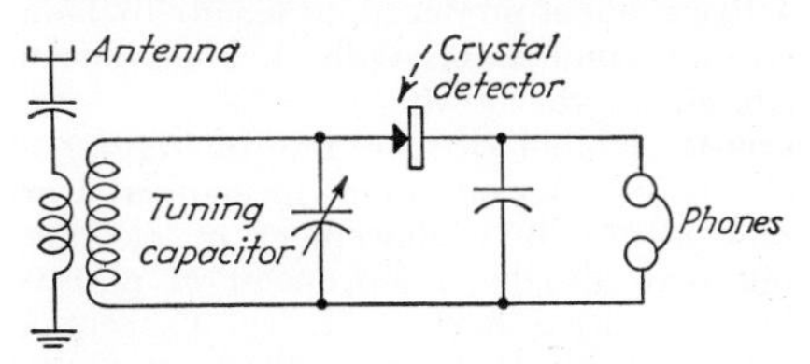

Crystal set circuit. A long outdoor antenna is usually required.

**crystal slab** A relatively thick piece of crystal from which crystal blanks are cut.

**crystal spectrometer** *Bragg spectrometer.*

**crystal-stabilized transmitter** A transmitter employing automatic frequency control, in which the reference frequency is that of a crystal oscillator.

**crystal system** One of the seven main categories to which a crystal may be assigned according to the symmetry of its external form or internal structure. The systems are cubic, tetragonal, hexagonal, trigonal, orthorhombic, monoclinic, and triclinic.

**crystal unit** A complete assembly of one or more quartz plates in a crystal holder.

**crystal video receiver** A broad-tuning radar or other microwave receiver consisting only of a crystal detector and a video or audio amplifier.

**C scan** *C display.*

**C scope** A radarscope that produces a C display.

**CT** Symbol used on diagrams to indicate a center tap.

**cu** Abbreviation for *crosstalk unit.*

**cue circuit** A one-way communication circuit used to convey program control information.

**cumulative dose** The total dose resulting from repeated exposures to radiation.

**cumulative ionization** *Avalanche.*

**cuprous oxide** [symbol $Cu_2O$] A semiconductor material formed on copper by heat. Electrons flow readily only in the direction from the metallic copper toward the oxide layer on the surface. This effect is utilized for rectification in copper-oxide rectifiers.

**curie** [abbreviated c] A unit of radioactivity, defined by international agreement in 1953 as that quantity of any radioactive nuclide which has $3.700 \times 10^{10}$ distintegrations per second. Before 1953, the curie was defined as the quantity of radon that is in equilibrium with 1 gram of radium.

**Curie point** The temperature above which a ferromagnetic material becomes substantially nonmagnetic.

**curium** [symbol Cm] A transuranic radioactive element produced artificially by bombarding plutonium with helium nuclei. Atomic number is 96.

**current** [symbol *I*] The rate of transfer of electricity from one point to another. Current is usually a movement of electrons but may also be a movement of positive ions, negative ions, or holes. Current is measured in amperes, milliamperes, and microamperes. Also called electric current and juice (slang).

**current amplification** The ratio of output signal current to input signal current for an electron tube, transistor, or magnetic amplifier, the multiplier section of a multiplier phototube, or any other amplifying device. Often expressed in decibels by multiplying the common logarithm of the ratio by 20.

**current amplifier** An amplifier capable of delivering considerably more signal current than is fed in.

**current antinode** A point at which current is a maximum along a transmission line, antenna, or other circuit element having standing waves. Also called current loop.

**current attenuation** The ratio of input signal circuit for a transducer to the current in a specified load impedance connected to the transducer. Often expressed in decibels.

**current-carrying capacity** The maximum current that can be continuously carried without causing permanent deterioration of electrical or mechanical properties of a device or conductor.

**current density** 1. The current per unit cross-sectional area of a conductor. 2. A vector whose component perpendicular to a surface in a nuclear reactor equals the net number of particles crossing that surface per unit area and unit time. Often referred to as current, as in neutron current.

**current drain** The current taken from a voltage source by a load. Also called drain.

**current feed** Feed to a point where current is a maximum, as at the center of a half-wave antenna.

**current feedback** Feedback introduced in series with the input circuit of an amplifier.

**current generator** A two-terminal circuit element whose terminal current is independent of the voltage between its terminals.

**current limiter** A device that restricts the flow of current to a certain amount, regardless of applied voltage.

**current-limiting resistor** A resistor inserted in an electric circuit to limit the flow of current to some predetermined value. Used chiefly to protect tubes and other components during warmup.

**current loop** *Current antinode.*

**current node** A point at which current is zero along a transmission line, antenna, or other circuit element having standing waves.

**current regulator** A device that maintains the output current of a voltage source at a predetermined essentially constant value despite changes in load impedance.

**current relay** A relay that operates at a specified current value rather than at a specified voltage value.

**current saturation** *Anode saturation.*

**current transformer** An instrument transformer intended to have its primary winding connected in series with a circuit carrying the current to be measured or controlled. The current is measured across the secondary winding.

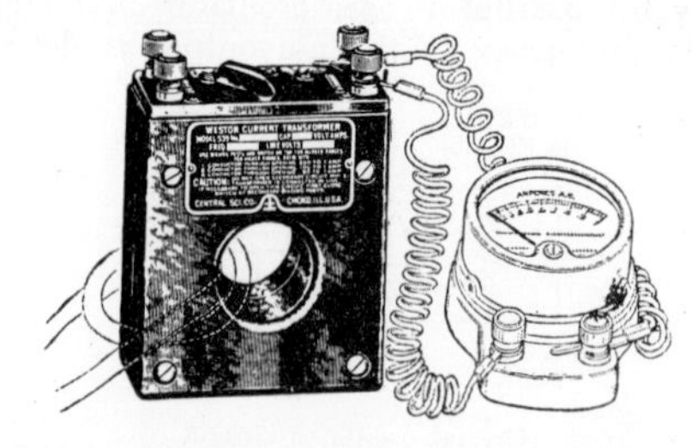


Current transformer, showing meter connections and method of looping current-carrying line through transformer.

**current-type telemeter** A telemeter in which the magnitude of a single current is the translating means.

**cursor** A clear or amber-colored filter that can be placed over a radar screen and rotated until an etched diameter line on the filter passes through a target echo. The bearing from radar to target can then be read accurately on a stationary 360° scale surrounding the filter. Another type of cursor for ground radar gives the bearing of an enemy aircraft with respect to a friendly aircraft being directed by radio from the ground.

**cursor target bearing** Target bearing as measured by a cursor on a ppi radar display.

**curtain** A thin shield, usually cadmium, used in a nuclear reactor to shut off a flow of slow neutrons.

**curtain array** An antenna array consisting of vertical wire elements stretched between two suspension cables. It may be backed by a second curtain serving as reflector. The active elements are usually half-wave dipoles.

**curtain rhombic antenna** A multiple-wire rhombic antenna having a constant input impedance over a wide frequency range. Two or more conductors join at the feed and terminating ends but are spaced apart vertically from 1 to 5 feet at the side poles.

**Curtis-winding resistor** A wirewound resistor in which residual inductance and capacitance are reduced by reversing the direction of alternate turns. This is achieved by passing the wire through a diametral slot in the coil form after each complete turn.

**cut** 1. A section of a crystal having two parallel major surfaces. Cuts are specified by their orientation with respect to the axes of the natural crystal, such as X cut, Y cut, BT cut, and AT cut. 2. The fraction that is removed as product or advanced to the next separative element in an isotope separation process. 3. An order to stop an action, turn off a television camera, or disconnect all microphones in a radio studio.

**cutie pie** A radiation dose-rate meter having a pistol grip, a plastic cylinder or barrel containing an ionization chamber, and an indicating meter mounted above the grip.

**Cutler feed** A resonant cavity that transfers r-f energy from the end of a waveguide to the reflector of a radar spinner assembly.

**cutoff** 1. The minimum value of negative grid bias that will prevent the flow of anode current in an electron tube. 2. *Cutoff frequency.*

**cutoff bias** The d-c bias voltage that must be applied to the grid of an electron tube to stop the flow of anode current.

**cutoff field** *Critical field.*

**cutoff frequency** A frequency at which the attenuation of a device begins to increase sharply, such as the limiting frequency below which a traveling wave in a given mode cannot be maintained in a waveguide, or the frequency above which an electron tube loses efficiency rapidly. Theoretical cutoff frequency and effective cutoff frequency are more specific terms. Also called critical frequency and cutoff.

**cutoff voltage** 1. The electrode voltage value that reduces the dependent variable of an electron-tube characteristic to a specified low value. 2. *Critical voltage.*

**cutoff wavelength** 1. The ratio of the velocity of electromagnetic waves in free space to the cutoff frequency in a uniconductor waveguide. 2. The wavelength corresponding to the cutoff frequency.

**cut paraboloidal reflector** A paraboloidal reflector that is not symmetrical with respect to its axis.

**cut-set** A set of branches of a network such that the cutting of all the branches of the set increases the number of separate parts of the network, but the cutting of all the branches except one does not.

**cutter** An electromagnetic or piezoelectric device that converts an electric input to a mechanical output, used to drive a stylus that cuts a wavy groove in the highly polished wax surface of a disk recording. Also called cutting head, head, and recording head.

**cutting angle** The angle between the vertical cutting face of the stylus and the surface of the record. It should ordinarily be 90°. Deviation from this value is sometimes specified as dig-in angle or drag angle.

**cutting head** *Cutter.*

**cutting stylus** A recording stylus with a sharpened tip that removes material to produce a groove in the recording medium.

**c-w** Abbreviation for *continuous wave.*

**cybernetics** A comparative study of the methods of automatic control and communication that are common to man and machines. Used in analyzing and improving the efficiency of communication systems, information-handling machines, and feedback control systems.

**cycle** 1. One complete sequence of values of an alternating quantity, including a rise to a maximum in one direction, a return to zero, a rise to a maximum in the opposite direction, and a return to zero. The number of cycles occurring in 1 second is called the frequency. 2. A set of operations that is repeated as a unit. 3. A computer shift in which digits dropped off at one end of a word are returned at the other end of the word. Also called cyclic shift. 4. To run a machine through an operating cycle. 5. *Cycle per second.*

**cycle-matching loran** *Low-frequency loran.*

**cycle per second** [abbreviated cps] A unit

of frequency. Commonly shortened to cycle, as in 15,750 cycles. Also called hertz in Europe. Also abbreviated c.

**cycle timer** A timer that opens or closes circuits according to a predetermined time schedule.

**cyclically magnetized** Under the influence of a magnetizing force varying between two specific limits long enough so the magnetic induction has the same value for corresponding points in successive cycles.

**cyclic binary code** *Gray code.*

**cyclic shift** *Cycle.*

**cycling** A periodic change of the controlled variable from one value to another in an automatic control system. Also called oscillation.

**cyclophon** A vacuum tube in which a beam of electrons serves as a switching element.

**cyclotron** An accelerator in which charged particles are successively accelerated by a constant-frequency alternating electric field that is synchronized with movement of the particles on spiral paths in a constant magnetic field normal to their path.

**cyclotron frequency** The frequency at which an electron traverses an orbit in a steady, uniform magnetic field and zero electric field.

**cyclotron-frequency magnetron** A magnetron whose frequency of operation depends on synchronism between the a-c electric field and the electrons oscillating in a direction parallel to this field. An example is a split-anode magnetron having a resonator between the anodes.

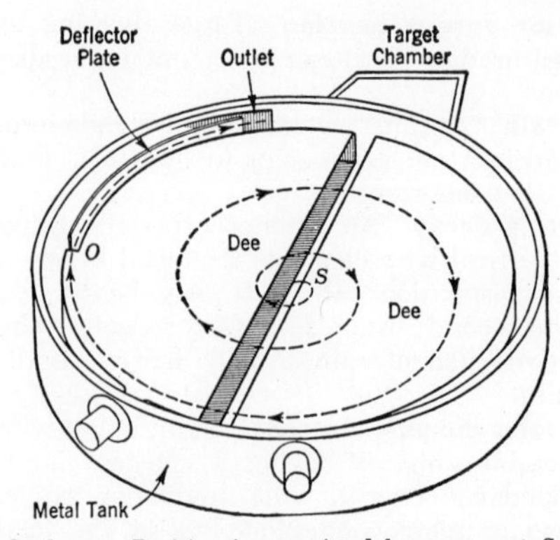


Cyclotron. Positive ions emitted by source at *S* in center are accelerated in steps between D-shaped hollow electrodes and bent on half-circles by an axial magnetic field. Ions leave the left dee at *O* and hit target in chamber at top.

**cyclotron wave** A wave associated with the electron beam of a traveling-wave tube.

**cylindrical antenna** An antenna in which hollow cylinders serve as radiating elements.

**cylindrical reflector** A reflector that is a portion of a cylinder. This cylinder is usually parabolic.

**cylindrical wave** A wave whose equiphase surfaces form a family of coaxial cylinders.

**cytac** A system in which hyperbolic lines of position are determined by measuring the time relationship between two synchronized radio signals. It is similar to loran but capable of higher accuracy and greater range.

# D

**D** Symbol for *deuterium.*

**d** 1. Symbol for *deuterium.* 2. Symbol for *deuteron.*

**damage-risk criterion** The maximum sound pressure levels of a noise, as a function of frequency, to which people can be exposed for a specified time without risk of significant hearing loss.

**damp** To make an oscillating indicator or other device come to rest.

**damped oscillation** Oscillation in which the amplitude of the oscillating quantity decreases with time.

**damped wave** A wave in which the amplitudes of successive cycles progressively diminish at the source.

**damper** A diode used in the horizontal deflection circuit of a television receiver to make the sawtooth deflection current decrease smoothly to zero instead of oscillating at zero. The diode conducts each time the polarity is reversed by a current swing below zero.

**damper tube** A diode, pentode, or other vacuum tube used in a damper circuit.

**damping** 1. Any action or influence that extracts energy from a vibrating system in order to suppress the vibration or oscillation. 2. Reducing or eliminating reverberation in a room by placing sound-absorbing materials on the walls and ceiling. Also called soundproofing.

**damping factor** The ratio of the amplitude of any one of a series of damped oscillations to that of the following one. Also called decrement.

**damping magnet** A permanent magnet used in conjunction with a disk or other moving conductor to produce a force that opposes motion of the conductor and thereby provides damping.

**danger coefficient** The change in reactivity caused by inserting a substance in a nuclear reactor. It depends on the amount and distribution of the substance inserted.

**daraf** [farad spelled backward] The unit of elastance, which is the reciprocal of capacitance in farads.

**dark conduction** Residual conduction in a photosensitive substance that is not illuminated.

**dark current** The current that flows through a photoelectric device in a given circuit in total darkness.

**dark-current pulse** A phototube dark-current excursion that can be resolved by the system employing the phototube.

**dark discharge** An invisible electric discharge in a gas.

**dark resistance** The resistance of a selenium cell or other photoelectric device in total darkness.

**dark space** A region in a glow discharge that produces little or no light.

**dark-trace tube** A cathode-ray tube having a bright face that does not necessarily luminesce, on which signals are displayed as dark traces or dark blips where the potassium chloride screen is hit by the electron beam. When the screen is illuminated externally by powerful lamps, the image may be enlarged greatly by optical lenses and projected onto a conventional rear-projection glass screen. In World War II this screen was usually mounted horizontally at table height and a translucent map of the area under radar survey was placed over it for viewing by many controllers during naval actions. Also called skiatron.

**d'Arsonval movement** Deprecated term for *permanent-magnet moving-coil instrument.*

**Dart** An Army surface-to-surface guided missile having a range of about 2 miles, intended for use by front-line troops against enemy tanks.

**data acquisition** The phase of data handling that begins with the sensing of variables and ends with a magnetic recording or other record of raw data. It may include a complete radio telemetering link.

**data-handling capacity** The maximum number of unit situations that can be handled by an electronic navigation facility within a specified period without deteriorating the performance below certain minimum values.

**data-handling system** Automatically operated equipment used to interpret data gathered by instrument installations. Also called data reduction system.

**data processing** Changing the form, meaning, appearance, location, or other characteristics of data. Processing includes data handling and data reduction.

**data processor** A machine for handling data in a sequence of operations.

**data reduction** The process of converting recorded data into desired useful forms.

**data-reduction system** *Data-handling system.*

**data stabilization** Stabilization of the display of radar signals with respect to a selected reference regardless of changes in radar-carrying vehicle attitude, as in azimuth-stabilized ppi.

**data transcription** Conversion of data from one recorded form to another, as from magnetic tape to punched cards.

**data-transmission system** A system used to transmit data from one instrument to another.

**data-transmitting element** The element in a fire-control system that transmits data between elements of the system that are located some distance apart.

**datum line** A reference line from which calculations or measurements are made.

**daughter** *Decay product.*

**db** Abbreviation for *decibel.*

**dba** Abbreviation for *adjusted decibel.*

**dbk** Abbreviation for *decibels above 1 kilowatt.*

**dbm** Abbreviation for *decibels above 1 milliwatt,* a unit used in specifying input levels.

**db meter** *Decibel meter.*

**dbp** Abbreviation for *decibels above 1 picowatt.*

**dbrn** Abbrevation for *decibels above reference noise.*

**dbx** Abbreviation for *decibels above reference coupling.*

**d-c** Abbreviation for *direct current.*

**d-c amplifier** Abbreviation for *direct-current amplifier.*

**d-c component** The average value of a signal. In television it represents the average luminance of the picture being transmitted. In radar it is the level from which the transmitted and received pulses rise.

**d-c convergence** *Static convergence.*

**d-c coupling** That type of coupling in which the zero-frequency term of the Fourier series representing the input signal is transmitted.

**d-c dump** The removal of all d-c power from a computer system or component intentionally, accidentally, or conditionally. In some types of storage, this results in loss of stored information.

**d-c electron-stream resistance** The electron-stream potential divided by the d-c component of electron-stream current in an electron tube.

**d-c erase** Use of direct current to energize an erasing head.

**d-c erasing head** A magnetic head that uses direct current to produce a unidirectional magnetic field for erasing magnetic tape.

**d-c generator** A rotating electric machine that converts mechanical power into d-c power.

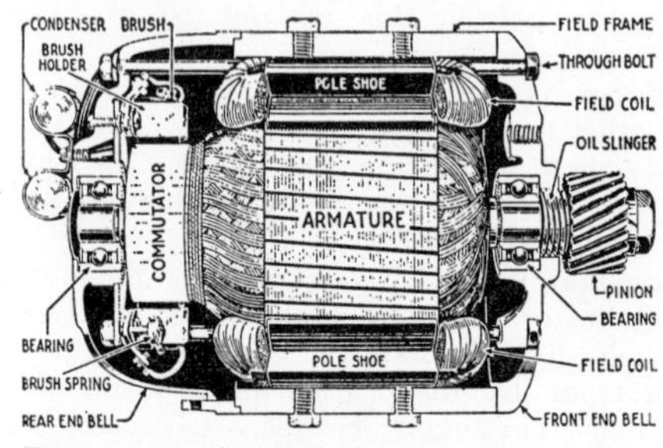


D-c generator for wind charger, with housing cut away to show construction.

**d-c inserter** A television transmitter stage that adds to the video signal a d-c component known as the pedestal level.

**d-c magnetic biasing** Magnetic biasing by means of direct current in magnetic recording.

**d-c motor control** *Electronic motor control.*

**d-c picture transmission** Television transmission in which the signal contains a d-c component that represents the average illumination of the entire scene.

**d-c power supply** A power supply that provides one or more d-c output voltages, such as a d-c generator, rectifier-type power supply, converter, or dynamotor.

**d-c receiver** A radio receiver designed to operate directly from a 115-volt d-c power line.

**d-c reinsertion** *Clamping.*

**d-c resistance** *Resistance.*

**d-c restoration** *Clamping.*

**d-c restorer** A clamping circuit used in television receivers to restore the d-c component of the video signal after a-c amplification. The resulting d-c voltage serves as the bias voltage for the grid of the picture tube, to make average reproduced brightness correspond to the average brightness of the scene being transmitted. Also called clamper, reinserter, and restorer.

**d-c self-synchronous system** A system for

transmitting angular position or motion, in which an arrangement of resistors serves as a transmitter that furnishes a receiver with two or more voltages which are functions of transmitter shaft position. The receiver has two or more stationary coils that set up a magnetic field which causes a rotor to take an angular position corresponding to the angular position of the transmitter shaft.

**d-c tachometer** A d-c generator operating with negligible load current and with constant field flux provided by a permanent magnet, so its d-c output voltage is proportional to speed.

**d-c transmission** Television transmission in which the d-c component of the picture signal is still present. The true level of background illumination is thus maintained at all times.

**d-c voltage** *Direct voltage.*

**d-c working volts** [abbreviated dcwv] The maximum continuously applied d-c voltage for which a capacitor is rated.

**dcwv** Abbreviation for *d-c working volts.*

**D display** A radarscope C display in which targets appear as bright spots that are vertically elongated in rough proportion to range. Also called D scan.

**deaccentuator** A circuit used in an f-m receiver to offset the preemphasis of higher audio frequencies introduced at the f-m transmitter.

**dead band** The range of values over which a measured variable can change without affecting the output of a magnetic amplifier or automatic control system. Also called dead zone and neutral zone.

**deadbeat** Coming to rest without vibration or oscillation, as when the pointer of a meter moves to a new position without overshooting.

**dead end** 1. The end of a sound studio that has the greater sound-absorbing characteristics. 2. The portion of a tapped coil through which no current is flowing at a particular switch position.

**dead-end effect** Absorption of energy by unused portions of a tapped coil.

**dead-end switch** A switch used to short-circuit unused portions of a tapped coil to prevent dead-end effects.

**dead reckoning** Determination of the approximate position of a vehicle by combining vectors for speed, direction, and other factors with the last known position. Usually allows for wind in air navigation, but excludes wind and currents in marine navigation; allowance for wind and current then gives estimated position at sea. The term comes from deduced reckoning, which was abbreviated ded. reckoning.

**dead-reckoning tracer** A mechanical device used to produce a continuous plot of all ships within range of a ship's radar.

**dead room** *Anechoic room.*

**dead short** A short-circuit path that has extremely low resistance.

**dead spot** 1. A geographic location in which signals from a radio, television, or radar transmitter are received poorly or not at all. 2. A portion of the tuning range of a receiver in which stations are heard poorly or not at all, due to improper design of tuning circuits.

**dead time** 1. The time interval, after a response to one signal or event, during which a system is unable to respond to another. For a radiation counter it is the interval after the start of a count during which the counter is insensitive to further ionizing events. For a transponder it is the interval after the start of a pulse during which no new pulse can be received or produced. Also called insensitive time. 2. The time interval between a change in the input signal to a process control system and the response to the signal.

**dead-time correction** A correction applied to an observed counting rate to allow for the probability of the occurrence of events within the dead time. Also called coincidence correction.

**dead zone** *Dead band.*

**deathnium center** An imperfection in the arrangement of atoms in a semiconductor crystal. It facilitates the generation and recombination of electron-hole pairs.

**death ray** A ray that can kill living cells. Ultraviolet rays can kill bacteria, radio waves of certain frequencies can kill insects, and x-rays can kill small animals.

**de Broglie wavelength** The wavelength ascribed by wave or quantum mechanics to a particle having a given momentum. This wavelength is equal to Planck's constant divided by the momentum of the particle.

**debugging** 1. The process of eliminating from a newly designed system the components and circuits that cause early failures. 2. Removing mistakes from a computer program.

**debunching** A tendency for electrons in a beam to spread out both longitudinally and transversely due to mutual repulsion. The effect is a drawback in velocity modulation tubes.

**decade** 1. A group of 10. 2. The interval

between any two quantities having the ratio of 10 to 1.

**decade box** An assembly of precision resistors, coils, or capacitors whose individual values vary in submultiples and multiples of 10. Each section contains 10 equal-value components connected in series. The

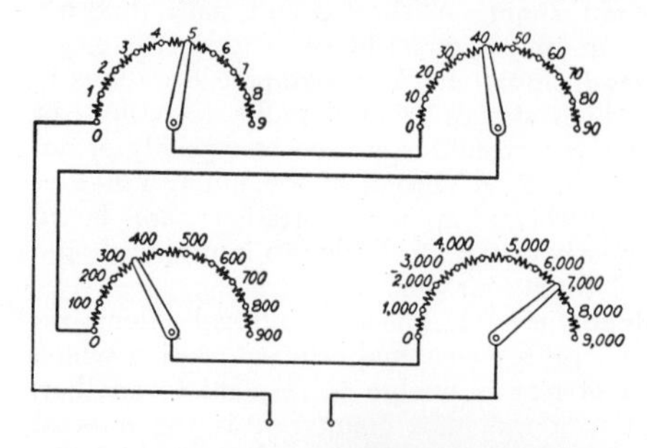


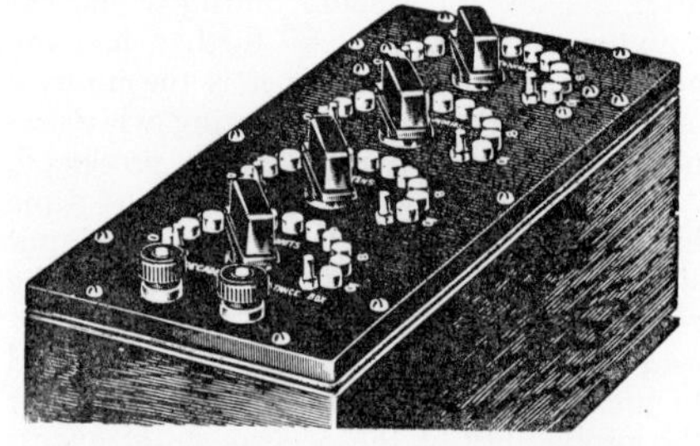

Decade box using precision resistors.

total value of each section is 10 times that of the preceding section. By appropriately setting the 10-position selector switch for each section, the decade box can be set to any desired value within its range.

**decade counter** *Decade scaler.*

**decade counter tube** 1. *Beam-switching tube.* 2. *Cold-cathode counter tube.* 3. *Counter tube.*

**decade glow counting tube** *Cold-cathode counter tube.*

**decade scaler** A scaler that produces one output pulse for every ten input pulses. Also called decade counter and scale-of-ten circuit.

**decametric wave** British term for a radio wave in the range from 10 to 100 meters in wavelength.

**decay** 1. Gradual reduction in the magnitude of a quantity, as of current, magnetic flux, a stored charge, or phosphorescence. 2. *Radioactive decay.*

**decay chain** *Radioactive series.*

**decay characteristic** *Persistence characteristic.*

**decay constant** The constant in the equation that relates the mean rate of decay of a radioactive species to the number of atoms present at a given time. Also called disintegration constant, radioactive decay constant, and transformation constant.

**decay curve** A graph showing how the activity of a radioactive sample varies with time. Alternatively, it may show the amount of radioactive material remaining at any time.

**decay family** *Radioactive series.*

**decay product** A nuclide resulting from the radioactive disintegration of a radionuclide. The nuclide is formed directly or is the result of successive transformations in a radioactive series. A decay product may be either radioactive or stable. Also called daughter.

**decay series** *Radioactive series.*

**decay time** The time taken by a quantity to decay to a stated fraction of its initial value. The fraction is commonly $1/e$, where $e$ is the base of natural logarithms. Also called storage time (deprecated).

**decca** A continuous-wave hyperbolic radio aid to navigation in which a receiver measures and indicates the relative phase difference between signals received from two or more synchronized ground stations. The system provides differential distance information from which position can be determined.

**decelerating electrode** An electrode whose potential provides an electric field to decrease the velocity of the beam electrons in an electron-beam tube.

**deci-** A prefix representing $10^{-1}$.

**decibel** [abbreviated db] A unit used to express the magnitude of a change in signal or sound level. One decibel is the change in level of a pure sine-wave sound that is just barely detectable by the average human ear. The difference in decibels between two signals is 10 times the common logarithm of their ratio of powers or 20 times the common logarithm of their ratio of voltages or currents. One decibel is one-tenth of a bel.

**decibel meter** An instrument used to measure directly the power level of a signal in decibels above or below an arbitrary reference level. Also called db meter.

**decibels above 1 kilowatt** [abbreviated dbk] A power level equal to 10 times the common logarithm of the ratio of a given power in watts to 1,000 watts.

**decibels above 1 milliwatt** [abbreviated dbm] A power level equal to 10 times the common logarithm of the ratio of a given power in watts to 0.001 watt. A negative value, such as −2.7 dbm, means decibels below 1 milliwatt.

**decibels above 1 picowatt** [abbreviated dbp] A power level equal to 10 times the common logarithm of the ratio of a given power in watts to 1 picowatt, which is equal to 1 micromicrowatt or $10^{-12}$ watt.

**decibels above reference coupling** [abbreviated dbx] A measure of the coupling between two circuits, expressed in relation to a reference value of coupling that gives a specified reading on a specified noise-measuring set when a test tone of 90 dba is impressed on one circuit.

**decibels above reference noise** [abbreviated dbrn] A unit used to show the relationship between the interfering effect of a noise frequency, or band of noise frequencies, and a fixed amount of noise power commonly called reference noise. A 1,000-cps tone having a power level of −90 dbm was originally selected as the reference noise power. This reference level was later changed to −85 dbm and the new unit called adjusted decibel, abbreviated dba.

**decimal digit** One of the 10 digits, 0, 1, 2, 3, 4, 5, 6, 7, 8, and 9, used in the scale of 10. Two of these digits, 0 and 1, also serve as binary digits in the scale of 2.

**decimal notation** A system of notation that uses the scale of 10.

**decimal number system** A system of positional notation in which the successive digits relate to successive powers of the base 10.

**decimal point** The point that marks the place between integral and fractional powers of 10 in a decimal number.

**decimal-to-binary conversion** The mathematical process of converting a number written in the scale of 10 into the same number written in the scale of 2.

**decimetric wave** An electromagnetic wave having a wavelength between 0.1 and 1 meter.

**decineper** One-tenth of a neper.

**decision element** A circuit that performs a logical operation such as "and," "or," "not," or "except" on one or more binary digits of input information representing "yes" or "no," and expresses the result in its output.

**deck** A magnetic tape transport mechanism.

**deck motion predictor** A device that predicts the roll and fore-and-aft motion of a ship, to permit firing a missile at a desired deck angle.

**deck switch** *Gang switch.*

**declassify** To remove the security classification from a document or piece of equipment.

**declination** The angle between the horizontal component of the earth's magnetic field and true north.

**decoder** 1. A matrix network in which a combination of inputs produces a single output. Used for converting digital information to an analog form. 2. A circuit that responds to a particular coded signal while rejecting others. 3. *Matrix.* 4. *Tree.*

**decommutation** The process of recovering a signal from the composite signal previously created by a commutation process.

**decommutator** The section of a telemetering system that extracts analog data from a time-serial train of samples representing a multiplicity of data sources transmitted over a single r-f link.

**decontamination** The removal of unwanted radioactive material.

**decontamination factor** The absolute ratio of initial specific radioactivity to final specific radioactivity resulting from a separation process.

**decontamination index** The logarithm of the ratio of initial specific radioactivity to final specific radioactivity resulting from a separation process.

**decoupling network** Any combination of resistors, coils, and capacitors placed in power-supply leads or other leads that are common to two or more circuits, to prevent unwanted interstage coupling.

**decoy** A countermeasure device intended to divert a guided missile from its target.

**decoy transponder** A transponder that returns a strong signal when triggered directly by a radar pulse. When used for electronic countermeasures, the transponders produce large and misleading target signals on enemy radar screens.

**decrement** *Damping factor.*

**decrypt** To convert a cryptogram or series of electronic pulses into plain text by electronic means.

**dectra** A British radio navigation aid that provides coverage over a specific section of a long ocean route, using equipment and techniques similar to decca. A master and slave station are located at each end of the route.

**dee** A hollow accelerating electrode in a cyclotron.

**dee line** A structural member that supports the dee of a cyclotron and acts with the dee to form the resonant circuit.

**deemphasis** A process for reducing the relative strength of higher audio frequencies before reproduction, to complement and thereby offset the preemphasis that was previously introduced to help these components override noise or reduce distortion. Used chiefly in frequency-modulated and phase-modulated receivers and in phonograph amplifiers. Also called post-emphasis and post-equalization.

**deemphasis network** An RC filter inserted in a system to restore preemphasized signals to their original form.

**deenergize** To disconnect from the source of power.

**deerhorn antenna** A dipole antenna whose ends are swept back to reduce wind resistance when mounted on an airplane.

**definition** 1. The fidelity with which a television or facsimile receiver forms an image. 2. The extent to which the fine-line details of a printed circuit correspond to the master drawing.

**deflection** The displacement of an electron beam from its straight-line path by an electrostatic or electromagnetic field.

**deflection center** The intersection of the forward projection of the electron path prior to deflection and the backward projection of the electron path in the field-free space after deflection in a cathode-ray tube.

**deflection coil** One of the coils in a deflection yoke.

**deflection defocusing** Defocusing that becomes greater as deflection is increased in a cathode-ray tube, because the beam hits the screen at an increasingly greater slant and its spot becomes increasingly more elliptical as it approaches the edges of the screen.

**deflection electrode** An electrode whose potential provides an electric field that deflects an electron beam. Also called deflection plate.

**deflection factor** The reciprocal of the deflection sensitivity in a cathode-ray tube. Deflection factor is usually expressed in amperes per inch for electromagnetic deflection and volts per inch for electrostatic deflection.

**deflection plane** A plane perpendicular to the cathode-ray-tube axis containing the deflection center.

**deflection plate** *Deflection electrode.*

**deflection sensitivity** The displacement of the electron beam at the target or screen of a cathode-ray tube per unit of change in the deflection field. Usually expressed in inches per volt applied between deflection electrodes or inches per ampere in a deflection coil. Deflection sensitivity is the reciprocal of deflection factor.

**deflection voltage** The voltage applied between a pair of deflection electrodes to produce an electric field.

**deflection yoke** An assembly of one or more electromagnets that is placed around the neck of an electron-beam tube to produce a magnetic field for deflection of one or more electron beams. Also called yoke.

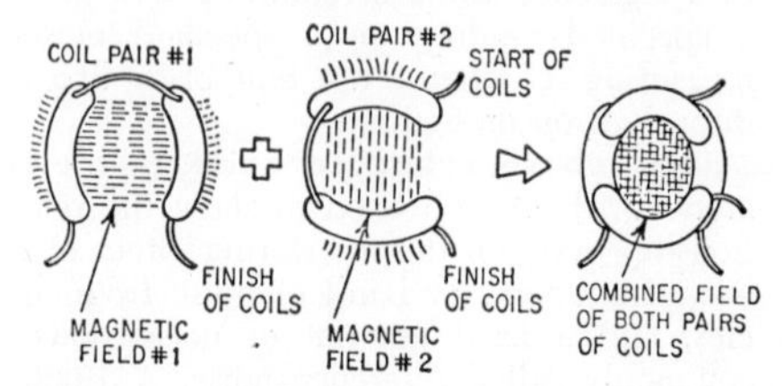

Deflection yoke for television picture tube contains four separate coils.

**deflection-yoke pullback** 1. In a color picture tube, the distance between the maximum possible forward position of the yoke and the position of the yoke that gives optimum color purity. 2. In a monochrome picture tube, the maximum distance the yoke can be moved back along the tube axis without producing neck shadow.

**defocus-dash mode** A mode of cathode-ray-tube storage of binary digits in which the writing beam is initially defocused so as to excite a small circular area on the screen. For one kind of binary digit it remains defocused. For the other kind it is suddenly focused to a concentric dot and drawn out into a dash during the time interval before the beam is cut off and moved to the next position.

**defocus-focus mode** A variation of the defocus-dash mode in which the focused dot is not drawn out into a dash.

**defruiting** The process of eliminating asynchronous returns in a radar beacon system.

**degassing** The process of driving out and exhausting the gases occluded in the internal parts of an electron tube, generally by heating during evacuation.

**degaussing** A method of neutralizing the magnetic field of a ship by placing a cable around its hull and sending through it a direct current of the correct value to neutralize the magnetic effect of the hull. The current adjustment is made at a degaussing station equipped with underwater equipment that indicates when the resultant magnetic field has been sufficiently weakened so it will not actuate a magnetic mine.

**degaussing control** A control that automatically varies the current in degaussing coils as a ship changes heading or rolls and pitches.

**degeneracy** The condition wherein two or more modes in a resonant device have the same resonant frequency.

**degenerate modes** A set of modes having the same resonant frequency or propagation constant.

**degenerate semiconductor** A semiconductor in which the number of electrons in the conduction band approaches that of a metal.

**degeneration** *Negative feedback.*

**degradation** *Moderation.*

**deionization** The recombination of ions with electrons in a glow or arc discharge to form neutral atoms and molecules.

**deionization potential** The potential at which ionization of the gas in a gas-filled tube ceases and conduction stops.

**deionization time** The time required for a gas tube to regain its preconduction characteristics after interruption of anode current, so the grid regains control. Also called recontrol time.

**deionizing grid** A grid used in a gas tube to speed up deionization of the gas.

**Dekatron** Trademark of Baird-Atomic for their cold-cathode counting tubes.

**delay circuit** A circuit in which the output signal is delayed by a specified time interval with respect to the input signal.

**delay coincidence circuit** A coincidence circuit that is actuated by two pulses, one of which is delayed by a specified time interval with respect to the other.

**delay distortion** Phase distortion in which the rate of change of phase shift with frequency of a circuit or system is not constant over the frequency range required for transmission. Also called envelope delay distortion.

**delayed alpha particle** An alpha particle emitted by an excited nucleus that was formed an appreciable time after a beta disintegration process.

**delayed automatic gain control** An automatic gain control system that does not operate until the signal exceeds a predetermined magnitude. Weaker signals thus receive maximum amplification. Also called biased automatic gain control, delayed automatic volume control, and quiet automatic volume control.

**delayed automatic volume control** *Delayed automatic gain control.*

**delayed coincidence** Occurrence of a count in one detector at a short but measurable time later than a count in another detector, the two counts being due to successive events in the same nucleus.

**delayed critical** The condition in which a nuclear reactor is critical because of delayed neutrons alone, without requiring the contribution of prompt neutrons.

**delayed fission neutron** A fission neutron emitted by fission products.

**delayed neutron** A neutron emitted by excited nuclei formed during beta disintegration, an appreciable time after the fission. Since neutron emission itself is prompt, the observed half-life is that of the preceding beta emitter.

**delayed ppi** A plan position indicator in which initiation of the time base is delayed a fixed time after each transmitted pulse, to give expansion of the range scale for distant targets so they show more clearly on the screen.

**delayed sweep** A sweep whose beginning is delayed for a definite time after the pulse that initiates the sweep.

**delay equalizer** A corrective network used to make the phase delay or envelope delay of a circuit or system substantially constant over a desired frequency range.

**delay line** A device utilizing the time of wave propagation to produce a time delay of a signal. A transmission line using either lumped or distributed constants gives a delay determined by the electrical length of the line. In one form, the inductance is increased by winding a helix on a flexible powdered iron core to serve as the inner conductor. High capacitance is achieved by using only a few layers of thin paper to separate the inner conductor from the stranded outer conductor. An ultrasonic delay line gives a delay determined by the length of the path taken by acoustic waves through the medium. Also called artificial delay line.

**delay-line memory** *Delay-line storage.*

**delay-line storage** A computer storage or memory device consisting of a delay line and means for regenerating and reinserting information into the delay line. Also called delay-line memory.

**delrac** A British radio navigation system designed to provide worldwide coverage by using 21 pairs of master-slave stations, with a 3,000-mile range for each pair of stations. Frequencies used are in the band from 10 to 14 kc. Decca indicating equipment can be used with delrac.

**delta** The difference between a partial-select output of a magnetic cell in a one state and a partial-select output of the same cell in a zero state.

**delta connection** A combination of three components connected in series to form a triangle like the Greek letter delta. Also called mesh connection.

**delta impulse function** An extremely narrow pulse having such great amplitude that the product of height and duration is unity.

**delta-matched antenna** A single-wire antenna, usually one half-wavelength long, to which the leads of an open-wire transmission line are connected in the shape of a Y. The flared parts of the Y serve to match

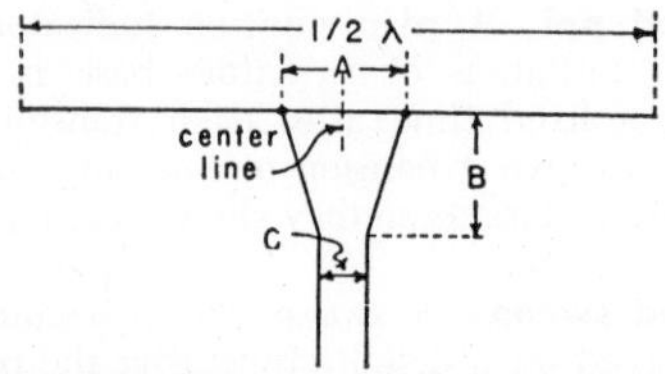


Delta-matched antenna. Dimensions A, B, and C determine matching to transmission line.

the transmission line to the antenna. As the top of the Y is not cut, the matching section has the triangular shape of the Greek letter delta. Also called Y antenna.

**delta-matching transformer** The Y-shaped matching section of a delta-matched antenna.

**delta network** A set of three branches connected in series to form a mesh.

**delta pulse code modulation** A modulation system that converts audio signals into corresponding trains of digital pulses to give greater freedom from interference during transmission over wire or radio channels.

**delta ray** An electron or proton ejected by recoil when a rapidly moving alpha particle or other primary ionizing particle passes through matter.

**delta wave** A brain wave having a frequency below 9 cps.

**demagnetization curve** A portion of the hysteresis loop of a magnetic material, showing the peak value of residual induction and the manner in which magnetization reduces to zero.

**demagnetizing force** A magnetizing force applied in the direction that reduces the residual induction in a magnetized object.

**demodulation** *Detection.*

**demodulator** *Detector.*

**demountable tube** An electron tube that can be taken apart for repair or replacement of electrodes.

**Dempster mass spectrograph** A mass spectrograph in which ions are accelerated through a slit by an electric field, then deflected by a magnetic field so all ions of the same charge-to-mass ratio pass through a second slit.

**demultiplexer** A device used to separate two or more signals that were previously combined by a compatible multiplexer and transmitted over a single channel.

**demultiplexing circuit** A circuit used to separate the signals that have been combined for transmission by multiplex.

**denaturant** A nonfissionable isotope that can be added to fissionable material to make it unsuitable for use in atomic weapons without extensive processing. An example is $U^{233}$ which, when added in sufficient quantity, makes $U^{235}$ and $U^{238}$ unsuitable for weapon use.

**densitometer** 1. An instrument used to measure the optical density of a material. 2. An instrument used to measure the amount of darkening of film badges to determine the radiation dosage received by the wearer.

**density** 1. Weight per unit volume. 2. A measure of the light-transmitting properties of an area. It is expressed as the common logarithm of the ratio of incident light to transmitted light. 3. Amount per unit cross-sectional area, as for current, magnetic flux, or electrons in a beam.

**density modulation** Modulation of an electron beam by making the density of the electrons in the beam vary with time.

**density packing** The number of units of useful information that can be stored within a given linear dimension on a single track of a magnetic tape or drum by a single head.

**depleted material** Material in which the amount of one or more isotopes of a constituent has been reduced by an isotope separation process or by a nuclear reaction.

**depletion** The percentage reduction in the quantity of fissionable atoms in the fuel assemblies or fuel mixture that occurs during operation of a nuclear reactor.

**depletion layer** An electric double layer formed at the surface of contact between a metal and a semiconductor having different work functions. Electrons diffuse from the substance having the lower work function toward the other, leaving equivalent positive charges at the layer in the first substance. This action occurs because the mobile carrier charge density is insufficient to neutralize the fixed charge density of donors and acceptors. Also called space-charge layer. Formerly called barrier layer and blocking layer.

**depletion-layer capacitance** The capacitance of the imaginary capacitor formed by the charges of a depletion layer. This capacitance is a function of reverse voltage. Also called barrier-layer capacitance (deprecated).

**depletion-layer transistor** A transistor that relies directly on motion of carriers through depletion layers, such as a spacistor.

**depolarization** Prevention of polarization in an electric cell or battery.

**deposit dose** The residual radioactivity deposited on the surface after a nuclear explosion, as by water falling as rain from the base surge of an underwater atomic explosion.

**depression deviation indicator** A sonar depth-of-target indicator used with a tilting transducer to show whether the transducer is aimed at, above, or below the target. Depth is automatically computed from transducer angle and echo range after the transducer is aligned with the target.

**depth dose** The radiation dose delivered at a particular depth beneath the surface of a body. It is usually expressed as per cent of surface dose or per cent of air dose.

**depth finder** *Fathometer.*

**depth of heating** The depth below the surface of a material in which effective dielectric heating can be confined when the applicator electrodes are applied adjacent to one surface only.

**depth of modulation** The ratio of the difference in field strength of the two lobes of a directional antenna system to the field strength of the greater at a given point in space. Used to determine direction in a radio guidance system.

**depth of penetration** The thickness of a layer, extending inward from the surface of a flat conductor, that has the same resistance to direct current as the conductor as a whole has to alternating current of a given frequency in induction heating.

**depth of velocity modulation** The ratio of the peak amplitude of the velocity modulation of an electron stream, expressed in equivalent volts, to the potential of the electron stream.

**depth sounder** *Fathometer.*

**derate** To reduce the rating of a device to improve reliability or to permit operation at high ambient temperatures.

**derivative action** Control action in which the speed at which a correction is made depends on how fast the system error is increasing. Also called rate action.

**desired track** *Course.*

**Destriau effect** Sustained emission of light by suitable phosphor powders that are embedded in an insulator and subjected only to the action of an alternating electric field.

**destruct** The deliberate action of exploding a missile or vehicle after it has been launched but before it has completed its course. Destructs are executed by radio when the missile gets off its plotted course or otherwise becomes a hazard.

**destructive testing** Intentional operation of equipment until it fails, to reveal design weaknesses.

**detail** The extent to which image elements that are close together can be individually distinguished. Detail requires contrast for its recognition.

**detection** The process of recovering the modulating wave from a modulated carrier. Also called demodulation.

**detection probability** The probability, expressed as a percentage, that a single sonar ping will return a recognizable echo at various distances and directions from a sonar transmitter.

**detectophone** An audio system used to listen to conversations secretly. A sensitive microphone is concealed in the room and connected to an audio amplifier feeding headphones or a magnetic tape recorder in a nearby room. Alternatively, the microphone may feed a wired radio transmitter that sends the signals over power lines to a more remote listening location.

**detector** The stage in a receiver at which demodulation takes place. In a superheterodyne receiver this is called the second detector. Also called demodulator.

**detent** A mechanism used on a multiposition control to hold it firmly in each position. One common type consists of a spring-loaded ball that falls into equally spaced indentations on a plate as the control shaft is rotated.

**detune** To change the inductance or capacitance of a tuned circuit so its resonant frequency is different from the incoming signal frequency.

**detuning stub** A quarter-wave stub used to match a coaxial line to a sleeve-stub antenna. The stub detunes the outside of the coaxial feed line while tuning the antenna itself.

**deuterium** [symbol d, D, or $H^2$] The hydrogen isotope having mass number 2. It is one form of heavy hydrogen, the other being tritium.

**deuterium oxide** *Heavy water.*

**deuteron** [symbol d] The nucleus of a deuterium atom. It is believed to consist of a neutron and a proton. The symbol d is used for the deuteron as a projectile.

**deviation** 1. The difference between the actual value of a controlled variable and the desired value corresponding to the set point. 2. *Frequency deviation.*

**deviation distortion** Distortion in an f-m receiver caused by inadequate bandwidth, inadequate amplitude-modulation rejection, or inadequate discriminator linearity.

**deviation from pulse flatness** The difference between the maximum and minimum amplitudes of a pulse divided by the maximum amplitude, all taken between the first and last knees of the pulse.

**deviation ratio** The ratio of the maximum possible frequency deviation to the maximum audio modulating frequency in an f-m system.

**deviation sensitivity** 1. The rate of change of course indication with respect to the change of displacement from the course line in a navigation indicator. 2. The lowest frequency deviation that produces a specified output power in an f-m receiver.

**dew line** [Distant Early-Warning line] A line of radar stations extending for about 3,000 miles from Alaska to Greenland along the Arctic circle, to give the earliest possible warning of the approach of enemy airplanes or missiles. Automatic target-recognition circuits eliminate the need for constant attention by human observers.

**d-f** Abbreviation for *direction finder.*

**diagnostic routine** A routine designed to locate a computer malfunction or a mistake in coding.

**dial** A separate scale or other device for indicating the value to which a control is set.

**dial cable** Braided cord or flexible wire cable used to make a pointer move over a dial when a separate control knob is rotated, or used to couple two shafts together mechanically.

**dial cord** A braided cotton, silk, or glass fiber cord used as a dial cable.

**dial lamp** A small lamp used to illuminate a dial. When used to indicate that a circuit is energized, it is called a pilot lamp.

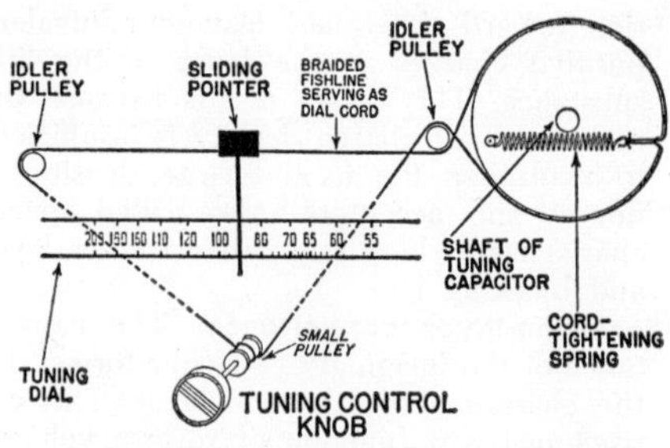

Dial-cord layout for typical slide-rule dial.

**diamagnetic material** A material that has a magnetic permeability less than 1, such as bismuth and antimony. Diamagnetic materials are repelled by a magnet, and therefore tend to position themselves at right angles to magnetic lines of force.

**diameter equalization** Increasing the high-frequency response in proportion to decreasing diameter of a disk recording.

**diamond antenna** *Rhombic antenna.*

**Diamondback** A Navy air-to-air guided missile similar to the Sidewinder, using infrared homing.

**diamond stylus** A stylus having a carefully ground diamond as its point.

**diaphragm** 1. A thin flexible sheet that can be moved by sound waves, as in a microphone, or can produce sound waves when moved, as in a loudspeaker. 2. An adjustable opening used in television cameras to reduce the effective area of a lens so as to increase the depth of focus. 3. *Iris.*

**diathermy** Therapeutic use of r-f energy to produce heat within some part of the body. Also called radiothermy.

**diathermy interference** Television interference caused by diathermy equipment. It produces a herringbone pattern in a dark horizontal band across the picture.

**diathermy machine** An r-f oscillator, sometimes followed by r-f amplifier stages, used to generate high-frequency currents that produce heat within some part of the body for therapeutic purposes.

**dichroic mirror** A glass surface coated with a special metal film that reflects certain colors of light while allowing other colors to pass through. Used in some color television cameras.

**Dicke radiometer** A radiometer-type receiver that detects weak signals in noise by modulating or switching the incoming signal before it is processed by conventional receiver circuits. After amplification and detection the modulation frequency is recovered. The product of this signal and a reference waveform from the input modulator is smoothed in a low-pass filter, and the filter output is used to drive a display device that indicates the presence of a signal.

**didymium** A mixture of neodymium and praseodymium, both of which are rare-earth metals.

**dielectric** A material that can serve as an insulator because it has poor electric conductivity. A dielectric such as air, mica, paper, or plastic film is used between the metal-foil plates of a capacitor to separate the plates electrically and store electric energy. A dielectric undergoes electric polarization when subjected to an electric field.

**dielectric absorption** The persistence of electric polarization in certain dielectrics after removal of the electric field. The effect may last for several years in certain mixtures of wax that are allowed to harden in a strong electric field. Electrets are based on this effect.

**dielectric amplifier** An amplifier using a ferroelectric capacitor whose capacitance varies with applied voltage in such a way as to give signal amplification.

**dielectric antenna** An antenna in which a dielectric is the major component used to produce a desired radiation pattern.

**dielectric constant** The property of a material that determines how much electrostatic energy can be stored per unit volume when unit voltage is applied. In effect, it is the ratio of the capacitance of a capacitor filled with a given dielectric to that of the same capacitor having only a vacuum as dielectric. Also called permittivity.

**dielectric current** The current flowing at any instant through a surface of a dielectric that is located in a changing electric field.

**dielectric diode** A capacitor in which the negative electrode can emit electrons into the normally insulating region between the plates. The charge stored on the capacitor is thus continuously in transit between the electrodes, to give current flow in one direction. Cadmium sulfide crystals can serve as the dielectric.

**dielectric dissipation factor** The cotangent of the dielectric phase angle of a dielectric material.

**dielectric fatigue** The property of some dielectrics in which resistance to breakdown decreases after a voltage has been applied for a considerable time.

**dielectric heating** The heating of a nominally insulating material by placing it in a high-frequency electric field. The heat results from internal losses during the rapid reversal of polarization of molecules in the dielectric material.

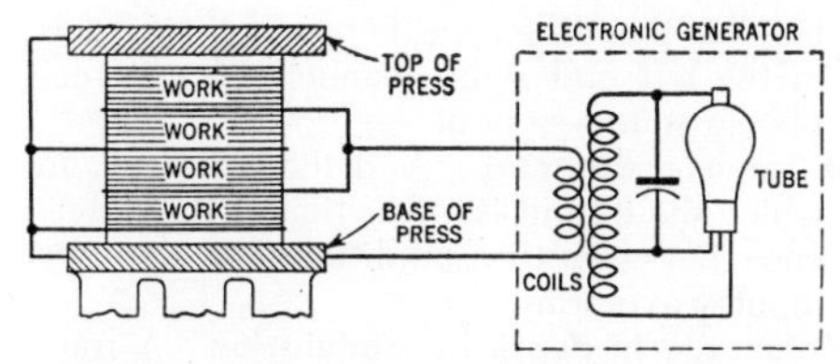


Dielectric heating setup for setting glue in plywood.

**dielectric hysteresis** A lagging of the electric field in a dielectric with respect to the alternating voltage applied to the dielectric. This effect causes a dielectric hysteresis loss comparable to that produced by magnetic hysteresis in a ferrous material.

**dielectric lens** A lens made of dielectric material in such a way that it refracts radio waves in the same manner that an optical lens refracts light waves. Used with microwave antennas.

**dielectric loss** The electric energy that is converted into heat in a dielectric subjected to a varying electric field.

**dielectric phase angle** The angular difference in phase between the sinusoidal alternating voltage applied to a dielectric and the resulting alternating current.

**dielectric power factor** The cosine of the dielectric phase angle.

**dielectric strength** The maximum potential gradient that a material can withstand without rupture. Usually specified in volts per millimeter of thickness. Also called electric strength.

**dielectric test** A test involving application of a voltage higher than the rated value for a specified time, to determine the margin of safety against later failure of insulating materials.

**dielectric waveguide** A waveguide consisting of a dielectric cylinder surrounded by air. Electromagnetic waves travel through the solid dielectric in much the same manner that sound waves travel through a speaking tube.

**dielectric wedge** A wedge-shaped piece of dielectric used in one waveguide to match its impedance to that of a different waveguide.

**dielectric wire** A dielectric waveguide used to transmit uhf radio waves short distances between parts of a circuit.

**difference amplifier** *Differential amplifier.*

**difference channel** An audio channel that

handles the difference between the signals in the left and right channels of a stereophonic sound system.

**difference detector** A detector circuit in which the output is a function of the difference between the amplitudes of the two input waveforms.

**difference in depth of modulation** A fraction obtained by subtracting the percentage of modulation of the smaller signal from that of the larger signal in a directive antenna system using overlapping lobes, then dividing the difference by 100. An instrument low-approach system is an example in which this rating applies.

**difference limen** The increment in a stimulus that is barely noticed in half of the trials. Also called difference threshold.

**difference number** *Neutron excess.*

**difference of potential** The voltage between two points.

**difference threshold** *Difference limen.*

**difference transfer function** The transfer function that relates a loop difference signal to the corresponding input signal of a feedback control loop.

**differential** The difference between levels for turn-on and turn-off operation in a control system.

**differential absorption ratio** The ratio of the concentration of a radioisotope in a given organ or tissue to the concentration that would be obtained if the same administered quantity of this isotope were uniformly distributed throughout the body.

**differential amplifier** An amplifier whose output is proportional to the difference between the voltages applied to its two inputs. Also called difference amplifier.

**differential analyzer** An analog computer designed and used for integrating and solving differential equations.

**differential capacitor** A two-section variable capacitor having one rotor and two stators so arranged that as capacitance is reduced in one section, it is increased in the other.

**differential cross section** The cross section for scattering of particles or photons in a small solid angle in a specified direction. Integration over all angles from 0° to 180° gives the ordinary cross section for a nuclear process.

**differential discriminator** A discriminator that passes only pulses whose amplitudes are between two predetermined values, neither of which is zero.

**differential gain** The amount that unity is exceeded by the ratio of the output amplitudes of a small high-frequency sine-wave signal at two stated levels of a low-frequency signal on which it is superimposed in a video transmission system. Differential gain is expressed in percent by multiplying the above difference by 100, and expressed in decibels by multiplying the common logarithm of the ratio itself by 20.

**differential gain control** *Sensitivity-time control.*

**differential ionization chamber** A two-section ionization chamber in which electrode potentials are such that output current is equal to the difference between the separate ionization currents of the two sections.

**differential microphone** *Double-button carbon microphone.*

**differential-mode signal** A signal that is applied between the two ungrounded terminals of a balanced three-terminal system.

**differential null detector** A null indicator in which a differential transformer delivers an output voltage that is proportional to the vector difference between two input voltages. One input is a 1,000-cps source voltage and the other is the unknown signal, applied through an amplifier and phase shifter that can be adjusted so the difference between the two voltages is zero.

**differential permeability** The slope of the magnetization curve for a magnetic material.

**differential phase** The difference in output phase of a small high-frequency sine-wave signal at two stated levels of a low-frequency signal on which it is superimposed in a video transmission system.

**differential pressure pickup** An instrument that measures the difference in pressure between two pressure sources and translates this difference into a change in inductance, resistance, voltage, or some other electrical quality.

**differential pulse-height discriminator** *Pulse-height selector.*

**differential relay** A two-winding relay that operates when the difference between the currents in the two windings reaches a predetermined value.

**differential scattering cross section** The cross section for scattering of a particle from its initial velocity to a new velocity, per unit solid angle per unit speed at the new velocity.

**differential transformer** A transformer

used to join two or more sources of signals to a common transmission line.

**differential winding** A winding whose magnetic field opposes that of a nearby winding.

**differentiating circuit** A circuit whose output voltage is proportional to the rate of change of the input voltage. The output waveform is then the time derivative of the input waveform, and the phase of the output waveform leads that of the input by 90°. An RC circuit gives this differentiating action. Also called differentiating network and differentiator.

**differentiating network** *Differentiating circuit.*

**differentiator** 1. A device, usually of the analog type, whose output is proportional to the derivative of an input signal in a computer. 2. *Differentiating circuit.*

**diffracted wave** A wave whose front has been changed in direction by an obstacle or other nonhomogeneity in a medium, other than by reflection or refraction.

**diffraction** The bending of a wave as it passes the edges of an object or opening. It is caused by interference between wave components scattered by different parts of the object.

**diffraction grating** A polished metal or glass surface having closely spaced parallel reflecting grooves that produce a spectrum by interference between different colors of light when white light arrives at a certain angle. There may be as many as 50,000 lines per inch. Also called grating.

**diffraction instrument** *Diffractometer.*

**diffraction pattern** The pattern produced on film exposed in an x-ray diffraction camera, consisting of portions of circles having various spacings depending on the material being examined.

**diffractometer** An instrument used to study the structure of matter or the properties of radiation by means of the diffraction of x-rays, electrons, neutrons, or other waves. Also called diffraction instrument.

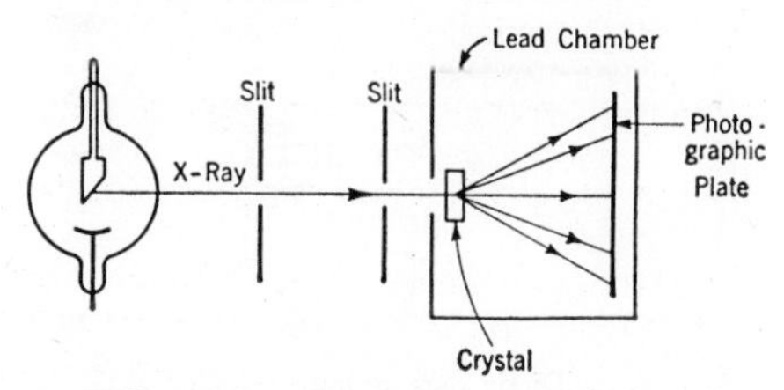


Diffractometer using x-rays to study crystals.

**diffused-alloy transistor** A transistor in which the semiconductor wafer is subjected to gaseous diffusion to produce a nonuniform base region, after which alloy junctions are formed in the same manner as for an alloy-junction transistor. It may also have an intrinsic region, to give a p-n-i-p unit. Also called drift transistor.

**diffused-base transistor** A transistor in which a nonuniform base region is produced by gaseous diffusion. The collector-base junction is also formed by gaseous diffusion, while the emitter-base junction is a conventional alloy junction.

**diffused emitter-collector transistor** A transistor in which both the emitter and collector are produced by diffusion.

**diffused junction** A junction that has been formed by diffusing an impurity a small controlled amount within a semiconductor crystal wafer without heating.

**diffused-junction rectifier** A semiconductor diode in which the p-n junction is produced by diffusion.

**diffused-junction transistor** A transistor in which the emitter and collector electrodes have been formed by diffusion of an impurity metal into the semiconductor wafer without heating.

**diffuse reflection** Reflection of light, sound, or radio waves from a surface in all directions according to the cosine law.

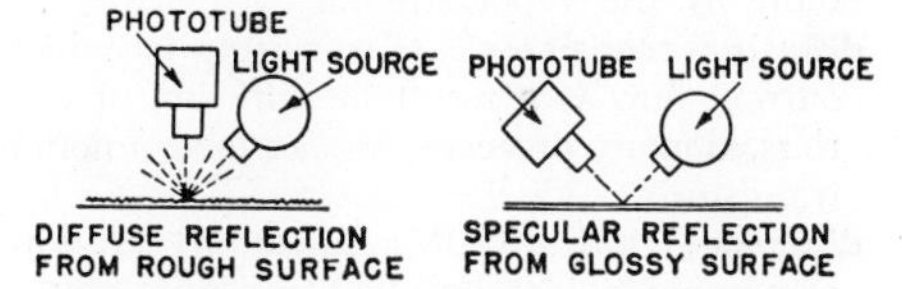


Diffuse-reflection measurement, as compared to direct or specular reflection.

**diffuse sound** Sound that has uniform energy density in a given region and is such that all directions of energy flux at all parts of the region are equally probable.

**diffuse transmission** Transmission in which all the emergent radiation is observed.

**diffuse transmission density** The value of the photographic transmission density obtained when light flux impinges normally on the sample and all the transmitted flux is collected and measured.

**diffusion** 1. The passage of particles through matter in such circumstances that the probability of scattering is large. 2. The movement of carriers, donors, or acceptors in a semiconductor. 3. The migration of atoms and vacancies of one material into another.

**diffusion coefficient** The constant of proportionality in Fick's law.

**diffusion constant** The diffusion current density in a homogeneous semiconductor divided by the charge carrier concentration gradient. The resulting constant is also equal to the product of the drift mobility and the average thermal energy per unit charge of carriers.

**diffusion equation** An equation applying to the operation of a homogeneous nuclear reactor.

**diffusion kernel** A Green's function of the elementary diffusion equation for a nuclear reactor. Also called Yukawa kernel.

**diffusion length** 1. The average distance to which minority carriers diffuse between generation and recombination in a homogeneous semiconductor. 2. The average distance traveled by a thermal neutron from its point of formation to its point of absorption.

**diffusion process** A method of producing a junction by diffusing an impurity metal into a semiconductor at a high temperature.

**diffusion pump** A vacuum pump in which a stream of heavy molecules, such as mercury vapor, carries gas molecules out of the volume being evacuated. Also used for separating isotopes according to weight, the lighter molecules being pumped preferentially by the vapor stream.

**diffusion transistor** A transistor in which current flow is a result of diffusion of carriers, donors, or acceptors, as in a junction transistor.

**digicom** [DIGItal COMmunication] A wire communication system that transmits speech signals in the form of corresponding trains of pulses, and transmits digital information directly from computers, radar, tape readers, teleprinters, and telemetering equipment. Delta pulse code modulation is used to convert the audio signals to digital signals. The system includes automatic switching circuits that locate the fastest message path from origin to destination.

**dig-in angle** A stylus cutting angle such that the point is driving into the coating. It is the opposite of drag angle.

**digit** A character that stands for zero or for a positive integer smaller than the base of an ordinary number system used in a digital computer.

**digital-analog decoder** A device used to convert digital information into a form suitable for use by an analog device.

**digital computer** A computer that processes information in digital form. Electronic digital computers generally use binary or decimal notation and solve problems by repeated high-speed use of the fundamental arithmetic processes of addition, subtraction, multiplication, and division.

**digital differential analyzer** A digital computer device that uses built-in integrators to perform integration.

**digital recorder** A recorder that records information as discrete numerically defined points, usually in binary code.

**digital transducer** A transducer that measures physical quantities and transmits the information as coded digital signals rather than as continuously varying currents or voltages.

**digitize** To convert an analog measurement of a quantity into a numerical value.

**dihedral reflector** A corner reflector having two sides meeting at a line.

**diheptal base** A tube base having 14 pins or possible pin positions. Used chiefly on television cathode-ray tubes.

**dilution** Reducing the intensity of a color by adding white.

**dina** An airborne radar jamming transmitter operating in the band from 92 to 210 mc with an output of 30 watts, radiating noise in one sideband for spot or barrage jamming. The carrier and the other sideband are suppressed.

**diode** 1. A two-electrode electron tube containing an anode and a cathode. 2. *Crystal diode.*

**diode amplifier** A parametric amplifier that uses a special diode in a cavity to amplify signals at frequencies as high as 6,000 mc. Also called cavity-type diode amplifier.

**diode characteristic** The composite electrode characteristic of an electron tube when all electrodes except the cathode are connected together.

**diode demodulator** A demodulator using one or more crystal or electron-tube diodes to provide a rectified output whose average value is proportional to the original modulation. Also called diode detector.

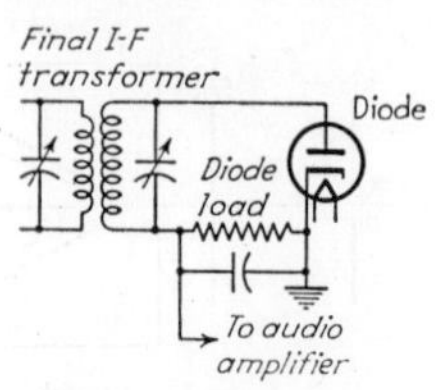


Diode demodulator circuit.

**diode detector** *Diode demodulator.*

**diode limiter** A peak-limiting circuit employing a diode that becomes conductive when signal peaks exceed a predetermined value.

**diode mixer** A mixer that uses a crystal or electron-tube diode. It is generally small enough to fit directly into an r-f transmission line.

**diode modulator** A modulator using one or more diodes to combine a modulating signal with a carrier signal. Used chiefly for low-level signaling because of inherently poor efficiency.

**diode-pentode** An electron tube having a diode and a pentode in the same envelope.

**diotron** A computer circuit using an emission-limited diode.

**dip brazing** A brazing process in which the parts to be joined are clamped together with appropriate preforms of nonferrous filler metal and dipped in a molten chemical or metal bath. The filler metal is distributed in the joints by capillary attraction. A metal bath may also provide the filler metal. Used for assembling waveguide components.

**dip coating** Application of a protective plastic coating to a transformer or other device by dipping.

**diplexer** A coupling system that allows two different transmitters to operate simultaneously or separately from the same antenna.

**diplex radio transmission** The simultaneous transmission of two signals using a common carrier wave.

**diplex reception** Simultaneous reception of two signals having some feature in common, such as a single receiving antenna or a single carrier frequency.

**dipole** 1. An antenna approximately one half-wavelength long, split at its electrical center for connection to a transmission line. The impedance of the antenna is about 72 ohms. The radiation pattern is a maximum at right angles to the axis of the antenna. Also called dipole antenna, doublet, doublet antenna, and half-wave dipole. 2. Two nuclear particles a very small distance apart, having opposite electric charges or opposite magnetic polarities. Also called doublet and electric doublet.

**dipole antenna** *Dipole.*

**dipole moment** A term used in specifying mathematically the field due to a given distribution of electric or magnetic charges.

**dip soldering** Soldering of printed wiring boards or other assemblies by immersion in a pool of molten solder.

**direct-acting recorder** A recorder in which the marking device is mechanically con-

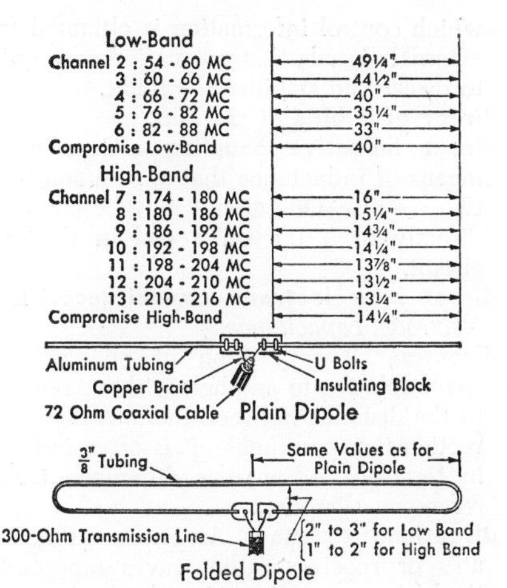

Dipole lengths for specific channels in television bands.

nected to, or directly operated by, the primary detector.

**direct command guidance** Control of a missile or drone entirely from the launching site, by radio or by signals sent over a wire.

**direct-coupled amplifier** [always spell out, to avoid confusion with the abbreviation for direct-current amplifier] A direct-current amplifier in which a resistor or a direct connection provides the coupling between stages, so small changes in direct currents can be amplified.

**direct-coupled attenuation** The insertion loss of a tr, pre-tr, or attenuator tube, measured with the resonant gaps (or their functional equivalent) short-circuited.

**direct coupling** Coupling of two circuits by means of a nonfrequency-sensitive device such as a wire, resistor, or battery, so both direct and alternating current can flow through the coupling path.

**direct current** [abbreviated d-c] An electric current that flows in one direction.

**direct-current amplifier** [abbreviated d-c amplifier] An amplifier that is capable of amplifying d-c voltages and slowly varying voltages. The basic types of d-c amplifiers are the direct-coupled amplifier, the carrier amplifier, and combinations of direct-coupled and carrier amplifiers. The commonest type of carrier amplifier is the chopper amplifier; when the chopper amplifier is used with a direct-coupled amplifier, the combination is called a chopper-stabilized amplifier.

**directed reference flight** A control system for guided missiles or target aircraft, in which control information is obtained from external signals that are varied as required to direct and stabilize the flight.

**direct grid bias** *Grid bias.*

**direct inductive coupling** Coupling by means of inductance that is common to the two circuits. One circuit may be connected directly to a tap on a coil in the other circuit.

**direct interelectrode capacitance** *Interelectrode capacitance.*

**direction** The position of one point in space relative to another, without reference to the distance between them. Although direction is not an angle, it is often indicated in terms of its angular difference from a reference direction.

**directional antenna** An antenna that radiates or receives radio waves more effectively in some directions than others.

**directional beam** A radio or radar wave that is concentrated in a given direction.

**directional characteristic** The variation in the behavior of a transducer or other device with respect to direction.

**directional control** Control exercised over an aircraft so as to move it in a desired direction.

**directional counter** A counter that is more sensitive to nuclear radiation from some directions than from others.

**directional coupler** A device that couples a secondary system only to a wave traveling in a particular direction in a primary transmission system, while completely ignoring a wave traveling in the opposite direction. The amount of coupling is ordinarily expressed in decibels of attenuation that the signal undergoes in passing through the coupling to the secondary line. Coaxial lines use loop-type directional couplers or two-hole directional couplers. For waveguide directional couplers the arrangement is comparable to a two-hole coupler.

**directional filter** A low-pass, bandpass, or high-pass filter that separates the bands of frequencies used for transmission in opposite directions in a carrier system.

**directional gain** *Directivity index.*

**directional gyroscope** A gyroscope that holds its position in azimuth and indicates deviation from a desired heading.

**directional homing** Homing in which the relative bearing is maintained constant.

**directional hydrophone** A hydrophone whose response varies significantly with the direction of sound incidence.

**directional microphone** A microphone

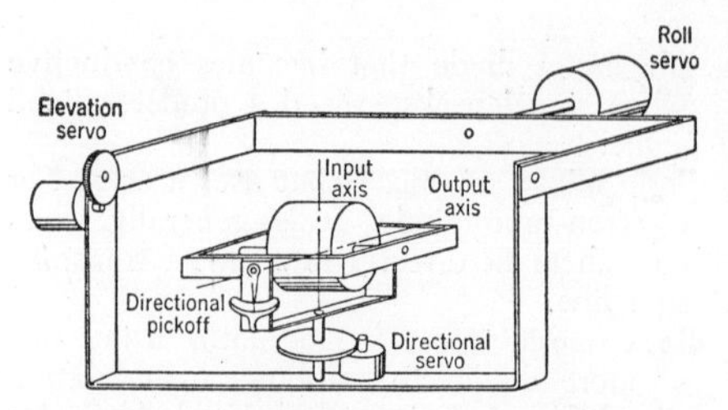


Directional gyroscope mounted on platform.

whose response varies significantly with the direction of sound incidence.

**directional phase shifter** A passive phase shifter in which the phase change for transmission in one direction differs from that for transmission in the opposite direction.

**directional response pattern** *Directivity pattern.*

**directional stabilizer** A directional gyro used to maintain direction automatically, as in an autopilot or bombsight.

**direction finder** [abbreviated d-f] *Radio direction finder.*

**direction-finder bearing indicator** An instrument used with an airborne radio direction finder to indicate the relative, magnetic, or true bearing of a station from an aircraft or the reciprocals of these bearings.

**direction-finder deviation** The difference between the observed radio bearing obtained with a direction finder and the true bearing of the transmitter.

**direction-finding station** A radio station having equipment for determining the direction of arrival of radio waves. Two or more such stations working together can determine the location of the transmitter by triangulation.

**direction of propagation** The direction of time-average energy flow at any point in a homogeneous isotropic medium. For a linearly polarized wave it is the direction of the electric vector. In a uniform waveguide the direction of propagation is often taken along the axis.

**direction rectifier** A rectifier that supplies a d-c voltage whose magnitude and polarity vary with the magnitude and relative polarity of an a-c synchro error voltage.

**directive gain** An antenna rating equal to $4\pi$ times the ratio of the radiation intensity in a given direction to the total power radiated by the antenna.

**directivity** 1. The value of the directive gain of an antenna in the direction of its maximum value. The higher the directivity value, the narrower is the beam in which the radiated energy is concentrated. 2. The ratio of the power measured at the forward wave sampling terminals of a directional coupler, with only a forward wave present in the transmission line, to the power measured at the same terminals when the direction of the forward wave in the line is reversed. The ratio is usually expressed in db. For a perfect coupler it would be infinitely high.

**directivity factor** 1. The ratio of radiated sound intensity at a remote point on the principal axis of a loudspeaker or other transducer to the average intensity of the sound transmitted through a sphere passing through the remote point and concentric with the transducer. The frequency must be stated. 2. The ratio of the square of the voltage produced by sound waves arriving parallel to the principal axis of a microphone or other receiving transducer to the mean square of the voltage that would be produced if sound waves having the same frequency and mean-square pressure were arriving simultaneously from all directions with random phase. The frequency must be stated. Directivity factor in acoustics is equivalent to directivity as applied to antennas.

**directivity index** The directivity factor expressed in decibels. It is 10 times the logarithm to the base 10 of the directivity factor. Also called directional gain.

**directivity pattern** A graphical or other description of the response of a transducer used for sound emission or reception. It is presented as a function of the direction of the transmitted or incident sound waves in a specified plane and at a specified frequency. Also called beam pattern and directional response pattern.

**directly heated cathode** *Filament.*

**director** A parasitic element placed a fraction of a wavelength ahead of a dipole receiving antenna to increase the gain of the array in the direction of the major lobe. It is usually a rod slightly shorter than the receiving dipole, with no connection to the lead-in.

**direct pickup** The transmission of television images without intermediate photographic or magnetic recording.

**direct radiator loudspeaker** A loudspeaker in which the radiating element acts directly on the air, without a horn.

**direct recording** Recording in which a visible record is produced immediately, without subsequent processing, in response to received signals.

**direct reflection** Reflection of light, sound, radar, or radio waves in accordance with the laws of optical reflection, as by a mirror. Also called mirror reflection, regular reflection, and specular reflection.

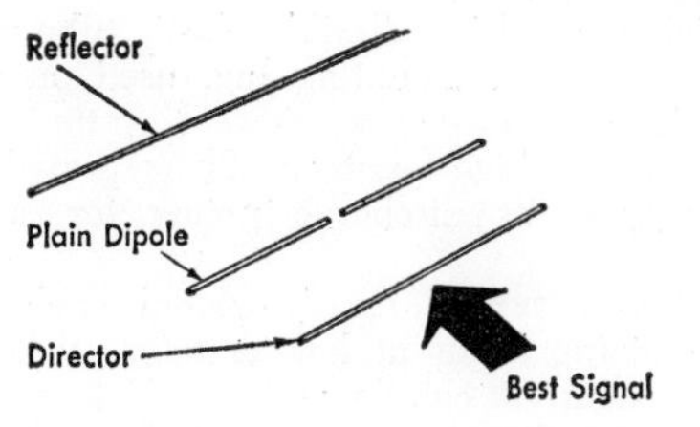


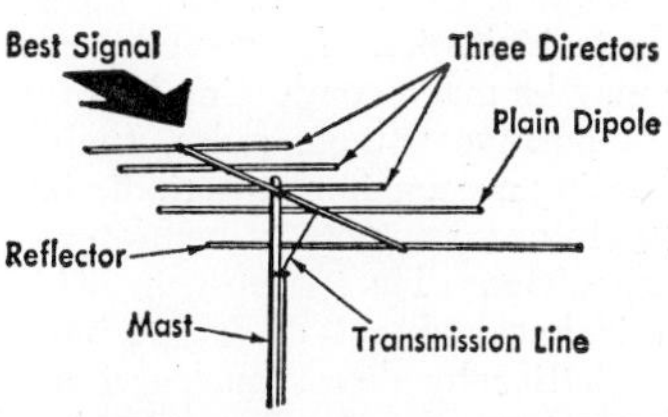


Director used with plain dipole and reflector, and example of commercial plain-dipole Yagi antenna using three directors.

**direct scanning** A scanning method in which the subject is illuminated at all times and only one elemental area of the subject is viewed at a time by the television camera. Used in live television broadcasting, whereas indirect or flying-spot scanning is sometimes used in industrial television systems.

**direct voltage** A voltage that forces electrons to move through a circuit in the same direction continuously, thereby producing a direct current. Also called d-c voltage.

**direct wave** A radio wave that is propagated directly through space from transmitter to receiver without being refracted by the ionosphere.

**disadvantage factor** The ratio of the average neutron flux in the moderator of a nuclear reactor to the average neutron flux in the fissionable material.

**discharge** 1. The passage of electricity through a gas, usually accompanied by a glow, arc, spark, or corona. 2. To remove a charge from a battery, capacitor, or other electric energy storage device.

**discharge lamp** A lamp in which light is produced by an electric discharge between electrodes in a gas or vapor at low or high pressure. Examples include fluorescent lamps, mercury-vapor lamps, neon tubing, and sodium-vapor lamps. Also called gas discharge lamp.

**discharger** A silver-impregnated cotton

wick encased in a flexible plastic tube with an aluminum mounting lug, used on aircraft to reduce precipitation static. The many fine high-resistance fibers provide a multitude of discharge points for static electricity.

**discharge tube** An evacuated enclosure containing a gas at low pressure, through which current can flow when sufficient voltage is applied between metal electrodes in the tube. The resulting current flow is due chiefly to ionization of a gas or vapor. The tube may be made conducting by triggering with a positive voltage pulse, such as for discharging a capacitor at periodic intervals to generate a sawtooth waveform.

**discomposition** The process in which an atom is knocked out of its position in a crystal lattice by direct nuclear impact, as by fast neutrons or by fast ions that have been previously knocked out of their lattice positions. The atom so displaced eventually comes to rest at an interstitial position or at a lattice edge.

**discone antenna** A biconical antenna in which one of the cones is spread out to 180° to form a disk. The center conductor of the coaxial line terminates at the center of the disk, and the cable shield terminates at the vertex of the cone. Both the input impedance and radiation pattern remain essentially constant over a wide frequency range. The disk is normally parallel to the earth, giving an omnidirectional radiation pattern in a horizontal plane.

**disconnect** To open a circuit by removing wires or connections, as distinguished from opening a switch to stop current flow.

**discontinuity** A break in the continuity of a medium or material, at which a reflection of wave energy can occur.

**discrete sampling** Sampling in which the individual samples are of such long duration that the frequency response of the channel is not deteriorated by the sampling process.

**discrete sentence intelligibility** The percent intelligibility obtained when the speech units under consideration are sentences, usually of simple form and content.

**discrete variable** A quantity that may assume any one of a number of individually distinct or separate values.

**discrete word intelligibility** The percent intelligibility obtained when the speech units under consideration are words, usually presented so as to minimize the contextual relation between them.

**discrimination** *Conditional jump.*

**discriminator** A circuit in which magnitude and polarity of the output voltage depend on how an input signal differs from a standard or from another signal. Thus, a frequency discriminator converts frequency

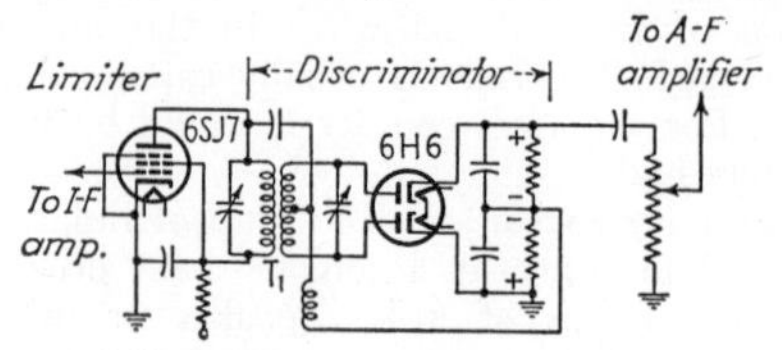

Discriminator circuit for f-m receiver.

deviations from a carrier frequency into corresponding amplitude variations. A pulse-height discriminator delivers an output voltage only for pulses that exceed a predetermined height. A phase discriminator converts phase variations into corresponding amplitude variations.

**discriminator transformer** A transformer designed to be used in a stage where f-m signals are converted directly to a-f signals or in a stage where frequency changes are converted to corresponding voltage changes.

**disengage** To break the contact between two objects.

**dish** Slang for *parabolic reflector.*

**disintegration** 1. Radioactive decay in which energy is emitted from the nuclei of atoms. 2. The transformation of one nuclide into one or more different nuclides by bombardment with high-energy particles such as alpha particles or helium ions, deuterons, protons, neutrons, or gamma rays.

**disintegration chain** *Radioactive series.*

**disintegration constant** *Decay constant.*

**disintegration energy** The energy released during radioactive decay. It is distributed among the decay products. Also called *Q* value.

**disintegration family** *Radioactive series.*

**disintegration rate** 1. The absolute rate of decay of a radioactive substance, usually expressed in terms of disintegrations per unit of time. 2. The absolute rate of transformation of a nuclide under bombardment.

**disintegration series** *Radioactive series.*

**disintegration voltage** The lowest anode voltage at which destructive positive-ion bombardment of the cathode occurs in a hot-cathode gas tube.

**disk** *Phonograph record.*

**disk recorder** A mechanical recorder that uses a disk as the recording medium.

**disk recording** *Phonograph record.*

**disk-seal tube** An electron tube having

disk-shaped electrodes arranged in closely spaced parallel layers, to give low interelectrode capacitance along with high power output up to 2,500 mc. The edges

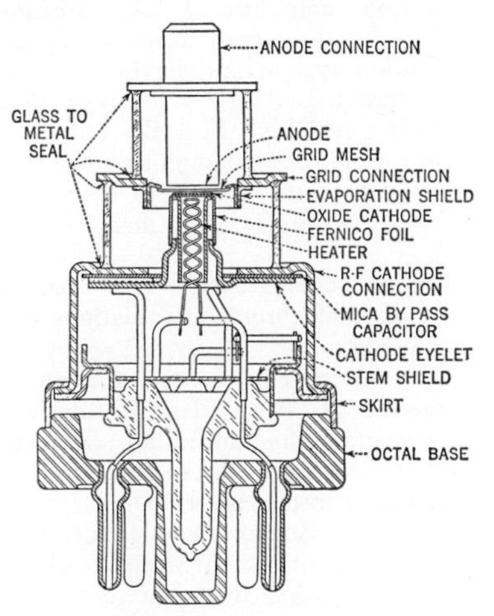


Disk-seal tube construction.

of the disk electrodes are fused into and project through the glass or ceramic envelope, to serve as external contacts. Also called lighthouse tube and megatron.

**dispenser cathode** An electron-tube cathode having provisions for continuously replacing evaporated electron-emitting material.

**dispersal gettering** Absorption of gas during dispersal of a getter in an electron tube.

**dispersion** 1. The process of separating radiation into components having different frequencies, energies, velocities, or other characteristics. A prism or diffraction grating disperses white light into its component colors. A magnetic field disperses or sorts electrons according to their velocities. 2. Scattering of microwave radiation by an obstruction. 3. A distribution of finely divided particles in a medium.

**displacement gyroscope** A gyroscope that senses, measures, and transmits angular displacement data.

**displacement kernel** A kernel that is a function of the distance between two points in a medium.

**displacement law** *Radioactive displacement law.*

**display** A visual presentation of output information, as on the cathode-ray tube screen of a radar set or navigation system.

**display loss** *Visibility factor.*

**display primaries** The television receiver primary colors that, when mixed in proper proportions, serve to produce other desired colors. The three primaries usually used are red, green, and blue. Also called receiver primaries.

**display tube** *Character-writing tube.*

**disruptive discharge** A sudden and large increase in current through an insulating medium due to complete failure of the medium under electrostatic stress.

**dissector tube** *Image dissector.*

**dissipation** An undesired loss of energy, generally by conversion into heat. Thus, the collector dissipation rating in watts for a transistor is the maximum amount of energy that can be lost as heat at the collector electrode without damage to the transistor.

**dissipation factor** The reciprocal of *Q*.

**dissipation line** A length of stainless steel or Nichrome wire used as a noninductive terminating impedance for a rhombic transmitting antenna when several kilowatts of power must be dissipated.

**dissolve** The merging of two television camera signals in such a way that as one scene disappears, another slowly appears.

**dissonance** An unpleasant combination of harmonics heard when certain musical tones are played simultaneously.

**dissymmetrical network** *Dissymmetrical transducer.*

**dissymmetrical transducer** A transducer whose input and output image impedances are not equal. Also called dissymmetrical network.

**distance mark** A mark produced on a radar display by a special signal generator. When the mark is moved to a target position on the screen, the range to that target can be read directly on the calibrated dial of the signal generator.

**distance-measuring equipment** [abbreviated dme] A radio aid to navigation that provides distance information by measuring total round-trip time of transmission from an airborne interrogator to a ground-based transponder and return.

**distance resolution** The minimum radial distance by which targets must be separated to be separately distinguishable by a particular radar. Also called range discrimination (deprecated) and range resolution (deprecated).

**distortion** An undesired change in waveform.

**distortion and noise meter** An instrument that uses a null network to remove

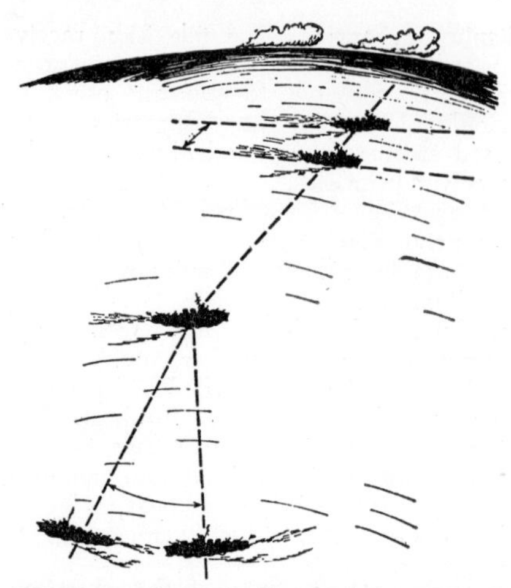

Distance resolution is illustrated by upper two ships, and bearing resolution by lower two ships, for radar on ship at center.

the fundamental frequency of a sinusoidal signal. The remainder, due to distortion and noise, is then measured.

**distortion meter** An instrument that provides a visual indication of the harmonic content of an audio-frequency wave.

**distress frequency** A frequency allotted to distress calls, generally by international agreement. For ships at sea and aircraft over the sea, it is 500 kc.

**distress signal** The international signal used when a ship, aircraft, or other vehicle is threatened by grave and imminent danger and requests immediate assistance. In radiotelegraphy, it consists of three dots, three dashes, and three dots transmitted as a single signal in which the dashes are emphasized. In radiotelephony, it is the word mayday.

**distributed amplifier** A wideband amplifier in which tubes are distributed along artificial delay lines made up of coils acting with the input and output capacitances of the tubes. Gain can be increased indefinitely by adding more tubes.

**distributed capacitance** The capacitance that exists between adjacent turns in a coil or between adjacent conductors in a cable. Also called self-capacitance.

**distributed constant** A circuit parameter that exists along the entire length of a transmission line. For a transverse electromagnetic wave on a two-conductor transmission line, the distributed constants are series resistance, series inductance, shunt conductance, and shunt capacitance per unit length of line.

**distributed inductance** The inductance that exists along the entire length of a conductor, as distinguished from inductance concentrated in a coil.

**distribution amplifier** 1. An a-f power amplifier used to feed a speech or music distribution system and having sufficiently low output impedance so changes in load do not appreciably affect the output voltage. 2. An r-f power amplifier used to feed television or radio signals to a number of receivers, as in an apartment house or hotel.

**distribution coefficients** The tristimulus values of monochromatic radiations having equal power.

**distribution control** *Linearity control.*

**disturbance** 1. An undesired interference or noise signal affecting radio, television, or facsimile reception. 2. An undesired command signal in a control system.

**disturbed-one output** A one output of a magnetic cell to which partial-read pulses have been applied since that cell was last selected for writing.

**disturbed-zero output** A zero output of a magnetic cell to which partial-write pulses have been applied since that cell was last selected for reading.

**dither** A force having a controlled amplitude and frequency, applied continuously to a device driven by a servomotor so the device is constantly in small-amplitude motion and cannot stick at its null position. Also called buzz.

**divergence** 1. The spreading of a cathode-ray stream due to repulsion of like charges (electrons). 2. A scalar quantity used in computations involving vectors.

**divergence loss** The portion of the transmission loss in an acoustic system that is due to the divergence or spreading of sound rays.

**diversity radar** A radar that uses two or more transmitters and receivers, each pair operating at a slightly different frequency but sharing a common antenna and video display, to obtain greater effective range and reduce susceptibility to jamming.

**diversity receiver** A radio receiver designed for space or frequency diversity reception.

**diversity reception** Radio reception in which the effects of fading are minimized by combining two or more sources of signal energy carrying the same modulation. Space diversity takes advantage of the fact that fading does not occur simultaneously for antennas spaced several wavelengths apart. Frequency diversity takes advantage of the fact that signals differing slightly in frequency do not fade simultaneously.

**dividing network** *Crossover network.*

**D layer** The lowest layer of ionized air above the earth, occurring in the D region only in the daytime hemisphere. It reflects frequencies below about 50 kc and partially absorbs higher-frequency waves.

**dme** Abbreviation for *distance-measuring equipment.*

**dme-cotar** A navigation system that gives complete trajectory information at a single-site ground station by combining the range indication of dme with the direction indication of cotar.

**D neutron** Obsolete term for a neutron of such energy that it is strongly absorbable by rhodium covered with cadmium.

**doghouse** A small enclosure placed at the base of a transmitting antenna tower to house antenna tuning equipment.

**Doherty amplifier** A linear r-f power amplifier that is divided into two sections whose inputs and outputs are connected by quarter-wave networks. Operating parameters are so adjusted that, for all values of input signal voltage up to one-half maximum amplitude, section No. 1 delivers all the power to the load. Above this level, section No. 2 comes into operation. At maximum signal input, both sections are operating at peak efficiency and each section is delivering half the total output power to the load.

**dollar** A unit of reactivity, equal to the difference between the reactivities for delayed critical and prompt critical conditions in a given nuclear reactor. One dollar is about 0.0078 in reactivity, which is equal to the contribution of delayed neutrons to reactivity. A smaller unit is the cent, equal to 1/100th of a dollar.

**dolly** A wheeled platform on which a television camera or other apparatus is mounted.

**domain** A region in a ferroelectric or ferromagnetic crystal in which the electric or magnetic polarization is saturated in a single direction and varies only with temperature.

**dome** An enclosure for a sonar transducer, projector, or hydrophone and associated equipment. It is designed to have minimum effect on sound waves traveling under water.

**domestic induction heater** A cooking utensil heated by current (usually of commercial power-line frequency) induced in it by a primary inductor.

**dominant mode** *Fundamental mode.*

**dominant wave** The electromagnetic wave that has the lowest cutoff frequency in a given uniconductor waveguide. It is the only wave that will carry energy when the excitation frequency is between the lowest cutoff frequency and the next higher cutoff frequency.

**dominant wavelength** The single wavelength of light that, when combined in suitable proportions with a reference standard light, matches the color of a given sample.

**donor** An impurity that is added to a pure semiconductor material to increase the number of free electrons. Because conduction is then due chiefly to movements of electrons, which are negative charges, the resulting alloy is called an n-type semiconductor. Antimony, arsenic, and phosphorus are frequently used as donors for germanium. Also called donor impurity.

**donor impurity** *Donor.*

**donor level** An intermediate energy level close to the conduction band in the energy diagram of an extrinsic semiconductor.

**donut** *Doughnut.*

**donutron** An all-metal tunable magnetron.

**doorknob capacitor** A high-voltage plastic-encased capacitor resembling a doorknob in size and shape. Sheet plastic insulation is used. A typical voltage rating is 20 kilovolts.

**doorknob tube** An ultrahigh-frequency electron tube having the approximate size and shape of a doorknob. Electrodes are small and closely spaced, and leads are brought out directly through the glass envelope. The Western Electric 316A is one example.

**doped junction** A junction produced by adding an impurity to the melt during growing of a semiconductor crystal.

**doping** The addition of impurities to a semiconductor to achieve a desired characteristic, such as to produce an n-type or p-type material.

**doping agent** An impurity element added to semiconductor materials used in crystal diodes and transistors. Common doping agents for germanium and silicon include aluminum, antimony, arsenic, gallium, and indium.

**doping compensation** The addition of donor impurities to a p-type semiconductor or of acceptor impurities to an n-type semiconductor.

**doppler broadening** Frequency-spreading

that occurs in single-frequency radiation when the radiating atoms, molecules, or nuclei do not all have the same velocity. Each radiating particle may then give rise to a different doppler shift.

**doppler effect** The change in the observed frequency of a wave due to relative motion of source and observer. When the distance between source and observer is decreasing, the observed frequency is higher than the source frequency. When the distance is increasing, the observed frequency is lower. The effect occurs for sound waves as well as radio waves. Named after Christian Doppler (1803–1853), a German mathematician.

**doppler frequency** *Doppler shift.*

**doppler navigation system** An airborne navigation system utilizing the doppler effect to determine drift and ground speed. Four beams of pulsed microwave energy are beamed toward the earth along the corners of an imaginary pyramid whose peak is at the aircraft. Echoes from the front-pointing beams undergo upward doppler shift, while echoes from the rearward beams undergo downward doppler shift. Similarly, drift causes doppler shift of echoes from beams on one side with respect to beams on the other side of the aircraft. Comparison of the doppler shifts in a computer provides complete information for navigation even in zero-visibility weather, at all altitudes, without reference to ground stations.

**doppler radar** A radar that makes use of the doppler shift of an echo due to relative motion of target and radar. This shift in frequency permits differentiation between fixed and moving targets. The velocity of a moving target can be determined with high accuracy by measuring the frequency shift. A doppler radar may be either continuous-wave or pulsed.

**doppler radar guidance** Missile guidance that makes use of a doppler navigation system built into the missile to determine surface velocity and drift angle.

**doppler shift** The amount of the change in the observed frequency of a wave due to doppler effect, expressed in cycles. Also called doppler frequency.

**doran** [DOppler RANge] A doppler ranging system that uses phase comparison of three different modulation frequencies on the carrier wave, such as 0.01 mc, 0.1 mc, and 1 mc, to obtain missile range data with high accuracy. Some aspects of doran are similar to dovap.

**dose** The amount of ionizing radiation delivered to a specified volume, measured in rads, reps, or rems. For x-rays or gamma rays having quantum energies up to 3 million electron-volts, the roentgen unit may be used. Also called absorbed dose and dosage.

**dosage** *Dose.*

**dosemeter** *Dosimeter.*

**dose rate** The rate at which nuclear radiation is delivered.

**dose-rate meter** An instrument that measures radiation dose rate.

**dosimeter** An instrument that measures the total dose of nuclear radiation received in a given period. Also called dosemeter.

**dot** *Button.*

**dot generator** A signal generator that produces a dot pattern on the screen of a three-gun color television picture tube, for use in convergence adjustments. When convergence is out of adjustment, the dots occur in groups of three, one for each of the receiver primary colors. When convergence is correct, the three dots of each group converge to form a single white dot.

**dot-sequential color television** A color television system in which the red, blue, and green primary-color dots are formed in rapid succession along each scanning line.

**double-base diode** *Unijunction transistor.*

**double-base junction diode** *Unijunction transistor.*

**double-base junction transistor** A tetrode transistor that is essentially a junction

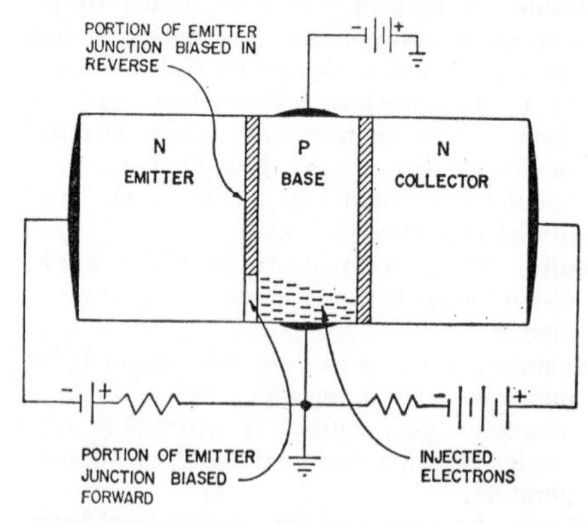


Double-base junction transistor construction and circuit.

triode transistor having two base connections on opposite sides of the central region of the transistor. Also called tetrode junction transistor.

**double-beam cathode-ray tube** A cathode-ray tube having two beams and capable of producing two independent traces that may overlap. The beams may be produced by splitting the beam of one gun or by using two guns.

**double bridge** *Kelvin bridge.*

**double-button carbon microphone** A carbon microphone having two carbon-filled button-like containers, one on each side of the diaphragm, so as to give twice the resistance change obtainable with a single button. Also called differential microphone.

**double-channel duplex** A method that provides for simultaneous communication between two stations through use of two r-f channels, one in each direction.

**double-channel simplex** A method that provides for nonsimultaneous communication between two stations through use of two r-f channels, one in each direction.

**double-diffused transistor** A transistor in which two p-n junctions are formed in the semiconductor wafer, by gaseous diffusion of both p-type and n-type impurities. An intrinsic region can also be formed.

**double-diode** *Duodiode.*

**double-doped transistor** The original grown-junction transistor, formed by successively adding p-type and n-type impurities to the melt during growing of the crystal.

**double-doublet antenna** Two half-wave doublet antennas crisscrossed at their center, with one being shorter than the other so as to give broader frequency coverage.

**double-hump response** *Bandpass response.*

**double image** A television picture consisting of two overlapping images, due to reception of the signal over two paths that differ in length so signals arrive at slightly different times. The longer path generally involves reflection of the signal by a hill, building, or large metal structure. The later-arriving reflected signal is often called a ghost because it is usually weaker than the direct signal and produces a phantom-like image to the right of the regular image.

**double-length number** A number having twice as many digits as are ordinarily used in a given computer. Also called double-precision number.

**double limiter** *Cascade limiter.*

**double-moding** Undesirable shifting of a magnetron from one frequency to another at irregular intervals.

**double modulation** A method of modulation in which a subcarrier is first modulated with the desired intelligence, and the modulated subcarrier is then used to modulate a second carrier having a higher frequency.

**double-pole double-throw** [abbreviated dpdt] A six-terminal switch or relay contact arrangement that simultaneously connects one pair of terminals to either of two other pairs of terminals.

**double-pole-piece magnetic head** A magnetic head having two pole pieces, opposite in polarity and mounted on opposite sides of the magnetic recording medium. Either or both may have an energizing winding.

**double-pole single-throw** [abbreviated dpst] A four-terminal switch or relay contact arrangement that simultaneously opens or closes two separate circuits or both sides of the same circuit.

Double-pole single-throw and double-pole double-throw knife switches.

**double-precision computation** Use of twice as many digits as are normally handled in a digital computer, by keeping track of the numerical fragments that go beyond the computer capacity.

**double-precision number** *Double-length number.*

**double-pulsing station** A loran station that belongs to two pairs and emits pulses at two pulse rates.

**doubler** 1. *Frequency doubler.* 2. *Voltage doubler.*

**double-sideband transmission** The transmission of a modulated carrier wave accompanied by both of the sidebands resulting from modulation. The upper sideband corresponds to the sum of the carrier and modulation frequencies, whereas the lower sideband corresponds to the difference between the carrier and modulation frequencies.

**double-sided mosaic** An array of photosensitive elements insulated one from the other and mounted in a television camera tube in such a way that an image can be projected optically on one side of the mosaic. The corresponding electric signal is obtained by electronically scanning the other side of the mosaic.

**double-spot tuning** The reception of a given station by a superheterodyne receiver at two different dial settings, one where the local oscillator is above the station frequency by the i-f value and the other where the local oscillator is below the station frequency by the i-f value. This effect can occur only if the receiver has poor selectivity. Also called repeat-point tuning.

**double-stream amplifier** A traveling-wave amplifier in which amplification occurs through interaction of two electron beams having different average velocities.

**double-stub tuner** A tuner consisting of two stubs, usually 3/8 wavelength apart, connected in parallel with a transmission line. Used for impedance-matching.

**double superheterodyne** A superheterodyne receiver in which the incoming carrier frequency is lowered to the first i-f value by beating with one oscillator frequency in the first mixer, then reduced to the final i-f value by beating with another oscillator frequency in a second mixer. Used in uhf receivers to obtain higher gain without instability, along with greater suppression of image frequencies and higher adjacent-channel selectivity. Also called dual-conversion receiver.

**double-surface transistor** A point-contact transistor in which the emitter and collector whiskers are in contact with opposite sides of the base.

**doublet** *Dipole.*

**doublet antenna** *Dipole.*

**double-track tape recorder** A tape recorder with a recording head that covers half the tape width, so two parallel tracks can be recorded on one tape. After one track is recorded, the reels are turned over and reversed in position, for recording the second track in the opposite direction on the tape. Also called dual-track tape recorder and half-track recorder.

**double-track tape recording** Magnetic recording in which two adjacent tracks are placed on the tape either for stereophonic reproduction or for doubling the recording capacity of the tape.

**double triode** An electron tube having two triodes in the same envelope.

**doublet trigger** A trigger signal consisting of two pulses spaced a predetermined amount for coding purposes.

**double-tuned circuit** A circuit that is resonant to two adjacent frequencies, so there are two approximately equal values of peak response, with a dip in between.

**double-tuned detector** A type of f-m discriminator in which the limiter output transformer has two secondaries, one tuned above the resting frequency and the other tuned an equal amount below. Without modulation, both diodes conduct equally at the resting frequency and the a-f output is zero. Signal frequency deviation makes one diode conduct more than the other, giving a-f output.

**doubling time** The time required for a breeding reactor to double its fuel inventory.

**doughnut** 1. The toroidal vacuum chamber in which electrons are accelerated in a betatron or synchrotron. It is generally made of glass or ceramic material and is placed between magnet poles. Also called donut and toroid. 2. An assembly of enriched fissionable material, often doughnut-shaped, used in a thermal reactor to provide a local increase in fast neutron flux for experimental purposes.

**dovap** [DOppler Velocity And Position] A phase-coherent tracking system used to determine velocity and position of missiles and space vehicles. A ground transmitter radiates a signal to a transponder on the missile and to receivers at several ground stations. The transponder doubles the frequency of the signal and retransmits it to the ground stations. The doppler frequency shift, separated by comparing the received signal with twice the original transmitted signal frequency, is then proportional to missile velocity. Position is determined by combining the readings of two or more ground stations.

**Dove** A Navy air-to-surface guided missile using radar homing.

**down-doppler** The sonar situation wherein the target is moving away from the transducer, so the frequency of the echo is less than the frequency of the reverberations received immediately after the end of the outgoing ping. Opposite of up-doppler.

**down-lead** *Lead-in.*

**downrange** Any area along the flight course of a missile. Downrange tracking stations report on missile flight behavior and receive telemetered data from the missile.

**down time** The time during which a piece of equipment is not in operation because of a breakdown.

**downward modulation** Modulation in which the instantaneous amplitude of the modulated wave is never greater than the amplitude of the unmodulated carrier.

**Dow oscillator** *Electron-coupled oscillator.*

**dpdt** Abbreviation for *double-pole double-throw.*

**dpst** Abbreviation for *double-pole single-throw.*

**drag angle** A stylus cutting angle such that the point drags in the coating during recording on a disk. It is the opposite of dig-in angle.

**drag-cup motor** An induction motor having a cup-shaped rotor of conducting material, inside of which is a stationary magnetic core. It can be reversed by reversing connections to one phase of the two-phase stator winding. The light-weight rotor permits quick starts, high speed, quick stops, and sudden reversals.

**drain** *Current drain.*

**D region** The region of the ionosphere up to about 60 miles above the earth, below the E and F regions.

**drift** 1. A slow change in some characteristic of a device, such as frequency, balance current, direction (as in a gyro), or desired course of travel. Temperature variations are a common cause of frequency drift and unbalance in circuits. 2. The movement of current carriers in a semiconductor under the influence of an applied voltage.

**drift angle** The angular difference between the course and the course made good. Also called course error (deprecated).

**drift correction angle** The angular difference between course and heading. Also called crab angle.

**drift mobility** The average drift velocity of carriers per unit electric field in a homogeneous semiconductor.

**drift rate** The rate at which the voltage drop of a voltage regulator tube or reference tube changes with time under constant operating conditions. It is equal to the slope of the smoothed curve of tube voltage drop plotted against time.

**drift space** A space in an electron tube that is substantially free of externally applied alternating fields, in which repositioning of electrons takes place. In a klystron the velocity-modulated electrons form bunches in this space.

**drift transistor** 1. A transistor having two plane parallel junctions, with a resistivity gradient in the base region between the junctions to improve the high-frequency response. 2. *Diffused-alloy transistor.*

**drift tunnel** A piece of metal tubing, held at a fixed potential, that forms the drift space in a microwave tube or linear accelerator.

**drift velocity** The average velocity of an electron that is moving under the influence of an electric field. Drift velocity corresponds to the net current in an electron tube or a semiconductor device.

**drive** *Excitation.*

**drive control** *Horizontal drive control.*

**driven array** An antenna array consisting of a number of driven elements, usually half-wave dipoles, fed in phase or out of phase from a common source. Examples include broadside, collinear, and end-fire arrays.

**driven element** An antenna element that is directly connected to the transmission line.

**drive pattern** An undesired pattern of density variations caused by periodic errors in the position of the recording spot in a facsimile system. When caused by drive gears the effect is called gear pattern.

**drive pin** An upward-projecting rod near the center pin of a turntable, used with a two-hole disk record to prevent slippage during recording.

**drive-pin hole** An off-center hole in a disk record, positioned to mate with the turntable drive pin.

**drive pulse** A pulsed magnetomotive force applied to a magnetic cell from one or more sources.

**driver** 1. The amplifier stage preceding the output stage in a receiver or transmitter. Also called driver stage. 2. The portion of a horn loudspeaker that converts electric energy into acoustic energy and feeds the acoustic energy to the small end of the horn.

**driver stage** *Driver.*

**driving-point admittance** The complex ratio of alternating current to applied alternating voltage for an electron tube, network, or other transducer.

**driving-point impedance** The complex ratio of applied alternating voltage to the resulting alternating current in an electron tube, network, or other transducer.

**driving power** The power supplied to the grid circuit of a tube in which the grid swings positive and draws current for a part of each cycle of the input signal.

**driving signal** A signal that times horizontal or vertical scanning at a television transmitter. Driving signals are usually provided by a central sync generator at the transmitter.

**drone** A remotely controlled pilotless aircraft. It is generally controlled by radio, either from the ground or from a mother plane. Used on missions too hazardous for a human pilot, such as for target practice, probing the cloud of a nuclear explosion, taking photographs at low level over enemy territory, or carrying a television camera and transmitter over enemy territory.

**drop** *Voltage drop.*

**dropout current** The maximum current at which a relay or other magnetically operated device will release to its deenergized position.

**dropout voltage** The maximum voltage at which a relay or other magnetically operated device will release to its deenergized position.

**dropping resistor** A resistor used in series with a load to decrease the voltage applied to the load.

**dropsonde** A radiosonde dropped by parachute from a high-flying aircraft to measure weather conditions and report them back to the aircraft. Used over water or other areas in which no ground station can be maintained.

**drum recorder** A facsimile recorder in which the record sheet is mounted on a rotating drum or cylinder.

**drum speed** The number of revolutions per minute made by a facsimile transmitter or recorder drum.

**drum transmitter** A facsimile transmitter in which the subject copy is mounted on a rotating drum or cylinder.

**dry battery** A battery made up of a series, parallel, or series-parallel arrangement of dry cells in a single housing to provide desired voltage and current values.

**dry cell** A voltage-generating cell having an immobilized electrolyte. The commonest form has a positive electrode of carbon and a negative electrode of zinc in an electrolyte of sal ammoniac paste.

**dry-charged battery** A storage battery in which the electrolyte is drained from the battery after the plates are formed. The battery can be stored for several years in this dry condition. To place in service, it is filled with electrolyte and charged for only a few minutes.

**dry circuit** A relay circuit in which open-circuit voltages are very low and closed-circuit currents extremely small, so there is no arcing to roughen the contacts. As a result, an insulating film prevents closing of the circuit when the contacts are brought together mechanically by the relay.

**dry-disk rectifier** *Metallic rectifier.*

**dry electrolytic capacitor** An electrolytic capacitor in which the electrolyte is a paste rather than a liquid. The dielectric is a thin film of gas formed on one of the plates by chemical action.

**D scan** *D display.*

**D scope** A radarscope that produces a D display.

**dual automatic radio compass** A combination of two automatic radio compasses feeding an azimuth indicator having two pointers, each indicating direction to one of the two radio stations being used to obtain a radio fix. Complete data for a fix is thus visible to the pilot at all times, eliminating the need for tuning first to one station and then to the other.

**dual-channel amplifier** An a-f amplifier having two separate amplifiers for the two channels of a stereophonic sound system, usually operating from a common power supply mounted on the same chassis.

**dual-channel television sound system** A television receiver sound system in which the sound i-f signal is taken off ahead of the video detector, for handling in a separate sound channel. Used chiefly in older television receivers, and replaced now by the simpler intercarrier sound system.

**dual-conversion receiver** *Double superheterodyne.*

**dual-diversity receiver** A diversity radio receiver in which the two antennas feed separate r-f systems, with mixing occurring after the converter. An automatic selection system may connect the output to whichever channel is stronger at each instant.

**dual-frequency induction heater** An induction heater in which the charge receives energy by induction, simultaneously or successively, from a work coil or coils operating at two different frequencies.

**dual modulation** The process of modulating a common carrier wave or subcarrier with two different types of modulation, each conveying separate information.

**dual networks** *Structurally dual networks.*

**dual-track tape recorder** *Double-track tape recorder.*

**dub** To transfer recorded material from one recording to another, with or without the addition of new sounds, background music, or sound effects.

**dubbing** Combining two or more sources of sound into a complete recording, at least one of the sources being a recording.

**duct** 1. An atmospheric condition that makes possible abnormally long-range microwave signal propagation in the troposphere. Temperature inversions cause abnormal refractions that make the beam refract up and down between the two air layers

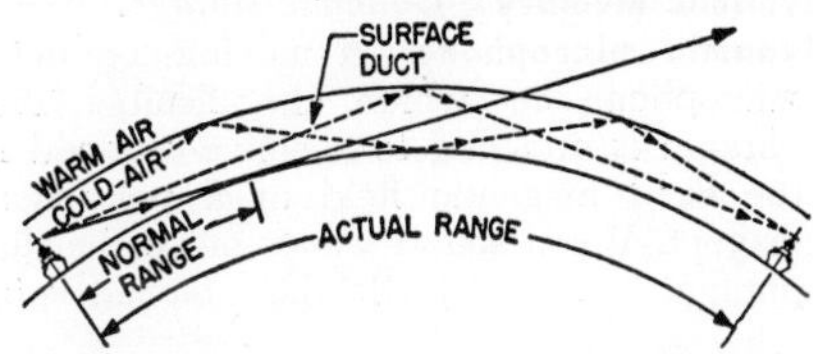


Duct of cold air under warm air makes radio waves follow curvature of earth to ship at right.

forming the duct, so microwave signals travel 10 or more times farther than the normal line-of-sight limit. Also called tropospheric duct. 2. An enclosed runway for cables.

**dumbbell marker** An improved version of the fan marker used for air navigation, radiating a dumbbell-shaped signal pattern about 1½ miles wide and 12 miles across.

**dumbbell slot** A dumbbell-shaped hole in a wall or diaphragm of a waveguide, designed to serve as a slot radiator.

**dummy** An artificial address, instruction, or other unit of information inserted in a digital computer solely to fulfill prescribed conditions (such as word-length or block-length) without affecting operations.

**dummy antenna** A device that has the impedance characteristic and power-handling capability of an antenna but does not radiate or receive radio waves. Used chiefly for testing transmitters. Also called artificial antenna.

**dummy load** A dissipative device used at the end of a transmission line or waveguide to convert transmitted energy into heat, so essentially no energy is radiated outward or reflected back to its source.

**dummy tube** An electron-tube substitute that can be plugged into a socket to make normal connections between socket terminals for alignment or test purposes, without providing any of the additional functions of the tube.

**dump** 1. To withdraw all power from a computer accidentally or intentionally. 2. In digital computer programing, to transfer all or part of the contents of one section of computer memory into another section.

**dump check** A computer check that usually consists of adding all the digits during dumping, and verifying the sum when retransferring.

**dunking sonar** Sonar designed to be lowered into the water from a hovering aircraft by means of a cable.

**duo-cone loudspeaker** A loudspeaker having two concentric conical diaphragms, the larger serving for low audio frequencies and the smaller for high audio frequencies.

**duodiode** An electron tube having two diodes in the same envelope, with either a common cathode or separate cathodes. Also called double diode.

**duodiode-pentode** An electron tube having two diodes and a pentode in the same envelope, generally with a common cathode.

**duodiode-triode** An electron tube having two diodes and a triode in the same envelope, generally with a common cathode.

**duplex** *Duplex operation.*

**duplex cable** Two insulated stranded conductors twisted together. They may or may not have a common insulating covering.

**duplex cavity** A radar tr-tube cavity.

**duplexer** A switching device used in radar to permit alternate use of the same antenna for both transmitting and receiving. It contains the tr tube that blocks out the receiver when the transmitter is operating. Other forms of duplexers serve for two-way radio communication using a single antenna at lower frequencies. Also called duplexing assembly.

**duplexing** *Duplex operation.*

**duplexing assembly** *Duplexer.*

**duplex operation** The operation of associated transmitting and receiving apparatus concurrently as in ordinary telephones, without manual switching between talking and listening periods. A separate frequency band is required for each direction of transmission. Also called duplex and duplexing.

**duplication check** A computer check which requires that the results of two independent performances (either concurrently on duplicate equipment or at a later time on the same equipment) of the same operation be identical.

**dust core** *Ferrite core.*

**duty cycle** 1. The time intervals devoted to starting, running, stopping, and idling when a device is used for intermittent duty. 2. The ratio of working time to total time for an intermittently operating device, usually expressed as a per cent. Also called duty factor. 3. The product of pulse duration and pulse repetition frequency, equal to the time per second that pulse power is applied. Also called duty factor. 4. The ratio of pulse width to the interval between like portions of successive pulses, usually expressed as a per cent. Also called duty factor. 5. Deprecated term for *duty ratio* in radar.

**duty factor** *Duty cycle.*

**duty ratio** In a pulse radar or similar system, the ratio of average to peak pulse power. Also called duty cycle (deprecated).

**dx** Abbreviation for distance reception, used in connection with reception of, or communication with, distant radio stations.

**dynamic analogies** Analogies that make it possible to convert the differential equations for mechanical and acoustic systems to equivalent electrical equations that can be represented by electric networks and solved by circuit theory.

**dynamic braking** The braking of an electric motor by using it as a generator that feeds energy to a heat-dissipating resistance load or back into the power system. Also called regenerative braking and resistance braking.

**dynamic characteristic** *Load characteristic.*

**dynamic convergence** The process whereby the locus of the point of convergence of electron beams in a color television or other multibeam cathode-ray tube is made to fall on a specified surface during scanning. Without dynamic convergence that varies with beam angle, the locus would be a spherical surface at a constant radius from the center of deflection of the beam.

**dynamic demonstrator** A large schematic circuit diagram that has been cemented to a board, with all components mounted near or on their symbols and connected together to give a working circuit of a radio, television receiver, or other electronic apparatus. Used for training purposes in classrooms and laboratories.

**dynamic focusing** The process of varying the focusing electrode voltage for a color picture tube automatically so the electron-beam spots remain in focus as they sweep over the flat surface of the screen. Without dynamic focusing, part of the image would be out of focus at all times.

**dynamic loudspeaker** A loudspeaker in which the moving diaphragm is attached to a current-carrying voice coil that interacts with a constant magnetic field to give the in-and-out motion required for the production of sound waves. These waves correspond to the audio-frequency current flowing through the voice coil. In a permanent-magnet loudspeaker the constant magnetic field is produced by a permanent magnet, while in an excited-field loudspeaker it is produced by a field coil. Also called dynamic speaker and moving-coil loudspeaker.

**dynamic memory** *Dynamic storage.*

**dynamic microphone** A moving-conductor microphone in which the flexible diaphragm is attached to a coil positioned in the fixed magnetic field of a permanent magnet. When sound waves move the diaphragm back and forth, the attached voice

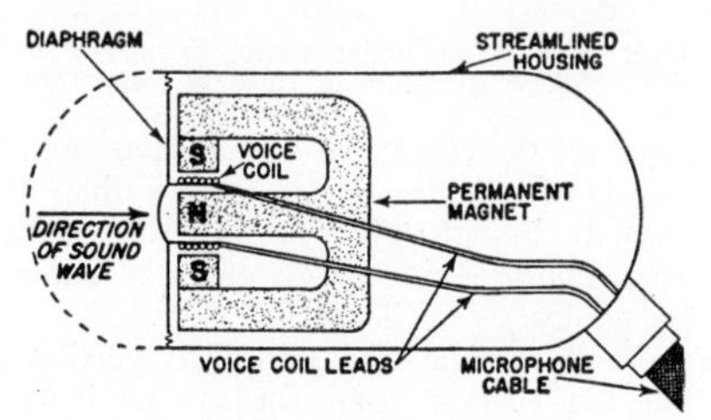


Dynamic microphone.

coil cuts magnetic lines of force. The desired a-f voltage is thus induced in the coil. Also called moving-coil microphone.

**dynamic noise suppressor** An a-f filter circuit that automatically adjusts its band-pass limits according to signal level, generally by means of reactance tubes. At low signal levels, when noise becomes more noticeable, the circuit reduces the low-frequency response and sometimes also reduces the high-frequency response.

**dynamic pickup** *Moving-coil pickup.*

**dynamic range** The ratio of the specified maximum signal level capability of a system or component to its noise level. Usually expressed in decibels.

**dynamic reproducer** *Moving-coil pickup.*

**dynamic sensitivity** The alternating component of phototube anode current divided by the alternating component of incident radiant flux.

**dynamic speaker** *Dynamic loudspeaker.*

**dynamic storage** Computer storage in which information at a certain position is not always available instantly because it is moving, as in an acoustic delay line or magnetic drum. Also called dynamic memory.

**dynamo** *Generator.*

**dynamoelectric** Pertaining to the conversion of mechanical energy to electric energy, or vice versa.

**dynamoelectric amplifier** A generator that serves as a power amplifier at low frequencies or direct current. The input signal is applied to the stationary field to change the excitation, and the amplified output is taken from the rotating armature.

**dynamometer-type instrument** An instrument in which current, voltage, or power is measured by the force between a fixed coil and a moving coil.

**dynamotor** A rotating electric machine having two or more windings on a single armature containing a commutator for d-c operation and slip rings for a-c operation. When one type of power is fed in for motor

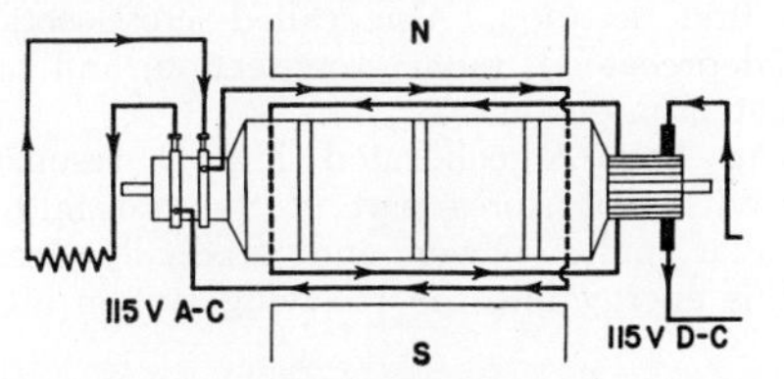


Dynamotor operating principle. Here d-c input power drives motor section, and generator section feeds a-c power to resistor load.

operation, the other type of power is delivered by generator action. Used chiefly as an inverter for converting d-c to a-c to operate electronic equipment from a storage battery in an automobile, boat, or other vehicle or location. Also called converter, rotary converter, and synchronous inverter.

**dynatron oscillator** An oscillator in which secondary emission of electrons from the anode of a screen-grid tube causes the

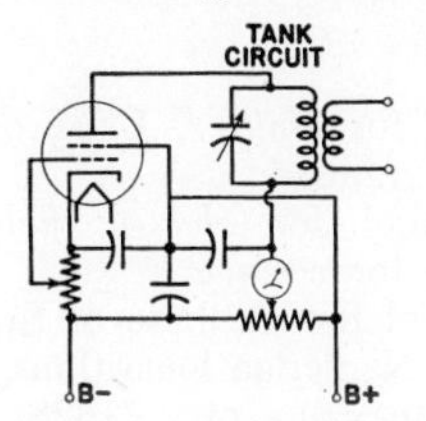


Dynatron oscillator circuit.

anode current to decrease as anode voltage is increased, giving the negative resistance characteristic required for oscillation. This action is achieved by making the screen grid more positive than the anode.

**dynode** An electrode whose primary function is secondary emission of electrons. Used in multiplier phototubes and some types of television camera tubes.

**dysprosium** [symbol Dy] A rare-earth element. Atomic number is 66.

**Dzus fastener** Trademark of Dzus Fastener Co. for a lock-type fastener used on airplane cowlings, in which the stud locks on a spring permanently attached to the mounting.

# E

**E** Symbol for *emitter.* Used on transistor circuit diagrams.

***E*** 1. Symbol for *electric field strength.* 2. Symbol for *voltage.*

**e** 1. Symbol for the base of the system of natural or Napierian logarithms, having the approximate value of 2.71828. 2. Symbol for the instantaneous value of an alternating voltage.

**early-warning radar** A long-range search radar used near the periphery of a defended area to detect approaching aircraft.

**earphone** A small, lightweight electroacoustic transducer that fits inside the ear, to function the same as a headphone or telephone receiver. Used chiefly with hearing aids.

**earphone coupler** A shaped cavity used for testing earphones. A microphone is mounted inside to measure pressures developed in the cavity.

**earth** British term for *ground.*

**earthed** British term for *grounded.*

**earth satellite** A body that orbits about the earth. When placed in orbit by man, it is called an artificial earth satellite.

**earth-space service** A radio communication service between earth stations and space stations.

**earth station** A radio communication station in the earth-space service, located either on the earth's surface or on an object that is limited to flight between points on the earth's surface.

**E bend** *E-plane bend.*

**eccentric circle** *Eccentric groove.*

**eccentric groove** A locked groove whose center is different from that of a disk record, to provide an in-and-out motion of the pickup arm for actuating the trip mechanism of an automatic record changer at the end of the record. Also called eccentric circle.

**Eccles-Jordan circuit** *Flip-flop circuit.*

**ecg** Abbreviation for *electrocardiogram.*

**echo** 1. A wave that has been reflected or otherwise returned with sufficient delay and magnitude to be perceived in some manner as a wave distinct from that directly transmitted. 2. The signal reflected by a radar target, or the trace produced by this signal on the screen of the cathode-ray tube in a radar receiver. Also called radar echo and return. 3. A sound wave reflected from a hard surface. 4. *Ghost signal.*

**echo area** The radar target area that effectively returns echoes to the transmitter location. Also called cross-section (deprecated), radar cross-section, and target cross-section.

**echo box** A calibrated high-*Q* resonant cavity that stores part of the transmitted radar pulse power and gradually feeds this energy into the receiving system after completion of the pulse transmission. Used to provide an artificial target signal for test and tuning purposes. Also called phantom target.

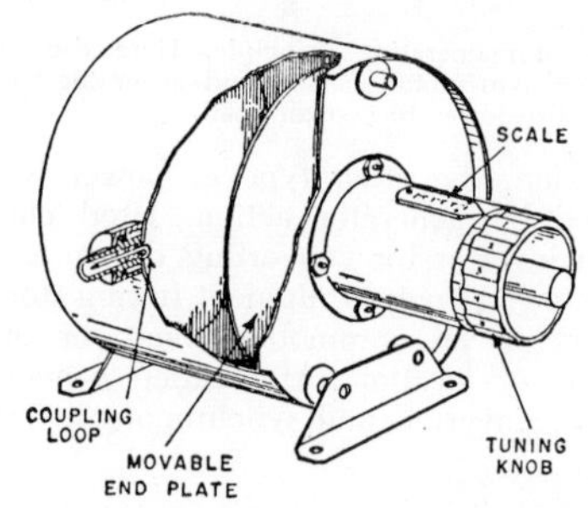


Echo box for coaxial cable feed from 10-cm radar.

**echo chamber** A reverberant room or enclosure used in a studio to add echo effects to sounds for radio or television programs.

**echo checking** A checking system in which the transmitted information is reflected back to the transmitter and compared with that which was transmitted.

**echo depth sounder** *Fathometer.*

**echoing area** *Backscattering coefficient.*

**echo matching** Rotating a radar antenna or antenna array to a position at which the two echoes corresponding to the two directions of an echo-splitting radar are equal.

**echo-ranging sonar** Active sonar, in which underwater sound equipment generates bursts of ultrasonic sound and picks up echoes reflected from submarines, fish, and other objects within range, to determine both direction and distance to each target.

**echo-repeater target** An electronic target used to simulate a submarine for testing sonar-equipped homing torpedoes. It consists of two transducers lowered over the side of a small target boat to the desired test depth. One transducer picks up the sonar pulse from the torpedo. After amplification, this signal is fed to the other transducer to send out an echo approximating that which would be reflected from an actual submarine at that location.

**echo-splitting radar** Radar in which the echo is split by special circuits associated with the antenna lobe-switching mechanism, to give two echo indications on the screen of the radarscope. When the two echo indications are equal in height, the target bearing is read from a calibrated scale.

**echo suppressor** 1. A circuit that desensitizes electronic navigation equipment for a fixed period after the reception of one pulse, for the purpose of rejecting delayed pulses arriving from indirect reflection paths. 2. A relay or other device used on a transmission line to prevent a reflected wave from returning to the sending end of the line.

**ecm** Abbreviation for *electronic countermeasure.*

**eco** Abbreviation for *electron-coupled oscillator.*

**eddy current** A circulating current induced in a conducting material by a varying magnetic field. These currents are undesirable in most instances because they represent loss of energy and cause heat. Laminations are used for the iron cores of transformers, filter chokes, and a-c relays to shorten the paths for eddy currents and thus keep eddy-current losses at a minimum. At radio frequencies, eddy-current paths must be broken up still more by using powdered iron cores.

**eddy-current heating** *Induction heating.*

**eddy-current loss** Energy loss due to undesired eddy currents circulating in a magnetic core.

**edge effect** An outward-curving distortion of lines of force near the edges of two parallel metal plates that form a capacitor. A correction must be made for this effect when computing capacitance from the geometry of a structure, or a special guard ring must be used to eliminate the effect.

**edgewise bend** A bend in a rectangular waveguide such that the longitudinal axis remains in a plane that is parallel to the wide side of the waveguide.

**Edison base** The standard screw-thread base used for ordinary electric lamps in the United States.

**Edison effect** The emission of electrons from hot bodies. The rate of emission increases rapidly with temperature. Discovered by Edison in 1883, when a current flow was obtained between the filament of an incandescent lamp and an auxiliary electrode inside the lamp. Also called Richardson effect.

**Edison storage battery** An alkaline storage battery that produces an open-circuit voltage of 1.2 volts per cell. The active material on the negative plates is an iron alloy, whereas that on the positive plates is nickel oxide.

**E display** A rectangular radarscope display in which targets appear as bright spots, with target distance indicated by the horizontal coordinate and target elevation by the vertical coordinate. Also called E scan.

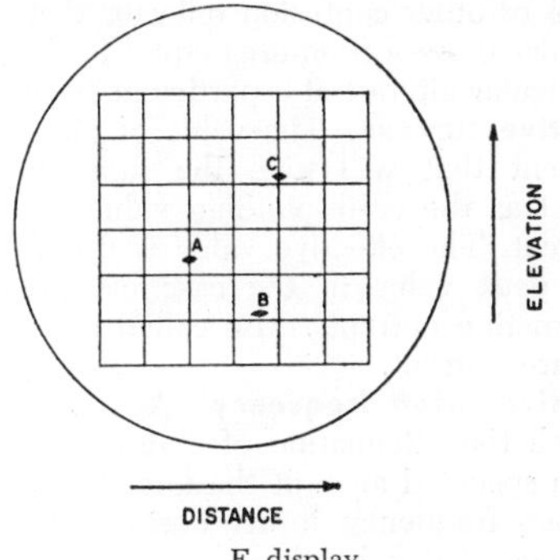


E display.

**edit** To arrange or rearrange digital computer output information before printing it out. Editing may involve deletion of unwanted data, selection of pertinent data, insertion of invariant symbols such as page numbers and typewriter characters, and the application of standard processes such as zero-suppression.

**eeg** Abbreviation for *electroencephalogram.*

**effective acoustic center** The point on or near a loudspeaker or other acoustic generator from which spherically divergent sound waves appear to diverge.

**effective antenna length** The electrical length of an antenna, as distinguished from its physical length.

**effective area** A directional antenna rating, equal to the square of the wavelength multiplied by the power gain (or directive gain) of an antenna in a specified direction, and the result divided by $4\pi$.

**effective atomic number** A number calculated from the composition and atomic numbers of a compound or mixture.

**effective band** The frequency band of a facsimile signal wave equal in width to that between zero frequency and maximum keying frequency.

**effective bandwidth** The bandwidth of an assumed rectangular bandpass filter having the same transfer ratio at a reference frequency as a given actual bandpass filter, and passing the same mean-square value of a hypothetical current having even distribution of energy throughout that bandwith.

**effective bunching angle** The transit angle that would be required in a hypothetical reflex klystron drift space in which the potentials vary linearly over the same range as in the given drift space and in which the bunching action is the same as in the given drift space.

**effective confusion area** The amount of chaff or other confusion reflector that gives a radar cross-section area equal to that of a particular aircraft at a particular frequency.

**effective current** The value of alternating current that will give the same heating effect as the corresponding value of direct current. The effective value is 0.707 times the peak value in the case of sine-wave alternating currents. Also called root-mean-square current.

**effective cutoff frequency** A frequency at which the attenuation of a device exceeds by a specified amount the loss at some reference frequency in its normal operating or transmission band.

**effective echoing area of target** The area of a hypothetical, perfect radar target, perpendicular to the incident beam, that would produce at the receiver a signal equal to that produced by the actual target. For an average aircraft it is generally from 1 to 10 square meters.

**effective half-life** The half-life of a radioisotope in a biological organism, resulting from a combination of radioactive decay and biological elimination.

**effective height** The height of the center of radiation of a transmitting antenna above the effective ground level.

**effective particle velocity** The root-mean-square value of the instantaneous particle velocities at a point. Also called root-mean-square particle velocity.

**effective percentage modulation** The ratio, for a single sinusoidal input component, of the peak value of the fundamental component of the envelope to the d-c component in the modulated condition, expressed as a percentage.

**effective radiated power** The product of antenna input power and antenna power gain, expressed in kilowatts.

**effective radius of earth** A radius value used in place of the geometric radius to correct for atmospheric refraction when the index of refraction in the atmosphere changes linearly with height. Under conditions of standard refraction the effective radius of the earth is 8,500,000 meters, or 4/3 the geometric radius.

**effective resistance** *High-frequency resistance.*

**effective sound pressure** The root-mean-square value of the instantaneous sound pressures at a point during a complete cycle, expressed in dynes per square centimeter. Also called pressure, root-mean-square sound pressure, and sound pressure.

**effective value** *Root-mean-square value.*

**effective wavelength** The wavelength of a monochromatic x-ray that undergoes the same percentage attenuation in a specified filter as the heterogeneous x-ray beam under consideration.

**efficiency** 1. The ratio of useful output of a device to total input, generally expressed as a percentage. 2. The probability that a count will be produced in a counter tube by a specified particle or quantum incident.

**ehf** Abbreviation for *extremely high frequency.*

**eht** Abbreviation for *extra-high tension.*

**E-H tee** A waveguide junction composed of a combination of E-plane and H-plane tee junctions having a common point of intersection with the main waveguide.

**E-H tuner** An E-H tee used for impedance transformation, having two arms terminated in adjustable plungers.

**EIA** Abbreviation for *Electronic Industries Association.*

**EIA color code** One of the systems of color markings developed by the Electronic Industries Association for specifying electrical values and terminal connections of resistors, capacitors, and other components. Formerly called RETMA color code and RMA color code.

**einsteinium** [symbol Es] A transuranic radioactive element produced artificially by bombardment of uranium or plutonium with nuclei. Atomic number is 99.

**ekg** Abbreviation for *electrocardiogram.*

**elastance** The reciprocal of capacitance, measured in darafs.

**elastic collision** A collision of nuclear particles in which the physical contents and energies of the colliding systems are unchanged, although the directions of their relative motions may be changed.

**elastic scattering** Scattering in which the kinetic energy of neutron plus nucleus is unchanged by the collision, and the nucleus itself is unchanged.

**E layer** A layer of ionized air occurring at various heights in the E region of the ionosphere, capable of bending radio waves back to earth. Also called Heaviside layer and Kennelly-Heaviside layer.

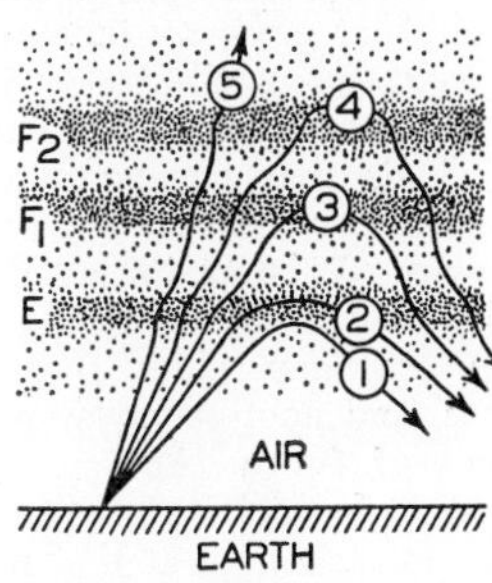


E layer, closest to earth, reflects signals along paths 1 and 2. F layers give reflection for paths 3 and 4. Signals along path 5 pass through all layers into outer space.

**elbow** *Corner.*

**electra** A continuous-wave radio navigation aid that uses special radio beacons to provide a number (usually 24) of equisignal zones. Electra is similar to sonne except that in sonne the equisignal zones as a group are periodically rotated in bearing.

**electret** A permanently polarized piece of dielectric material, produced by heating the material and placing it in a strong electric field during cooling. Some barium titanate ceramics can be polarized in this way, as also can carnauba wax and mixtures of certain other organic waxes. The electric field of an electret corresponds to the magnetic field of a permanent magnet.

**electric** Containing, producing, arising from, or actuated by electricity. Examples are electric energy, electric lamp, and electric motor. Often used interchangeably with electrical.

**electrical** Related to or associated with electricity, but not containing it or having its properties or characteristics. Examples are electrical engineer, electrical handbook, and electrical rating. Often used interchangeably with electric.

**electrical angle** The angle that specifies a particular instant in an a-c cycle. Usually expressed in degrees. One cycle is equal to 360°, hence a quarter-cycle is 90°. The phase difference between two alternating quantities is expressed as an electrical angle.

**electrical center** The point that divides a component into two equal electrical values.

**electrical degree** One 360th of a cycle of an alternating quantity.

**electrical distance** The distance between two points, expressed in terms of the duration of travel of an electromagnetic wave in free space between the two points. A convenient unit is the light-microsecond, which is approximately 983 feet or 300 meters.

**electrical engineer** An engineer whose training includes a degree in electrical engineering from an accredited college or university (or comparable knowledge and experience), to prepare him for dealing with the generation, transmission, and utilization of electric energy.

**electrical length** The length of a conductor expressed in wavelengths, radians, or degrees.

**electrical zero** A standard reference position from which rotor angles are measured in synchros and other rotating devices.

**electric anesthesia** Anesthesia produced by electric means, as with interrupted direct current.

**electric arc** A discharge of electricity through a gas, normally characterized by a voltage drop approximately equal to the ionization potential of the gas. Also called arc.

**electric axis** The *x* axis in a quartz crystal. There are three in a crystal, each parallel to one pair of opposite sides of the hexagon. All pass through and are perpendicular to the optical or *z* axis.

**electric cell** *Cell.*

**electric charge** *Charge.*

**electric circuit** A path or group of interconnected paths capable of carrying electric currents.

**electric conduction** The conduction of electricity by means of electrons, ionized atoms, ionized molecules, or semiconductor holes.

**electric contact** A physical contact that permits current flow between conducting parts.

**electric control** The control of a machine or device by switches, relays, or rheostats, as contrasted to electronic control by electron tubes or devices that do the work of electron tubes.

**electric controller** A device that governs

in some predetermined manner the electric power delivered to apparatus.

**electric coupling** A rotating machine in which the torque is transmitted or controlled by electric or magnetic means.

**electric current** *Current.*

**electric dipole** A pair of equal and opposite charges an infinitesimal distance apart.

**electric-discharge machining** A metal-cutting process in which high-frequency discharges from a negatively charged metal tool remove metal from the work piece by electro-erosion. There is no electrolyte, but the work is submerged in oil to flush away eroded particles and to delay each spark until peak energy is built up.

**electric displacement** *Electric flux density.*

**electric displacement density** *Electric flux density.*

**electric doublet** *Dipole.*

**electric dynamometer** An electric generator or motor equipped with means for indicating torque.

**electric eye** Slang term for *photocell.*

**electric field** 1. The region around an electrically charged body in which other charged bodies are acted on by an attracting or repelling force. 2. The electric component of the electromagnetic field associated with radio waves and with electrons in motion.

**electric field strength** [symbol $E$] The magnitude of the electric field vector.

**electric field vector** The force on a stationary positive charge per unit charge at a point in an electric field. Usually measured in volts per meter. Also called electric vector.

**electric flux** *Electric line of force.*

**electric flux density** A vector whose magnitude is equal to the charge per unit area that would appear on one face of a thin metal plate which is placed at a point in an electric field and oriented for maximum charge. The vector is perpendicular to the plate and directed from the negative to the positive face. Also called electric displacement and electric displacement density.

**electric forming** The process of applying electric energy to a semiconductor device to modify permanently its electrical characteristics.

**electric guitar** A guitar in which a contact microphone placed under the strings picks up the acoustic vibrations for amplification and reproduction by a loudspeaker. Volume and tone controls are usually provided.

**electric hygrometer** An instrument for indicating by electric means the humidity of the ambient atmosphere. Usually based on the relation between the electric conductance of a film of hygroscopic material and its moisture content.

**electric image** An array of electric charges, either stationary or moving, in which the density of charge is proportional at each point to the light values at corresponding points in an optical image to be reproduced.

**electricity** A fundamental quantity in nature, consisting of electrons and protons at rest or in motion. Electricity at rest has an electric field that possesses potential energy and can exert force. Electricity in motion (an electric current) has both electric and magnetic fields that possess potential energy and can exert force.

**electric lamp** A lamp in which light is produced by electricity, as in the incandescent lamp, arc lamp, glow lamp, mercury-vapor lamp, and fluorescent lamp.

**electric line of force** An imaginary line, each segment of which represents the direction of the electric field at that point. Also called electric flux.

**electric megaphone** *Electronic megaphone.*

**electric motor** *Motor.*

**electric noise** Unwanted electric energy in a receiver or transmission system, other than crosstalk.

**electric phonograph** A phonograph in which the turntable is driven by an electric motor and the output of the pickup is fed to an a-f amplifier and loudspeaker.

**electric precipitation** The collecting of dust or other finely divided particles of matter by charging the particles inductively with an electric field, then attracting them to highly charged collector plates. Also called electrostatic precipitation.

**electric scanning** Scanning in which the required changes in radar beam direction are produced by variations in phase and/or amplitude of the currents fed to the various elements of the antenna array.

**electric sheet** Special sheet iron or steel, generally alloyed with silicon, from which laminations are punched for transformers and other iron-core devices.

**electric shock** *Shock.*

**electric shock therapy** *Electroshock therapy.*

**electric storm** 1. A meteorological disturbance in which the air is highly charged with electricity, occurring in fine weather, without clouds or rain, and often accompanied by dry, dusty winds. 2. A sudden change in the pattern of earth currents, causing interference with radio reception.

**electric strength** *Dielectric strength.*

**electric tachometer** An instrument for measuring rotational speed by measuring the output voltage of a generator driven by the rotating shaft.

**electric telemetering** *Telemetering.*

**electric thermometer** An instrument that utilizes electric means to measure temperature, such as a thermocouple or resistance thermometer.

**electric transcription** *Transcription.*

**electric transducer** A transducer in which all of the waves concerned are electric.

**electric tuning** Tuning a receiver to a desired station by switching a set of preadjusted trimmer capacitors or coils into the tuning circuits.

**electric twinning** A defect occurring in natural quartz crystals, in which adjacent regions of quartz have their electric axes oppositely poled. Each type of axis is usable, but both cannot be in the same plate. During manufacture, the crystal is cut apart on the dividing line.

**electric vector** *Electric field vector.*

**electric wind** *Convective discharge.*

**electrification** The process of establishing a charge in an object.

**electroacoustic** Pertaining to a device involving both electricity and acoustics, such as a loudspeaker or microphone.

**electroacoustic transducer** A transducer that receives waves from an electric system and delivers waves to an acoustic system, or vice versa.

**electrocardiogram** [abbreviated ecg or ekg] A graphic record made by an electrocardiograph, showing how the voltages due to heart action vary with time. Also called cardiogram.

**electrocardiograph** An instrument for recording the changes in voltage that occur in the human body in synchronism with heartbeats.

**electrochemical recording** Recording by means of a chemical reaction brought about by the passage of signal-controlled current through the sensitized portion of the record sheet.

**electrocoagulation** The coagulation of tissue by means of a high-frequency electric current.

**electrode** 1. A conducting element that performs one or more of the functions of emitting, collecting, or controlling the movements of electrons or ions in an electron tube, or the movements of electrons or holes in a semiconductor device. 2. A terminal or surface at which electricity passes from one material or medium to another, as at the electrodes of a battery, electrolytic capacitor, or welder. 3. One of the terminals used in dielectric heating or diathermy for applying the high-frequency electric field to the material being heated.

**electrode admittance** An admittance value for a particular electrode of an electron tube.

**electrode capacitance** A capacitance value for a particular electrode of an electron tube.

**electrode characteristic** The relation between voltage and current for an electrode of an electron tube, all other electrode voltages being maintained constant.

**electrode conductance** The in-phase or real part of electrode admittance.

**electrode current** The current passing to or from an electrode of an electron tube through the interelectrode space, such as cathode current, grid current, and anode current.

**electrode dark current** The electrode current that flows when there is no radiant flux incident on the photocathode in a phototube or camera tube.

**electrode dissipation** The power dissipated as heat by an electrode as a result of electron and/or ion bombardment in an electron tube.

**electrode impedance** The reciprocal of electrode admittance.

**electrodeless discharge** A luminous discharge produced by a high-frequency electric field in a gas-filled glass tube having no internal electrodes.

**electrodeposition** The process of depositing a substance on an electrode by electroplating or electroforming. Also called electrolytic deposition.

**electrode potential** 1. The instantaneous voltage of an electrode with respect to the cathode of an electron tube. 2. The voltage existing between an electrode and the solution or electrolyte in which it is immersed.

**electrode radiator** A metal structure, often of large area, that is used as an external extension of an electrode of an electron tube to facilitate the dissipation of heat.

**electrode reactance** The imaginary component of electrode impedance.

**electrode resistance** The reciprocal of electrode conductance. This is the effective parallel resistance and is not the real component of electrode impedance.

**electrode voltage** The voltage between

an electrode and the cathode or a specified point on a filamentary cathode of an electron tube.

**electrodynamic instrument** An instrument that depends for its operation on the reaction between the current in one or more movable coils and the current in one or more fixed coils.

**electrodynamic loudspeaker** *Excited-field loudspeaker.*

**electroencephalogram** [abbreviated eeg] A graphic record of how brain voltages vary with time.

**electroencephalograph** An instrument for recording the waveforms of voltages developed in the brain, using electrodes applied to the scalp.

**electroflor** A material that changes color when electrically activated, but does not radiate light.

**electroforming** 1. The electrodeposition of metal on a conducting mold in sufficient thickness to make a desired metal object, such as a complex waveguide structure. The mold is often of graphite-coated wax so it can be removed by melting. 2. Production of a p-n junction in a point-contact diode or transmitter by passing a large current pulse through the semiconductor material.

**electroless deposition** Chemical deposition of a metal on a material, without electrolytic or electroplating action.

**electroluminescence** Luminescence produced in a gas by electric discharge, or in a solid by an electric field, electron beam, x-rays, or other radiation.

**electrolysis** The production of chemical changes by passing current from an electrode to an electrolyte, or vice versa, as in electroplating, electroforming, or electropolishing. Used also in separating isotopes, as in the concentration of deuterium (heavy water) by electrolysis of ordinary water.

**electrolyte** A liquid, paste, or other conducting medium in which the flow of electric current takes place by migration of ions.

**electrolytic capacitor** A capacitor consisting of two electrodes separated by an electrolyte. A dielectric film, usually a thin layer of gas, is formed on the surface of one electrode.

**electrolytic cleaning** Alkaline cleaning during which a current is passed through the cleaning solution and the metal to be cleaned.

**electrolytic deposition** *Electrodeposition.*

**electrolytic iron** A pure iron having excellent magnetic properties, produced by an electrolytic process.

**electrolytic recording** Electrochemical recording in which the chemical change is made possible by the presence of an electrolyte.

**electrolytic rectifier** A rectifier consisting of metal electrodes in an electrolyte, in which rectification of alternating current is accompanied by electrolytic action. A polarizing film formed on one of the electrodes permits current flow in one direction but not the other.

**electrolytic switch** A switch having two electrodes projecting into a chamber containing a precisely measured quantity of a conductive electrolyte, leaving an air bubble of predetermined width. When the switch is tilted from true horizontal, the bubble shifts position and changes the amount of electrolyte in contact with the electrodes, thereby changing the amount of current passed by the switch. Used as a leveling switch in gyro systems.

**electrolytic tank** A tank in which voltages are applied to an enlarged scale model of a tube electrode system immersed in a poorly conducting liquid. The equipotential lines between electrodes are traced with measuring probes, as an aid to electron-tube design.

**electromagnet** A magnet consisting of a coil wound around a soft iron or steel core. The core is strongly magnetized when current flows through the coil, and is almost completely demagnetized when the current is interrupted. Used for attracting a movable external iron object such as the armature of a relay. In a solenoid, the iron core itself is movable.

**electromagnetic** Pertaining to the combined electric and magnetic fields associated with movements of electrons through conductors.

**electromagnetic cathode-ray tube** A

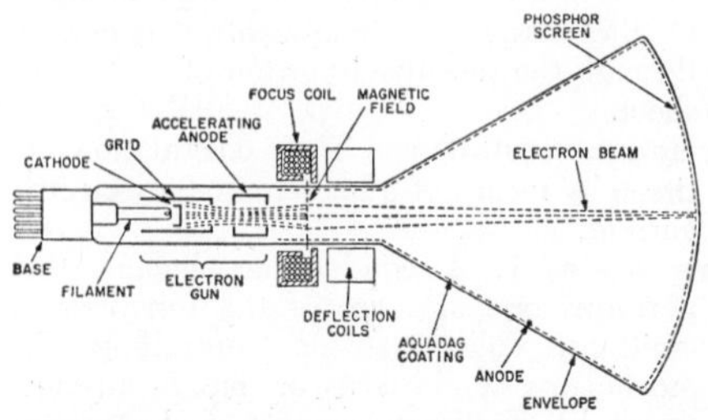


Electromagnetic cathode-ray tube with electromagnetic focus coil.

cathode-ray tube in which electromagnetic deflection is used to deflect the electron beam.

**electromagnetic constant** The speed of propagation of electromagnetic waves in a vacuum. Latest measurements, using microwave techniques, give a value of 299,793 kilometers per second.

**electromagnetic crack detector** An instrument that detects cracks in iron or steel objects by applying a strong magnetizing force to the object and measuring the resulting magnetic flux through the object. When a flawed portion passes through the magnetizing coil, the magnetic flux drops.

**electromagnetic deflection** Deflection of an electron stream by means of a magnetic field. In a television picture tube, the magnetic fields for horizontal and vertical deflection of the electron beam are produced by sending sawtooth currents through coils in a deflection yoke that goes around the neck of the picture tube.

**electromagnetic energy** Energy associated with radio waves, heat waves, light waves, x-rays, and other types of electromagnetic radiation.

**electromagnetic field** The field associated with electromagnetic radiation, consisting of a moving electric field and a moving magnetic field acting at right angles to each other and at right angles to their direction of motion.

**electromagnetic flowmeter** A flowmeter that offers no obstruction to liquid flow. Two coils produce an electromagnetic field

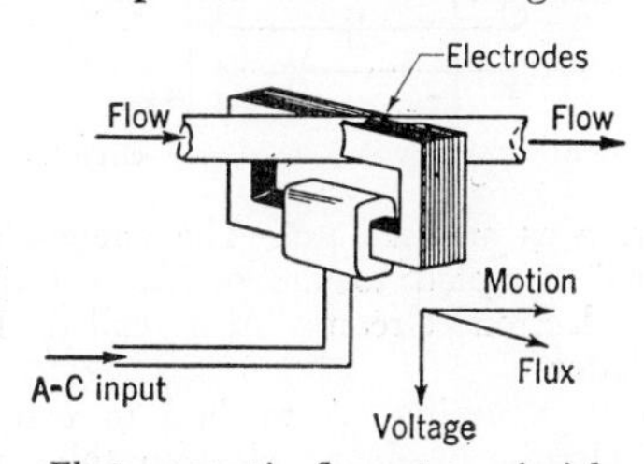


Electromagnetic flowmeter principle.

in the conductive moving fluid. The current induced in the liquid, detected by two electrodes, is directly proportional to the rate of flow.

**electromagnetic focusing** Focusing the electron beam in a television picture tube by means of a magnetic field parallel to the beam, produced by sending an adjustable value of direct current through a focusing coil mounted on the neck of the tube.

**electromagnetic horn** A horn-shaped antenna structure used to provide highly directional radiation characteristics. Signal power is fed to the horn by a waveguide or by an exciting dipole or loop at the input end of the horn.

**electromagnetic induction** The production of a voltage in a coil by a change in the number of magnetic lines of force passing through the coil.

**electromagnetic loudspeaker** *Magnetic-armature loudspeaker.*

**electromagnetic plane wave** A transverse electric wave, transverse electromagnetic wave, or transverse magnetic wave.

**electromagnetic pump** A pump in which a conductive liquid is made to move through a pipe by sending a large current transversely through the liquid. This current reacts with a magnetic field that is at right angles to the pipe and to current flow, to move the current-carrying liquid conductor just as a solid conductor is moved in an electric motor. Used for moving liquid sodium metal through the pipes of nuclear reactor cooling systems.

**electromagnetic radiation** Radiation associated with a periodically varying electric and magnetic field that is traveling at the speed of light, including radio waves, light waves, x-rays, and gamma radiation.

**electromagnetic relay** A relay in which current flow through a coil produces a magnetic field that results in contact actuation.

**electromagnetic separation** Separation of ions of varying mass by a combination of electric and magnetic fields. In the most common application an isotopic mixture of ions is produced either by electron bombardment of a gas or by thermionic emission. The ionized particles are accelerated and collimated into a beam by a system of electrodes, and the beam is projected into a magnetic field where the paths of the ions depend on their mass-to-charge ratio. Properly located collectors can be placed to receive ions of specified masses, as in the mass spectrograph.

**electromagnetic spectrum** The total range of wavelengths or frequencies of electromagnetic radiation, extending from the longest radio waves to the shortest known cosmic rays. Also called spectrum.

**electromagnetic units** [abbreviated emu] An obsolete system of electrical units.

**electromagnetic wave** A wave of electromagnetic radiation, characterized by variations of electric and magnetic fields.

**electromagnetism** Magnetism produced by an electric current rather than by a permanent magnet.

**electromechanical recording** Recording by means of a signal-actuated mechanical device.

**electromechanical transducer** A transducer for receiving waves from an electric system and delivering waves to a mechanical system, or vice versa.

**electrometer** An instrument for measuring voltage without drawing appreciable current. Older electrometer designs are based on the electrostatic force exerted between bodies that are charged with the voltage to be measured, such as suspended parallel strips of gold leaf. Modern vacuum-tube electrometers are essentially voltage-measuring amplifiers having such a high input resistance (usually above $10^{10}$ ohms) that they draw negligible current.

**electrometer tube** A high-vacuum electron tube having a high input impedance (low control-electrode conductance) to facilitate measurement of extremely small direct currents or voltages.

**electromotance** *Electromotive force.*

**electromotive force** [abbreviated emf] The force that tends to produce an electric current in a circuit. Usually called voltage. Also called electromotance.

**electromotive series** An arrangement of the metal elements in the order of the amount of electromotive force (voltage) set up between metal and solution when the metal is placed in a normal solution of any of its salts. Each metal is negative to those ahead of it in the list, and positive to those following it. The series for the more common metals is: sodium, magnesium, aluminum, manganese, zinc, chromium, iron, cadmium, cobalt, nickel, tin, lead, antimony, copper, silver, mercury, platinum, and gold.

**electromyograph** An instrument for recording muscular action currents or physical movements during muscular contractions.

**electron** An elementary negative particle having a mass of $9.107 \times 10^{-28}$ gram and a charge of $4.802 \times 10^{-10}$ statcoulomb. It is the smallest electric charge that can exist. Also called negatron. An electron may also have a positive charge, in which case it is usually called a positron. The term electron was first used in an article by George J. Stoney in the July 1891 issue of *The Scientific Transactions of the Royal Dublin Society,* "On the Cause of Double Lines and of Equidistant Satellites in Spectra of Gases."

**electronarcosis** *Electroshock therapy.*

**electron beam** A narrow stream of electrons moving in the same direction, all having about the same velocity.

**electron-beam magnetometer** A magnetometer that depends for its operation on the change in intensity or direction of an electron beam that passes through the magnetic field to be measured.

**electron-beam tube** An electron tube whose performance depends on the formation and control of one or more electron beams.

**electron binding energy** *Ionization voltage.*

**electron capture** A radioactive transformation of a nuclide in which a bound electron merges with its nucleus. This decreases the atomic number by 1 but leaves the mass number unchanged in the new nuclide. A proton is transformed to a neutron within the nucleus, a bound electron is taken up, and a neutrino emerges. Examples are K-electron, L-electron, and M-electron capture.

**electron collector** *Collector.*

**electron-coupled oscillator** [abbreviated eco] An oscillator employing a multigrid tube in which the cathode and two grids

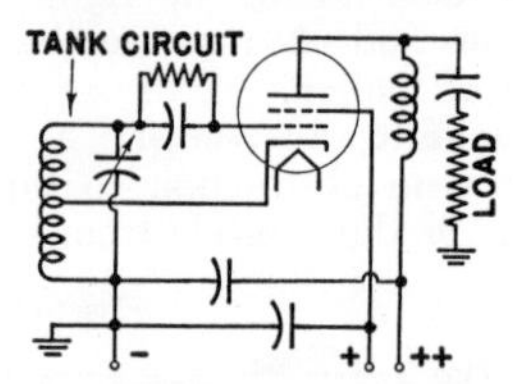

Electron-coupled oscillator circuit.

operate as an oscillator. The anode-circuit load is coupled to the oscillator through the electron stream. Also called Dow oscillator.

**electron coupling** A method of coupling two circuits inside an electron tube, used principally with multigrid tubes. The electron stream passing between electrodes in one circuit transfers energy to electrodes in the other circuit.

**electron device** A device in which conduction is principally by electrons moving through a vacuum, gas, or semiconductor, as in a crystal diode, electron tube, transistor, or selenium rectifier.

**electron diffraction camera** A camera used to obtain a photographic record of the position and intensity of the diffracted beams produced when a specimen is irradiated by a beam of electrons.

**electron emission** The liberation of electrons from an electrode into the surrounding space, usually under the influence of heat, light, or a high voltage. Also called emission.

**electron flow** A current produced by the movement of free electrons toward a positive terminal. The direction of electron flow is opposite to that of current.

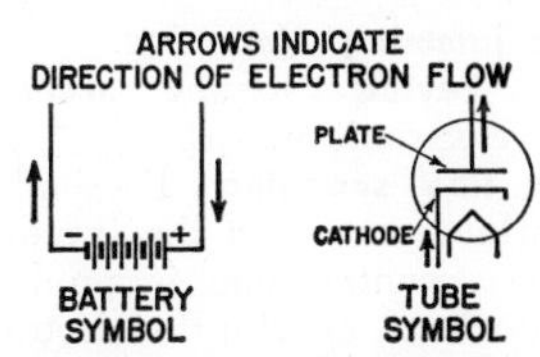

Electron flow in battery and tube.

**electron gun** An electrode structure that produces and may control, focus, deflect, and converge one or more electron beams in an electron tube. Also called gun.

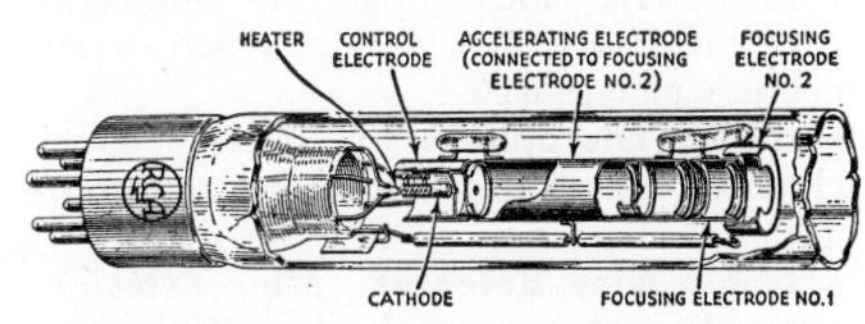

Electron gun of cathode-ray tube.

**electron-gun density multiplication** The ratio of the average current density at any specified aperture through which the stream passes to the average current density at the cathode surface.

**electronic** Pertaining to electron devices or to circuits or systems utilizing electron devices, including electron tubes, magnetic amplifiers, transistors, and other devices that do the work of electron tubes.

**electronically tuned oscillator** An oscillator in which the operating frequency is changed by changing an electrode voltage or current.

**electronic altimeter** *Radio altimeter.*

**electronic Bohr magneton** A unit of magnetic moment equal to $9.27 \times 10^{-21}$ erg gauss$^{-1}$.

**electronic bug** A bug that automatically produces both dots and dashes having the correct lengths and spacings for international Morse code signals.

**electronic calculating punch** A card-handling machine that reads a punched card, performs a number of sequential operations, and punches the result on the card.

**electronic carillon** A carillon that uses electric and electronic circuits to generate, amplify, and reproduce musical tones approximating those of bells.

**electronic chimes** A set of tubular chimes actuated by strikers electromagnetically controlled from a keyboard, with the resulting sounds being picked up, amplified, and reproduced by loudspeakers.

**electronic circuit** A circuit containing one or more electron tubes, transistors, magnetic amplifiers, or other devices providing comparable functions.

**electronic clock** A clock that uses a ferrite rod and coil to pick up the electromagnetic field of 60-cps power line wiring in a home. This 60-cps voltage is amplified by a transistor amplifier and used to control a transistor oscillator that drives a tiny permanent-magnet synchronous clock motor. Two mercury cells provide power for the transistors.

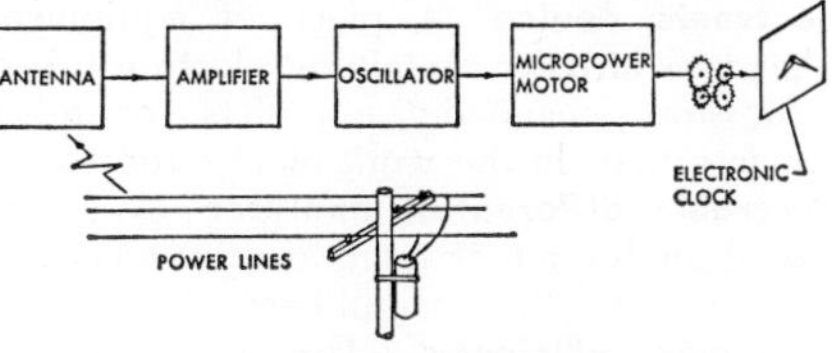

Electronic clock.

**electronic commutator** An electron-tube or transistor circuit that switches one circuit connection rapidly and in succession to many other circuits, without the wear and noise of mechanical switches. An example is the radial-beam tube, in which a rotating magnetic field causes an electron beam to sweep over one anode after another and produce the desired switching action.

**electronic computer** A computer that uses electron devices.

**electronic control** The control of a machine or process by circuits using electron tubes, transistors, magnetic amplifiers, or other devices having comparable functions.

**electronic controller** An electric controller in which some or all of the basic functions are performed by electron devices.

**electronic counter** A circuit using electron tubes or equivalent devices for counting electric pulses.

**electronic countermeasure** [abbreviated ecm] An offensive or defensive tactic using electronic and reflecting devices to reduce the military effectiveness of enemy equipment involving electromagnetic radiations.

**electronic-countermeasure reconnaissance** Aerial reconnaissance flown by aircraft equipped with electronic devices capable of locating enemy radar stations, determining their area coverage, and making radarscope photographs for combat mission folders. Also called electronic reconnaissance.

**electronic curve tracer** A photoelectric instrument in which a spot of light automatically traces along an inked line. It can be used to measure the area within a closed curve or to control a cutting torch for duplicating an irregular design.

**electronic deception** An electronic countermeasure in which electromagnetic waves are radiated or reradiated in such a way as to mislead the enemy or decoy enemy missiles away from their targets.

**electronic device** A piece of equipment that uses circuits containing electron tubes, transistors, magnetic amplifiers, or other devices that do the work of electron tubes.

**electronic differential analyzer** A differential analyzer that integrates by means of high-gain feedback amplifiers.

**electronic efficiency** The ratio of the power delivered by the electron stream to an oscillator or amplifier circuit at the desired frequency to the average power supplied to the stream in an electron tube.

**electronic flash tube** *Flash tube.*

**electronic fuel injection** The forced injection of fuel under pressure into an engine, under electronic control.

**electronic fuze** A fuze, such as the radio proximity fuze, that is set off by an electronic circuit incorporated in the fuze.

**electronic gap admittance** The difference between the gap admittance of an electron tube with the electron stream traversing the gap and the gap admittance with the stream absent.

**electronic generator** A high-power oscillator used to generate r-f energy for electronic heating. Also called electronic heater.

**electronic heater** *Electronic generator.*

**electronic heating** Heating by means of r-f current produced by an electron-tube oscillator or an equivalent r-f power source. The two types of electronic heating are induction heating for metals and dielectric heating for nonmetals. Also called high-frequency heating and r-f heating.

**Electronic Industries Association** [abbreviated EIA] A trade association made up chiefly of electronic component and equipment manufacturers. Its functions include standardization of sizes, specifications, and terminology for electronic products. Known as Radio Manufacturers Association (RMA) 1924–1950, Radio-Television Manufacturers Association (RTMA) 1950–1953, and Radio-Electronics-Television Manufacturers Association (RETMA) 1953–1957.

**electronic intelligence** Intelligence regarding the location, volume, direction, and type of enemy electronic devices, obtained chiefly by intercepting enemy signals.

**electronic jamming** *Jamming.*

**electronic keying** Keying accomplished solely by electronic means.

**electronic line scanning** Facsimile scanning in which motion of the scanning spot along the scanning line is produced by electronic means, as with the electron beam of a cathode-ray tube.

**electronic locator** *Metal detector.*

**electronic megaphone** A megaphone consisting of a microphone, audio amplifier, and horn loudspeaker built as a single unit. Also called electric megaphone.

**electronic micrometer** An electronic instrument for measuring and indicating small linear distances in air or across nonmetallic materials.

**electronic microphone** A microphone in which vibrations or sound waves act on one of the electrodes in an electron tube.

**electronic mine detector** *Mine detector.*

**electronic motor control** A control circuit used to vary the speed of a d-c motor operated from an a-c power line. Thyra-

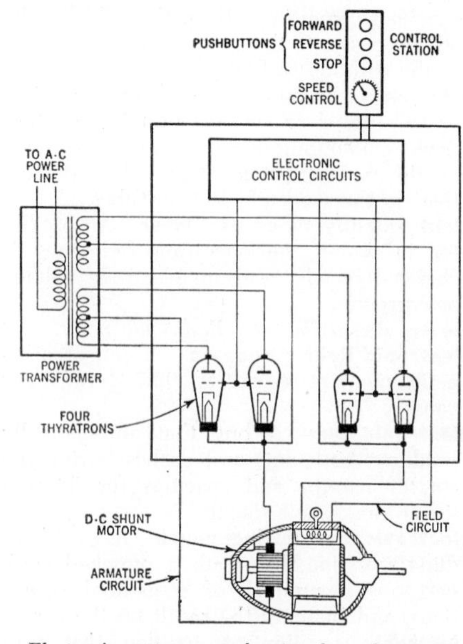


Electronic motor control using four thyratrons.

trons or ignitrons are used to rectify the a-c voltage and vary the field current of the motor. Also called d-c motor control and motor control.

**electronic music** Music consisting of tones originating in electronic sound and noise generators used alone or in conjunction with electroacoustic shaping means and sound-recording equipment. The resulting sounds may or may not resemble those of conventional musical instruments.

**electronic musical instrument** A musical instrument in which an audio signal is produced by a pickup or audio oscillator and amplified electronically to feed a loudspeaker, as in an electric guitar, electronic carillon, electronic organ, or electronic piano.

**electronic organ** A musical instrument that uses electronic circuits to produce music similar to that of a pipe organ.

**electronic photometer** *Photoelectric photometer.*

**electronic piano** A piano without a sounding board, in which vibrations of each string affect the capacitance of a capacitor microphone and thereby produce audio-frequency signals that are amplified and reproduced by a loudspeaker.

**electronic pickup** A phonograph pickup in which the output signal is produced by causing the needle to move an electrode of an electron tube.

**electronic pilotage** A method of pilotage in which landmark images appearing on a radarscope are used for guidance.

**electronic potentiometer** A potentiometer circuit that is continuously balanced by an electronic servosystem.

**electronic profilometer** An electronic instrument for measuring surface roughness. The stylus of a pickup is moved over the surface to be examined, and the resulting varying voltage is amplified, rectified, and measured with a meter calibrated to read directly in microinches of deviation from smoothness.

**electronic range** A range that uses dielectric heating to cook food.

**electronic raster scanning** Facsimile scanning in which motion of the scanning spot in both dimensions is accomplished by electronic means, as by the electron beam of a cathode-ray tube.

**electronic reconnaissance** 1. Reconnaissance with the aid of electronic devices. 2. *Electronic-countermeasure reconnaissance.*

**electronic rectifier** A rectifier in which electron tubes or equivalent selenium, silicon, copper-oxide and other semiconductor devices are used as rectifying elements.

**electronic relay** An electronic circuit that provides the function of a relay but has no moving parts.

**electronics** 1. The branch of science and technology that deals with electron devices, including electron tubes, magnetic amplifiers, transistors, and other devices that do the work of electron tubes. 2. Pertaining to the field of electronics, as in electronics consultant, electronics course, electronics engineer, electronics laboratory, and electronics training.

**electronic scanning** 1. Scanning a television image with an electron beam in a cathode-ray television camera tube, as distinguished from mechanical scanning. 2. Scanning a region in space with a stationary multielement antenna so designed that the beam can be electronically aimed by changing the current feeds to the various elements.

**electronics engineer** An engineer whose training includes a degree in electronic engineering from an accredited college or university, a degree in electrical engineering with a major in electronics, or comparable knowledge and experience as required for working with electronic circuits and devices.

**electronic sewing** The use of dielectric heating for uniting thermoplastic sheet materials.

**electronic sextant** A sextant using a highly directional radio receiving system or a photoelectric system for determining position with respect to a selected member of the solar system.

**electronics serviceman** A serviceman who is qualified to repair and maintain electronic equipment.

**electronics technician** A technician who is qualified to work under the direction of an electronics engineer in assembling, testing, and repairing electronic equipment, generally in factories and laboratories.

**electronic switch** An electronic circuit used to perform the function of a high-speed switch. Applications include switching a cathode-ray oscilloscope back and forth between two inputs at such high speed that both input waveforms appear simultaneously on the screen.

**electronic television** A television system employing cathode-ray tubes to scan the scene at the transmitter and to reconstruct it at the receiver, as used today.

**electronic timer** A timer that uses an electronic circuit to operate a relay at a predetermined interval of time after the circuit is energized, as in timing exposures for a photographic printer or in controlling an electronic generator.

**electronic tonometer** *Tonometer.*

**electronic tube** *Electron tube.*

**electronic tuning** The process of changing the operating frequency of a system by changing the characteristics of a coupled electron stream. In a reflex klystron this is done by changing the repeller voltage.

**electronic tuning range** The frequency range of continuous tuning, between two operating points of specified minimum power output, for an electronically tuned oscillator.

**electronic tuning sensitivity** The rate of change in oscillator frequency with changes in electrode voltage or current for an electronically tuned oscillator.

**electronic viewfinder** A television camera viewfinder using a small cathode-ray picture tube to show the image being picked up.

**electronic voltmeter** A voltmeter that uses the rectifying and amplifying properties of electron devices and their associated circuits to secure desired characteristics, such as high input impedance, wide frequency range, and peak indications. Called vacuum-tube voltmeter when the electron devices are vacuum tubes.

**electronic warfare** Warfare in which electronic instruments are used, as in radio communications, guided-missile control, and target detection.

**electronic wattmeter** A wattmeter using two matched electronic voltmeters to give a reading proportional to the product of two voltages, on a scale calibrated to read power values directly. One voltage is that appearing across the load, and the other is obtained across a resistor in series with the line.

**electron image** 1. An image formed in a stream of electrons. The electron density in a cross-section of the stream is at each point proportional to the brightness of the corresponding point in an optical image. 2. A pattern of electric charges on an insulating plate, with the magnitude of the charge at each point being proportional to the brightness of the corresponding point in an optical image.

**electron image tube** *Image tube.*

**electron injector** The electron gun used to inject a beam of electrons into the vacuum chamber of a mass spectrometer, betatron, or other large electron accelerator.

**electron lens** An arrangement of electrodes, with or without magnetic focusing coils, used to control the size of a beam of electrons in an electron tube.

**electron microscope** A microscope in which a beam of electrons focused by an electron lens in a vacuum chamber is sent through a thin sample of the material being examined. The sample absorbs electrons in proportion to the density at each point, so the emerging beam is an electron image of the sample. This image is magnified thousands of times by another electron lens, then made visible on a fluorescent screen or recorded on photographic film.

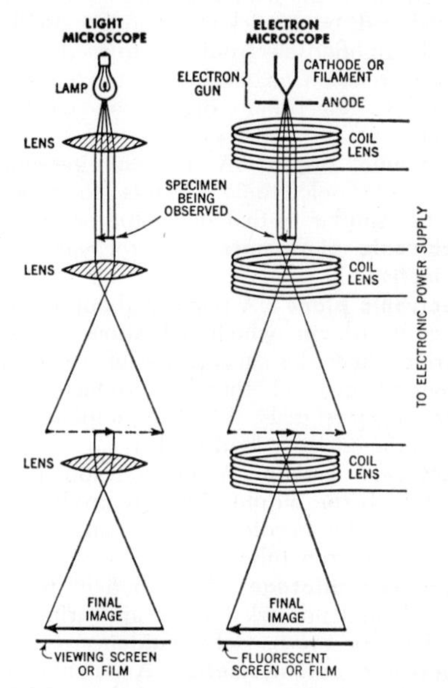

Electron microscope using electromagnetic lenses, as compared to ordinary optical microscope.

**electron multiplier** An electron-tube structure that employs secondary electron emission from solid reflecting electrodes (dynodes) to produce current amplification. The electron beam containing the desired signal current is reflected from each dynode surface in turn. At each reflection, an impinging electron releases two or more secondary electrons, so the beam builds up in strength. A typical arrangement of nine dynodes can give an amplification of several million. Used in multiplier phototubes and television camera tubes. Also called multiplier and secondary-electron multiplier.

**electron-multiplier phototube** *Multiplier phototube.*

**electron-multiplier tube** An electron tube having an electron multiplier.

**electron optics** The branch of electronics that deals with the control of electron beams in a vacuum by means of electron lenses using electric or magnetic fields, or both.

**electron paramagnetic resonance** Paramagnetic resonance involving conduction electrons in a metal or semiconductor.

**electron-positron pair** The electron and positron simultaneously created by the process of pair production.

**electron radius** The classical value of $2.82 \times 10^{-13}$ centimeter for the radius of an electron. It is obtained by equating the rest-mass energy of the electron to its electrostatic self-energy.

**electron-ray indicator** *Cathode-ray tuning indicator.*

**electron shell** The arrangement of electrons in a given orbital outside the nucleus of an atom. All electrons in a shell have the same energy level.

**electron spin** The rotation of an electron about its own axis, contributing to the total angular momentum of the electron.

**electron-spin resonance** Interaction of electric and magnetic fields with the spin of an electron about its own axis.

**electron-stream potential** The time average of the voltage between a point in an electron stream and the electron-emitting surface at which the stream originates.

**electron synchrotron** A synchrotron designed to accelerate electrons. The electron beam is allowed to strike an internal target, producing high-energy gamma rays that are used outside the machine.

**electron telescope** A telescope in which an infrared image of a distant object is focused on the photosensitive cathode of an image converter tube. The resulting electron image is enlarged by electron lenses and made visible by a fluorescent screen. An electron telescope can be used in complete darkness. The sniperscope and snooperscope are examples of early military versions.

**electron trajectory** The path of one electron in an electron tube.

**electron tube** An electron device in which conduction of electricity is provided by electrons moving through a vacuum or gaseous medium within a gastight envelope. A tube may provide rectification, amplification, modulation, demodulation, oscillation, limiting, and a variety of other functions.

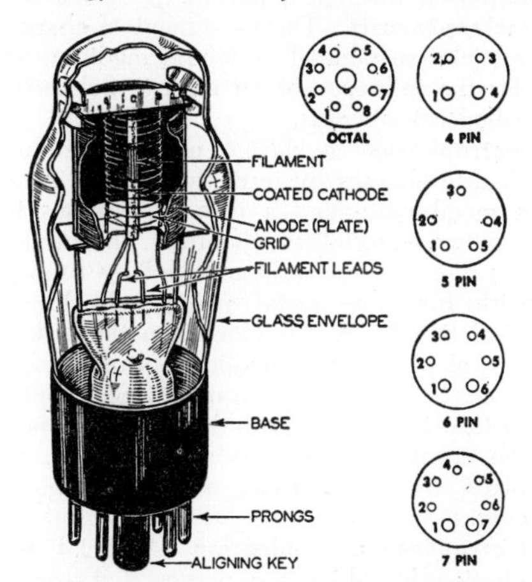


Electron-tube construction, showing typical heater-type triode with octal base, and bottom views of five different tube bases. On octal tubes, one or more base pins are sometimes omitted.

Examples include cathode-ray tubes, gas tubes, phototubes, and vacuum tubes. Also called electronic tube, radio tube, tube, and valve (British).

**electron-tube amplifier** An amplifier in which electron tubes provide the required increase in signal strength.

**electron-tube coupler** A coupler specifically designed to be inserted between an electron tube and an input or output device, as between a magnetron and a transmission line.

**electron-tube generator** A generator in which d-c energy is converted to r-f energy by an electron tube in an oscillator circuit.

**electron-volt** [abbreviated ev] A unit of energy equal to the energy acquired by an electron when it passes through a potential difference of 1 volt in a vacuum. One electron-volt is equal to $1.602 \times 10^{-12}$ erg.

**electron-wave tube** An electron tube in which mutually interacting streams of electrons having different velocities cause a signal modulation to change progressively along the length of the electron streams.

**electro-optical shutter** A shutter that uses a Kerr cell to modulate a beam of light.

**electro-osmosis** The movement of fluids through diaphragms as a result of the application of an electric current.

**electrophonic effect** The sensation of

hearing produced when an alternating current of suitable frequency and magnitude is passed through a person.

**electrophoresis** The movement of charged particles suspended in a fluid medium, under the influence of an electric field. Also called cataphoresis.

**electrophorus** A device used to produce electric charges by induction, consisting of a metal plate and a disk of resinous insulating material. In operation, the insulating disk is negatively charged by rubbing with fur. The metal plate, held by an insulating handle, is placed on the disk so it is charged by induction (bottom surface positive and top surface negative). The top surface is touched with a finger to remove the negative charge. When lifted off, the plate now has a strong positive charge all over.

**electrophrenic respiration** Artificial respiration in which the nerves that control breathing are stimulated electrically through appropriately placed electrodes. The equipment needed is commercially available in portable form and is used by many rescue squads in preference to manual methods of artificial respiration.

**electroplating** The electrodeposition of an adherent metal coating on a conductive object for protection, decoration, or other purposes. The object to be plated is placed in an electrolyte and connected to one terminal of a d-c voltage source. The metal to be deposited is similarly immersed and connected to the other terminal. Ions of the metal provide transfer of metal as they make up the current flow between the electrodes.

**electropolishing** The process of producing a smooth, lustrous surface on a metal by making it the anode in an electrolytic solution and preferentially dissolving the minute protuberances.

**electropositive** Having a positive electric polarity.

**electrorefining** The process of dissolving a metal from an impure anode by means of electrodeposition, and redepositing it in a purer state on a cathode.

**electroretinogram** A graphic record of the manner in which the voltage developed by the retina of the eye varies with time.

**electroscope** An instrument for detecting an electric charge by means of the mechanical forces exerted between electrically charged bodies. In one form, two narrow strips of gold leaf suspended in a glass jar spread apart when charged. The angle between the strips is then related to the charge.

**electrosensitive recording** Recording in which the image is produced by passing electric current through the record sheet.

**electroshock therapy** The use of an electric current, generally passed through the brain, to induce convulsions in the treatment of certain types of mental disorders. Also called electric shock therapy, electronarcosis, and shock therapy.

**electrostatic** Pertaining to electricity at rest, such as an electric charge on an object.

**electrostatic accelerator** *Electrostatic generator.*

**electrostatic actuator** An auxiliary external electrode used to apply a known electrostatic force to the diaphragm of a microphone for calibration purposes.

**electrostatically focused traveling-wave tube** *Estiatron.*

**electrostatic cathode-ray tube** A cathode-ray tube in which electrostatic deflection is used to deflect the electron beam.

**electrostatic deflection** The deflection of an electron beam by means of an electrostatic field produced by electrodes on opposite sides of the beam. Used chiefly in cathode-ray tubes for oscilloscopes. The electron beam is attracted to a positive electrode and repelled by a negative electrode.

**electrostatic field** An electric field having constant intensity, such as that produced by stationary charges.

**electrostatic focusing** A method of focusing an electron beam by the action of an electric field, as in the electron gun of a cathode-ray tube, so that the beam will have the required small area at a screen or other surface.

**electrostatic generator** A high-voltage generator in which electric charges are generated by friction or induction, then transferred mechanically to an insulated electrode to build up a voltage that may be as high as 9,000,000 volts. Examples include the Van de Graaff generator and the Wimshurst machine. Also called electrostatic accelerator and electrostatic machine.

**electrostatic induction** The process of charging an object electrically by bringing it near another charged object.

**electrostatic instrument** A meter that depends for its operation on the forces of attraction and repulsion between electrically charged bodies.

**electrostatic lens** A lens consisting of coaxial metal cylinders and pierced diaphragms operated at potentials that produce electrostatic focusing of an electron beam directed along the axis of the lens.

**electrostatic loudspeaker** A loudspeaker in which the mechanical forces are produced by the action of electrostatic fields. In one type, the fields are produced between a thin metal diaphragm and a rigid metal plate. Also called capacitor loudspeaker.

**electrostatic machine** *Electrostatic generator.*

**electrostatic memory** A memory in which information is stored in the form of the presence or absence of electrostatic charges at specific spot locations, generally on the screen of a special type of cathode-ray tube known as a storage tube. Also called electrostatic storage.

**electrostatic microphone** *Capacitor microphone.*

**electrostatic painting** A painting process that uses the particle-attracting property of electrostatic charges. A d-c voltage of about 100,000 volts is applied to a grid of wires through which the paint is sprayed, to charge each particle. The metal objects to be sprayed are connected to the opposite terminal of the high-voltage circuit, so that they attract the particles of paint and thereby minimize waste of paint.

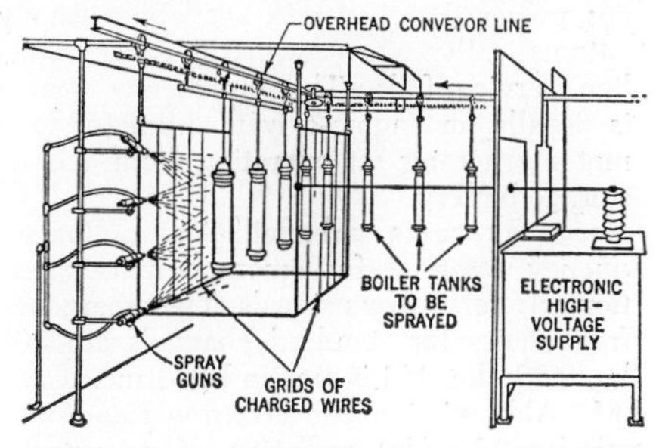


Electrostatic painting setup for hot-water boiler tanks.

**electrostatic precipitation** *Electric precipitation.*

**electrostatic precipitator** *Precipitator.*

**electrostatic radius** The nuclear radius as deduced from an analysis of nuclear binding energies, especially of mirror nuclides.

**electrostatic recording** Recording by means of a signal-controlled electrostatic field.

**electrostatics** The science that deals with electricity at rest, such as with charged objects and constant-intensity electric fields.

**electrostatic scanning** Scanning that involves electrostatic deflection of an electron beam.

**electrostatic separator** A separator in which a finely pulverized mixture falls through a powerful electric field between two electrodes. Materials having different specific inductive capacitances are deflected by varying amounts and fall into different sorting chutes.

**electrostatic shield** A grounded metal screen, sheet, or enclosure placed around a device or between two devices to prevent electric fields from acting through the shield. Often used to prevent interaction between the electric fields of adjacent parts on a chassis.

**electrostatic storage** *Electrostatic memory.*

**electrostatic storage tube** *Storage tube.*

**electrostatic tweeter** A tweeter loudspeaker in which a flat metal diaphragm is driven directly by a varying high voltage applied between the diaphragm and a fixed metal electrode.

**electrostatic unit** [abbreviated esu] A unit used in the centimeter-gram-second electrostatic system, in which the statcoulomb is the charge that repels an exactly similar charge in a vacuum with a force of 1 dyne.

**electrostatic voltmeter** A voltmeter in which the voltage to be measured is applied between fixed and movable metal vanes. The resulting electrostatic force deflects the movable vane against the tension of a spring. An attached pointer moving over a scale indicates the voltage of the circuit. Usually used for measuring high values of d-c voltage.

**electrostriction** The change in dimensions that occurs in some dielectric materials when placed in an electric field. The change is independent of the polarity of the electric field. The reverse effect does not take place, whereas in piezoelectricity the effect is reversible.

**electrotherapy** The use of electricity in treating disease.

**electrothermal instrument** An instrument that depends for its operation on the heating effect of a current.

**electrothermal recording** Recording in which the marking on the record sheet is produced principally by signal-controlled thermal action.

**electrotonus** A change in the characteristics of a nerve or muscle during or after the passage of an electric current.

**element** 1. One of the 102 known substances that cannot be divided into simpler substances by chemical means. 2. A part of an electron tube, semiconductor device, or antenna array that contributes directly to the electrical performance. 3. *Component.*

**elemental area** *Picture element.*

**elementary charge** The unit charge of electricity, corresponding to the charge on a single electron, and equal to about $4.802 \times 10^{-10}$ electrostatic unit.

**elementary particle** A term sometimes applied to an electron, proton, neutron, positron, meson, V particle, neutrino, and photon. Also called fundamental particle.

**elevated duct** A tropospheric radio duct having both its upper and lower boundaries above the ground.

**elevation** 1. *Altitude.* 2. *Elevation angle.*

**elevation angle** The angle that a radio, radar, or other beam of radiation makes with the horizontal. Also called elevation.

**elevation deviation indicator** An indicator that presents visually the relationship between a target and the elevation angle of a radar or radio beam.

**elevation indicator** A component that presents visually the angle between a fixed reference point and a target in the same vertical plane.

**elliptically polarized wave** An electromagnetic wave for which the component of the electric vector in a plane normal to the direction of propagation describes an ellipse at a given frequency.

**elliptical polarization** Polarization in which the magnitude of the vector representing the wave varies as the radius of an ellipse while the vector rotates about a point.

**ellipticity** Deprecated term for *axial ratio.*

**elmint** [ELectroMagnetic INTelligence] An Air Force system that obtains and processes electromagnetic intelligence.

**elongation** The extension of the envelope of a signal due to delayed arrival of multipath components.

**elsse** [ELectronic Sky Screen Equipment] A range safety system for ground-launched missiles, consisting of a single pair of antennas located on a baseline that passes through the launch point. As long as the missile remains in the vertical plane passing through the baseline, the direction cosine indication is zero.

**elsse cotar** A passive range instrumentation system using an elsse system on the ground to receive transmitted data from any airborne transponder, such as from an f-m/f-m telemetering transmitter in a missile.

**emanation** 1. [symbol Em] A name sometimes given to radioactive element 86, having mass number 222 and a half-life of 3.8 days. Also called emanon and radon. 2. *Radioactive emanation.*

**emanon** *Emanation.*

**embedding** The process of molding an insulating plastic around a component or assembly to form a solid block having only the leads or terminals exposed.

**embossed-foil printed circuit** A printed circuit formed by indenting the desired pattern of metal foil into an insulating base, then mechanically removing the remaining unwanted raised portions.

**embossed-groove recording** Disk recording in which a comparatively blunt stylus pushes aside the material in the modulated groove. No material is removed from the disk.

**embossing stylus** A recording stylus with a rounded tip that forms a groove by displacing material in the recording medium.

**emergency power supply** A source of 60-cps power that becomes available, usually automatically, when normal 60-cps power-line service fails. The emergency source is usually an engine-driven alternator or a motor-generator set operating from a large storage battery.

**emergency radio channel** Any radio frequency reserved for emergency use, particularly for distress signals. The emergency frequencies for standard channels are 500 kc, 8,364 kc, 121.5 mc, and 243 mc.

**emf** Abbreviation for *electromotive force.*

**emission** 1. Any radiation of energy by means of electromagnetic waves, as from a radio transmitter. 2. *Electron emission.*

**emission bandwidth** The band of frequencies comprising 99% of the total radiated power, extended to include any other discrete frequency at which the power is over 0.25% of the total radiated power.

**emission characteristic** The relation between the emission and a factor controlling the emission, such as temperature, voltage, or current of the filament or heater in an electron tube.

**emission spectrum** The spectrum produced by radiation from any emitting source, such as the spectrum of radiation from an invisible infrared source.

**emissivity** The ratio of the radiation emitted by a surface to the radiation emitted by a perfect blackbody radiator at the same temperature.
**emitron** A television camera tube similar to an iconoscope, made in Great Britain.
**emittance** The power radiated per unit area of a radiating surface.
**emitter** [symbol E] A transistor electrode from which a flow of carriers enters the interelectrode region. The emitter roughly corresponds to the cathode of an electron tube.
**emitter bias** A bias voltage applied to the emitter electrode of a transistor.
**emitter follower** A grounded-collector transistor amplifier whose operation is similar to a cathode follower using a vacuum tube.
**emitter junction** A junction between emitter and base of a semiconductor device, normally biased in the low-resistance direction to inject minority carriers into an interelectrode region.
**$E_{m,n}$ mode** British term for *$TM_{m,n}$ mode.*
**$E_{m,n}$ wave** British term for *$TM_{m,n}$ wave.*
**emphasis** *Preemphasis.*
**emphasizer** *Preemphasis network.*
**empire cloth** Cotton cloth coated with insulating varnish.
**empirical** Based on actual measurement, observation, or experience, rather than on theory.
**empty band** A band of possible energy levels in an atom, none of which correspond to the energy of any electron in the given substance in the given state.
**emu** Abbreviation for *electromagnetic unit.*
**enabling gate** A circuit that initiates the start and determines the length of a generated pulse.
**enabling pulse** A pulse that prepares a circuit for some subsequent action.
**enameled wire** Wire coated with an insulating layer of baked enamel.
**encapsulating** The process of placing a heavy protective coating on a component or assembly by dipping it in a thick insulating plastic fluid or other insulating material.
**encephalogram** A radiograph made by passing x-rays through the brain to a sensitive film.
**encephalography** Radiography of the brain.
**enclosure** A housing for loudspeakers or other electronic equipment.
**encode** 1. To express given information by means of a code. 2. To prepare a routine in machine language for a specific computer.
**encoder** *Matrix.*
**end-around carry** A computer carry signal that is sent directly to the least significant digit place when generated in the most significant digit place.
**end effect** The effect of capacitance at the ends of an antenna. It requires that the actual length of a half-wave antenna be about 5% less than a half-wavelength.
**end-fed vertical antenna** *Series-fed vertical antenna.*
**end-fire array** A linear array whose direction of maximum radiation is along the axis of the array. It may be either unidirectional or bidirectional. The elements of the array are parallel and in the same plane, as in a fishbone antenna.
**end instrument** A pickup used in telemetering to convert a physical quantity to a change of inductance, resistance, voltage, or other electrical quantity that can be transmitted over wires or by radio.
**endothermic** Involving absorption of heat or other energy.
**end-point control** Quality control through continuous, automatic analysis of the final product. In automatic control the necessary process or machine corrections are made automatically by a controller.
**end product** The final product of a reaction or process, such as the stable nuclide that is the final member of a radioactive series.
**end shield** A shield placed at each end of the interaction space of a magnetron to prevent electrons from bombarding the end seals.
**energize** To apply rated voltage.
**energized** Electrically connected to a voltage source. Also called alive, hot, and live.
**energy** Ability to do work. Energy may be transferred from one form to another, but cannot be created or destroyed.
**energy band** The sets of discrete but closely adjacent energy levels, equal in number to the number of atoms, that arise from each of the quantum states of the atoms of a substance when the atoms condense to a solid state from a nondegenerate gaseous condition. Also called Bloch band. For a semiconductor the highest energy level is the conduction band, containing only the excess electrons resulting from crystal impurities. The next highest level is the valence band, usually completely filled with electrons. In between these bands is the forbidden band, which is

wider for an insulating material than for a semiconductor and vanishes in a conducting material.

**energy dependence** The characteristic response of a radiation detector to a given range of radiation energies or wavelengths, as compared to the response of a standard open-air chamber.

**energy diagram** *Energy-level diagram.*

**energy gap** The energy range between the bottom of the conduction band and the top of the valence band in a semiconductor.

**energy level** A constant-energy state for particles in an atom. Only a limited number of electrons can exist at each energy level. An electron radiates energy when moving to a lower energy level, and absorbs energy when moving to a higher energy level.

**energy-level diagram** A diagram in which the energy levels of the particles of a quantized system are indicated by distances of horizontal lines from a zero energy level. Also called energy diagram.

**energy product curve** A curve obtained by plotting the product of the values of magnetic induction $B$ and magnetic field strength $H$ for each point on the demagnetization curve of a permanent-magnet material.

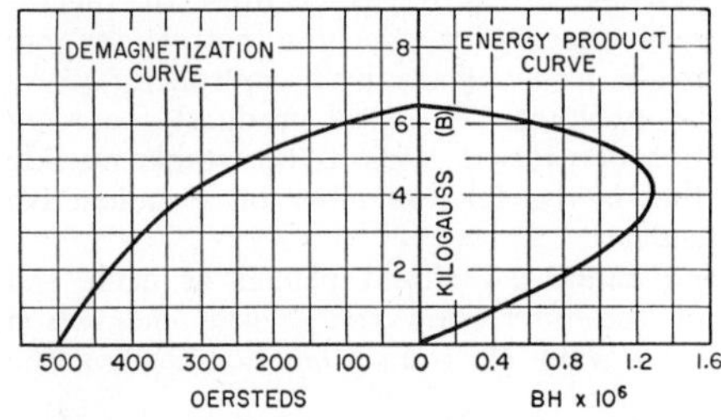


Energy product curve at right is drawn by obtaining values of $B$ and $H$ from demagnetization curve at left, multiplying them for each point, then plotting the $BH$ products against $B$.

**energy unit** Proposed unit for dosage of ionizing radiation in tissue. When the energy absorbed at a point is 93 ergs per gram, the dosage is 1 energy unit.

**engineer** A person having the training and experience required to perform professional duties in a branch of engineering. An engineer generally designs, constructs, operates, or supervises, whereas a scientist seeks to uncover new knowledge, new principles, or new materials.

**engineering** A profession in which a knowledge of the mathematical and physical sciences, gained by study, experience, and practice, is applied with judgment to the utilization of the materials and forces of nature.

**enhanced-carrier demodulation** An amplitude demodulation system in which a synchronized local carrier of proper phase is fed into the demodulator to reduce demodulation distortion.

**eniac** [Electronic Numerical Integrator And Calculator] A digital computer built for the Army Ordnance Department.

**enriched material** Material in which the amount of one or more isotopes of a constituent has been increased above normal.

**enriched reactor** A nuclear reactor in which the fuel is an enriched material.

**enriched uranium** Uranium in which the natural proportion of $U^{235}$ has been increased artificially.

**enrichment** A process that changes the isotopic ratio in a material. For uranium, the process increases the ratio of $U^{235}$ to $U^{238}$ by separation of isotopes. Enrichment processes include thermal diffusion, gaseous barrier diffusion, centrifugal separation, and electromagnetic separation.

**enrichment factor** The ratio of the isotopic abundance after enrichment to that before enrichment.

**envelope** 1. The glass or metal housing of an electron tube. Also called bulb. 2. A curve drawn to pass through the peaks of a graph showing the waveform of a modulated r-f carrier signal.

**envelope delay** The time required for the envelope of a wave to travel between two points in a system.

**envelope delay distortion** *Delay distortion.*

**envelope demodulator** A diode detector whose output is shunted by a capacitor, so as to make the output proportional to the peaks of the rectified amplitude-modulated carrier. Commonly used as the second detector in a television receiver.

**environment** The aggregate of all the conditions and influences that affect the operation of equipment and components. Natural uncontrolled environments include temperature, humidity, rain, snow, ice, sleet, hail, fog, wind, lightning, sand, dust, fungi, radiation, pressure, sun, salt spray, and static electricity. Environments caused by operation or location include vibration, shock, noise, acceleration, aerodynamic heating, erosion in flight, electromagnetic effects, and force.

**epicadmium** Energy above the cadmium

cutoff level of about 0.3 electron-volt in a nuclear reactor.

**epithermal** Above thermal in a nuclear reactor.

**epithermal neutron** A neutron having an energy in the range immediately above the thermal range, or between about 0.02 and 100 electron-volts.

**epithermal reactor** A nuclear reactor in which a substantial fraction of fissions is induced by neutrons having more than thermal energy.

**E-plane bend** A rectangular waveguide bend in which the longitudinal axis of the waveguide remains in a plane parallel to the electric field vector throughout the bend. Also called E bend.

**E-plane tee junction** A rectangular waveguide tee junction in which the electric field vector of the dominant wave of each arm is parallel to the plane of the longitudinal axes of the waveguides.

**epoxy resin** A good insulating plastic having high strength and low shrinkage during curing. Widely used in encapsulating and embedding electronic assemblies.

**eppi** Abbreviation for *expanded plan position indicator.*

**epsilon** [symbol $\epsilon$] A Greek letter used to represent the number 2.71828. . ., which is the base of the natural system of logarithms.

**Eput meter** Trademark of Beckman/Berkeley Division for an electronic instrument that counts and indicates the number of cycles, pulses, or other events occurring in a predetermined unit of time.

**equal-energy source** A light source for which the time rate of emission of energy per unit of wavelength is constant throughout the visible spectrum.

**equal-energy white** The light produced by a source that radiates equal energy at all visible wavelengths.

**equalization** The effect of all frequency-discriminative means employed in transmitting, recording, amplifying, or other signal-handling systems to obtain a desired overall frequency response. Also called frequency-response equalization.

**equalization curve** A curve showing the frequency response needed in a high-fidelity sound-reproducing system to compensate for preemphasis introduced at the broadcasting or recording studio.

**equalizer** A network designed to compensate for an undesired amplitude-frequency or phase-frequency response of a system or component. It is usually some combination of coils, capacitors, and resistors.

**equalizing pulse** One of the pulses occurring just before and just after the vertical synchronizing pulses in a television signal and serving to minimize the effect of line-frequency pulses on interlace. The equalizing pulses occur at twice the line frequency and serve to make each vertical deflection start at the correct instant for proper interlace.

**equally tempered scale** A series of notes selected by dividing the octave into 12 equal intervals.

**equilibrium orbit** *Stable orbit.*

**equilibrium time** Start-up time in an isotope separation plant.

**equiphase surface** Any wave surface over which the field vectors at the same instant are in phase or 180° out of phase.

**equiphase zone** The region in space within which a difference in phase of two radio signals is indistinguishable.

**equipment** One or more assemblies capable of performing a complete function.

**equipotential** Having the same potential at all points.

**equipotential cathode** *Indirectly heated cathode.*

**equipotential line** An imaginary line in space or in a medium, having the same potential at all points.

**equipotential surface** An imaginary surface in space, or in a medium, on which all points have the same potential.

**equisignal localizer** An aircraft guidance localizer in which the localizer on-course line is centered in a zone of equal amplitude of two transmitted signals. Deviations from this zone are detectable as unbalance in the levels of the two signals. Also called equisignal radio-range beacon and tone localizer.

**equisignal radio-range beacon** *Equisignal localizer.*

**equisignal sector** *Equisignal zone.*

**equisignal zone** The region in space within which the difference in amplitude of two radio signals (usually emitted by a single station) is indistinguishable. Also called bisignal zone and equisignal sector.

**equivalent absorption** The area of perfectly absorbing surface that will absorb sound energy at the same rate as the given surface or object under the same conditions. The acoustical unit of equivalent absorption is the sabin.

**equivalent binary digits** The number of binary digits that is equivalent to a given number of decimal digits or other characters. When a decimal number is converted into a binary number, the number of binary digits needed is in general about 3.3 times the number of decimal digits. In coded decimal notation, the number of binary digits needed is ordinarily four times the number of decimal digits.

**equivalent circuit** A circuit that is electrically equivalent to a more complex circuit or device. Used to simplify circuit analysis.

**equivalent constant potential** The constant potential that must be applied to an x-ray tube to produce radiation having an absorption curve in a given material closely similar to that of the beam under consideration.

**equivalent dark-current input** The incident luminous flux required to give a signal output current equal to the dark current in a phototube.

**equivalent diode** An imaginary diode consisting of the cathode of a triode or multigrid tube and a virtual anode to which is applied a composite controlling voltage such that the diode cathode current is the same as in the more complex tube under consideration.

**equivalent loudness level** *Loudness level.*

**equivalent noise conductance** A quantitative representation of the spectral density of a noise-current generator at a specified frequency, expressed in conductance units such as mhos.

**equivalent noise resistance** A quantitative representation of the spectral density of a noise-voltage generator at a specified frequency, expressed in resistance units such as ohms.

**equivalent roentgen** *Roentgen equivalent physical.*

**equivalent stopping power** 1. *Relative stopping power.* 2. *Stopping equivalent.*

**erase** 1. To remove recorded material from magnetic tape by passing the tape through a strong, constant magnetic field (d-c erase) or through a high-frequency alternating magnetic field (a-c erase). 2. To change all the binary digits in a digital computer storage device to binary zeros. 3. To eliminate previously stored information in a charge-storage tube by charging or discharging all storage elements.

**erase oscillator** The oscillator used in a magnetic recorder to provide the high-frequency signal needed to erase a recording on magnetic tape. The bias oscillator usually serves also as the erase oscillator.

**erasing head** A magnetic head used to obliterate material previously recorded on magnetic tape.

**erasing speed** The rate of erasing successive storage elements in a charge-storage tube.

**erbium** [symbol Er] A rare-earth element. Atomic number is 68.

**E region** The region of the ionosphere extending from about 60 to 90 miles above the earth, between the D and F regions.

**erg** The absolute centimeter-gram-second unit of energy and work. It is the work done when a force of 1 dyne is applied through a distance of 1 centimeter. One foot-pound is equal to 13,560,000 ergs.

**error** 1. The difference between the true value and a calculated or observed value. 2. *Malfunction.*

**error-detecting code** *Self-checking code.*

**error-rate damping** A type of damping in which servo control is accomplished by two voltages—one proportional to the error, and the other proportional to the rate at which the error changes. Also called proportional plus derivative control.

**error signal** 1. A voltage that depends on the signal received from the target in a tracking system, having a polarity and magnitude dependent on the angle between the target and the center of the scanning beam. 2. *Error voltage.*

**error voltage** A voltage, usually obtained from a selsyn, that is proportional to the difference between the angular positions of the input and output shafts of a servosystem. This voltage acts on the system to produce a motion that tends to reduce the error in position. Also called error signal.

**erythema** A skin condition caused by ultraviolet radiation from an atomic blast, characterized by reddened skin and painful blisters.

**erythmal flux unit** The amount of radiant flux that will give the same erythmal effect as 10 microwatts of radiant power at 2,967 angstroms. Also called erytheme and E viton.

**erytheme** *Erythmal flux unit.*

**Esaki diode** *Tunnel diode.*

**E scan** *E display.*

**escape velocity** The velocity required for an object to escape from the gravitational influence of the earth.

**E scope** A radarscope that produces an E display.

**escutcheon** An ornament used around a dial, window, control knob, or other panel-mounted part.

**Estiatron** Trademark of RCA for an electrostatically focused traveling-wave tube. No permanent magnets are needed. Bifilar helices provide the required beam focusing,

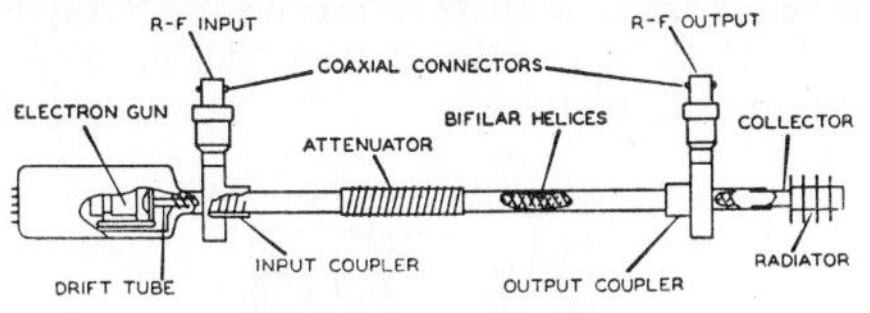

Estiatron construction.

along with their usual r-f functions. Also called electrostatically focused traveling-wave tube.

**esu** Abbreviation for *electrostatic unit.*

**etched printed circuit** A printed circuit formed by chemical etching or chemical and electrolytic removal of unwanted portions of a layer of conductive material bonded to an insulating base.

**Ettingshausen effect** When a metal strip is placed with its plane perpendicular to a magnetic field and an electric current is sent longitudinally through the strip, corresponding points on opposite edges of the strip will have different temperatures.

**E unit** A method of designating radar signal-to-noise ratio. E-1 is a 1-to-1 ratio, barely perceptible; E-2 is 2 to 1, weak; E-3 is 4 to 1, good; E-4 is 8 to 1, strong; E-5 is 16 to 1, very strong or saturating.

**eureka** The ground radar beacon of the rebecca-eureka navigation system, operating at spot frequencies between 215 mc and 235 mc. Range is about 20 to 40 miles, depending on the altitude of the aircraft carrying interrogating rebecca equipment.

**europium** [symbol Eu] A rare-earth element. Atomic number is 63.

**eutectic** The liquid alloy composition having the lowest freezing point. For lead-tin solder, the eutectic is 63% tin and 37% solder, giving a freezing point of 361°F.

**ev** Abbreviation for *electron-volt.*

**evacuate** To remove gases and vapors from an enclosure. Also called exhaust.

**evacuation** The removal of gases and vapors from the envelope of an electron tube during manufacture.

**evanescent mode** The mode of oscillation in which the amplitude diminishes along a waveguide without change of phase.

**evaporative-cooled** *Vaporization-cooled.*

**E vector** A vector that represents the electric field of an electromagnetic wave. In free space it is perpendicular to the H vector and to the direction of propagation.

**even-even nuclei** Nuclei that contain an even number of protons and an even number of neutrons.

**even-odd nuclei** Nuclei that contain an even number of protons and an odd number of neutrons.

**E viton** *Erythmal flux unit.*

**E wave** British term for *transverse magnetic wave* (TM wave).

**exalted-carrier receiver** A receiver that counteracts selective fading by maintaining the carrier at a high level at all times. The receiver may have a local oscillator that is synchronized by the incoming carrier, or may amplify the incoming carrier separately and recombine it with the sidebands. Also called reconditioned-carrier receiver.

**except** A logical operator which has the property that if P and Q are two statements, then the statement "P except Q" is true only when P alone is true. It is false for the other three combinations (P false Q false, P false Q true, and P true Q true). The "except" operator is equivalent to "and not."

**excess conduction** Conduction by excess electrons in a semiconductor.

**excess electron** An electron in excess of the number needed to complete the bond structure in a semiconductor, generally resulting from donor impurities.

**excess-three code** A number code in which the decimal digit $n$ is represented by the four-bit binary equivalent of $n + 3$, so decimal digits 0 through 9 become 0011, 0100, 0101, 0110, 0111, 1000, 1001, 1010, 1011, and 1100 respectively.

**exchange force** *Nuclear force.*

**excitation** 1. The signal voltage that is applied to the control electrode of an electron tube. Also called drive. 2. Application of signal power to a transmitting antenna. 3. The transfer of a nuclear system from its ground state to an excited state by adding energy. 4. The application of voltage to field coils to produce a magnetic field, as required for the operation of an excited-field loudspeaker or a generator.

**excitation anode** An anode used to maintain a cathode spot on a pool cathode of a gas tube when output current is zero.

**excitation curve** A curve showing the relative yield of a specified nuclear reaction as a function of the energy of the incident particles or photons. Also called excitation function.

**excitation energy** The minimum energy required to change a system from its ground state to a particular excited state.

**excitation function** 1. The cross section for a specified nuclear reaction, expressed as a function of the energy of the incident particle or photon. 2. *Excitation curve.*

**excitation purity** *Purity.*

**excitation winding** The magnetic amplifier winding that applies a unidirectional magnetomotive force to the core.

**excited** Having higher energy than the ground state of a nucleus, atom, or molecule.

**excited-field loudspeaker** A dynamic loudspeaker in which the steady magnetic field is produced by an electromagnet called the field coil, through which a direct current is sent. Also called electrodynamic loudspeaker.

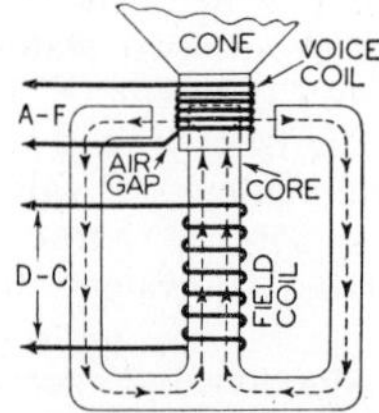


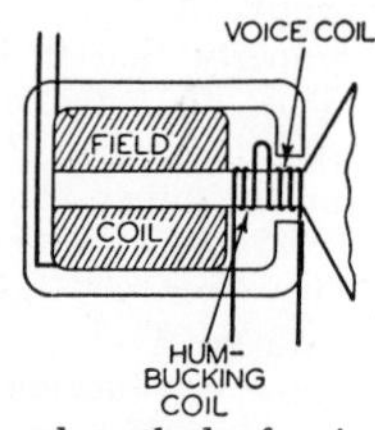


Excited-field loudspeaker, and method of using hum-bucking coil in series with voice coil.

**exciter** 1. The portion of a directional transmitting antenna system that is directly connected to the transmitter. 2. A crystal oscillator or self-excited oscillator used to generate the carrier frequency of a transmitter. 3. A small auxiliary generator that provides field current for an a-c generator. 4. A loop or probe extending into a resonant cavity or waveguide. 5. *Exciter lamp.*

**exciter lamp** A bright incandescent lamp having a concentrated filament. Used in variable-area sound-on-film recording, in reproducing photographic sound tracks on film, and in illuminating subject copy in facsimile transmitters. The lamp serves to excite a phototube or photocell. Also called exciter.

**exciting current** *Magnetizing current.*

**exciting interval** The portion of the a-c supply cycle of a magnetic amplifier in which the entire supply voltage sends current through the gate winding.

**exciton** The combination of an electron and a hole in a semiconductor that is in an excited state. In attracting the electron, the hole acts as a positive charge.

**excitron** A single-anode mercury pool tube provided with means for maintaining a continuous cathode spot.

**exclusion principle** *Pauli exclusion principle.*

**executive routine** A digital computer routine designed to process and control other routines.

**exhaust** *Evacuate.*

**exhaust tube** A glass or metal tube through which air is evacuated from the envelope of an electron tube.

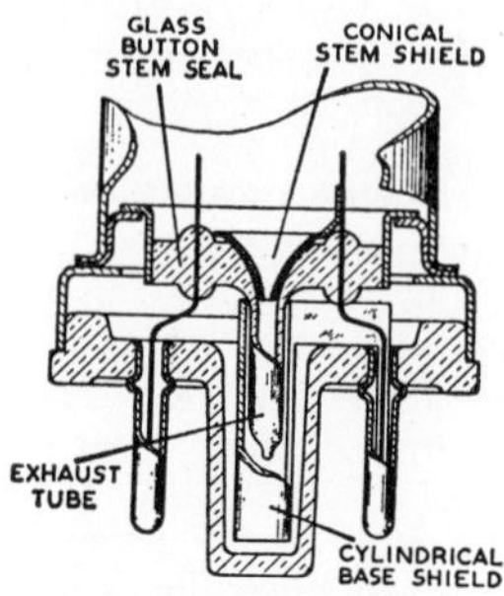


Exhaust tube of metal tube is protected by molded plastic aligning plug in base.

**exit dose** The dose of radiation at the surface from which an irradiating beam emerges after passing through a body.

**exoergic** *Exothermic.*

**exothermic** Involving the evolution of heat or other energy. All radioactive processes are exothermic. Also called exoergic.

**exp** An abbreviation sometimes used in equations to indicate that the quantity immediately following in brackets is to be considered as the exponent of $\epsilon$, which is 2.718, the base of the natural system of logarithms.

**expanded-center ppi display** A modified ppi display in which each sweep starts about ¼ inch out from the center of the tube screen. Bearings of nearby objects can then be determined more accurately. Also called center expansion.

**expanded plan position indicator** [abbreviated eppi] A special high-precision indicator used in some ground-controlled approach equipment to display azimuth, elevation, and range information simultaneously. Also called azel indicator.

**expanded scope** A magnified portion of a cathode-ray presentation.

**expanded sweep** A cathode-ray sweep in which the movement of the electron beam across the screen is speeded up during a selected portion of the sweep time.

**expander** A transducer that, for a given input amplitude range, produces a larger output range. One type of expander increases the amplitude range as a linear function of the envelope of speech waves.

**expandor** The part of a compandor that

is used at the receiving end of a circuit to return the compressed signal to its original form. It attenuates weak signals and amplifies strong signals.

**expansion** A process in which the effective gain of an amplifier is varied as a function of signal magnitude, the effective gain being greater for large signals than for small signals. The result is greater volume range in an audio amplifier and greater contrast range in facsimile.

**expansion chamber** *Cloud chamber.*

**experimental breeder reactor** A fast heterogeneous nuclear reactor used for research and breeding. Its core consists of enriched $U^{235}$ surrounded by a blanket of natural uranium.

**Explorer** The first United States earth satellite placed in orbit, on January 31, 1958. It contained an 11-pound package of instruments and radio transmitters.

**exploring coil** A small coil used to measure a magnetic field or to detect changes produced in a magnetic field by a hidden object. The coil is connected to an indicating instrument either directly or through an amplifier. Also called magnetic test coil and search coil.

**exponential absorption** The removal of particles or photons from a beam at an exponential rate as the beam passes through matter.

**exponential decay** Decay of radiation, charge, signal strength, or some other quantity at an exponential rate.

**exponential experiment** A nuclear experiment involving a subcritical assembly of fissionable and moderator material. Used to determine the neutron properties of reactor cores. The decay of neutron flux is usually exponential in one direction.

**exponential flareout** *Altimetric flareout.*

**exponential horn** A horn whose cross-sectional area increases exponentially with axial distance.

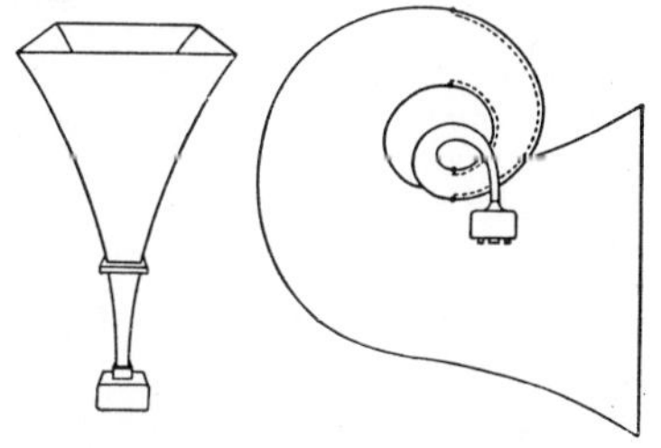

Exponential horns, straight and curled.

**exponential reactor** A nuclear reactor designed specifically for exponential experiments, as well as for determining critical size.

**exponential transmission line** A two-conductor transmission line whose characteristic impedance varies exponentially with electrical length along the line.

**exponential well** A potential well having an exponentially changing value.

**exposure** The total quantity of radiation at a given point, measured in air.

**exposure dose** A measure of x-ray or gamma radiation at a point, based on its ability to produce ionization. The unit of exposure dose is the roentgen.

**exposure-dose rate** The exposure dose per unit time. The unit of exposure-dose rate is the roentgen per unit time.

**exposure meter** An instrument used to measure the intensity of light reflected from an object, for the purpose of determining proper camera exposure. Modern exposure meters consist of a photovoltaic cell connected to an indicating meter.

**extended-interaction tube** A microwave tube in which a moving electron stream interacts with a traveling electric field in a long resonator. Bandwidth is between that of klystrons and traveling-wave tubes.

**extended-play record** A 7-inch, 45-rpm record having closely spaced grooves giving up to eight minutes of playing time per side, as compared to three to five minutes per side with conventional 45-rpm microgroove records.

**extension cord** A line cord having a plug at one end and an outlet at the other end.

**extensive shower** A cosmic-ray shower occurring over an area as large as 300 feet in diameter, presumably initiated high in the atmosphere by a single cosmic ray having energy as high as $10^{17}$ electron-volts. Also called Auger shower and giant air shower.

**externally quenched counter tube** A radiation-counter tube that requires an external quenching circuit to inhibit reignition.

**external memory** A memory that is separate from a digital computer but holds information stored in language acceptable to the machine, such as on recorded magnetic tape or punched cards.

**extinction voltage** The lowest anode voltage at which a discharge is sustained in a gas tube.

**extract** 1. To form a new computer word by extracting and putting together selected segments of given words. 2. To remove from a computer register or memory all items that meet a specified condition.

**extra-high tension** [abbreviated eht] British term for the high d-c voltage applied to the second anode in a cathode-ray tube, ranging from about 4,000 to 50,000 volts in various sizes of tubes.

**extraordinary component** The component of light that is plane-polarized and passes through a Nicol prism. The ordinary component is totally reflected by the prism.

**extraordinary-wave component** The magneto-ionic wave component in which the electric vector rotates in the opposite sense to that for the ordinary-wave component. Also called X wave.

**extrapolate** To extend the range of known values by estimating. Interpolate means to estimate missing values between those that are known.

**extrapolated boundary** The surface outside a nuclear reactor on which the neutron flux would be zero if extrapolated linearly from the flux a few mean free paths inside the medium.

**extrapolated range** The range of a given charged particle in matter as obtained by extrapolation of the absorption curve.

**extrapolation ionization chamber** An ionization chamber so designed that volume, electrode separation, or some other factor can be varied in suitable steps for measurement purposes. The resulting measured values are plotted in appropriate form and the desired result is obtained by extrapolation of the curve.

**extremely high frequency** [abbreviated ehf] A Federal Communications Commission designation for the band from 30,000 to 300,000 mc in the radio spectrum.

**extrinsic properties** The properties of a semiconductor as modified by impurities or imperfections within the crystal.

**extrinsic semiconductor** A semiconductor whose electrical properties are dependent on impurities added to the semiconductor crystal, in contrast to an intrinsic semiconductor whose properties are characteristic of an ideal pure crystal.

# F

**F** 1. Symbol for *filament.* 2. Symbol for *fuse.* 3. Abbreviation for *Fahrenheit.*

**F+** *A+.*

**F−** *A−.*

**f** Abbreviation for *farad.*

**f** Symbol for *frequency.*

**face** *Faceplate.*

**faceplate** The transparent or semitransparent glass front of a cathode-ray tube, through which the image is viewed or projected. The inner surface of the face is coated with fluorescent chemicals that emit light when hit by an electron beam. Also called face.

**facom** A radio navigation system using phase comparison techniques to give range accuracy within 1% at ranges up to 3,000 miles, even under adverse propagation and noise conditions. Range is obtained by phase comparison of received and locally generated signals for two positions of the receiver; if the signals are in phase at one position, the phase difference at the other position will be directly proportional to the range. Only two stations are needed to give a fix.

**facsimile** [abbreviated fax] A system of communication in which a photograph, map, or other fixed graphic material is scanned and the information converted into signal waves for transmission by wire or radio to a facsimile receiver at a remote point. Also called phototelegraphy, radiophoto, telephoto, telephotography, and wirephoto.

**facsimile converter** A device that changes the type of modulation in a facsimile system.

**facsimile receiver** The receiver used to translate the facsimile signal from a wire or radio communication channel into a facsimile record of the subject copy.

**facsimile recorder** The section of a facsimile receiver that performs the final conversion of electric picture signals to an image of the subject copy on the record medium. The image may be produced by a modulated light beam focused on photographic printing paper mounted on a rotating drum, by sending electric current through special paper mounted on the drum, or by using a pressure stylus with appropriate pressure-sensitive paper on the drum.

**facsimile signal** The picture signal produced by scanning the subject copy in a facsimile transmitter.

**facsimile-signal level** The maximum facsimile signal power or voltage (root-mean-square or d-c) measured at any point in a facsimile system.

**facsimile system** An integrated assembly of the elements used for facsimile.

**facsimile transmission** The transmission of signal waves produced by scanning fixed graphic material, including pictures, for reproduction in record form.

**facsimile transmitter** The apparatus used to translate the subject copy into facsimile signals suitable for delivery to a communication system. In most transmitters, the subject copy is placed on a rotating drum driven by a synchronous motor, for scanning by a photoelectric system mounted on a carriage that moves along the length of the drum. Alternatively, the rotating drum may move axially past a stationary photoelectric system.

**facsimile transmitting converter** A converter that changes amplitude modulation to frequency-shift modulation in a facsimile system.

**fade** To change signal strength gradually.

**fade in** To increase signal strength gradually, as at the start of a radio or television program or when changing to a new scene. This makes sound volume and picture brightness increase gradually.

**fade out** To reduce signal strength gradually in a sound or television channel. Often done at the end of a program to make a picture go dark gradually, with volume of background music decreasing correspondingly.

**fadeout** A gradual and temporary loss of a received radio or television signal due to magnetic storms, atmospheric disturbances, or other conditions along the transmission path. A blackout is a fadeout that may last several hours or more at a particular frequency.

**fader** A multiple-unit volume control used

for gradual changeover from one microphone, audio channel, or television camera to another. In each case, the level is held essentially constant because one section of the fader increases signal level in its channel while the other fader section reduces signal level a corresponding amount.

**fading** Variations in the field strength of a radio signal, usually gradual, that are caused by changes in the transmission medium.

**fading margin** An attenuation allowance made in radio system planning so that anticipated fading will still keep the signal above a specified minimum signal-to-noise ratio.

**Fahnestock clip** A spring-type terminal to which a temporary connection can readily be made.

Fahnestock clip.

**Fahrenheit** [abbreviated F] A temperature scale in which the freezing point of water is 32° and the boiling point of water is 212° at normal atmospheric pressure. Absolute zero is −459.6°F. To change degrees Fahrenheit to degrees centigrade, subtract 32 and then multiply the result by 5/9. Fahrenheit and centigrade scales agree at one point: −40°F = −40°C.

**fail-safe control** A control so designed that control-circuit failure cannot cause a dangerous condition under any circumstances.

**fairing** A contoured structure mounted on a sonar dome that projects from the hull of a ship, to reduce the water resistance around the dome when the ship is in motion.

**fairlead** A plastic, wood, or metal tube with a funnel-shaped end through which a trailing antenna is reeled on an aircraft.

**Falcon** An Air Force air-to-air guided missile having either radar or infrared homing guidance, a speed of about mach 2, and a range of about 5 miles. It can be carried in quantity by interceptor aircraft.

**fallback** The material that drops back to the earth or water at the site of a surface or subsurface nuclear explosion.

**fallout** The material that descends to the earth or water well beyond the site of a surface or subsurface nuclear explosion.

**fall time** The time for the voltage at the trailing edge of a pulse to change from 90% to 10% of the original negative or positive amplitude.

**false course** A spurious additional course-line indication produced by a navigation aid, due to undesired reflections or maladjusted equipment.

**false echo** A misleading extra echo indication on the plan position indicator of a ship radar set. On rivers, shore structures may cause false echoes. A target may also be hit by a side lobe of the beam, and the echo reflected back into the main beam path by a mast or funnel.

**fan beam** 1. A radio beam having an elliptically shaped cross-section in which

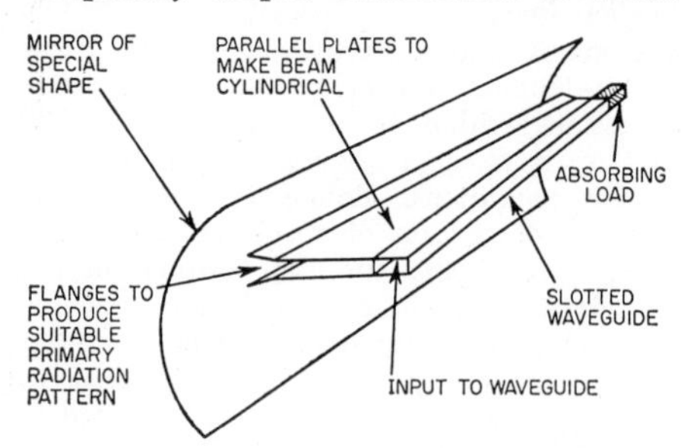

Fan beam is produced by special waveguide feed to mirror that is section of cylinder.

the ratio of the major to minor axes usually exceeds 3 to 1. The beam is broad in the vertical plane and narrow in the horizontal plane. 2. A radar beam having the shape of a fan.

**fan marker** *Fan marker beacon.*

**fan marker beacon** A vhf radio facility having a vertically directed fan beam intersecting an airway to provide a fix. When used near an airport as part of an instrument landing system, it may be called a boundary, middle, or outer marker beacon. Also called fan marker and radio fan marker beacon.

**fanned-beam antenna** A unidirectional antenna so designed that transverse cross-sections of the major lobe are approximately elliptical. Used to produce a fan beam.

**farad** [abbreviated f] The basic unit of capacitance. A capacitor has a capacitance of one farad when a voltage change of one volt per second across it produces a current of one ampere. The farad is too large a unit for practical work, so two smaller units are generally used. The microfarad is equal to one-millionth of a farad, and the micro-microfarad is equal to one-millionth of a microfarad.

**faraday** A unit of quantity of electricity, equal to 96,500 coulombs.

**Faraday dark space** The relatively nonluminous region that separates the negative glow from the positive column in a cold-cathode glow-discharge tube.

**Faraday effect** When a plane-polarized beam of light passes through certain transparent substances in a direction parallel to the lines of a strong magnetic field, the plane of polarization is rotated a certain amount. The same effect governs the action of a ferrite rotator in a waveguide.

**Faraday rotation isolator** An isolator used in circular waveguides to pass a wave traveling in one direction but not in the other direction. A ferrite rod is positioned

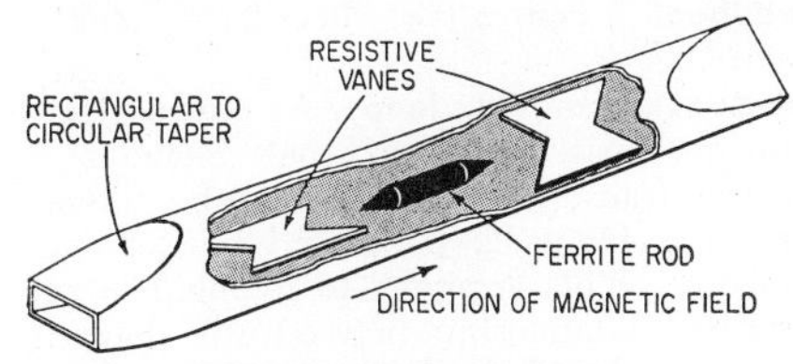


Faraday rotation isolator.

between resistive vanes to rotate the plane of polarization of the waves 45°. With appropriate rectangular-to-circular transitions, the isolator may be used with rectangular waveguides.

**Faraday shield** An electrostatic shield made of a series of parallel wires connected at one end like a comb. The common point is grounded. The shield provides electrostatic shielding while passing electromagnetic waves.

**Faraday's law** The voltage induced in a circuit is proportional to the rate at which the magnetic flux linkages of the circuit are changing. Also called law of electromagnetic induction.

**faradic current** An intermittent and nonsymmetrical alternating current like that obtained from the secondary winding of an induction coil. Used in electrobiology.

**faradization** Use of a faradic current to stimulate muscles and nerves.

**far-end crosstalk** Crosstalk that travels along the disturbed circuit in the same direction as desired signals in that circuit. When it occurs at carrier telephone repeater stations, the output signals of one repeater go out also over the output line for the other repeater.

**far field** The radiation field in the Fraunhofer region surrounding a transmitting antenna.

**Farnsworth image dissector tube** *Image dissector.*

**far region** *Fraunhofer region.*

**far zone** *Fraunhofer region.*

**fast breeder reactor** An experimental nuclear reactor that uses a uranium-plutonium alloy as fuel and sodium as coolant, with no moderator.

**fast effect** The reactivity change (increase in neutrons) due to fissions caused by fast neutrons in a thermal reactor.

**fastener** A device for holding two or more parts together.

**fast fission** Fission caused by fast neutrons in a thermal reactor.

**fast forward** High-speed forward travel of magnetic tape, used to reach another part of the tape when intervening recorded material is not desired.

**fast groove** An unmodulated spiral groove having a pitch that is much greater than that of the recorded grooves in a disk recording. Also called fast spiral.

**fast multiplication factor** The number of fission-source neutrons due to all fissions, divided by the number due to fission by thermal neutrons in a nuclear reactor.

**fast neutron** A neutron having energy much greater than some arbitrary lower limit that may be only a few thousand electron-volts.

**fast reactor** A nuclear reactor in which most of the fissions are produced by fast neutrons.

**fast spiral** *Fast groove.*

**fast-time-constant circuit** [abbreviated ftc circuit] A circuit having a short time-constant, used in radar to emphasize short-duration signals and discriminate against low-frequency components of clutter. Also used to suppress rain and snow echoes.

**fathometer** A sonar-type instrument used to measure ocean depth and locate underwater objects such as schools of fish. A sound pulse is transmitted vertically downward by a piezoelectric or magnetostriction transducer mounted on the hull of the ship. The time required for the pulse to return, after reflection from the bottom of the sea or an intervening object, is measured electronically and converted into a depth indication on a dial or recorder chart. Also called acoustic depth finder, depth finder, depth sounder, echo depth sounder, and sonic depth finder.

**fatigue** The decrease of efficiency of a

luminescent or light-sensitive material as a result of excitation.

**fault** A defect, such as an open, short-circuit, or ground in a circuit, component, or line.

**fault electrode current** The current to an electrode under fault conditions, such as during arcbacks and load short-circuits.

**fax** Abbreviation for *facsimile.*

**fay surface** A surface to which a soldered connection is to be made.

**FCC** Abbreviation for *Federal Communications Commission.*

**F display** A rectangular radarscope display in which a target appears as a centralized bright spot when the radar antenna is aimed at it. Horizontal and vertical aiming errors are respectively indicated by the horizontal and vertical displacement of the spot. Also called F scan.

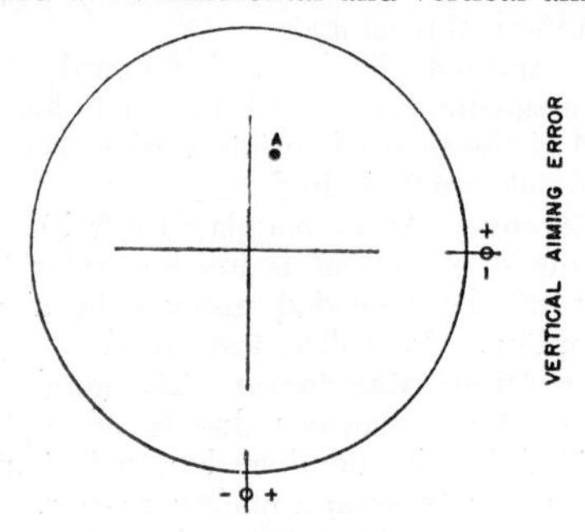

F display.

**fdm** Abbreviation for *frequency-division multiplex.*

**FDS law** Abbreviation for *Fermi-Dirac-Sommerfield velocity-distribution law.*

**Feather analysis** A technique for determining the range in aluminum of the beta rays of a species by comparing the absorption curve of that species with the absorption curve of a reference species.

**Federal Communications Commission** [abbreviated FCC] A board of seven commissioners appointed by the President under the Communications Act of 1934, having the power to regulate all electric communication systems originating in the United States, including radio, television, facsimile, telegraph, telephone, and cable systems.

**feed** 1. To supply a signal to the input of a circuit, transmission line, or antenna. 2. The part of a radar antenna that is connected to or mounted on the end of the transmission line, and serves to radiate r-f energy to the reflector or receive energy therefrom.

**feedback** The return of a portion of the output of a circuit or device to its input. With positive feedback, the signal fed back is in phase with the input and increases amplification, but may cause oscillation. With negative feedback, the signal fed back is 180° out of phase with the input and decreases amplification, but stabilizes circuit performance and tends to minimize noise and distortion.

**feedback admittance** The short-circuit transadmittance from the output electrode to the input electrode of an electron tube.

**feedback amplifier** An amplifier in which a passive network is used to return a portion of the output signal to its input in such a way as to change the performance characteristics of the amplifier.

**feedback controller** *Feedback control system.*

**feedback control loop** A closed transmission path or loop that includes an active transducer and consists of a forward path, a feedback path, and one or more mixing points arranged to maintain a prescribed relationship between the loop input signal and the loop output signal.

**feedback control system** A control system, comprising one or more feedback control loops, in which functions of the controlled signals are combined with functions of the commands to tend to maintain prescribed relationships between the commands and the controlled signals. Also called closed-loop control system and feedback controller.

**feedback cutter** A disk recording cutter that has an auxiliary feedback coil in the magnetic field. Signals exciting the cutter are induced into the feedback coil and fed back to the input of the cutter amplifier. This feedback gives a substantially uniform frequency response.

**feedback oscillator** An oscillating circuit, including an amplifier, in which the output is fed back in phase with the input. Oscillation is maintained at a frequency determined by the values of the components in the amplifier and the feedback circuits.

**feedback path** The transmission path from the loop output signal to the loop feedback signal in a feedback control loop.

**feedback regulator** A feedback control system that tends to maintain a prescribed relationship between certain system signals and other predetermined quantities. Some of the system signals in a regulator are adjustable reference signals. Under certain methods of operation, a feedback regulator may also be a servomechanism.

**feedback signal** *Primary feedback.*

**feedback transfer function** The transfer function of the feedback path in a feedback control loop.

**feedback winding** A winding to which feedback connections are made in a magnetic amplifier.

**feeder** A transmission line used between a transmitter and an antenna.

**feed reel** The reel on a tape recorder that supplies the magnetic tape to the recording or playback head.

**feedthrough** A conductor that connects patterns on opposite sides of a printed-circuit board. Also called interface connection.

**feedthrough capacitor** A feedthrough insulator that provides a desired value of capacitance between the feedthrough conductor and the metal chassis or panel through which the conductor is passing. Used chiefly for bypass purposes in uhf circuits.

**feedthrough insulator** An insulator designed for mounting in a hole in a panel, wall, or bulkhead, with a conductor in the center of the insulator to permit feeding electricity through the partition.

**feet per second** [abbreviated fps] A unit used in specifying the speed at which sound waves travel through a medium. In air at standard sea-level conditions, the speed of sound is about 1,080 feet per second.

**female connector** A connector having one or more contacts set into recessed openings. Jacks, sockets, and wall outlets are examples of female connectors.

**Fermat's principle** An electromagnetic wave will take a path that involves the least travel time when propagating between two points.

**Fermi age** The value calculated for the slowing-down area used in the Fermi-age model.

**Fermi-age model** A model used in studying the slowing down of neutrons by elastic collisions. It is assumed that the slowing down takes place by a very large number of very small energy changes.

**Fermi constant** A universal constant, introduced in beta disintegration theory, that expresses the strength of the interaction between the transforming nucleon and the electron-neutrino field. Its value lies between $10^{-48}$ and $10^{-49}$ gm cm$^5$ sec$^{-2}$.

**Fermi-Dirac distribution function** A function having a value in the range from zero to unity, specifying the probability that an electron in a semiconductor will occupy a certain quantum state of energy when thermal equilibrium exists. The energy for which the value of the function is 0.5 is called the Fermi level.

**Fermi-Dirac-Sommerfield velocity-distribution law** [abbreviated FDS law] An equation giving the number of particles in a quantized system in equilibrium.

**Fermi-Dirac statistics** Quantum statistics in which no more than one of a set of identical particles may occupy a particular quantum state.

**Fermi level** The level of the electron energy at which the Fermi-Dirac distribution function has a value of 0.5.

**fermion** A particle that obeys Fermi-Dirac statistics. All known fermions have total angular moments of $(n + \frac{1}{2})\ h$, where $n$ is zero or an integer and $h$ is the quantum or unit of orbital angular momentum.

**Fermi plot** *Kurie plot.*

**Fermi selection rules** A set of selection rules for allowed beta transitions.

**fermium** [symbol Fm] A transuranic radioactive element produced artificially by bombardment of einsteinium with alpha particles or neutrons. Atomic number is 100.

**fernico** An iron-nickel-cobalt alloy used for metal-to-glass seals. Not a trademark.

**ferret** An aircraft equipped to detect, locate, record, and analyze electromagnetic radiation.

**ferric oxide** [symbol $Fe_2O_3$] A magnetic iron oxide (red) used as a coating on magnetic recording tapes.

**ferrimagnetic amplifier** A microwave amplifier using ferrites.

**ferrimagnetism** A type of magnetism in which the magnetic moments of neighboring ions tend to align antiparallel to each other. The moments are of different magnitudes, however, so there can be a large resultant magnetization. Observed in the ferrites and similar compounds.

**ferristor** A miniature two-winding saturable reactor that operates at a high carrier frequency and may be connected as a coincidence gate, current discriminator, free-running multivibrator, oscillator, or ring counter.

**ferrite** A powdered, compressed, and sintered magnetic material having high resistivity, consisting chiefly of ferric oxide combined with one or more other metals.

The high resistance makes eddy current losses extremely low at high frequencies. Examples of ferrite compositions include nickel ferrite, nickel-cobalt ferrite, manganese-magnesium ferrite, yttrium-iron garnet, and single-crystal yttrium-iron garnet. Also called ferrospinel.

**ferrite core** A magnetic core made of ferrite material. Also called dust core and powdered iron core.

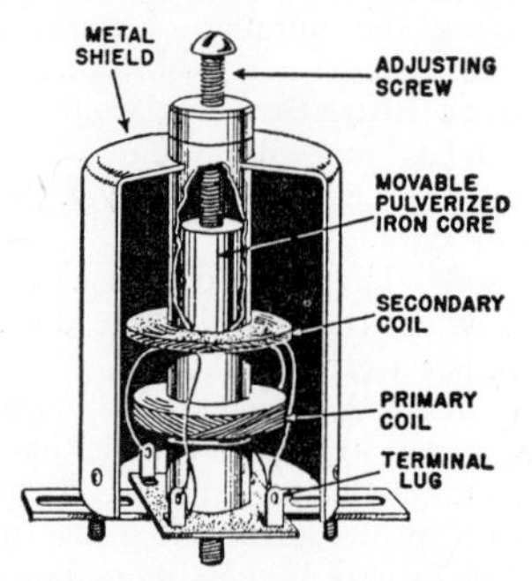


Ferrite core used to change coupling between primary and secondary of i-f transformer.

**ferrite-core memory** A magnetic memory consisting of read-in and read-out wires threaded through a matrix of tiny toroidal cores molded from a square-loop ferrite. Some cores of this type are only 0.05 inch in diameter.

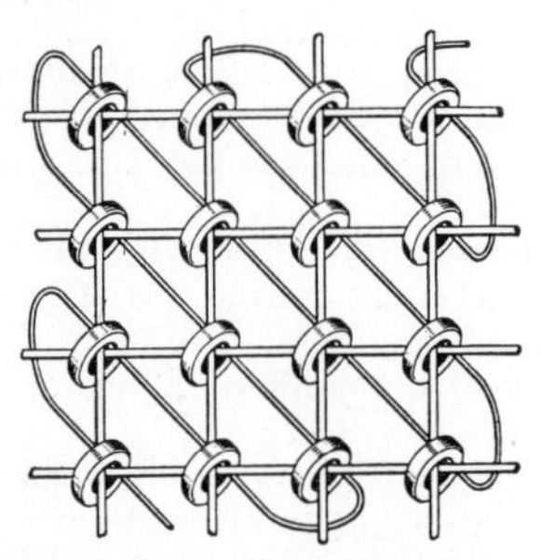

Ferrite-core memory construction. A single matrix for a computer may have over 50,000 such cores.

**ferrite isolator** An isolator that passes energy in one direction through a waveguide while making possible the absorption of energy from the opposite direction. A ferrite rod, centered on the axis of a short length of circular waveguide, is located between rectangular-waveguide end sections displaced 45° with respect to each other. A signal in the desired direction is rotated 45° by the ferrite rod to make it correct for the output waveguide. A backward signal is rotated 45° in the wrong direction and its energy is absorbed by resistance cards.

**ferrite phase-differential circulator** A combination microwave duplexer and load isolator that functions as a switching device between a high-power radar magnetron, a radar receiver, and a radar antenna. It consists of a folded magic tee, a ferrite section, and a short-slot hybrid coupler with a dual adapter that connects a matched load and the antenna to the hybrid. The ferrite section consists of a double waveguide unit with a wall between the two waveguide lines. Ferrite slabs are placed in the waveguide lines to produce the desired differential phase shift.

**ferrite-rod antenna** An antenna consisting of a coil wound on a rod of ferrite. Used in place of a loop antenna in radio receivers. The coil generally serves as the tuning inductance for the first stage of the receiver. Also called ferrod and loopstick antenna.

**ferrite rotator** A gyrator consisting of a ferrite cylinder surrounded by a ring-type permanent magnet, inserted in a waveguide to rotate the plane of polarization of the electromagnetic wave passing through

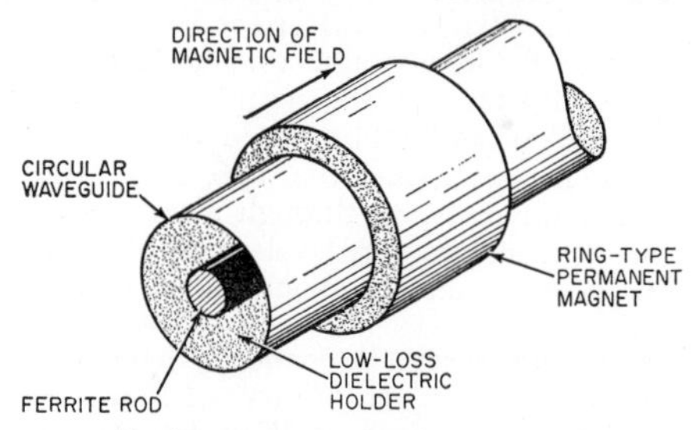


Ferrite rotator in circular waveguide.

the waveguide. The dimensions of the ferrite cylinder and the magnetization of the ferrite determine the amount of rotation.

**ferrite switch** A ferrite device that blocks the flow of energy through a waveguide by rotating the electric field vector 90°. The switch is energized by sending direct current through its magnetizing coil. The rotated energy is then reflected from a reactive mismatch or absorbed in a resistive card.

**ferrod** *Ferrite-rod antenna.*

**ferrodynamic instrument** An electrodynamic instrument in which the forces are materially augmented by the presence of ferromagnetic material.

**ferroelectric converter** A converter that transforms thermal energy into electric energy by utilizing the change in the dielectric constant of a ferroelectric material when heated beyond its Curie temperature. A large capacitor using a ferroelectric dielectric such as barium strontium titanate, initially at its Curie temperature, is charged by a battery through a diode, then heated beyond the Curie temperature. The dielectric constant then drops and capacitance drops correspondingly, so that capacitor voltage goes up. After this voltage is discharged through a load, the capacitor is cooled and the cycle is repeated. Solar radiation on spacecraft could provide the required heat.

**ferroelectric material** A nonlinear dielectric material in which electric dipoles line up spontaneously by mutual interaction, just as magnetic dipoles line up in a magnetic material. Examples of ferroelectric materials include barium titanate, potassium dihydrogen phosphate, and Rochelle salt. Used in ceramic capacitors, acoustic transducers, and dielectric amplifiers.

**ferromagnetic amplifier** A parametric amplifier based on the nonlinear behavior of ferromagnetic resonance at high r-f power levels. In one version, microwave pumping power is supplied to a garnet or other ferromagnetic crystal mounted in a

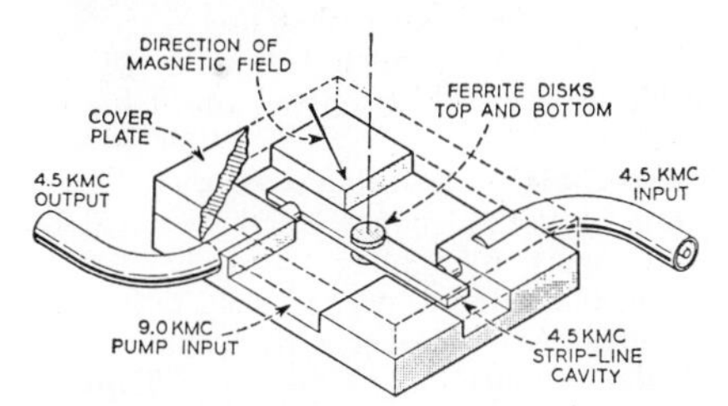


Ferromagnetic amplifier for 4,500-mc signals.

cavity containing a strip line. A permanent magnet provides sufficient field strength to produce gyromagnetic resonance in the garnet at the pumping frequency. The input signal is applied to the crystal through the strip line, and the amplified output signal is extracted from the other end of the strip line. Sometimes incorrectly called a garnet maser, but the operating principle differs from that of masers.

**ferromagnetic material** A magnetic material having a permeability that is considerably greater than the permeability of a vacuum and varies with the magnetizing force. The various forms of iron, steel, cobalt, nickel, and their alloys are examples.

**ferromagnetic resonance** Resonance at which the apparent permeability of a magnetic material at microwave frequencies reaches a sharp maximum. This resonance occurs in the presence of a steady transverse magnetic field when the microwave frequency equals the precession frequency of the electron orbits in the atoms of the magnetic material. The resonance frequency depends on the strength of the transverse field.

**ferromagnetics** The science that deals with storage of information and control of pulse sequences by means of the magnetic polarization properties of materials.

**ferromagnetic tape** A tape made of magnetic material for use in winding closed magnetic cores of toroids and transformers.

**ferrometer** An instrument used to make permeability and hysteresis tests of iron and steel.

**ferroresonant circuit** A resonant circuit in which a saturable reactor provides nonlinear characteristics, with tuning being accomplished by varying circuit voltage or current.

**ferrospinel** *Ferrite.*

**ferrule terminal** A cylindrical end terminal sometimes used on resistors, cartridge fuses, and other parts to permit quick insertion and removal from holders having corresponding spring contacts.

**fertile** Capable of being transformed into a fissionable material by capture of a neutron. Thorium $Th^{232}$ and uranium $U^{238}$ are examples.

**Fiberglas** Trademark of Owens-Corning Fiberglas Corp. for their glass fiber materials.

**Fick's law** Diffusion current density is proportional to the negative of the gradient of neutron density.

**fidelity** The degree to which a system accurately reproduces at its output the essential characteristics of the signal impressed on its input.

**field** 1. One of the equal parts into which a frame is divided in interlaced scanning for television. A field includes one complete scanning operation from top to bottom of the picture and back again. In the present U. S. television broadcasting system there are two fields per frame, with each field taking 1/60th second and including 262.5 lines. 2. A region containing electric or magnetic lines of force, or both. 3. A set of one or more columns reserved for a particular type of information on a punched card.

**field coil** A coil used to produce a constant-strength magnetic field in an electric motor, generator, or excited-field loudspeaker.

**field-displacement isolator** A ferrite isolator that can be used directly in a rectangular waveguide, eliminating the rectangular-to-circular transitions needed with a Faraday rotation isolator.

**field-effect transistor** *Unipolar transistor.*

**field-effect varistor** A passive two-terminal nonlinear semiconductor device that maintains constant current over a wide voltage range.

**field emission** The liberation of electrons from an unheated solid or liquid by a strong electric field at the surface.

**field-enhanced photoelectric emission** The increased photoelectric emission resulting from the action of a strong electric field on the emitter.

**field-enhanced secondary emission** The increased secondary emission resulting from the action of a strong electric field on the emitter.

**field-free emission current** The electron current emitted by a cathode when the electric field at the surface of the cathode is zero.

**field frequency** The number of fields transmitted per second in television. In the United States, it is 60 fields per second. The field frequency is equal to the frame frequency multiplied by the number of fields that make up one frame. Also called field repetition rate.

**field intensity** *Field strength.*

**field ion emission microscope** A high-magnification microscope in which an intense electric field is applied to a sharp metal point to make movements of atoms visible on the point.

**fieldistor** *Unipolar transistor.*

**field-neutralizing coil** A coil used around the faceplate of a color television picture tube. Direct current is sent through this coil to produce a constant magnetic field that offsets the effect of the earth's magnetic field on the electron beams.

**field-neutralizing magnet** A permanent magnet mounted near the edge of the faceplate of a color picture tube to serve the same function as a field-neutralizing coil. Also called rim magnet.

**field of force** A region in space in which force is exerted on electric charges by other stationary or moving charges.

**field pattern** *Radiation pattern.*

**field period** The time required to transmit one television field, equal to 1/60th of a second in the United States.

**field pole** A structure of magnetic material on which a field coil of a loudspeaker, motor, generator, or other electromagnetic device may be mounted.

**field quantum** The fundamental field particle that is the result of quantizing a field.

**field repetition rate** *Field frequency.*

**field-sequential color television** A color television system in which the individual red, blue, and green primary colors are associated with successive fields.

**field-simultaneous system** A color television system in which a complete full-color field is presented simultaneously as a unit. The eyes then see a succession of full-color images rather than a succession of primary color fields.

**field strength** The strength of an electric, magnetic, or electromagnetic field at a point. For electromagnetic radiation, it is generally expressed in volts, millivolts, or microvolts per meter of effective antenna height. Also called field intensity.

**field-strength meter** A calibrated radio receiver used to measure the field strength of radiated electromagnetic energy from a radio transmitter.

**field telephone** A portable telephone designed for field or combat use.

**field wire** An insulated flexible wire or cable used in field telephone and telegraph systems.

**figure-of-eight radiation pattern** A radiation pattern having equal broad lobes 180° apart, resembling the numeral 8.

**figure of merit** A performance rating that governs the choice of a device for a particular application. Thus, the figure of merit of a magnetic amplifier is the ratio of usable power gain to the control time constant.

**filament** [symbol F] A cathode made of resistance wire or ribbon, through which an electric current is sent to produce the high temperature required for emission of electrons in a thermionic tube. Also called directly heated cathode, filamentary cathode, and filament-type cathode. A filament emits electrons directly, whereas a heater merely provides heat for a separate cathode.

**filamentary cathode** *Filament.*

**filamentary transistor** *Unijunction transistor.*

**filament current** The current supplied to the filament of an electron tube for heating purposes.

**filament emission** Liberation of electrons from a heated filament wire in an electron tube.

**filament resistance** The resistance in ohms of the filament of an electron tube. For metal filaments, the resistance increases with temperature, so the hot resistance is many times the cold resistance.

**filament saturation** *Temperature saturation.*

**filament transformer** A small transformer used exclusively to supply filament or heater current for one or more electron tubes.

**filament-type cathode** *Filament.*

**filament-type tube** An electron tube in which electron emission is produced directly by a heated filament.

**filament voltage** The voltage applied to the terminals of the filament in an electron tube.

**filament winding** The secondary winding of a power transformer that furnishes a-c heater or filament voltage for one or more electron tubes.

**filled band** An energy band in which each energy level is occupied by an electron.

**filler** A finely divided inert material that is added to the plastic composition for a disk record to improve its working properties, permanence, strength, and other qualities.

**film** 1. A thin sheet or coating of material. 2. The layer adjacent to the valve metal in an electrochemical valve, in which is located the high voltage drop when current flows in the direction of high impedance.

**film badge** A badge containing one or more pieces of unexposed x-ray film, worn on the person as a means of measuring radiation exposure. The films may have different sensitivities, and some of the film may be shielded from certain types of radiation by a filter. The films are removed from time to time, developed, and the resulting emulsion density measured to determine exposure.

**film pickup** A special motion-picture film projector combined with a television camera.

**film reproducer** Equipment in which sprocket-hole film is the medium from which a magnetic recording is reproduced.

**film resistor** A fixed resistor in which the resistance element is a thin layer of conductive material on an insulated form. The conductive material does not contain either binders or insulating material.

**film ring** A small film badge mounted on a finger ring.

**film scanning** The process of converting motion-picture film into corresponding electric signals that can be transmitted by a television system.

**film sound recorder** Equipment that uses oxide-coated sprocket-hole film as the medium for magnetic recording.

**filter** A selective device that transmits a desired range of matter or energy while substantially attenuating all other ranges. Thus, an electric filter is a network that transmits alternating currents of desired frequencies while substantially attenuating all other frequencies. An acoustic filter transmits only desired sound frequencies. An optical filter transmits desired wavelength ranges in the visible, ultraviolet, and infrared spectrums.

**filter capacitor** A capacitor used in a power-supply filter system to provide a low-reactance path for alternating currents and thereby suppress ripple currents, without affecting direct currents. Electrolytic capacitors are generally used for this purpose.

**filter center** An information center at which all radar and other observed information concerning movements of friendly and enemy planes within a certain sector is screened and disseminated.

**filter choke** An iron-core coil used in a power-supply filter system to pass direct current while offering high impedance to pulsating or alternating currents.

**filtration** Removal of some components of a heterogeneous beam of x-ray and other radiation by inserting in the beam path a sheet of material such as aluminum or copper.

**final amplifier** The transmitter stage that feeds the antenna.

**finder** An optical or electronic device that shows the field of action covered by a television camera.

**fine chrominance primary** The chrominance primary that is associated with the greater transmission bandwidth in the two-primary U. S. system of color television. The fine chrominance primary is the I signal, and has frequency components up to 1.5 mc. The coarse chrominance primary is the Q signal, and has a bandwidth of only 0.5 mc.

**fine structure** The occurrence of a spectral line in spectroscopy as a doublet, triplet, or other multiple, due to interaction between the spin angular momentum and the orbital angular momentum of the electrons in the emitting atoms.

**fine tuning control** A control used to make small changes in the frequency of the r-f oscillator in a television tuner, usually by means of an adjustable capacitor, after switching to a desired channel.

**finished crystal blank** The finished crystal product after the completion of all processes, including application of electrodes to the faces of the crystal. Also called piezoid.

**finite** Having fixed and definite limits or magnitudes.

**finite clipping** Clipping in which the threshold level is large but is below the peak input signal amplitude.

**fin waveguide** A waveguide containing a thin longitudinal metal fin that serves to increase the wavelength range over which the waveguide will transmit signals efficiently. Usually used with circular waveguides.

**Firebee** A high-speed drone used chiefly as a target.

**fire control** Control of the aiming and firing of guns or rockets.

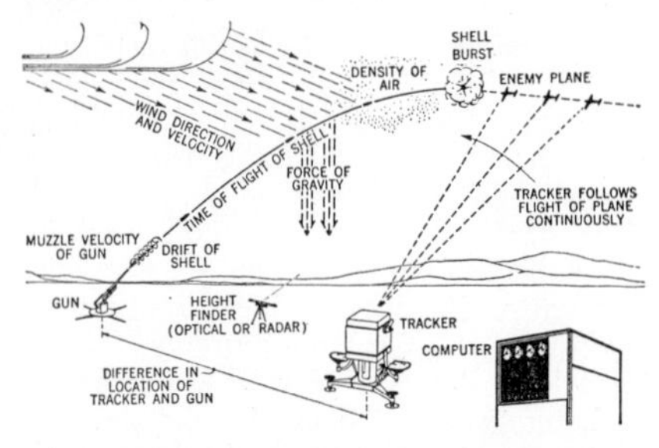


Fire control of antiaircraft gun.

**fire-control radar** Radar equipment used in a fire-control system.

**fire-control system** A system that determines the azimuth and elevation at which a gun or missile launcher must be pointed, the instant at which the missile must be launched, and the time for which the fuze must be set if using a timed-fuze missile. A typical system includes position-finding, tracking, predicting, ballistic, correcting, data-transmitting, compensating, pointing, and fuze-setting elements.

**fired tube** A tr, atr, or pre-tr tube in which a radio-frequency glow discharge exists at either the resonant gap, resonant window, or both.

**firing** 1. The transition from the unsaturated to the saturated state of a saturable reactor. 2. The gas ionization that initiates current flow in a gas discharge tube. 3. Excitation of a magnetron or tr tube by a pulse.

**first detector** *Mixer.*

**first Fresnel zone** The Fresnel zone that is centered on the line-of-sight path between two microwave antennas and bounded by all paths whose lengths are one half-wavelength longer than the direct path.

**first harmonic** *Fundamental frequency.*

**first-order servo** A servo that has zero static error but has a finite steady following error for a velocity input.

**first quantum number** *Main quantum number.*

**fishbone antenna** An end-fire array in which the elements are arranged in a plane along both sides of a transmission line, as in the skeleton of a fish.

**fishpaper** A type of fiber used in sheet form for insulating purposes where high mechanical strength is required, as in insulating transformer windings from the transformer core.

**fishpole antenna** *Whip antenna.*

**fissile** British term for *fissionable.*

**fission** *Nuclear fission.*

**fissionable** Having the property of being able to capture neutrons and thereupon split into two particles having high kinetic energy. Also called active and fissile (British).

**fissionable material** An element that will undergo fission when it absorbs a slow neutron.

**fission bomb** *Atomic bomb.*

**fission chamber** An ionization chamber used to detect slow neutrons. The inside wall of the chamber has a thin coating of uranium. A slow neutron produces a fission in the uranium, and the resulting highly ionizing fission fragments produce a count in the chamber.

**fission cross section** The cross section for the reaction of neutron absorption followed by fission.

**fission fragments** The nuclear species first produced when an atom such as $U^{238}$ or $Pu^{239}$ undergoes fission. In slow neutron fission the fragments are generally divided between a heavier group having masses around 140 and a lighter group having masses around 95.

**fission gammas** Gamma rays resulting from a fission process.

**fissioning distribution** The distribution of fission-producing neutrons throughout the energy spectrum.

**fission neutron** A neutron emitted as a result of nuclear fission. Fission neutrons have a continuous spectral distribution in energy, with a maximum at about 1 million electron-volts.

**fission poison** A fission fragment that has appreciable capture cross section for neutrons. An example is $Xe^{135}$, which has an absorption cross section of 3.5 million barns for slow neutrons.

**fission product** A nuclide produced by the fission of a heavy-element nuclide such as $U^{235}$ or $Pu^{239}$.

**fission recoil** A fission fragment at the instant of separation.

**fission spectrum** The energy distribution of neutrons arising from fission.

**fission yield** The per cent of fissions that gives a particular nuclide. The sum of all nuclide yields is within experimental error equal to 200%, indicating binary fission.

**fix** A determination of position without reference to any former position.

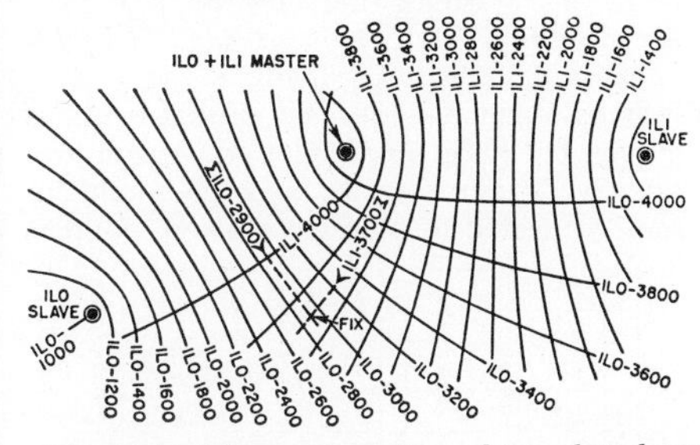


Fix obtained from one master and two slave loran stations. Pair 1L0 gave time difference of 2,900 microseconds and pair 1L1 gave time difference of 3,700 microseconds, so that location is at intersection of loran lines 1L0-2900 and 1L1-3700.

**fixed attenuator** *Pad.*

**fixed bias** A constant value of bias voltage, independent of signal strength. For an electron-tube circuit it may be provided by a battery or other d-c voltage source between cathode and ground or between grid and ground, or may be obtained from a voltage divider connected across the anode voltage source.

**fixed capacitor** A capacitor having a definite capacitance value that cannot be adjusted.

**fixed contact** A relatively immovable contact that is engaged and disengaged by a moving contact to make and break a circuit, as in a switch or relay.

**fixed-cycle operation** A computer operation that is performed in a fixed amount of time, under synchronous or clock-type control.

**fixed echo** An echo indication that remains stationary on a radar ppi display, indicating the presence of a fixed target.

**fixed-frequency iff** A type of iff equipment that responds immediately to every interrogation, permitting display of the response on a ppi indicator.

**fixed-frequency transmitter** A transmitter designed for operation on a single carrier frequency.

**fixed-loop radio compass** An aircraft radio compass having a loop antenna that is fixed in position in such a way that the pointer of the left-right indicator is centered when the aircraft is headed directly toward or away from the transmitting station to which the compass receiver is tuned.

**fixed-point calculation** A computer calculation in which a fixed location of the decimal point or binary point in each number is used or assumed.

**fixed-point system** A point system of positional notation in which the location of the point is assumed to remain fixed with respect to one end of the numerical expressions.

**fixed resistor** A resistor that has no provision for varying its resistance value.

**fixed service** A radio communication service between specific fixed points.

**fixing** The determination of position without reference to any former position.

**fixture** A device that holds a chassis or other part in a desired position during assembly operations but does not guide the tools performing the operations.

**flag** 1. A large sheet of metal or fabric used to shield television camera lenses from light when not in use. 2. A small metal tab that holds the getter during assembly of an electron tube.

**flag alarm** A semaphore-type indicator used in electronic navigation instruments to warn that the indications are unreliable.

**flame attenuation** Attenuation of radio signals by the ionized gases in the exhaust flame of a guided missile when the transmission path between the missile and the ground station passes through the flared-out flame.

**flange** A fitting used at the end of a waveguide for the purpose of bolting it to a microwave component or to another waveguide. Also called waveguide flange.

**flap attenuator** A waveguide attenuator in which a contoured sheet of dissipative material is moved into the guide through a nonradiating slot to provide a desired amount of power absorption. Also called vane attenuator.

**flare** 1. A radar screen target indication having an enlarged and distorted shape due to excessive brightness. 2. British term for *horn antenna.*

**flareout** That portion of the approach path of an aircraft in which the vertical component is modified to lessen the impact of landing.

**flash arc** A sudden increase in the emission of large thermionic vacuum tubes, probably due to irregularities in the cathode surface. It sometimes causes complete breakdown.

**flashback voltage** The peak inverse voltage at which ionization occurs in a gas tube.

**flashing** Application of a high-frequency electromagnetic field to an electron tube to flash its getter during evacuation or to test the quality of the vacuum. A glow is seen inside a tube if there is enough gas left to cause ionization.

**flash magnetization** Magnetization of a ferromagnetic object by a current impulse of such short duration that magnetization does not penetrate beyond a shallow surface layer of the material. Sometimes used in electromagnetic crack detectors.

**flashover** An electric discharge around or over the surface of an insulator.

**flash radiography** Radiography in which the exposure time is extremely short, such as 1 microsecond.

**flash test** A method of testing insulation by applying momentarily a voltage much higher than the rated working voltage.

**flash tube** A gas discharge tube used in a photoflash unit to produce high-intensity, short-duration flashes of light for photography. Also called electronic flash tube and photoflash tube.

**flat coaxial transmission line** *Strip transmission line.*

**flat fading** Fading in which all components of the received radio signal fluctuate simultaneously in the same manner.

**flat leakage power** The peak r-f power transmitted through a tr or pre-tr tube after establishment of the steady-state r-f discharge.

**flat response** Uniform amplification or reproduction of a specified band of frequencies.

**flat-top antenna** An antenna having two or more lengths of wire parallel to each other and in a plane parallel to the ground, each fed at or near its midpoint.

**flat-top response** *Bandpass response.*

**F layer** One of the layers of ionized air occurring at various heights in the F region of the ionosphere, capable of reflecting radio waves back to earth at frequencies up to about 50 mc. Also called Appleton layer. In the daytime hemisphere there are normally two F layers, called the $F_1$ and $F_2$ layers.

**$F_1$ layer** The lower of the two ionized layers normally existing in the F region in the daytime hemisphere. It is usually somewhere between 90 and 150 miles above the earth, and chiefly affects frequencies from about 1.5 to about 25 mc.

**$F_2$ layer** The single ionized layer normally existing in the F region in the nighttime hemisphere. In the daytime hemisphere it is the higher of the two F layers.

**flection-point emission current** The value of diode current for which the second derivative of the current with respect to voltage has its maximum negative value. This current corresponds to the upper flection point of the diode characteristic.

**Fleming's rule** 1. *Left-hand rule.* 2. *Right-hand rule.*

**Fletcher-Munson curves** Equal-loudness curves for pure tones, plotted against frequency. They show the average sound intensity needed to produce a given loudness sensation throughout the audio-frequency range.

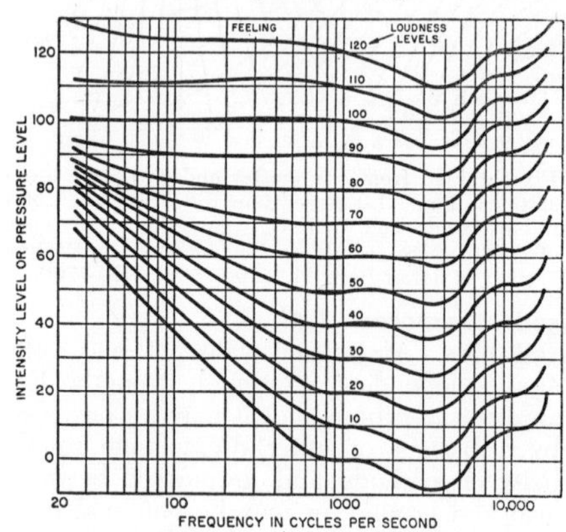


Fletcher-Munson curves of equal-loudness contours.

**flexible coupling** 1. A coupling designed to allow a limited angular movement between the axes of two waveguides. 2. A coupling that connects two shafts end to end and permits rotation even though the shafts are not exactly aligned.

**flexible resistor** A wirewound resistor having the appearance of a flexible lead. It is made by winding Nichrome resistance wire around a length of asbestos or other heat-resistant cord, then covering the winding with asbestos and a braided insulating covering. This covering is generally color-coded to indicate the resistor value.

**flexible shaft** A shaft that transmits rotary motion at any angle up to about 90°. Used in electronic equipment to permit mounting adjustable controls at optimum positions with respect to other parts while still securing desirable groupings of controls on the panel.

**flexible waveguide** A waveguide that can be bent or twisted without appreciably changing its electrical properties.

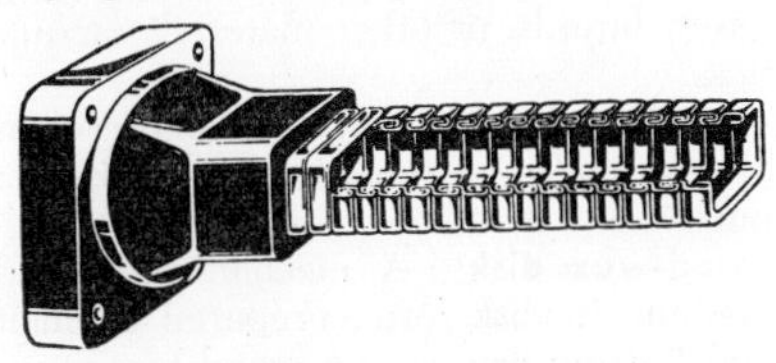

Flexible waveguide.

**flicker** A visual sensation produced by periodic fluctuations in light at rates ranging from a few cycles per second to a few tens of cycles per second. In U. S. television, interlaced scanning eliminates the flicker that might otherwise be noticed at 30 frames per second.

**flicker effect** Random variations in the output current of an electron tube having an oxide-coated cathode, due to random changes in cathode emission.

**flicker noise** Electric noise produced by the flicker effect in an electron-tube circuit. It is a low-frequency noise that is in excess of shot noise.

**flight indicator** *Artificial horizon.*

**flight instrument** Any aircraft instrument that indicates the altitude, attitude, airspeed, drift, or direction of an aircraft.

**flight path** A line in space planned or used as the path for an aircraft or missile.

**flight-path angle** The acute angle between the horizontal and the flight path of an aircraft or missile during ascent or descent.

**flight-path computer** A computer that includes all of the functions of a course-line computer and also provides means for controlling the altitude of an aircraft in accordance with a desired plan of flight.

**flight-path deviation indicator** An instrument providing a visual indication of deviation from a planned flight path.

**flight-path-reference flight** Stabilized flight in which control information is obtained from a navigation system capable of providing heading or altitude guidance, or both, with respect to a desired flight path. An example is flight in which a vhf omnirange, distance-measuring equipment, or instrument landing system feeds an automatic pilot.

**flight-path selector** An instrument used with a flight-path computer to preset the values defining the flight path to a way point.

**flight simulator** Equipment that simulates any or all of the conditions of actual flight. Used chiefly for training purposes. Also called flight trainer.

**flight track** The path in space actually traced by a vehicle. Flight track is the three-dimensional equivalent of track.

**flight trainer** *Flight simulator.*

**flip coil** A small coil used to measure the strength of a magnetic field. It is placed in the field, connected to a ballistic galvanometer or other instrument, and suddenly flipped over 180°. The resulting generated electric energy is a measure of magnetic field strength. Alternatively, the coil may be held stationary and the magnetic field reversed.

**flip-flop circuit** A two-stage multivibrator circuit having two stable states. In one state, the first stage is conducting and the second is cut off. In the other state, the second stage is conducting and the first stage is cut off. A trigger signal changes

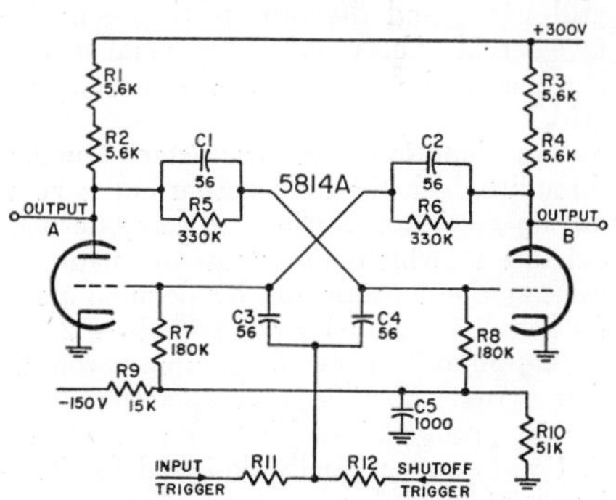


Flip-flop circuit, using preferred circuit design by National Bureau of Standards, provides positive or negative 180-volt gate pulse in response to 25-volt input trigger pulse. Gate pulse is terminated by 25-volt shutoff trigger pulse. Values of R11 and R12 in input circuit can be from 5,000 to 50,000 ohms each.

the circuit from one state to the other, and the next trigger signal changes it back to the first state. For counting and scaling purposes, a flip-flop can be used to deliver one output pulse for each two input pulses. Also called bistable multivibrator, Eccles-Jordan circuit, and trigger circuit.

**floating action** Controller action in which there is a predetermined relation between the deviation and the speed of a final control element. A neutral zone, in which no motion of the final control element occurs, is often used in floating action.

**floating address** *Symbolic address.*

**floating-average-position action** Floating action in which there is a predetermined relation between deviation of the controlled variable and the rate of change of the time-average position of a final control element that is moved periodically from one of two fixed positions to the other.

**floating battery** A storage battery connected permanently in parallel with another power source. The battery normally handles only small charging or discharging currents, but takes over the entire load upon failure of the main supply.

**floating-carrrier modulation** *Controlled-carrier modulation.*

**floating charge** Application of a constant voltage to a storage battery, sufficient to maintain an approximately constant state of charge while the battery is idle or on light duty.

**floating grid** An electron-tube grid that is not connected to a circuit. The grid assumes a negative potential with respect to the cathode, due to electrons hitting the grid wires, and the tube is then sensitive to external effects such as movement of a hand near the envelope. Also called free grid.

**floating junction** A transistor junction through which the average current is zero.

**floating-point calculation** A computer calculation in which provisions are made for varying the location of the decimal point (if base 10) or binary point (if base 2).

**floating-point routine** A computer routine that permits floating-point operation for a specific problem. Usually used in computers that were not originally designed for floating-point operation.

**floating-point system** A point system of positional notation in which the position of the point is regularly recalculated and may be moved. A floating-point system usually locates the point by expressing a power of the base, and involves the use of two sets of digits. For floating decimal notation the base is 10, so 6,200,000 would be 6.2, 6. For floating binary notation the base is 2, so 88 would be 11, 3.

**float switch** A switch actuated by a float at the surface of a liquid.

**flock** Finely divided felt used on phonograph turntable surfaces, underneath microphone stands, and in similar locations where a nonscratching surface is desired. It is sifted over a layer of cement applied to the surface.

**floodlighting** Covering a wide area with radar waves.

**flood projection** An optical method of facsimile scanning in which the subject copy is floodlighted and the scanning spot is defined by an aperture that moves in the path of the reflected or transmitted light.

**flow** The movement of electric charges, gases, liquids, or other materials or quantities.

**flow diagram** A graphical representation of a program or routine for a digital computer.

**flowed-wax disk** A mechanical recording medium in disk form, prepared by melting and flowing wax onto a metal base.

**flowmeter** An instrument that measures and indicates the rate of flow of a liquid or gas.

**flow transmitter** A device used to measure the flow of liquids in pipe lines and convert the results into proportional electric signals that can be transmitted to distant receivers or controllers.

**fluctuation noise** *Random noise.*

**fluidized reactor** A nuclear reactor in which the fuel has been given the properties of a quasi-fluid, such as by suspension of fine fuel particles in a carrying gas or liquid.

**Fluon** Trademark of Imperial Chemical Industries, England, for their polytetrafluoroethylene resin that is comparable to Teflon.

**fluorescence** Emission of light or other electromagnetic radiation by a material exposed to another type of radiation or to a beam of particles, with the luminescence ceasing within about $10^{-8}$ second after irradiation is stopped. Certain minerals fluoresce in characteristic colors during exposure to ultraviolet radiation. A cathode-ray screen fluoresces when hit by the electron beam in the tube. Other materials give off characteristic x-rays when irradiated by higher-frequency x-rays.

**fluorescence yield** The probability that an excited atom will emit an x-ray photon rather than an Auger electron in the first transition. The yield is a number between zero and one, depending on the element and its state of excitation.

**fluorescent lamp** A tubular discharge lamp in which ionization of mercury vapor produces radiation that activates the fluorescent coating on the inner surface of the glass.

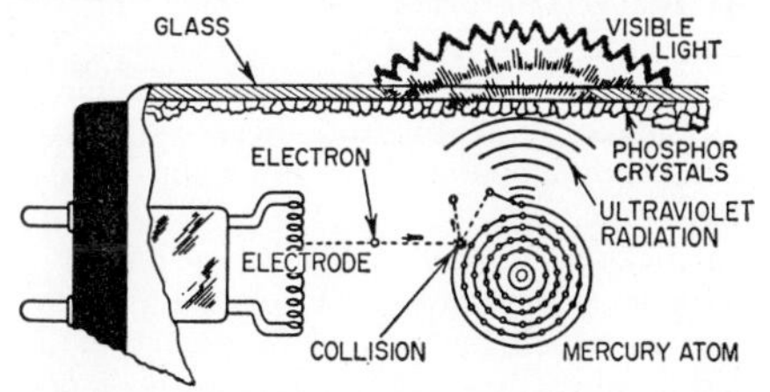


Fluorescent lamp construction and operation. Electron emitted by electrode at one end collides with one of electrons in outer ring of mercury atom, producing ultraviolet radiation, which in turn acts on phosphor crystals inside glass wall to produce visible light.

**fluorescent screen** A sheet of material coated with a fluorescent substance that emits visible light when irradiated with ionizing radiation such as x-rays or electron beams. In a cathode-ray tube the fluorescent screen is a coating on the inside surface of the tube face.

**fluorine** [symbol F] A highly corrosive and poisonous gaseous element. Atomic number is 9.

**fluorocarbon resin** General term for a family of plastics having excellent electrical insulating qualities and relatively high service temperatures. Examples include polychlorotrifluoroethylene resin, marketed as Kel-F, and polytetrafluoroethylene resin, marketed as Teflon and Fluon.

**fluorography** Photography of an image produced on a fluorescent screen.

**fluorometer** An instrument that measures the intensity of x-rays and other radiation by measuring the intensity of the fluorescence produced.

**fluoroscope** A fluorescent screen designed for use with an x-ray tube to permit direct visual observation of x-ray shadow images of objects interposed between the x-ray tube and the screen. The screen serves to transform invisible x-ray radiation into visible light.

**fluoroscopy** Use of a fluoroscope for x-ray examination.

**flush-printed circuit** A printed circuit in which the outer surface of the reproduced pattern is flush with the outer surface of the base.

**flutter** 1. Distortion that occurs in sound reproduction as a result of undesired speed variations during the recording, duplicating, or reproducing process. The variations in speed and hence pitch occur at a much higher rate than for wow. 2. A fast-changing variation in received signal strength, such as may be caused by antenna movements in a high wind or interaction with a signal on another frequency.

**flutter echo** A radar echo consisting of a rapid succession of reflected pulses resulting from a single transmitted pulse.

**flux** 1. The electric or magnetic lines of force in a region. 2. The rate of flow of particles or photons through a unit area. 3. The product of the number of particles per unit volume and their average velocity. 4. A material used to remove oxide films from the surfaces of metals in preparation for soldering, brazing, or welding. Rosin is widely used as a flux for soldering electronic circuits.

**flux density** *Magnetic induction.*

**flux gate** A detector that gives an electric signal whose magnitude and phase are proportional to the magnitude and direction of the external magnetic field acting along its axis. A flux gate may consist of three magnetic cores having appropriate excitation windings and load windings to give perfect balance in the absence of external magnetic fields.

**flux-gate compass** *Gyro flux-gate compass.*

**flux-gate magnetometer** A magnetometer in which the degree of saturation of the core by an external magnetic field is used as a measure of the strength of the field.

**flux guide** A shaped piece of magnetic material used to guide electromagnetic flux in desired paths in induction heating. The guide may be used either to direct flux to preferred locations or to prevent the flux from spreading beyond definite regions.

**flux leakage** Magnetic flux that does not pass through an air gap or other part of a magnetic circuit where it is required.

**flux linkage** The product of the number of turns in a coil and the number of magnetic lines of force passing through the turns. Also called linkage.

**fluxmeter** An instrument for measuring magnetic flux. It is usually calibrated to read either in maxwells or webers.

**flyback** The time interval in which the electron beam of a cathode-ray tube returns to its starting point after scanning one line or one field of a television picture or after completing one trace in an oscilloscope. In television the beam is blanked out during flyback. Also called retrace and return trace.

**flyback power supply** A high-voltage power supply used to produce the d-c voltage of about 10,000 to 25,000 volts required for the second anode of a cathode-ray tube in a television receiver or oscilloscope. The sudden reversal of horizontal deflection-coil current in the horizontal output transformer during each flyback induces a voltage pulse that is increased to the required higher value by autotransformer action, then rectified and filtered. Also called kickback power supply.

**flyback transformer** *Horizontal output transformer.*

**flying-spot scanner** A television scanner in which a simple phototube replaces a more complex iconoscope or other pickup tube at the camera. A moving spot of light, controlled either mechanically or electrically, scans the image field to be transmitted. The light reflected from or transmitted

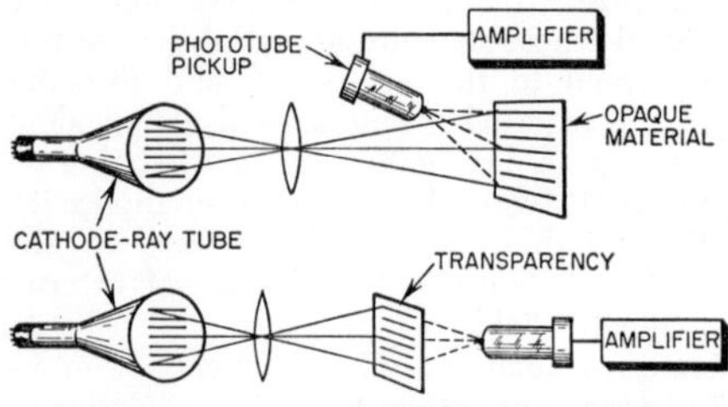


Flying-spot scanner in which light source is spot that traces raster pattern on screen of cathode-ray tube. Position of phototube depends on whether material being televised is opaque or transparent.

by the image field is picked up by the phototube to generate the video signal. Used today chiefly for film and slide transmission. A high-intensity cathode-ray tube is generally used as the flying-spot light source. Also called light-spot scanner.

**flywheel effect** The ability of a resonant circuit to maintain oscillation at an essentially constant frequency when fed with short pulses of energy at constant frequency and phase.

**flywheel synchronization** Automatic frequency control of a scanning system by using the average timing of the incoming sync signals, rather than by making each pulse trigger the scanning circuit. Used in high-sensitivity television receivers designed for fringe-area reception, where noise pulses might otherwise trigger the sweep circuit prematurely.

**f-m** 1. Abbreviation for *frequency-modulated.* 2. Abbreviation for *frequency modulation.*

**f-m broadcast band** Abbreviation for *frequency-modulation broadcast band.*

**f-m broadcast channel** Abbreviation for *frequency-modulation broadcast channel.*

**f-m cyclotron** *Synchrocyclotron.*

**f-m/f-m telemetering** A telemetering system used in tests of guided missiles. The subcarrier bands are frequency-modulated on subcarrier waves and these are in turn frequency-modulated on the main r-f carrier wave.

**f-m pickup** *Variable-capacitance pickup.*

**f-m/p-m telemetering** A telemetering system in which the several frequency-modulated subcarriers are used to phase-modulate the main r-f carrier.

**f-m radar** Abbreviation for *frequency-modulated radar.*

**f-m radio altimeter** Abbreviation for *frequency-modulated radio altimeter.*

**f-m receiver** Abbreviation for *frequency-modulation receiver.*

**f-m scanning sonar** Abbreviation for *frequency-modulated scanning sonar.*

**f-m transmitter** Abbreviation for *frequency-modulated transmitter.*

**f-m tuner** Abbreviation for *frequency-modulation tuner.*

**f number** A lens rating obtained by dividing the focal length of the lens by the effective maximum diameter of the lens. The lower the f number, the shorter is the exposure required or the lower is the illumination needed for satisfactory results with a television camera or ordinary camera. An f number of 3.5, for example, is usually expressed as f/3.5.

**foamed plastic** A resinous material that has been expanded into a multicellular structure having low density and relatively high strength.

**focal length** The distance between the optical center of a lens and the television camera screen or photographic camera film when the camera is focused on a distant object.

**focal spot** The small area on the target of an x-ray tube that gives off x-rays when hit by the electron stream.

**focus** 1. The point at which rays of light or electrons of a beam converge to form a minimum-diameter spot. 2. To move a lens or adjust a voltage or current in order to obtain a focus.

**focus control** A control that adjusts spot size at the screen of a cathode-ray tube, to give the sharpest possible image. It may vary the current through a focusing coil or change the position of a permanent magnet.

**focus-defocus mode** A mode of storage of binary digits in which the writing beam of a cathode-ray storage tube is initially focused. For one type of binary digit it remains focused, while for the other type it is suddenly defocused to a small concentric circular area, in the time interval before the beam is cut off and moved to the next position.

**focusing** 1. The process of controlling convergence or divergence of the electron paths within one or more beams to obtain a desired image or a desired current density distribution in the beam. 2. The process of moving an optical lens toward or away from a screen or film to obtain the sharpest possible image of a desired object.

**focusing and switching grille** A color-selecting electrode system used in one type of color picture tube. An array of wires includes at least two mutually insulated sets of conductors. Switching is accomplished by changing the voltage between the sets of wires. Focusing is accomplished by maintaining the proper average voltages on the array and on the phosphor screen of the color picture tube.

**focusing anode** An anode used in a cathode-ray tube to change the size of the electron beam at the screen. Varying the voltage on this anode alters the paths of electrons in the beam and thus changes the position at which they cross or focus.

**focusing coil** A coil that produces a magnetic field parallel to an electron beam, for the purpose of focusing the beam. The coil is usually mounted on the neck of a cathode-ray picture tube, and carries a direct current whose value can be adjusted by a focus control rheostat.

**focusing electrode** An electrode to which a potential is applied to control the cross-sectional area of the electron beam in a cathode-ray tube.

**focusing magnet** A permanent magnet used to produce a magnetic field for focusing an electron beam.

**fog chamber** *Cloud chamber.*

**foil** A flexible sheet of thin aluminum, lead, or tin, widely used in fixed capacitors.

**folded cavity** A cavity used in some klystrons to make the incoming wave act on the electron stream from the cathode at several places to produce a cumulative effect.

**folded-dipole antenna** A dipole antenna whose outer ends are folded back and joined together at the center. The impedance is about 300 ohms, as compared to 70 ohms for a single-wire dipole. Widely used with television and f-m receivers.

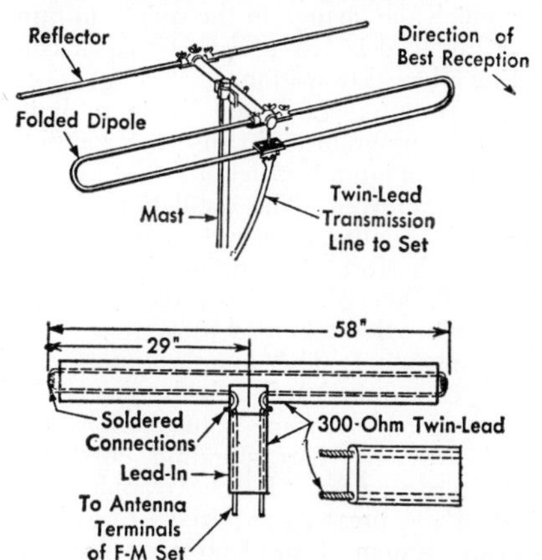


Folded-dipole antenna with reflector mounted on mast for television reception, and folded-dipole antenna made from twin-line for f-m reception.

**folded horn** An acoustic horn in which the path from throat to mouth is folded or curled to give the longest possible path in a given volume.

**foldover** Picture distortion seen as a white line on either side, top, or bottom of a television picture. Generally caused by nonlinear operation in either the horizontal or vertical deflection circuits of a receiver.

**foot-candle** [abbreviated ft-c] A unit of illuminance used when the foot is taken as the unit of length. It is the illuminance on a surface one square foot in area on which there is a uniformly distributed flux of one lumen, or the illuminance at a surface all points of which are at a distance of one foot from a uniform source of one candle.

**foot-lambert** [abbreviated ft-l] A unit of luminance equal to $1/\pi$ candle per square foot, or to the uniform luminance of a perfectly diffusing surface emitting or reflecting light at the rate of one lumen per square foot. A foot-candle is a unit of incident light, while a foot-lambert is a unit of emitted or reflected light. For a perfectly reflecting and perfectly diffusing surface, the number of foot-candles is equal to the number of foot-lamberts.

**forbidden band** An energy band in which there can be no electrons in a given substance.

**forbidden-combination check** A test for the occurrence of a nonpermissible code expression in a computer. Used to detect computer errors.

**forbidden transition** A transition between two states of a quantum-mechanical system for which the change in the quantum numbers involved is less probable than a competing allowed transition.

**force** To intervene manually in a digital computer program and cause the computer to execute a jump instruction.

**forced oscillation** The oscillation of some physical quantity of a system when external periodic forces determine the period of the oscillation. Also called forced vibration.

**forced vibration** *Forced oscillation.*

**force factor** 1. The complex quotient of the force required to block the mechanical system of an electromechanical transducer, divided by the corresponding current in the electric system. 2. The complex quotient of the pressure required to block the acoustic system of an electroacoustic transducer, divided by the corresponding current in the electric system.

**forcing** The application of control impulses that are larger than warranted by the error in a system, to achieve a greater rate of correction.

**forcing function** A mathematical description of the characteristics of a source of excitation.

**forcing resistance** A resistance used in series with a control winding of a magnetic amplifier or dynamoelectric amplifier to give higher response speed.

**foreign-body locator** A device for locating foreign metallic bodies in tissue by use of suitable probes that generate a magnetic field. The presence of a magnetic body within this field is indicated by a meter or sound signal.

**fork beat** *Carrier beat.*

**fork oscillator** An oscillator that uses a tuning fork as the frequency-determining element.

**form factor** 1. The ratio of the effective value of an alternating quantity to the average value during a half-cycle. The form factor is about 1.11 for a pure sine wave, and is equal to the ratio of the readings obtained for a given a-c quantity on root-mean-square and rectifier-type meters. 2. A factor that takes the shape of a coil into account when computing its inductance. Also called shape factor.

**Formica** Trademark of Synthane Corp. for their line of plastics and plastic products.

**forming** Application of voltage to an electrolytic capacitor, electrolytic rectifier, or semiconductor device to produce a desired permanent change in electrical characteristics as a part of the manufacturing process.

**form-wound coil** A coil that is formed or bent to an irregular shape, as for a deflection yoke.

**forsterite** A ceramic having low dielectric loss over a wide range of temperatures up to about 1,000°C. Good at high frequencies. Used in making insulators.

**fortuitous telegraph distortion** Distortion that results in departure, for one occurrence of a particular signal pulse, from the average combined effects of bias and characteristic telegraph distortion.

**forty-five/forty-five pickup** *Stereo pickup.*

**forty-five rpm record** A 7-inch-diameter disk recorded and reproduced at a speed of 45 rpm, having a center hole 1.5 inches in diameter and grooves designed for a stylus having a point radius of 1 mil.

**forward bias** A bias voltage that is applied to a p-n junction in the direction that causes a large current flow. Used in some semiconductor diode circuits.

**forward current** The current that flows through a rectifying junction in the conducting direction.

**forward direction** The direction of lesser resistance to current flow through a rectifier.

**forward path** The transmission path from the loop actuating signal to the loop output signal in a feedback control loop.

**forward recovery time** The time required for the forward current of a semiconductor diode to reach a specified value after a forward bias is instantaneously applied.

**forward-scatter propagation** *Scatter propagation.*

**forward transadmittance** The transfer admittance of a klystron or other microwave tube. It is equal to the fundamental component of the short-circuit current induced in the second of any two gaps, divided by the fundamental component of the voltage across the first gap.

**forward transfer function** The transfer function of the forward path in a feedback control loop.

**forward voltage** A voltage having the correct polarity to send current through a rectifier in the forward direction.

**forward wave** A wave whose group velocity is in the same direction as the electron-stream motion in a traveling-wave tube.

**Foster-Seeley discriminator** *Phase-shift discriminator.*

**four-course radio range** A radio range that beams on-course signals in four different directions for aircraft guidance. The Adcock radio range is an example.

**Fourier analysis** The process of determining the amplitude, frequency, and phase of each of the sinusoidal components in a given waveform.

**Fourier series** A mathematical expression by which any periodic function can be represented as a combination of sine and cosine terms that are integral multiples of a fundamental frequency.

**Fourier transform** A mathematical expression relating the energy in a transient to that in a continuous energy spectrum of adjacent frequency components.

**four-pole network** *Two-terminal pair network.*

**four-wire circuit** A circuit in which communication signals are transmitted in one direction on one path and in the other direction on the other path. Multiplexing methods such as frequency division or time division may be used to reduce the actual number of wires required.

**four-wire repeater** A repeater that provides amplification in opposite directions on two transmission paths.

**fps** Abbreviation for *feet per second.*

**fractional distillation** A distillation process involving the use of a fractionating tower to separate isotopes in liquid form by evaporation and recondensation. An upward-directed stream of vapor and a downward-directed stream of liquid are constantly exchanging molecules.

**fractional-horsepower motor** Any motor built into a frame smaller than that for a motor having an open construction and a continuous rating of 1 horsepower at 1,800 revolutions per minute.

**fractional isotopic abundance** The ratio of the number of atoms of a specified isotope to the total number of atoms making up a particular element. Commonly expressed as a percentage.

**Frahm frequency meter** *Vibrating-reed frequency meter.*

**frame** 1. One complete coverage of a television picture. In the United States a frame contains 525 horizontal scanning lines, repeated at the rate of 30 frames per second. Each frame is scanned in two interlaced fields, one covering the 262.5 odd scanning lines and the other the remaining even scanning lines. In Great Britain a 405-line picture is scanned 25 times per second in two interlaced 202.5-line fields. 2. A single complete picture on motion-picture film. For 35-mm film the standard rate of projection is 24 frames per second. This means that a special projector is required to convert this to 30 frames per second for U. S. television. 3. A rectangular area representing the size of copy handled by a facsimile system. The width of a facsimile frame is the available line width, and the length is determined by the service requirements.

**frame frequency** The number of times per second that the frame is completely scanned in television. It is 30 frames per second in the United States and 25 frames per second in Great Britain. Also called picture frequency.

**frame of reference** A set of lines or surfaces used as references for coordinates defining a moving or stationary point.

**frame period** A time interval equal to the reciprocal of the frame frequency. In the United States the frame period is 1/30th second.

**framer** A device for adjusting facsimile equipment so that the start and end of a recorded line are the same as on the corresponding line of the subject copy.

**framing** 1. Adjusting a television picture to a desired position on the screen of the picture tube. 2. Adjusting a facsimile picture to a desired position in the direction of line progression. Also called phasing.

**framing control** 1. A control that adjusts the centering, width, or height of the image on a television receiver screen. 2. A control that shifts a received facsimile picture horizontally.

**framing signal** A signal used to adjust a facsimile picture to a desired position in the direction of line progression.

**francium** [symbol Fr] The heaviest alkali metal element. Atomic number is 87. Formerly called virginium.

**Franklin antenna** An antenna several half-wavelengths long, having nonradiating phasing coils between half-wave sections.

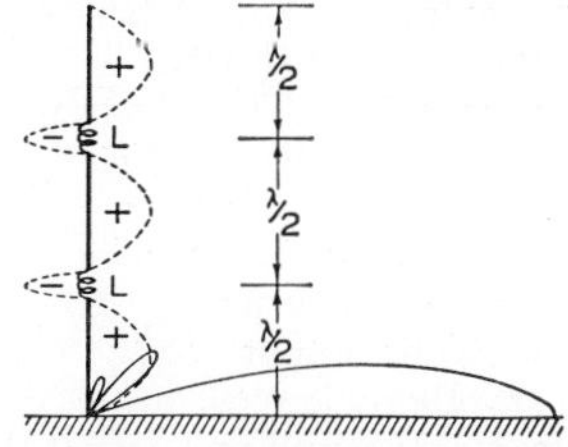

Franklin antenna, showing vertical radiation pattern for all directions (solid curves) and current distribution curves (dotted).

**Fraunhofer diffraction** Diffraction that occurs when radiation passes the edges of one or more apertures.

**Fraunhofer lines** Dark lines in the spectrum of sunlight, as obtained with a spectroscope. They are due to absorption of certain wavelengths by gases and vapors in the solar atmosphere.

**Fraunhofer region** The region in which energy flow from an antenna proceeds essentially as though coming from a point source located in the vicinity of the antenna. It is beyond the Fresnel region, and begins at a distance equal to about twice the square of antenna length divided by wavelength. Also called far region and far zone.

**free-air ionization chamber** An air-filled ionization chamber in which a sharply defined beam of radiation passes between the electrodes without striking them or other internal parts of the equipment. The observed ionization current is then due entirely to ions and electrons resulting from the action of radiation on the air. Used for x-ray dosimetry. Also called open-air ionization chamber.

**free electron** An electron that is not constrained to remain in a particular atom. It is therefore able to move freely in matter or in a vacuum, when acted on by external electric or magnetic fields.

**free field** A field in which the effects of boundaries are negligible over the region of interest. An object in a free sound field will have the same disturbing effect as a boundary, however, unless the acoustic impedance of the object matches the acoustic impedance of the medium.

**free-field current response** The ratio of the short-circuit output current of a microphone or other electroacoustic transducer to the free-field sound pressure at the transducer location. Usually expressed in decibels, as 20 times the logarithm to the base 10 of the quotient of the observed ratio divided by a reference ratio that is usually 1 ampere per microbar. Also called receiving current sensitivity.

**free-field emission** The electron emission that occurs when the electric field at the surface of an emitter is zero.

**free-field room** *Anechoic room.*

**free-field voltage response** The ratio of the open-circuit output voltage of a microphone or other electroacoustic transducer to the free-field sound pressure at the transducer location. Usually expressed in decibels, as 20 times the logarithm to the base 10 of the quotient of the observed ratio divided by a reference ratio that is usually 1 volt per microbar. Also called receiving voltage sensitivity.

**free grid** *Floating grid.*

**free gyro** A gyroscope mounted in two or more gimbal rings so that it is free to maintain a fixed orientation in space, with no external means for changing its normal precession.

**free impedance** The impedance at the input of a transducer when the impedance of its load is made zero.

**free motional impedance** The complex impedance remaining after the blocked impedance of a transducer has been subtracted from the free impedance.

**free network** A radio communication network in which any station may communicate with any other station in the network without permission from the network control station.

**free oscillation** Oscillation that continues in a circuit or system after the applied force has been removed. The frequency is determined by the parameters in the system or circuit. Also called free vibration and shock-excited oscillation.

**free-point tester** An adapter that permits transferring a tube from its operating socket to a socket on a test panel at which voltage or current measurements for any tube terminal can readily be made. Connections are made to the original socket with a cable and plug.

**free progressive wave** A wave in a medium free from boundary effects. Also called free wave.

**free-running** Operating without external synchronizing pulses.

**free-running frequency** The frequency at which a normally synchronized oscillator operates in the absence of a synchronizing signal.

**free-running multivibrator** *Astable multivibrator.*

**free space** A region high enough so that the radiation pattern of an antenna is not affected by surrounding objects such as buildings, trees, hills, and the earth.

**free-space field intensity** The field intensity that would exist at a point in the absence of waves reflected from the earth or other reflecting objects.

**free-space loss** The theoretical radiation loss, depending only on frequency and distance, that would occur if all variable factors were disregarded when transmitting energy between two antennas.

**free-space radiation pattern** The radiation pattern that an antenna would have in free space.

**free vibration** *Free oscillation.*

**free wave** *Free progressive wave.*

**F region** The region of the ionosphere between about 90 miles and 250 miles above the earth, above the D and E regions.

**frequency** [symbol $f$] The number of complete cycles per unit of time for a periodic quantity such as alternating current, sound waves, or vibrating objects. Frequency is expressed in cycles, kilocycles, megacycles, and kilomegacycles per second. Frequency in cycles per second is equal to 300,000,000 divided by wavelength in meters. Frequency in megacycles is equal to 300 divided by wavelength in meters.

**frequency allocation** Assignment of available frequencies in the radio spectrum to specific stations and for specific purposes, to give maximum utilization of frequencies with minimum interference between stations. Allocations in the United States are made by the Federal Communications Commission.

**frequency band** A continuous range of frequencies extending between two limiting frequencies. A band may include a number of channels; thus, the broadcast band extends from 535 to 1,605 kc.

**frequency bridge** A bridge in which the balance varies with frequency in a known manner, such as the Wien bridge. Used to measure frequency.

**frequency calibrator** An instrument that generates a highly accurate signal at one or more fixed frequencies, for use in calibrating other frequency sources.

**frequency changer** *Frequency converter.*

**frequency characteristic** *Amplitude-frequency response.*

**frequency compensation** *Compensation.*

**frequency constant** The number relating a natural vibrational frequency to a linear dimension of a piezoid (finished crystal blank).

**frequency conversion** The process of converting the carrier frequency of a received signal from its original value to the intermediate-frequency value in a superheterodyne receiver.

**frequency converter** A circuit, device, or machine that changes an alternating current from one frequency to another, with or without a change in voltage or number of phases. In a superheterodyne receiver, the oscillator and mixer-first detector stages together serve as the frequency converter. Also called frequency changer.

**frequency cutoff** The frequency at which the current gain of a transistor drops 3 db below the low-frequency gain value.

**frequency departure** The amount of variation of a carrier frequency or center frequency from its assigned value. Use of the term frequency deviation for this meaning is deprecated because frequency deviation is used for frequency and phase modulation.

**frequency deviation** The peak difference between the instantaneous frequency of a frequency-modulated wave and the carrier frequency. Also called deviation.

**frequency-deviation meter** An instrument that indicates the number of cycles a broadcast transmitter has drifted from its assigned carrier frequency.

**frequency discriminator** A discriminator circuit that delivers an output voltage which is proportional to the deviations of a signal from a predetermined frequency value. Used in frequency-modulated receivers and automatic frequency control circuits. Examples include the double-tuned detector, locked-in oscillator, and ratio detector.

**frequency distortion** Nonlinear distortion in which the relative magnitudes of the different frequency components of a wave are changed during transmission or amplification. Frequency distortion occurs when an audio amplifier cannot amplify equally well all the frequencies present in the input signal. Also called amplitude distortion, amplitude-frequency distortion, and waveform-amplitude distortion.

**frequency diversity** Diversity reception involving the use of carrier frequencies separated 500 cps or more and having the same modulation, to take advantage of the fact that fading does not occur simultaneously on different frequencies. The receiver minimizes the effects of fading by using at each instant the frequency having the higher signal strength.

**frequency-diversity radar** A radar that uses two or more frequencies simultaneously, with means for combining the resulting echoes or selecting the strongest of the echoes from the target at each instant.

**frequency divider** A harmonic conversion transducer in which the frequency of the output signal is an integral submultiple of the input frequency.

**frequency-division multiplex** [abbreviated fdm] A multiplex system for transmitting two or more signals over a common path by using a different frequency band for each signal.

**frequency doubler** An amplifier stage whose resonant anode circuit is tuned to the second harmonic of the input frequency. The output frequency is then twice the input frequency. Also called doubler.

**frequency drift** A gradual change in the frequency of an oscillator or transmitter, due to temperature or other changes in the circuit components that determine frequency.

**frequency frogging** Interchanging of frequency allocations for carrier channels to prevent singing, reduce crosstalk, and reduce the need for equalization. Modulators in each repeater translate a low-frequency group to a high-frequency group, and vice versa.

**frequency-hour** One frequency used for one hour, as in international short-wave broadcasting.

**frequency hysteresis** Failure of an oscillator to change frequency smoothly when continuously tuned, due to use of a long transmission line between oscillator and load.

**frequency interlace** Interlace of interfering signal frequencies with the spectrum of harmonics of scanning frequencies in television, to minimize the effect of interfering signals by altering the appearance of their pattern on successive scans.

**frequency keying** Keying in which the carrier frequency is shifted alternately between two predetermined values.

**frequency meter** An instrument for measuring the frequency of an alternating current.

**frequency-modulated** [abbreviated f-m] Pertaining to frequency modulation.

**frequency-modulated cyclotron** *Synchrocyclotron.*

**frequency-modulated pickup** *Variable-capacitance pickup.*

**frequency-modulated radar** [abbreviated f-m radar] A continuous-wave radar in which the carrier frequency is alternately increased and decreased at a predetermined rate. The frequency of the beat between the returning echo and the wave transmitted at the instant of echo arrival is proportional to range.

**frequency-modulated radio altimeter** [abbreviated f-m radio altimeter] A radio altimeter used in aircraft to give accurate absolute altitude indications from a few feet above the surface up to a limit of about 5,000 feet. The frequency of the downward-radiated radio wave is varied back and forth continuously at some cyclic rate. Some of the energy from the transmitter is also fed to the input of the receiver for mixing with the energy reflected from the surface. The beat between the direct and reflected waves, equal to the change in transmitter frequency during the time of travel of the wave to the surface and back, is a direct indication of altitude. Also called low-altitude radio altimeter and terrain-clearance indicator.

**frequency-modulated scanning sonar** [abbreviated f-m scanning sonar] Scanning sonar in which the frequency of the transmitter is electronically decreased at a constant rate for several seconds, then suddenly increased to its original value and the process repeated. The tone of the echo is then related to target position. The chief drawback is doppler error when there is relative motion of transmitter and target.

**frequency-modulated transmitter** [abbreviated f-m transmitter] A radio transmitter that transmits a frequency-modulated wave.

**frequency modulation** [abbreviated f-m] Modulation in which the instantaneous frequency of the modulated wave differs from the carrier frequency by an amount proportional to the instantaneous value of the modulating wave. The amplitude of the modulated wave is constant. A frequency-modulation broadcast system is practically immune to atmospheric and manmade interference.

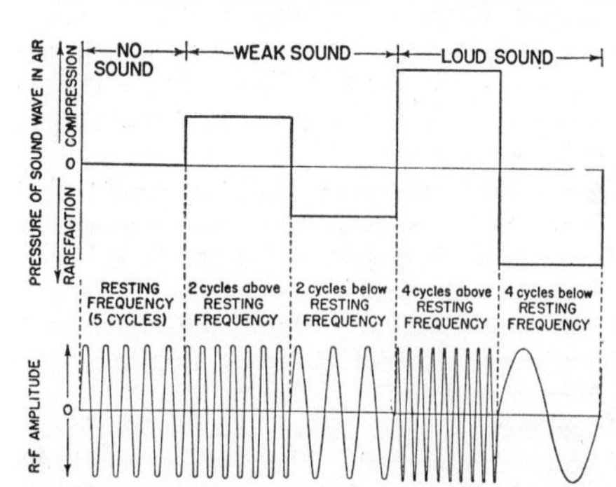

Frequency modulation, showing how carrier or resting frequency varies with strength of audio modulation.

**frequency-modulation broadcast band** [abbreviated f-m broadcast band] The band of frequencies extending from 88 to 108 mc, used for frequency-modulation radio broadcasting in the United States.

**frequency-modulation broadcast channel** [abbreviated f-m broadcast channel] A band of frequencies 200 kc wide in the frequency-modulation broadcast band, designated by its center frequency. Assigned center frequencies begin at 88.1 mc and continue in successive steps of 0.2 mc up to 107.9 mc in the United States.

**frequency-modulation receiver** [abbreviated f-m receiver] A complete radio receiver that receives frequency-modulated waves and delivers corresponding sound waves.

**frequency-modulation tuner** [abbreviated f-m tuner] A tuner containing an r-f amplifier, converter, i-f amplifier, and demodulator for frequency-modulated signals, used to feed a low-level a-f signal to a separate a-f amplifier and loudspeaker.

**frequency modulator** A circuit or device for producing frequency modulation.

**frequency monitor** An instrument for indicating the amount of deviation of the carrier frequency of a transmitter from its assigned value.

**frequency multiplier** A harmonic conversion transducer in which the frequency of the output signal is an exact integral multiple of the input frequency. Also called multiplier.

**frequency overlap** The portion of a 6-mc television channel that is common to both the monochrome and chrominance signals in a color television system. Frequency overlap is a form of bandsharing.

**frequency pulling** A change in the frequency of an oscillator due to a change in load impedance.

**frequency pulsing** Oscillator operation at two different frequencies alternately. Build-up of oscillation at one frequency creates conditions favorable for oscillation at the second frequency, and vice versa.

**frequency range** The range of frequencies over which a transmission system or device may be considered useful when used with different circuits under a variety of operating conditions. In contrast, bandwidth is a measure of useful frequency range with fixed circuits and fixed operating conditions.

**frequency record** A recording of various known frequencies at known amplitudes, usually for testing or measuring.

**frequency regulator** A device that maintains the frequency of an a-c generator at a predetermined value.

**frequency response** *Amplitude-frequency response.*

**frequency-response equalization** *Equalization.*

**frequency run** A series of tests made to determine the amplitude-frequency response characteristic of a transmission line, circuit, or device.

**frequency separator** The circuit that separates the horizontal and vertical synchronizing pulses in a monochrome or color television receiver.

**frequency shift** A change in the frequency of a radio transmitter or oscillator.

**frequency-shift converter** A device that converts a received frequency-shift signal to an amplitude-modulated signal or a direct-current signal.

**frequency-shift keying** [abbreviated fsk] A form of frequency modulation in which the modulating wave shifts the output frequency between predetermined values corresponding to the frequencies of correlated sources. When frequency-shift keying is used for code transmission, operation of the keyer shifts the carrier frequency back and forth between two distinct frequencies to designate mark and space. When frequency-shift keying is used for facsimile transmission, one carrier frequency represents picture black and another, generally 800 cps away, represents picture white.

**frequency stability** The ability of an oscillator to maintain a desired frequency. Usually expressed as per cent deviation from the assigned frequency value.

**frequency stabilization** The process of controlling the center frequency of an oscillator so it does not differ from that of a reference source by more than a prescribed amount.

**frequency standard** A stable oscillator, usually controlled by a crystal or tuning fork, that is used primarily for frequency calibration.

**frequency swing** 1. The instantaneous departure of the frequency of an emitted wave from the center frequency during frequency modulation. 2. The difference between the maximum and minimum design values of the instantaneous frequency in a frequency-modulation system.

**frequency tolerance** The extent to which the carrier frequency of a transmitter may be permitted to depart from the assigned frequency.

**frequency tripler** An amplifier or other device that delivers output voltage at a frequency equal to three times the input frequency.

**frequency-type telemeter** A telemeter that

employs the frequency of a signal as the translating means.

**fresnel** *Megamegacycle.*

**Fresnel diffraction** Diffraction of microwaves that permits reception behind obstacles, with no focusing by the obstacles. The Fresnel diffraction region of an antenna is the near-field region in which the radiation pattern varies with distance.

**Fresnel lens** A thin lens constructed with stepped setbacks so as to have the optical properties of a much thicker lens.

**Fresnel region** The region between the near field of an antenna and the Fraunhofer region. The boundary between the two is generally considered to be at a radius equal to twice the square of antenna length divided by wavelength.

**Fresnel zone** [after Augustin Jean Fresnel, 1788–1827] One of the conical zones that exist between microwave transmitting and receiving antennas due to cancellation of some portions of the wavefront by other

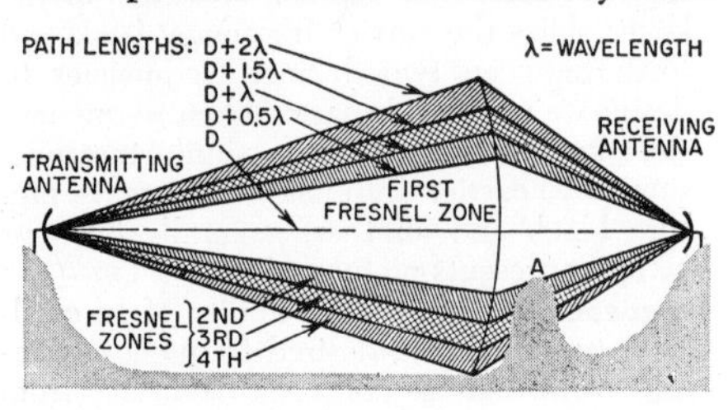


Fresnel zones between microwave antennas. Hill at A cuts off interfering waves, making received signal stronger than if hill did not exist. Hill may be higher, to give clearance of 0.6 of first Fresnel zone, without appreciably reducing signal strength.

portions that travel different distances. The boundary of the first Fresnel zone includes all paths ½ wavelength longer than the line-of-sight path. The outer boundaries of the second, third, and fourth Fresnel zones are formed by paths 1, 1½, and 2 wavelengths longer than the direct path.

**friction tape** Cotton tape impregnated with a sticky moisture-repelling compound. Used chiefly to hold rubber-tape insulation in position over a joint or splice. Friction tape itself has relatively poor insulating qualities.

**fringe area** An area just beyond the limits of the reliable service area of a television transmitter, in which signals are weak and erratic. High-gain directional receiving antennas and high-sensitivity receivers are generally required for satisfactory fringe-area reception.

**fringe effect** The extension of the electrostatic field of an air capacitor outside the space between its plates.

**fringe howl** A howl or squeal heard when some circuit in a radio receiver is on the verge of oscillation.

**frogging repeater** A carrier repeater having provisions for frequency frogging to permit use of a single multipair voice cable without having excessive crosstalk.

**front contact** The stationary contact of the normally open contacts on a relay.

**front end** The tuner of a television receiver, containing one or more r-f amplifier stages, the local oscillator, and the mixer, along with all channel-tuning circuits.

**front porch** The portion of a composite picture signal that lies between the leading edge of the horizontal blanking pulse and the leading edge of the corresponding sync pulse. The duration of the front porch is 1.27 microseconds in the standard U. S. television signal.

**front projection** A projection television system that uses a nontranslucent reflecting screen. Rear projection uses a translucent screen, with viewers and the projector on opposite sides of the screen.

**front-to-back ratio** *Front-to-rear ratio.*

**front-to-rear ratio** The ratio of the effectiveness of a directional antenna, loudspeaker, or microphone toward the front and toward the rear. Usually expressed in decibels. Also called front-to-back ratio.

**fruit** Deprecated term for *fruit pulse.*

**fruit pulse** An unsynchronized pulse reply resulting from interrogation of a transponder by interrogators not associated with the responsor in question. Also called fruit (deprecated).

**F scan** *F display.*

**F scope** A radarscope that produces an F display.

**fsk** Abbreviation for *frequency-shift keying.*

**ft-c** Abbreviation for *foot-candle.*

**ftc circuit** Abbreviation for *fast-time-constant circuit.*

**ft-l** Abbreviation for *foot-lambert.*

**fuel** Fissionable material having reasonably long life, suitable for producing energy in a nuclear reactor.

**fuel assembly** A combination of fuel and structural materials, used in some nuclear reactors to facilitate assembly of the core.

**fuel cell** A cell that converts chemical energy directly into electric energy, with electric power being produced as a part of a chemical reaction between the electrolyte and a fuel such as kerosene or industrial

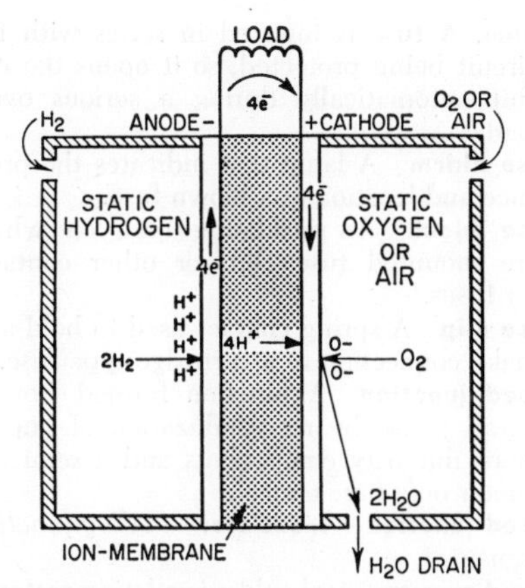


Fuel cell connected to inductive load. Hydrogen gas, serving as fuel, is fed into cell at upper left as required to meet power requirements of load. Cell shown is ion-membrane type, with no liquid electrolyte. Byproduct is water, which drains off at lower right.

fuel gas. The fuel is fed into the cell from an outside source.

**fuel rod** A long rod-shaped fuel assembly. Short fuel rods are called slugs.

**full-duplex operation** Operation of telegraph or other signal systems in opposite directions simultaneously.

**full integrand** An integrand that contains the maximum absolute capacity, either positive or negative.

**full load** The greatest load that a circuit or piece of equipment is designed to carry under specified conditions. Any additional load is an overload.

**full shot noise** The fluctuation in the current of charge carriers passing through a surface at statistically independent times.

**full-wave gas rectifier** A cold-cathode rectifier in which the cathode is much smaller than the two anodes, so as to give rectification.

**full-wave rectification** Rectification in which output current flows in the same direction during both half-cycles of the alternating input voltage.

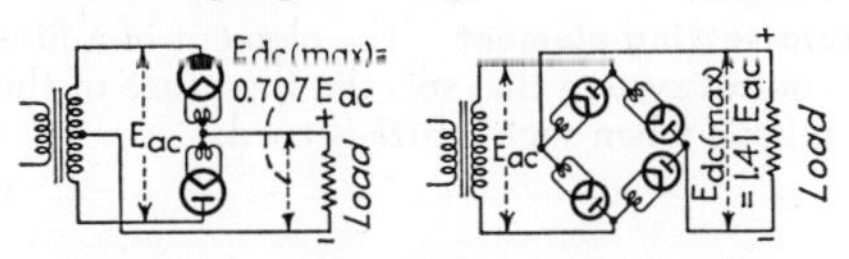


Full-wave rectification with conventional full-wave rectifier circuit (left) and with bridge rectifier circuit that gives twice as much d-c output voltage for a given a-c input voltage.

**full-wave rectifier** A double-element rectifier that provides full-wave rectification. One element functions during positive half-cycles and the other during negative half-cycles.

**full-wave vibrator** A vibrator having an armature that moves back and forth between two fixed contacts, so as to change the direction of direct-current flow through a transformer at regular intervals and thereby permit voltage step-up by the transformer. Used in battery-operated power supplies for mobile and marine radio equipment.

**fully active homing** Homing in which the missile generates its own radar signals and carries a computer that provides guidance signals for lock-on to give a collision course.

**function** A quantity whose value depends upon the value of one or more other quantities.

**function digit** A computer instruction word digit that determines the arithmetical or logical operation to be performed.

**function generator** 1. An analog computer device that indicates the value of a given function as the independent variable is increased. 2. A signal generator that delivers a choice of a number of different waveforms, with provisions for varying the frequency over a wide range. In an arbitrary function generator such as a photoformer, a mask having the desired waveform is inserted in a photoelectric scanning system.

**function multiplier** An analog computer device that takes in the changing values of two functions and puts out the changing value of their product as the independent variable is changed.

**function switch** A network having a number of inputs and outputs so connected that input signals expressed in a certain code will produce output signals that are a function of the input information but in a different code.

**function table** 1. A table that lists the values of a function for various values of the variable. 2. Sets of computer information arranged so an entry in one set selects one or more entries in the other sets. 3. A computer device that converts multiple inputs into a single output or encodes a single input into multiple outputs.

**fundamental** *Fundamental frequency.*

**fundamental component** *Fundamental frequency.*

**fundamental field particle** The field quantum that is the result of quantizing a field.

**fundamental frequency** 1. The lowest frequency component of a complex vibration, sound, or electric signal. Used as the basis for harmonic analysis of a wave. The fundamental frequency is the reciprocal of the period of a wave. Also called first harmonic, fundamental, and fundamental component. The frequency that is twice the fundamental frequency is called the second harmonic. 2. The first order or lowest frequency of an intended mode of vibration for a quartz plate or other vibrating object. Also called fundamental mode.

**fundamental-frequency magnetic modulator** A magnetic modulator in which the output is at the fundamental frequency of the supply.

**fundamental mode** 1. The waveguide mode having the lowest critical frequency. Also called dominant mode and principal mode. 2. *Fundamental frequency.*

**fundamental particle** *Elementary particle.*

**fundamental scanning frequency** *Maximum keying frequency.*

**fundamental tone** The component tone of lowest pitch in a complex tone.

**fundamental unit** An arbitrarily defined unit that serves as the basis of a system of units. All other units in the system are derived from a set of fundamental units. The dimensions of any physical quantity may be expressed as combinations of fundamental units.

**fundamental wavelength** The wavelength corresponding to the fundamental frequency.

**fungiproofing** Application of a protective chemical coating that inhibits growth of fungi on electronic equipment in humid tropical regions.

**fuse** [symbol F] A protective device containing a short length of special wire that melts when the current through it exceeds the rated value for a definite period of time. A fuse is inserted in series with the circuit being protected, so it opens the circuit automatically during a serious overload.

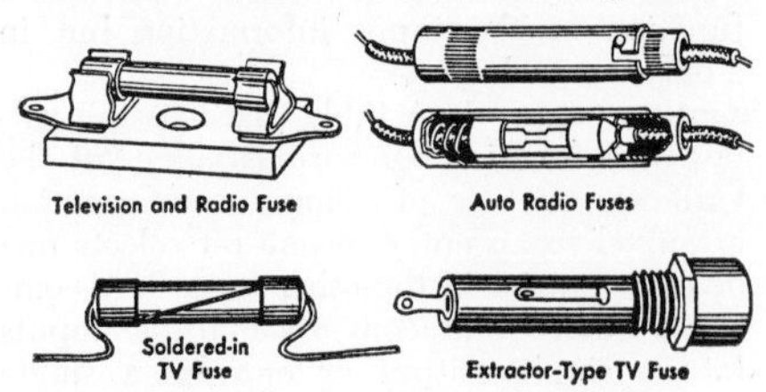

Fuses used in electronic equipment.

**fuse alarm** A lamp that indicates the presence and location of a blown fuse.

**fuse block** An insulating base on which are mounted fuse clips or other contacts for fuses.

**fuse clip** A spring contact used to hold and make connection to a cartridge-type fuse.

**fused junction** A junction formed from a liquid phase by recrystallization of one or more impurity components and a semiconductor on a base crystal.

**fused-junction transistor** *Alloy-junction transistor.*

**fused quartz** A glasslike insulating material made by melting crushed crystals of natural quartz or a certain type of quartz sand.

**fuse wire** Wire made from an alloy that melts at a relatively low temperature and overheats to this temperature when carrying a particular value of overload current.

**fusible resistor** A resistor designed to protect a circuit against overload. Its resistance limits current flow and thereby protects against surges when power is first applied to a circuit. Its fuse characteristic opens the circuit when current drain exceeds design limits. Generally used in series with anode supply circuits.

**fusion** *Nuclear fusion.*

**fusion bomb** *Hydrogen bomb.*

**fuze** A device used to detonate an explosive charge automatically when the proper conditions of impact, elapsed time, external command, or proximity are achieved in a bomb, missile, torpedo, or mine.

**fuze chronograph** A chronograph that accurately measures and records the time interval between firing of a time-fuzed projectile and the air burst as detected by a photoelectric pickup aimed at the expected point of burst. The recorder may be started by the passage of a magnetized projectile through a coil mounted on the muzzle of the gun.

**fuze-setting element** The element in a fire-control system that sets the time fuze of the missile when such a fuze is used.

# G

**G** 1. Symbol for *grid.* 2. Symbol for *generator.*

**g** Abbreviation for *gravitational force.*

**G** Symbol for *conductance.*

**GaAs** Symbol for *gallium arsenide.*

**gadolinium** [symbol Gd] A rare-earth metallic element. Atomic number is 64.

**gage** A measuring device or measuring instrument.

**gain** 1. The increase in signal power that is produced by an amplifier. Generally expressed in decibels. Also called transmission gain. 2. *Antenna gain.*

**gain-bandwidth product** The midband gain of an amplifier stage multiplied by the bandwidth in megacycles. The bandwidth is the difference between the two frequencies at which the power output is a specified fraction (usually ½) of the midband value.

**gain control** A device for adjusting the gain of a system or component.

**gain margin** The amount of increase in gain that would cause oscillation in a feedback control system.

**gain-time control** *Sensitivity-time control.*

**gain turndown control** An automatic receiver gain control incorporated in a transponder to protect the transmitter from overload.

**galaxy noise** Noise signals similar to thermal noise, coming from the direction of the Milky Way.

**galena** A crystalline form of lead sulfide, once used in crystal detectors.

**gallium** [symbol Ga] A metallic element. Atomic number is 31.

**gallium arsenide** [symbol GaAs] A semiconductor having a forbidden band gap of 1.4 electron-volts and a maximum operating temperature of 400°C when used in a transistor.

**gallium phosphide** [symbol GaP] A semiconductor having a forbidden band gap of 2.4 electron-volts and a maximum operating temperature of 870°C when used in a transistor.

**galvanic** Pertaining to electricity flowing as a result of chemical action.

**galvanic cell** An electrolytic cell that is capable of producing electric energy by electrochemical action.

**galvanometer** An instrument for indicating or measuring a small electric current by means of a mechanical motion derived from electromagnetic or electrodynamic forces produced by the current.

**galvanometer constant** A number by which a particular relative reading of a galvanometer must be multiplied to obtain the current value.

**galvanometer recorder** A sound recorder in which the audio signal voltage is applied to a coil suspended in a magnetic field. The resulting movements of the coil cause a tiny attached mirror to move a reflected light beam back and forth across a slit in front of a moving photographic film. This provides a variable-area photographic record of the signal.

**galvanometer shunt** A resistor connected in parallel with a galvanometer to reduce its sensitivity under certain conditions. It allows only a known fraction of the current to pass through the galvanometer.

**gamma** A numerical indication of the degree of contrast in a television or photographic image. It is equal to the slope of the straight-line portion of the H and D curve for the emulsion or screen.

**gamma correction** Correction of the effective value of gamma by introducing a nonlinear output-input characteristic.

**gamma emitter** An atom whose radioactive decay process involves the emission of gamma rays.

**gamma radiation** Radiation of gamma rays.

**gamma radiography** Radiography by means of gamma rays.

**gamma ray** A quantum of electromagnetic radiation emitted by a nucleus as the result of a quantum transition between two energy levels of the nucleus. Gamma rays have energies usually between 10 kilo-electron-volts and 10 million electron-volts, with shorter wavelengths than x-rays. They are more penetrating than alpha and beta particles and are not affected by magnetic fields.

**gamma-ray source** A quantity of radioactive material that emits gamma radiation and is in a form convenient for radiology.

**gamma-ray spectrometer** An instrument that measures the energy distribution of gamma rays.

**gammate** To apply a nonlinear transfer characteristic to a television signal in order to achieve a desired value of gamma.

**gamma uranium** An allotropic modification of uranium that is stable above approximately 770°C.

**Gamow barrier radius** *Coulomb barrier radius.*

**Gamow-Teller selection rules** [abbreviated GT selection rules] A set of selection rules for allowed beta transitions.

**gang capacitor** A combination of two or more variable capacitors mounted on a common shaft to permit adjustment by a single control.

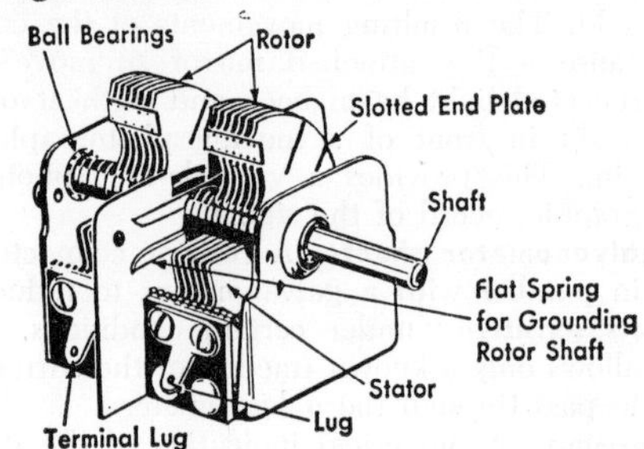


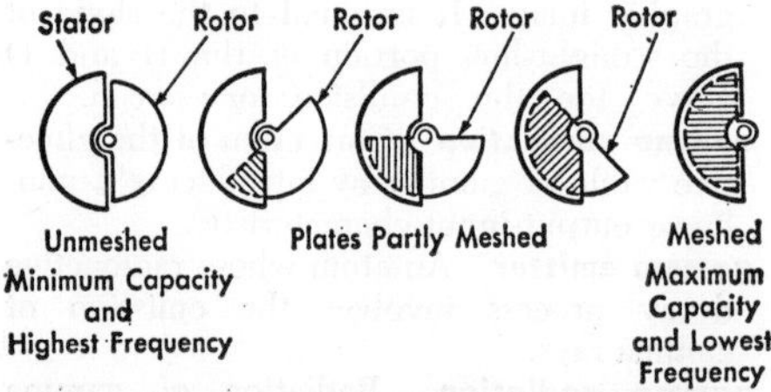


Gang capacitor construction and tuning action

**ganged tuning** Simultaneous tuning of two or more circuits with a single control knob.

**ganged volume control** A combination of two or more volume controls, one for each channel of a stereophonic sound system, mounted on a common shaft to permit changing the volume of all loudspeakers simultaneously without changing their balance with respect to each other.

**ganging** A mechanical means of operating two or more controls with one control knob.

**gang switch** A combination of two or more switches mounted on a common shaft to permit operation by a single control. Also called deck switch.

**gang tuning capacitor** A gang capacitor used for tuning purposes in a receiver.

**GaP** Symbol for *gallium phosphide.*

**gap** 1. A break in a closed magnetic circuit, containing only air or filled with a nonmagnetic material. 2. The spacing between two electric contacts. 3. A region not adequately covered by the main lobes of a radar antenna.

**gap admittance** The admittance of a microwave circuit at a gap in the absence of an electron stream. More specifically called circuit gap admittance.

**gap coding** A method of communicating information by interrupting the transmission of an otherwise regular signal in such a way that the interruptions form a telegraphic-type message. Used in some radar beacons.

**gap factor** The ratio of peak energy gained (in electron-volts) to peak resonator voltage in a traveling-wave tube.

**gap-filler** An auxiliary radar antenna used to cover gaps in the main radar antenna pattern.

**gap length** The physical distance between adjacent surfaces of the poles of a longitudinal magnetic recording head.

**gap loading** The electronic gap admittance that results from the movement of electrons in the gap of a microwave tube. The three types are multipactor gap loading, primary transit-angle gap loading, and secondary-electron gap loading.

**garble** To alter a message intentionally or unintentionally so that it is difficult to understand.

**garnet maser** Term sometimes used incorrectly for *ferromagnetic amplifier.*

**gas amplification** The ratio of the charge collected to the charge liberated by the initial ionizing event in a radiation-counter tube.

**gas amplification factor** The ratio of radiant or luminous sensitivities with and without ionization of the contained gas in a gas phototube.

**gas capacitor** A capacitor consisting of two or more electrodes separated by a gas, other than air, that serves as a dielectric.

**gas-cooled reactor** A nuclear reactor in which a gas such as air, carbon dioxide, or helium is used as a coolant.

**gas counter** A counter in which the radioactive sample is prepared in the form of a gas and introduced into the counter tube.

**gas current** A positive-ion current produced by collisions between electrons and residual gas molecules in an electron tube. Also called ionization current.

**gas diode** A tube having a hot cathode and an anode in an envelope containing a small amount of an inert gas or vapor. When the anode is made sufficiently positive, the electrons flowing to it collide with gas atoms and ionize them. As a result, anode current is much greater than that for a comparable vacuum diode.

**gas discharge** Conduction of electricity in a gas, due to movements of ions produced by collisions between electrons and gas molecules.

**gas discharge lamp** *Discharge lamp.*

**gaseous diffusion process** A method of separating isotopes in which an isotopic mixture of gases is allowed to diffuse through a porous wall. The lighter molecules pass through the porous wall more readily than the heavier molecules.

**gas-filled cable** A coaxial or other cable containing gas under pressure to serve as insulation and keep out moisture.

**gas-filled radiation-counter tube** A gas tube used to detect radiation by means of gas ionization.

**gas-filled rectifier** *Cold-cathode rectifier.*

**gas-flow counter tube** A radiation-counter tube in which an appropriate atmosphere is maintained by a flow of gas through the tube.

**gas focusing** A method of concentrating an electron beam by utilizing the residual gas in a tube. Beam electrons ionize the gas molecules, forming a core of positive ions along the path of the beam. This core of ions attracts beam electrons and thereby makes the beam more compact. Also called ionic focusing.

**gas maser** A maser in which the microwave electromagnetic radiation interacts with the molecules of a gas such as ammonia. Use is limited chiefly to highly stable oscillator applications, as in atomic clocks.

**gas noise** Electric noise due to random ionization of gas molecules in gas tubes and partially evacuated vacuum tubes.

**gas phototube** A phototube into which a quantity of gas has been introduced after evacuation, usually to increase its sensitivity. In a vacuum phototube, no appreciable gas is present.

**gas ratio** The ratio of the ion current in a tube to the electron current that produces it.

**gassy tube** A vacuum tube that has not been fully evacuated or has lost part of its vacuum due to release of gas by the electrode structure during use, so that enough gas is present to impair operating characteristics appreciably. Also called soft tube.

**gas tetrode** *Thyratron.*

**gaston** A modulator that uses a gas tube as a noise source for producing a random-noise modulation signal that can be fed into any standard aircraft communication transmitter for jamming purposes.

**gas triode** *Thyratron.*

**gas tube** An electron tube in which the contained gas or vapor performs the primary role in the operation of the tube.

**gas-tube generator** A power source comprising a gas-tube oscillator, a power supply, and associated control equipment.

**gas x-ray tube** An x-ray tube in which the emission of electrons from the cold cathode is produced by positive-ion bombardment when the applied cathode-anode voltage is made sufficiently high.

**gate** 1. A circuit having an output and a multiplicity of inputs so designed that the output is energized when and only when a certain combination of pulses is present at the inputs. An and-gate delivers an output pulse only when every input is energized simultaneously in a specified manner. An or-gate delivers an output pulse when any one or more of the input pulses meet the specified conditions. Used in digital computers. 2. A circuit in which one signal, generally a square wave, serves to switch another signal on and off. Used in radar to block the receiver, except for brief instants when echo signals may be returning from a target. 3. A movable barrier of shielding material used for closing a hole in a nuclear reactor.

**gate circuit** A circuit that admits and amplifies or passes a signal only when a gating pulse is present.

**gate current** The alternating or pulsating direct current that flows through the gate winding of a magnetic amplifier.

**gated-beam tube** A pentode electron tube having special electrodes that form a sheet-shaped beam of electrons. This beam may be deflected away from the anode by a relatively small voltage applied to a control electrode, thus giving extremely sharp cutoff of anode current. Used in some f-m detector circuits.

**gate generator** A circuit used to generate gate pulses. In one form it consists of a multivibrator having one stable and one unstable position.

**gate impedance** The impedance of one of

the gate windings in a magnetic amplifier.

**gate voltage** The voltage across the terminals of the gate winding in a magnetic amplifier.

**gate winding** A winding used in a magnetic amplifier to produce on-off action of load current.

**gating** The process of selecting those portions of a wave that exist during one or more selected time intervals or that have magnitudes between selected limits. Usually achieved by applying a pulsed voltage to a normally cut-off electron tube, transistor, or magnetic amplifier to make the device conductive (open the gate) for the duration of the pulse. The opposite action, wherein the device is cut off for the duration of the pulse, is used to black out a television cathode-ray tube during retraces and is called blanking.

**gating pulse** A pulse that modifies the operation of a circuit for the duration of the pulse.

**gausitron** A mercury-pool rectifier in which starting is achieved by applying a high-voltage pulse between an insulated probe and the mercury pool.

**gauss** [plural gauss] The cgs electromagnetic unit of magnetic induction $B$.

**Gaussian distribution** A distribution of random variables comparable to that found in nature, characterized by a symmetrical and continuous distribution decreasing gradually to zero on either side of the most probable value. Also called normal distribution.

**Gaussian noise generator** A signal generator that produces a random noise signal whose frequency components have a Gaussian distribution centered on a predetermined frequency value. A gas thyratron may be used as the primary source of noise.

**Gaussian well** A potential well whose value varies according to a Gaussian distribution.

**gaussmeter** A magnetometer provided with a scale graduated in gauss or kilogauss. One version is used to measure the magnetic field strength between the pole faces of magnetic structures for magnetrons, ranging from 1,200 to 9,600 gauss.

**gca** Abbreviation for *ground-controlled approach.*

**gci** Abbreviation for *ground-controlled interception.*

**G display** A rectangular radarscope display in which a target appears as a laterally centralized bright spot when the radar antenna is aimed at it in azimuth. Wings appear to grow on the spot as the distance to the target is diminished. Horizontal and vertical aiming errors are respectively indicated by horizontal and vertical displacement of the spot, just as in an F display. Also called G scan.

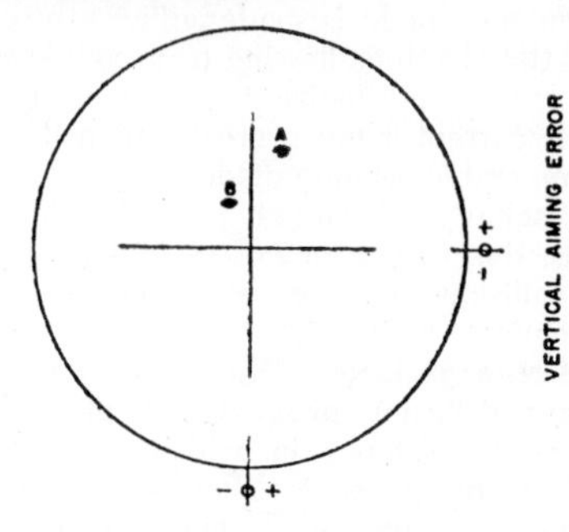


G display.

**gearmotor** A motor combined with a set of speed-reducing gears.

**gee** [Ground Electronics Engineering] A vhf radio navigation system in which three or more ground stations transmit synchronized pulses. Hyperbolic lines of position are determined by measuring the differences in the times of arrival of these pulses. Developed in Great Britain, and similar to loran.

**gee-H system** A combination of gee and H systems of navigation, used to give greater precision for bombing.

**Geiger counter** A radiation counter that uses a Geiger counter tube in appropriate circuits for detecting and counting ionizing particles, such as cosmic-ray particles. Each particle ionizes the gas in the tube in such a way that the total ionization per event is independent of the energy of the ionizing particle. Ionization results in electron flow from the tube wall to the center electrode, producing one output voltage pulse for each particle. Also called Geiger-Mueller counter. 2. *Geiger counter tube.*

**Geiger counter tube** A radiation-counter tube operated in the Geiger region. It usually consists of a gas-filled cylindrical metal chamber containing a fine-wire anode at its axis. Also called Geiger counter,

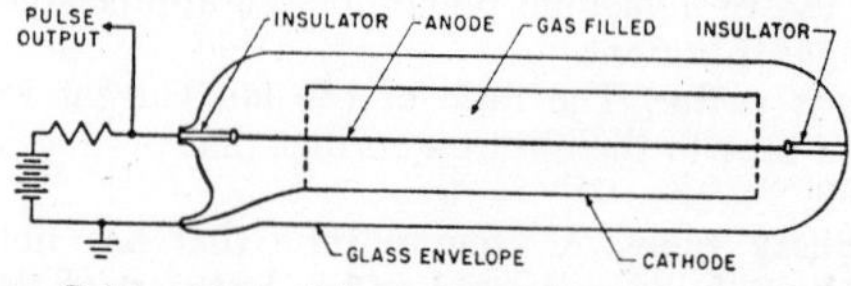


Geiger-counter-tube construction and circuit.

Geiger-Mueller counter tube, and Geiger-Mueller tube.

**Geiger-Mueller counter** [abbreviated G-M counter] *Geiger counter.*

**Geiger-Mueller counter tube** *Geiger counter tube.*

**Geiger-Mueller region** *Geiger region.*

**Geiger-Mueller threshold** *Geiger threshold.*

**Geiger-Mueller tube** *Geiger counter tube.*

**Geiger region** The range of applied voltage in which the charge collected per isolated count is independent of the charge liberated by the initial ionizing event of a radiation-counter tube. Also called Geiger-Mueller region.

**Geiger threshold** The lowest applied voltage at which the charge collected per isolated tube count in a radiation-counter tube is substantially independent of the nature of the initial ionizing event. Also called Geiger-Mueller threshold.

**Geissler tube** An experimental two-electrode discharge tube used to show the luminous effects of electric discharges through various gases at low pressures. The electrodes are at opposite ends of the tube, just as in modern neon signs.

**Genemotor** Trademark of Carter Motor Co. for their line of dynamotors.

**general routine** A routine expressed in computer coding for solving an entire class of problems.

**generate** To produce, trace out, originate, or otherwise create, as, for example, to generate a needed computer subroutine from parameters and skeletal coding.

**generating electric field meter** A device in which a flat conductor is alternately exposed to the electric field to be measured and then shielded from it. The resulting current to the conductor is rectified and used as a measure of the potential gradient at the conductor surface. Also called gradient meter.

**generating voltmeter** A device in which a capacitor is connected across the voltage to be measured, and its capacitance is varied cyclically. The resulting current in the capacitor is rectified and used as a measure of the voltage. Also called rotary voltmeter.

**generation rate** The time rate of creation of electron-hole pairs in a semiconductor.

**generation time** The mean time required for neutrons arising from a fission to produce new fissions.

**generator** [symbol G] 1. A machine that converts mechanical energy into electric energy. In its commonest form, a large number of conductors are mounted on an armature that is rotated in a magnetic field produced by field coils. Also called dynamo. 2. A vacuum-tube oscillator or any other nonrotating device that generates an alternating voltage at a desired frequency when

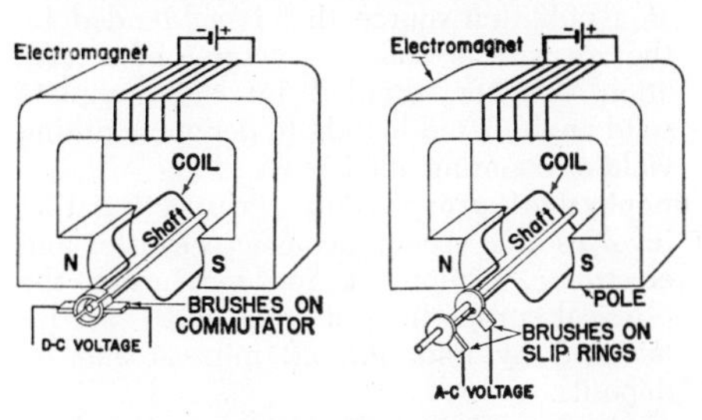


Generator principles, showing use of commutator for d-c generator and slip rings for a-c generator.

energized with d-c power or low-frequency a-c power. Such generators are used to produce large amounts of r-f power, such as for high-frequency heating and ultrasonic cleaning. 3. A circuit that generates a desired repetitive or nonrepetitive waveform, such as a pulse generator.

**Genie** An Air Force air-to-air missile having a nuclear warhead.

**geodesic** The shortest line between two points, as measured on any specified surface that includes the points, such as the surface of the earth.

**geomagnetic electrokinetograph** An instrument that can be suspended from the side of a ship to measure the direction and speed of ocean currents while the ship is under way. Electrodes suspended in the conductive sea water are connected to a potentiometer recorder to measure the voltage induced in the moving conductive sea water by the magnetic field of the earth. Two runs at right angles to each other are generally made to determine the direction and rate of drift due to local ocean currents.

**geometric distortion** An aberration that causes a reproduced picture to be geometrically dissimilar to the perspective plane projection of the original scene.

**geometric error** 1. A systematic error due to calibrating a navigation system on the basis of a spherical earth rather than an oblate spheroidal earth. 2. *Ionospheric height error.*

**geometric factor** The ratio of the change in a navigation coordinate to the change in distance, taken in the direction of maximum navigation-coordinate change.

**geometric mean** The square root of the product of two quantities.

**geometry** The physical arrangement of a device or assembly, such as a neutron counter.

**geometry factor** The average solid angle at a radiation source that is subtended by the aperture or sensitive volume of a radiation detector, divided by the complete solid angle. Used loosely to denote counting yield or counting efficiency.

**geophysical prospecting** Prospecting that involves the use of acoustic, electric, and electronic equipment for measuring the physical properties of the earth as influenced by underground mineral and oil deposits.

**geophysics** A branch of science that deals with factors affecting the structure of the earth.

**georef** [GEOgraphic REFerence] Abbreviation for *world geographic reference system.*

**georef grid** [GEOgraphical REFerence grid] The grid system used on U. S. Air Force aeronautical charts for identifying the location of any point or area in the world. The world chart is divided into 24 parallel north-south strips 15° wide lettered from A through Z (I and O omitted), beginning at the 180th meridian, and into 12 parallel east-west strips 15° wide lettered from A through M (I omitted), beginning at the South Pole. Each quadrangle is subdivided into 15 lettered units eastward and 15 lettered units northward. These are lettered from A through Q (I and O omitted). Each 1° quadrangle is subdivided into 60 numbered minute units. Minute units may be subdivided further into decimal parts.

**germanium** [symbol Ge] A brittle, grayish-white metallic element having semiconductor properties. Widely used in transistors and crystal diodes. Atomic number is 32.

**germanium diode** A semiconductor diode that uses a germanium crystal pellet as the rectifying element.

**getter** A special metal that is placed in a vacuum tube during manufacture and vaporized after the tube has been evacuated. When the vaporized metal condenses, it absorbs residual gases. Metals used as getters include barium, calcium, magnesium, potassium, sodium, and strontium. Some getters leave a silvery film on the inside of the glass envelope.

**gettering** Removal of residual gas from an electron tube, after evacuation, by evaporation of a getter. An ionizing electron discharge is sometimes passed through the gas during the gettering process, to accelerate absorption of the residual gas by the getter.

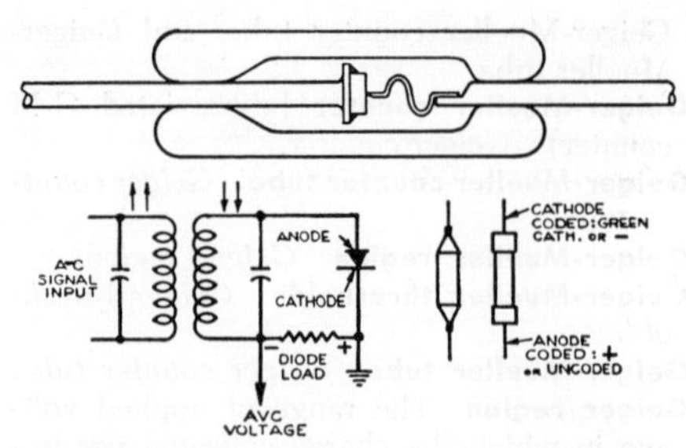


Germanium diode mounted in glass, typical circuit showing diode symbol, and polarity markings.

**gev** Abbreviation for *giga-electron-volt.*

**ghost** *Ghost image.*

**ghost image** 1. An undesired duplicate image at the right of the desired image on a television receiver. It is due to multipath effect, wherein a reflected signal traveling over a longer path arrives slightly later than the desired signal. Hills and large buildings on either side of the direct path between transmitter and receiver are the commonest cause of ghost images. Also called ghost. 2. An undesired or unknown echo indication on the screen of a radar indicator. Also called ghost.

**ghost mode** A waveguide mode having a trapped field associated with an imperfection in the wall of the waveguide. A ghost mode can cause trouble in a waveguide operating close to the cutoff frequency of a propagation mode.

**ghost pulse** *Ghost signal.*

**ghost signal** 1. The reflection-path signal that produces a ghost image on a television receiver. Also called echo. 2. Any signal appearing on a loran or gee display at other than the basic repetition rate being observed. Also called ghost pulse.

**giant air shower** *Extensive shower.*

**Gibson girl** Slang term for one type of portable, hand-operated radio transmitter carried on an airplane, for use in cranking out distress signals when down at sea.

**giga-** A prefix representing $10^9$.

**gigacycle** *Kilomegacycle.*

**giga-electron-volt** [abbreviated gev] Billion electron-volt, equal to $10^9$ electron-volts in the United States and France.

**gilbert** The unit of magnetomotive force in the cgs electromagnetic system. The magnetomotive force in gilberts is equal to the line integral of the magnetic field strength in oersteds around the magnetic circuit. One gilbert is equivalent to 0.7956 ampere-turn. One gilbert per centimeter is equal to 1 oersted.

**Gill-Morrell oscillator** An oscillator of the retarding-field type in which the frequency of oscillation is dependent on associated circuit parameters as well as on electron-transit time within the tube.

**gimbal lock** Alignment and locking of gimbals that are normally at right angles to each other in a two-axis gyroscope. This catastrophic malfunction occurs when the precession angle reaches 90°, usually as a result of excessive angular motion of the aircraft or missile in which the gyroscope is mounted.

**gimbals** A mounting having two mutually perpendicular and intersecting axes of rotation. A body mounted on gimbals is free to incline in any direction.

**gimmick** A small capacitance formed by twisting two insulated wires together.

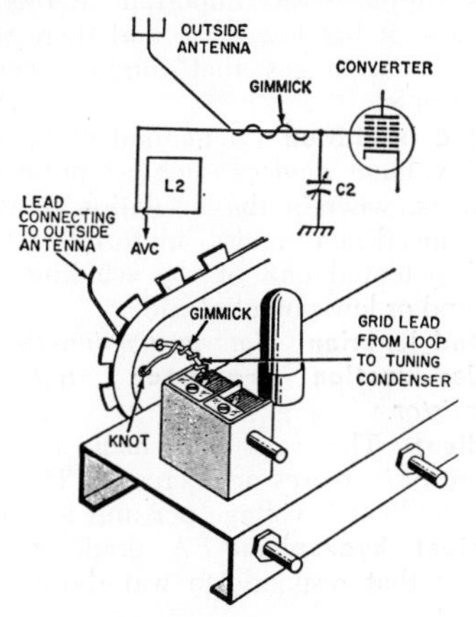


Gimmick in radio circuit.

**Giorgi system** *Meter-kilogram-second system.*

**glass-bonded mica** An insulating material made by compressing a mixture of powdered glass and powdered natural or synthetic mica at high temperature. Examples of commercial products include Mycalex, Mykroy, and Synthamica.

**glass fiber** A glass thread less than a thousandth of an inch thick, used loosely or in woven form as an acoustic, electric, or thermal insulating material and as a reinforcing material in laminated plastics.

**glassine** An insulating paper used between layers of iron-core transformer windings.

**glass-to-metal seal** An airtight seal between glass and metal parts of an electron tube, made by fusing together a special glass and special metal alloy having nearly the same temperature coefficients of expansion.

**glass-type tube** An electron tube having a glass envelope.

**glide bomb** A bomb fitted with airfoils to provide lift. It is carried and released in the direction of a target by an airplane, and may or may not be remotely controlled.

**glide path** The three-dimensional path used by an aircraft in airport approach procedures, as defined by the radio transmitters of an instrument landing system.

**glide-path station** A directional radio beacon used with an instrument landing system to provide aircraft guidance in the vertical plane during an instrument landing.

**glide slope** An inclined surface that includes a glide path, generated by an instrument landing system.

**glide-slope angle** *Slope angle.*

**glide-slope deviation** The difference between the actual vertical-plane path and the planned slope of an aircraft making an instrument landing, expressed in angular or linear measurements.

**glide-slope facility** The radio transmitter and associated equipment used at an airport to provide a glide slope.

**glide-slope sector** The sector containing the glide slope in an instrument landing system. The sector is bounded above and below by radial lines from the glide-slope transmitter, along each of which there is a specified difference in depth of modulation.

**glide-slope transmitter** A transmitter used to feed the antennas that radiate the beam for the glide slope of an instrument landing system at an airport.

**G line** A single dielectric-coated round wire used for transmitting microwave energy.

**glint** A pulse-to-pulse variation in the amplitude and apparent origin of a reflected radar signal, due to reflection of the radar beam from different surfaces of a rapidly moving target or to reflections from a propeller on the target. Also called glitter.

**glitter** *Glint.*

**glomb** [GLide bOMB] A glider adapted for use as a glide bomb. It is towed to the target area, released, and guided to its target by radio from a mother airplane.

**glossmeter** An instrument, often photoelectric, for measuring the ratio of the light reflected from a surface in a definite direction to the total light reflected in all directions.

**glow discharge** A discharge of electricity through a gas in an electron tube, characterized by a cathode glow and a voltage drop, in the vicinity of the cathode, that is much higher than the ionization voltage of the gas.

**glow-discharge cold-cathode tube** *Glow-discharge tube.*

**glow-discharge rectifier** A glow-discharge tube used as a rectifier.

**glow-discharge tube** A gas tube that depends for its operation on the properties of a glow discharge. Also called glow-discharge cold-cathode tube and glow tube.

**glow lamp** A two-electrode electron tube in which light is produced by a negative glow close to the negative electrode when voltage is applied between the two electrodes. The envelope contains a small quantity of an inert gas such as neon or argon. With an a-c voltage both electrodes appear to glow because each is negative half the time. Neon glow lamps, having an orange-red glow, are the commonest example. Less-used argon glow lamps have a blue-violet glow. Wattages of glow lamps range from 1/25 watt to 3 watts.

**glow switch** An electron tube containing contacts that are operated thermally by a glow discharge. Used as a starter in some fluorescent lamp circuits. The heat of the glow discharge causes a bimetallic strip to close the contacts.

**glow tube** *Glow-discharge tube.*

**glow voltage** The voltage at which a glow discharge begins in a gas tube.

**glue-line heating** Dielectric heating in which the electrodes are designed to give preferential heating to a thin film of glue or other relatively high-loss material located between layers of relatively low-loss material such as wood.

$g_m$ Symbol for *transconductance.*

**G-M counter** Abbreviation for *Geiger-Mueller counter.*

**GMT** Abbreviation for *Greenwich mean time.*

**gnd** Abbreviation for *ground.*

**gnomonic projection** An aeronautical chart made by projecting the surface of a sphere onto a plane touching the sphere at one point. The projection is made by means of radials from the center of the sphere. Used in determining great-circle courses for aircraft and for tracking aircraft electronically.

**gold** [symbol Au] A precious-metal element. Atomic number is 79. Used for electroplating electronic components that must withstand severe corrosive conditions such as exist in the tropics.

**goniometer** An instrument for measuring angles. Used to calculate and resolve mathematical problems or electrical functions, as well as to establish directional phase difference between two transmitted or received signals. In one form it has two fixed windings crossed at 90° to each other, along with a rotatable third winding. In another form it is used for measuring the angles between the reflecting surfaces of a crystal or prism. An x-ray version serves for measuring the angular positions of the axes of a quartz crystal.

**gpi** Abbreviation for *ground-position indicator.*

**graded filter** A power-supply filter in which connections to the output stage of a receiver or amplifier are made at or near the filter input to obtain maximum d-c voltage. Ripple is less important at this stage because it has low gain, and there are no subsequent stages that might accentuate the ripple.

**graded insulation** A method of insulating high-voltage devices such as pulse transformers, wherein the insulation to ground is reduced more or less uniformly from the high-potential end of the winding to the ground or low-potential end.

**graded junction** *Rate-grown junction.*

**graded-junction transistor** *Rate-grown transistor.*

**gradient** The rate at which a variable quantity increases or decreases. Thus, voltage gradient is voltage per unit length.

**gradient hydrophone** A gradient microphone that responds to waterborne sound waves.

**gradient meter** *Generating electric field meter.*

**gradient microphone** A microphone whose output corresponds to a gradient of the sound pressure. A pressure microphone is a gradient microphone of zero order. A velocity microphone is a first-order gradient microphone.

**gradiometer** A flux-gate magnetometer whose output is proportional to the gradient of the magnetic field being measured.

**grain** A small particle of metallic silver remaining in a photographic emulsion after developing and fixing. These grains together form the dark areas of a photographic image.

**graininess** Visible coarseness in a photographic image under specified conditions, due to silver grains.

**gramophone** British term for *phonograph.*

**gram-rad** A unit of integral absorbed dose of radiation, equal to 100 ergs per gram.

**gram-roentgen** A unit of energy conversion, equal to a dose of 1 roentgen delivered to 1 gram of air. One gram-roentgen is about 84 ergs. Used to describe total energy absorption by a patient undergoing radiation therapy. The quantity obtained by integrating gram-roentgens throughout a region is called the integral dose.

**graph** A line drawing showing the relation between two variable quantities.

**graphechon** *Storage tube.*

**graphite** A form of carbon used in resistor elements and as a moderator in nuclear reactors.

**grass** Clutter due to circuit noise in a radar receiver, seen on an A scope as a pattern resembling a cross-section of turf. Also called hash.

**grating** 1. An arrangement of fine parallel wires used in waveguides to pass only a certain type of wave. 2. An arrangement of crossed metal ribs or wires that acts as a reflector for a microwave antenna and offers minimum wind resistance. 3. *Diffraction grating.*

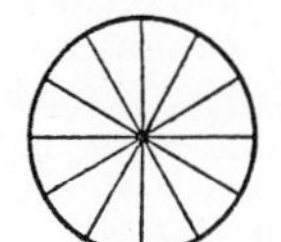
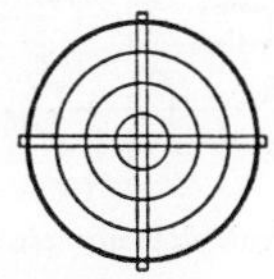
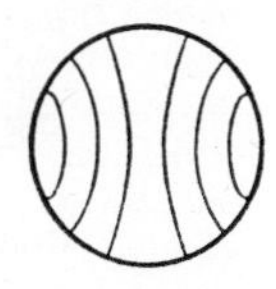


Gratings for circular waveguides. The grating of radial wires blocks all types of transverse electric waves, while the other two block certain types of transverse magnetic waves.

**grating converter** A wave converter consisting of a double grating positioned just ahead of a coaxial sheet grating in a circular waveguide. One grating conforms to the pattern of the arriving wave and the other to the pattern of the converted wave.

**grating reflector** An openwork metal structure designed to provide a good reflecting surface for microwave antennas.

**Gratz rectifier** A three-phase full-wave rectifying circuit utilizing six rectifying elements.

**gravitational force** [abbreviated g] The gravitational pull of the earth, or the comparable force required to accelerate or decelerate a freely movable body at the rate of approximately 32.16 feet per second per second.

**graybody** A body whose spectral emissivity remains constant through the spectrum and is less than that of a blackbody radiator at the same temperature.

**Gray code** A modified binary code used for analog-to-digital conversion, such as for changing an angle to a binary alphabet value. Also called cyclic binary code and reflective code.

**gray filter** *Neutral-density filter.*

**gray scale** A series of achromatic tones having varying proportions of white and black, to give a full range of grays in between white and black. A gray scale is usually divided into 10 steps.

**green gain control** A variable resistor used in the matrix of a three-gun color television receiver to adjust the intensity of the green primary signal.

**green gun** The electron gun whose beam strikes phosphor dots emitting the green primary color in a three-gun color television picture tube.

**Green Quail** A decoy missile containing radar countermeasure equipment, designed to be launched from aircraft.

**green restorer** The d-c restorer for the green channel of a three-gun color television picture tube circuit.

**Green's function** A kernel in which the integral operator is the inverse of a differential operator.

**green video voltage** The signal voltage output from the green section of a color television camera, or the signal voltage between the receiver matrix and the green-gun grid of a three-gun color television picture tube.

**Greenwich mean time** [abbreviated GMT] The mean solar time at the Greenwich meridian, now reckoned from midnight in both the United States and Great Britain.

**grenz ray** An x-ray produced at the long-wavelength end of the x-ray spectrum, involving wavelengths of the order of 1 to 10 angstroms, by using special tubes that operate at voltages from only 5,000 to 15,000 volts. Used in skin therapy and radiography of insects and botanical specimens.

**grenz tube** A low-voltage x-ray tube having a special glass window capable of transmitting x-ray wavelengths ranging from 1 to 10 angstroms, which are blocked by ordinary glass.

**grid** [symbol G] 1. An electrode located between the cathode and anode of an electron tube and having one or more openings through which electrons or ions can pass under certain conditions. A grid serves to control the flow of electrons from cathode to anode. 2. A network of equally spaced lines forming squares, used for determining permissible locations of holes on a printed-circuit board or a chassis. 3. *Potter-Bucky grid.*

**grid-anode transconductance** *Transconductance.*

**grid bearing** A bearing in which the reference line is grid north.

**grid bias** The d-c voltage applied between the control grid and cathode of an electron tube to establish the desired operating point. Also called bias, C bias, and direct grid bias.

**grid-bias cell** *Bias cell.*

**grid blocking capacitor** *Grid capacitor.*

**grid cap** A top-cap terminal for the control grid of an electron tube.

**grid capacitor** A small capacitor used in the grid circuit of an electron tube to pass signal current while blocking the d-c anode voltage of the preceding stage. Also called grid blocking capacitor.

**grid circuit** The circuit connected between the grid and cathode of an electron tube.

**grid control** The control of anode current of an electron tube by varying the voltage of the control grid with respect to the cathode.

**grid-controlled mercury-arc rectifier** A mercury-arc rectifier in which one or more electrodes are employed exclusively to control the starting of the discharge. Also called grid-controlled rectifier.

**grid-controlled rectifier** *Grid-controlled mercury-arc rectifier.*

**grid-control tube** A mercury-vapor tube having external grid control.

**grid current** Electron flow to a positive grid in an electron tube.

**grid detection** Detection in the grid circuit of a vacuum tube, as in a grid-leak detector.

**grid-dip meter** An absorption wavemeter in which the resonance indicator is a vacuum-tube oscillator having in its grid circuit a sensitive current-indicating meter. The meter dips (reads lower grid current) when energy is drawn from the oscillator by a calibrated tuned circuit that is lightly coupled to the source whose frequency is being measured.

**grid dissipation** The power lost as heat at the grid of an electron tube.

**grid-drive characteristic** A relation between electric or light output of an electron tube and the control-electrode voltage as measured from cutoff.

**grid driving power** The average product of the instantaneous value of the grid current and the alternating component of the grid voltage of an electron tube over a complete cycle.

**grid emission** Electron or ion emission from a grid of an electron tube.

**grid-glow tube** A glow-discharge cold-cathode tube in which one or more control electrodes initiate but do not limit the anode current except under certain operating conditions.

**grid leak** A resistor used in the grid circuit of an electron tube to provide a discharge path for the grid capacitor and for charges built up on the control grid.

**grid-leak capacitor** A small capacitor connected in parallel with a grid leak. The two parts together serve to provide a grid-bias voltage.

**grid-leak detector** A detector in which the desired a-f voltage is developed across a grid leak and grid capacitor by the flow

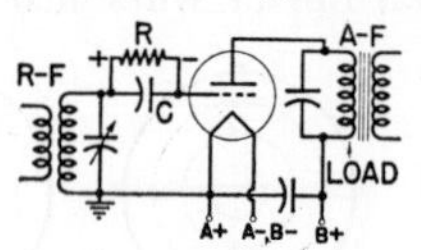


Grid-leak detector circuit.

of modulated r-f grid current. The circuit provides square-law detection on weak signals and linear detection on strong signals, along with amplification of the a-f signal.

**grid limiting** Limiting action achieved by placing a high-value resistor in series with the grid of a vacuum tube. The voltage drop across this resistor increases with input signal strength, giving a varying negative grid bias that serves to level input signals that are above a certain value.

**grid modulation** Modulation produced by feeding the modulating signal to the control-grid circuit of any electron tube in which the carrier is present.

**grid neutralization** Neutralization in which a portion of the grid-cathode a-c voltage of a vacuum tube is shifted 180° and applied to the anode-cathode circuit through a neutralizing capacitor.

**grid north** An arbitrary reference direction used in connection with the grid system of navigation. The reference direction is the top of a grid which, for polar navigation, consists of rectangular coordinates superimposed over the polar regions. One line on this grid coincides with the Greenwich meridian. On this grid, north is usually the direction of the North Pole from Greenwich, England. Also called polar grid.

**grid-pool tank** A grid-pool tube having a heavy metal envelope resembling a tank.

**grid-pool tube** An electron tube having a mercury-pool cathode, one or more anodes, and a control electrode or grid that controls the start of current flow in each cycle. The excitron and ignitron are examples.

**grid pulse modulation** Modulation produced in an amplifier or oscillator by applying one or more pulses to a grid circuit.

**grid pulsing** Pulsing of an r-f oscillator by biasing the grid sufficiently negative to block oscillation, and applying a positive pulse to the grid to remove this bias.

**grid resistor** A resistor used to limit grid current.

**grid return** The portion of a grid circuit that completes the electrical path from the grid to the cathode of an electron tube.

**grid swing** The total variation in grid-cathode voltage from the positive peak to the negative peak of the applied signal voltage.

**grid transformer** A transformer used to apply a signal or a-c voltage to the grid circuit of an electron tube.

**grid voltage** The voltage between a grid and the cathode of an electron tube.

**grille** An arrangement of wood, metal, or plastic bars placed across the front of a loudspeaker in a cabinet for decorative and protective purposes.

**grille cloth** A loosely woven cloth stretched across the front of a loudspeaker to keep out dust and provide protection without appreciably impeding sound waves.

**grivation** [GRId VAriaTION] The variation between grid north and magnetic north, usually expressed as an angle.

**grivation computer** A computer that calculates the angle between grid north and magnetic north as a function of longitude convergence in the polar regions of the earth.

**grommet** A circumferentially grooved self-retaining insulating washer used to cover sharp edges of a hole through which conductors are run.

Grommets made from rubber and from plastics.

**groove** The track inscribed in a record by the cutting or embossing stylus during mechanical recording, including undulations or modulations caused by the vibration of the stylus.

**groove angle** The angle between the two walls of an unmodulated disk-recording groove.

**groove shape** The contour of the groove in a disk recording.

**groove speed** The linear speed of the groove with respect to the stylus in disk recording.

**ground** [abbreviated gnd] 1. A conducting path, intentional or accidental, between an electric circuit or equipment and the earth, or some conducting body serving in place of the earth. Also called earth (British). 2. The lowest energy state of a nucleus, atom, or molecule. All other states are excited.

**ground clamp** A clamp used to connect a grounding conductor to a grounded object.

**ground clutter** Clutter on a ground or airborne radar due to reflection of signals from the ground or objects on the ground. Also called ground flutter, ground return, land return, and terrain echoes.

**ground control** Control of an aircraft or missile in flight by a person on the ground.

**ground-controlled approach** [abbreviated gca] A ground radar system used at an airport to provide information by which aircraft approaches for landings may be directed from the ground by radio during conditions of poor visibility and low cloud ceiling. It consists of an airport surveillance radar for guiding the aircraft to the start of the final approach path, and a precision approach radar for showing the exact position of the aircraft on its final approach path. Also called talk-down system.

**ground-controlled interception** [abbreviated gci] A radar system by means of which a controller at the ground or ship radar may direct an aircraft by radio to make an interception of another aircraft.

**ground distance** The mean sea-level great-circle component of distance from one point to another.

**grounded** Connected to earth or to some conducting body that serves in place of the earth. Also called earthed (British).

**grounded-anode amplifier** *Cathode follower.*

**grounded-base amplifier** An amplifier that uses a transistor in a grounded-base connection.

**grounded-base connection** A transistor circuit in which the base electrode is common to both the input and output circuits. The base need not be directly connected to circuit ground. Also called common-base connection.

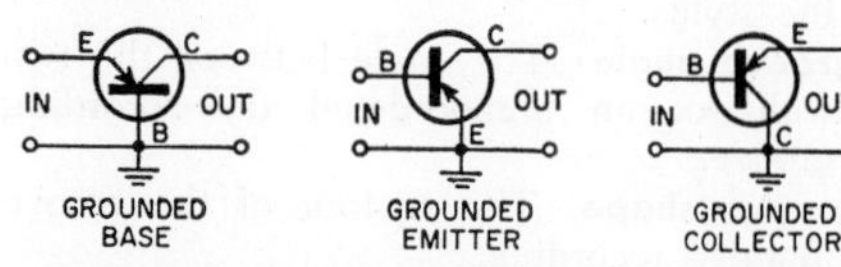


Grounded-base, grounded-emitter, and grounded-collector connections for p-n-p junction transistor.

**grounded-cathode amplifier** A conventional electron-tube amplifier circuit. The cathode is at ground potential at the operating frequency. The input is applied between control grid and ground. The output load is connected between anode and ground.

**grounded-collector amplifier** An amplifier that uses a transistor in a grounded-collector connection.

**grounded-collector connection** A transistor circuit in which the collector electrode is common to both the input and output circuits. The collector need not be directly connected to circuit ground. Also called common-collector connection.

**grounded-emitter amplifier** An amplifier that uses a transistor in a grounded-emitter connection.

**grounded-emitter connection** A transistor circuit in which the emitter electrode is common to both the input and output circuits. The emitter need not be directly connected to circuit ground. Also called common-emitter connection.

**grounded-grid amplifier** An electron-tube amplifier circuit in which the control grid is at ground potential at the operating frequency. The input signal is applied between cathode and ground, and the output load is connected between anode and ground. Chief advantages are low input impedance and freedom from oscillation due to feedback.

**ground-equalizer coil** A coil having relatively low inductance, placed in one or more of the circuits connected to the grounding points of an antenna to distribute the current to the various points in a desired manner.

**ground fault** Accidental grounding of a conductor.

**ground flutter** *Ground clutter.*

**ground-guided missile** A missile guided by radio control from the ground.

**ground influence mine** An aerial mine designed to rest on the ground under water, and detonated by magnetic or other influence.

**grounding outlet** An outlet having, in addition to the current-carrying contacts, one grounded contact that can be used for grounding portable appliances and equipment.

**grounding plate** An electrically grounded metal plate on which a person stands to discharge static electricity picked up by his body.

**ground noise** The residual system noise in the absence of the signal in recording and reproducing. Usually caused by inhomogeneity in the recording and reproducing media, but may also include tube noise and noise generated in resistive elements in the amplifier system.

**ground-plane antenna** A vertical antenna combined with turnstile or other horizontal antenna elements to lower the angle of radiation. A concentric base support and

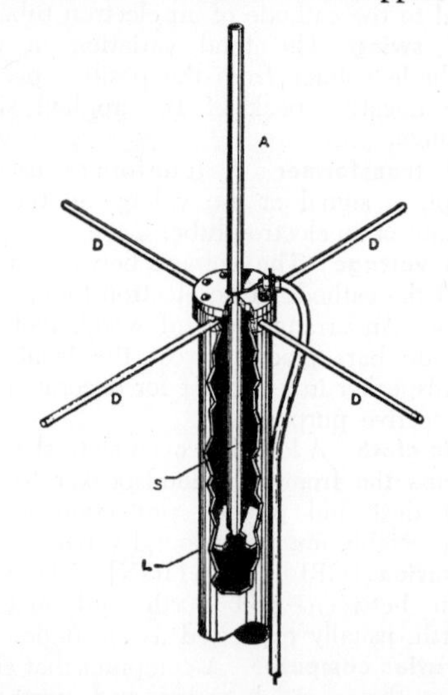


Ground-plane antenna, formed by adding turnstile elements D to high-frequency vertical antenna A having metallic support L and center conductor S acting as an inductance. The combination is fed by a coaxial transmission line running outside the support.

center conductor together serve to place the antenna at ground potential even though it may be located several wavelengths above ground. The horizontal radiation pattern is nondirectional.

**ground position** A position on the ground directly under an airborne aircraft.

**ground-position indicator** [abbreviated gpi] A computer similar to an air-position indicator but having provision for taking account of drift so as to give the actual ground position with respect to some fixed point on the ground.

**ground potential** Zero voltage with respect to the ground or to a chassis that serves in place of a ground connection.

**ground-reflected wave** A radio wave that is reflected from the ground somewhere along the transmission path.

**ground return** 1. An echo received from the ground by an airborne radar set. 2. Use of the earth or a chassis as the return path for a transmission line. 3. Use of a chassis as a return path for a circuit. 4. *Ground clutter.*

**ground rod** A rod that is driven into the earth to provide a good ground connection.

**ground speed** The speed of a vehicle along its track. For an aircraft, it is the speed relative to the earth's surface.

**ground-speed recorder** A recorder that makes a permanent record of the horizontal speed of a moving object with respect to the earth.

**ground state** The lowest energy state of a nucleus, atom, or molecule. Also called normal state.

**ground-state beta disintegration** The total energy released in a beta transition between isobars in their ground states. It includes the energies of any gamma and associated radiations following the beta process.

**ground-state disintegration** The disintegration energy that is present when all reactant and product nuclei are in their ground states.

**ground surveillance radar** A surveillance radar operated at a fixed point for observation and control of the position of aircraft or other vehicles in the vicinity.

**ground system** The portion of an antenna that is closely associated with an extensive conducting surface, which may be the earth itself.

**ground-to-air communication** One-way radio communication from ground stations to aircraft.

**ground vector** A vector representing the track and ground speed of an aircraft.

**ground wave** A radio wave that is propagated over the earth and is ordinarily affected by the presence of the ground and the troposphere. The ground wave includes all components of a radio wave over the earth except ionospheric and tropospheric waves. The ground wave is refracted because of variations in the dielectric constant of the troposphere, including the condition known as a surface duct. Also called surface wave.

**ground wire** A conductor used to connect electric equipment to a ground rod or other grounded object.

**ground zero** A point on the surface of land or water that is directly below or above an atomic explosion.

**grouped-frequency operation** Use of different frequency bands for channels in opposite directions in a two-wire carrier system.

**grouping** 1. Periodic error in the spacing of recorded lines in a facsimile system. 2. Nonuniform spacing between the grooves of a disk recording.

**group modulation** The process by which a number of channels, already separately modulated to a specific frequency range, are modulated upon a new carrier so as to shift the entire group to another frequency range.

**group velocity** The velocity of propagation of the envelope of a plane wave occupying a frequency band over which the envelope delay is approximately constant. Group velocity differs from phase velocity in a medium in which the phase velocity varies with frequency.

**growler** An electromagnetic device consisting essentially of two field poles arranged as in a motor, used for locating short-circuited coils in the armature of a generator or motor and for magnetizing or demagnetizing objects. A growling noise indicates a short-circuited coil.

**grown-diffused transistor** A junction transistor in which the final junctions are formed by diffusion of impurities near a grown junction.

**grown junction** A junction produced by changing the types and amounts of donor and acceptor impurities that are added during the growth of a semiconductor crystal from a melt.

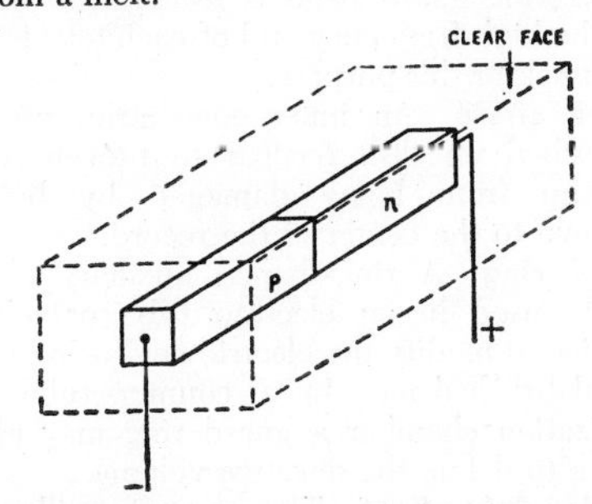


Grown-junction photocell mounted in transparent plastic block.

**grown-junction photocell** A photodiode consisting of a bar of semiconductor material having a p-n junction at right angles to its length and an ohmic contact at each end of the bar.

**grown-junction transistor** A transistor having a grown junction.

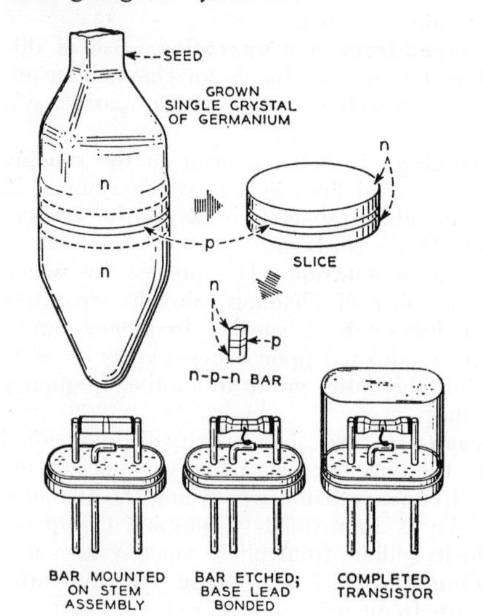


Grown-junction transistor construction procedure.

**growth curve** A curve showing how a quantity increases with time.

**G scan** *G display.*

**G scope** A radarscope that produces a G display.

**GT selection rules** Abbreviation for *Gamow-Teller selection rules.*

**guard** To monitor constantly a specific radio frequency or channel.

**guard band** A narrow frequency band provided between adjacent channels in certain portions of the radio spectrum to prevent interference between stations. In television, a 0.25-mc guard band is usually provided at the high-frequency end of each television channel for this purpose.

**guard circle** An inner concentric groove inscribed on disk records to prevent the pickup from being damaged by being thrown to the center of the record.

**guard ring** A ring-shaped auxiliary electrode used in an electron tube or other device to modify the electric field or reduce insulator leakage. In a counter tube or ionization chamber a guard ring may also serve to define the sensitive volume.

**Gudden-Pohl effect** The momentary illumination produced when an electric field is applied to a phosphor previously excited by ultraviolet radiation.

**guidance** Control of the path of a missile along part or all of its path.

**guidance system** Any system that controls the path of a missile from launch to target. Examples are homing guidance and beam-rider guidance.

**guidance tape** A magnetic or punched paper tape that is placed in a missile or its computer to program desired events during flight.

**guide** *Waveguide.*

**guided** Controlled or controllable as to direction by preset mechanisms, radio commands, or built-in self-reacting devices.

**guided aircraft missile** A guided missile designed for launching from an aircraft in flight.

**guided aircraft rocket** A rocket-powered guided missile designed to be launched from an aircraft in flight.

**guided ballistic missile** A ballistic missile that is guided only during the powered portion of its trajectory.

**guided bomb** An aerial bomb that is guided in range or azimuth, or both, during its drop.

**guided missile** A missile that is directed to its target by a preset or self-reacting device within the missile or by radio command outside the missile. Also called missile.

**guided rocket** A guided missile having rocket propulsion.

**guided wave** A wave whose energy is concentrated near a boundary or between substantially parallel boundaries separating materials of different properties, and whose direction of propagation is effectively parallel to these boundaries. Waveguides transmit guided waves.

**guide field** The magnetic flux that holds particles in a stable circular orbit during the accelerating period in a betatron or synchrotron.

**guide wavelength** The wavelength in a waveguide.

**Guillemin line** A network or artificial line used in high-level pulse modulation to generate a nearly square pulse, with steep rise and fall. Used in radar sets to control pulse width.

Guillemin line.

**gun** 1. *Electron gun.* 2. *Soldering gun.*

**gun-laying radar** Radar equipment specifically designed to determine range, azimuth, and elevation of a target and sometimes also to automatically aim and fire antiaircraft artillery or other guns.

**gutta-percha** A natural vegetable gum similar to rubber, used principally as insulation for wires and cables.

**guy wire** A wire used to hold a pole or tower in an upright position.

**gyrator** A waveguide component that uses a ferrite section to give zero phase shift for one direction of propagation and 180° phase shift for the other direction. Also called microwave gyrator.

**gyro** *Gyroscope.*

**gyrocompass** A compass that uses a gyroscope to provide a reference direction.

**gyro flux-gate compass** A compass in which a flux gate, horizontally stabilized by a gyroscope, senses the horizontal component of the earth's magnetic field. Being fixed with respect to the aircraft, the compass reacts to each change in heading with a change in current. This current is amplified and used to actuate the dial of a master indicator. Also called flux-gate compass.

**gyrofrequency** The natural frequency of rotation of a charged particle under the influence of a constant magnetic field, such as the magnetic field of the earth.

**gyro horizon** *Artificial horizon.*

**gyromagnetic** Pertaining to the magnetic properties of rotating electric charges, such as spinning electrons moving within atoms.

**gyromagnetic compass** A magnetic compass in which gyroscopic stabilization is used to indicate direction.

**gyromagnetic ratio** The ratio of the magnetic moment to the angular momentum for a charged particle moving in a closed orbit.

**gyroscope** A wheel mounted and driven at high speed in such a way that its spinning axis is free to rotate about either of two other axes perpendicular to itself and to each other. Used to maintain a stable equilibrium such as is required for autopilots. Also called gyro.

**gyroscopic horizon** A gyroscopic instrument that indicates the lateral and longitudinal attitude of an aircraft by simulating the natural horizon.

**gyro sight** A gun sight that uses a gyroscope.

**gyro stabilizer** A stabilizer that uses a gyroscope to compensate for the roll and pitch of a ship.

**gyrotron** A device that detects motion of a system by measuring the phase distortion that occurs when a vibrating tuning fork is moved. The altered fork frequency is compared to a reference frequency that was originally in phase with the fork frequency.

**G — Y signal** The green-minus-luminance color-difference signal used in color television. It is combined with the luminance signal in a receiver to give the green color-primary signal.

# H

**H** 1. Symbol for *heater.* 2. *H system.*

**$H^2$** Symbol for *deuterium.*

**$H^3$** Symbol for *tritium.*

**h** Abbreviation for *henry.*

***H*** Symbol for *magnetic field strength.*

***h*** Symbol for *Planck's constant.*

**hafnium** [symbol Hf] A metallic element used as a neutron-absorbing shield in nuclear reactors. Atomic number is 72.

**halation** An area of glow surrounding a bright spot on a fluorescent screen, due to scattering by the phosphor or to multiple reflections at front and back surfaces of the glass faceplate.

**half-adder** A digital computer circuit having two input and two output channels for binary digit signals, related in such a way that two half-adders can be combined to give one binary adder.

**half-cycle** The time interval corresponding to half a cycle or 180° at the operating frequency of a circuit or device.

**half-cycle transmission** A data transmission and control system that uses synchronized sources of 60-cycle power at the transmitting and receiving ends. Either of two receiver relays can be actuated by choosing the appropriate half-cycle polarity of the 60-cycle transmitter power supply.

**half-duplex operation** Operation of a telegraph system in either direction over a single channel, but not in both directions simultaneously.

**half-life** The average time required for half the atoms of a sample of a radioactive substance to lose their radioactivity by decaying into stable atoms. Also called half-value period.

**half-power frequency** One of the two values of frequency, on the sides of an amplifier response curve, at which the voltage is 70.7% of a midband or other reference value.

**half-power width** An angular rating for antenna beam width. In a plane containing the direction of the maximum of the radiation lobe, it is the full angle between the two directions in that plane in which the radiation intensity is half the maximum value of the lobe.

**half-rhombic antenna** A long-wire antenna array in which the lengths of the sides are several wavelengths long at the lowest frequency of operation. The radiating sides form half of a rhombic antenna, with the lines connected to opposite ends. When terminated in a resistor, the radiation pattern is unidirectional; when unterminated, it is bidirectional.

**half-step** *Semitone.*

**halftone characteristic** A relation between the density of the recorded copy and the density of the subject copy in a facsimile system.

**half-track recorder** *Double-track tape recorder.*

**half-value layer** *Half-value thickness.*

**half-value period** *Half-life.*

**half-value thickness** The thickness of a given substance which, when introduced in the path of a given beam of radiation, will reduce its intensity to half of the initial value. Also called half-value layer.

**half-wave** 1. Having an electrical length of one half-wavelength. 2. Pertaining to half of one cycle of a wave.

**half-wave antenna** An antenna whose electrical length is half the wavelength being transmitted or received.

**half-wave dipole** *Dipole.*

**half-wavelength** The distance correspond-

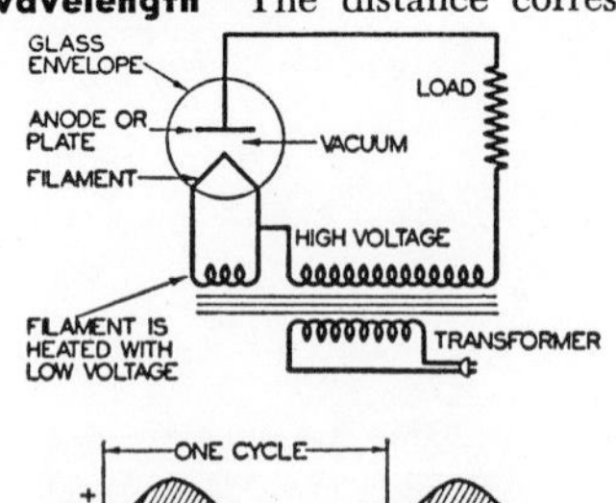


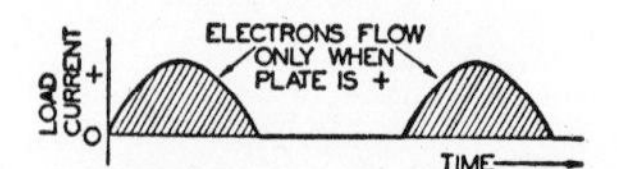


Half-wave rectifier circuit and waveforms.

ing to an electrical length of half a wavelength at the operating frequency of a transmission line, antenna element, or other device.

**half-wave rectification** Rectification in which current flows only during alternate half-cycles.

**half-wave rectifier** A rectifier that provides half-wave rectification.

**half-wave vibrator** A vibrator having only one pair of contacts. It interrupts the flow of direct current through the primary of a power transformer but does not reverse the current, whereas a full-wave vibrator changes the direction of current flow twice per operating cycle.

**Hall constant** The constant of proportionality in the equation for a current-carrying conductor in a magnetic field. The constant is equal to the transverse electric field (Hall field) divided by the product of the current density and the magnetic field strength. The sign of the majority carrier can be inferred from the sign of the Hall constant.

**Hall effect** The development of a voltage between the two edges of a current-carrying metal strip whose faces are perpendicular to a magnetic field.

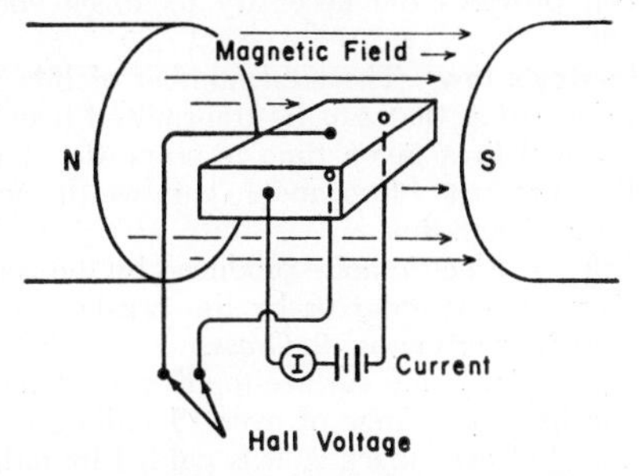


Hall effect, in which voltage is developed between top and bottom surfaces of current-carrying metal strip in magnetic field.

**Hall mobility** The product of conductivity and the Hall constant for a conductor or semiconductor. It is a measure of the mobility of the electrons or holes in a semiconductor.

**Hallwachs effect** The ability of ultraviolet radiation to discharge a negatively charged body in a vacuum.

**halo** An undesirable bright or dark ring surrounding an image on the fluorescent screen of a television cathode-ray tube. Generally due to overloading or maladjustment of the camera tube.

**ham** Slang for *amateur.*

**H and D curve** A characteristic curve of a photographic emulsion, obtained by plotting film density against the logarithm of the exposure. Also called Hurter and Driffield curve.

**Handie-Talkie** Trademark of Motorola, Inc., for their line of portable two-way radio communication units.

**hand-reset** Requiring manual resetting after an operation, as with a circuit breaker.

**handset** A hand-held combination telephone-type receiver and telephone-type transmitter, as commonly used in telephone sets.

**handwheel** A wheel used in place of a knob on a control shaft.

**hangover** 1. Faulty reproduction of bass notes by a loudspeaker that is poorly damped or improperly mounted. 2. *Tailing.*

**hard copy** A message in page form, produced as a result of a transmission.

**hard cosmic ray** A cosmic radiation component that penetrates a moderate thickness of an absorber, such as 4 inches of lead. The hard component consists chiefly of mesons, but has some fast protons and fast electrons.

**hard-drawn copper wire** Copper wire that has been drawn to size through several dies without being annealed, to increase its hardness and tensile strength.

**hard magnetic material** A magnetic material that is not easily demagnetized. Used in permanent magnets.

**hardness** 1. The penetrating ability of x-rays. The shorter the wavelength, the harder are the rays and the greater is their penetrating ability. 2. The degree of evacuation in an x-ray tube or other vacuum tube. The harder the tube, the better is its vacuum.

**hard rubber** Rubber that has been vulcanized at high temperatures and pressures, to give hardness. Once used extensively as an insulating material.

**hard tube** A vacuum tube that has been evacuated to a high degree, approaching a perfect vacuum. A tube containing an appreciable amount of gas is called a soft tube.

**hardware** 1. Electronic components, subassemblies, and finished pieces of equipment, as contrasted to a drawing-board design. 2. Bolts, nuts, fasteners, brackets, handles, and similar small structural parts used in assembling electronic equipment.

**hard x-ray** An x-ray having high penetrating power.

**harmonic** A sinusoidal component of a periodic wave, having a frequency that is an integral multiple of the fundamental frequency. The frequency of the second harmonic is twice that of the fundamental frequency or first harmonic. Also called harmonic component and harmonic frequency.

**harmonic analysis** 1. Any method of identifying and evaluating the harmonics that make up a complex waveform of voltage, current, or some other varying quantity. 2. The expression of a given function as a series of sine and cosine terms that are approximately equal to the given function, such as a Fourier series.

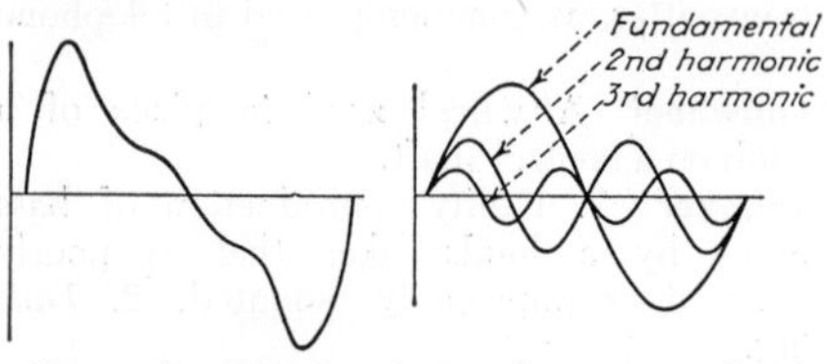


Harmonic analysis of complex wave, showing fundamental and harmonic components.

**harmonic analyzer** An analyzer that measures the strength of each frequency component in a complex wave. Also called harmonic wave analyzer.

**harmonic antenna** An antenna whose electrical length is an integral multiple of a half-wavelength at the operating frequency of the transmitter or receiver.

**harmonic attenuation** Attenuation of an undesired harmonic component in the output of a transmitter, as by use of a pi network whose shunt reactances are tuned to have zero impedance at the harmonic frequency to be suppressed.

**harmonic component** *Harmonic.*

**harmonic content** The components remaining after the fundamental frequency has been removed from a complex wave.

**harmonic conversion transducer** A conversion transducer in which the output signal frequency is a multiple or submultiple of the input frequency, as in frequency dividers and frequency multipliers.

**harmonic distortion** Nonlinear distortion in which undesired harmonics of a sinusoidal input signal are generated because of circuit nonlinearity.

**harmonic filter** A filter that is tuned to suppress an undesired harmonic in a circuit.

**harmonic frequency** *Harmonic.*

**harmonic generator** A generator operated under conditions such that it generates strong harmonics along with the fundamental frequency.

**harmonic interference** Interference due to the presence of harmonics in the output of a radio station.

**harmonic leakage power** The total r-f power transmitted through a fired tr or pre-tr tube in its mount at frequencies other than the fundamental frequencies generated by the transmitter.

**harmonic-mode crystal unit** *Overtone crystal unit.*

**harmonic wave analyzer** *Harmonic analyzer.*

**harness** An assembly of insulated wires of various lengths, bent to a pattern and tied together before installation in a piece of equipment.

**hartley** A unit of information content, equal to the designation of one of ten possible and equally likely values or states of anything used to store or convey information. One hartley is equal to $\log_2 10$ bits or 3.323 bits.

**Hartley oscillator** A vacuum-tube oscillator in which the parallel-tuned tank circuit is connected between grid and anode. The tank coil has an intermediate tap at cathode potential, so the grid-cathode portion of the coil provides the necessary feedback voltage.

**Hartley's law** The total number of bits of information that can be transmitted over a channel in a given time is proportional to the product of channel bandwidth and transmission time.

**hash** 1. Electric noise produced by the contacts of a vibrator or by the brushes of a generator or motor. 2. *Grass.*

**Hawk** An Army surface-to-air guided missile having a range of over 15 miles and a speed of over mach 2. It is guided by radio command to attack low-flying enemy airplanes, and may also have radar homing.

**Hay bridge** A four-arm a-c bridge used to measure inductance in terms of capacitance, resistance, and frequency. Bridge balance depends on frequency.

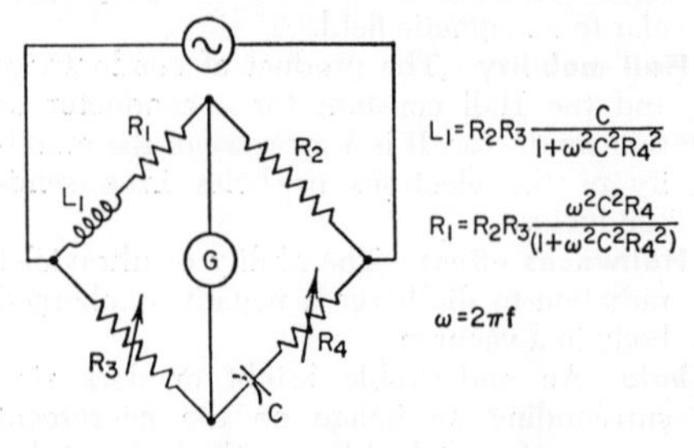


Hay bridge circuit and equations.

**H bend** *H-plane bend.*
**H-bomb** *Hydrogen bomb.*
**H display** A radarscope B display modified to include indication of angle of elevation. The target appears as two closely spaced

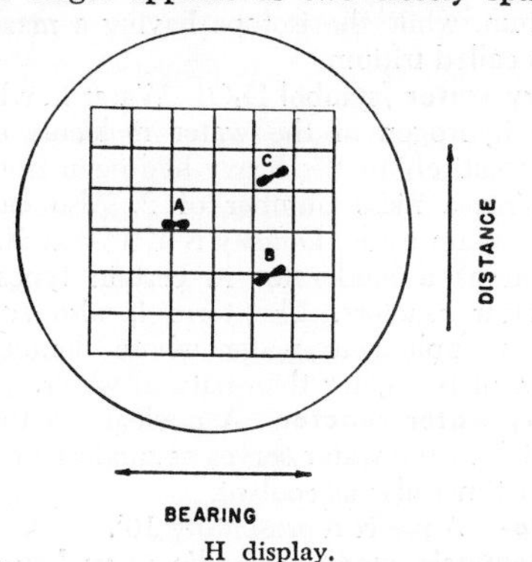

H display.

bright spots that approximate a short bright line, the slope of which is proportional to the sine of the angle of target elevation. Also called H scan.
**HDO** Symbol for water in which one of the hydrogen atoms is replaced by a deuterium atom.
**head** 1. The photoelectric unit that converts the sound track on motion-picture film into corresponding audio signals in a motion-picture projector. 2. *Cutter.* 3. *Magnetic head.*
**head amplifier** 1. An amplifier that is mounted close to the head that serves as its signal source, for amplifying the weak signal before it is fed through a cable to the main amplifier. 2. British term for the video amplifier that is mounted close to the pickup tube in a television camera.
**head demagnetizer** A device that eliminates residual magnetism built up in a magnetic-tape recording head.
**header** A mounting plate through which the insulated terminals or leads are brought out from a hermetically sealed relay, transformer, transistor, tube, or other device.
**heading** The horizontal direction in which a vehicle is directed, expressed as an angle between a reference line and the line extending in the direction the vehicle is headed, usually measured clockwise from the reference line. Heading is the instantaneous actual direction in which an aircraft, ship, or other vehicle is pointed, whereas course is the intended direction of travel. Also called relative heading (deprecated).
**heading marker** A line of light produced on a marine radar ppi display at the instant when the rotating radar antenna is at the position corresponding to the ship's heading, to indicate whether objects ahead of the ship are located to port or starboard.
**headlight antenna** A radar antenna small enough to be housed within the thickness of an airplane wing in the manner of an automobile headlight, yet producing a beam that can be directed much like a searchlight.
**headphone** An electroacoustic transducer designed to be held against an ear by a clamp passing over the head, for private listening to the audio output of a communication, radio, or television receiver or other

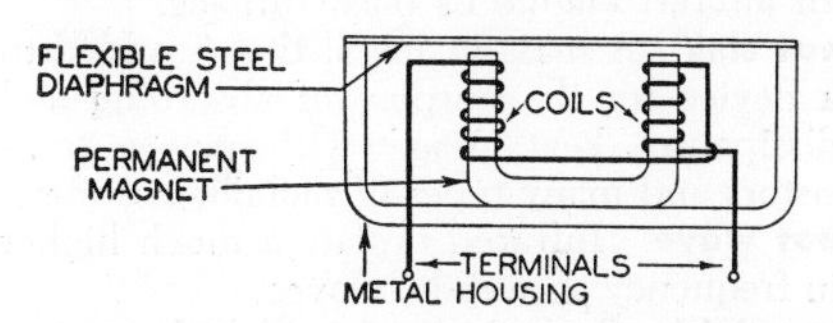

Headphone construction. Audio-frequency currents flowing through diaphragm produce magnetic field that alternately aids and opposes fixed magnetic field of permanent magnet, causing diaphragm to move and produce sound waves.

source of a-f signals. Usually used in pairs, one for each ear, with the clamping strap holding both in position. Also called headset and phone.
**headset** *Headphone.*
**health physics** The protection of personnel from harmful effects of ionizing radiation by such means as routine radiation surveys, area and personnel monitoring, protective equipment, and protective procedures.
**hearing aid** A miniature portable sound amplifier for persons with impaired hearing, consisting of a microphone, audio amplifier, earphone, and batteries.
**hearing loss** The ratio of the threshold of audibility of a particular human ear to the normal threshold, expressed in decibels.
**heat** *Thermal radiation.*
**heater** [symbol H] An electric heating element for supplying heat to an indirectly heated cathode in an electron tube.
**heater current** The current flowing through a heater for an indirectly heated cathode in an electron tube.
**heater-type cathode** *Indirectly heated cathode.*
**heater voltage** The voltage applied to the terminals of the heater in an electron tube.
**heating element** A coiled or other arrangement of resistance wire, supported by one or more ceramic insulators, through which current is sent to produce heat.

**heat loss** Power loss that is due to the conversion of electric energy into heat.

**heat reactor** A nuclear reactor designed primarily to supply heat for industrial purposes.

**heat run** A series of temperature measurements made on an electric device during operating tests under various conditions.

**heat-seal** To bond or weld a flexible plastic film or sheet of material to itself or to another material by heat alone. Dielectric heating is frequently used for this purpose.

**heatseeker** A guided missile incorporating an infrared device for homing on heat-radiating machines or installations, such as an aircraft engine or blast furnace.

**heat sink** A mass of metal that is added to a device for the purpose of absorbing and dissipating heat. Used with power transistors and many types of metallic rectifiers.

**heat wave** Infrared radiation much higher in frequency than radio waves.

**Heaviside-Campbell mutual-inductance bridge** A Heaviside mutual-inductance bridge in which one of the inductive arms

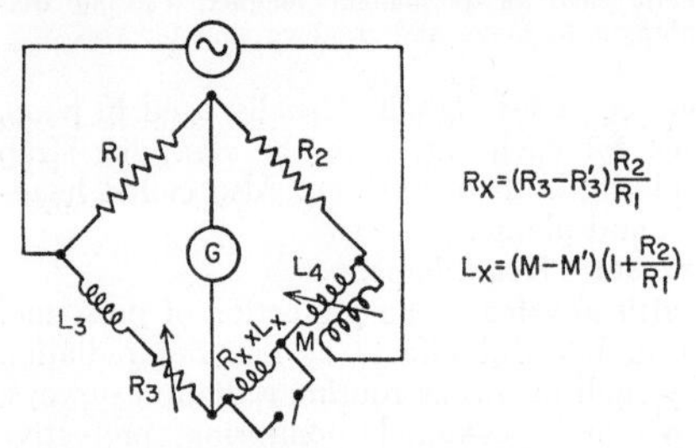


Heaviside-Campbell mutual-inductance bridge and equations.

contains a separate inductor that is included in the bridge arm during the first of a pair of measurements and is short-circuited during the second. Bridge balance is independent of frequency.

**Heaviside layer** *E layer.*

**Heaviside mutual-inductance bridge** An a-c bridge used to compare self-inductances

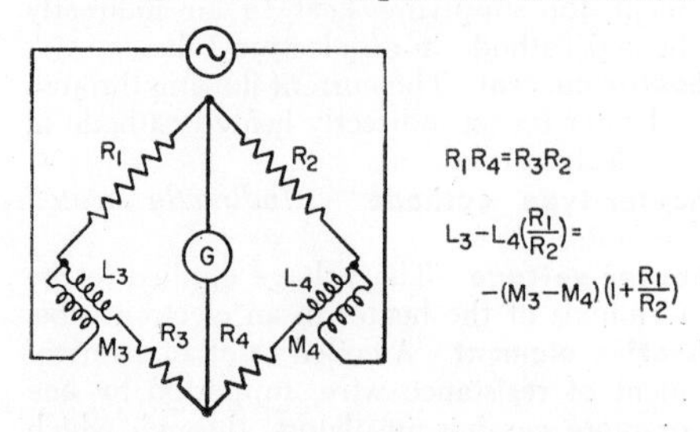


Heaviside mutual-inductance bridge and equations.

and mutual inductances. Bridge balance is independent of frequency.

**heavy hydrogen** An isotope of hydrogen having a mass number of either 2 or 3. The isotope having a mass of 2 is called deuterium, while the isotope having a mass of 3 is called tritium.

**heavy water** [symbol $D_2O$] Water in which the hydrogen of the water molecule consists entirely of the heavy-hydrogen isotope having a mass number of 2. Also called deuterium oxide. Density is 1.1076 at 20°C. Used as a moderator in certain types of nuclear reactors. The term is also sometimes applied to water whose deuterium content is greater than natural water.

**heavy-water reactor** A nuclear reactor in which heavy water serves as moderator and sometimes also as coolant.

**hecto-** A prefix representing $10^2$.

**hectometric wave** A radio wave between the wavelength limits of 100 and 1,000 meters, corresponding to the frequency range of 3,000 to 300 kc.

**height** The vertical distance to a target from a horizontal plane passing through the location of the height computer.

**height control** The television receiver control that adjusts picture height.

**height effect** *Antenna effect.*

**height error** Deprecated term for *ionospheric height error.*

**height finder** A radar set that measures and determines the height of an airborne object.

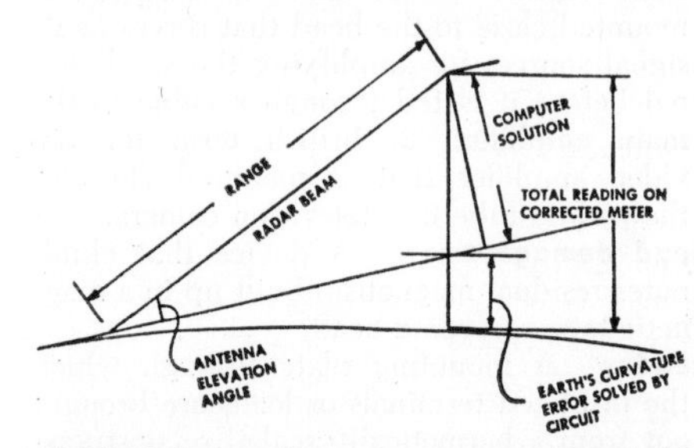


Height finder requires computer to combine earth's curvature with range and elevation data provided by radar set.

**height marker** A type of calibration marker used on some radar displays.

**height-range indicator** A radarscope that gives both height and range of a target.

**Heising modulation** *Constant-current modulation.*

**helical antenna** An antenna having the form of a helix. When the helix circumference is much smaller than one wavelength, the antenna radiates at right angles to the axis of the helix. When the helix circumference is of the order of one wavelength, maximum radiation is along the helix axis. Also called helix antenna.

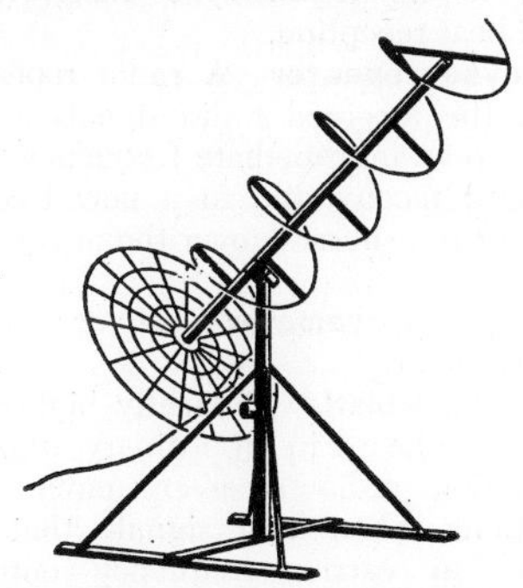

Helical antenna used with telemetering receivers.

**helical potentiometer** A multiturn precision potentiometer in which a number of complete turns of the control knob are required to move the contact arm from one end of the helically wound resistance element to the other end.

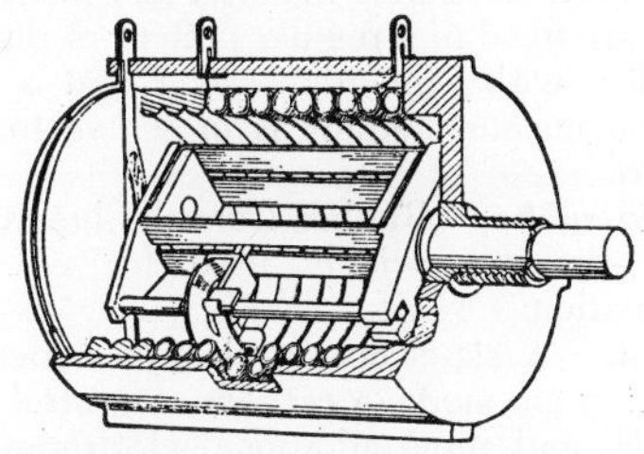

Helical potentiometer having ten-turn resistance element.

**helical scanning** 1. A method of facsimile scanning in which a single-turn helix rotates against a stationary bar to give horizontal movement of an elemental area. 2. A method of radar scanning in which the antenna beam rotates continuously about the vertical axis while the elevation angle changes slowly from horizontal to vertical, so that a point on the radar beam describes a distorted helix.

**helium** [symbol He] A light gaseous element that exists as a monatomic gas except at extremely low temperatures, is chemically inert, and has practically no absorption cross section for neutrons. Atomic number is 2.

**heliumtight** *Hermetic.*

**helix antenna** *Helical antenna.*

**helix recorder** A recorder that uses helical scanning.

**helix waveguide** A waveguide consisting of closely wound turns of insulated copper wire covered with a lossy jacket.

**Helmholtz resonator** An acoustic enclosure having a small opening of such dimensions that the enclosure resonates at a single frequency determined by the geometry of the resonator.

**HEM wave** Abbreviation for *hybrid electromagnetic wave.*

**henry** [abbreviated h; plural henrys] The centimeter-gram-second unit of inductance or mutual inductance. The inductance of a circuit is 1 henry when a current change of 1 ampere per second induces 1 volt. One millihenry is one-thousandth of a henry. One microhenry is one-millionth of a henry.

**heptode** A seven-electrode electron tube containing an anode, a cathode, a control electrode, and four additional electrodes that are ordinarily grids.

**herald** [Harbor Echo Ranging And Listening Device] A sensitive ultrasonic underwater sound-detection system used at harbor entrances to detect submerged submarines running at silent speed. The operator is able to listen to the sounds as well as determine range and bearing for their source.

**Hermes** A surface-to-surface rocket-powered guided missile using beam-rider guidance. Range is 50 miles, speed is mach 2, and maximum altitude is 100,000 feet.

**hermetic** Permanently sealed by fusion, soldering, or other means, to prevent the transmission of air, moisture vapor, and all other gases. Also called heliumtight, leaktight, and vacuumtight.

**hermetically sealed crystal unit.** A crystal unit sealed in its glass or metal holder, usually by soldering, for protection against all external conditions except vibration and temperature.

**hermetically sealed relay** A relay that is permanently sealed in its metal, glass, or ceramic housing by fusion or soldering.

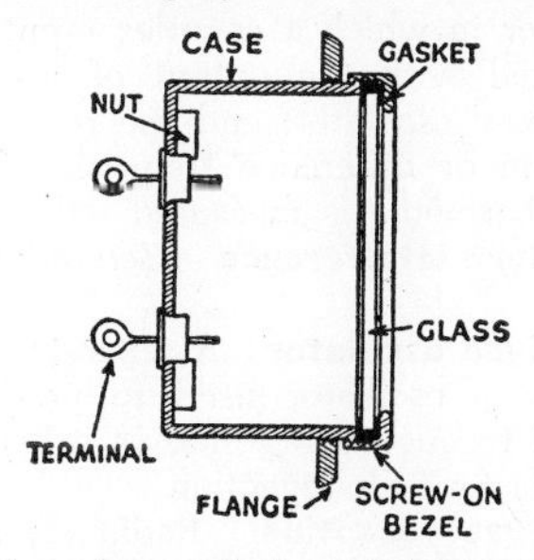


Hermetic seals around glass window and terminals of meter housing.

**hermetic seal** A seal that prevents passage of air, water vapor, and all other gases.

**herringbone pattern** An interference pattern sometimes seen on television receiver screens, consisting of a horizontal band of closely spaced V- or S-shaped lines.

**hertz** [abbreviated Hz] European term for *cycle per second.*

**Hertz antenna** An ungrounded half-wave antenna.

**Hertzian wave** *Radio wave.*

**Hertz vector** A vector used to specify the electromagnetic field of a radio wave.

**heterodyne** To mix two alternating-current signals of different frequencies in a nonlinear device for the purpose of producing two new frequencies, corresponding respectively to the sum of and the difference between the two original frequencies. This action is the basis of all superheterodyne receivers.

**heterodyne conversion transducer** *Converter.*

**heterodyne detector** A detector in which an unmodulated carrier frequency is combined with the signal of a local oscillator having a slightly different frequency, to provide an audio-frequency beat signal that can be heard with a loudspeaker or headphones. Used chiefly for code reception.

**heterodyne frequency** Either of the two new frequencies resulting from heterodyne action. One is the sum of the two input frequencies, while the other is the difference between the input frequencies.

**heterodyne frequency meter** A frequency meter in which a known adjustable frequency is heterodyned with an unknown frequency until a zero beat is obtained. Alternatively, the unknown frequency may be heterodyned with a fixed known frequency to produce a lower-frequency signal, usually in the audio-frequency range, whose value is measured by other means. Also called heterodyne wavemeter.

**heterodyne harmonic analyzer** A harmonic analyzer in which a complex input voltage is mixed with the output of a variable-frequency oscillator, and the magnitude of the sum or difference frequency for each input harmonic is measured with a meter.

**heterodyne interference** *Heterodyne whistle.*

**heterodyne oscillator** A separate variable-frequency oscillator used to produce the second frequency required in a heterodyne detector for code reception.

**heterodyne reception** Radio reception in which the incoming r-f signal is combined with a locally generated r-f signal of different frequency, followed by detection. The resulting beat frequency may be audible or may be at a higher intermediate frequency, as in a superheterodyne receiver. Also called beat reception.

**heterodyne repeater** A radio repeater in which the received radio signals are converted to an intermediate frequency, amplified, and reconverted to a new frequency band for transmission over the next repeater section.

**heterodyne wavemeter** *Heterodyne frequency meter.*

**heterodyne whistle** A steady high-pitched audio tone heard in an ordinary amplitude-modulation radio receiver under certain conditions when two signals that differ slightly in carrier frequency enter the receiver and heterodyne to produce an audio beat. Also called heterodyne interference.

**heterogeneous radiation** Radiation having a number of different frequencies, different particles, or different particle energies.

**heterogeneous reactor** A nuclear reactor in which fissionable material and moderator are arranged in a regular pattern of discrete bodies with dimensions such that a non-homogeneous medium is presented to neutrons.

**hevea rubber** Rubber from the hevea brasiliensis tree, widely used for electrical insulation.

**hexode** A six-electrode electron tube containing an anode, a cathode, a control electrode, and three additional electrodes that are ordinarily grids. Also called six-electrode tube.

**h-f** Abbreviation for *high frequency.*

**HgTe** Symbol for *mercuric telluride.*

**hi-fi** *High fidelity.*

**high-altitude radio altimeter** *Radar altimeter.*

**high band** The television band extending from 174 to 216 mc, which includes channels 7 to 13.

**high boost** *High-frequency compensation.*

**high-confidence countermeasure** A countermeasure that is very difficult for the enemy to overcome and may therefore be expected to retain its usefulness for a number of years.

**high-contrast image** An image in which the contrast between black and white is great and intermediate tones are poor.

**high definition** The television equivalent of high fidelity, in which the reproduced image contains such a large number of accurately reproduced elements that picture details approximate those of the original scene.

**high fidelity** Fidelity of audio reproduction so perfect that listeners hear exactly what they would have heard if present at the original performance. Also called hi-fi.

**high-fidelity receiver** A radio receiver that reproduces audio frequencies with high fidelity, so as to duplicate faithfully the original sound picked up by the microphone.

**high-flux reactor** A nuclear reactor operating at a high power density, so that a high flux results from a high rate of fission per unit volume.

**high frequency** [abbreviated h-f] A Federal Communications Commission designation for the band from 3 to 30 mc in the radio spectrum.

**high-frequency compensation** Increasing the amplification at high frequencies with respect to that at low and middle frequencies in a given band, such as in a video band or an audio band. Also called high boost.

**high-frequency furnace** An induction furnace in which the heat is generated within the charge, within the walls of the containing crucible, or in both, by currents induced by high-frequency magnetic flux produced by a surrounding coil. Also called coreless-type induction furnace.

**high-frequency heating** *Electronic heating.*

**high-frequency induction heater** An induction heater that produces electric current flow in a charge to be heated, at a frequency higher than that of the a-c power line.

**high-frequency resistance** The total resistance offered by a device in an a-c circuit, including the d-c resistance and the resistance due to eddy current, hysteresis, dielectric, and corona losses. Also called a-c resistance, effective resistance, and r-f resistance.

**high-frequency trimmer** A trimmer capacitor that controls the calibration of a tuning circuit at the high-frequency end of a tuning range in a superheterodyne receiver.

**high-frequency welding** Welding with high-frequency current obtained from an electronic generator.

**high-gamma tube** 1. A television camera tube in which the voltage output increases uniformly with the intensity of the light on the image. 2. A television picture tube in which the light intensity on the screen is directly proportional to the control-grid voltage.

**high-level detector** A power detector or linear detector, for which the voltage-current characteristic is essentially a straight line or two intersecting straight lines extending up to high input voltage levels.

**high-level firing time** The time required to establish an r-f discharge in a switching tube after r-f power is applied.

**high-level modulation** Modulation produced in the anode circuit of the last stage of a transmitter.

**high-level radio-frequency signal** A radio-frequency signal of sufficient power to cause a tr, atr, or pre-tr tube to become fired.

**high-level vswr** The voltage standing wave ratio due to a fired switching tube in its mount, located between a generator and a matched termination in the waveguide.

**highlight** A bright area in a television image.

**highlight brilliance** The maximum brilliance in a television image, corresponding to the area having highest brilliance in the original scene.

**high-mu tube** A vacuum tube having a high amplification factor.

**high-pass filter** A filter that transmits all frequencies above a given cutoff frequency and substantially attenuates all others.

**high-pressure cloud chamber** A cloud chamber in which the gas is maintained at high pressure to reduce the range of high-energy particles and thereby increase the probability of observing events.

**high-pressure laminate** A laminated plastic that is generally produced at pressures above 1,000 pounds per square inch.

**high-pressure mercury-vapor lamp** A discharge tube containing an inert gas and a small quantity of liquid mercury. The initial glow discharge through the gas heats and vaporizes the mercury, after which the discharge through mercury vapor produces an intensely brilliant light.

**high Q** A characteristic wherein a component has a high ratio of reactance to effective resistance, so that its $Q$ factor is high.

**high-recombination-rate contact** A semiconductor-to-semiconductor or metal-to-semiconductor contact at which thermal-equilibrium carrier densities are maintained substantially independent of current density.

**high-resistance voltmeter** A voltmeter having a resistance considerably higher than 1,000 ohms per volt, so that it draws little current from the circuit in which a measurement is made.

**high-speed carry** A carry that bypasses the

normal adding circuit in a computer. Also called standing-on-nines carry.

**high-speed memory** *Rapid memory.*

**high-speed relay** A relay specifically designed for short operate time, short release time, or both.

**high-temperature reactor** A power reactor in which the temperature is high enough for efficient generation of mechanical power.

**high tension** 1. High voltage, of the order of thousands of volts. 2. British term for *anode voltage,* of the order of several hundred volts.

**high vacuum** A degree of vacuum at which essentially no gases or vapors are present, so that ionization cannot occur.

**high-velocity scanning** The scanning of a target with electrons of such velocity that the secondary-emission ratio is greater than unity, so that more than one secondary electron leaves the target for every electron hitting it.

**hill-and-dale recording** *Vertical recording.*

**hiss** Random noise in the audio-frequency range, similar to prolonged sibilant sounds.

**histogram** A frequency-distribution graph, in which the number of quantities having each particular range of values is represented by a vertical bar on the graph.

**$H_{m,n}$ mode** British term for *$TE_{m,n}$ mode.*

**$H_{m,n}$ wave** British term for *$TE_{m,n}$ wave.*

**H network** An attenuation network composed of five branches. Two are connected in series between an input terminal and an output terminal. Two are connected in

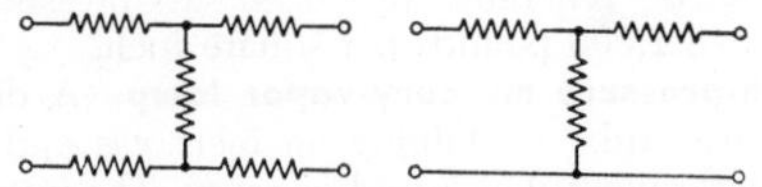

H network has five impedance branches, while pi network at right has only three branches.

series between another input terminal and output terminal. The fifth is connected from the junction point of the first two branches to the junction point of the second two branches. Also called H pad.

**hodoscope** An array of small Geiger counters used in tracing the paths of cosmic rays.

**hoghorn antenna** British term for *horn antenna.*

**hold** 1. To maintain storage elements at equilibrium voltages in a charge storage tube by electron bombardment. 2. To retain information in a computer storage device for further use after it has been initially utilized.

**hold control** A manual control that changes the frequency of the horizontal or vertical sweep oscillator in a television receiver, so that the frequency more nearly corresponds to that of the incoming synchronizing pulses. The two controls used for this purpose are called the horizontal hold control and the vertical hold control. Also called speed control.

**holder** A device that mechanically and electrically accommodates one or more crystals, fuses, or other components in such a way that they can readily be inserted or removed.

**holding anode** A small auxiliary anode used in a mercury-pool rectifier to keep a cathode spot energized during the intervals when the main anode current is zero.

**holding beam** A diffused beam of electrons used to regenerate the charges stored on the dielectric surface of a cathode-ray storage tube.

**holding coil** A separate relay coil that is energized by contacts which close when a relay pulls in, to hold the relay in its energized position after the original operating circuit is opened.

**hold time** The time during which pressure is applied to a welded joint after welding current ceases to flow.

**hole** A mobile vacancy in the electronic valence structure of a semiconductor. It is equivalent to a positive charge. A hole exists when an atom has less than its normal number of electrons.

**hole-and-slot anode** A magnetron anode in which the cavity resonators are circular

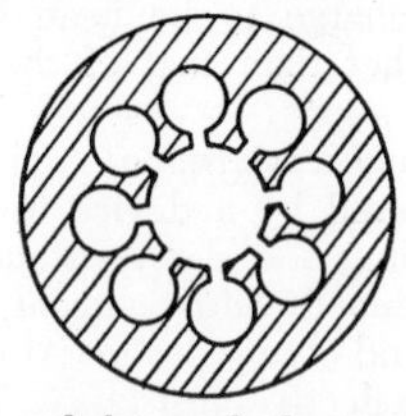

Hole-and-slot anode for magnetron.

holes connected by slots to the space between the central cathode and the anode.

**hole conduction** Conduction occurring in a semiconductor when electrons move into holes under the influence of an applied voltage, and thereby create new holes. The apparent movement of such holes is toward the more negative terminal, and is hence equivalent to a flow of positive charges in that direction.

**hole injection** The production of holes in an n-type semiconductor when voltage is applied to a sharp metal point in contact with the surface of the material.

**hole mobility** The ability of a hole to travel readily through a semiconductor.

**holmium** [symbol Ho] A rare-earth element. Atomic number is 67.

**home** 1. To fly toward a radiation-emitting source, using the radiated waves as a guide. 2. To travel to a target by guidance of heat radiation, radar echoes, radio waves, sound waves, or other phenomena reflected from or originating in the target, as in homing missiles and homing torpedoes.

**homer** 1. A ground-based direction-finding station that utilizes radio transmissions from aircraft to determine their bearing, then guides the aircraft toward the station by voice communication. 2. *Target seeker.*

**homing** 1. The process of approaching a desired point by maintaining constant some indicated navigation parameter other than altitude. 2. Use of radiation from a target to establish a collision course in missile guidance. The three types of homing are active, semiactive, and passive homing. 3. Flying toward a radio or radar transmitter by using the transmitted radiation for navigation guidance. 4. Returning to the starting position, as in a stepping relay or tuning motor.

**homing adapter** A device attached to an aircraft radio receiver to permit homing the aircraft on a radio beacon or other radio transmitting station. The adapter indicates direction to the transmitting station by aural or visual signals.

**homing aid** A system designed to guide an aircraft to an airport or carrier.

**homing-and-busting aircraft** Armed aircraft capable of homing on and destroying sources of electromagnetic radiation.

**homing antenna** A directional antenna array used when it is desired to fly directly to a target that is emitting or reflecting radio or radar waves.

**homing beacon** A radio beacon, either airborne or on the ground, toward which an aircraft can fly if equipped with a radio compass or homing adapter. Also called radio homing beacon.

**homing device** 1. A transmitter, receiver, or adapter used for homing aircraft or used by aircraft for homing purposes. 2. A device incorporated in a guided missile or the like to home it on a target. 3. A control device that automatically starts in the correct direction of motion or rotation to achieve a desired change, as in a remote-control tuning motor for a television receiver. A nonhoming device may first go to the end of its travel in the wrong direction.

**homing guidance** A guidance system in which a missile directs itself to a target by means of a self-contained mechanism that reacts to a particular characteristic of the target. Such characteristics include heat, light, sound, a reflected radar echo or other electromagnetic radiation, ionization of air, or air pollution by the target's exhaust gases. Homing guidance may be active, semiactive, or passive.

**homing range** The maximum distance from a target or homing beacon at which a homing device is effective.

**homing station** A station at which a beacon emits signals that may be used for homing.

**homing torpedo** A torpedo having homing guidance designed for homing on a surface vessel or a submerged submarine.

**homodyne reception** A system of radio reception for suppressed-carrier systems of radiotelephony, in which the receiver generates a voltage having the original carrier frequency and combines it with the incoming signal. Also called zero-beat reception.

**homogeneous nuclear reactor** A nuclear reactor in which fissionable material and moderator (if used) are intimately mixed to form an effectively homogeneous medium for neutrons. Uranyl sulfate dissolved in water is generally used as the fuel.

**homogeneous radiation** Radiation having an extremely narrow band of frequencies, or a beam of monoenergetic particles of a single type, so that all components of the radiation are alike.

**homopolar generator** A d-c generator in which the poles presented to the armature are all of the same polarity, so that the voltage generated in active conductors has the same polarity at all times. A pure direct current is thus produced without commutation. Also called acyclic machine, homopolar machine, and unipolar machine.

**homopolar machine** *Homopolar generator.*

**honeycomb coil** A coil wound in a crisscross manner to reduce distributed capacitance. Also called lattice-wound coil.

**hood** An opaque shield placed above the screen of a cathode-ray tube to eliminate extraneous light.

**hook transistor** A transistor having four alternating p-type and n-type layers, with one layer floating between the base layer and the collector layer. This arrangement gives high emitter-input current gains. A p-n-p-n transistor has a p-type floating layer, while an n-p-n-p transistor has an n-type floating layer.

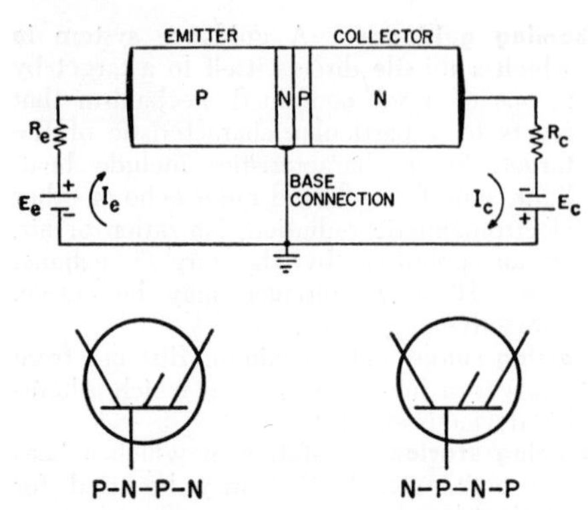


Hook transistor construction and circuit for p-n-p-n type, and symbols for both types.

**hookup** An arrangement of circuits and apparatus for a particular purpose.

**hookup wire** Tinned and insulated solid or stranded soft-drawn copper wire used in making circuit connections. The size is usually No. 18, 20, or 22.

**hop** A single reflection of a radio wave from the ionosphere back to the earth in traveling from one point to another. Usually used in such expressions as single-hop, double-hop, and multihop. The number of hops is called the order of reflection.

**Hopi** A Navy air-to-surface medium-range missile designed for launching from carrier aircraft, and capable of carrying a nuclear warhead.

**horizon** The apparent junction of earth and sky as seen from a transmitting antenna site. The horizon bounds that part of the earth's surface which is reached by the direct wave of a radio station.

**horizon sensor** A passive infrared device that detects the thermal discontinuity between the earth and space, for use in establishing a stable vertical reference for control of the attitude or orientation of a missile or satellite in space. Thermistors serve as the infrared detectors. Also called wide-angle horizon sensor.

**horizontal blanking** Blanking of a television picture tube during the horizontal retrace.

**horizontal blanking pulse** The rectangular pulse that forms the pedestal of the composite television signal in between active horizontal lines. This pulse causes the beam current of the picture tube to be cut off during retrace. Also called line-frequency blanking pulse.

**horizontal centering control** The centering control provided in a television receiver or cathode-ray oscilloscope to shift the position of the entire image horizontally in either direction on the screen.

**horizontal convergence control** The control that adjusts the amplitude of the horizontal dynamic convergence voltage in a color television receiver.

**horizontal definition** *Horizontal resolution.*

**horizontal deflection electrodes** The pair of electrodes that moves the electron beam horizontally from side to side on the fluorescent screen of a cathode-ray tube employing electrostatic deflection.

**horizontal deflection oscillator** The oscillator that produces, under control of the horizontal synchronizing signals, the sawtooth voltage waveform that is amplified to feed the horizontal deflection coils on the picture tube of a television receiver. Also called horizontal oscillator.

**horizontal drive control** The control in a television receiver, usually at the rear of the set, that adjusts the output of the horizontal oscillator. Also called drive control.

**horizontal flyback** Flyback in which the electron beam of a television picture tube returns from the end of one scanning line to the beginning of the next line. Also called horizontal retrace and line flyback.

**horizontal frequency** *Line frequency.*

**horizontal hold control** The hold control that changes the free-running period of the horizontal deflection oscillator in a television receiver, so that the picture remains steady in the horizontal direction.

**horizontal linearity control** A linearity control that permits narrowing or expanding the width of the left-hand half of a television receiver image, to give linearity in the horizontal direction so that circular objects appear as true circles. Usually mounted at the rear of the receiver.

**horizontal line frequency** *Line frequency.*

**horizontally polarized wave** A linearly polarized wave whose electric field vector is horizontal.

**horizontal oscillator** *Horizontal deflection oscillator.*

**horizontal output stage** The television receiver stage that feeds the horizontal deflection coils of the picture tube through the horizontal output transformer. It may also include a part of the second-anode power supply for the picture tube.

**horizontal output transformer** A transformer used in a television receiver to provide the horizontal deflection voltage, the high voltage for the second-anode power supply of the picture tube, and the filament

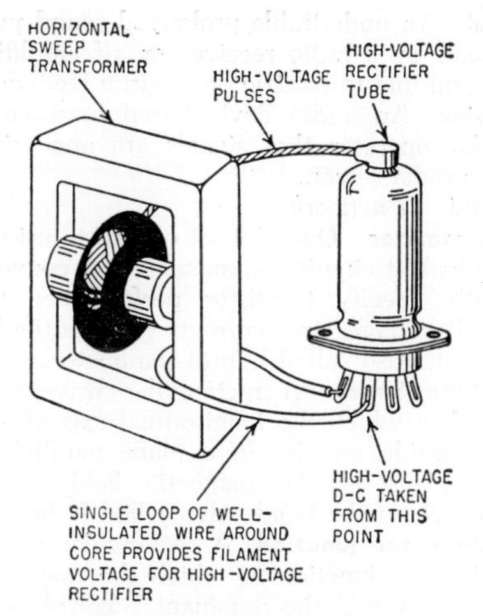

Horizontal output transformer.

voltage for the high-voltage rectifier. Also called flyback transformer and horizontal sweep transformer.

**horizontal polarization** Transmission of radio waves in such a way that the electric lines of force are horizontal, while the magnetic lines of force are vertical. With this polarization, transmitting and receiving dipole antennas are placed in a horizontal plane. The U. S. television system uses horizontal polarization, whereas the British television system uses vertical polarization.

**horizontal resolution** The number of individual picture elements or dots that can be distinguished in a horizontal scanning line of a television or facsimile image. Usually determined by observing the wedge of fine vertical lines on a test pattern. Also called horizontal definition.

**horizontal retrace** *Horizontal flyback.*

**horizontal-ring induction furnace** An induction furnace in which a magnetic core links the output transformer primary winding with a ring-shaped horizontal trough in which is placed the metal to be melted.

**horizontal sweep** The sweep of the electron beam from left to right across the screen of a cathode-ray tube.

**horizontal sweep transformer** *Horizontal output transformer.*

**horizontal synchronizing pulse** The rectangular pulse transmitted at the end of each line in a television system, to keep the receiver in line-by-line synchronism with the transmitter. Also called line synchronizing pulse.

**horizon tracker** A device for establishing a vertical reference in a navigation system by precisely tracking the visible horizon.

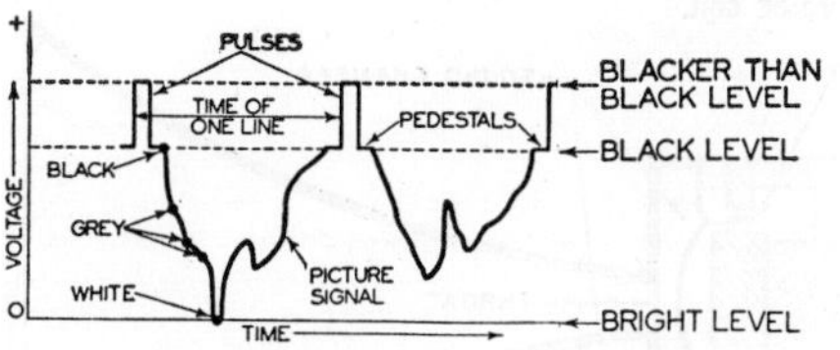

Horizontal synchronizing pulse occurs at end of each line of picture signal.

**horn** 1. An acoustic device used with a loudspeaker driving element to improve the radiation of sound and to achieve desired directional characteristics. The cross-sectional area increases progressively from the throat to the mouth. Also called acoustic horn. 2. *Horn antenna.*

**horn antenna** A microwave antenna produced essentially by flaring out the end of a circular or rectangular waveguide into the shape of a horn, for radiating radio waves directly into space. In a rectangular horn

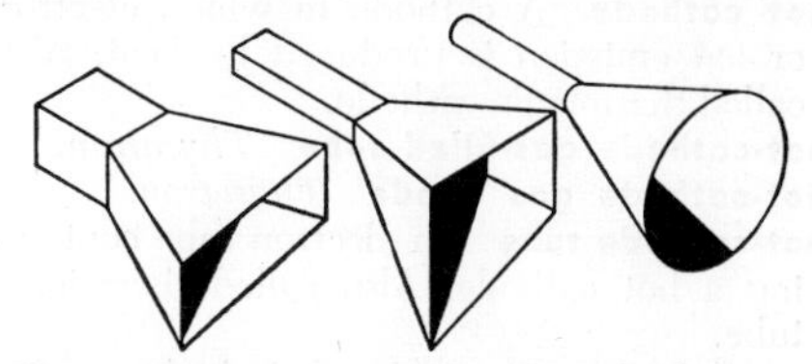
Horn antennas—rectangular, pyramidal, and conical.

antenna, either one or both transverse dimensions increase linearly from the small end or throat to the mouth. Also called flare (British), hoghorn antenna (British), horn, and horn radiator.

**horn arrester** A lightning arrester in which the spark gap has thick wire horns that spread outward and upward. The arc forms at the narrowest bottom part of the gap, travels upward, and extinguishes itself when it reaches the widest part of the gap.

**horn feed** A horn antenna used to feed a parabolic reflector in a radar antenna system.

**horn loudspeaker** A loudspeaker in which the radiating element is coupled to the air or another medium by means of a horn.

**horn mouth** The end of a horn that has the larger cross-sectional area.

**horn radiator** *Horn antenna.*

**horn throat** The end of a horn that has the smaller cross-sectional area.

**horseshoe magnet** A permanent magnet or electromagnet in which the core is horse-

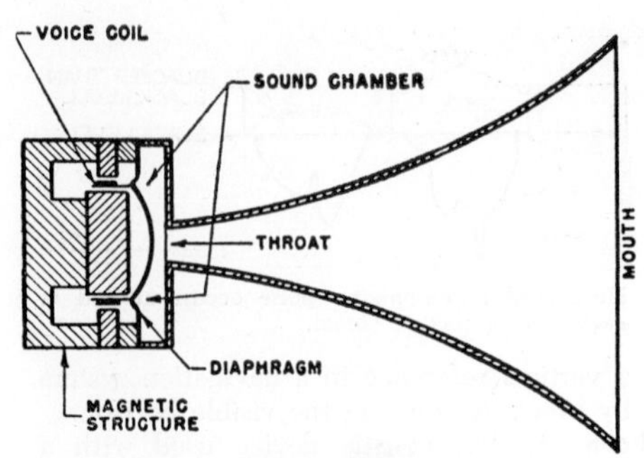


Horn loudspeaker.

shoe-shaped or has parallel sides like a U, to bring the two poles near each other.

**hot** 1. Highly radioactive. 2. *Energized.*

**hot atom** An atom that has high energy as a result of a nuclear process such as beta decay or neutron capture.

**hot-atom chemistry** The chemical reactions and properties of atoms that are in a high state of excitation or possess high kinetic energy as a result of nuclear processes.

**hot cathode** A cathode in which electron or ion emission is produced by heat. Also called thermionic cathode.

**hot-cathode gas-filled tube** *Thyratron.*

**hot-cathode gas triode** *Thyratron.*

**hot-cathode tube** An electron tube containing a hot cathode. Also called thermionic tube.

**hot-cathode x-ray tube** A high-vacuum x-ray tube in which the cathode is heated by current flow through a heater.

**hot laboratory** A laboratory designed for research with radioactive materials having such high strengths that special handling precautions are required.

**hot-wire ammeter** An ammeter in which alternating or direct current is measured by sending it through a fine wire. The resulting expansion or sag of the wire due to heat is used to deflect the meter pointer. Also called thermal ammeter.

**hot-wire instrument** An instrument that depends for its operation on the expansion by heat of a wire carrying a current.

**hot-wire microphone** A velocity microphone that depends for its operation on the change in resistance of a hot wire as the wire is cooled or heated by varying particle velocities in a sound wave.

**Hound Dog** An Air Force air-to-surface guided missile having a range of several hundred miles, capable of carrying a nuclear warhead, and intended for launching from long-range bombers of the Strategic Air Command.

**howl** An undesirable prolonged sound produced by a radio receiver or a-f amplifier system due to electric or acoustic feedback.

**howler** An audio device used to warn a radar operator that signals are appearing on a radar screen.

**H pad** *H network.*

***h* parameter** One of a set of four transistor equivalent-circuit parameters that conveniently specify transistor performance for small voltages and currents in a particular circuit. Also called hybrid parameter.

**H-plane bend** A rectangular waveguide bend in which the longitudinal axis of the waveguide remains in a plane parallel to the plane of the magnetic field vector throughout the bend. Also called H bend.

**H-plane tee junction** A rectangular waveguide tee junction in which the magnetic field vector of the dominant wave of each arm is parallel to the plane of the longitudinal axes of the waveguides.

**H scan** *H display.*

**H scope** A radarscope that produces an H display.

**H system** A radar navigation system that uses two ground radar beacons in conjunction with airborne equipment that gives the direction and distance to each beacon. The principal H systems are gee-H, micro-H, rebecca-H, and shoran. Also called H.

**HTO** Symbol for water that has been labeled by replacing one of the hydrogen atoms with a tritium atom.

**hue** The name of a color, such as red, yellow, green, blue, or purple, corresponding to the dominant wavelength. White, black, and gray are not considered as being hues.

**hue control** A control that varies the phase of the chrominance signals with respect to that of the burst signal in a color television receiver, in order to change the hues in the image. Also called phase control.

**huff-duff** [from pronunciation of first letters of High-Frequency Direction Finder] A direction finder that indicates the direction of arrival of high-frequency radio signals on the screen of a cathode-ray tube.

**hum** 1. An electrical disturbance occurring at the power-supply frequency or its harmonics, usually 60 or 120 cps in the United States and 50 or 100 cps in Great Britain. An example is ripple. 2. A sound produced by an iron core of a transformer due to loose laminations or magnetostrictive effects. The frequency of the sound is twice the power-line frequency.

**human engineering** Engineering involving adaptation of equipment designs to best meet the physical and psychological requirements of a human operator.

**hum balancer** An adjustable resistor connected between the heater leads of a tube, with the adjustable center tap connected to ground.

**hum bar** A dark horizontal band extending across a television picture, due to excessive hum in the video signal applied to the input of the picture tube.

**hum-bucking coil** A coil wound on the field coil of an excited-field loudspeaker and connected in series opposition with the voice coil, so that hum voltage induced in the voice coil is canceled by that induced in the hum-bucking coil.

**humidity detector** A detector that opens or closes a switch when the amount of moisture in the atmosphere reaches a preset value. The sensing element may be moisture-absorbing paper, stretched human

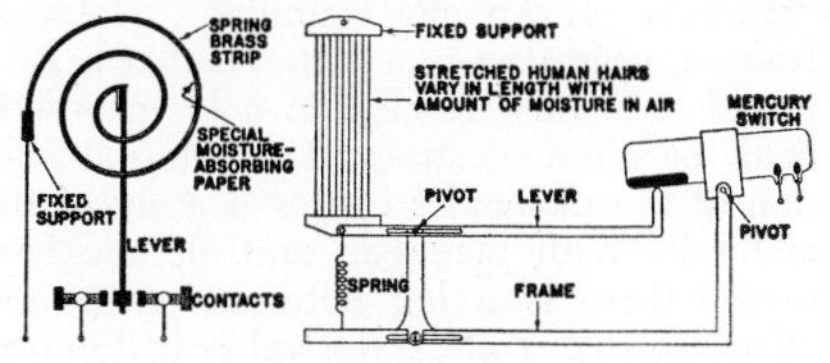


Humidity detectors.

hairs, stretched nylon fibers, or any other material having a characteristic that changes with humidity in a known and repeatable manner.

**hum modulation** Modulation of an r-f signal or detected a-f signal by hum. This type of hum is heard in a radio receiver only when a station is tuned in.

**hum slug** A copper ring placed around the core of an excited-field loudspeaker adjacent to the voice coil, to serve as a single shorted turn for hum currents induced by the field coil.

**hunting** Undesirable oscillation of an automatic control system, wherein the controlled variable swings on both sides of the desired value.

**Hurter and Driffield curve** *H and D curve.*

**Huygens principle** Every point on an advancing wavefront acts as a source that sends out new waves. The combined effect is propagation of the wave as a whole.

**H vector** A vector that represents the magnetic field of an electromagnetic wave. In free space it is perpendicular to the E vector and to the direction of propagation.

**H wave** British term for *transverse electric wave* (TE wave).

**hybrid** *Hybrid junction.*

**hybrid coil** *Hybrid transformer.*

**hybrid electromagnetic wave** [abbreviated HEM wave] An electromagnetic wave having components of both the electric and magnetic field vectors in the direction of propagation.

**hybrid junction** A transformer, resistor, or waveguide circuit or device that has four pairs of terminals so arranged that a signal entering at one terminal pair will divide

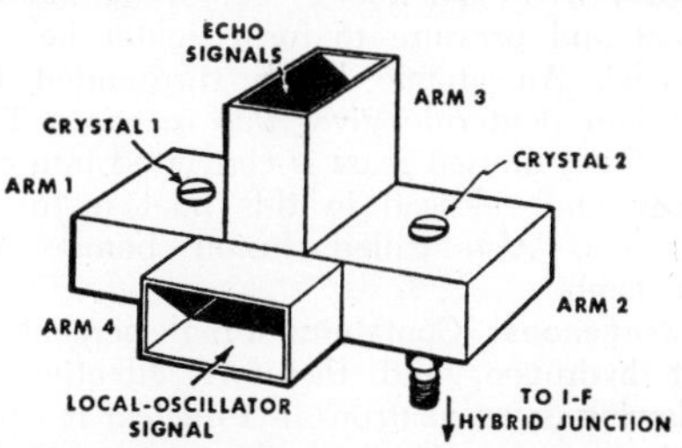


Hybrid junction for rectangular waveguides, serving as balanced mixer for radar receiver.

and emerge from the two adjacent terminal pairs, but will be unable to reach the opposite terminal pair. Also called bridge hybrid and hybrid.

**hybrid parameter** *H parameter.*

**hybrid repeater** *Hybrid transformer.*

**hybrid ring** A doughnut-shaped waveguide serving as a hybrid junction for four waveguide sections. Coaxial lines are sometimes used in place of waveguides. Also called rat race.

**hybrid set** Two or more transformers interconnected to form a hybrid junction. Also called transformer hybrid.

**hybrid tee** A microwave hybrid junction composed of an E-H tee with internal matching elements. It is reflectionless for a wave propagating into the junction from any arm when the other three arms are match-terminated. Also called magic tee.

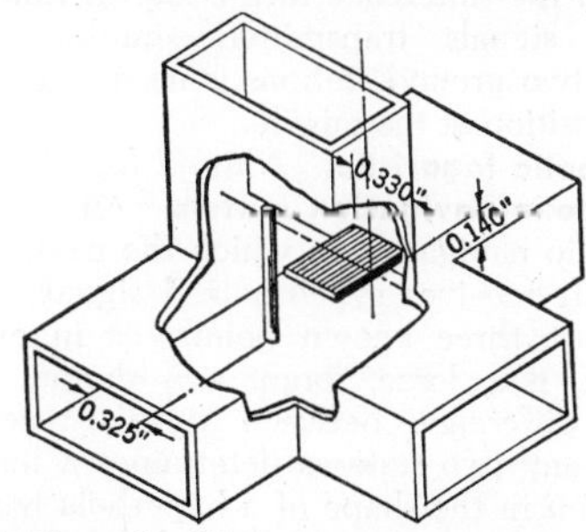


Hybrid tee for ½- by ¼-inch waveguide, using a post and iris inside for matching. Often called a magic tee.

**hybrid transformer** A single transformer that performs the essential function of a hybrid set. Also called hybrid coil and hybrid repeater.

**hydrogen** [symbol H] A gaseous element having the simplest known atom, consisting of only one proton and one electron. Atomic number is 1. Its three isotopes are ordinary or light hydrogen, deuterium, and tritium.

**hydrogen bomb** A nuclear bomb in which heavy hydrogen nuclei fuse under intense heat and pressure to form lighter helium nuclei. An atomic bomb surrounded by lithium deuteride gives this reaction. The resulting unused mass is converted into energy and released in this nuclear fusion process. Also called fusion bomb and H-bomb.

**hydrogenous** Containing a high percentage of hydrogen, and therefore effective in slowing down neutrons in a nuclear reactor.

**hydrogen thyratron** A thyratron containing hydrogen, used in radar pulse circuits to provide high peak currents at high anode voltages. Use of hydrogen instead of mercury vapor gives freedom from effects of changes in ambient temperature.

**hydrophone** An electroacoustic transducer that responds to waterborne sound waves and delivers essentially equivalent electric waves. Used chiefly to detect the approach of submarines and other craft. Examples include moving-coil hydrophones and moving-conductor hydrophones.

**hydrostatic pressure** *Static pressure.*

**hygrometer** An instrument that measures the humidity of the atmosphere.

**hygroscopic** Tending to absorb moisture.

**hyperbolic flareout** A flareout obtained by changing the glide slope from a straight line to a hyperbolic curve at an appropriate distance from touchdown at an airport.

**hyperbolic guidance** Missile guidance in which the difference in the arrival times of radio signals transmitted simultaneously from two ground stations is used to control the position of the missile.

**hyperbolic logarithm** *Natural logarithm.*

**hyperbolic navigation system** Any system of radio navigation in which the navigating aircraft receives synchronized signals from at least three known points, as in cytac, decca, gee, lorac, loran, and shoran. The time difference between signals received from any two stations determines a line of position in the shape of a hyperbola having the two transmitters at its foci. Crossing this line of position with another line of position, obtained by using at least one other transmitting station, establishes a fix.

**hyperdirective antenna** An antenna or antenna system so energized as to have a more compact directional pattern than naturally corresponds to its extent as measured in wavelengths.

**hyperfine structure** A set of closely spaced lines making up a spectral line. Also called isotope structure.

**hyperfragment** An unstable neutral hyperon particle produced in a high-energy collision.

**hyperon** An unstable particle whose mass is between that of a neutron and a deuteron.

**hypersonic** Having a velocity greater than five times the speed of sound in air (greater than mach 5).

**hysteresigraph** An instrument that automatically measures and draws the hysteresis curve for a specimen of magnetic material.

**hysteresis** 1. An effect similar to internal friction, occurring in a material that is subjected to a varying electric field, magnetic field, or physical strain. The internal friction of the molecules causes heating of the material. With magnetic and electric hysteresis there is a lag between cause and effect, so that a measured value is different for a cause that increases to a final value than for a cause that decreases to the same final value. 2. A temporary change in the counting-rate–voltage characteristic of a radiation-counter tube, caused by its previous operation. 3. An oscillator effect wherein a given value of an operating parameter may result in multiple values of output power and/or frequency.

**hysteresis curve** A curve that shows the steady-state relation between the magnetic induction in a material and the steady-state alternating magnetic intensity that produces it.

**hysteresis distortion** Distortion that occurs in circuits containing magnetic components, due to nonlinearity caused by hysteresis.

**hysteresis heater** An induction heater in which a ferrous charge or charge container

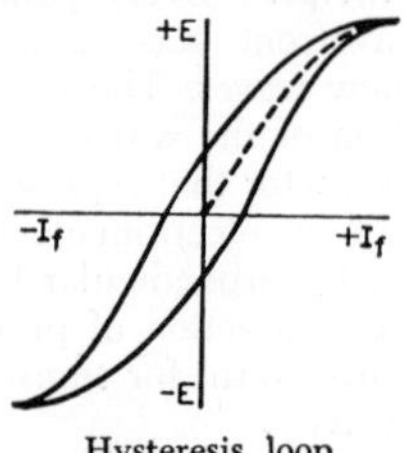


Hysteresis loop.

is heated principally by hysteresis losses due to varying magnetic flux in the magnetic material. In normal induction heating, the heat is due to eddy-current losses.

**hysteresis loop** A curve showing, for each value of magnetizing force, two values of the magnetic flux density in a cyclically magnetized material—one when the magnetizing force is increasing, the other when it is decreasing.

**hysteresis loss** The power loss in an iron-core transformer or other a-c device due to magnetic hysteresis.

**hysteresis motor** A small synchronous motor sometimes used for light constant-speed duty, as for phonograph motors. It starts by virtue of the hysteresis losses induced in its hardened steel secondary member by the revolving field of the primary.

**Hz** Abbreviation for *hertz.*

# I

**I** [from intensity] Symbol for *current.*

**iagc** Abbreviation for *instantaneous automatic gain control.*

**IC** Symbol for *internal connection.* Used on tube-base diagrams.

**icbm** Abbreviation for *intercontinental ballistic missile.*

**ICI** Deprecated abbreviation for *Commission Internationale de l'Eclairage.*

**iconoscope** A television camera tube in which a beam of high-velocity electrons scans a photoemissive mosaic that is capable of storing an electric charge pattern corresponding to an optical image focused on the mosaic. The mosaic consists of globules of light-sensitive material on a mica sheet having a conducting film on its back surface. A small value of capacitance thus exists between each globule and the metal film. Each globule emits electrons in proportion to the light on it, thus producing charges on the capacitances between globules. The electron beam discharges each capacitance in turn during scanning of the mosaic. The resulting variations in the current taken from the metal film form the desired output signal. Also called storage-type camera tube.

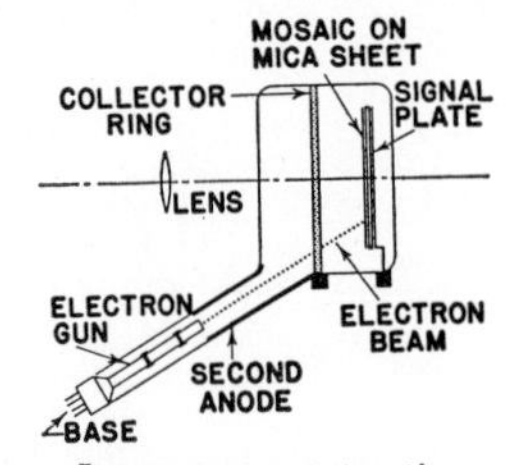

Iconoscope construction.

**icw** Abbreviation for *interrupted continuous wave.*

**ideal bunching** A theoretical condition in which the bunching of electrons in a velocity-modulation tube is such that all electrons in a bunch have the same velocity and phase, corresponding to an infinitely large current peak.

**ideal noise diode** A diode that has an infinite internal impedance and in which the current exhibits full shot-noise fluctuations.

**ideal permeability** The value of permeability obtained by superimposing a large alternating magnetizing force on the desired magnetizing force, then slowly reducing the alternating component to zero.

**ideal rectifier** A rectifier in which the back conductance, forward resistance, and capacitance are zero.

**ideal transducer** A hypothetical passive transducer that transfers the maximum possible power from the source to the load, and therefore has no losses.

**ideal transformer** A hypothetical transformer that neither stores nor dissipates energy, has unity coefficient of coupling, and has pure inductances of infinitely great value.

**I demodulator** The demodulator in which the chrominance signal and the color-burst oscillator signal are combined to recover the I signal in a color television receiver.

**identification** The process of determining the identity of a particular displayed target or determining which blip represents a specific target in radar.

**identification, friend or foe** [abbreviated iff] A system used for identifying an aircraft or ship that is picked up by radar, wherein a coded challenging transmission received by a friendly craft causes automatic transmission of a coded identification signal.

**I display** A radarscope display in which a target appears as a complete circle when the radar antenna is correctly pointed at it, the radius of the circle being proportional to target distance. When the antenna is not aimed at the target, the circle reduces to

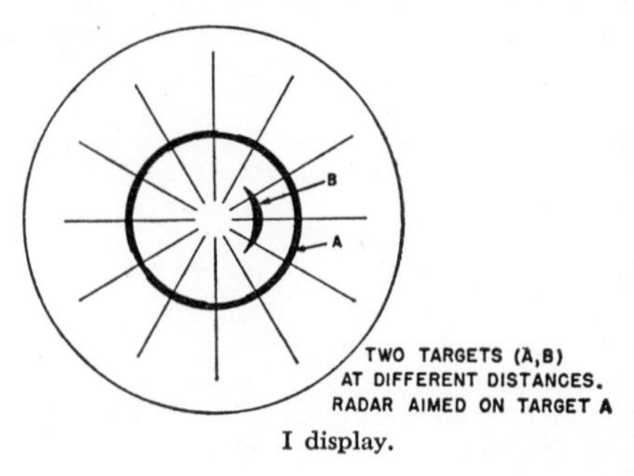

I display.

a segment of a circle. The segment length is then inversely proportional to the magnitude of the pointing error, and its angular position is reciprocal to the direction of the pointing error. Also called I scan.

**idler** An intermediate rubber-tired wheel used to transmit power from a motor shaft to a driven shaft, as from a phonograph drive motor shaft to the rim of the turntable.

**i-f** Abbreviation for *intermediate frequency.*

**iff** Abbreviation for *identification, friend or foe.*

**ifr** Abbreviation for *instrument flight rules.*

**ifr clearance** An aircraft clearance for a flight under instrument flight rules.

**ifr conditions** Weather conditions below the minimum specified for visual flight rules, in which instrument flight rules apply.

**ignition control** Control of the starting instant of current flow in the anode circuit of a gas tube.

**ignition interference** Interference due to the spark discharges in an automotive or other ignition system.

**ignitor** 1. An electrode used to initiate and sustain the ignitor discharge in a switching tube. Also called keep-alive electrode (deprecated). 2. A pencil-shaped electrode

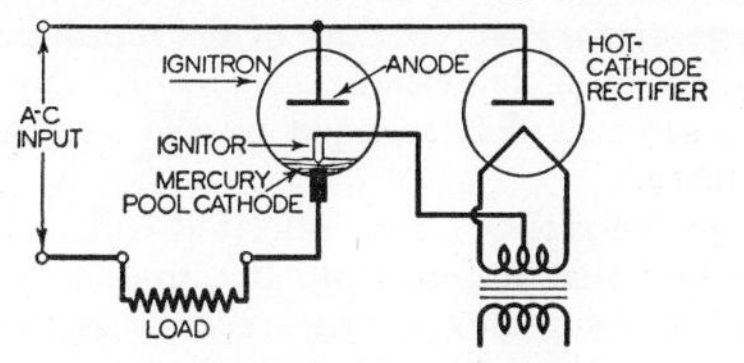


Ignitor circuit for ignitron.

made of carborundum or some other conducting material that is not wetted by mercury, partly immersed in the mercury-pool cathode of an ignitron and used to initiate conduction at the desired point in each a-c cycle.

**ignitor-current temperature drift** The variation in ignitor electrode current caused by a change in ambient temperature of a tr, pre-tr, or attenuator tube.

**ignitor discharge** A d-c glow discharge, between the ignitor electrode and a suitably located electrode of a switching tube, used to facilitate r-f ionization.

**ignitor firing time** The time interval between the application of a direct voltage to the ignitor electrode and the establishment of the ignitor discharge in a switching tube.

**ignitor interaction** The difference between the insertion loss measured at a specified ignitor current and that measured at zero ignitor current in a tr, pre-tr, or attenuator tube.

**ignitor leakage resistance** The insulation resistance of a switching tube, measured in the absence of an ignitor discharge, between the ignitor electrode terminal and the adjacent r-f electrode.

**ignitor oscillation** A relaxation-oscillator type of oscillation occurring in the ignitor circuit of a tr, pre-tr, or attenuator tube.

**ignitor voltage drop** The direct voltage between the cathode and the anode of the ignitor discharge in a switching tube at a specified ignitor current.

**ignitron** A single-anode pool tube in which an ignitor electrode is employed to initiate the cathode spot on the surface of the mercury pool before each conducting period.

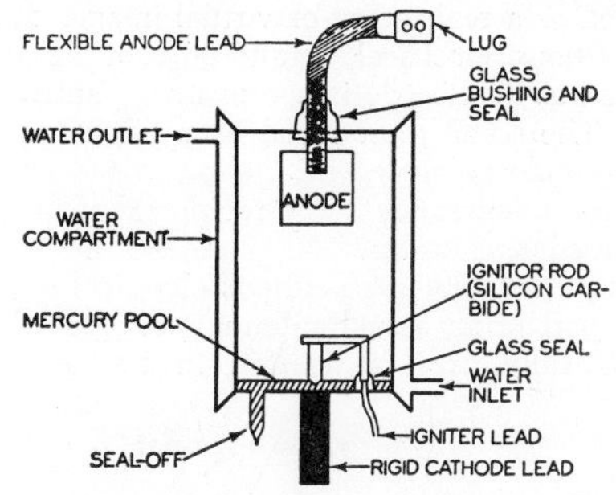


Ignitron construction, showing mercury pool that serves as cathode.

**ignitron rectifier** A high-power rectifier consisting of a number of large metal-tank ignitrons, used in industrial plants such as aluminum refineries and steel mills.

Ignitron rectifiers used to change three-phase a-c power to d-c power.

**ignore** A punched-tape code or computer instruction which indicates that no action should be taken by a computer or other device. In the Teletype and Flexowriter codes, punching of all holes is an ignore.

**IGY** Abbreviation for *International Geophysical Year.*

**illinium** Former name for *promethium.*

**illuminance** The density of the luminous flux on a surface. It is equal to the flux divided by the area of the surface when the latter is uniformly illuminated. Also called illumination.

**illuminant C** The reference white of color television. It closely matches average daylight.

**illumination** *Illuminance.*

**illumination sensitivity** The signal output current divided by the incident illumination on a camera tube or phototube.

**illuminometer** An instrument for measuring illumination on a surface directly in footcandles.

**ils** Abbreviation for *instrument landing system.*

**image** 1. An optical counterpart of an object, as a real image or virtual image. 2. A fictitious electrical counterpart of an object, as an electric image or image antenna. 3. The scene reproduced by a television or facsimile receiver.

**image admittance** The reciprocal of image impedance.

**image antenna** A fictitious electrical counterpart of an actual antenna, acting mathematically as if it existed in the ground directly under the real antenna and served as the direct source of the wave that is reflected from the ground by the actual antenna.

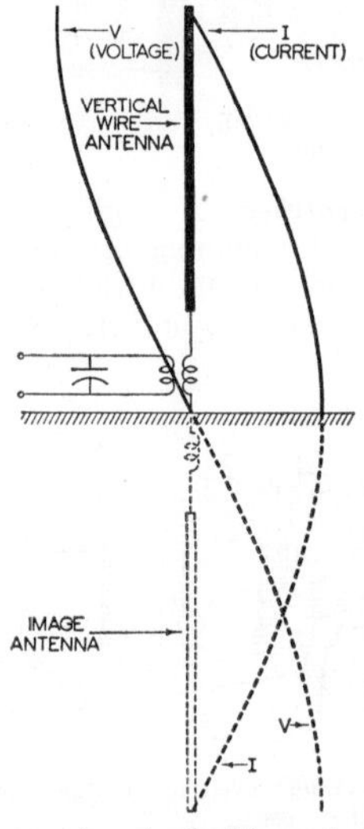


Image antenna for simple Marconi vertical-wire antenna, with voltage and current distribution curves.

**image attenuation constant** The real part of the transfer constant.

**image burn** *Retained image.*

**image converter tube** *Image tube.*

**image dissector** A television camera tube in which the scene to be transmitted is focused on a light-sensitive surface, each point of which emits electrons in proportion to incident light. The resulting broad beam of electrons is drawn down the tube by a positive anode. Magnetic fields produced by coils keep the electron image in focus as they sweep it in a scanning motion past an aperture opening into an electron multiplier. The output voltage of the electron multiplier is thus proportional at each instant to the brightness of an elemental area of the scene being scanned in orderly sequence. Also called dissector tube and Farnsworth image dissector tube.

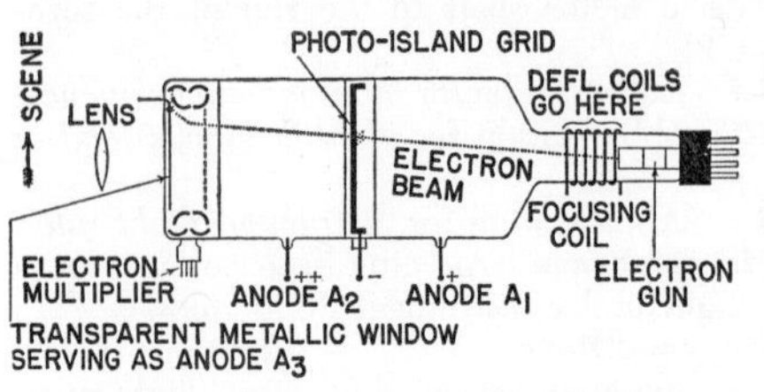


Image dissector construction details.

**image distortion** Failure of the reproduced image in a television receiver to appear the same as that scanned by the television camera.

**image frequency** An undesired carrier frequency that differs from the frequency to which a superheterodyne receiver is tuned by twice the intermediate frequency. Thus, if the oscillator frequency is higher than that of the desired incoming carrier, the image frequency will be the same amount lower than the oscillator frequency. If the receiver has poor selectivity, a strong image-frequency signal can get through its tuning circuits to beat with the local oscillator and produce the correct intermediate frequency.

**image-frequency rejection ratio** The ratio of the response of a superheterodyne receiver at the desired frequency to the response at the image frequency.

**image iconoscope** A camera tube in which an optical image is projected on a semitransparent photocathode, and the resulting electron image emitted from the other side of the photocathode is focused on a separate storage target. The target is scanned on the same side by a high-velocity electron beam, neutralizing the elemental charges in sequence to produce the camera output signal at the target. Sensitivity is much higher than for an iconoscope. Also called superemitron camera (British).

**image impedance** One of the impedances that, when connected to the input and output of a transducer, will make the impedances in both directions equal at the input terminals and at the output terminals. The load impedance and the equivalent internal impedance of the transducer are then images of each other. This condition gives maximum power transfer. When the two image impedances are equal, their value is the same as the characteristic impedance.

**image intensifier** An electron tube used in fluoroscopy, containing a special screen from which electrons are released by x-rays. These electrons are accelerated and focused onto a fluorescent screen, giving an image much brighter than would be produced by direct action of the x-ray beam on the fluorescent screen.

**image interference** Interference occurring in a superheterodyne receiver when a station broadcasting on the image frequency is received along with the desired station. Image interference can occur when the receiver circuits do not have sufficient selectivity to reject the image-frequency signal. For a standard radio broadcast receiver, the image frequency would be 910 kc higher than that of the desired station.

**image-interference ratio** A superheterodyne receiver rating indicating the effectiveness of the preselector in rejecting signals at the image frequency.

**image orthicon** A camera tube in which an electron image is produced by a photoemitting surface and focused on one side of a separate storage target that is scanned on its opposite side by an electron beam, usually consisting of low-velocity electrons. It has such high sensitivity that images can be picked up even in semidarkness.

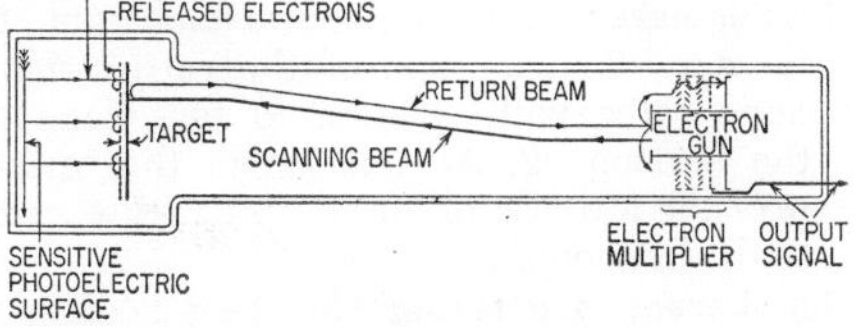

Image orthicon construction and operation. Camera lens produces optical image shown as feathered arrow on photoelectric surface at left.

**image pattern** The pattern of charged particles on an insulating surface in a cathode-ray tube used in a television camera or for storage in a computer.

**image phase constant** The imaginary part of the transfer constant.

**image ratio** The ratio of the field strength at the image frequency to the field strength at the desired frequency, each field being applied in turn to the receiving antenna of a superheterodyne receiver, under specified conditions, to produce equal receiver outputs.

**image reactor** A nuclear reactor for which the criticality equations have been solved by using theoretical images.

**image rejection** Suppression of signals at the image frequency in a superheterodyne receiver.

**image response** The response of a superheterodyne receiver to an undesired signal at its image frequency.

**image tube** An electron tube that reproduces on its fluorescent screen an image of the optical image or other irradiation pattern incident on its photosensitive surface. Also called electron image tube and image converter tube.

**immittance** [IMpedance and adMITTANCE] A term used to denote both impedance and admittance, as commonly applied to transmission lines, networks, and certain types of measuring instruments.

**impedance** [symbol Z] The total opposition offered by a component or circuit to the flow of an alternating or varying current. Impedance Z is expressed in ohms and is a combination of resistance $R$ and reactance $X$, computed as the square root of the sum of the squares of $R$ and $X$. Impedance is also computed as $Z = E/I$, where $E$ is the applied a-c voltage and $I$ is the resulting alternating current flow in the circuit. In computations, impedance is handled as a complex ratio of voltage to current.

**impedance characteristic** A graph in which the impedance of a circuit is plotted against frequency.

**impedance coupling** Coupling of two circuits with an impedance, such as a tuned circuit or choke.

**impedance match** The condition in which the impedance of a connected load is equal to the internal impedance of the source, or to the surge impedance of a transmission line, thereby giving maximum transfer of energy from source to load, minimum reflection, and minimum distortion.

**impedance-matching network** A network of two or more resistors, coils, and/or capacitors used to couple two circuits in such a manner that the impedance of each circuit will be equal to the impedance into which it looks.

**impedance-matching transformer** A transformer used to obtain an impedance match between a given signal source and a load having a different impedance than the source.

**impedance triangle** A diagram consisting of a right-angle triangle with sides proportional to the resistance and reactance, respectively, of an a-c circuit. The hypotenuse then represents the impedance of the circuit. The cosine of the angle between resistance and impedance lines is equal to the power factor of the circuit.

**impedor** A term sometimes used in place of impedance to describe a circuit element having impedance.

**imperfection** Any deviation in structure from that of an ideal crystal.

**implant** A quantity of radioactive material in a suitable container, intended to be embedded in a tissue for therapeutic purposes.

**implode** To burst inward.

**implosion** The inward collapse of an evacuated container, such as the glass envelope of a cathode-ray tube.

**importance factor** The probability that failure or malfunction of a given equipment or portion of a system will cause failure of the intended mission.

**importance function** A measure of the importance to the chain reaction in a nuclear reactor of a neutron at a given position and with a given velocity.

**impregnated** Having spaces filled with a substance having good insulating properties.

**impregnated cathode** A dispenser cathode made by impregnating a porous tungsten body with a suitable molten barium compound. Evaporated electron-emitting surface barium is continuously replenished from within the cathode body.

**impregnated coil** A coil in which the spaces between the turns of insulated wire are filled with an insulating varnish or plastic material.

**impregnated tape** *Magnetic powder-impregnated tape.*

**improvement threshold** The value of carrier-to-noise ratio below which the signal-to-noise ratio decreases more rapidly than the carrier-to-noise ratio.

**impulse** *Pulse.*

**impulse excitation** *Pulse excitation.*

**impulse generator** *Pulse generator.*

**impulse noise** *Pulse noise.*

**impulse test** *Pulse test.*

**impurity** An atom that is foreign to the crystal in which it exists. In a semiconductor crystal, an impurity can produce either excess electrons or holes.

**impurity level** An energy level that is due to the presence of impurity atoms.

**impurity semiconductor** A semiconductor whose properties are due to impurity levels produced by foreign atoms.

**InAs** Symbol for *indium arsenide.*

**in-band signaling** The transmission of signaling tones within the channel normally used for voice transmission.

**incandescent lamp** An electric lamp in which light is produced by sending electric current through a filament of resistance material so as to heat it to incandescence.

**inches per second** [abbreviated ips] A magnetic tape speed rating. Commonly used speed values include 1⅞, 3¾, 7½, 15, and 30 inches per second.

**inching** *Jogging.*

**incidence angle** The angle between an approaching beam of radiation and the perpendicular (normal) to the surface that is in the path of the beam.

**incidental radiation device** A device that radiates radio-frequency energy during normal operation, although not intentionally designed to generate such energy.

**incident light** The direct light that falls on a surface.

**incident wave** 1. A wave that impinges on a discontinuity or on a medium having different propagation characteristics. 2. A current or voltage wave that is traveling through a transmission line in the direction from source to load.

**inclination** The angle that a line, surface, vector, or aircraft makes with the horizontal.

**inclinometer** 1. An instrument used to measure the direction of the earth's magnetic force with relation to the plane of the horizon. 2. An instrument that measures the attitude of an aircraft with respect to the horizontal.

**incoherent scattering** Scattering of particles or photons in which the scattering elements act independently of one another, so that there are no definite phase relationships among the different parts of the scattered beam.

**increment** A small change in the value of a variable.

**incremental frequency shift** A method of superimposing incremental intelligence on another intelligence by shifting the center frequency of an oscillator a predetermined amount.

**incremental permeability** The ratio of a small cyclic change in magnetic induction to the corresponding cyclic change in magnetizing force when the average magnetic induction is greater than zero.

**incremental sensitivity** The smallest change in a quantity being measured that can be detected by a particular instrument.

**incremental tuner** A television tuner in which the antenna, r-f amplifier, and r-f oscillator tuning coils are continuous or in small sections connected in series. Rotary switches make connections to the required portions of the total inductance required for a given channel, or short-circuit all of an inductance except that required for a given channel.

**independent firing** Initiating conduction of an ignitron by obtaining a firing pulse for the ignitor from a circuit independent of the anode circuit of the ignitron.

**independent-particle model** A nuclear model in which each proton and neutron moves independently in the field corresponding to the average positions of the other protons and neutrons. Also called individual-particle model, nuclear model, shell model, and single-particle model.

**independent variable** The independent quantity or condition that, through the action of the control system of an automatic controller, directs the change in the controlled variable according to a predetermined relationship.

**index of cooperation** The product of the total length of a facsimile scanning or recording line and the number of scanning or recording lines per unit length.

**index of refraction** The ratio of the velocity of a wave in a vacuum to that in a specified medium.

**index register** A computer register whose contents are used to automatically modify addresses incorporated in instructions just prior to their execution. The original instruction remains intact and unmodified in the memory.

**indicated course error** An instrumental error resulting in a discrepancy between the actual line of position offered by a navigation facility and the intended line of position.

**indicating instrument** An instrument in which the present value of the quantity being measured is visually indicated.

**indicator** A cathode-ray tube or other device that presents information transmitted or relayed from some other source, as from a radar receiver.

**indicator gate** A rectangular voltage applied to the grid or cathode circuit of an indicator cathode-ray tube to sensitize it during the desired portion of a radar operating cycle.

**indicator tube** An electron-beam tube in which useful information is conveyed by the variation in cross-section of the electron beam at a luminescent target.

**indirectly controlled variable** The quantity or condition that is controlled by virtue of its relation to the controlled variable and is not directly measured for control in a feedback control system.

**indirectly heated cathode** A cathode to which heat is supplied by an independent heater element in a thermionic tube. As a result, this cathode has the same potential on its entire surface, whereas the potential along a directly heated filament varies from one end to the other. Also called equipotential cathode, heater-type cathode, and unipotential cathode.

**indirect scanning** Scanning in which a narrow beam of light is moved across the area being televised. Used in flying-spot scanning of films, where the light transmitted by each illuminated elemental area in turn is picked up by one or more phototubes.

**indium** [symbol In] A metallic element. Atomic number is 49.

**indium antimonide** [symbol InSb] An intermetallic compound having semiconductor properties.

**indium arsenide** [symbol InAs] An intermetallic compound having semiconductor properties.

**indium phosphide** [symbol InP] An intermetallic compound having semiconductor properties.

**individual-particle model** *Independent-particle model.*

**indoor antenna** A receiving antenna located entirely inside a building but outside the receiver.

**induced charge** An electrostatic charge produced on an object by an electric field.

**induced current** A current produced in a conductor by a time-varying electromagnetic field, as in induction heating.

**induced nuclear disintegration** *Induced nuclear reaction.*

**induced nuclear reaction** A reaction in which a nucleus interacts with another nucleus, an elementary particle, or a photon to produce one or more other nuclei and possibly neutrons, mesons, neutrinos, and photons. The reaction occurs in about $10^{-12}$

second or less. Also called induced nuclear disintegration.

**induced radioactivity** *Artificial radioactivity.*

**induced radionuclide** A radionuclide that has a geologically short lifetime and is a product of nuclear reactions occurring currently or recently in nature. Examples are $C^{14}$ (natural radiocarbon) produced by cosmic-ray neutrons in the atmosphere and $Pu^{239}$ produced in uranium minerals by neutron capture.

**induced voltage** A voltage produced in a circuit by a change in the number of magnetic lines of force passing through a coil in the circuit.

**inductance** [symbol *L*] 1. The property of a circuit or circuit element that opposes a change in current flow. Inductance thus causes current changes to lag behind voltage changes. Inductance is measured in henrys, millihenrys, and microhenrys. 2. *Coil.*

**inductance bridge** An instrument similar to a Wheatstone bridge, used to measure an unknown inductance by comparing it with a known inductance.

**inductance-capacitance** [abbreviated *LC*] Containing both inductance and capacitance, as provided by coils and capacitors.

**inductance standard** *Standard inductor.*

**induction** The process by which a voltage, electrostatic charge, or magnetic field is produced in an object by lines of force.

**induction accelerator** *Betatron.*

**induction brazing** An electric brazing process in which the heat is produced by induced current flowing through the resistance of the joint being brazed.

**induction coil** A device for changing direct current into high-voltage alternating current. Its primary coil contains relatively few turns of heavy wire, and its secondary coil, wound over the primary, contains many turns of fine wire. Interruption of the direct current in the primary by a vibrating-contact arrangement induces a high voltage in the secondary.

**induction-conduction heater** A heating device in which electric current is conducted through a charge but is restricted by induction to a preferred path.

**induction coupling** A coupling in which torque is transmitted by the interaction of the magnetic field produced by magnetic poles on one rotating member and induced currents in the other rotating member.

**induction field** 1. The portion of the electromagnetic field of a transmitting antenna that acts as if it were permanently associated with the antenna. The radiation field leaves the transmitting antenna and travels through space as radio waves. 2. The electromagnetic field of a coil carrying alternating current, responsible for the voltage induced by that coil in itself or in a nearby coil.

**induction furnace** A furnace in which electric energy is transformed into heat by electromagnetic induction.

**induction hardening** A process of hardening a ferrous alloy by using induction heating.

**induction heater** A generator and associated equipment used for induction heating. At low frequencies, down to power-line frequencies, rotating equipment is generally used. At higher frequencies, more suitable for surface heating, electronic generators are used.

**induction heating** Heating of a conducting material by placing the material in a varying electromagnetic field. The resulting eddy currents induced in the material produce heat, just as current flowing through

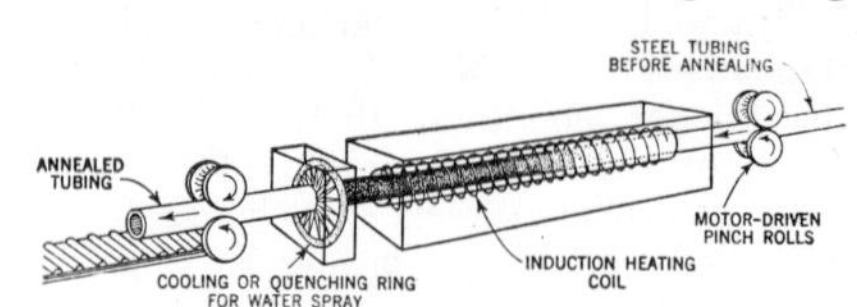


Induction heating setup used to heat steel tubing for annealing.

a resistor produces heat. The frequencies used range from 60 to over 500,000 cps, depending on the size and shape of the object to be heated. Also called eddy-current heating.

**induction instrument** An instrument that depends for its operation on the reaction between magnetic flux set up by currents in fixed windings and other currents set up by electromagnetic induction in movable conducting parts.

**induction loudspeaker** A loudspeaker in which the audio-frequency current that reacts with the steady magnetic field is induced in the moving member.

**induction motor** An a-c motor in which a primary winding on one member (usually the stator) is connected to the power source, and a polyphase secondary winding or a squirrel-cage secondary winding on the other member (usually the rotor) carries induced current.

**induction potentiometer** A resolver-type

synchro that delivers a polarized voltage whose magnitude is directly proportional to angular displacement from a reference position and whose phase indicates the direction of shaft rotation from the reference position. The linear range can be as great as 85°.

**induction radio** A carrier communication system for railroads, in which the short-range induction field of a loop antenna on a train is used to transfer signals to wayside telegraph wires connected to carrier

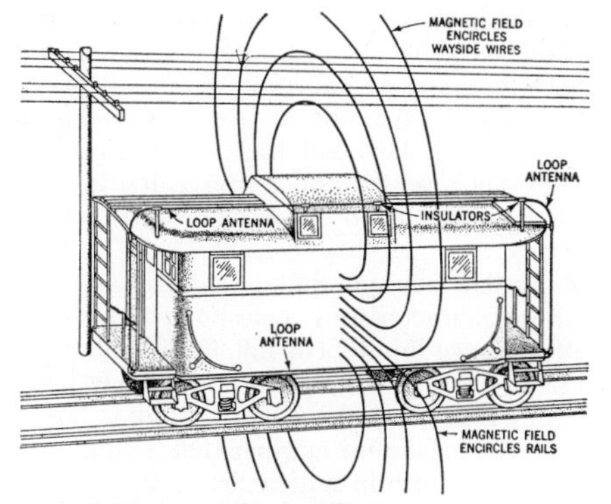


Induction radio installation on caboose.

equipment. Used to provide two-way communication between moving trains and wayside stations. Alternatively, the rails may be used in place of overhead wires to transmit the carrier signals.

**induction ring heater** A form of core-type induction heater used for heating electrically conducting charges in the form of a ring or loop, the core being open or separable to facilitate linking the charge.

**induction voltage regulator** A type of transformer having a primary winding connected in parallel with a circuit and a secondary winding in series with the circuit. The relative positions of the primary and secondary windings are changed to vary the voltage or phase relations in the circuit.

**inductive** 1. Pertaining to inductance. 2. Pertaining to the inducing of a voltage through mutual inductance. 3. Pertaining to the inducing of an electric charge by electrostatic induction.

**inductive circuit** A circuit containing a higher value of inductive reactance than capacitive reactance.

**inductive coupling** Coupling of two circuits by means of the mutual inductance provided by a transformer. Coupling by means of self-inductance that is common to two circuits is called direct inductive coupling. Also called transformer coupling.

**inductive diaphragm** A resonant window used in a waveguide to provide the equivalent of inductive reactance at the frequency being transmitted.

**inductive feedback** Feedback of energy from the anode circuit of an electron tube to its grid circuit through an inductance or by means of inductive coupling.

**inductive load** A load that is predominantly inductive, so that the alternating load current lags behind the alternating voltage of the load. Also called lagging load.

**inductive neutralization** A method of neutralizing an amplifier whereby the feedback susceptance due to the anode-to-grid capacitance is canceled by the equal and opposite susceptance of a coil. Also called coil neutralization and shunt neutralization.

**inductive-output tube** A tube in which output energy is obtained from the electron stream by electric induction between a cylindrical output electrode and the electron stream that flows through but does not touch this electrode.

**inductive post** A metal post or screw extending across a waveguide parallel to the E field, so as to add inductive susceptance in parallel with the waveguide for tuning or matching purposes.

**inductive reactance** [symbol $X_L$] Reactance due to the inductance of a coil or circuit. Inductive reactance is measured in ohms and is equal to 6.28 $fL$, where $f$ is the frequency in cycles per second and $L$ is the inductance in henrys.

**inductive tuning** Tuning involving use of a variable inductance.

**inductive window** A conducting diaphragm extending into a waveguide from one or both sidewalls of the waveguide, to give the effect of an inductive susceptance in parallel with the waveguide.

**inductometer** *Variable inductance.*

**inductor** *Coil.*

**inductor generator** An a-c generator in which all the windings are fixed and the flux linkages are varied by rotating an appropriately toothed ferromagnetic rotor. Sometimes used for generating high power at frequencies up to several thousand cycles per second for induction heating.

**inductuner** *Spiral tuner.*

**industrial heating equipment** Equipment that uses an r-f generator to produce energy for dielectric or induction heating operations in a manufacturing process.

**industrial radiography** Radiography of castings, welded joints, and other industrial products for quality control and flaw detection.

**industrial, scientific, and medical equipment** Equipment that uses radio waves for purposes other than communication.

**industrial television** [abbreviated itv] Closed-circuit television used for remote viewing of industrial processes and operations.

**industrial tube** An electron tube designed specifically for industrial electronic applications.

**inelastic collision** A collision in which at least one system gains internal excitation energy at the expense of the total kinetic energy of the center-of-gravity motion of the colliding systems.

**inelastic scattering** Scattering in which the kinetic energy of the neutron-plus-nucleus combination is decreased and the nucleus is left in an excited state.

**inertance** The acoustical equivalent of inductance.

**inert gas** A chemically inert gas, such as argon, helium, krypton, neon, and xenon.

**inertial-celestial guidance** An inertial guidance system in which position is checked automatically from time to time by means of celestial guidance.

**inertial-gravitational guidance system** A guidance system that is independent of all outside information except gravity.

**inertial guidance** A self-contained electronic guidance system that automatically follows a given course toward a ground target whose precise geographical position has been set into the computer of the system along with the starting point or a check-point position. It responds to inertial effects resulting from each change of course or speed, and makes appropriate corrections under control of its built-in computer. It is independent of interference and is not subject to jamming. Used in missiles, aircraft, and long-range nuclear-powered submarines.

**inertia switch** A switch that is actuated by an abrupt change in the velocity of the item on which it is mounted.

**I neutron** Obsolete term for a neutron having an energy of about 36 electron-volts, so that it is strongly absorbable in iodine.

**infinite attenuation** Attenuation so great that a voltage applied to the input terminals of the filter produces no output voltage. The term is used to specify a frequency at which infinite attenuation would be produced by a filter if its coils and capacitors had zero loss.

**infinite baffle** A loudspeaker baffle in which no acoustic energy travels from the front of the loudspeaker diaphragm to the back. It is usually a large sealed enclosure with the loudspeaker mounted in a hole cut into one side.

**infinite clipping** Clipping in which the threshold level is very small, so the output waveform is essentially rectangular.

**infinite-impedance detector** A detector in which the input circuit has infinite impedance. The load resistor is connected between cathode and ground, and is paralleled by a capacitor to bypass r-f signals. Distortion is low, but the circuit is seldom used because it cannot conveniently provide an automatic volume control voltage.

**infinite line** A hypothetical transmission line whose characteristics correspond to those of an ordinary line that is infinitely long.

**infinity** [symbol $\infty$] 1. An indefinitely large number or amount. 2. Any number larger than the maximum number that a computer is able to store in any register. When such a number is calculated, the computer usually stops and signals an alarm indicating an overflow.

**inflection point** A point at which a curve takes a definite change in direction.

**inflection-point emission current** The value of current that corresponds to the inflection point on the characteristic of a diode. It is approximately equal to the maximum space-charge-limited emission current.

**influence fuze** *Proximity fuze.*

**information gate** A gate that passes only the information contained on one telemetering channel.

**infradyne receiver** A superheterodyne receiver in which the intermediate frequency is higher than the signal frequency, so as to obtain high selectivity.

**infrared absorption spectrum** A spectrum produced by molecular absorption of infrared radiation.

**infrared beacon** A source of infrared radiation used to establish a geographical reference point, the bearing of which may be determined.

**infrared communication set** The components required to operate a two-way electronic system utilizing infrared radiation to carry intelligence.

**infrared detector** A device used to determine the presence and/or bearing of an object by measuring infrared radiation from its surfaces.

**infrared guidance system** A missile guidance system that uses an infrared detector with amplifiers and control units as required to detect and home on a heat-emitting enemy target.

**infrared homing** Homing in which the target is tracked by means of the infrared radiation that it emits.

**infrared image converter** An electron tube that converts an invisible infrared-illuminated scene into a visible image on a fluorescent screen. An infrared lens focuses the desired scene on a photocathode at the input end of the tube. The resulting stream

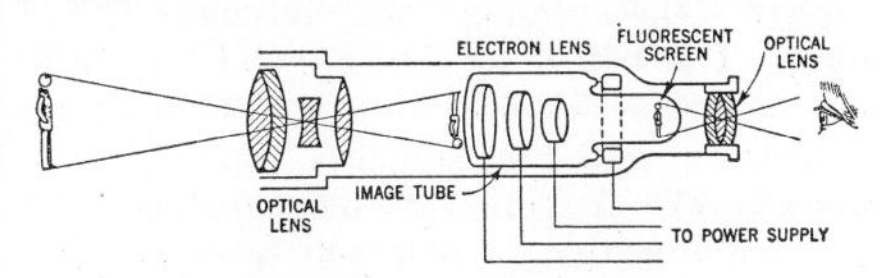


Infrared image converter.

of electrons emitted from the back of the photocathode, proportional to illumination at each point, passes through electron lenses that focus the stream on a fluorescent screen at the other end of the tube to give a visible image. Used in sniperscopes, snooperscopes, and other infrared viewing equipment.

**infrared mapping** Mapping in which a sensitive infrared detector is mounted on a motor-driven scanner like that used for sector-scan radar antennas, to scan the field of view line by line much as in television. The resulting thermal image is translated into varying shades of gray on photographic film, to give a line-pattern image much like that seen on a television screen. Used for mapping over enemy territory at night.

**infrared maser** An optical maser that uses an infrared frequency as the pumping frequency, to permit signal radiation and detection at millimeter wavelengths. Also called iraser and laser.

**infrared radiation** Electromagnetic radiation in the infrared spectrum, ranging in wavelength from about 0.75 to 1,000 microns. Also called nancy ray.

**infrared receiver** A device that intercepts and/or demodulates infrared radiations that may carry intelligence. Also called nancy receiver (U. S. Navy term).

**infrared spectroscope** A spectroscope designed to make measurements in the infrared spectrum.

**infrared spectrum** The portion of the electromagnetic spectrum between visible red light (about 0.75 micron) and the shortest microwaves (about 1,000 microns). All bodies that are not at absolute-zero temperature radiate in this range, making pos-

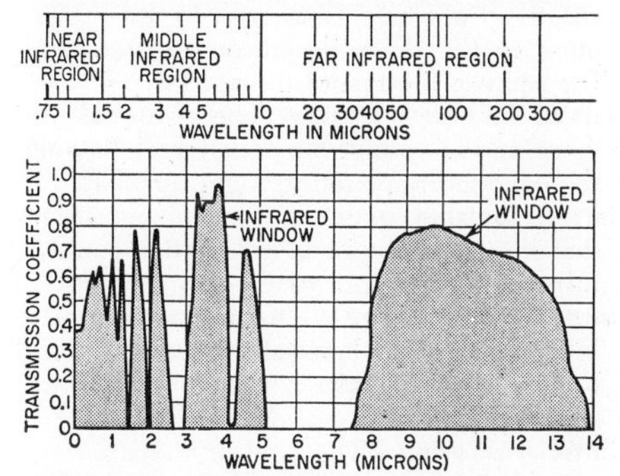


Infrared spectrum, extending between visible light at left end and microwave region starting at about 1,000 microns. Below is transmission spectrum of the earth's atmosphere for the near and middle infrared regions.

sible their detection in the dark by infrared systems. The infrared region is sometimes subdivided according to wavelength as follows: near infrared–0.75–1.5 microns; middle infrared–1.5–10 microns; far infrared–10–1,000 microns. Also called black light (deprecated).

**infrared transmitter** A transmitter that emits energy in the infrared spectrum. It may be modulated with intelligence signals.

**infrasonic** Pertaining to signals, equipment, or phenomena involving frequencies below the range of human hearing, hence below about 15 cps. Also called subsonic (deprecated).

**inherent filtration** Filtration introduced by the wall of an x-ray tube and any permanent tube enclosure.

**inherent regulation** Regulation such that equilibrium is reached after a disturbance, without the aid of a control system.

**inherited error** An error inherited from previous steps in a computer process.

**inhibiting input** A gate input that blocks an output which might otherwise occur in a computer.

**inhibition gate** A gate circuit placed in parallel with the circuit being controlled, for use as a switch.

**inhibit pulse** A drive pulse that tends to prevent flux reversal of a magnetic cell by certain specified drive pulses.

**inhour** [INverse HOUR] A unit of reactivity of a reactor. One inhour is the reactivity that will give the reactor a period of one hour. For a reactor that is very nearly critical, the reactivity in inhours is the inverse of the reactor period measured in hours.

**inhour equation** An equation relating the reactivity of a nuclear reactor to the parameters of the delayed-neutron emitters and the neutron lifetime of the reactor.

**initial inverse voltage** The value of inverse anode voltage immediately following the conducting period in a gas rectifier.

**initial ionizing event** An ionizing event that initiates a count in a radiation-counter tube. Also called primary ionizing event.

**initial permeability** The normal permeability that exists when both the magnetizing force and the magnetic induction approach zero.

**initial radiation** The nuclear radiation accompanying an atomic explosion and emitted from the resultant fireball. It includes the neutrons and gamma rays given off at the instant of the explosion, along with the alpha, beta, and gamma rays emitted by the rising fireball and the column of smoke.

**injection grid** A vacuum-tube grid used to feed the local oscillator signal into the mixer stage in some superheterodyne receivers.

**injector** An electrode through which charge carriers (holes or electrons) are forced to enter the high-field region in a spacistor.

**ink-mist recording** *Ink-vapor recording.*

**ink recorder** A recorder that employs an ink-filled pen or capillary tube to produce the graphic record.

**ink-vapor recording** A type of recording in which vaporized ink particles are directly deposited on the record sheet. Also called ink-mist recording.

**in-line assembly machine** An assembly machine that inserts components into a wiring board one at a time as the board is moved from station to station by a conveyor or other transport mechanism.

**in-line heads** Two magnetic-tape heads mounted so that their gaps are in exact vertical alignment. This is now the standard arrangement for stereophonic tape players. Also called stacked heads.

**in-line stereophonic tape** Magnetic stereophonic tape that has been recorded by using in-line heads. At any point on the tape the signal on one track is in line with the signal on the other. Also called stacked stereophonic tape.

**inner bremsstrahlung** 1. A beta-disintegration process resulting in the emission of a photon whose energy is between zero and the maximum energy available in the transition. 2. The radiation resulting from the inner-bremsstrahlung process.

**inner marker beacon** *Boundary marker beacon.*

**InP** Symbol for *indium phosphide.*

**in phase** Having waveforms that are of the same frequency and pass through corresponding values at the same instant.

**in-phase rejection** *Common-mode rejection.*

**in-phase signal** *Common-mode signal.*

**in-pile test** An irradiation test in which the effects of radiation are measured while the specimen is being subjected to radiation and neutrons in a nuclear reactor.

**in-port** The entrance for a network.

**input** The power or signal that is fed into an electronic device, or the terminals to which the power or signal is applied.

**input block** A section of internal storage in a computer that is generally reserved for the receiving and processing of input information.

**input capacitance** The short-circuit transfer capacitance that exists between the input terminal and all other terminals of an electron tube (except the output terminal) connected together. This quantity is equal to the sum of the interelectrode capacitances between the input electrode and all other electrodes except the output electrode.

**input equipment** The equipment used for feeding data into a computer.

**input gap** An interaction gap used to initiate a variation in an electron stream. In a velocity-modulated tube this gap is in the buncher resonator.

**input impedance** The impedance that exists between the input terminals of an amplifier or transmission line when the source is disconnected.

**input level** The ratio in decibels of audio input signal power to a reference power level of 1 milliwatt when the signal is working into a given impedance. Commonly expressed as dbm.

**input resonator** *Buncher resonator.*

**input transformer** A transformer used to provide a correct impedance match between a signal source and the input of a circuit or device.

**InSb** Symbol for *indium antimonide.*

**insensitive time** *Dead time.*

**insert earphone** A small earphone that fits partially inside the ear.

**insertion gain** The ratio of the power delivered to a part of the system following an inserted amplifier, to the power delivered to that same part before insertion of the amplifier. Usually expressed in decibels.

**insertion head** A mechanism used to insert a component in a printed-wiring board. The mechanism may also include automatic tools for cutting, forming, and clinching the leads of each component.

**insertion loss** The loss in load power due to the insertion of a component or device at some point in a transmission system. Generally expressed as the ratio in decibels of the power received at the load before insertion of the apparatus, to the power received at the load after insertion.

**insertion voltage gain** The complex ratio of the alternating component of voltage across the output terminals of a system when an amplifier is inserted between source and output, to the output voltage when the source is connected directly to the output termination.

**inside spider** A flexible device placed inside a voice coil to center it accurately with respect to the pole pieces of a dynamic loudspeaker.

**instantaneous automatic gain control** [abbreviated iagc] The portion of a radar system that automatically adjusts the gain of an amplifier for each pulse so as to obtain a substantially constant output-pulse peak amplitude with different input-pulse peak amplitudes. The adjustment is fast enough to act during the time a pulse is passing through the amplifier.

**instantaneous companding** Companding in which the effective gain variations are made in response to instantaneous values of the signal wave.

**instantaneous frequency** The time rate of change of the angle of a wave whose phase angle is a function of time.

**instantaneous particle velocity** The total particle velocity at a point minus the steady velocity at that point.

**instantaneous power output** The rate at which energy is delivered to a load at a particular instant.

**instantaneous recording** A recording intended for direct reproduction without further processing.

**instantaneous sampling** The process of obtaining a sequence of instantaneous values of a wave.

**instantaneous sound pressure** The total instantaneous pressure at a point minus the static pressure that exists at that point when no sound waves are present. The commonly used unit is the microbar.

**instantaneous speech power** The rate at which sound energy is being radiated by a speech source at any given instant.

**instantaneous value** The value of a sinusoidal or otherwise varying quantity at a particular instant.

**instantaneous volume velocity** The total instantaneous volume velocity at a point minus the static volume velocity at that point.

**Institute of Radio Engineers** [abbreviated IRE] A nonprofit professional organization of engineers and scientists, established for the advancement of the theory and practice of radio and electronics, including allied branches of engineering and the related arts and sciences.

**instruction** A set of characters, with or without one or more addresses, that defines a computer operation. Also called order. The signal that initiates the operation is called a command.

**instruction code** An artificial language for describing or expressing the instructions that can be carried out by a digital computer. Each instruction word usually contains a part specifying the operation to be performed and one or more addresses that identify a particular location in storage or serve some other purpose.

**instrument** 1. A device for measuring and sometimes also recording and controlling the value of a quantity under observation. The term "instrument" is usually applied to combinations of a meter with associated electron-tube circuits. 2. *Meter.*

**instrumental error** The error due to calibration, limited course sensitivity, and other inaccuracies introduced in any portion of a navigation system by the mechanism that translates path-length differences into navigation coordinate information.

**instrumental straggling** Additional straggling of particles due to such instrumental effects as noise, gain instability, source thickness, and poor geometry.

**instrument approach** An approach to a landing that is made by use of navigation instruments and radio guidance, with visual reference to the landing area only after the aircraft breaks through the overcast.

**instrument approach system** An aircraft navigation system that furnishes guidance in the vertical and horizontal planes to aircraft during descent from an initial-approach altitude to a point near the landing area. Completion of a landing requires guidance to touchdown by visual or other means.

**instrumentation** The use of measuring devices to determine the values of varying quantities, often for the purpose of controlling those quantities within prescribed limits.

**instrumentation package** The portion of a missile or artificial satellite that contains measuring and telemetering equipment along with a power source.

**instrument bombing** Bombing by the use of radar or other instruments, without visual reference to the ground.

**instrument conditions** Weather conditions in which instrument flying is mandatory, or in which continued flight is possible only by using instruments.

**instrument flight rules** [abbreviated ifr] Regulations governing flying when weather conditions are below the minimum for visual flight rules.

**instrument flying** Flying in which navigation is carried out by the use of flight and navigation instruments, including radio and radar equipment, without visual reference to the ground. Also called blind flying.

**instrument landing** An aircraft landing made with the aid of an instrument landing system.

**instrument landing system** [abbreviated ils] A radio system that provides in an aircraft the directional, longitudinal, and vertical guidance necessary for landing. It usually employs uhf ground transmitters and fixed directional antennas to define a beam that laterally localizes the runway extension and defines a slope plane at some angle between 2° and 5° leading to the optimum point of touchdown on the runway.

**instrument multiplier** A highly accurate resistor used in series with a voltmeter to extend its voltage range. Also called voltage multiplier and voltage-range multiplier.

**instrument panel** A panel or board containing indicating meters.

**instrument range** 1. The total range of values that an instrument is capable of measuring. 2. An intermediate range of reaction rate in a nuclear reactor, above the counter range and below the power range, in which the neutron flux can be measured by permanently installed control instruments such as ion chambers.

**instrument shunt** A resistor designed to be connected in parallel with an ammeter to extend its current range.

**instrument takeoff** A takeoff using aircraft instruments or other aids, without visual reference to the ground.

**instrument transformer** A transformer that transfers primary current, voltage, or phase values to the secondary circuit with sufficient accuracy to permit connecting an instrument to the secondary rather than the primary. Used so only low currents or low voltages are brought to the instrument. With a current transformer, the primary winding is inserted in the circuit carrying the current to be measured or controlled. With a voltage transformer, the primary winding is connected across the circuit whose voltage is to be measured or controlled.

**instrument-type relay** A relay constructed like a meter, with one adjustable contact mounted on the scale and the other contact mounted on the pointer. Also called contact-making meter.

**insulated** Separated from other conducting surfaces by a nonconducting material.

**insulated carbon resistor** A carbon resistor encased in a fiber, plastic, or other insulating housing.

**insulated wire** A conductor covered with a nonconducting material.

**insulating oil** A mineral oil used in transformers as an insulating and cooling medium.

**insulating tape** Tape impregnated with insulating material, usually adhesive. Used to cover joints in insulated wires or cables.

**insulating varnish** A varnish having good insulating qualities. Applied to coils and windings to improve their insulation and sometimes also to improve mechanical rigidity.

**insulation** A material having high electric resistance, and therefore suitable for separating adjacent conductors in an electric circuit or preventing possible future contact between conductors.

**insulation resistance** The electric resistance between two conductors separated by an insulating material.

**insulator** 1. A device having high electric resistance, used for supporting or separating conductors so as to prevent undesired flow of current from the conductors to other objects. 2. A substance in which the normal energy band is full and is separated from the first excitation band by a forbidden band that can be penetrated only by an electron having an energy of several electron-volts, sufficient to disrupt the substance.

**integer** A whole number.

**integral absorbed dose** *Integral dose.*

**integral action** A control action in which the rate of change of the correcting force is proportional to the deviation.

**integral dose** The total energy imparted to

an irradiated body by an ionizing radiation. Usually expressed in gram-rads or gram-roentgens. Also called integral absorbed dose and volume dose.

**integrating accelerometer** An accelerometer that measures the forces of acceleration along the longitudinal axis, records the velocity, and totalizes the distance traveled. When installed in a missile, it may be preset to switch off the fuel flow when the required speed is reached.

**integrating circuit** *Integrating network.*

**integrating dosimeter** An ionization chamber and measuring system used to measure the total radiation administered during an exposure. A device may be included to terminate the exposure when it reaches a desired value.

**integrating filter** A filter in which successive pulses of applied voltage cause cumulative buildup of charge and voltage on an output capacitor.

**integrating gyroscope** A gyroscope that senses the rate of angular displacement and measures and transmits the time integral of this rate.

**integrating ionization chamber** An ionization chamber whose collected charge is stored in a capacitor for subsequent measurement.

**integrating meter** An instrument that totalizes electric energy or some other quantity consumed over a period of time.

**integrating network** A circuit or network whose output waveform is the time integral of its input waveform. Also called integrating circuit and integrator.

**integrating-sphere densitometer** A photoelectric densitometer in which a beam of interrupted light is directed through the specimen into a sphere having a white inside surface. The light reflected from any part of the sphere is proportional to the total light entering the sphere. A phototube inserted in the wall of the sphere converts the reflected light into a corresponding signal that can be amplified and fed to a meter calibrated in density ratings.

**integrating timer** A timer that totalizes a number of small time intervals.

**integrator** 1. A computer device that approximates the mathematical process of integration. 2. *Integrating network.*

**intelligence** Data, information, or messages that are to be transmitted.

**intelligence signal** Any signal that conveys information, such as code, facsimile diagrams and photographs, music, television scenes, and spoken words.

**intelligibility** *Articulation.*

**intensification modulation** Modulation of the intensity of an electron beam in such a way that the sweep of the beam across the screen is normally almost invisible, and is made visible in proportion to the strength of the signal being displayed. Used in radar, where the echo pulse causes the beam to brighten at a point corresponding to the range of the target.

**intensifier electrode** An electrode used to increase the velocity of electrons in a beam near the end of their trajectory, after deflection of the beam. Also called post-accelerating electrode and post-deflection accelerating electrode.

**intensifier ring** A metallic ring-shaped coating on the inside of the glass envelope of a cathode-ray tube near the fluorescent screen. When a high positive voltage is applied to this ring, it increases the velocity of the electrons in the beam and thereby increases the intensity of the picture on the screen.

**intensify** To increase the brilliance of an image on the screen of a cathode-ray tube.

**intensifying screen** A thin screen coated with a substance that fluoresces readily under the influence of x-rays. It is placed next to the emulsion of x-ray film to increase the effect of x-rays on the film. In industrial radiography a sheet of thin lead is sometimes used for this purpose; here the secondary electrons and x-rays emitted by the lead produce the intensifying action.

**intensitometer** An instrument for determining relative x-ray intensities during radiography, to control exposure time.

**intensity** The strength or amount of a quantity, as of current, magnetization, radiation, or radioactivity. The symbol $I$ for current comes from this word.

**intensity control** *Brightness control.*

**intensity level** A term used in acoustics to specify the relation of one sound intensity to another. The intensity level is expressed in decibels, and is equal to 10 times the common logarithm of the ratio of the intensities. Also called sound-energy flux density level and specific sound-energy flux level.

**intensity modulation** Modulation of electron-beam intensity in a cathode-ray tube in accordance with the magnitude of the received signal. The luminance of the trace on the screen then varies with signal strength. Used in television and radar. Also called *z*-axis modulation.

**interaction-circuit phase velocity** The phase velocity of a wave traveling through the interaction gap of a traveling-wave tube in the absence of electron flow.

**interaction crosstalk** Crosstalk resulting from mutual coupling between two paths by means of a third path.

**interaction gap** An interaction space between electrodes in a microwave tube.

**interaction impedance** A measure of the r-f field strength at the electron stream of a traveling-wave tube for a given power in the interaction circuit.

**interaction space** A region of an electron tube in which electrons interact with an alternating electromagnetic field.

**intercarrier beat** An interference pattern that appears on television pictures when the 4.5-mc beat frequency of an intercarrier sound system gets through the video amplifier to the video input circuit of the picture tube.

**intercarrier noise suppressor** *Noise suppressor.*

**intercarrier sound system** A television receiver arrangement in which the television picture carrier and the associated sound carrier are amplified together by the video i-f amplifier and passed through the second detector, to give the conventional video signal plus a frequency-modulated sound signal whose center frequency is the 4.5-mc difference between the two carrier frequencies. The new 4.5-mc sound signal is then separated from the video signal for further amplification before going to the frequency-modulation detector stage.

**intercept** 1. To meet or interrupt the course of a moving vessel, aircraft, or missile. 2. To tap or tune to a telephone or radio message not intended for the listener.

**intercept bearing** A bearing taken by electronic means on an enemy radio signal to determine the location of the station.

**interceptor missile** A surface-to-air or air-to-air guided missile used to intercept enemy aircraft or guided missiles.

**intercom** *Intercommunication system.*

**intercommunication system** An audio-frequency amplifier system that provides two-way voice communication between two or more locations that are usually in the same structure. Each station contains a dynamic loudspeaker that also serves as a microphone. The amplifier may be at a central station, or each station may have its own amplifier. Connections between stations may be by means of wires or by means of carrier signals traveling over electric wiring in the building. Widely used also on ships and large aircraft. Also called intercom.

**intercontinental ballistic missile** [abbreviated icbm] A missile flying a ballistic trajectory after guided powered flight, usually over ranges in excess of 4,000 miles.

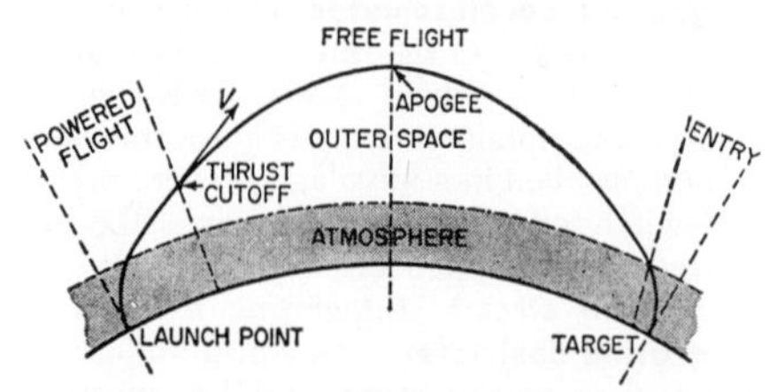

Intercontinental ballistic missile trajectory.

**intercontinental missile** A missile designed for travel from one continent to another, such as an intercontinental ballistic missile.

**interdigital magnetron** A magnetron having axial anode segments around the cathode, with alternate segments connected together at one end. The remaining segments are connected together at the opposite end.

**interelectrode capacitance** The capacitance between one electrode of an electron tube and the next electrode on the anode side. Also called direct interelectrode capacitance.

**interelectrode transadmittance** The short-circuit transfer admittance from one electrode of an electron tube to the next electrode on the anode side.

**interelectrode transconductance** The real part of interelectrode transadmittance.

**interface** A surface that forms the boundary between two types of material.

**interface connection** *Feedthrough.*

**interference** Any undesired energy that tends to interfere with the reception of desired signals. Manmade interference is generated by improperly operating electric devices, with the resulting interference signals either being radiated through space as electromagnetic waves or traveling over power lines. Radiated interference may also be due to atmospheric phenomena such as lightning. Radio transmitters themselves may interfere with each other in certain locations. Also called radio interference.

**interference blanker** A device that permits simultaneous operation of two or more pieces of radio or radar equipment without confusion of intelligence, or suppresses undesired signals when used with a single receiver.

**interference control** Monitoring of radio frequencies assigned to missile ranges, to detect signals that might interfere with missiles.

**interference eliminator** *Interference filter.*

**interference filter** 1. A filter used to attenuate manmade interference signals entering a receiver through its power line. Also called interference eliminator. 2. A filter used to attenuate unwanted carrier-frequency signals in the tuned circuits of a receiver.

**interference generator** A generator designed to produce r-f signals that are amplitude-modulated or frequency-modulated by random frequencies having erratic amplitudes, to simulate atmospheric static. Also called interference unit.

**interference guard band** One of the two bands of frequencies that border the authorized communication band and frequency tolerance of a station, provided to minimize interference between stations on adjacent channels.

**interference threshold** The minimum signal-to-noise ratio required for essentially error-free message transmission and reception.

**interference unit** *Interference generator.*

**interferometer** An apparatus used to produce and show interference between two or more wave trains.

**interferometer homing** A homing guidance system in which target direction is determined by comparing the phase of the echo signal as received at two antennas precisely spaced a few wavelengths apart.

**interlace** 1. To assign successive memory location numbers to physically separated locations on a storage tape or magnetic drum of a computer, usually to reduce access time. 2. *Interlaced scanning.*

**interlaced scanning** A scanning process in which the distance from center to center of successively scanned lines is two or more times the nominal line width, so that adjacent lines belong to different fields. In U. S. television, double interlace is used, wherein 262.5 alternate lines are scanned in one field and the remaining 262.5 lines in the next field. Used to minimize flicker of the picture. Also called interlace, interlacing, and line interlace.

**interlacing** *Interlaced scanning.*

**interlock** A switch or other device that prevents activation of a piece of equipment when a protective door is open or some other hazard exists.

**interlock circuit** A circuit in which one action cannot occur until one or more other actions have first taken place. The interlocking action is generally obtained with relays.

**interlock relay** A relay composed of two or more coils, each with its own armature and associated contacts, so arranged that movement of one armature or the energizing of its coil is dependent on the position of the other armature.

**interlock switch** A switch designed for mounting on a door, drawer, or cover in such a way that it opens automatically when the door or other part is opened.

**intermediate frequency** [abbreviated i-f] The frequency produced by combining the received signal with that of the local oscillator in a superheterodyne receiver.

**intermediate-frequency amplifier** The section of a superheterodyne receiver that amplifies signals after they have been converted to the fixed i-f value by the frequency converter. It is located between the frequency converter and second detector.

**intermediate-frequency harmonic interference** Interference due to acceptance of harmonics of an i-f signal by r-f circuits in a superheterodyne receiver.

**intermediate-frequency interference ratio** *Intermediate-frequency response ratio.*

**intermediate-frequency response ratio** The ratio of the field strength at a specified frequency in the i-f band to the field strength at the desired frequency, each field being applied in turn, under specified conditions, to produce equal outputs. Also called intermediate-frequency interference ratio.

**intermediate-frequency signal** A modulated signal whose carrier frequency is the i-f value of the receiver. This value is usually 455 kc for a broadcast-band radio receiver, 45.75 mc for a television receiver picture channel, and 41.25 mc for a television picture sound channel.

**intermediate-frequency stage** One of the stages in the i-f amplifier of a superheterodyne receiver.

**intermediate-frequency strip** A receiver subassembly consisting of the i-f amplifier stages mounted on an insulating strip, installed or replaced as a unit.

**intermediate-frequency transformer** The transformer used at the input and output

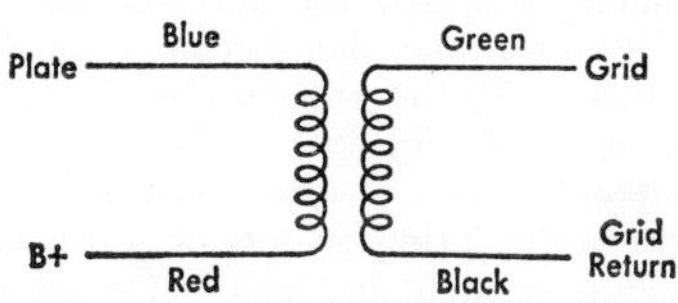


Intermediate-frequency transformer color code. For full-wave transformer feeding two diodes, one secondary lead is black, the other is green-black, and the center tap is black.

of each i-f amplifier stage in a superheterodyne receiver for coupling purposes and to provide selectivity.

**intermediate horizon** A hill, ridge, or building that is similar to the radar horizon but is not the most distant. An intermediate horizon may screen a valley between it and a mountain range that is the actual radar horizon.

**intermediate neutron** A neutron having energy in a range from about 100 to 100,000 electron-volts.

**intermediate-range ballistic missile** [abbreviated irbm] A missile flying a ballistic trajectory after guided powered flight and having a range of approximately 1,500 miles.

**intermediate reactor** A nuclear reactor in which fission is induced predominantly by neutrons having energies greater than thermal but much less than those of fission neutrons (from about 0.5 electron-volt to 100,000 electron-volts).

**intermediate subcarrier** A carrier that may be modulated by one or more subcarriers and used as a modulating wave to modulate another carrier.

**intermetallic compound** A semiconductor that consists only of metallic atoms held together by metallic bonds, to give a basic crystal structure consisting of two different metallic elements. An intermetallic compound is semiconducting when the two metals together contribute just enough electrons to fill the valence band, as in bismuth telluride, gallium arsenide, gallium phosphide, indium antimonide, indium arsenide, indium phosphide, and mercuric telluride.

**intermittent defect** A defect that is not continuously present.

**intermittent-duty rating** An output rating based on operation of a device for specified intervals of time rather than continuous duty.

**intermittent-duty relay** A relay that must be deenergized at intervals to avoid overheating of its coil.

**intermittent reception** A radio receiver complaint in which the receiver operates normally for a time, then becomes defective for a time, with the process repeating itself at regular or irregular intervals.

**intermittent service area** The area surrounding the primary service area of a broadcast station, in which the ground wave is received but is subject to some interference and fading.

**intermodulation** Modulation of the components of a complex wave by each other, producing new waves whose frequencies are equal to the sums and differences of integral multiples of the component frequencies of the original complex wave.

**intermodulation distortion** Nonlinear distortion characterized by the appearance of output frequencies equal to the sums and differences of integral multiples of the input frequency components. Harmonic components also present in the output are usually not included as part of the intermodulation distortion.

**intermodulation interference** Interference that occurs when the signals from two undesired stations differ by exactly the i-f value of a superheterodyne receiver and both signals are able to pass through the preselector due to poor selectivity. The undesired signals combine in the mixer stage to give an undesired i-f signal that interacts with the desired i-f signal. As a result, squeals and garbled sounds interfere with reception of all desired stations.

**internal connection** [symbol IC] An electron-tube connection, brought out to a base pin, that serves only for manufacturing purposes, such as for flashing the getter during evacuation.

**internal conversion** The ejection of an orbital electron by direct coupling with an excited nucleus. The ejected electron is called a conversion electron. Subsequent filling of the vacancy in the shell produces characteristic x-rays.

**internal conversion coefficient** *Conversion fraction.*

**internal memory** The total memory or storage that is accessible automatically to a computer without human intervention. Also called internal storage.

**internal resistance** The resistance of a voltage source, acting in series with the source.

**internal shield** [symbol IS] A metallic coating placed on the inner surface of the glass envelope of an electron tube for shielding purposes, with a connection being brought out to a base pin.

**internal storage** *Internal memory.*

**international broadcast station** A broadcast station employing frequencies allocated to the broadcasting service between 5,950 and 26,100 kc, whose transmissions are intended to be received directly by the general public in foreign countries.

**International Commission on Illumination** [abbreviated ICI] British version of Commission Internationale de l'Eclairage.

**International Geophysical Year** [abbreviated IGY] A period beginning July 1, 1957, and ending Dec. 31, 1959, scheduled by the world's scientists for a cooperative effort to advance scientific knowledge of the world itself.

**international Morse code** The code universally used for radiotelegraphy.

**international radio silence** A 3-minute period of radio silence on the international distress frequency of 500 kc only, commencing 15 and 45 minutes after each hour, during which radio stations may listen on that frequency for distress signals of ships and aircraft.

**international standard atmosphere** A standardized atmosphere, adopted internationally, used for comparing performance of aircraft and missiles. It is a pressure of 1,013.2 millibars at a mean sea-level temperature of 15°C, and a lapse rate of 6.5°C per kilometer of altitude up to 11 kilometers, above which the temperature is assumed constant at −56.5°C.

**International Telecommunications Union** [abbreviated ITU] An international civil organization established to provide standardized communication procedures, including frequency allocations and radio regulations, on a worldwide basis.

**international unit** A unit accepted internationally between 1908 and 1950 as a legal standard.

**interphone** An intercommunication system using headphones and microphones for communication between adjoining or nearby studios or offices, or between crew locations on an aircraft, vessel, tank, or other vehicle. Also called talk-back circuit.

**interpolate** To estimate missing values between those which are known. Extrapolate means to estimate values outside the known range.

**interpreter** A computer executive routine that translates a stored program expressed in some machine-like pseudo-code into machine code and performs the indicated operations by means of subroutines as they are translated. Also called interpretive routine.

**interpretive routine** *Interpreter.*

**interrogation** Transmission of a signal intended to trigger a transponder, racon, or iff system. Also called challenge and challenging signal.

**interrogator** The transmitting section of an interrogator-responsor. Also called challenger.

**interrogator-responsor** A transmitter and receiver combined, used for sending out pulses to interrogate a radar beacon and for receiving and displaying the resulting replies.

**interrupted continuous wave** [abbreviated icw] A continuous wave that is interrupted at a constant audio-frequency rate high enough to give several interruptions for each keyed code dot. The technique permits reception of code signals by receivers without beat oscillators, but radiates more sideband frequencies than other methods.

**interrupter** An electric, electronic, or mechanical device that periodically interrupts the flow of a direct current so as to produce pulses.

**interrupting capacity** The highest current that a device can interrupt at its rated voltage.

**interstage** Between stages.

**interstage transformer** A transformer used to provide coupling between two stages.

**interstation noise suppressor** *Noise suppressor.*

**interval** The spacing in pitch or frequency between two sounds. The frequency interval is the ratio of the frequencies or the logarithm of this ratio.

**interval timer** An instrument for signaling the expiration of a predetermined time. It may also actuate a switch at the end of the time interval. Also called timer.

**intrinsic angular momentum** The angular momentum associated with axial rotation of an elementary particle.

**intrinsic coercive force** The magnetizing force required to bring to zero the intrinsic induction of a magnetic material that is in a symmetrically and cyclically magnetized condition.

**intrinsic conduction** Conduction associated with movement of electron-hole pairs in a semiconductor under the influence of an electric field.

**intrinsic induction** The additional magnetic induction that exists in a given magnetic medium, above that which would exist at the same location for the same magnetizing force if the medium were a vacuum.

**intrinsic-junction transistor** A four-layer transistor having an i-type semiconductor layer between the base and collector layers, as in p-n-i-p, n-p-i-n, p-n-i-n, and n-p-i-p transistors.

**intrinsic mobility** The mobility of the electrons in an intrinsic semiconductor.

**intrinsic noise** Noise due to a device or

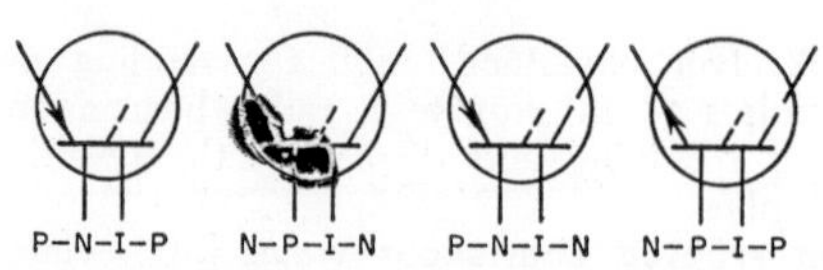

Intrinsic-junction transistor symbols. Unconnected short broken line represents intrinsic region. Two parallel broken lines indicate that intrinsic region is between similar regions.

transmission path, independent of modulation.

**intrinsic permeability** The ratio of intrinsic induction to the corresponding magnetizing force.

**intrinsic properties** The properties of a semiconductor that are characteristic of the pure ideal crystal.

**intrinsic Q** *Unloaded Q.*

**intrinsic region** A semiconductor region in which current flow is made up of approximately equal numbers of electrons and positive holes.

**intrinsic semiconductor** A semiconductor whose electrical properties are essentially characteristic of the pure ideal crystal. Current flow is made up of both electrons and positive holes in approximately equal numbers, hence the material is neither n-type nor p-type. Also called i-type semiconductor.

**intrusion alarm** A photoelectric, capacitance-controlled, electric, acoustic, or other system for setting off an alarm that announces the presence of an intruder at the boundaries of a protected area or inside that area.

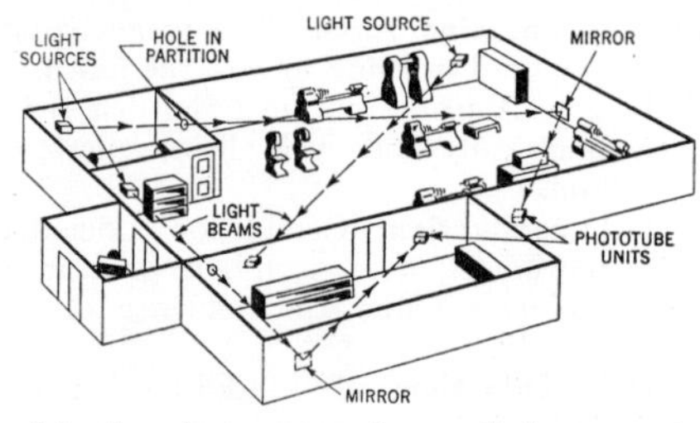

Intrusion alarm system for small factory, using three photoelectric systems.

**Invar** Trademark of Carpenter Steel Co. for their alloy of nickel and iron, containing about 36% nickel, that remains essentially constant in length over a wide range of temperature. Used for tuning forks and microwave cavities.

**inverse current** The current resulting from an inverse voltage in a contact rectifier.

**inverse direction** The direction of greater resistance in a rectifier, going from the positive electrode to the negative electrode. It is the opposite of the conducting direction. Also called reverse direction.

**inverse electrode current** The current flowing through an electrode of an electron tube in the direction opposite to that for which the tube is designed. Thus, for inverse anode current the electrons would flow from the anode to the cathode.

**inverse feedback** *Negative feedback.*

**inverse-feedback filter** A tuned filter circuit used at the output of a high-selectivity amplifier having negative feedback. The filter is adjusted so that the negative-feedback output is zero at the desired resonant frequency, but increases rapidly to reduce the amplification as the signal frequency departs from this value.

**inverse function** The function that would be obtained if the dependent and independent variables of a given function were interchanged.

**inverse limiter** A limiter whose output is constant for instantaneous input values within a specified range. Above and below that range it is linear or is some other prescribed function of the input. Used to remove the low-level portions of signals from an output wave, such as to eliminate the annoying effects of crosstalk.

**inverse-parallel connection** A connection of two rectifying elements such that the cathode of the first is connected to the anode of the second, and the anode of the first is connected to the cathode of the second.

**inverse peak voltage** 1. The peak value of the voltage that exists across a rectifier tube or x-ray tube during the half-cycle in which current does not flow. 2. The maximum instantaneous voltage value that a rectifier tube or x-ray tube can withstand in the inverse direction (with anode negative) without breaking down and becoming conductive.

**inverse photoelectric effect** The transformation of the kinetic energy of a moving electron into radiant energy at impact, as in the production of x-rays.

**inverse-square law** The intensity of light varies inversely as the square of the distance from the source.

**inverse voltage** An effective value of voltage that exists across a rectifier tube or x-ray tube during the half-cycle in which the anode is negative and current does not normally flow.

**inversion** 1. The process of scrambling speech for secrecy by beating the voice signal with a fixed higher audio frequency and using only the difference frequencies. The original low audio frequencies then become high audio frequencies, and vice versa. 2. A shallow layer of air in which temperature increases with altitude (temperature normally decreases with altitude). The resulting relatively sudden change in air density causes bending of radio waves.

**inverted-L antenna** An antenna consisting of a long horizontal wire, with the vertical lead-in wire connected to one end. Also called L antenna.

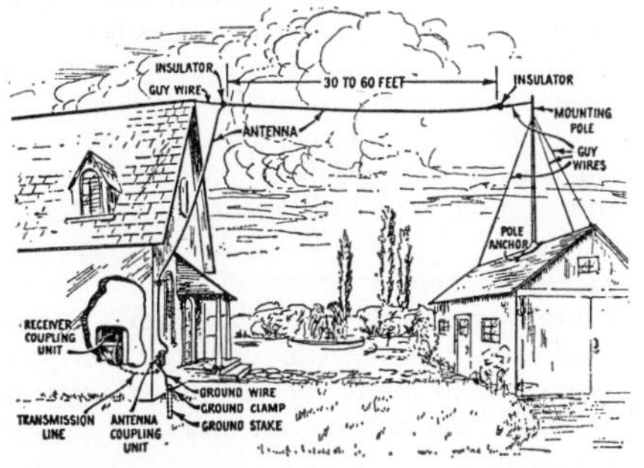


Inverted-L antenna installation for broadcast-band reception in rural areas.

**inverted speech** *Scrambled speech.*

**inverter** A device for converting direct current into alternating current. It may be electromechanical, as in a vibrator or synchronous inverter, or electronic, as in a thyratron inverter circuit.

**iodine** [symbol I] A nonmetallic element. Atomic number is 53.

**ion** A charged atom or group of atoms. A negative ion has gained one or more extra electrons, whereas a positive ion has lost one or more electrons.

**ion beam** A beam of ions drawn from a single source by a high voltage. In a cyclotron the voltages across two gaps reverse polarity at an r-f rate and the ions are acted on by a steady magnetic field, so the beam has a circular path.

**ion-beam scanning** The process of analyzing the mass spectrum of an ion beam in a mass spectrometer either by changing the electric or magnetic fields of the mass spectrometer or by moving a probe.

**ion burn** Deactivation and discoloration of a small area of phosphor at the center of the screen of a magnetically deflected cathode-ray tube, due to bombardment by heavy negative ions. An ion trap eliminates the effect by preventing the ions from leaving the electron gun.

**ion chamber** *Ionization chamber.*

**ion engine** A reaction engine designed for space travel, in which thrust is produced by a stream of positive ions obtained as a result of nuclear fission or fusion. The ions are accelerated by electrostatic fields, much as in a cathode-ray tube. To prevent buildup of space charges on the vehicle, electrons must be injected into the ion beam to neutralize the positive charge as the beam leaves the vehicle.

**ion gun** *Ion source.*

**ionic focusing** *Gas focusing.*

**ionic-heated cathode** A hot cathode that is heated primarily by ionic bombardment of the emitting surface.

**ionic-heated-cathode tube** An electron tube containing an ionic-heated cathode.

**ionium** [symbol Io] A naturally occurring radioisotope of thorium, having an atomic weight of 230.

**ionization** A process by which a neutral atom or molecule loses or gains electrons, thereby acquiring a net charge and becoming an ion. Ionization can be produced by collisions of particles, by radiation, and by other means.

**ionization by collision** Ionization produced by collisions of high-velocity electrons or ions with neutral atoms or molecules.

**ionization chamber** An enclosure containing two oppositely charged electrodes in a gas. When the chamber is exposed to nuclear radiation, the gas is ionized and each ion is drawn to the electrode of opposite polarity. The resulting current through the chamber is proportional to the intensity of the radiation. Also called ion chamber.

**ionization counter** An ionization chamber in which there is no internal amplification by gas multiplication. Used for counting ionizing particles.

**ionization cross section** The probability that a particle or photon passing through a particular gas or other form of matter will undergo an ionizing collision.

**ionization current** *Gas current.*

**ionization gage** A vacuum gage in which electrons accelerated between a hot cathode and a nearby positive electrode cause ionization of the residual gas. The resulting positive ions are attracted to another electrode to give a measurable current whose value bears a known relation to gas pressure.

**ionization instrument** An ionization chamber and associated equipment used to measure the intensity of gamma rays, x-rays, and other ionizing radiation.

**ionization path** The trail of ion pairs produced by an ionizing particle in its passage through matter. Also called ionization track.

**ionization spectrometer** *Bragg spectrometer.*

**ionization time** The time interval between the initiation of conditions for conduction in a gas tube and the establishment of conduction at some stated value of tube voltage drop.

**ionization track** *Ionization path.*

**ionization voltage** The energy per unit charge, usually expressed in volts, required to remove an electron from a particular kind of atom to an infinite distance. Also called electron binding energy.

**ionized layer** One of the atmospheric layers, such as the E layer and F layer, that reflect radio waves back to earth under certain conditions.

**ionizing energy** The average energy lost by an ionizing particle when producing an ion pair in a gas. For air the ionizing energy is about 32 electron-volts.

**ionizing event** An event in which one or more ions are produced.

**ionizing particle** A particle that produces ion pairs directly when it passes through a substance. The kinetic energy of the particle must be considerably greater than the ionizing energy of the medium.

**ionizing radiation** 1. Particles or photons that have sufficient energy to produce ionization directly in their passage through air. 2. Particles that are capable of nuclear interactions in which sufficient energy is released to produce ionization.

**ionophone** A high-frequency loudspeaker in which the a-f signal modulates the r-f supply to an arc maintained in a quartz tube, and the resulting modulated wave acts directly on ionized air to create sound waves.

**ionosphere** A region in the earth's outer atmosphere where ions and electrons are present in quantities sufficient to affect the propagation of radio waves. It begins about 30 miles above the earth and extends above 250 miles, with the height depending on the season of the year and the time of day. The chief regions of the ionosphere and their approximate heights are: D region—30 to 60 miles; E region—60 to 90 miles; F region—90 to 250 miles.

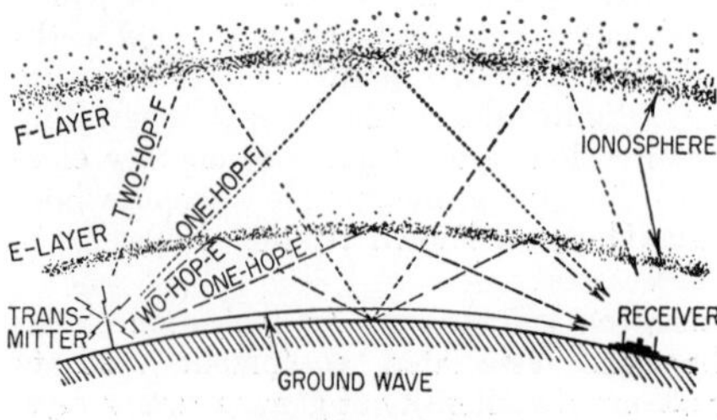


Ionosphere layers.

**ionospheric disturbance** A disturbance that makes the heights of the ionosphere layers go beyond the normal limits for a location, date, and time of day. Ionospheric storms and radio fadeouts are examples.

**ionospheric error** The total systematic and random error resulting from the reception of a navigation signal after ionospheric reflections. It may be due to variations in transmission paths, nonuniform height of the ionosphere, or nonuniform propagation within the ionosphere. Also called sky error (deprecated) and sky-wave error (deprecated).

**ionospheric height error** The component of total ionospheric error that is due to the difference in geometric configuration between ground paths and ionospheric paths. Also called geometric error and height error (deprecated).

**ionospheric scatter** A form of scatter propagation in which radio waves are scattered by the lower E layer of the ionosphere to permit communication over distances of from 600 to 1,400 miles when using the frequency range of about 25 to 100 mc.

**ionospheric storm** A turbulence in the F region of the ionosphere, usually due to a sudden burst of radiation from the sun. It is accompanied by a decrease in the density of ionization and an increase in the virtual height of the region. The higher frequencies in the band from 3 to 30 mc are most affected by the resulting radio blackouts. Also called auroral storm.

**ionospheric wave** *Sky wave.*

**ion pair** A positive ion and an equal-charge negative ion, usually an electron, that are produced by the action of radiation on a neutral atom or molecule.

**ion propulsion** A method of obtaining propulsion for spaceships by expelling ions and electrons from a combustion chamber. The recombination of electrons with ions outside the chamber prevents space charge effects that would counteract the thrust.

**ion sheath** A film of positive ions that forms on or near an electrode surface in a gas tube and limits the control action.

**ion source** A device in which gas ions are produced, focused, accelerated, and emitted as a narrow beam. Also called ion gun.

**ion spot** 1. A dark spot formed near the center of the screen of a cathode-ray tube due to ion burn. 2. A spurious signal resulting from bombardment of the target or photocathode of a camera tube or image tube by ions.

**ion trap** An arrangement whereby ions in the electron beam of a cathode-ray tube are prevented from bombarding the screen and producing an ion spot. Usually a part or all of the electron gun is tilted, and an external permanent magnet is used to bend the electron beam so it will pass through the tiny output aperture of the electron gun. The ions, being heavier and hence less affected by the magnetic field, are trapped harmlessly inside the gun.

**ion-trap magnet** One or more small permanent magnets with pole pieces, placed around the neck of a television picture tube

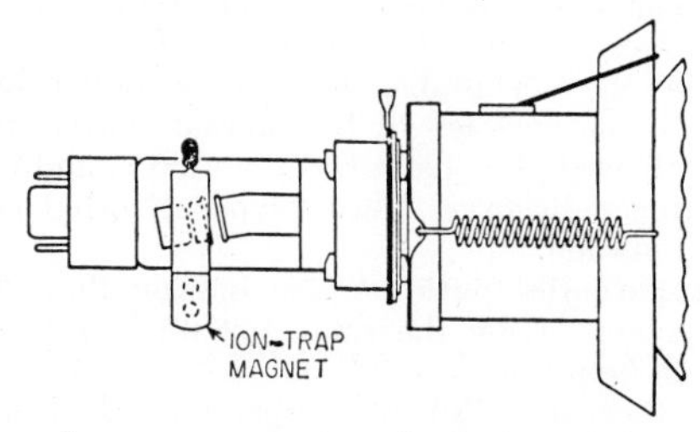


Ion-trap magnet on neck of picture tube.

to provide a magnetic field for ion-trap action in the electron gun. Also called beam bender.

**ion yield** The number of ion pairs produced per incident particle or quantum.

**ips** Abbreviation for *inches per second.*

**iraser** [InfraRed mASER] *Infrared maser.*

**irbm** Abbreviation for *intermediate-range ballistic missile.*

**irdome** [InfraRed DOME] A dome used to protect an infrared detector and its optical elements. Generally made from quartz, silicon, germanium, sapphire, calcium aluminate, or other material having high transparency to infrared energy.

***IR* drop** The voltage drop produced across a resistance *R* by the flow of current *I* through the resistance. Also called resistance drop.

**IRE** Abbreviation for *Institute of Radio Engineers.*

**iridium** [symbol Ir] An element. Atomic number is 77.

**iris** A conducting plate mounted across a waveguide to introduce impedance. When only a single mode can be supported, an iris acts substantially as a shunt admittance and may be used for matching the waveguide impedance to that of a load. Also called diaphragm.

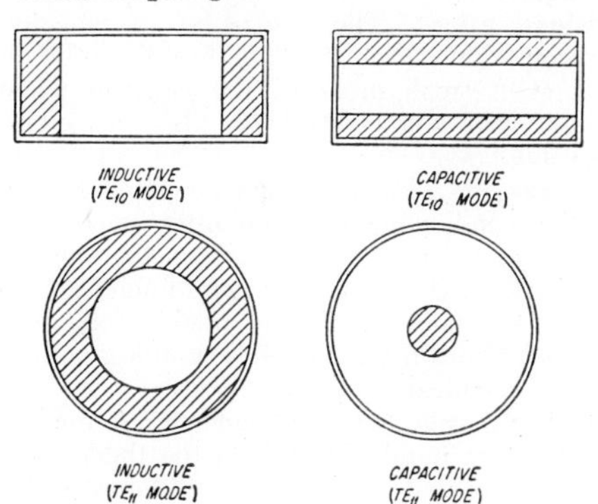


Iris designs for rectangular and circular waveguides.

**$I^2R$ loss** *Copper loss.*

**iron** [symbol Fe] A metallic element. Atomic number is 26.

**iron-core choke** *Iron-core coil.*

**iron-core coil** A coil in which solid or laminated iron or other magnetic material forms part or all of the magnetic circuit linking its winding. Also called iron-core choke.

**iron-core transformer** A transformer in which laminations of iron or other magnetic material make up part or all of the path for magnetic lines of force that link the transformer windings.

**iron-dust core** A core made by mixing finely powdered magnetic material with an insulating binder and molding under pressure to form a rod-shaped core that can be moved into or out of a coil or transformer to vary the inductance or degree of coupling for tuning purposes.

**iron loss** *Core loss.*

**iron-vane instrument** A measuring instrument in which the movable element is an iron vane.

**irradiance** *Radiant flux density.*

**irradiation** The exposure of a material, object, or patient to x-rays, gamma rays, ultraviolet rays, or other ionizing radiation.

**IS** Symbol for *internal shield.* Used on tube-base diagrams.

**I scan** *I display.*

**I scope** A radarscope that produces an I display.

**I signal** The in-phase component of the chrominance signal in color television, having a bandwidth of 0 to 1.5 mc. It consists of $+0.74(R-Y)$ and $-0.27(B-Y)$ where $Y$ is the luminance signal, $R$ is the red camera signal, and $B$ is the blue camera signal.

**island effect** The restriction of emission from the cathode of an electron tube to certain small areas of the cathode when the grid voltage is lower than a certain value.

**isobar** 1. A line that connects points having the same value of a quantity, such as a barometric pressure line on a meteorological chart. 2. One of two or more nuclides having the same number of nucleons in their nuclei but differing in their atomic numbers and chemical properties.

**isobaric spin quantum number** A nuclear quantum number based on the theory that a proton and a neutron are different states of a nucleon. The nucleon is assigned an isobaric spin quantum number of ½, and its two possible orientations, +½ and −½, are assigned to the neutron and proton, respectively. Also called isotopic spin quantum number and isotopic variable.

**isoclinic line** A line passing through points of equal magnetic inclination or dip on a magnetic map of the earth. Also called aclinic line.

**isodiaphere** One of several nuclides that have the same difference between the numbers of neutrons and protons in their nuclei.

**isoelectronic** Pertaining to atoms having the same number of electrons outside the nucleus of the atom.

**isogriv** A line that joins points of equal magnetic variation from grid north on a map or chart.

**isolation amplifier** An amplifier used to minimize the effects of a following circuit on the preceding circuit.

**isolation network** A network inserted in a circuit or transmission line to prevent interaction.

**isolation transformer** A transformer inserted in a system to separate one section of the system from undesired influences of other sections. Usually made with a 1-to-1 ratio of primary turns to secondary turns, to eliminate a direct connection without changing voltages.

**isolator** A passive attenuator in which the loss in one direction is much greater than that in the opposite direction. A ferrite isolator for waveguides is an example.

**isomagnetic line** A line passing through points having equal magnetic force but not necessarily the same deviation from vertical.

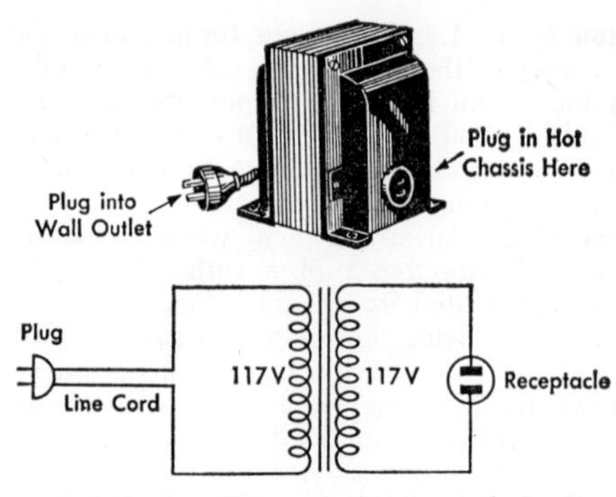


Isolation transformer construction and circuit.

**isomer** One of two or more nuclides having the same atomic and mass numbers but differing in other properties.

**isomeric transition** A radioactive transition from one nuclear isomer to another of lower energy. The deexcitation of the nuclei in the metastable state may occur by gamma emission or by internal conversion followed by emission of x-rays and/or Auger electrons. It is a type of forbidden transition.

**isopotential path** A line passing through points having the same potential or field strength.

**isothermal** Without temperature change.

**isotone** One of several nuclides having the same number of neutrons in their nuclei but differing in the number of protons.

**isotope** One of two or more nuclides having the same atomic number but differing in atomic weight and energy content because they have different numbers of neutrons. Thus, natural uranium consists of three isotopes, having atomic weights of 234, 235, and 238 and designated as $U^{234}$, $U^{235}$, and $U^{238}$, respectively.

**isotope shift** The slight difference in wavelength for a given spectral line of one isotope as compared with that of a related isotope.

**isotope structure** *Hyperfine structure.*

**isotopic abundance** The relative number of atoms of a particular isotope in a sample of an element.

**isotopic carrier** *Carrier.*

**isotopic indicator** *Isotopic tracer.*

**isotopic mass** Obsolete term for *atomic mass.*

**isotopic number** *Neutron excess.*

**isotopic spin quantum number** *Isobaric spin quantum number.*

**isotopic tracer** An isotope, usually radioactive, that is used as a chemical tracer for the element with which it is isotopic. Also called isotopic indicator.

**isotopic variable** *Isobaric spin quantum number.*

**isotron** A device for sorting isotopes, as of uranium, by accelerating all the ions to a given energy by applying a strong electric field. Ions of different mass then have different velocities. An r-f field is then applied to make the ions group themselves according to mass.

**isotropic antenna** *Unipole.*

**isotropic medium** A medium whose properties are the same in all directions.

**iterated fission expectation** The number of fissions produced per generation time by the daughter neutrons of a given neutron after a long period of operation of a critical assembly.

**iterative filter** A four-terminal filter that provides iterative impedance.

**iterative impedance** The impedance that, when connected to one pair of terminals of a four-terminal transducer, will cause the same impedance to appear between the other two terminals. The iterative impedance of a uniform transmission line is the same as the characteristic impedance. When a four-terminal transducer is symmetrical, the iterative impedances for the two pairs of terminals are equal and the same as the image impedances and the characteristic impedance.

**ITU** Abbreviation for *International Telecommunications Union.*

**itv** Abbreviation for *industrial television.*

**i-type semiconductor** *Intrinsic semiconductor.*

# J

**j** A complex operator that is mathematically equivalent to the square root of −1.

**jack** A connecting device into which a plug can be inserted to make circuit connections. The jack may also have contacts that open or close to perform switching functions when the plug is inserted or removed.

**jack box** A box designed to mount and protect one or more jacks and sometimes also one or more switches.

**jacket** A thin container for one or more fuel slugs, used to prevent the fuel from escaping into the coolant of a reactor. Also called can.

**jaff** Slang term for a combination of electronic and chaff jamming.

**jammer** A transmitter used to jam radio or radar transmissions.

**jamming** Radiation or reradiation of electromagnetic waves in such a way as to impair the usefulness of a specific segment of the radio spectrum that is being used by the enemy for communication or radar. Also called electronic jamming.

**JAN** Abbreviation for Joint Army-Navy, when referring to specifications and equipment issued by both services. Now generally superseded by the term MIL.

**Janet** A vhf point-to-point meteoric scatter communication system, based on forward scattering of signal bursts by the ionized trails of meteors.

**J antenna** A half-wave antenna that is end-fed by a parallel-wire quarter-wave section, so that the radiating elements somewhat resemble the letter J.

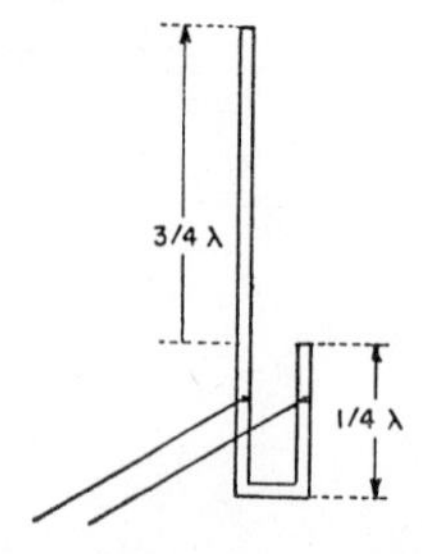


J antenna and method of connecting its quarter-wave matching section to an open-wire transmission line.

**Janus antenna array** An array providing both forward and backward beams, used in airborne doppler navigation systems. Named after the Roman god of doorways, who had a face on each side of his head.

**Janus technique** A technique of generating a doppler signal for an airborne navigation radar that is used to measure basic ground speed and drift angle. Microwave energy in the X band is radiated forward and backward from the aircraft toward the earth. The backscattered energy from both beams is detected, and the echo frequency of the aft beam is subtracted from that of the fore beam. The resulting low audio frequency is then a measure of ground speed.

**jar** An obsolete unit of capacitance used in the British Navy, equal to 1/900th microfarad.

**J display** A modified radarscope A display in which the time base is a circle. The target signal appears as an outward radial deflection from the time base. Also called J scan.

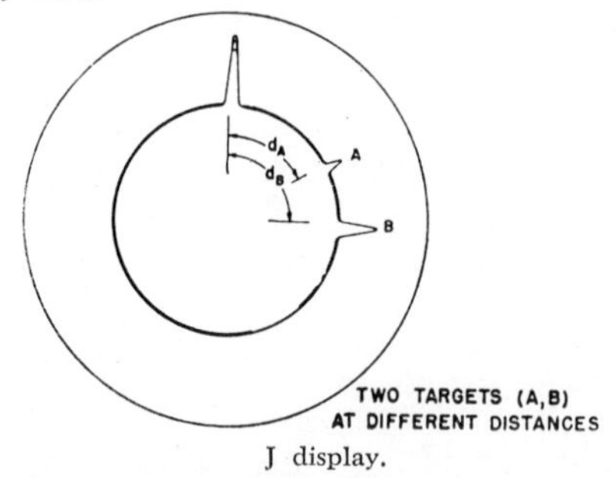


J display.

**JETEC** Abbreviation for *Joint Electron-Tube Engineering Council.*

**jewel bearing** A natural or synthetic jewel, usually sapphire, having a carefully ground conical depression that serves as a bearing for the pivot of a meter movement or as a bearing in other delicate instruments.

**jitter** Small, rapid variations in a waveform due to mechanical vibrations, fluctuations in supply voltages, control-system instability, and other causes. It can cause unsteadiness of a trace or picture on a cathode-ray screen or raggedness in received facsimile copy.

**j-j coupling** The interaction between two or more particles, each of which exhibits spin-orbit coupling.

**jogging** Quickly repeated opening and closing of a circuit to produce small movements of the driven machine. Also called inching.

**Johnson noise** *Thermal noise.*

**joint** A juncture of two wires or other conductive paths for current. Permanent joints are usually soldered, whereas temporary joints are generally held together by spring clips or screws.

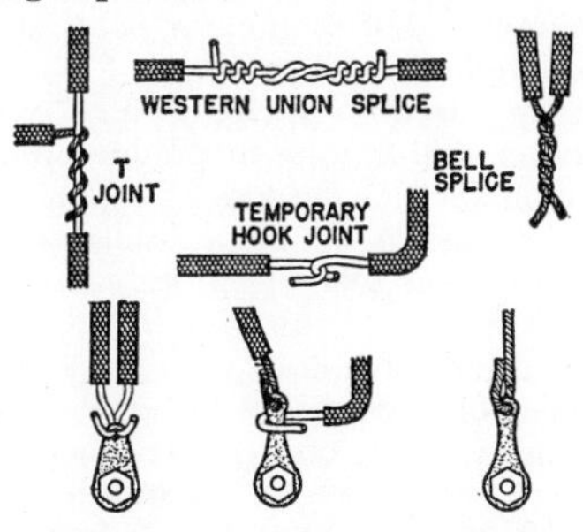


Joints used for soldered connections.

**Joint Electron-Tube Engineering Council** [abbreviated JETEC] A group established jointly by Electronic Industries Association and National Electrical Manufacturers Association to deal with industrywide electron-tube problems.

**joule** A unit of energy or work. One joule is equal to 1 watt-second.

**Joule effect** 1. The heating effect produced by the flow of current through a resistance. 2. *Magnetostriction.*

**Joule's law** The rate at which heat is produced in a constant-resistance electric circuit is proportional to the square of the current.

**J scan** *J display.*

**J scope** A radarscope that produces a J display.

**juice** Slang term for *current.*

**jukebox** An automatic phonograph that has labeled controls permitting a choice from as many as 200 records, played by depositing a coin in a slot.

**jump** A digital-computer programing instruction that conditionally or unconditionally specifies the location of the next instruction and directs the computer to that instruction. Used to alter the normal sequence of the computer. Also called transfer and transfer of control.

**jumper** A short length of conductor used to make a connection between two points or terminals in a circuit or to provide a path around a break in a circuit.

**junction** 1. A region of transition between two different semiconducting regions in a semiconductor device, such as a p-n junction. 2. A fitting used to join a branch waveguide at an angle to a main waveguide, as in a tee junction. Also called waveguide junction.

**junction box** An enclosure into which wires or cables are led and connected to form joints. It provides mechanical protection for the joints.

**junction diode** A semiconductor diode in which the rectifying characteristics occur at a diffused, grown, or alloyed junction between n-type and p-type semiconductor materials.

**junction point** *Branch point.*

**junction transistor** A transistor in which the base electrode is sandwiched between two or more junction electrodes.

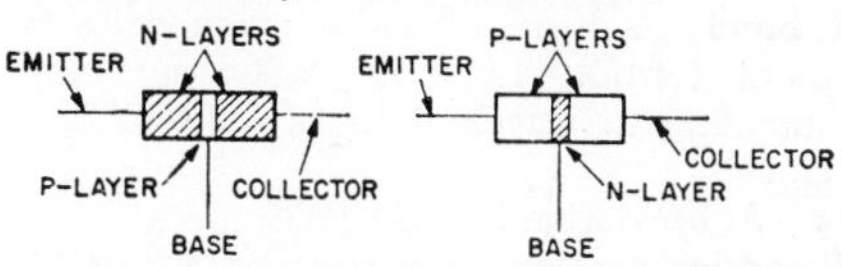


Junction transistor construction for n-p-n and p-n-p types.

**Jupiter** An Army surface-to-surface intermediate-range ballistic missile having an inertial guidance system, a range of about 1,500 miles, and a speed above mach 1. It can carry either conventional or nuclear warheads. A research version designated Jupiter-C was used to launch the first U.S. satellite into orbit January 31, 1958.

**just scale** A musical scale formed by taking three consecutive triads each having the ratio 4:5:6 or 10:12:15, with the highest note of one triad serving as the lowest note of the next.

# K

**K** 1. Symbol for *cathode.* 2. Abbreviation for *Kelvin.* 3. Abbreviation for *kilohm.* 4. Symbol for *relay.*

**kallitron oscillator** A negative-resistance oscillator using two triodes, with the tank circuit connected between the two anodes.

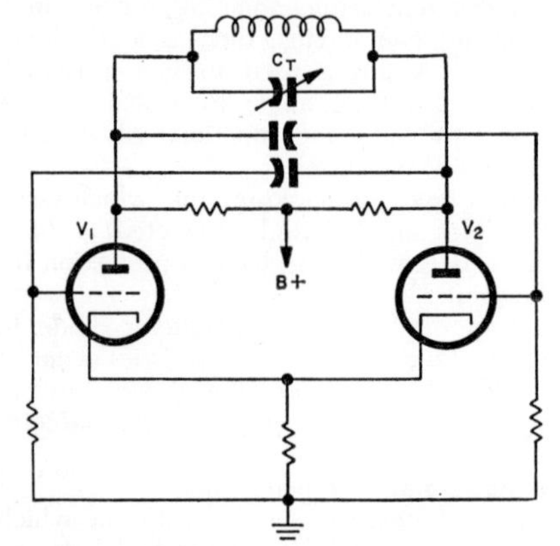


Kallitron oscillator circuit.

**K band** A band of radio frequencies extending from 11,000 to 33,000 mc, corresponding to wavelengths of 2.73 to 0.909 cm.

**kc** Abbreviation for *kilocycle.*

**K carrier system** A carrier system providing 12 telephone channels in a bandwidth of approximately 60 kc.

**K display** A modified radarscope A display in which a target appears as a pair of

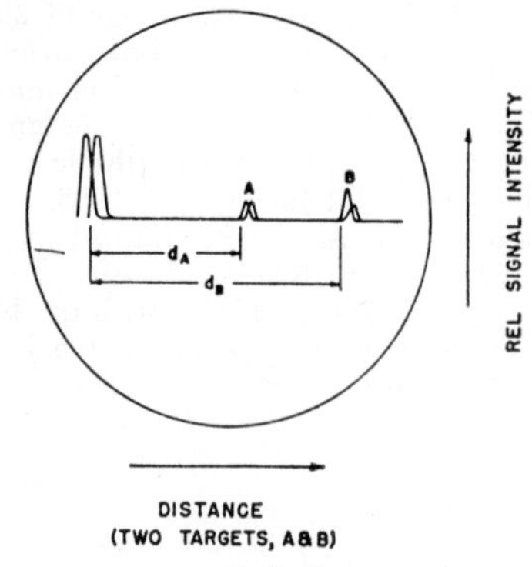


K display.

vertical deflections instead of as a single deflection. When the radar antenna is correctly pointed at the target in azimuth, the deflections are of equal height. When the antenna is not correctly pointed, the difference in pulse heights is an indication of direction and magnitude of azimuth pointing error. Also called K scan.

**kdp crystal** Abbreviation for *potassium dihydrogen phosphate crystal.*

**keep-alive circuit** A circuit used with a tr tube or anti-tr tube to produce residual ionization for the purpose of reducing the initiation time of the main discharge.

**keep-alive electrode** Deprecated term for *ignitor.*

**keeper** A bar of iron or steel placed across the poles of a permanent magnet to complete the magnetic circuit when the magnet is not in use, to avoid the self-demagnetizing effect of leakage lines. Also called magnet keeper.

**K electron** An electron having an orbit in the K shell, which is the first shell of electrons surrounding the atomic nucleus, counting out from the nucleus.

**K-electron capture** The radioactive decay process in which an orbital electron from the K shell of an atom is captured by the nucleus of that atom. It results in the production of x-rays characteristic of the daughter atom. Other examples of such electron capture are L-electron capture and M-electron capture.

**Kel-F** Trademark of M. W. Kellogg Co. for their fluorocarbon products based on polychlorotrifluoroethylene resin.

**Kelvin** [abbreviated K] An absolute temperature scale that uses centigrade degrees but with the entire scale shifted so that 0°K is at absolute zero. In this scale, water freezes at 273.16°K and boils at 373.16°K. Add 273.16 to a centigrade value to get the corresponding Kelvin value.

**Kelvin bridge** A seven-arm bridge used to compare the four-terminal resistances of two four-terminal resistors or networks. Also called double bridge and Thomson bridge.

**Kendall effect** A spurious pattern or other distortion in a facsimile record, caused by

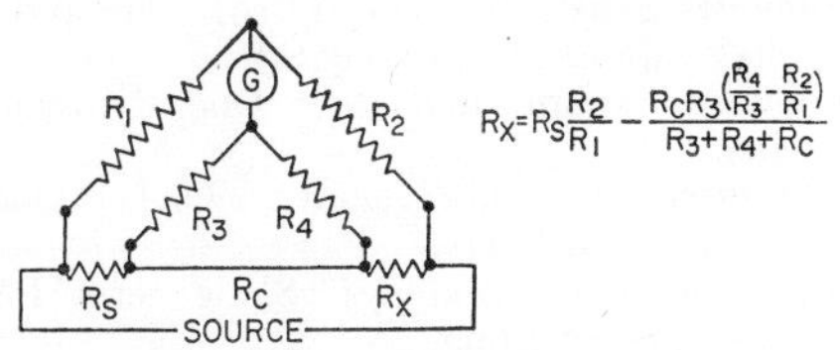

Kelvin bridge circuit and equation.

unwanted modulation products arising from the transmission of a carrier signal. It occurs principally when the width of one sideband is greater than half the facsimile carrier frequency.

**Kennelly-Heaviside layer** *E layer.*

**kenotron** A high-vacuum, high-voltage thermionic diode, used chiefly as a high-voltage rectifier.

**kernel** A function of two sets of variables used to define an integral operator in nuclear reactor theory. If the integral operator so defined is the inverse of a differential operator, the kernel is known as Green's function and belongs to the differential operator.

**Kerr cell** A cell containing a pair of electrodes in a dielectric liquid such as nitrobenzene. The dielectric becomes doubly refracting when under electric stress. If crossed Nicol prisms or Polaroid filters are

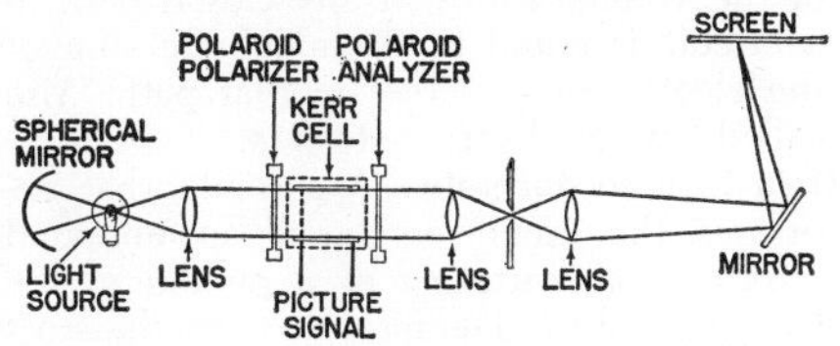

Kerr cell used in projection-type television receiver.

put before and after the Kerr cell, no light passes through the combination when no voltage is applied to the cell. When a signal voltage is applied to the cell, the plane-polarized light that enters the cell becomes elliptically polarized to an extent dependent on the voltage, and a proportional amount of light passes through the second prism. Used to modulate a beam of light or serve as a high-speed camera shutter.

**kev** Abbreviation for *kilo-electron-volt.*

**key** 1. A hand-operated switch used for transmitting code signals. Also called signaling key. 2. A special lever-type switch used for opening or closing a circuit only as long as the handle is depressed. Also called switching key.

**key click** A transient signal sometimes produced when a radiotelegraph sending key

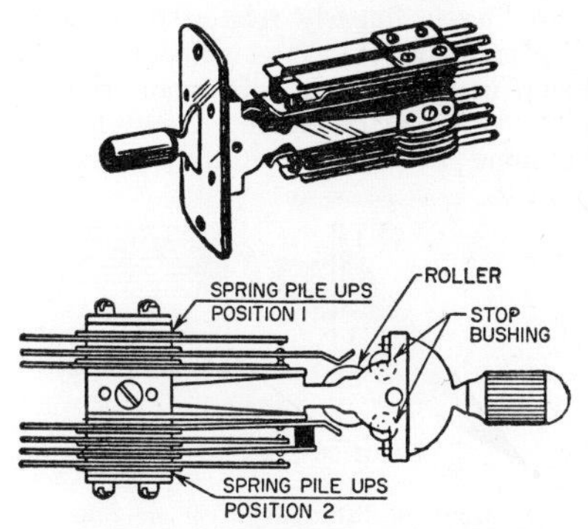

Key used for switching. Spring action returns lever handle to neutral position when it is released after being pushed up or down.

is opened or closed. It is heard in a loudspeaker or headphone as a click or chirp. Also called keying chirp.

**key-click filter** A filter that attenuates the surges produced each time the keying circuit of a transmitter is opened or closed by the key. Also called click filter.

**keyed automatic gain control** Automatic gain control in which an agc tube in a television receiver is biased to cutoff and is unblocked only when the peaks of positive horizontal sync pulses act on its grid. This technique prevents the agc voltage from being affected by noise pulses occurring in between sync pulses.

**keyed rainbow generator** A rainbow generator in which the colors of the spectrum are separated by black bars.

**keyer** A device that changes the output of a transmitter from one value of amplitude or frequency to another in accordance with the intelligence to be transmitted.

**keyer adapter** An adapter that detects a modulated signal and produces a corresponding d-c keying signal of varying amplitude for the frequency-shift exciter unit of a radio facsimile transmitter.

**keying** The forming of signals, such as for telegraph transmission, by modulating a d-c or other carrier between discrete values of some characteristic.

**keying chirp** *Key click.*

**keying frequency** The maximum number of times per second that a black line signal occurs when scanning the subject copy in facsimile.

**keying wave** *Marking wave.*

**key station** The station at which a network radio or television program originates.

**keystone distortion** Camera-tube distortion such that the length of a horizontal scan line is linearly related to its vertical displacement. It occurs when the electron beam in the camera tube scans the image plate at an acute angle. A system having keystone distortion distorts a rectangular

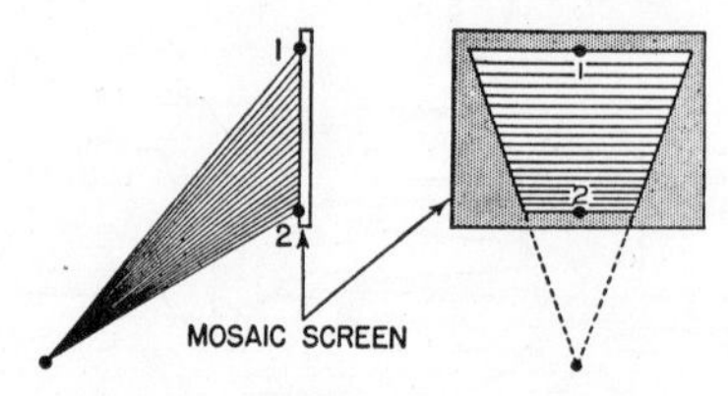

Keystone distortion in camera tube.

pattern into a trapezoidal pattern. The distortion is normally corrected by special transmitter circuits.

**kickback** The voltage developed across an inductance when current flow is cut off and the magnetic field collapses.

**kickback power supply** *Flyback power supply.*

**kick-sorter** British term for *pulse-height analyzer.*

**kilo-** A prefix representing $10^3$ or 1,000.

**kilocurie** One thousand curies.

**kilocycle** [abbreviated kc] One thousand cycles. Interpreted as meaning thousands of cycles per second.

**kilo-electron-volt** [abbreviated kev] One thousand electron-volts. It is the energy acquired by an electron that has been accelerated through a voltage difference of 1,000 volts.

**kilogauss** [plural kilogauss] One thousand gauss.

**kilohm** [abbreviated K] One thousand ohms. Thus, 15K is 15,000 ohms.

**kilomegacycle** [abbreviated kmc] One thousand megacycles. Also called billicycle and gigacycle.

**kilometric waves** British term for electromagnetic waves having wavelengths between 1,000 and 10,000 meters.

**kiloton** A unit used in specifying the yield of a fission bomb, equal to the explosive power of 1,000 tons of trinitrotoluene (TNT).

**kilovolt** [abbreviated kv] One thousand volts.

**kilovoltage** A voltage of the order of thousands of volts, such as the voltage applied to an x-ray tube.

**kilovolt-ampere** [abbreviated kva] One thousand volt-amperes.

**kilovoltmeter** A voltmeter whose scale is calibrated to indicate voltage in kilovolts.

**kilovolts peak** [abbreviated kvp] The peak voltage applied to an x-ray tube.

**kilowatt** [abbreviated kw] One thousand watts.

**kilowatthour** [abbreviated kwhr] One thousand watthours.

**kine** [pronounced kinny] Slang term for *kinescope recording.*

**kinescope** *Picture tube.*

**kinescope recording** A motion-picture film made by photographing images on the face of the picture tube in a television monitor or receiver, to permit repeating the same television program later and at different television stations. The sound portion of the program is usually recorded separately on magnetic tape or transcription disks. Also called kine (slang) and television recording.

**kinetic energy** Energy associated with motion.

**Kipp relay** A type of bistable multivibrator circuit that can be triggered by a short-duration pulse to transfer conduction from one tube to the other, and switched back by a similar pulse of opposite polarity.

**Kirchhoff's laws** The sum of the currents flowing to a given point in a circuit is equal to the sum of the currents flowing away from that point. The algebraic sum of the voltage drops in any closed path in a circuit is equal to the algebraic sum of the electromotive forces in that path. Also called laws of electric networks.

**Klein-Nishina formula** A formula that expresses the cross section of an unbound electron for scattering of a photon in the Compton effect. The formula gives the cross section as a function of the energy of the photon.

**K line** One of the characteristic lines in the x-ray spectrum of an atom. It is produced by excitation of the electrons of the K shell.

**klystron** An electron tube in which the electrons are periodically bunched by electric fields. The resulting velocity-modulated

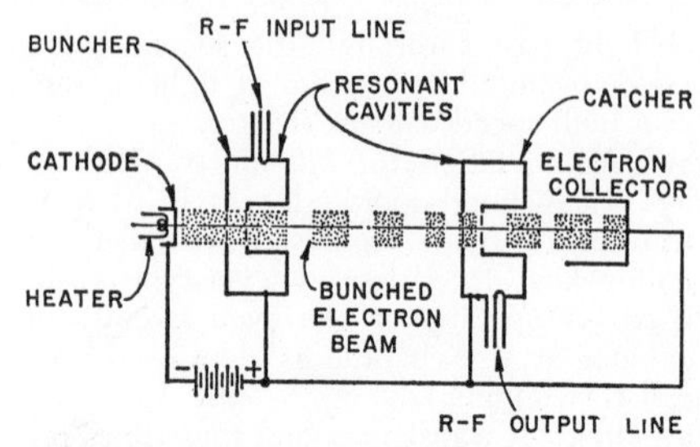

Klystron amplifier tube construction and operation.

electron beam is fed into a cavity resonator to sustain oscillations within the cavity at a desired microwave frequency. It is used as an oscillator or amplifier in uhf applications such as microwave relay and radar transmitters and receivers.

**klystron frequency multiplier** A two-cavity klystron in which the output cavity is tuned to a multiple of the fundamental frequency.

**klystron repeater** A microwave repeater consisting of a klystron inserted directly in a waveguide. Incoming waves velocity-modulate the electron stream emitted by the cathode of the tube. A second cavity converts the velocity-modulated beam back into waves, but with greatly increased amplitude, and feeds them into the output waveguide.

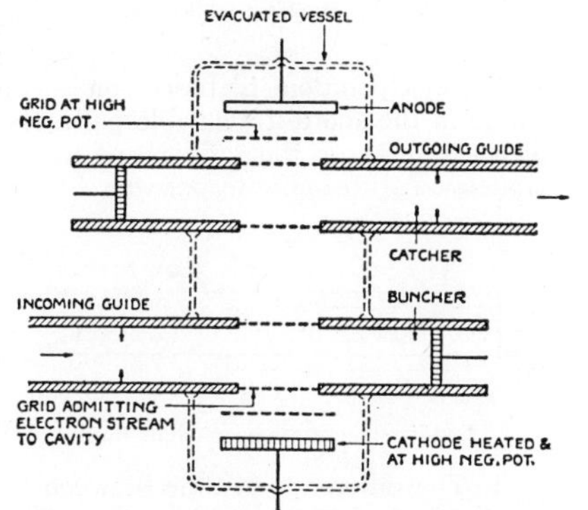

Klystron tube used as repeater in waveguide.

**kmc** Abbreviation for *kilomegacycle.*

**knee** The curve that joins two relatively straight portions of a characteristic curve.

**knife switch** A switch in which one or more hinged metal blades are manually pushed between spring contacts.

**knob** A component that is placed on a control shaft to facilitate manual rotation of the shaft. The knob sometimes has a pointer or markings to indicate shaft position.

**knot** A unit of speed equal to 1 nautical mile per hour or 1.15 miles per hour.

**Kovar** Trademark of Westinghouse Electric Corp. for an iron-nickel-cobalt alloy used in making metal-to-glass seals.

**krypton** [symbol Kr] A gaseous element. Atomic number is 36.

**K scan** *K display.*

**K scope** A radarscope that produces a K display.

**K shell** The innermost layer of electrons surrounding the atomic nucleus, having electrons characterized by the principal quantum number 1.

**Kurie plot** A graph of a beta-particle spectrum. Also called Fermi plot.

**kv** Abbreviation for *kilovolt.*

**kva** Abbreviation for *kilovolt-ampere.*

**kvp** Abbreviation for *kilovolts peak.*

**kw** Abbreviation for *kilowatt.*

**kwhr** Abbreviation for *kilowatthour.*

# L

**L** Symbol for *coil.*

***L*** Symbol for *inductance.*

**labeled molecule** A molecule containing one or more atoms of radioactive or stable isotopes that may be followed conveniently through biological, chemical, or physical processes.

**labile** *Unstable.*

**labile oscillator** An oscillator whose frequency is controlled from a remote location by wire or radio.

**laboratory system** A frame of reference that is attached to the observer's laboratory and hence usually is at rest relative to the surface of the earth. Used in nuclear physics.

**labyrinth** A loudspeaker enclosure having air chambers at the rear that absorb rearward-radiated acoustic energy, so as to prevent it from interfering with the desired forward-radiated energy.

**lacquer disk** A mechanical recording disk made of metal, glass, or paper coated with a lacquer compound that often contains cellulose nitrate. Also called cellulose nitrate disk.

**lacquer master** Deprecated term for *lacquer original.*

**lacquer original** An original recording made on a lacquer surface to serve for making a master. Also called lacquer master (deprecated).

**lacquer recording** Any recording made on a lacquer recording medium.

**Lacrosse** An Army surface-to-surface guided missile having a range of about 20 miles, controlled by radio command from a forward observation post. It uses conventional warheads and is employed tactically to replace and supplement conventional artillery.

**ladder attenuator** An attenuator having a series of symmetrical sections designed so the impedance remains essentially constant in both directions as the amount of attenuation is varied.

**ladder network** A network composed of a sequence of H, L, T, or pi networks connected in tandem.

**laddic** A multiaperture magnetic structure resembling a ladder, used to perform logic

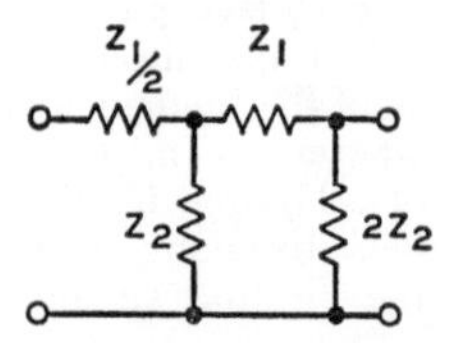

Ladder network, showing ratios of impedance values.

functions. Operation is based on a flux change in the shortest available path when adjacent rungs of the ladder are initially magnetized with opposite polarity.

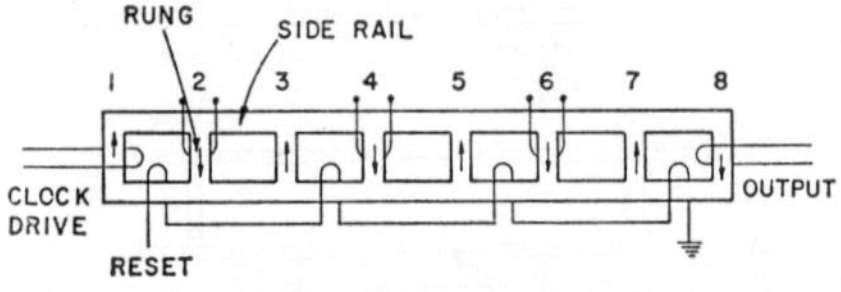

Laddic structure having eight rungs.

**lag** 1. The difference in time between two events or values considered together. Often expressed in degrees when comparing alternating quantities; thus, the current through

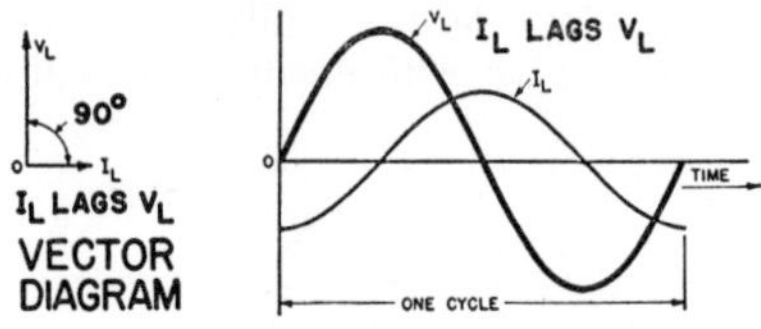

Lag of coil current behind coil voltage by 90°.

a perfect coil lags the applied voltage by 90°. 2. A persistence of the electric-charge image in a camera tube for a small number of frames.

**lagging load** *Inductive load.*

**lambda** [symbol λ] A Greek letter used to designate wavelength in meters.

**lambda particle** A hyperon having an extremely short life (about $3.7 \times 10^{-10}$ second) and a mass between that of neutrons and deuterons.

**lambert** A unit of luminance equal to $1/\pi$ candle per square centimeter. It corresponds to the uniform luminance of a perfectly diffusing surface that is emitting or reflecting light at the rate of 1 lumen per square centimeter.

**aminate** A product made by bonding together two or more layers of materials.

**aminated core** An iron core for a coil, transformer, armature, or other electromagnetic device, built up from laminations stamped from sheet iron or steel. The laminations are more or less insulated from each other by surface oxides and sometimes also by application of varnish. Laminated construction is used to minimize the effect of eddy currents.

**aminated plastic** A plastic material that is made by applying heat and pressure to particles or sheets of filler materials that have been impregnated with a thermosetting resin.

**aminated record** A mechanical recording medium composed of several layers of material, usually having a thin face of surface material on each side.

**amination** A single steel punching used to build up a magnetic circuit.

**aminography** Radiography of a particular layer of a body or object. In one method the x-ray tube and the film are moved simultaneously in opposite directions about a pivotal point in the plane of the layer. Only the plane of the layer then produces a well-defined image. Also called body-section radiography, planigraphy, and tomography.

**amp** A device that produces light, such as an electric lamp.

**amp bank** A number of incandescent lamps connected together in parallel or series to serve as a resistance load for full-load tests of electric equipment.

**amp cord** Two twisted or parallel insulated wires, usually No. 18 or No. 20, used chiefly for connecting electric equipment to wall outlets.

**anac** [LAminar Navigation AntiCollision] An aircraft radio navigation system consisting of airborne interrogator and ground transponder equipments with height-coding of the airborne interrogator pulses.

**and** 1. The record surface between two adjacent grooves of a mechanical recording. 2. *Terminal area.*

**anding aid** A lamp, searchlight, radio beacon, radar device, communicating device, or any system of such devices for aiding aircraft in an approach and landing.

**anding beacon** A radio beacon that produces a landing beam for aircraft guidance.

**anding beam** A radio beam, highly directional in both elevation and azimuth, that slants upward from the landing surface of an airport. It is produced by a landing beacon and serves as the glide path in an instrument landing system for aircraft.

**landline** A wire connection between two ground locations.

**landmine** A mine that is concealed below the surface of the earth and exploded by a fuze actuated by the weight of a vehicle or person.

**land return** *Ground clutter.*

**lane** The surface bounded by adjacent lines of position having the same value of the cyclic parameter for a navigation system.

**Langmuir dark space** A nonluminous region surrounding a negatively charged probe inserted in the positive column of a glow discharge.

**language** The characters, combining rules, and meanings used to express and process information for handling by computers and associated equipment.

**L antenna** *Inverted-L antenna.*

**lanthanide** One of the rare-earth elements having atomic numbers 58 to 71 inclusive. All have chemical properties similar to those of lanthanum.

**lanthanide contraction** The decreasing sequence of crystal radii of the tripositive rare-earth ions with increasing atomic numbers in the group from lanthanum through lutecium.

**lanthanum** [symbol La] A rare-earth element. Atomic number is 57.

**lap** 1. A rotating flat or curved disk, commonly of cast iron, used to grind quartz crystal plates, semiconductor blanks, flat optical objects, and other objects prior to polishing. 2. To grind a flat or curved surface with a lap.

**lap dissolve** Changeover from one television scene to another in such a way that the new picture appears gradually at the same rate at which the previous picture disappears.

**lapel microphone** A small microphone that can be attached to a lapel or pocket on the clothing of the user, to permit free movement while speaking.

**Laplace's law** The strength of the magnetic field at a given point, due to an element of a current-carrying conductor, is directly proportional to the strength of the current and the projected length of the element, and inversely proportional to the square of the distance of the element from the point in question.

**Laplace transform** A special case of a Fourier transform.

**Lark** A Navy rocket-powered surface-to-air guided missile having a radar homing guidance system. Used chiefly for training purposes.

**Larmor frequency** The angular frequency of precession of a charged particle rotating in a magnetic field. The frequency value is proportional to the strength of the magnetic field and to the gyromagnetic ratio.

**laser** [Light Amplification by Stimulated Emission of Radiation] *Infrared maser.*

**latching relay** *Locking relay.*

**latch-in relay** *Locking relay.*

**latency** The time a digital computer takes to deliver information from its memory. It may be the time spent waiting for the desired location on a magnetic drum to appear under a reading head. In a serial storage system, latency is the access time minus the word time.

**latent image** A stored image, as in the form of charges on a mosaic of small capacitances.

**latent period** The interval between irradiation and the appearance of results.

**lateral parity** Parity associated with an individual character, to indicate whether the number of holes punched across paper tape is even or odd. For magnetic tape the term applies to bits recorded in parallel tracks.

**lateral recording** A type of disk recording in which the groove modulation is parallel to the surface of the recording medium, so that the cutting stylus moves from side to side during recording.

**lattice** 1. A pattern of identifiable intersecting lines of position laid down in fixed positions with respect to the transmitters that establish the pattern for a navigation system. 2. A structure used in a nuclear reactor, made up of discrete bodies of fissionable and nonfissionable material arranged in a geometric pattern. 3. An orderly arrangement of atoms in a crystalline material.

**lattice imperfection** A deviation from a perfect homogeneous lattice in a crystal.

**lattice network** A network composed of four branches connected in series to form a mesh. Two nonadjacent junction points serve as input terminals, and the remaining two junction points serve as output terminals.

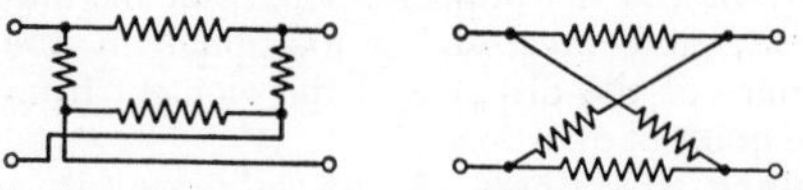

Lattice network as drawn in two different ways.

**lattice reactor** A heterogeneous nuclear reactor in which both fuel and moderator are in the form of long rods.

**lattice-wound coil** *Honeycomb coil.*

**Laue pattern** The characteristic photographic record obtained when x-rays from a pinhole or slit are sent through a single crystal that diffracts or bends the rays in all directions.

**launching** The process of transferring energy from a coaxial cable or transmission line to a waveguide.

**launching guidance** Navigation control of a missile during launching.

**Lauritsen electroscope** A rugged and sensitive electroscope in which a metallized quartz fiber is the sensitive element.

**lava** A natural fired stone consisting chiefly of magnesium silicate, used for insulators.

**law of electric charges** Like charges repel; unlike charges attract.

**law of electromagnetic induction** *Faraday's law.*

**law of electromagnetic systems** An electromagnetic system tends to change its configuration so that the flux of magnetic induction will be a maximum.

**law of electrostatic attraction** *Coulomb's law.*

**law of induced current** *Lenz's law.*

**law of magnetism** Like poles repel; unlike poles attract.

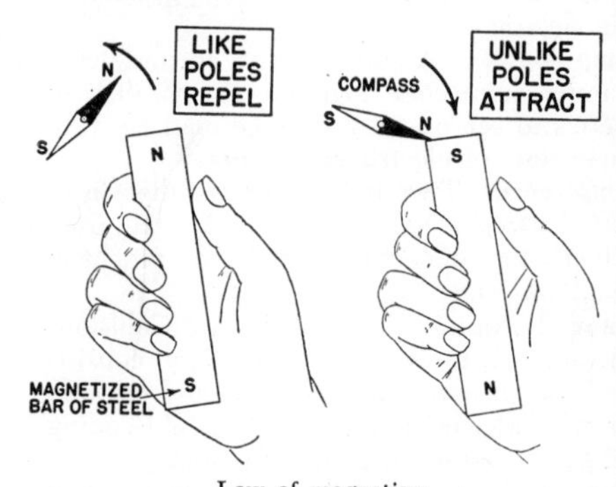


Law of magnetism.

**Lawrence tube** *Chromatron.*

**laws of electric networks** *Kirchhoff's laws.*

**layer winding** A coil-winding method in which adjacent turns are laid evenly side by side along the length of the coil form. Any number of additional layers may be wound over the first, usually with sheets of insulating material between the layers.

**layout** A diagram indicating the positions of various parts on a chassis or panel.

**lazy H antenna** An antenna array in which

two or more dipoles are stacked one above the other to obtain greater directivity.

**L band** A band of radio frequencies extending from 390 to 1,550 mc, corresponding to wavelengths of 76.9 to 10.37 cm.

**LC** Abbreviation for *inductance-capacitance.*

**L cathode** A dispenser cathode having a porous tungsten body covered with a layer of barium and oxygen atoms. Evaporated electron-emitting barium is replaced continuously from a compound located in a chamber behind the tungsten.

**LC product** The product of the inductance value $L$ in henrys and the capacitance value $C$ in farads.

**L/C ratio** The ratio of inductance $L$ in henrys to capacitance $C$ in farads for a resonant circuit.

**L display** A radarscope display in which the target appears as two horizontal pulses or blips, one extending to the right and one to the left from a central vertical time base. When the radar antenna is correctly aimed in azimuth at the target, both blips are of

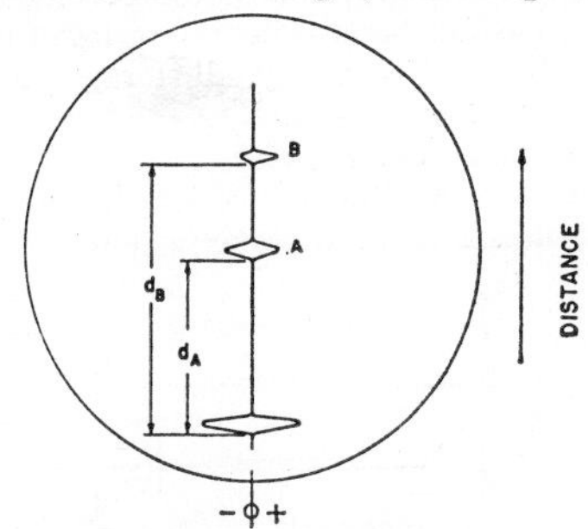


L display.

equal amplitude. When not correctly aimed, the relative blip amplitudes indicate the pointing error. The position of the signal along the baseline indicates target distance. The display may be rotated 90° when used for elevation aiming instead of azimuth aiming. Also called L scan.

**lead** [pronounced led; symbol Pb] A soft gray metallic element used as a shielding material in nuclear work and used with tin in solder. Atomic number is 82.

**lead** [pronounced leed] 1. The angle by which one alternating quantity leads another in time, expressed in degrees or in radians. Thus, the current through a perfect capacitor leads the applied voltage by 90°. 2. The distance between a moving target and the point at which a gun or missile is aimed. 3. A wire used to connect together two points in a circuit.

**lead-acid cell** The cell in an ordinary storage battery, in which the electrodes are grids of lead containing lead oxides that change in composition during charging and discharging. The electrolyte is dilute sulfuric acid.

**lead-covered cable** A cable whose conductors are protected from moisture and mechanical damage by a sheath of lead.

**lead equivalent** The thickness of lead that gives the same reduction in radiation dose rate as the material in question.

**leader cable** A cable used as a navigation aid, in which the path to be followed is defined by the magnetic field produced by current flowing through the cable.

**lead-in** A single wire used to connect a single-terminal outdoor antenna to a receiver or transmitter. Dipoles and other two-terminal antennas use transmission lines for this purpose. Also called down-lead.

**leading current** An alternating current that reaches its maximum value up to 90° ahead of the voltage that produces it. A leading current flows in any circuit that is predominantly capacitive.

**leading edge** The major portion of the rise of a pulse.

**leading-edge pulse time** The time at which the instantaneous amplitude of a pulse first reaches a stated fraction of the peak pulse amplitude.

**leading ghost** A ghost displaced to the left of the image on a television receiver screen.

**leading load** A load that is predominantly capacitive, so its current leads the alternating voltage applied to the load.

**lead-in groove** A blank spiral groove at the beginning of a disk recording, generally having a pitch much greater than that of the recorded grooves, provided to bring the pickup stylus quickly to the first recorded groove. Also called lead-in spiral.

**lead-in insulator** A tubular insulator inserted in a hole drilled through a wall, through which the lead-in wire can be brought into a building.

**lead-in spiral** *Lead-in groove.*

**leadout groove** A blank spiral groove at the end of a disk recording, generally of a pitch much greater than that of the recorded grooves, connected to either the locked or eccentric groove. Used to actuate the record-changing mechanism of an automatic record changer. Also called throwout spiral.

**leadover groove** A groove cut between

separate selections or sections on a disk recording to transfer the pickup stylus rapidly from one cut to the next. Also called crossover spiral.

**lead screw** [pronounced leed screw] A threaded shaft used to convert rotation to longitudinal motion. In a disk recorder it guides the cutter at a desired rate across the surface of an ungrooved disk during recording, so that the grooves will be appropriately spaced.

**lead storage battery** A storage battery that uses lead-acid cells.

**lead sulfide** [symbol PbS] The mineral galena, used as a crystal detector in the early days of radio and now used as an infrared detector.

**lead sulfide cell** A cell used to detect infrared radiation. Either its generated voltage or change of resistance may be used as a measure of the intensity of the radiation.

**lead telluride** [symbol PbTe] A compound of lead used in infrared detectors.

**lead time** [pronounced leed time] The time allowed or required to initiate and develop a piece of equipment that must be ready for use at a given time.

**leakage** Undesired and gradual escape or entry of a quantity, such as loss of neutrons by diffusion from the core of a nuclear reactor, escape of electromagnetic radiation through joints in shielding, flow of electricity over or through an insulating material, and flow of magnetic lines of force beyond the region in which useful work is performed.

**leakage current** 1. Undesirable flow of current through or over the surface of an insulating material or insulator. 2. The flow of direct current through a poor dielectric in a capacitor. 3. The alternating current that passes through a rectifier without being rectified.

**leakage flux** Magnetic lines of force that go beyond their intended path and do not serve their intended purpose.

**leakage inductance** Self-inductance due to leakage flux in a transformer.

**leakage power** The r-f power transmitted through a fired tr or pre-tr tube.

**leakage radiation** Radiation from anything other than the intended radiating system. A common example is electromagnetic radiation that escapes through joints or defects in shielding.

**leakage reactance** Inductive reactance due to leakage flux that links only the primary winding of a transformer.

**leakage resistance** The resistance of the path over which leakage current flows. It is normally a high value.

**leakage spectrum** A spectrum that shows the energy distribution of neutrons leaving a reactor by leakage.

**leaktight** *Hermetic.*

**leaky** Having leakage. Often applied to a capacitor in which the resistance has dropped far below its normal value so that excessive leakage current flows.

**leaky-pipe antenna** A microwave antenna consisting of a length of waveguide having one or more holes or slots in a wall, of such size and position as to give a desired radiation pattern for the electromagnetic energy that leaks out through the holes.

**leaky waveguide** A waveguide having a narrow longitudinal slot through which energy leaks out continuously.

**leapfrog test** A computer test using a special program that performs a series of arithmetical or logical operations on one group of storage locations, transfers itself to another group, checks the correctness of the transfer, then begins the series of operations again. Eventually, all storage positions will have been tested.

**Lecher line** *Lecher wires.*

**Lecher-line oscillator** A Hartley oscillator that uses Lecher wires as a tank circuit.

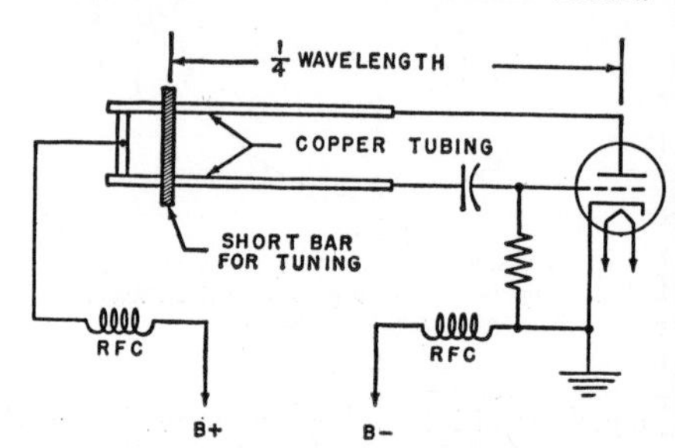


Lecher-line oscillator circuit.

**Lecher wires** Two parallel wires that are several wavelengths long and a small fraction of a wavelength apart, used to measure the wavelength of a microwave source that is connected to one end of the wires. A sliding shorting bar is moved along the wires, the positions of standing-wave nodes are noted, and the distance between nodes is then used to determine wavelength or frequency. Also called Lecher line.

**Leclanche cell** The common dry cell, which is a primary cell having a carbon positive electrode and a zinc negative electrode in an electrolyte of sal ammoniac and a depolarizer.

**Leduc current** An asymmetrical alternating current obtained from or similar to that obtained from the secondary winding of an induction coil. Used in electrobiology.

**left-hand polarized wave** An elliptically polarized transverse electromagnetic wave in which the rotation of the electric field vector is counterclockwise for an observer looking in the direction of propagation. Also called counterclockwise polarized wave.

**left-hand rule** 1. For a current-carrying wire: If the fingers of the left hand are placed around the wire in such a way that the thumb points in the direction of electron flow, the fingers will be pointing in the direction of the magnetic field produced by the wire. For conventional current flow

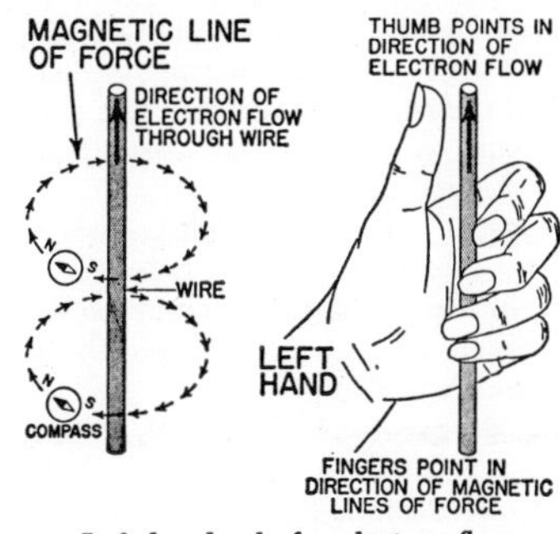


Left-hand rule for electron flow.

(the opposite of electron flow), the right hand is used. 2. For a movable current-carrying wire or an electron beam in a magnetic field: If the thumb, first, and second fingers of the left hand are extended at right angles to one another, with the first finger representing the direction of magnetic lines of force and the second finger representing the direction of electron flow, the thumb will be pointing in the direction of motion of the wire or beam. Also called Fleming's rule.

**left-hand taper** A taper in which there is greater resistance in the counterclockwise half of the operating range of a rheostat or potentiometer (looking from the shaft end) than in the clockwise half.

**leg** *Radio-range leg.*

**L electron** An electron having an orbit in the L shell, which is the second shell of electrons surrounding the atomic nucleus, counting out from the nucleus.

**L-electron capture** A mode of electron capture similar to K-electron capture except that the electrons are initially in the L shell of the atom.

**Lenard rays** Cathode rays produced in air by a Lenard tube.

**Lenard tube** An early experimental electron-beam tube having a thin glass or metallic foil window at the end opposite the cathode, through which the electron beam can pass into the atmosphere.

**lens** 1. A dielectric or metallic structure that is transparent to radio waves and can bend these waves to produce a desired radiation pattern. Used with antennas for radar and microwave relay systems. 2. One or more pieces of precisely contoured glass or other transparent material, used to focus light rays and form images by refraction. 3. An arrangement of electrodes used to produce an electric field that serves to focus electrons into a beam. 4. An arrangement of coils or permanent magnets used to produce a magnetic field that serves to focus electrons. Sometimes used in combination with an electric field. 5. A structure used for concentrating sound waves by refraction.

**lens antenna** A microwave antenna in which a dielectric lens is placed in front of the dipole or horn radiator to concentrate the radiated energy into a narrow beam. Also used to focus received energy on the receiving dipole or horn.

**Lenz's law** The current induced in a circuit due to its motion in a magnetic field or to a change in its magnetic flux is in such a direction as to exert a mechanical force opposing the motion or to oppose the change in flux. Also called law of induced current.

**lepton** A particle having a small mass, such as an electron, positron, neutrino, or antineutrino.

**lethargy** A neutron energy rating. The lower the energy of a neutron, the more lethargic it is and the higher is its lethargy.

**level** 1. The difference between a quantity and an arbitrarily specified reference quantity, usually expressed as the logarithm of the ratio of the quantities. In audio and communication work, a common reference level for power is 1 milliwatt, and power levels are expressed in dbm (decibels above 1 milliwatt). Thus, 0.01 milliwatt is −20 dbm, 1 milliwatt is 0 dbm, and 1 watt is 30 dbm. 2. A specified position on an amplitude scale applied to a signal waveform, such as reference white level and reference black level in a standard television signal. 3. A charge value that can be stored in a given storage element of a charge storage tube and distinguished in the output from other charge values. 4. Volume of sound.

5. A single bank of contacts, as on a stepping relay.

**level above threshold** The pressure level of a sound in decibels above its threshold of audibility for the individual observer. Also called sensation level.

**level indicator** 1. An instrument that indicates liquid level. 2. An indicator that shows the audio voltage level at which a recording is being made. It may be a volume-unit meter, neon lamp, or cathode-ray tuning indicator.

**Leyden jar** An early type of capacitor, consisting simply of metal foil sheets on the inner and outer surfaces of a glass jar.

**l-f** Abbreviation for *low frequency.*

**l-f loran** Abbreviation for *low-frequency loran.*

**lie detector** An instrument that indicates or records one or more functional variables of a body that may change when a person undergoes the emotional stress associated with a lie. These variables include blood pressure, heart action, and skin resistance. Also called psychointegroammeter.

**life test** A test in which a device is operated under conditions that simulate a normal lifetime of use, to obtain an estimate of life expectancy.

**lifetime** *Mean life.*

**light** Electromagnetic radiation having wavelengths capable of causing the sensation of vision, ranging approximately from 4,000 angstroms (extreme violet) to 7,700 angstroms (extreme red). The velocity of light is about 186,300 miles per second. Also called light wave.

**light-beam pickup** A phonograph pickup in which a beam of light is a coupling element of the transducer.

**light carrier injection** A method of introducing the carrier in a facsimile system by periodic variation of the scanner light beam, the average amplitude of which is varied by the density changes of the subject copy. Also called light modulation.

**light chopper** A rotating fan or other mechanical device used to interrupt a light beam that is aimed at a phototube, to permit a-c amplification of the phototube output and to make its output independent of strong, steady ambient illumination.

**light flux** *Luminous flux.*

**lighthouse tube** *Disk-seal tube.*

**light-microsecond** The distance a light wave travels in free space in one-millionth of a second. Used as a unit of electrical distance, equal to approximately 983 feet.

**light modulation** *Light carrier injection.*

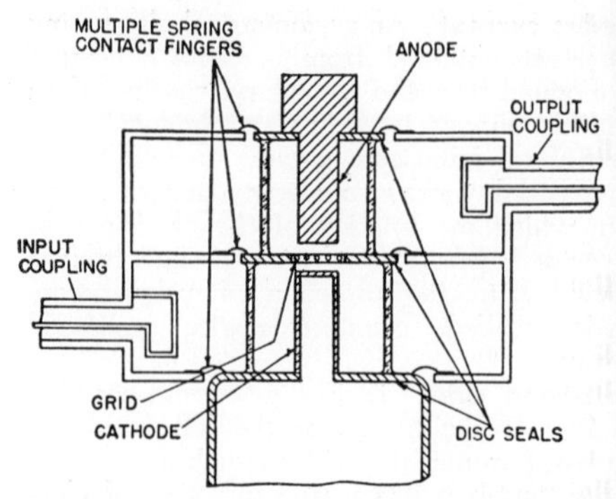


Lighthouse tube, showing close spacing of anode, grid, and cathode. Lower portion of tube, not shown, contains conventional octal base and heater that projects up inside cathode.

**light modulator** The combination of a source of light, an appropriate optical system, and a means for varying the resulting light beam to produce a sound track on motion-picture film.

**light-negative** Having negative photoconductivity, hence decreasing in conductivity (increasing in resistance) when exposed to light. Selenium sometimes exhibits this property.

**lightning arrester** A device that provides a discharge path to ground for lightning striking an antenna or transmission line. It generally contains a spark gap that has high resistance at normal circuit voltages. This gap breaks down to become a low resistance when acted on by the high-voltage surge of lightning. Also called arrester.

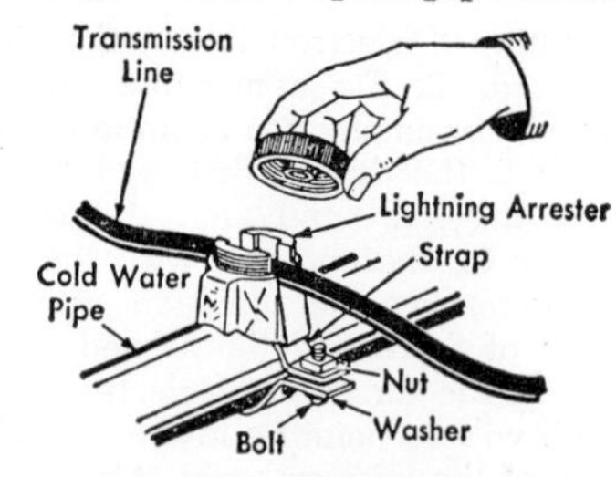


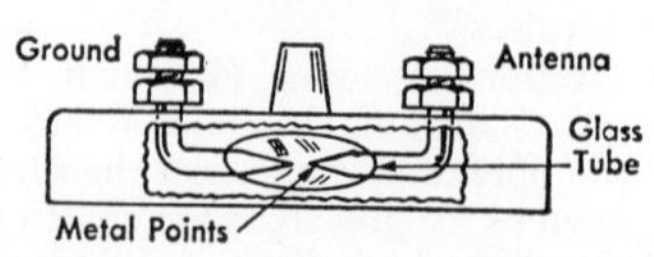


Lightning arrester designed for mounting on cold-water pipe and connecting to television twin line, and cross-section of typical lightning arrester.

**light pipe** A solid transparent plastic rod that transmits light from one end to the other even when bent. Used for coupling a scintillator to a phototube. Unfocused transmission and reflection minimize photon losses. Also used to distribute light more uniformly over a photocathode.

**light-positive** Having positive photoconductivity, hence increasing in conductivity (decreasing in resistance) when exposed to light. Selenium ordinarily has this property.

**light ray** A beam of light having a small cross-section.

**light relay** *Photoelectric relay.*

**light-sensitive** Having photoconductive, photoemissive, or photovoltaic characteristics. Also called photosensitive.

**light-sensitive cell** A cell that changes its electrical characteristics when exposed to light, such as a photovoltaic cell and a photoconductive cell.

**light source** A lamp used to supply radiant energy, as for a photoelectric control system.

**light-spot scanner** *Flying-spot scanner.*

**light valve** A device whose light transmission can be made to vary in accordance with an externally applied electrical quantity, such as voltage, current, electric field, magnetic field, or an electron beam. Used in exposing the sound track on motion-picture film.

**light wave** *Light.*

**limen** *Threshold.*

**limit bridge** A form of Wheatstone bridge used for rapid routine electrical tests of manufactured products. No attempt is made to balance the bridge for each test; instead, all products that produce deflections within limits corresponding to permissible tolerance are passed.

**limited-proportionality region** *Region of limited proportionality.*

**limited signal** A radar signal that is intentionally limited in amplitude by the dynamic range of the radar system.

**limited stability** *Conditional stability.*

**limiter** A circuit that limits the amplitude of its output signal to some predetermined threshold level. It may act on positive amplitude swings, on negative swings, or on both. Used after the i f amplifier in some f-m receivers to remove all amplitude variations from the frequency-modulated signal. For infinite limiting, the threshold level is very small and the output has an essentially rectangular waveform. Also called automatic peak limiter, clipper, peak clipper, and peak limiter.

**limiting** A desired or undesired amplitude-limiting action performed on a signal by a limiter. Also called clipping.

**limit switch** A switch designed to cut off power automatically at or near the limit of travel of a moving object that is actuated by electric means.

**Lindemann glass** A lithium borate-beryllium oxide glass having no element higher in atomic number than oxygen, used as window material for low-voltage x-ray tubes because it will pass x-rays of extremely long wavelength, such as grenz rays.

**line** 1. A transmission line or power line. 2. A production line used in mass-production assembly of electronic equipment. 3. The path covered by the electron beam of a television picture tube in one sweep from left to right across the screen. 4. One horizontal scanning element in a facsimile system. 5. *Trace.*

**line amplifier** An audio amplifier that feeds a program to a transmission line at a specified signal level. Also called program amplifier.

**linear** Having an output that varies in direct proportion to the input.

**linear absorption coefficient** The fractional decrease in intensity of a beam of photons or particles per unit distance traversed.

**linear accelerator** An accelerator having ring-shaped electrodes arranged in a straight line. When the electrode potentials are properly varied in amplitude at an ultrahigh frequency, particles passing through the electrodes receive successive increments of energy and are accelerated along an essentially linear path. Also called linear electron accelerator.

**linear actuator** An actuator that converts electric energy into linear mechanical motion.

**linear amplifier** An amplifier in which changes in output current are directly proportional to changes in applied input voltage.

**linear array** An antenna array in which the dipole or other half-wave elements are arranged end to end on the same straight line. Maximum radiation is produced in a plane perpendicular to the axis of the array. Also called collinear array.

**linear control** A rheostat or potentiometer having uniform distribution of resistance along the entire length of its resistance element.

**linear detector** A detector in which the output signal voltage is directly proportional to the changes in input carrier amplitude for amplitude modulation or to the changes in input carrier frequency for frequency modulation.

**linear electron accelerator** *Linear accelerator.*

**linear energy transfer** The linear rate of loss of energy by an ionizing particle traversing a material medium, expressed in electron-volts per micron of thickness.

**linear feedback control system** A feedback control system in which the relationships between the pertinent measures of the system signals are linear.

**linearity** 1. The condition wherein the change in the value of one quantity is directly proportional to the change in the value of another quantity. 2. Uniformity of distribution of the lines and elements of an image on a television picture tube, so that straight lines in a scene are straight in the image.

**linearity control** A television receiver control that varies the amount of correction applied to the sawtooth scanning wave in order to provide the desired linear scanning of lines, so that lines appear straight and round objects appear as true circles. Separate linearity controls, known as the horizontal linearity control and the vertical linearity control, are usually provided for the horizontal and vertical sweep oscillators. Also called distribution control.

**linearity region** The region, for an aircraft instrument approach and guidance system, in which the deviation sensitivity remains within specified values.

**linearly polarized wave** A transverse electromagnetic wave whose electric field vector is always along a fixed line.

**linear modulation** Modulation in which the amplitude of the modulation envelope (or the deviation from the resting frequency) is directly proportional to the amplitude of the intelligence signal at all modulation frequencies.

**linear power amplifier** A power amplifier in which the signal output voltage is directly proportional to the signal input voltage.

**linear pulse amplifier** A pulse amplifier in which the peak amplitude of the output pulses is directly proportional to the peak amplitude of the corresponding input pulses.

**linear rectifier** A rectifier whose output current or voltage contains a wave having a form identical to that of the envelope of an impressed signal wave.

**linear scan** A radar scan in which the beam is oscillated back and forth over a fixed angle in a given plane, as in sector scanning.

**linear stopping power** The energy loss per unit distance when a charged particle passes through a medium.

**linear sweep** A cathode-ray sweep in which the beam moves at constant velocity from one side of the screen to the other, then suddenly snaps back to the starting side.

**linear taper** A taper that gives the same change in resistance with rotation over the entire range of a potentiometer.

**linear transducer** A transducer for which the pertinent measures of all the waves concerned are linearly related.

**linear varying-parameter network** A linear network in which one or more parameters vary with time.

**line balance** Balance in which the conductors of a transmission line have the same electrical characteristics with respect to each other, to other conductors, and to ground.

**line-balance converter** *Balun.*

**line cord** A two-wire cord terminating in a two-prong plug at one end and connected permanently to a radio receiver or other appliance at the other end. Used to make connections to a source of power. Also called power cord. The wire alone is commonly called lamp cord. A third wire and prong are sometimes included in a line cord, for making a safety connection to ground.

**line-cord resistor** An asbestos-enclosed wirewound resistor incorporated in a line cord along with the two regular wires.

**line coupling** The coupling capacitors, line-tuning circuits, and lead-in circuits that together provide a connection between power or telephone lines and the transmitter-receiver assembly of a carrier communication system.

**line diffuser** An oscillator used in a television monitor or receiver to produce small vertical oscillations of the spot on the screen so as to make the line structure of the image less noticeable at short viewing distances.

**line drop** The voltage drop existing between two points on a power line or transmission line, due to the impedance of the line.

**line equalizer** An equalizer containing inductance and/or capacitance, inserted in a transmission line to modify the frequency response of the line.

**line filter** 1. A filter inserted between a power line and a receiver, transmitter, or other unit of electric equipment to prevent passage of noise signals through the power line in either direction. Also called power-line filter. 2. A filter inserted in a transmission line or high-voltage power line for carrier communication purposes.

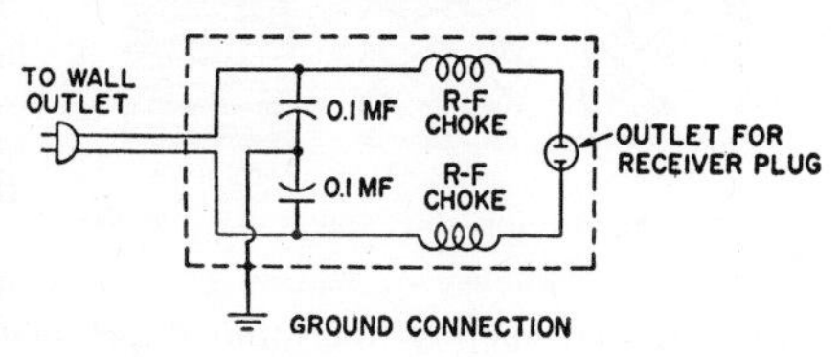


Line-filter circuit for use between radio receiver and power line.

**line flyback** *Horizontal flyback.*

**line-focus tube** An x-ray tube in which the focal spot is roughly a line. This gives an essentially square beam of x-rays at one angle of reflection from the target.

**line frequency** The number of times per second that the scanning spot sweeps across the screen in a horizontal direction in a television system. In the United States, with 525 lines and 30 complete pictures per second, it is 15,750 sweeps per second, including the horizontal scans made during the vertical return intervals. In Great Britain, with 405 lines and 25 pictures per second, it is 10,125 sweeps per second. Also called horizontal frequency and horizontal line frequency.

**line-frequency blanking pulse** *Horizontal blanking pulse.*

**line hydrophone** A directional hydrophone consisting of one straight-line element, an array of suitably phased elements mounted in line, or the acoustic equivalent of such an array.

**line impedance** The impedance measured across the terminals of a transmisson line.

**line interlace** *Interlaced scanning.*

**line microphone** A highly directional microphone consisting of a single straight-line element or an array of small parallel tubes of different lengths, with one end of each abutting a microphone element, so as to give highly directional characteristics. Also called machine-gun microphone.

**line noise** Noise originating in a transmission line, from such causes as poor joints and inductive interference from power lines.

**line of force** An imaginary line in an electric or magnetic field, each segment of which represents the direction of the field at that point.

**line of position** [abbreviated lop] The intersection of two surfaces of position, used to establish a position or fix in air navigation.

**line of sight** The straight, unobstructed path between two points.

**line-of-sight distance** The distance from a transmitter to the horizon, normally representing the range limit of a radio or radar station. Usually under 200 miles. Under certain conditions, atmospheric refraction may extend the range.

**line-of-sight path** The direct, essentially straight path taken by a radio wave from a transmitting antenna to a receiving antenna.

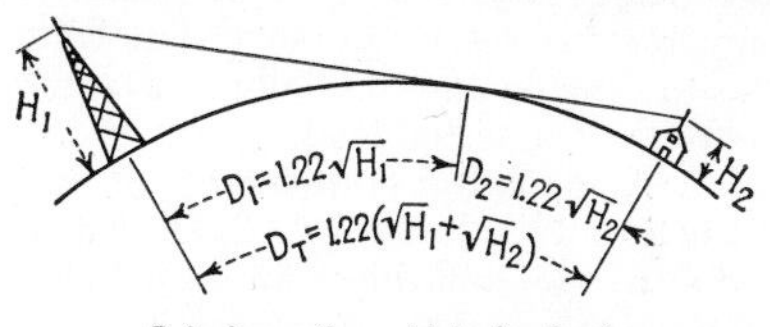


Line-of-sight path, and formula for computing path length D when heights of transmitting and receiving antennas are known.

**line-of-sight stabilization** The stabilization of a radar antenna mounted on a ship or aircraft, to compensate for roll and pitch, by changing the elevation angle of the antenna. For horizontal scanning, the beam would then be aimed at the horizon at all times.

**line pad** A pad inserted between a program amplifier and a transmission line, to isolate the amplifier from impedance variations of the line.

**line printer** A high-speed printer that prints out an entire line of characters simultaneously across a page.

**line-sequential color television** A color television system in which an entire line is one color, with the colors changing from line to line in a red, blue, and green sequence.

**line spectrum** A spectrum whose components have discrete frequencies or other discrete values.

**line-stabilized oscillator** An oscillator in which a section of high-Q transmission line is used as a frequency-controlling element.

**line stretcher** A section of waveguide or rigid coaxial line whose physical length is variable. A telescoping design is commonly used for this purpose.

**line synchronizing pulse** *Horizontal synchronizing pulse.*

**line transformer** A transformer inserted in a system for such purposes as isolation, impedance matching, or additional circuit derivation.

**line trap** A filter consisting of a series inductance shunted by a tuning capacitor, inserted in series with the power or telephone line for a carrier-current system to minimize the effects of variations in line attenuation and reduce carrier energy loss.

**line voltage** The voltage provided by a power line at the point of use. In the United States, it is usually between 110 and 125 volts at outlets in homes, with 117 volts as an average.

**line-voltage regulator** A regulator that counteracts variations in power-line voltage, so as to provide an essentially constant voltage for the connected load.

**link** 1. A radio transmitter-receiver system connecting two locations. 2. A flat strip serving as a removable connector between two terminal screws.

**linkage** *Flux linkage.*

**link coupling** Coupling consisting of two coils connected together by a short length of transmission line, with each coil inductively coupled to the coil of a separate tuned circuit.

**link neutralization** Neutralization by means of link coupling between the output and input tuned circuits.

**lin-log receiver** A radar receiver having a linear amplitude response for small-amplitude signals and a logarithmic response for large-amplitude signals.

**lip microphone** A contact microphone designed for use against the upper lip of a person. An acoustic balancing arrangement cancels sounds originating at a distance. Used where noise level is extremely high, as in military tanks. Sound waves from the person's lips enter through only one aperture and act on the microphone, whereas sound waves from a distance enter through both apertures and act on opposite sides of the diaphragm simultaneously, so that their effects cancel.

**liquid-metal-fuel reactor** A nuclear reactor that uses a solution of uranium in molten bismuth as fuel, circulating through a graphite structure that serves as moderator.

**Lissajous figure** The pattern appearing on an oscilloscope screen when sine waves are applied simultaneously to the horizontal and vertical deflection plates.

**listening station** A radio or radar receiving station that is continuously manned for various purposes, such as for radio direction-finding or for gaining information about enemy electronic devices.

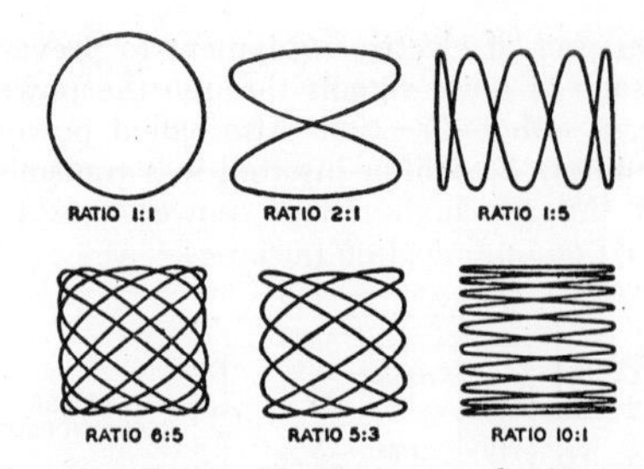


Lissajous figures for sine waves having various frequency ratios.

**lithium** [symbol Li] An alkali metal element having characteristics similar to those of sodium. Sometimes used on the cathodes of phototubes because it gives high response at the extreme violet end of the light spectrum. Atomic number is 3.

**litz wire** Wire consisting of a number of separately insulated strands woven together so each strand successively takes up all possible positions in the cross-section of the entire conductor, to reduce skin effect and thereby reduce r-f resistance. Sometimes used for winding coils.

**live** 1. Broadcast directly at the time of production, instead of from recorded or filmed program material. 2. *Energized.*

**live end** The end of a radio studio that gives almost complete reflection of sound waves.

**live room** A room having a minimum of sound-absorbing material.

**L line** One of the characteristic lines in the x-ray spectrum of an atom. It is produced by excitation of the electrons of the L shell.

**L/M ratio** The ratio of the number of internal conversion electrons from the L shell to the number from the M shell, emitted in the deexcitation of a nucleus.

**ln** Abbreviation for *natural logarithm.*

**L network** A network composed of two branches in series, with the free ends connected to one pair of terminals. The junction point and one free end are connected to another pair of terminals.

**load** 1. The device that receives the useful signal output of an amplifier, oscillator, or other signal source. 2. A device that consumes electric power. 3. The amount of electric power that is drawn from a power line, generator, or other power source. 4. The material to be heated by an induction heater or dielectric heater. Also called work.

**load characteristic** The relation between the instantaneous values of a pair of variables for an electron tube in a specified operating circuit, such as electrode voltage and current, when all direct electrode supply voltages are maintained constant. Also called dynamic characteristic.

**load circuit** The complete circuit required to transfer power from a source to a load, such as the coupling network, leads, and load material connected to the output terminals of an induction heater.

**load coil** A coil that delivers a-c energy by induction to a charge to be heated in induction heating. Also called work coil.

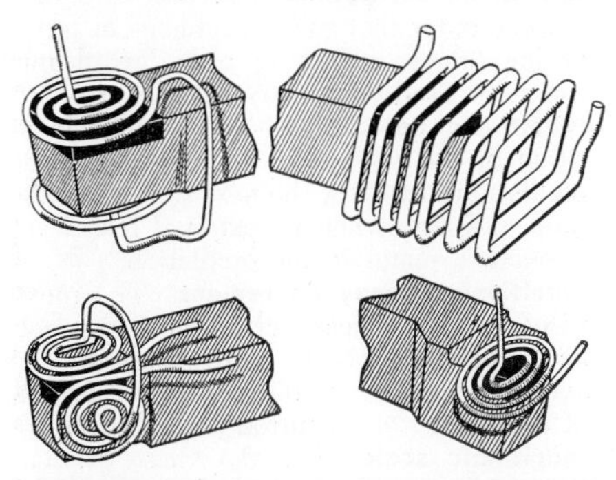

Load coils used to produce localized heat for brazing tungsten carbide tips on cutting tools.

**loaded antenna** An antenna having extra inductance in series, to increase its electrical length.

**loaded impedance** The impedance at the input of a transducer when the output is connected to its normal load.

**loaded motional impedance** *Motional impedance.*

**loaded Q** The $Q$ of a switching tube when modified by the coupled impedances.

**load impedance** The complex impedance presented to a transducer by its load.

**load impedance diagram** A diagram showing how the performance of an oscillator is affected by variations in load impedance.

**loading** The addition of inductance to a transmission line to improve its transmission characteristics throughout a given frequency band.

**loading coil** 1. An iron-core coil connected into a telephone line or cable at regular intervals to lessen the effect of line capacitance and reduce distortion. Also called Pupin coil and telephone loading coil. 2. A coil inserted in series with a radio antenna to increase its electrical length and thereby lower the resonant frequency.

**loading disk** A circular metal piece mounted at the top of a vertical antenna to increase its natural wavelength.

**loading machine** A machine used to introduce fuel into a nuclear reactor.

**load line** A straight line drawn across a series of tube or transistor characteristic curves to show how output signal current will change with input signal voltage when a specified load resistance is used.

**load matching** Matching the load circuit impedance to that of the source so as to give maximum transfer of energy, as desired in induction and dielectric heating.

**load-matching network** A network for load matching in induction and dielectric heating.

**load-matching switch** A switch used in a load-matching network to compensate for a sudden change in load characteristics, such as that which occurs when passing through the Curie point of the load material in induction and dielectric heating.

**load transfer switch** A switch used to connect a generator or power source optionally to either of two load circuits.

**lobe** One of the three-dimensional portions of the radiation pattern of a directional antenna. The direction of maximum radiation coincides with the axis of the major lobe. All other lobes in the pattern are called minor lobes.

**lobe switch** A switch used for systematically shifting the radiation pattern of an antenna.

**lobe switching** A method of determining the exact direction to a target by periodically shifting the beam of a radar antenna slightly to the left and to the right of the dead-ahead position, using electronic or mechanical means. While comparing received signal strengths, the entire antenna is turned until equal signals are received from both lobes. The antenna is then accurately aimed, without using an impracticably narrow beam width. Used also in radio direction finding. Also called beam switching.

**local channel** A standard broadcast channel in which several stations may operate with powers not in excess of 250 watts.

**local control** Control of a transmitter directly at the transmitter, rather than by remote control.

**localizer** A radio facility that provides signals for use in lateral guidance of aircraft with respect to a runway centerline.

**localizer on-course line** A line in a vertical plane passing through a localizer, on either side of which the received indications have opposite sense.

**localizer sector** The sector included between two radial equisignal localizer lines having the same specified difference in depth of modulation.

**local oscillator** The oscillator in a superheterodyne receiver, whose output is mixed with the incoming modulated r-f carrier signal in the mixer to give the frequency conversion needed to produce the i-f signal.

**local oscillator tube** The electron tube used in a heterodyne conversion transducer (such as a local oscillator) to provide the local oscillator signal for the mixer tube.

**location** *Storage register.*

**locator** A radar or other device designed to detect and locate airborne aircraft.

**lock** To fasten onto and automatically follow a target by means of a radar beam.

**locked groove** A blank and continuous groove placed at the end of the modulated grooves on a disk recording to prevent further travel of the pickup. Also called concentric groove.

**locked-in oscillator** A type of discriminator that does not react to amplitude modulation and hence requires no limiter preceding it. The circuit uses a pentagrid tube with three tank circuits, each tuned to the signal resting frequency and so located that average anode current changes only with signal frequency.

**locked-rotor current** The current drawn by a stalled electric motor.

**lock-in base** *Loktal base.*

**locking relay** A relay that locks into position when its coil is energized momentarily. It may have a manual release lever or a second coil that is energized to release the relay. Also called latching relay and latch-in relay.

**lock-in range** The frequency range over which an oscillator may be synchronized by means of a synchronization signal.

**lock-on** The instant at which a radar begins to track its target automatically.

**loctal base** *Loktal base.*

**lodar** A direction finder used to determine the direction of arrival of loran signals, free of night effect, by observing the separately distinguishable ground and sky-wave loran signals on a cathode-ray oscilloscope and positioning a loop antenna to obtain a null indication of the component selected to be most suitable. Also called lorad.

**log** 1. A written record of radio and television station operating data, required by law. 2. Abbreviation for *logarithm.*

**logarithm** [abbreviated log] The power to which a number, called the base, must be raised in order to equal the original number. The common system of logarithms uses 10 as a base; here, the logarithm of 1,000 to the base 10 is 3, because $10^3$ is 1,000. Another commonly used base is 2.71828, designated by the Greek letter $\epsilon$ (epsilon) or by $e$ and known as the hyperbolic, Napierian, or natural logarithm; here the notation $\log_\epsilon N$ is often abbreviated as $\ln N$.

**logarithmic amplifier** An amplifier whose output signal is a logarithmic function of the input signal.

**logarithmic computer** A section of a digital computer that solves problems in terms of logarithmic values or as a logarithmic function. Used in gunnery calculations.

**logarithmic decrement** The natural logarithm of the ratio of the amplitude of one oscillation to that of the next which has the same polarity, when no external forces are applied to maintain the oscillation.

**logarithmic energy decrement** The mean value of the increase in lethargy per collision of particles. It corresponds to the average decrease in the natural logarithm of the energy of a neutron per collision.

**logarithmic scale** A scale whose graduations are spaced logarithmically rather than linearly.

**logger** A recorder that automatically scans measured quantities at specified times and records or logs their values on a chart.

**logic** 1. The basic principles and applications of truth tables, interconnections of on-off circuit elements, and other factors involved in mathematical computation in a computer. 2. *Logical design.*

**logical comparison** The operation of comparing two items in a computer and producing a one output if they are equal or alike, and a zero output if not alike.

**logical design** The preliminary design that concentrates on the logical and mathematical interrelationships required in a computer or data-processing system. It precedes the detailed engineering design work. Also called logic.

**logical diagram** A diagram representing the logical elements of a computer and their interconnections, without necessarily showing construction or engineering details.

**logical element** The smallest building block that can be represented by an operator in an appropriate system of symbolic logic for a computer or data-processing system. Typical logical elements are the and-gate and the flip-flop.

**logical function** A means of expressing a definite state or condition in magnetic amplifier, relay, and computer circuits. Examples include: (a) the "and" function, where an output is produced only when a number of input signals are present and combined; (b) the "or" function, where an output is obtained when any one of a number of input signals is applied; (c) the "not" function, where an output is obtained only when there is no input signal.

**logical operation** A nonarithmetical operation in a computer, such as comparing, selecting, making references, matching, sorting, and merging, where logical yes-or-no quantities are involved.

**logical switch** A diode matrix or other switching arrangement that is capable of directing an input signal to one of several outputs.

**logical symbol** A graphic symbol representing the means of performing some specified simple computer operation, such as coincidence gating.

**loktal base** A tube base having a grooved metal center post designed to lock firmly in a corresponding eight-pin loktal socket. The tube pins are sealed directly into the glass envelope. Also called lock-in base and loctal base.

**loktal tube** An electron tube having a loktal base.

**lone electron** An electron that is alone on an energy level.

**long-distance navigation aid** A navigation aid that is usable at distances beyond about 200 miles, or beyond the radio line-of-sight limit.

**longitudinal current** A current that flows in the same direction in both wires of a pair, and uses the earth or other conductors for a return path.

**longitudinal fuze** An electronic fuze that has maximum sensitivity at right angles to the sides of the missile or other explosive, rather than in the straight-ahead direction.

**longitudinal heating** Dielectric heating in which the electrodes are so positioned that the electric field is parallel to the layers of a laminated material being heated.

**longitudinal magnetization** Magnetization of a magnetic recording medium in a direction essentially parallel to the line of travel.

**longitudinal parity** Parity associated with bits recorded on one track in a data block, to indicate whether the number of recorded bits in the block is even or odd.

**longitudinal wave** A wave in which the direction of displacement at each point of the medium is perpendicular to the wavefront.

**long-line effect** An effect occurring when an oscillator is coupled to a transmission line with a bad mismatch. Two or more frequencies may then be equally suitable for oscillation, and the oscillator jumps from one of these frequencies to another as its load changes.

**long-persistence screen** A fluorescent screen containing phosphorescent compounds that increase the decay time, so a pattern may be seen for several seconds after it is produced by the electron beam.

**long-play record** [abbreviated lp record] A 10-inch or 12-inch phonograph record that operates at a speed of 33⅓ rpm and has closely spaced grooves, to give playing times up to 30 minutes for one 12-inch side. Designed for use with a 1-mil stylus, whose point radius is 0.001 inch. Also called microgroove record.

**long-range alpha particle** An alpha particle that is produced directly from the excited states of nuclei during beta disintegration.

**long-range radar** A radar set having a range of between 300 and 800 miles on a reflecting target 1 meter square and perpendicular to the signal path, provided that line of sight exists between the radar set and the target.

**long-tail pair** A two-tube circuit having a common cathode resistance that gives strong negative feedback.

**long wave** An electromagnetic wave having a wavelength longer than the longest broadcast band wavelength of about 545 meters, corresponding to frequencies below about 550 kc.

**long-wire antenna** An antenna whose length is a number of times greater than its operating wavelength, so as to give a directional radiation pattern.

**look-through** 1. Interruption of a jamming transmission for extremely short periods at irregular intervals to permit monitoring of the victim signal and determine whether jamming is still needed. 2. Observing or monitoring a desired signal that is being jammed by the enemy, during interruptions in the jamming signals.

**loop** 1. A curved conductor that connects the ends of a coaxial line or other transmission line and projects into a resonant cavity for coupling purposes. 2. A closed curve on a graph, such as a hysteresis loop. 3. A closed path or circuit over which a signal can circulate, as in a feedback control system. 4. Repetition of a sequence of operations in a computer routine. 5. *Antinode.* 6. *Loop antenna* 7. *Mesh.*

**loop actuating signal** The signal derived by mixing the loop input signal with the loop feedback signal of a control system.

**loop antenna** An antenna consisting of one or more complete turns of a conductor, usually tuned to resonance by a variable capacitor connected to the terminals of the loop. The radiation pattern is bidirectional, with maximum radiation or pickup in the plane of the loop and minimum radiation at right angles to the loop. Also called loop.

**loop control** *Photoelectric loop control.*

**loop difference signal** A type of loop actuating signal that is produced at a summing point of a feedback control loop, when a particular loop input signal is applied to that summing point.

**loop error** The desired value of the loop output signal minus the actual value in a control system.

**loop error signal** The loop actuating signal in those cases in which it is the loop error.

**loop feedback signal** The signal derived as a function of the loop output signal and fed back to the mixing point for control purposes.

**loop gain** The product of the gain values acting on a signal passing around a closed-loop path. In a feedback control system it is the forward gain multiplied by the feedback network gain. In a repeater, carrier terminal, or complete closed system the loop gain is the maximum gain that can be used without causing oscillation or singing, and the loop gain here may therefore be less than the product of the gain values.

**loop input signal** An external signal applied to a feedback control loop.

**loop output signal** The signal extracted from a feedback control loop.

**loop return signal** A type of loop input signal that is returned through a feedback control loop to a summing point in response to a loop input signal applied to that summing point. It subtracts from the applied loop input signal.

**loopstick antenna** *Ferrite-rod antenna.*

**loop test** A telephone or telegraph line test that is made by connecting a faulty line to good lines in such a way as to form a loop in which measurements can be made to determine the position of the fault.

**loop transfer function** The transfer function of the transmission path formed by opening and properly terminating a feedback loop.

**loop transfer ratio** The transfer ratio of a loop return signal to the corresponding loop difference signal.

**loop-type directional coupler** A directional coupler for coaxial lines, using a loop projecting into the main line in such a way that a wave traveling in one direction transfers a part of its energy to the secondary coaxial line connected to the loop, while rejecting energy traveling in the opposite direction. One end of the secondary line must be terminated in its characteristic impedance.

**loop-type radio range** An A-N radio range using two separate loop antennas fed by a single transmitter.

**loose coupling** Coupling that is considerably less than critical coupling, so there is very little transfer of energy. Also called weak coupling.

**lop** Abbreviation for *line of position.*

**lorac** [LOng-Range-ACcuracy radar system] A long-range navigation system that determines a position fix by the intersection of lines of position. Each line is defined by the phase angle between two beat-frequency waves; one wave is the beat between the continuous-wave signals from two widely spaced transmitters, and the other is the reference wave of the same frequency, obtained by beating the same two continuous-wave signals at a fixed location and transmitting the beat to the navigation receiver over a second r-f channel.

**lorad** *Lodar.*

**loran** [LOng-RAnge Navigation] A long-distance radio navigation system for aircraft and ships, utilizing synchronized pulses transmitted simultaneously by widely spaced transmitting stations. Hyperbolic lines of position are determined by measuring the difference in the time of arrival of these pulses. The intersection of two of these lines of position, obtained from either three or four stations, gives a position fix. Standard loran operates on frequencies between 1,800 and 2,000 kc. Low-frequency loran and sky-wave-synchronized loran are other types.

**loran C** *Low-frequency loran.*

**loran chain** A chain of four or more loran stations, forming three or more pairs of stations for loran navigation.

**loran chart** A chart showing loran lines of position along with a limited amount of topographic detail.

**loran fix** A fix obtained by determining

the intersection of two loran lines of position.

**loran guidance** Missile guidance in which a loran receiver in the missile receives signals continuously from three or more fixed ground stations. The resulting missile position data is fed into a computer that in turn produces the control signals required to guide the missile to its intended fixed ground target.

**loran indicator** An indicator that displays the pulse signals from two loran ground stations simultaneously and also shows the time difference between reception of the two signals.

**loran line** A line of position on a loran chart. Each line is the locus of points whose distances from two fixed stations differ by a constant amount.

**loran set** A receiving set or indicator that displays the pulses from loran transmitting stations.

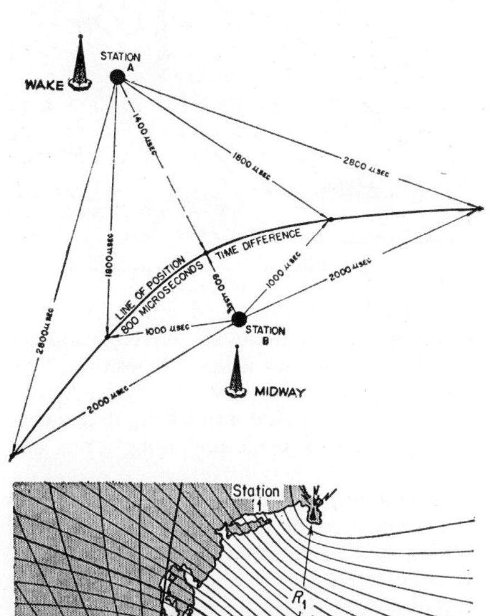


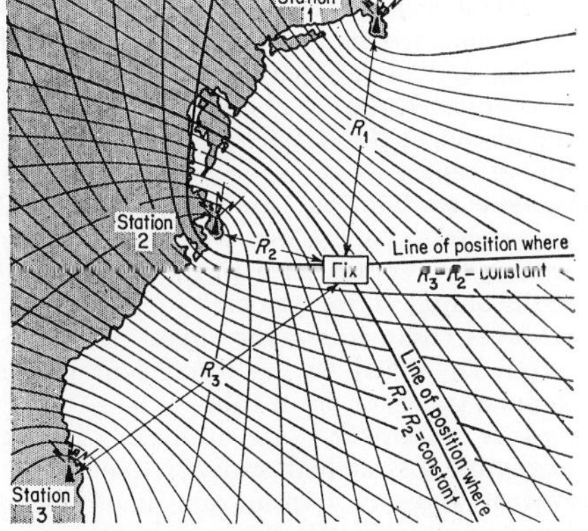


Loran station pair, showing line of position for a time difference of 800 microseconds, and example of fix on loran chart for east coast of United States.

**loran station** A transmitting station in a loran system.

**loran triplet** A combination of three loran stations, with one of the stations forming a pair with each of the others.

**Lorenz instrument landing system** A continuous-wave instrument landing system used in continental Europe and Great Britain, consisting of a runway localizing beacon and two radio marker beacons.

**lorhumb line** A navigation course line in a lattice such that the derivative of one coordinate with respect to the other coordinate constantly equals the ratio of the difference of the coordinates at the beginning and end points of the course line.

**loss** 1. Power that is dissipated in a device or system without doing useful work. 2. *Transmission loss.*

**losser** A material or element that dissipates energy, used intentionally in a circuit to prevent oscillation and for other purposes.

**Lossev effect** Radiation resulting from recombination of charge carriers injected in a p-n or p-i-n junction that is biased in the forward direction.

**loss factor** The power factor of a material multiplied by its dielectric constant. The loss factor varies with frequency and determines the amount of heat generated in a material.

**loss modulation** *Absorption modulation.*

**loss tangent** A measure of the amount of power lost as heat when a dielectric or semiconductor material is subjected to a high-frequency electric or electromagnetic field.

**lossy** Having losses, generally intentional as applied to a dielectric material.

**lossy line** A transmission line having intentionally high attenuation per unit length. High loss is sometimes achieved by using Nichrome for the center conductor.

**loudness** The intensity characteristic of an auditory sensation, in terms of which sound may be described on a scale extending from soft to loud.

**loudness contour** A curve that shows the related values of sound pressure level and frequency required to produce a given loudness sensation for a typical listener.

**loudness control** A combination volume and tone control that boosts bass frequencies when the control is set for low volume, to compensate automatically for the reduced response of the ear to low frequencies at low volume levels. Some loudness controls also provide the same automatic compensation for treble frequencies. Also called compensated volume control.

**loudness level** The sound pressure level, in decibels relative to 0.0002 microbar, of a pure 1,000-cps tone that is judged by listeners to be equivalent in loudness to the sound under consideration. Also called equivalent loudness level. One decibel of change under these conditions is called a phon.

**loudspeaker** [abbreviated spkr] An electroacoustic transducer that converts audio-frequency electric power into acoustic power and radiates the acoustic power effectively at a distance in air. Also called speaker.

**loudspeaker dividing network** *Crossover network.*

**loudspeaker impedance** The impedance rating of the voice coil of a loudspeaker, corresponding to the impedance of the amplifier output terminals to which the loudspeaker should be connected to obtain its rated performance. Common impedance values are 4, 8, and 16 ohms.

**loudspeaker system** A combination of one or more loudspeakers with associated baffles, horns, and dividing networks, arranged to work together as a coupling means between the driving electric circuit and the air or between the circuit and another acoustic medium.

**loudspeaker voice coil** *Voice coil.*

**louver** 1. An arrangement of concentric or parallel slats or equivalent grille members used to conceal and protect a loudspeaker while allowing sound waves to pass. Often molded integrally with a plastic radio cabinet. 2. An arrangement of fixed or adjustable slot-like openings provided in a cabinet for ventilation.

**low-altitude radio altimeter** *Frequency-modulated radio altimeter.*

**low band** The band that includes television channels 2 to 6, extending from 54 to 88 mc.

**low-definition television** Television involving less than about 200 scanning lines per complete image.

**lower sideband** The sideband containing all frequencies below the carrier-frequency value that are produced by an amplitude-modulation process.

**lowest useful high frequency** The lowest high frequency that is effective at a specified time for ionospheric propagation of radio waves between two specified points. The frequency value is determined by such factors as absorption, transmitter power, antenna gain, receiver characteristics, type of service, and noise conditions.

**low-flux reactor** A nuclear reactor having a relatively low neutron flux.

**low frequency** [abbreviated l-f] A Federal Communications Commission designation for the band from 30 to 300 kc in the radio spectrum.

**low-frequency compensation** Compensation that serves to extend the frequency range of a broadband amplifier to lower frequencies.

**low-frequency induction heater** An induction heater in which current flow at the commercial power-line frequency is induced in the charge to be heated.

**low-frequency loran** [abbreviated l-f loran] A modification of standard loran, operating in the low-frequency range of approximately 100 to 200 kc to increase range over land and during daytime. Whereas ordinary loran matches the envelopes of the r-f pulses to obtain a line of position, low-frequency loran matches the cycles within the pulses to provide a much more accurate fix. Also called cycle-matching loran and loran C.

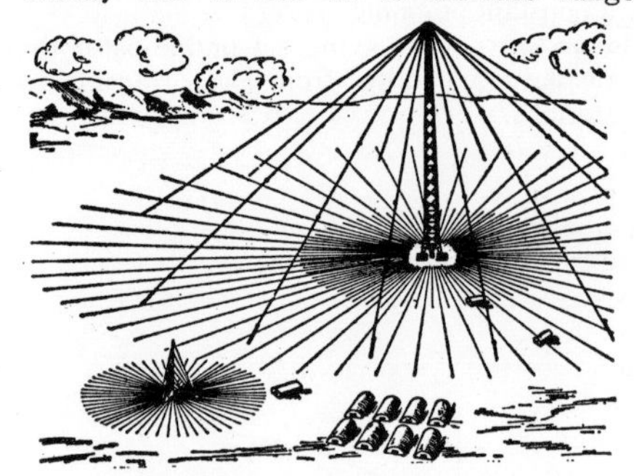
Low-frequency loran station.

**low-frequency padder** A trimmer capacitor connected in series with the oscillator tuning coil of a superheterodyne receiver. It is adjusted during alignment to calibrate the circuit at the low-frequency end of the tuning range.

**low-level modulation** Modulation produced at a point in a system where the power level is low compared with the power level at the output of the system.

**low-level radio-frequency signal** A radio-frequency signal having insufficient power to fire a tr, atr, or pre-tr tube.

**low-loss insulator** An insulator having negligible loss at high radio frequencies.

**low-loss line** A transmission line having low power dissipation per unit length.

**low-pass filter** A filter that transmits alternating currents below a given cutoff frequency and substantially attenuates all other currents.

**low-pressure cloud chamber** A cloud chamber in which the gas is maintained at low pressure to increase the range or decrease the scattering of particle tracks.

**low-pressure laminate** A laminated plastic that is produced and cured at pressures well below 1,000 pounds per square inch.

**low-temperature reactor** A nuclear reactor designed to operate at a relatively low temperature.

**low tension** British term for low voltage, generally as applied to filament and heater voltages of tubes.

**low vacuum** A degree of vacuum at which so much gas or vapor is still present that ionization can occur in an electron tube.

**low-velocity scanning** The scanning of a target with electrons of velocity less than the minimum velocity needed to give a secondary-emission ratio of unity. Used in the emitron, image orthicon, and vidicon types of camera tubes.

**L pad** A volume control having essentially the same impedance at all settings. It consists essentially of an L network in which both elements are adjusted simultaneously.

**lp record** Abbreviation for *long-play record.*

**L scan** *L display.*

**L scope** A radarscope that produces an L display.

**L shell** The second layer of electrons surrounding the nucleus of an atom, having electrons whose principal quantum number is 2.

**lubber line** A line placed on a compass or cathode-ray indicator parallel to the longitudinal axis of a ship or aircraft, to serve as a reference for determining the heading.

**lucero** A British interrogator-responsor used both for iff and rebecca-eureka systems.

**Lucite** Trademark of Du Pont for their transparent acrylic resins.

**lug** A stamped metal strip used as a terminal to which wires can be soldered. Also called soldering lug.

**lumen** The unit of luminous flux. It is equal to the flux on a unit surface all points of which are at unit distance from a uniform point source of 1 candle.

**lumerg** [LUMen-ERG] A unit of luminous flux, equal to 1 erg of radiant energy emitted by a source having a luminous efficiency of 1 lumen per watt.

**luminance** The luminous intensity of any surface in a given direction per unit of projected area of the surface, as viewed from that direction. Units of luminance are candles per square foot and candles per square meter. Formerly called brightness.

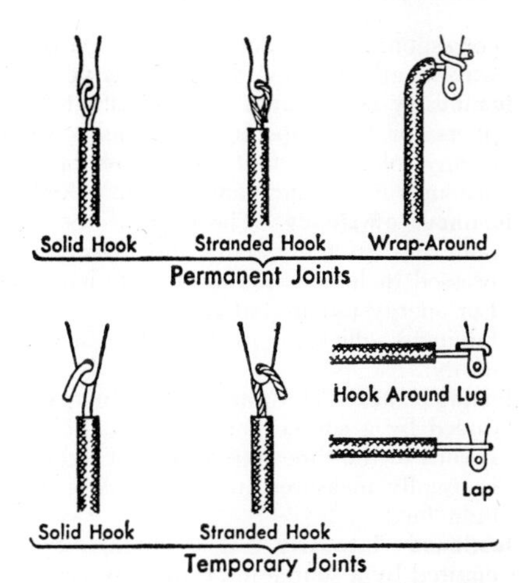


Lug connections for permanent and temporary soldered joints.

**luminance channel** A path intended primarily for the luminance signal in a color television system.

**luminance flicker** Flicker that results from fluctuation of luminance only.

**luminance primary** One of the three transmission primaries whose amount determines the luminance of a color in a color television system.

**luminance signal** The color television signal that is intended to have exclusive control of the luminance of the picture. It is made up of 0.30 red, 0.59 green, and 0.11 blue and is capable of producing a complete monochrome picture. Also called Y signal.

**luminescence** Emission of light by a material at lower than incandescent temperatures, as a result of chemical action, electrical action, physiological processes, exposure to certain types of radiation, or other nonthermal processes.

**luminescence threshold** The lowest frequency of radiation that is capable of exciting a luminescent material. Also called threshold of luminescence.

**luminescent** Capable of exhibiting luminescence.

**luminescent screen** The screen in a cathode-ray tube, which becomes luminous when bombarded by an electron beam and maintains its luminosity for an appreciable time.

**luminosity** The ratio of luminous flux to the

corresponding radiant flux at a particular wavelength. Expressed in lumens per watt.

**luminosity coefficients** The constant multipliers for the respective tristimulus values of any color, such that the sum of the three products is the luminance of the color.

**luminous efficiency** The ratio of the luminous flux to the radiant flux. Usually expressed in lumens per watt of radiant flux. For energy radiated at a single wavelength, luminous efficiency is the same as luminosity.

**luminous flux** The total visible energy produced by a source per unit time. It corresponds to the time rate of flow of light, and is usually measured in lumens. Also called light flux.

**luminous intensity** The luminous flux emitted by a source in an infinitesimal solid angle, divided by the solid angle.

**luminous sensitivity** The output current of a phototube or camera tube divided by the incident luminous flux.

**lumped constant** A single constant that is electrically equivalent to the total of that type of distributed constant existing in a coil or circuit.

**lumped impedance** An impedance concentrated in a single component rather than distributed throughout the length of a transmission line.

**lunar probe** A space rocket or other vehicle intended for operation near or on the moon after being fired from the earth, for telemetering data on lunar phenomena back to earth by radio.

**lunar satellite** A lunar probe designed to orbit around the moon after being fired from the earth.

**Luneberg lens** An artificial type of lens used for focusing radiated electromagnetic energy at ultrahigh frequencies, to increase the gain of an antenna.

**lutecium** [symbol Lu] A rare-earth element. Atomic number is 71.

**lux** A unit of illumination in the metric system, equal to one lumen per square meter. Also called meter-candle.

**Luxemburg effect** Cross-modulation between two radio signals during their passage through the ionosphere, due to the nonlinearity of the propagation characteristics of free charges in space. Because of this effect, the program of a powerful station is sometimes heard when a receiver is tuned to a weaker station on a different frequency.

# M

**M** Symbol for *mutual inductance.*

**ma** Abbreviation for *milliampere.*

**Mace** An Air Force surface-to-surface guided missile developed from Matador but having a self-contained navigation system. Range is about 600 miles.

**machine check** An automatic check in a computer, or a programed check of machine functions.

**machine-gun microphone** *Line microphone.*

**machine language** A language involving a form and code that can be handled by a computer, such as code-punched paper tape.

**machine translation** *Mechanical translation.*

**machine word** The standard number of characters that a computer regularly handles in each transfer. As an example, a machine may regularly handle numbers or instructions in units of 36 binary digits.

**mach number** [pronounced mock] The ratio of flight speed to the speed of sound in the medium in which the object moves. At sea level, mach 1 is approximately 741 miles per hour at 32°F in dry air, whereas at 30,000 feet altitude it is about 675 miles per hour.

**macroscopic cross section** The cross section per unit volume.

**macroscopic property** A nuclear reactor property that can be treated independently of other factors.

**mad** Abbreviation for *magnetic airborne detector.*

**madt** Abbreviation for *microalloy diffused transistor.*

**magamp** Abbreviation for *magnetic amplifier.*

**magic number** The atomic number or neutron number of a nuclide that has greater-than-average stability and sometimes other exceptional properties. Some magic numbers are 2, 8, 20, 28, 50, 82, and 126.

**magic tee** *Hybrid tee.*

**magnal base** A base having 11 pins, used for cathode-ray tubes.

**magnesium** [symbol Mg] An alkaline metallic element. Its compounds are sometimes used for cathodes of phototubes when maximum response is desired in the ultraviolet region. Atomic number is 12.

**magnesium cell** A primary cell in which the negative electrode is made of magnesium or one of its alloys.

**magnesium-copper sulfide rectifier** A metallic rectifier consisting of magnesium in contact with copper sulfide.

**Magnesyn** Trademark of Eclipse-Pioneer Division, Bendix Aviation Corp., for their line of synchros.

**magnet** An object that produces a magnetic field outside of itself. It has the property of attracting other magnetic objects, such as iron, and attracting or repelling other magnets. A permanent magnet produces a permanent magnetic field, whereas an electromagnet possesses magnetic properties only when current is flowing through its windings.

**magnet charger** A charger that restores or establishes the field strength of a permanent magnet by applying a strong magnetic field produced by a surge of direct current through a large electromagnet.

**magnetic airborne detector** [abbreviated mad] An airborne magnetometer used to locate submerged submarines at ranges up to about 400 feet. A direct current passes through a coil wound on a high-permeability core and balances out the effect of the earth's magnetic field on the core, while an alternating current saturates the core an equal amount on both positive and negative swings. The magnetic field of a submarine makes the swings unequal, thereby producing an output signal.

**magnetic amplifier** [abbreviated magamp] An amplifier that uses the nonlinear properties of saturable reactors, alone or in combination with other circuit elements, to provide amplification. The d-c or a-c input signal is applied to a control winding to change the degree of saturation of the core, thereby producing a larger change in alternating current in the output winding. Also called transductor.

**magnetic-armature loudspeaker** A loudspeaker in which the diaphragm is driven by a ferromagnetic armature that is alternately attracted and repelled by interaction between the field of a permanent magnet and the field produced in the armature by a coil that carries audio-frequency currents. Also called electromagnetic loudspeaker, magnetic loudspeaker, and moving-armature loudspeaker.

**magnetic bearing** The angle in the horizontal plane between the direction of magnetic north and the line along which an aircraft or vessel is pointing. Usually measured clockwise from magnetic north.

**magnetic biasing** Biasing of a magnetic recording medium during recording by superposing an additional magnetic field on that of the signal being recorded. Used to obtain a substantially linear relationship between the amplitude of the signal and the remanent flux density in the recording medium.

**magnetic biasing coil** A winding used on a saturable reactor to establish a basic magnetization of the core in both polarity and magnitude.

**magnetic bottle** A magnetic field used to confine a stream of plasma to minimum volume to produce the pinch effect in an electron tube.

**magnetic brake** A friction brake controlled by electromagnetic means.

**magnetic cartridge** *Variable-reluctance pickup.*

**magnetic cell** One unit of a magnetic memory, capable of storing one bit of ininformation as a zero state or a one state.

**magnetic circuit** A complete closed path for magnetic lines of force, having a reluctance that limits the amount of magnetic flux which can be sent through the circuit by the magnetomotive force.

**magnetic clutch** A clutch in which motion is transmitted from one rotating shaft to another by means of the attraction between magnetized poles.

**magnetic contactor** A contactor actuated by electromagnetic means.

**magnetic core** A quantity of ferrous material placed in a coil or transformer to provide a better path than air for magnetic flux, thereby increasing the inductance of the coil and increasing the coupling between the windings of a transformer. Also called core.

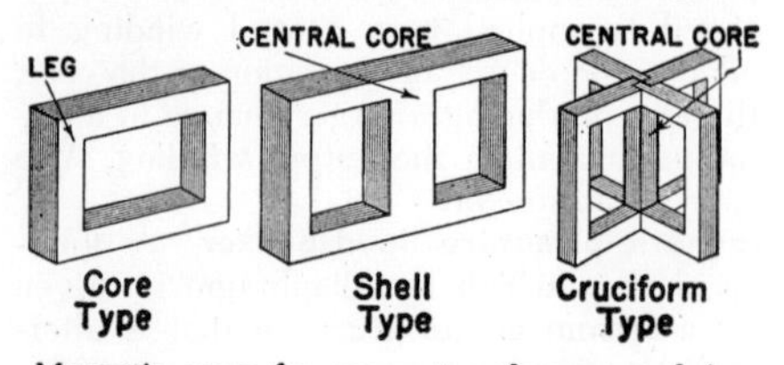


Magnetic cores for power transformers and iron-core chokes.

**magnetic-core storage** A computer storage system in which thousands of tiny doughnut-shaped magnetic cores each store one bit of information. Current pulses are sent through wires threading through the cores to record or read out data. Also called core memory.

**magnetic course** A course in which the direction of the reference line is magnetic north. Also called corrected compass course.

**magnetic cross-field reactor** A hollow toroidal iron core having an enclosed annular winding in the cavity and a conventional toroidal winding around the outside of the core. With an a-c signal flowing through the annular winding, the a-c output voltage developed in the toroidal winding will be proportional to the d-c bias voltage applied to that winding and will be twice the input frequency value.

**magnetic cutter** A cutter in which the mechanical displacements of the recording stylus are produced by the action of magnetic fields.

**magnetic damping** Damping of a mechanical motion by means of the reaction between a magnetic field and the current generated by the motion of a coil through the magnetic field.

**magnetic declination** The angle between true north (geographical) and magnetic north (the direction of the compass needle). This angle is different for different places and may vary from year to year.

**magnetic deflection** Deflection of an electron beam by the action of a magnetic field, as in a television picture tube.

**magnetic delay line** A delay line used for storage of data in a computer, consisting essentially of a metallic medium along which the velocity of propagation of magnetic energy is small compared to the speed of light. Storage is accomplished by recirculation of wave patterns containing information, usually in binary form.

**magnetic dip** The angle that the magnetic field of the earth makes with the horizontal at a particular location. Also called magnetic inclination.

**magnetic dipole** An elementary dipole associated with nuclear particles. It consists of two equal magnetic poles having opposite polarity, closely spaced together and so small that directive properties are independent of size and shape. Also called magnetic doublet.

**magnetic discriminator** A magnetic amplifier used with a transformer and other components to sense the polarity and magnitude of coded pulses for the purpose of producing output control voltages such as might be required for making rudder corrections during the flight of a guided missile.

**magnetic doublet** *Magnetic dipole.*

**magnetic drum** A rapidly rotating cylinder coated with magnetic material in which information is stored in the form of magnetized dipoles. The orientation or polarity of the dipoles is used to store binary information.

**magnetic field** Any space or region in which a magnetic force is exerted on moving electric charges. The magnetic field may be produced by a current-carrying coil or conductor, by a permanent magnet, or by the earth itself.

**magnetic field strength** [symbol $H$] The magnitude of the magnetic field vector. Usually expressed in oersteds or ampere-turns per meter. Also called magnetic force, magnetic intensity, and magnetizing force.

**magnetic flaw detector** A flaw detector in which a ferrous object is magnetized with an electromagnet or permanent magnet and sprayed with an ink containing fine iron particles. Surface or near-surface flaws then appear as black lines. These are even more visible if the surface being examined is first painted white.

**magnetic flip-flop** A flip-flop circuit in which one or more magnetic amplifiers are arranged to have two discrete levels of output. The circuit is changed from one output to the other by appropriately changing the control voltage or current.

**magnetic fluid clutch** A friction clutch that is engaged by magnetizing a liquid suspension of powdered iron located between pole-pieces mounted on the input and output shafts.

**magnetic flux** The magnetic lines of force produced by a magnet.

**magnetic flux density** *Magnetic induction.*

**magnetic focusing** Focusing an electron beam by using the action of a magnetic field.

**magnetic force** *Magnetic field strength.*

**magnetic friction clutch** A friction clutch in which the pressure between the friction surfaces is produced by magnetic attraction.

**magnetic gate** A gate circuit that uses a magnetic amplifier.

**magnetic hardness comparator** An instrument for comparing the hardness of steel parts. A sample part having the desired hardness is placed in one coil, and the parts to be tested are inserted one by one in a similar coil. If the two coils then have the same magnetic properties as displayed on a cathode-ray oscilloscope screen, the parts in the two coils have the same hardness.

**magnetic head** The electromagnet used for reading, recording, or erasing signals on a magnetic disk, drum, or tape. Also called head.

**magnetic heading** A heading in which the direction of the reference line is magnetic north. Also called corrected compass heading.

**magnetic hysteresis** Internal friction that occurs between the molecules of a magnetic material when subjected to a varying magnetic field. It results in heat loss, and makes the magnetic induction dependent on the previous state of magnetization of the material.

**magnetic inclination** *Magnetic dip.*

**magnetic induction** [symbol $B$] The number of magnetic lines of force per unit area at right angles to the lines. Generally expressed in gauss. Also called flux density and magnetic flux density.

**magnetic intensity** *Magnetic field strength.*

**magnetic leakage** Passage of magnetic flux outside the path along which it can do useful work.

**magnetic lens** A lens that uses an arrangement of electromagnets or permanent magnets to produce magnetic fields that serve to focus a beam of charged particles.

**magnetic line of force** An imaginary line, each segment of which represents the direction of the magnetic flux at that point.

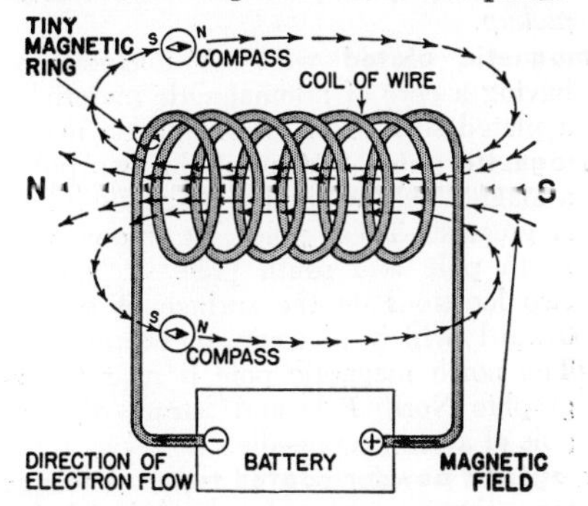


Magnetic lines of force around current-carrying coil make up magnetic field of coil.

**magnetic loudspeaker** *Magnetic-armature loudspeaker.*

**magnetic material** A material that shows magnetic properties. Ferromagnetic materials are strongly magnetic, while paramagnetic materials are feebly magnetic.

**magnetic memory** A computer memory that stores information in the form of varying degrees of magnetization of a magnetic material. The material may be in the form of magnetic cores, disks, drums, plates, or tapes. Also called magnetic storage.

**magnetic-memory plate** A magnetic memory consisting of a ferrite plate having a grid of small holes through which the read-in and read-out wires are threaded. Printed wiring may be applied directly to the plate in place of conventionally threaded wires, permitting mass production of plates having a high storage capacity.

**magnetic microphone** *Variable-reluctance microphone.*

**magnetic mine** An underwater mine that is detonated when the hull of a passing vessel changes the magnetic field at the mine.

**magnetic modulator** A modulator in which a magnetic amplifier serves as the modulating element for impressing an intelligence signal on a carrier.

**magnetic moment** The moment of a magnetic dipole. A magnetic moment is associated with the intrinsic spin of a particle and with the orbital motion of a particle in a system.

**magnetic north** The direction indicated by the north-seeking element of a magnetic compass when influenced only by the earth's magnetic field. Because magnetic meridians often follow zigzag lines, the compass needle at any given place does not necessarily point to the magnetic pole.

**magnetic pickup** *Variable-reluctance pickup.*

**magnetic plated wire** A magnetic wire having a core of nonmagnetic material and a plated surface of ferromagnetic material.

**magnetic pole** 1. One of the two poles of a magnet near which the magnetic intensity is greatest. These poles are known as the north pole and south pole. 2. Either of two locations on the surface of the earth toward which a compass needle points. The north magnetic pole is near the geographic North Pole and attracts the south pole of a compass needle.

**magnetic powder-coated tape** A magnetic tape that consists of a powdered ferromagnetic coating uniformly dispersed on a nonmagnetic base. Also called coated tape.

**magnetic powder-impregnated tape** A magnetic tape that consists of magnetic particles uniformly dispersed in a nonmagnetic material. Also called impregnated tape.

**magnetic printing** The permanent and usually undesired transfer of a recorded signal from one section of a magnetic recording medium to another when these sections are brought together, as on a reel of tape. Also called crosstalk and magnetic transfer.

**magnetic reading head** A magnetic head used to transform magnetic variations in magnetic tape or wire into corresponding voltage or current variations.

**magnetic recorder** A recorder that records audio-frequency signals on magnetic tape or wire as magnetic variations in the medium. It usually also contains provisions

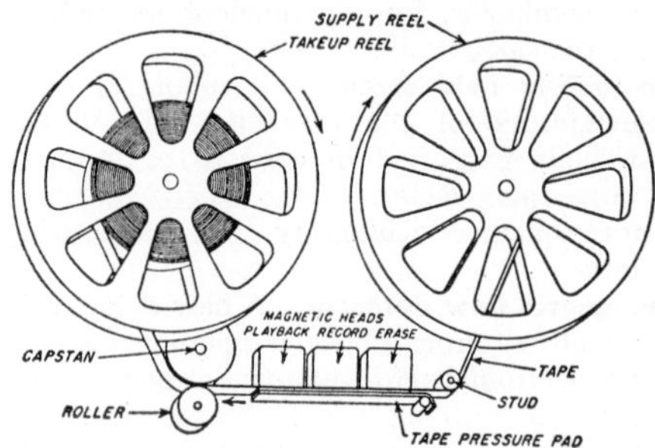

Magnetic recorder construction.

for playback, to convert the recorded magnetic variations back into electric variations, with or without amplification and reproduction as sound waves by a loudspeaker.

**magnetic recording** Recording by means of a signal-controlled magnetic field.

**magnetic recording head** A magnetic head used to transform electric variations into magnetic variations for storage on magnetic media.

**magnetic recording medium** A magnetizable material used in a magnetic recorder to retain the magnetic variations imparted during the recording process. It may have the form of a wire, tape, cylinder, card, or disk.

**magnetic reproducing head** A magnetic head used to convert magnetic variations on magnetic media into electric variations.

**magnetic resonance spectrum** A spectrum produced by varying the r-f electromagnetic field that is superimposed on a steady or slowly varying magnetic field about which the atoms of a material precess, to make molecules change their magnetic quantum numbers as they absorb or emit quanta of radio waves.

**magnetic rigidity** A measure of the momentum of a particle, equal to the product of the magnetic intensity perpendicular to the path of the particle and the resultant radius of curvature of the path of the particle.

**magnetics** The branch of science that deals with magnetic phenomena.

**magnetic saturation** The maximum possible magnetization of a magnetic substance. Also called saturation.

**magnetic sensitivity** The relationship between the current passing through the deflection coils of a cathode-ray tube and the physical distance that the electron beam is displaced at the screen.

**magnetic separator** An apparatus for separating powdered magnetic ores from nonmagnetic ores. An electromagnet deflects magnetic materials from the path taken by nonmagnetic materials.

**magnetic shield** An enclosure made from high-permeability magnetic material, used to protect instruments and electronic assemblies from the effects of stray magnetic fields.

**magnetic shift register** A shift register in which the pattern of settings of a row of magnetic cores is shifted one step along the row by each new input pulse. Diodes in the coupling loops between cores prevent backward flow of information.

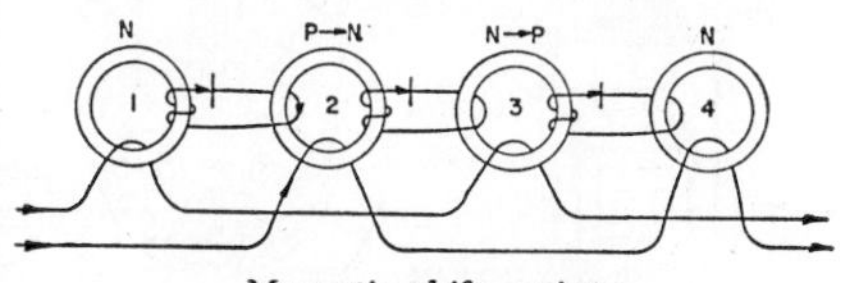

Magnetic shift register.

**magnetic shunt** A piece of magnetic material used to divert an adjustable amount of magnetic flux around an air gap, usually for calibration purposes.

**magnetic spectrograph** A spectrograph based on the action of a constant magnetic field on the paths of electrons or other charged particles. Used to separate particles having different velocities.

**magnetic storage** *Magnetic memory.*

**magnetic storm** A storm that causes rapid and erratic changes in the strength of the magnetic fields of the earth, affecting both radio and wire communications. It is believed to be due to sun-spot activity.

**magnetic tape** A plastic, paper, or metal tape that is coated or impregnated with magnetizable iron oxide particles. Also called tape.

**magnetic-tape core** A toroidal core formed by winding a strip of thin magnetic-core material around a form. Coils may be wound around the core with a toroidal winder, or the completed core may be sawed in half to permit insertion of prewound coils.

**magnetic-tape reader** A computer device that is capable of reading information recorded on magnetic tape and delivering the corresponding electric pulses.

**magnetic-tape recorder** A magnetic recorder that uses magnetic tape as the recording medium.

**magnetic test coil** *Exploring coil.*

**magnetic tracking** A method of coating the sound-track area of finished motion-picture film with iron oxide to form a magnetic track on which the sound accompaniment for the film can be recorded.

**magnetic transfer** *Magnetic printing.*

**magnetic-vane meter** An a-c meter in which the moving element is a metal vane pivoted inside a coil.

**magnetic wire** A wire made from magnetic material suitable for magnetic recording.

**magnetism** A property possessed by iron, steel, and certain other magnetic materials, wherein these materials can produce or conduct magnetic lines of force capable of interacting with electric fields or other magnetic fields.

**magnetization** 1. The degree to which a particular object is magnetized. 2. The process of magnetizing a magnetic material.

**magnetization curve** *B-H curve.*

**magnetizing current** The current that flows through the primary winding of a power transformer when no loads are connected to the secondary winding. This current establishes the magnetic field in the core and furnishes energy for the no-load power losses in the core. Also called exciting current.

**magnetizing force** *Magnetic field strength.*

**magnet keeper** *Keeper.*

**magneto** An a-c generator that uses one or more permanent magnets to produce its magnetic field. Also called magnetoelectric generator.

**magnetoelectric generator** *Magneto.*

**magnetohydrodynamics** The study of the effects of magnetic fields on ionized gases.

**magnetohydrodynamic system** A propulsion system based on the reaction of a weakly ionized arc-produced plasma or on the reaction of highly conductive nuclear plasma produced by controlled nuclear reactions.

**magneto-ionic wave component** Either of the two elliptically polarized wave components into which a linearly polarized wave incident on the ionosphere is separated because of the earth's magnetic field.

**magnetometer** An instrument for measuring the magnitude and sometimes also the direction of a magnetic field, such as the earth's magnetic field.

**magnetomotive force** The force that produces a magnetic field. It is the total magnetizing force acting around a complete closed magnetic circuit, and corresponds to voltage (electromotive force) in an electric circuit. If due to current in a coil, magnetomotive force is proportional to ampere-turns. The cgs unit of magnetomotive force is the gilbert, equal to about 0.8 ampere-turn.

**magneton** *Bohr magneton.*

**magnetoresistance** The change in resistance associated with a change in magnetization in some materials.

**magnetostriction** The change in the dimensions of a ferromagnetic object when placed in a magnetic field. Also called Joule effect.

**magnetostriction hydrophone** A magnetostriction microphone that responds to waterborne sound waves.

**magnetostriction loudspeaker** A loudspeaker in which the mechanical displacement is derived from the deformation of a material having magnetostrictive properties.

**magnetostriction microphone** A microphone that depends for its operation on the generation of an electromotive force by the deformation of a material having magnetostrictive properties. Used chiefly in ultrasonic and underwater sound applications.

**magnetostriction oscillator** An oscillator in which the anode circuit is inductively coupled to the grid circuit through a magnetostrictive element. The frequency of oscillation is determined by the magnetomechanical characteristics of the coupling element.

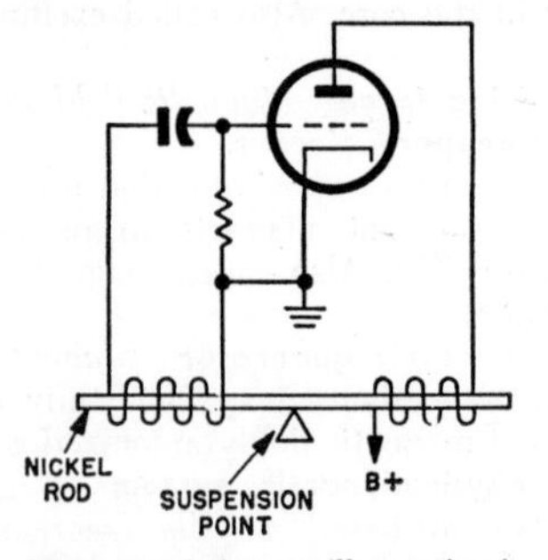


Magnetostriction oscillator circuit.

**magnetostriction transducer** A transducer used with sonar equipment to change an alternating current to sound energy at the same frequency and form the sound energy into a beam. A large number of tubes and coils are usually connected in a series-parallel arrangement. One end of each tube is attached to a diaphragm that is in contact with the sea water. The transducer acts also as a microphone for returning echoes.

**magnetostrictive** Changing in dimensions when placed in a magnetic field.

**magnetostrictive resonator** A ferromagnetic rod that can be made to vibrate at one or more definite resonant frequencies by applying an alternating magnetic field.

**magnetron** A two-electrode electron tube in which the flow of electrons to the anode is controlled by a combination of crossed steady electric and magnetic fields in such a way as to produce a-c power output.

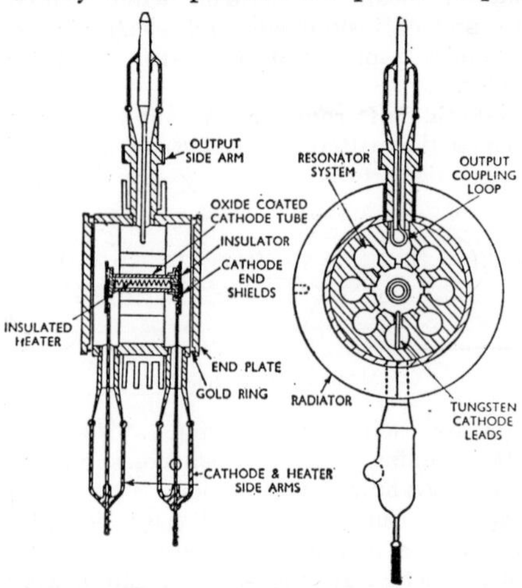


Magnetron construction, shown in side and top views. Coupling loop takes output microwave power from one cavity of hole-and-slot anode.

Used as an oscillator in microwave radio and radar transmitters. Basic types include interdigital, multicavity, multisegment, packaged, rising sun, split-anode, and traveling-wave magnetrons.

**magnetron amplifier** A magnetron used as an amplifier.

**magnetron arcing** Internal breakdown between the cathode and anode of a magnetron, usually due to a sudden release of gas.

**magnetron beam-switching tube** *Beam-switching tube.*

**magnetron oscillator** An oscillator circuit employing a magnetron.

**magnetron package** A complete magnetron assembly consisting of a magnetron, its permanent magnet, and its output-matching device.

**magnetron rectifier** A cold-cathode gas diode rectifier in which the electron stream is controlled by an external magnetic field.

**magnetron strapping** Connecting together alternate segments of a multiple-cavity resonator in a magnetron by means of copper straps, to improve efficiency and prevent frequency jumping.

**magnet steel** Steel having high retentivity, usually containing some combination of tungsten, cobalt, chromium, and manganese with steel. Used in permanent magnets.

**magnettor** A second-harmonic type of magnetic modulator, in which the second-harmonic output varies linearly with the d-c bias voltage over a wide range when the a-c input current to a winding on the saturable core is below the core saturation value.

**magnet wire** Insulated copper wire in any of the sizes commonly used for winding the coils of transformers, relays, and other electromagnetic devices.

**Magnistor** Trademark of Potter Instrument Co. for their line of toroidal saturable reactors, designed for such circuit applications as gating, switching, and counting at high frequencies.

**mag-slip** British term for *synchro.*

**main anode** An anode that conducts load current in a pool-cathode tube.

**main bang** The transmitted pulse in a radar system.

**main gap** The conduction path between a principal cathode and a principal anode of a glow-discharge tube.

**main quantum number** A positive integer that specifies the size of an electron orbit. Also called first quantum number.

**majorana particle** A neutrino that corresponds to an antineutrino. Double beta decay of such a particle gives emission and absorption of a neutrino.

**major apex face** One of the three large sloping faces extending to the apex or pointed end of a natural quartz crystal. The other three smaller sloping faces are the minor apex faces.

**major cycle** The time interval between successive appearances of a given storage position in a serial-access computer storage device.

**majority carrier** The type of carrier constituting more than half of the total number of carriers in a semiconductor.

**majority-carrier contact** The electric contact across which the ratio of majority-carrier current to applied voltage is substantially independent of the polarity of the voltage applied to a semiconductor. The ratio of minority-carrier current to applied voltage is not independent of the polarity of the voltage.

**majority emitter** An electrode from which a flow of majority carriers enters the interelectrode region of a transistor.

**major lobe** The radiation lobe containing the direction of maximum radiation.

**make-before-break contacts** Double-throw contacts so arranged that the moving contact establishes a new circuit before interrupting the old one.

**make contact** A normally open stationary contact on a relay. Its circuit is closed when the relay is energized.

**malfunction** A failure in the operation of a computer component. Also called error.

**manganese** [symbol Mn] A metallic element. Atomic number is 25.

**manganin** An alloy containing 84% copper, 12% manganese, and 4% nickel, used in making precision wirewound resistors because of its low temperature coefficient of resistance.

**Manhattan project** A project of the War Department, lasting from August 1942 to August 1946, in which the atomic bomb was developed.

**manipulated variable** The quantity or condition that is varied by the controller so as to change the value of the controlled variable in a feedback control ssytem.

**manipulator** A mechanical arm that is used to handle radioactive materials from a safe distance.

**manmade static** High-frequency noise signals created by sparking in an electric circuit. When picked up by radio receivers, manmade static causes buzzing and crashing sounds.

**manual direction finder** A rotatable-loop radio compass that is operated manually.

**manual telephone system** A telephone system in which telephone connections between customers are ordinarily established manually by telephone operators in accordance with orders given verbally by calling parties.

**manual tuning** Tuning in which a control

knob is rotated by hand to tune in a desired station on a radio or television receiver.

**many-one function switch** A function switch in which a combination of several inputs must be excited at one time to produce a corresponding single output.

**Marconi antenna** An antenna that is connected to ground at one end through the receiver or transmitter input coil and suitable tuning reactances.

**marginal checking** A preventive-maintenance checking procedure in which certain operating conditions, such as supply voltage or frequency, are varied about their normal values in order to detect and locate incipient defective units.

**mark** 1. The closed-circuit condition in telegraphic communication, during which the signal actuates the printer. It is the opposite of space. 2. A designation followed by a serial number, used to identify models of military equipment.

**marker** 1. A radio facility used in an instrument landing system to provide a signal that designates a small area immediately above the marker location. Also called marker beacon and radio marker beacon. 2. An electronic range or bearing indication on a radar indicator.

**marker beacon** *Marker.*

**marker generator** 1. An r-f generator used to inject one or more frequency-identifying pips on the pattern produced by a sweep generator on a cathode-ray oscilloscope screen. Used for adjusting response curves of tuned circuits, as when aligning f-m and television receivers. 2. An r-f generator used to generate pulses having precise amplitude, shape, duration, and recurrence characteristics, as required for producing reference indices on a radarscope for such quantities as target range, azimuth, and elevation.

**marker pip** An identifying mark on a cathode-ray display.

**marking wave** The emission that takes place in telegraphic communication while the active or mark portions of the code characters are being transmitted. Also called keying wave.

**mark-space ratio** The ratio of the duration of a single pulse to the interval between two successive recurrent pulses.

**mars** [Military Affiliated Radio System] A world-wide network of radio stations operated by amateurs both on and off military installations, but sponsored jointly by the Air Force and Army to provide an alternate communication system in case of emergency need.

**maser** [Microwave Amplification by Stimulated Emission of Radiation] A microwave amplifier in which amplification is achieved by raising atoms or molecules of a paramagnetic material to an unstable high energy level. A microwave input signal then triggers the radiation of the excess energy at a specific wavelength, with radiated energy greatly exceeding that in the signal. Examples include the gas maser, resonant-cavity maser, solid-state maser, and traveling-wave maser. Also called paramagnetic amplifier.

**mask** A frame used in front of a television picture tube to conceal the rounded edges of the screen.

**masking** 1. The amount by which the threshold of audibility of a sound is raised by the presence of another sound. The unit customarily used is the decibel. Also called audio masking and aural masking. 2. A programed procedure for eliminating radar coverage in areas where such transmissions may be of use to the enemy for navigation purposes, by weakening the beam in such directions or by use of additional transmitters on the same frequency at suitable sites to interfere with homing.

**masking audiogram** A graphical presentation of the masking due to a stated noise, plotted in decibels as a function of the frequency of the masked tone.

**mask microphone** A microphone designed for use inside an oxygen mask or other type of respiratory mask.

**mass conversion factor** *Atomic mass conversion factor.*

**mass defect** The difference between the atomic mass and the mass number of a nuclide.

**mass-energy equation** The equation developed by Albert Einstein for interconversion of mass and energy, written as $E = mc^2$, where $E$ is the energy in ergs, $m$ is mass in grams, and $c$ is the velocity of light in centimeters per second.

**mass formula** An equation giving the atomic mass of a nuclide as a function of its atomic number and mass number.

**mass number** The sum of the protons and neutrons in the nucleus of an atom or nuclide. The mass number is also equal to the sum of the atomic number and the neutron number. In the symbol for a nuclide, the mass number is usually shown as a superscript following the element symbol; thus, in $U^{235}$, the mass number is 235. Also called nuclear number and nucleon number.

**mass radiator** A radiator that is capable of generating and radiating frequencies in the ehf band and even higher, extending far into the infrared region. It consists of fine metal particles suspended in a liquid dielectric and subjected to a high voltage. The electromagnetic waves are generated by the sparking that occurs between the particles. Output power is very low.

**mass spectrograph** A mass spectrometer that provides a permanent record of the mass spectrum lines of a material on a photographic plate.

**mass spectrometer** A spectrometer designed for analyzing a substance in terms of the ratios of mass to charge of its components.

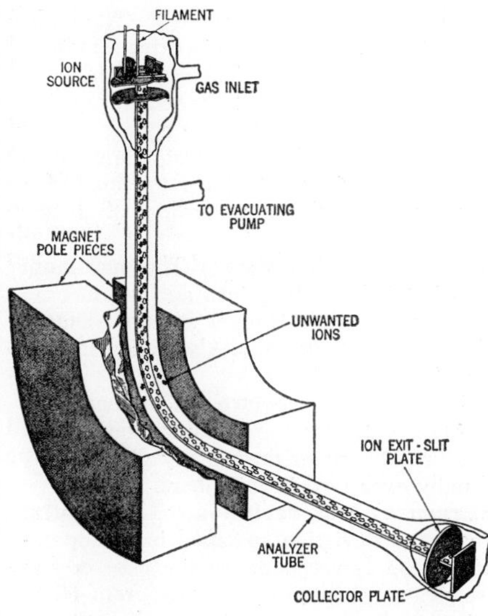


Mass spectrometer for analysis of gases.

Combined electric and magnetic fields are used to deflect the ions of the substance and focus each type in turn on an output electrode for detection and measurement.

**mass spectrum** A spectrum showing the distribution in mass or in mass-to-charge ratio of ionized atoms, molecules, or molecular fragments. The mass spectrum of an element shows the relative abundances of the isotopes of the element.

**mass stopping power** The energy loss per unit surface density traversed when a charged particle passes through a medium. Usually expressed in electron-volts per milligram per square centimeter.

**mass unit** *Atomic mass unit.*

**mast** A vertical metal pole serving as an antenna or antenna support.

**master** 1. The negative metal counterpart of a disk recording, produced by electroforming as one step in the production of phonograph records. 2. *Master station.*

**master brightness control** A variable resistor that adjusts the grid bias on all guns of a three-gun color picture tube simultaneously.

**master clock** The electronic or electric source of standard timing signals, often called clock pulses, required for sequencing the operation of a computer.

**master control** The control console that contains the main program controls for a radio or television transmitter or network.

**master gain control** 1. A variable resistor or potentiometer used in a stereo amplifier to control the gain of both audio channels simultaneously. 2. A control used in a radio, television, or recording studio to change the over-all audio output level without affecting the mixer controls that determine the balance of the microphones and other sound sources. Used also for fading out or fading in the sound volume.

**master oscillator** An oscillator that establishes the carrier frequency of the output of an amplifier or transmitter.

**master oscillator–power amplifier** [abbreviated mopa] An oscillator stage followed by an r-f power amplifier stage that serves also as a buffer.

**master routine** *Routine.*

**master station** The station of a synchronized group of radio stations to which the emissions of other stations of the group are referred. In a loran system it is the A station. Also called master.

**master synchronization pulse** A pulse distinguished from other telemetering pulses by amplitude and/or duration, used to indicate the end of a sequence of pulses.

**masurium** Former name for *technetium.*

**Matador** An Air Force surface-to-surface guided missile resembling an airplane, having a range of about 600 miles and capable of carrying conventional or nuclear weapons. It is an air-breather, hence altitude is limited to about 35,000 feet.

**matched load** A load having the impedance value that results in maximum absorption of energy from the signal source.

**matched power gain** The power gain obtained when the impedance of a load is matched to the effective output impedance of the amplifier to which it is connected.

**matched termination** A termination that

produces no reflected wave at any transverse section of a waveguide or other transmission line.

**matched transmission line** A transmission line having a matched termination.

**matched waveguide** A waveguide having a matched termination.

**matching** Connecting two circuits or parts together in such a way that their impedances are equal or are equalized by a coupling device, so as to give maximum transfer of energy.

**matching device** A device that matches unequal impedances, such as the output transformer of a radio receiver.

**matching diaphragm** A diaphragm consisting of a slit in a thin sheet of metal, placed transversely across a waveguide for matching purposes. The orientation of the slit with respect to the long dimension of the waveguide determines whether the diaphragm acts as a capacitive or inductive reactance.

**matching stub** A short length of two-wire transmission line connected at the antenna end or receiver end of a regular transmission line to add inductive or capacitive reactance for matching purposes. The shorting conductor is sometimes in the form of a slider that can be moved along the stub.

**matching transformer** A transformer used between unequal impedances for matching purposes, to give maximum transfer of energy.

**match-terminated** Terminated in a load equal to the characteristic impedance of the transmission line.

**materials-testing reactor** A nuclear reactor designed primarily for testing materials and equipment in strong radiation fields.

**mathematical check** A programed computer check of a sequence of operations, using the mathematical properties of that sequence.

**matrix** 1. A computer logical network consisting of a rectangular array of intersections of input-output leads, with diodes, magnetic cores, or other circuit elements connected at some of these intersections. The network usually functions as an encoder or decoder. 2. The section of a color television transmitter that transforms the red, green, and blue camera signals into color-difference signals and combines them with the chrominance subcarrier. Also called color coder, color encoder, and encoder. 3. The section of a color television receiver that transforms the color-difference signals into the red, green, and blue signals needed to drive the color picture tube. Also called color decoder and decoder. 4. A set of mathematical elements arranged in rows and columns, used in solving certain types of problems. 5. The precisely shaped form used as the cathode in electroforming. 6. *Translator.*

**matrixing** The process of performing a code conversion with a matrix, as in converting color television signal components from one form to another.

**mattress array** *Billboard array.*

**mavar** [Modulating Amplifier using VAriable Reactance] *Parametric amplifier.*

**maximum keying frequency** The frequency in cycles per second that is numerically equal to the spot speed divided by twice the horizontal dimension of the spot in a facsimile system. Also called fundamental scanning frequency.

**maximum modulating frequency** The highest picture frequency required for a facsimile transmission system. The maximum modulating frequency and the maximum keying frequency are not necessarily equal.

**maximum output** The greatest average output power delivered to the rated load of a receiver or amplifier, regardless of distortion.

**maximum retention time** The maximum time between writing into and reading an acceptable output from a storage element of a charge storage tube. Also called storage time (deprecated).

**maximum sound pressure** The maximum absolute value of the instantaneous sound pressure at a point during any given cycle. Usually expressed in microbars.

**maximum system deviation** The greatest frequency deviation specified in the operation of an f-m system. In the case of f-m broadcast systems in the range from 88 to 108 mc, the maximum system deviation is 75 kc.

**maximum undistorted output** The greatest average output power into the rated load of an amplifier at which distortion does not exceed a specified limit when the input is sinusoidal. Also called maximum useful output.

**maximum usable frequency** [abbreviated muf] The upper limit of the frequencies that can be used at a specified time for point-to-point radio transmission involving propagation by reflection from the regular ionized layers of the ionosphere. Higher frequencies may be transmitted only by sporadic and scattered reflections.

**maximum useful output** *Maximum undistorted output.*

**maxwell** The cgs electromagnetic unit of magnetic flux. It is equal to one gauss per square centimeter, or to one magnetic line of force.

**Maxwell-Boltzmann law** A law that gives the distribution of velocities among the molecules of a perfect steady-state gas.

**Maxwell-Boltzmann statistics** Statistics representing the distribution of particles among the various possible energy levels at such high temperatures that a large number of energy levels are excited.

**Maxwell bridge** A four-arm a-c bridge used to measure inductance (or capacitance) in terms of resistance and capacitance (or inductance). Bridge balance is independent of frequency.

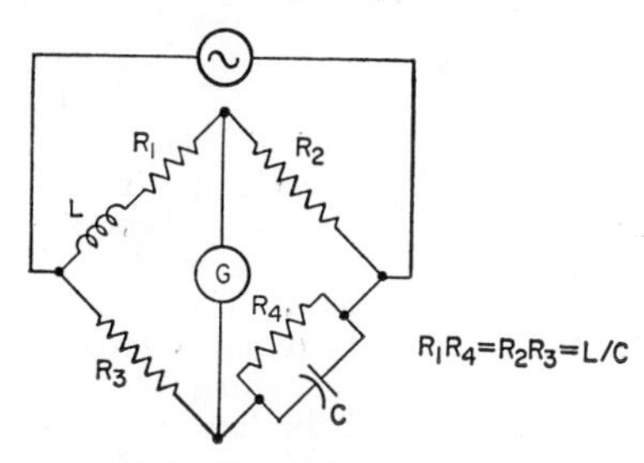


Maxwell bridge and equations.

**Maxwell d-c commutator bridge** A four-arm bridge normally used to measure capacitance in terms of resistance and time. One arm includes a commutator that alternately

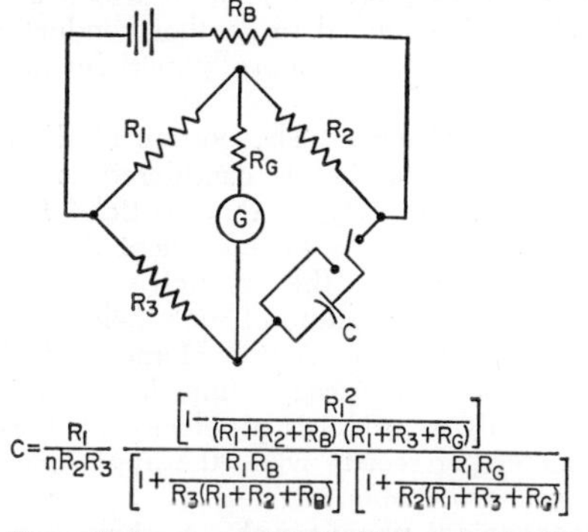


Maxwell d-c commutator bridge and equation.

connects the unknown capacitor in series with the bridge arm and then opens the bridge arm while short-circuiting the capacitor.

**Maxwellian distribution** The velocity distribution of the molecules of a gas in thermal equilibrium. This distribution is often assumed to hold for neutrons in thermal equilibrium with the moderator in a nuclear reactor.

**Maxwell inductance bridge** A four-arm a-c bridge used to compare inductances. Bridge balance is independent of frequency.

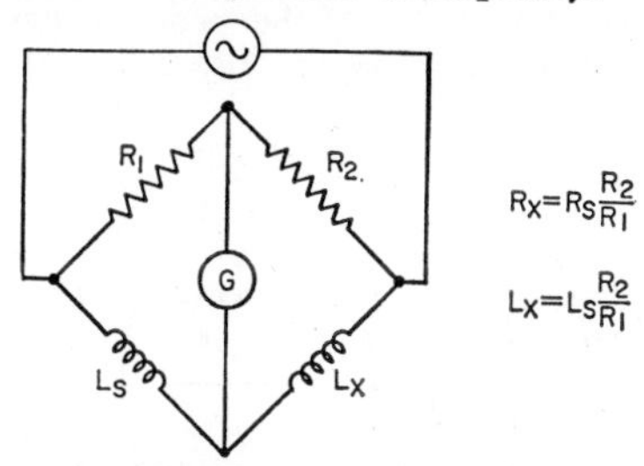


Maxwell inductance bridge circuit and equations.

**Maxwell mutual-inductance bridge** An a-c bridge used to measure mutual inductance in terms of self-inductance. Bridge balance is independent of frequency.

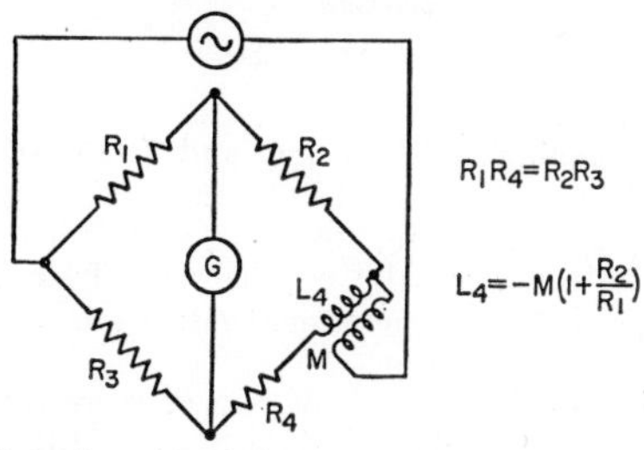


Maxwell mutual-inductance bridge circuit and equations.

**Maxwell triangle** The equilateral-triangle form of a chromaticity diagram, in which the primary colors are represented at the corners of the triangle.

**mayday** [French m'aider] The radiotelephone distress call, used throughout the world by international agreement.

**mc** 1. Abbreviation for *megacycle.* 2. Abbreviation for *millicurie.*

**McGill fence** *Mid-Canada line.*

**McNally tube** A single-cavity velocity-modulated microwave local oscillator tube, the frequency of which may be controlled over a wide range by electrical means.

**mcw** Abbreviation for *modulated continuous wave.*

**m-derived section** A T or pi network section so designed that when two or more sections are joined in a filter unit their impedances are matched at all frequencies, even though the sections may have different resonant frequencies.

**mdi** Abbreviation for *miss-distance indicator.*

**M display** A modified radarscope A display in which target distance is determined by moving an adjustable pedestal signal along the baseline until it coincides with the horizontal position of the target deflection. The control that moves the pedestal signal is calibrated in distance. Also called M scan.

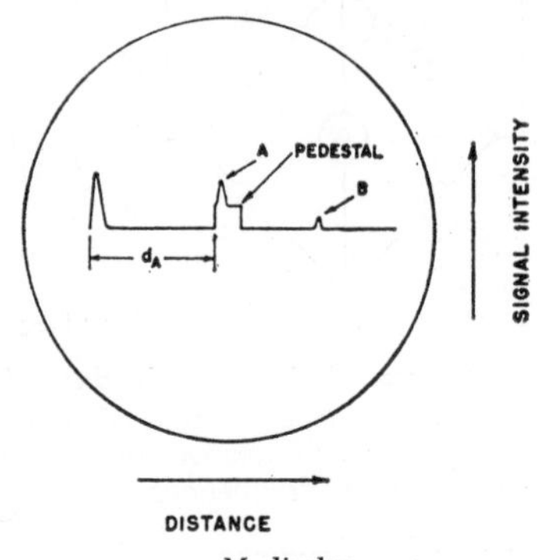

M display.

**meaconing** [MEAsuring and CONFusING] Measuring received radio navigation signals and broadcasting confusing signals instantly and automatically on the same frequency. Used to give enemy navigators a false indication of position.

**mean** The average of two or more quantities, obtained by adding the quantities and dividing by the number of quantities. Also called arithmetic mean.

**mean carrier frequency** The average carrier frequency of a transmitter, corresponding to the resting frequency in a frequency-modulation system.

**mean free path** 1. The average distance that a particle travels between successive collisions. 2. The average distance that sound waves travel between successive reflections in an enclosure.

**mean free time** The average time between successive collisions of a particle.

**mean lethal dose** *Median lethal dose.*

**mean life** The average time during which an atom or other system exists in a particular form. For a radionuclide, the mean life is the reciprocal of the disintegration constant. Mean life is 1.443 times the radioactive half-life. In a semiconductor, mean life is the time required for injected excess carriers to recombine with others of the opposite sign. Also called average life and lifetime.

**mean pulse time** The arithmetic mean of the leading-edge pulse time and the trailing-edge pulse time.

**mean range** The range that is exceeded by half of the particles under consideration.

**measurand** The physical quantity, property, or condition that is to be measured.

**mechanical axis** One of the $y$ axes in a quartz crystal. There are three, each perpendicular to one pair of opposite sides of the hexagon.

**mechanical bandspread** Bandspread in which a vernier tuning dial or other mechanical means provides greater angular rotation of a control knob for a given tuning range, to simplify tuning in crowded short-wave bands.

**mechanical compliance** The displacement of a mechanical element per unit of force, expressed in centimeters per dyne.

**mechanical filter** A filter consisting of shaped metal rods that act as coupled mechanical resonators when used with piezoelectric input and output transducers. Sometimes used in i-f amplifiers of highly selective superheterodyne receivers.

**mechanical impedance** The complex quotient of an applied alternating force divided by the resulting alternating linear velocity in the direction of the force at its point of application. The real component is mechanical resistance, and the imaginary component is mechanical reactance. The unit of mechanical impedance, reactance, and resistance is the mechanical ohm.

**mechanical jamming** *Passive jamming.*

**mechanical joint** A joint in which a conductor is clamped to another conductor or to a terminal mechanically, without the use of solder.

**mechanical mass** The portion of the mass of a particle that is considered to be an intrinsic property of that particle. It does not include the mass increment due to the interaction of the particle with itself through the medium of some field.

**mechanical ohm** A unit of mechanical resistance, reactance, or impedance. A force of 1 dyne produces a velocity of 1 centimeter per second when the opposition is 1 mechanical ohm.

**mechanical phonograph** A phonograph in which the playback stylus drives the diaphragm of an acoustic pickup that radiates acoustic energy.

**mechanical phonograph recorder** A recorder that transforms electric or acoustic signals into a corresponding mechanical motion and inscribes such motion in an appropriate medium by cutting or embossing. Also referred to as mechanical recorder.

**mechanical recorder** *Mechanical phonograph recorder.*

**mechanical rectifier** A rectifier in which rectification is accomplished by mechanical action, as in a synchronous vibrator.

**mechanical register** An electromechanical device that indicates a count of pulses.

**mechanical reproducer** *Acoustic pickup.*

**mechanical scanning** Scanning in which a beam of light controlled by a rotating scanning disk, rotating mirror, or other mechanical device is used to break up a scene or image into a rapid succession of narrow lines, as required for conversion into electric pulses.

**mechanical television system** A television system that uses mechanical scanning.

**mechanical translation** Automatic translation of one language into another by means of a computer or other machine that contains a dictionary lookup in its memory along with the programs needed to make logical choices from synonyms, supply missing words, and rearrange word order as required for the new language. The text to be translated must be converted into machine-readable form first, either manually by special typewriters, or automatically by photoelectric character readers. Also called machine translation.

**median** The value at the halfway point in a series, wherein there is the same number of values above the median as below the median.

**median energy of fission** The neutron energy above (or below) which half of the total number of fission-producing neutron absorptions occur.

**median lethal dose** The dose of radiation required to kill, within a specified period, 50% of the individuals in a large group of animals or organisms. Also called mean lethal dose.

**median lethal time** The time required, following administration of a specified dose of radiation, for death of 50% of the individuals in a large group of animals or organisms.

**medical electronics** A branch of electronics in which electronic instruments and equipment are used for such medical applications as diagnosis, therapy, research, anesthesia control, cardiac control, and surgery.

**medium frequency** [abbreviated m-f] A Federal Communications Commission designation for the band from 300 to 3,000 kc in the radio spectrum.

**medium-range radar** Radar whose maximum line-of-sight range is between 150 and 300 miles for a target having an area of 1 square meter perpendicular to the radar beam.

**meg** Abbreviation for *megohm.*

**mega-** A prefix representing $10^6$ or one million.

**megabar** An absolute unit of pressure equal to 1,000,000 bars. One megabar is almost exactly equal to normal atmospheric pressure.

**megacycle** [abbreviated mc] One million cycles.

**mega-electron-volt** *Million electron-volt.*

**megamegacycle** [abbreviated mmc] A unit of frequency equal to $10^{12}$ cycles per second or 1,000,000 megacycles. Also called fresnel and teracycle.

**megaton** 1. One million tons. 2. The explosive power of 1,000,000 tons of TNT, as in a megaton atomic bomb.

**megatron** *Disk-seal tube.*

**Megger** Trademark of James G. Biddle Co. for their high-range ohmmeter having a hand-driven d-c generator as its voltage source. It is used chiefly for measuring insulation resistance values in megohms.

**megohm** [abbreviated meg] One million ohms.

**Meissner oscillator** An oscillator in which the grid and anode circuits are inductively coupled through an independent tank circuit that determines the frequency of oscillation.

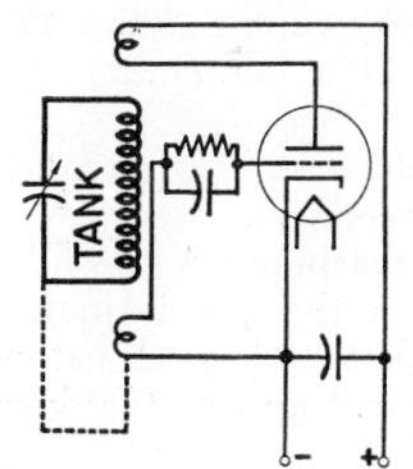

Meissner oscillator circuit.

**mel** A unit of pitch. By definition, a simple 1,000-cps tone 40 db above a listener's threshold produces a pitch of 1,000 mels. The pitch of any sound that is judged by the listener to be *n* times that of the 1-mel tone is *n* mels.

**M electron** An electron having an orbit in the M shell, which is the third shell of electrons surrounding the atomic nucleus, counting out from the nucleus.

**M-electron capture** A mode of electron capture similar to K-electron capture except that the electrons are initially in the M shell of the atom.

**meltback transistor** A junction transistor in which the junction is made by melting a properly doped semiconductor and allowing it to solidify again.

**melting channel** The restricted portion of the charge in a submerged horizontal-ring induction furnace, in which the induced currents are concentrated to heat and melt the charge.

**melt-quench transistor** A junction transistor that is made by suddenly cooling a melted-back region.

**memory** *Storage.*

**memory capacity** *Storage capacity.*

**memory tube** Deprecated term for *storage tube.*

**mendelevium** [symbol Mv] An unstable transuranic radioactive element produced artificially by bombardment of alpha particles. Atomic number is 101.

**mercuric telluride** [symbol HgTe] An intermetallic compound having characteristics similar to those of indium antimonide. It shows promise of being sensitive to about 12 microns as an infrared detector.

**mercury** [symbol Hg] A silvery white liquid metal that becomes a solid at −40°F. Used in mercury switches and electron tubes because the vapor of mercury ionizes readily and conducts electricity. The green line of mercury 198 very closely approaches pure monochromatic light. Atomic number is 80.

**mercury arc** An electric discharge through ionized mercury vapor, giving off a brilliant bluish-green light containing strong ultraviolet radiation.

**mercury-arc inverter** An inverter that uses mercury-arc rectifiers.

**mercury-arc rectifier** A gas-filled rectifier tube in which the gas is mercury vapor. Small sizes use a heated cathode, while larger sizes rated up to 8,000 kilowatts and higher use a mercury-pool cathode. Also called mercury-vapor rectifier.

Mercury-arc rectifiers having heated cathodes.

**mercury battery** A battery made up of mercury cells.

**mercury cell** A primary dry cell that delivers an essentially constant output voltage throughout its useful life by means of a chemical reaction between zinc and mercury oxide. Widely used in hearing aids. Also called mercury oxide cell.

**mercury delay line** An acoustic delay line in which mercury is the medium for sound transmission. Also called mercury memory and mercury storage.

**mercury-hydrogen spark-gap converter** A spark-gap generator that uses the oscillatory discharge of a capacitor through a coil and a spark gap as a source of r-f power. The spark gap consists of a solid electrode and a pool of mercury in a hydrogen atmosphere.

**mercury memory** *Mercury delay line.*

**mercury oxide cell** *Mercury cell.*

**mercury-pool cathode** A pool cathode in which the pool is liquid mercury.

**mercury-pool tube** *Pool tube.*

**mercury relay** A relay in which mercury, moved by a magnetic plunger, serves to connect the relay contacts together when the plunger is pulled into the mercury by the energized relay coil.

**mercury storage** *Mercury delay line.*

**mercury switch** A switch that is closed by making a large globule of mercury move up to the contacts and bridge them. The mercury is usually moved by tilting the entire switch.

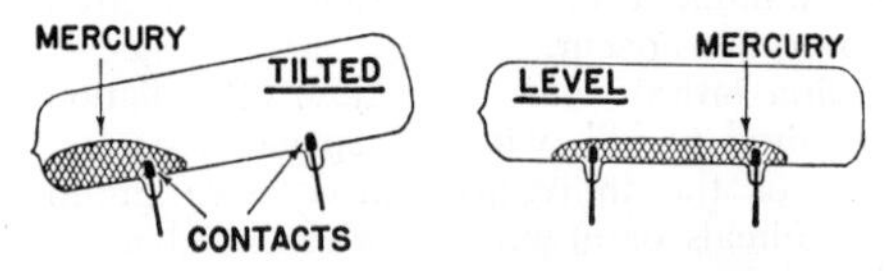


Mercury switch construction and operation.

**mercury-vapor lamp** A lamp in which light is produced by an electric arc between two electrodes in an ionized mercury-vapor atmosphere. It gives off a blue-green light that is rich in ultraviolet radiation.

**mercury-vapor rectifier** *Mercury-arc rectifier.*

**mercury-vapor tube** A gas tube in which the active gas is mercury vapor.

**merge** To collate two or more similarly ordered sets of values into one set having the same ordered sequence.

**mesa transistor** A transistor in which a germanium or silicon wafer is etched down in steps so the base and emitter regions appear as physical plateaus above the collector region.

**mesh** A set of branches forming a closed path in a network in such a way that if any one branch is omitted from the set, the remaining branches of the set do not form a closed path. Sometimes called loop.

**mesh connection** *Delta connection.*

**Mesny circuit** An ultrahigh-frequency oscillator employing a symmetrical arrangement of two vacuum tubes, with inductances in the grid and anode supply leads.

**meson** Any elementary particle having a rest mass intermediate in value between the mass of an electron and that of a proton. Mesons are found in cosmic rays and are produced in high-energy nuclear reactions. They may have a positive or negative charge or be neutral. Average lives are always shorter than a microsecond. They differ in masses and spins, and can in some cases change from one kind of meson to another. Known mesons include mu mesons, pi mesons, and tau mesons.

**mesothorium** A radioactive nuclide in the thorium series, occurring in two forms. Mesothorium I (symbol $MsTh_1$) has an atomic number of 88, while mesothorium II (symbol $MsTh_2$) has an atomic number of 89.

**mesotron** Obsolete term for *mu meson.*

**message** A group of words, variable in length, transported as a unit in a computer.

**metadyne** A type of rotating magnetic amplifier characterized by more than one brush per armature pole. In Great Britain, metadynes correspond to amplidynes.

**metal-backed screen** *Aluminized screen.*

**metal-ceramic** A ceramic mixed with a metal oxide, carbide, or nitride to obtain greater ductility and strength. Also called ceramet.

**metal-clad base material** A laminate that consists of metallic material bonded to one or both surfaces of an insulating base. Used for printed-wiring boards.

**metal-cone tube** A television picture tube having a metal cone rather than a glass cone between the glass faceplate and the glass neck of the tube.

**metal detector** An electronic device for detecting concealed metal objects, such as guns, knives, or buried pipe lines, generally by radiating a high-frequency electromagnetic field and detecting the change produced in that field by the ferrous or nonferrous metal object being sought. Also called electronic locator, metal locator, and radio metal locator.

**metallic antenna lens** A lens consisting of contoured parallel metal surfaces placed in front of an antenna to focus the beam. The vanes change the phase velocity in proportion to the distance traveled between metal surfaces at each part of the lens.

**metallic circuit** A complete circuit in which the earth or ground forms no part.

**metallic insulator** A shorted quarter-wave section of a transmission line. Such a line acts as an extremely high impedance at a frequency corresponding to its quarter-wavelength and may therefore be used as a mechanical support.

**metallic rectifier** A rectifier consisting of one or more disks of metal under pressure contact with semiconductor coatings or layers, such as a copper-oxide, selenium, or silicon rectifier. Also called contact rectifier, dry-disk rectifier, and semiconductor rectifier.

**metallic rectifier cell** An elementary metallic rectifier having one positive electrode, one negative electrode, and one rectifying junction.

**metallized paper capacitor** A paper capacitor in which a film of metal is deposited directly on the insulating paper to serve in place of a separate foil strip. The capacitor has self-healing characteristics.

**metallized resistor** A resistor made by depositing a thin film of high-resistance metal on the surface of a glass or ceramic rod or tube.

**metallized screen** *Aluminized screen.*

**metal locator** *Metal detector.*

**metal master** *Original master.*

**metal negative** *Original master.*

**metal positive** *Mother.*

**metal-tank mercury-arc rectifier** A mercury-arc rectifier in which the anodes and the mercury cathode are enclosed in a metal container or chamber.

**metal tube** A vacuum tube having a metal envelope, with electrode leads passing through glass beads fused into the metal housing.

**metascope** An infrared receiver used for converting pulsed invisible infrared rays into visible signals for communication purposes. Also used with an infrared source for reading maps in darkness.

**metastable** Capable of undergoing a quantum transition to a state of lower energy, but having a relatively long lifetime as compared with the most rapid quantum transitions of similar systems.

**metastable equilibrium** A condition of pseudoequilibrium in which the free energy

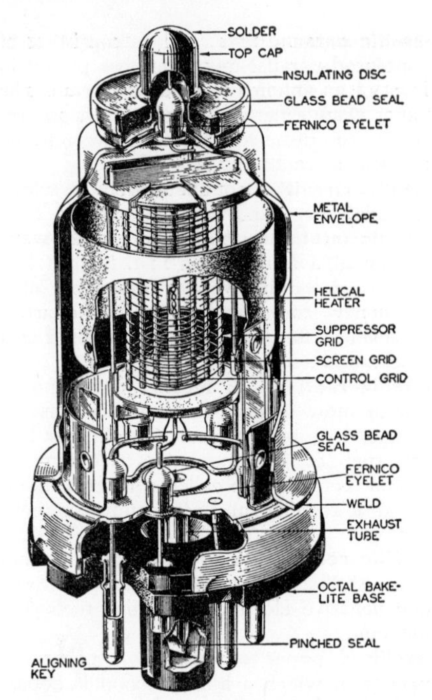


Metal tube, showing how leads pass through glass beads sealed into metal base.

of a system is at a minimum with respect to infinitesimal changes, but not with respect to major changes. Although a condition of greater thermodynamic stability exists, the system may remain in the metastable state because the transition to the more stable condition is extremely sluggish.

**metastable state** An excited state from which all quantum transitions to lower states are generally forbidden transitions.

**meteoric scatter** A form of scatter propagation in which meteor trails serve to scatter radio waves back to earth. Two radio links, working in opposite directions, are used. Any message transmitted on one link is sent back on the other, so the sender can verify satisfactory reception. The sender transmits only the first character of the message until it is received and returned as a result of scatter by a meteor. The desired message, previously recorded on magnetic tape, is then transmitted at high speed as a burst that may last from a fraction of a minute to several minutes, depending on the size and course of the meteor. One meteoric scatter system is known as Janet.

**meteorograph** An instrument that measures and records meteorological data such as air pressure, temperature, and humidity. When carried aloft, it is also called an aerograph. When used with a radio transmitter, it is called a radiosonde.

**meter** 1. A device for measuring the value of a quantity under observation. Also called instrument. The term meter is usually applied to an indicating instrument alone, such as a voltmeter, ammeter, ohmmeter, or wattmeter. 2. The basic unit of length in the metric system, equal to 39.37 inches or 3.28 feet. A meter is equal to 100 centimeters or 1,000 millimeters.

**meter-candle** *Lux.*

**meter-kilogram-second-ampere unit** [abbreviated mksa unit] A practical absolute electrical unit based on the meter, kilogram, second, and ampere as fundamental units.

**meter-kilogram-second system** [abbreviated mks system] An absolute system of units based on the meter, kilogram, and second as fundamental units. Also called Giorgi system.

**meter-type relay** A relay that uses a meter movement having a contact-bearing pointer which moves toward or away from a fixed contact mounted on the meter scale.

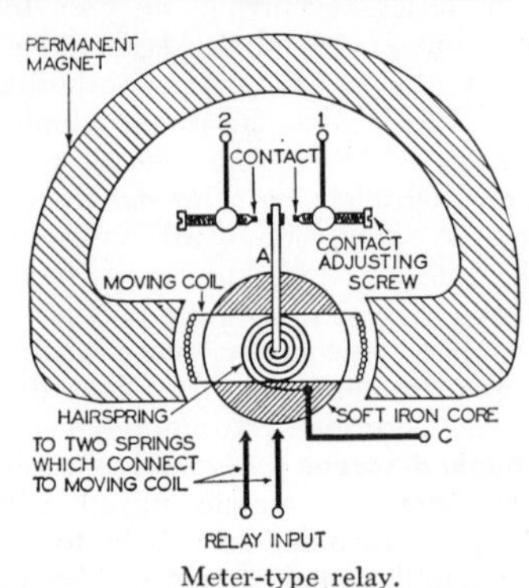


Meter-type relay.

**metric waves** British classification for wavelengths between 1 and 10 meters.

**mev** Abbreviation for *million electron-volt.*

**mew** Abbreviation for *microwave early warning.*

**m-f** Abbreviation for *medium frequency.*

**mf** Alternate abbreviation for *microfarad.*

**mfd** Alternate abbreviation for *microfarad.*

**mh** Abbreviation for *millihenry.*

**mho** The unit of conductance or admittance. It is the reciprocal of the ohm. A resistance or impedance of 1 ohm is equal to a conductance or admittance of 1 mho.

**mhometer** A meter used in a tube tester or other instrument to indicate conductance in mhos.

**mica** A transparent mineral that splits readily into thin sheets having excellent insulating and heat-resisting qualities. Used as the dielectric in mica capacitors and as electrode spacers in electron tubes.

**mica capacitor** A fixed capacitor that uses mica sheets as the dielectric.

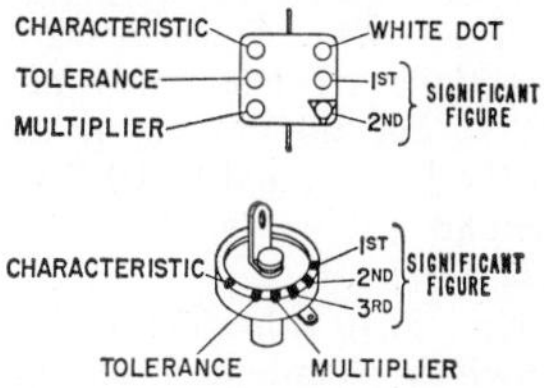

CHARACTERISTIC
WHITE DOT
TOLERANCE
1ST
SIGNIFICANT FIGURE
MULTIPLIER
2ND

CHARACTERISTIC
1ST
2ND
3RD
SIGNIFICANT FIGURE
TOLERANCE
MULTIPLIER

Mica capacitors—conventional molded plastic type and silvered mica button type—with methods of using color codes to give capacitance values in micromicrofarads and other significant characteristics. Many other color-marking methods are used also for mica capacitors.

**micro-** 1. A prefix representing $10^{-6}$ or one-millionth. Thus, 1 microvolt is 0.000001 volt. Often abbreviated $\mu$ (Greek letter mu), as in the abbreviation $\mu$v for microvolt. 2. A prefix indicating smallness, as in microwave. 3. A prefix indicating extreme sensitivity, as in microradiometer and microphone.

**microalloy diffused transistor** [abbreviated madt] A microalloy transistor in which the semiconductor wafer is first subjected to gaseous diffusion to produce a nonuniform base region.

**microalloy transistor** A transistor in which the emitter and collector electrodes are formed by etching depressions, then electroplating and alloying a thin film of the impurity metal to the semiconductor wafer, somewhat as in a surface-barrier transistor.

**microammeter** An ammeter whose scale is calibrated to indicate current values in microamperes.

**microampere** [abbreviated $\mu$a] One-millionth of an ampere.

**microbar** A unit of pressure commonly used in acoustics, equal to 1 dyne per square centimeter. The term bar properly denotes a pressure of $10^6$ dynes per square centimeter. Unfortunately, in acoustics the bar was used to mean 1 dyne per square centimeter. It is recommended, therefore, in respect to sound pressures that the less ambiguous terms microbar or dyne per square centimeter be used.

**microcurie** [abbreviated $\mu$c] One-millionth of a curie.

**microdensitometer** A high-sensitivity densitometer used in spectroscopy to detect spectrum lines too faint on a negative to be seen by the human eye.

**microfarad** [abbreviated $\mu$f] One-millionth of a farad. Alternate abbreviations are mf, mfd, and uf.

**microflash** A high-intensity, short-duration light source used in photography.

**microgroove record** *Long-play record.*

**microhenry** [abbreviated $\mu$h; plural microhenrys] One-millionth of a henry.

**microhm** One-millionth of an ohm.

**micro-H system** An H system of bombing and navigation radar that operates in the region of 10,000 mc.

**microlock** A phase-locking loop system for transmitting and receiving information by radio with reduced bandwidth. Used in tracking satellites and telemetering data to ground stations at line-of-sight distances as great as 3,000 miles.

**microlock network** A network of radio stations using microlock equipment to track missiles and satellites.

**micromho** One-millionth of a mho.

**micromicro-** [abbreviated $\mu\mu$] A prefix meaning one-millionth of one-millionth.

**micromicrofarad** [abbreviated $\mu\mu$f] One-millionth of a microfarad. Sometimes called picofarad (abbreviated pf) in Europe. Alternate abbreviations are mmf, mmfd, and uuf.

**micromicrowatt** A unit of power equal to $10^{-12}$ watt. Also called picowatt.

**microminiaturization** Miniaturization involving the design and construction of equipment several orders of magnitude smaller than with subminiature techniques, generally by combining, merging, or blending circuit elements with the device itself.

**micron** [abbreviated $\mu$] One-millionth of a meter, a unit used in specifying wavelengths of light. These range from about 0.38 micron for purple to 0.7 micron for red.

**microphone** An electroacoustic transducer that responds to sound waves and delivers essentially equivalent electric waves. Also called mike (slang).

**microphone boom** An overhead extension arm used to support a microphone within range of the sound to be picked up but outside the range of a television camera.

**microphone button** A button-shaped telescoping container filled with carbon particles and serving as the resistance element of a carbon microphone.

**microphone cable** A special shielded cable used to connect a microphone to an audio amplifier.

**microphone mixer** A mixer used to feed two or more microphones into the input of an a-f amplifier. Separate controls permit adjusting the output level of each microphone.

**microphone preamplifier** An a-f amplifier that amplifies the output of a microphone before the signal is sent through a transmission line to the main a-f amplifier. The preamplifier is often built into the microphone housing or stand.

**microphone shield** A protective covering used to protect a microphone diaphragm from condensed moisture originating from the operator's breath when in use, or to protect from other environmental conditions.

**microphone stand** A stand used to support a microphone in a desired position above the floor or on a table.

**microphone transformer** An iron-core transformer used for coupling certain types of microphones to a microphone preamplifier, to a transmission line, or to the main a-f amplifier.

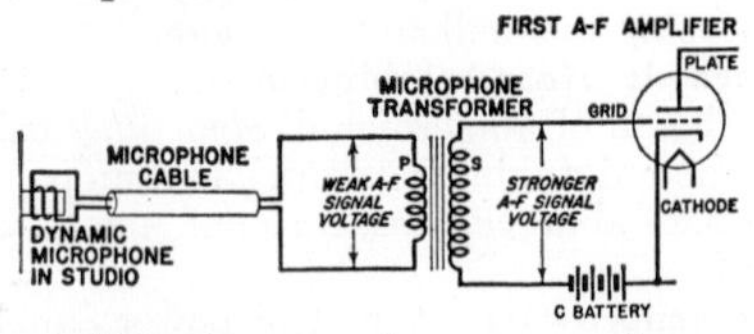

Microphone transformer connections.

**microphonic** Vulnerable to vibration that produces microphonics. A tube in a radio receiver is microphonic if a pinging sound is heard from the loudspeaker when the side of the tube is tapped with a finger.

**microphonics** Noise caused by mechanical vibration of the elements of an electron tube, component, or system. The vibration causes modulation of the signal currents flowing through or controlled by the vibrating device. Heard as noise in an a-f system and seen as an undesirable interference pattern in facsimile and television images. Also called microphonism.

**microphonism** *Microphonics.*

**microphotometer** A photometer that provides highly accurate illumination measurements. In one form, the changes in illumination are picked up by a phototube and converted into current variations that are amplified by vacuum tubes.

**microradiography** The radiography of small objects having details too fine to be seen by the unaided eye, with optical enlargement of the resulting negative.

**microradiometer** A radiometer used for measuring weak radiant power, in which a thermopile is supported on and connected directly to the moving coil of a galvanometer. Also called radiomicrometer (deprecated).

**microscopic cross section** The cross section per atom of isotope. The dimensions are those of an area. Since neutron reaction cross sections are frequently of the order of $10^{-24}$ sq cm, the commonly used unit of area is the barn, equal to $10^{-24}$ sq cm.

**microsecond** [abbreviated $\mu$sec] One-millionth of a second.

**microstrip** *Strip transmission line.*

**Micro Switch** Trademark of Micro Switch Division of Minneapolis-Honeywell Regulator Co. for their line of switches.

**microtron** A type of cyclotron in which electrons are accelerated.

**microvolt** [abbreviated $\mu$v] One-millionth of a volt.

**microvoltmeter** A voltmeter whose scale is calibrated to indicate voltage values in microvolts.

**microvolts per meter** A measure of the intensity of the signal produced by a radio transmitter at a given point. It is equal to the signal strength in microvolts at the receiving antenna divided by the effective height of the antenna in meters. Stronger signals are expressed in millivolts per meter.

**microwatt** [abbreviated $\mu$w] One-millionth of a watt.

**microwave** Pertaining to wavelengths in the microwave spectrum, ranging from about 30 to 0.3 cm.

**microwave circulator** *Circulator.*

**microwave early warning** [abbreviated mew] A high-power, long-range, early-warning radar with a number of indicators, giving high resolution and large traffic-handling capacity.

**microwave filter** A filter consisting of resonant cavity sections or other elements, built into a microwave transmission line to pass desired frequencies while rejecting or absorbing other frequencies.

**microwave frequencies** Frequencies of 890 mc and above, as used in FCC regulations.

**microwave gyrator** *Gyrator.*

**microwave radiometer** *Radiometer.*

**microwave relay system** *Microwave repeater.*

**microwave repeater** A radio repeater using highly directional radio beams at microwave frequencies to link towers spaced up to 50 miles apart. At each tower is a receiver and transmitter for picking up, amplifying, and passing on the signals in

Microwave repeater used for television networks.

either or both directions. Also called microwave relay system.

**microwaves** *Microwave spectrum.*

**microwave spectroscopy** Spectroscopy involving determination of the selective absorption of microwaves at various frequencies by solid or gaseous materials. Used in studying atomic, crystalline, and molecular structures.

**microwave spectrum** A spectrum of wavelengths located between the short-wave region and the far infrared region, and commonly considered to include wavelengths from about 30 cm (1,000 mc) to 0.3 cm (100,000 mc). Also called microwaves.

**microwave system** A radio system utilizing microwaves.

**microwave therapy** Therapeutic use of electromagnetic energy to generate heat within the body, using frequencies in the microwave spectrum.

**microwave transmission path** The actual path taken by r-f energy between two microwave antennas, as affected by reflectors, atmospheric refraction, and diffraction.

**microwave tube** An electron tube designed for operation at wavelengths in the range of about 30 to 0.3 cm.

**midas** [Missile Intercept Data Acquisition System] A system using three airborne telemetering transmitters and two bicotar ground sites to determine missile, target, and launch plane trajectories, as well as vector miss-distance, for an air-to-air firing range.

**mid-Canada line** A chain of radar stations along the 55th and 56th parallels in Canada. Also called McGill fence.

**midcourse guidance** The guidance applied to a missile between the launching phase and the target approach phase.

**middle marker** A fan marker beacon installed approximately 3,500 feet from the approach end of the runway on the localizer course of an instrument landing system.

**migration area** One-sixth the mean square distance that a neutron travels in a medium from its birth in fission until its absorption.

**migration length** The square root of the migration area.

**mike** Slang for *microphone.*

**MIL** Abbreviation for Military Specification, which replaces JAN specifications.

**mil** 1. A unit of angular measurement used in launching bombs and guided missiles. A true mil is the angle determined by an arc whose length is one-thousandth of the radius. For practical purposes, the mil is considered to be 1/6,400 (instead of 1/6,283) of 360°. 2. One-thousandth of an inch.

**Miller bridge** A bridge used to measure amplification factors of vacuum tubes.

**Miller effect** The increase in the effective grid-cathode capacitance of a vacuum tube due to the charge induced electrostatically on the grid by the anode through the grid-anode capacitance.

**Miller integrator** A resistor-capacitor charging network having a high-gain amplifier paralleling the capacitor. Used to produce a linear time-base voltage. Also called Miller time-base.

**Miller time-base** *Miller integrator.*

**milli-** A prefix representing $10^{-3}$ or one-thousandth.

**milliammeter** An ammeter whose scale is calibrated to indicate current values in milliamperes.

**milliampere** [abbreviated ma] One-thousandth of an ampere.

**millibar** One-thousandth of a bar. A bar is 1,000,000 dynes per sq cm.

**millicurie** [abbreviated mc] One-thousandth of a curie.

**millihenry** [abbreviated mh; plural millihenrys] One-thousandth of a henry.

**millilambert** One-thousandth of a lambert.

**millimeter** [abbreviated mm] One-thousandth of a meter.

**millimeter wave** An electromagnetic wave having a wavelength below 1 centimeter, corresponding to a frequency above 30,000 mc. An atmospheric window, with attenuation of only about 0.3 db per mile, occurs around 75,000 mc, whereas attenuation is almost 20 db per mile at 50,000 mc.

**millimicron** One-thousandth of a micron, hence $10^{-9}$ meter.

**milliohm** One-thousandth of an ohm.

**million electron-volt** [abbreviated mev] A unit of energy equal to 1,000,000 electron-volts. Also called mega-electron-volt.

**milliroentgen** One-thousandth of a roentgen.

**millisec** Abbreviation for *millisecond.*

**millisecond** [abbreviated millisec] One-thousandth of a second.

**millisone** A unit of loudness equal to 0.001 sone.

**millivolt** [abbreviated mv] One-thousandth of a volt.

**millivoltmeter** A voltmeter whose scale is calibrated to indicate voltage values in millivolts.

**millivolts per meter** A rating often used for signal intensities greater than 1,000 microvolts per meter.

**milliwatt** [abbreviated mw] One-thousandth of a watt.

**mine detector** A metal detector designed specifically to locate explosive mines that are buried or submerged. A higher-frequency version is used for locating nonmetallic antipersonnel mines. Also called electronic mine detector.

**miniature tube** A small electron tube having no base, with tube electrode leads projecting through the glass bottom in positions corresponding to those of pins for either a 7-pin or 9-pin tube base.

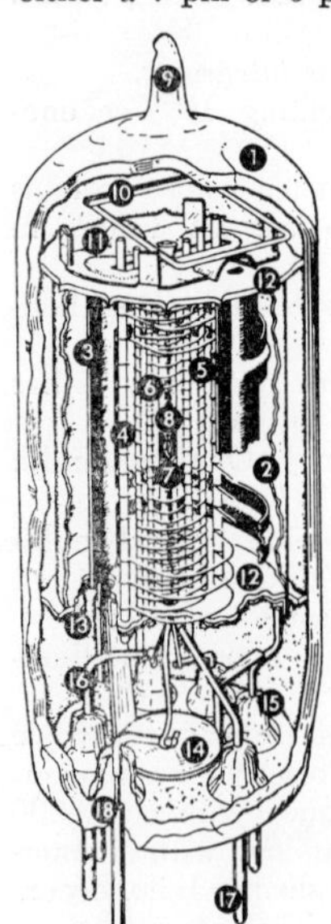


Miniature tube construction for pentode.

**miniaturization** Reduction in the size and weight of a system, package, or component by using small parts arranged for maximum utilization of space.

**minigroove record** A record having more lines per inch than the average 78-rpm phonograph record and yet not enough lines per inch to be called extended-play, long-play, or microgroove.

**minimum-access programing** The programing of a digital computer in such a way that minimum waiting time is required to obtain information out of the memory.

**minimum distance** The shortest distance at which a navigation system will function within its prescribed tolerances.

**minimum firing power** The minimum r-f power required to initiate an r-f discharge in a switching tube at a specified ignitor current.

**minimum ionization** The smallest possible value of the specific ionization that a charged particle can produce in passing through a particular substance. For singly charged particles in ordinary air, the minimum ionization is about 50 ion pairs per centimeter of path. In general, it is proportional to the density of the medium and to the square of the charge of the particle.

**minimum safe distance** The total distance required between desired ground zero of an atomic explosion and friendly positions to ensure troop safety.

**minitrack network** A network of minitrack radio stations placed at different points around the world to track the flight of an earth satellite.

**minitrack radio** A radio receiver that tracks an earth satellite or other object equipped with a miniature transmitter emitting telemeter-type signals.

**minor apex face** One of the three smaller sloping faces near but not touching the apex of a natural quartz crystal. The three larger sloping faces are the major apex faces.

**minor cycle** The time required for the transmission or transfer of one machine word, including the space between words, in a digital computer using serial transmission. Also called word time.

**minor face** One of the three longer sides of a natural hexagonal quartz crystal.

**minority carrier** The type of carrier that constitutes less than half of the total number of carriers in a semiconductor. The minority carriers are holes in an n-type semiconductor and electrons in a p-type semiconductor.

**minority emitter** An electrode from which a flow of minority carriers enters the interelectrode region of a transistor.
**minor lobe** Any lobe except the major lobe of a radiation pattern. Also called secondary lobe and side lobe.
**Minuteman** An Air Force intercontinental ballistic missile using solid fuel to permit almost instant firing when needed.
**miran** [MIssile RANging] A microwave omnidirectional pulse-type missile-tracking system that measures range only. A master interrogator at the master ground station triggers a missile beacon. Two or more slave stations measure the transit time of the resulting pulse from the beacon, for conversion to position data at the master station.
**mirror-backed screen** *Aluminized screen.*
**mirror galvanometer** A galvanometer having a small mirror attached to the moving element, to permit use of a beam of light as an indicating pointer.

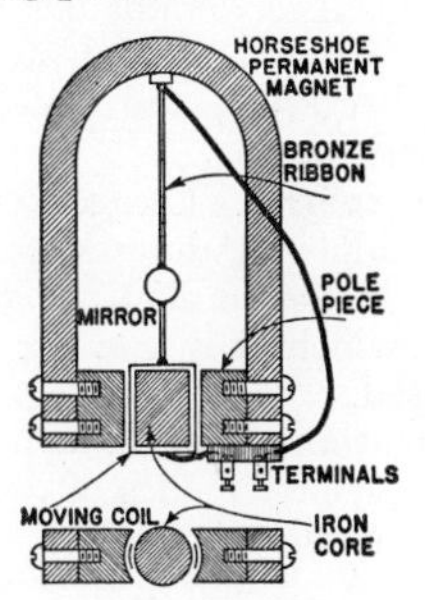


Mirror galvanometer construction, side view and top view.

**mirror nuclides** Pairs of nuclides, each member of which would be transformed into the other by exchanging all neutrons for protons and vice versa.
**mirror reflection** *Direct reflection.*
**mirror-reflection echo** A radar echo that undergoes multiple reflections, as by reflection from the side of an aircraft carrier or other large flat surface before being reflected from a nearby target.
**misch metal** An alloy of cerium, lanthanum, and didymium sometimes used on the cathodes of glow tubes.
**misfire** Failure to establish an arc between the main anode and the cathode of an ignitron or other mercury-arc rectifier during a scheduled conducting period.
**mismatch** The condition in which the impedance of a source does not match or equal the impedance of the connected load or transmission line.
**mismatch factor** *Reflection coefficient.*
**miss-distance indicator** [abbreviated mdi] An indicator system that uses radio transponders in a target and in each test-fired missile to transmit to receiving and recording equipment at a ground station the minimum separation between target and missile.
**missile** *Guided missile.*
**missile acquisition** A measuring system for providing angular data (azimuth and elevation angle) by phase comparison of radio-frequency signals received from a missile.
**missile master** A system of computers, communication equipment, and data-processing equipment used to collect data on approaching aircraft from all available sources, evaluate it, select targets, and direct fire of guided missile batteries automatically.
**missilier** A person skilled in launching and directing guided missiles.
**mistake** An error resulting from incorrect programing, coding, or some other manual operation in a computer.
**mixed-base notation** A computer number system in which a single base, such as 10 in the decimal system, is replaced by two number bases used alternately, such as 2 and 5. Biquinary notation is an example.
**mixed highs** The high-frequency signal components that are intended to be reproduced achromatically (without color) in a color television picture.
**mixer** 1. A device having two or more inputs, usually adjustable, and a common output. It is used to combine separate audio or video signals linearly in desired proportions to produce an output signal. 2. The stage in a superheterodyne receiver in which the incoming modulated r-f signal is combined with the signal of a local r-f oscillator to produce a modulated i-f signal. Crystal diodes are widely used as mixers in radar and other microwave equipment. Also called first detector and mixer-first detector. The mixer and oscillator together form the converter.
**mixer-first detector** *Mixer.*
**mixer tube** An electron tube that performs only the frequency-conversion function of a converter in a superheterodyne receiver. It is supplied with voltage or power by a separate local oscillator.
**mixing** Combining two or more signals, such as the outputs of several microphones.
**mixing amplifier** An amplifier having inputs to which two or more different signals

are applied, and a common output that delivers a composite signal.

**mksa unit** Abbreviation for *meter-kilogram-second-ampere unit.*

**mks system** Abbreviation for *meter-kilogram-second system.*

**M line** One of the characteristic lines in the spectrum of an atom. It is produced by excitation of the electrons of the M shell.

**mm** Abbreviation for *millimeter.*

**mmc** Abbreviation for *megamegacycle.*

**mmf** Alternate abbreviation for *micromicrofarad.*

**mmfd** Alternate abbreviation for *micromicrofarad.*

**mobile station** A radio station intended to be used while in motion or during halts at unspecified points.

**mobile telemetering** Telemetering between points that may have relative motion, using radio in place of interconnecting wires.

**mobile transmitter** A radio transmitter designed for installation in a vessel, vehicle, or aircraft, and normally operated while in motion.

**mobile unit** A truck or other vehicle equipped with television studio equipment. It is used for television pickups at remote locations. Picture and sound signals are sent back to the main transmitter either by a relay transmitter on the truck or through coaxial cables.

**mobility** Freedom of particles to move, either in random motion or under the influence of fields or forces.

**mode** 1. A state of a vibrating system that corresponds to a particular field pattern and one of the possible resonant frequencies of the system. The three common modes of vibration in a quartz plate are the extensional, flexural, and shear modes. 2. A form of propagation of guided waves that is characterized by a particular field pattern in a plane transverse to the direction of propagation. The field pattern is independent of position along the axis of the waveguide. For uniconductor waveguides the field pattern of a particular mode of propagation is also independent of frequency. The TE mode is the transverse electric mode, while the TM mode is the transverse magnetic mode. Also called transmission mode.

**mode changer** *Mode transducer.*

**mode filter** A selective device designed to pass energy along a waveguide in one or more modes of propagation and substantially reduce energy carried by other modes.

**mode jump** A sudden and irregular change in the oscillation frequency and power output of a magnetron, due to a change in the mode of operation from one pulse to the next.

**modem** [MOdulator DEModulator] A carrier terminal panel containing a modulator and demodulator, some circuits of which may be in common.

**mode purity** 1. The ratio of power present in the forward-traveling wave of a desired mode to the total power present in the forward-traveling waves of all modes. 2. The extent to which an atr tube in its mount is free from undesirable mode conversion.

**moder** *Coder.*

**moderation** The slowing down of a particle, usually a neutron, as a result of collisions with nuclei. Also called degradation.

**moderator** The material used in a nuclear reactor to moderate or slow down neutrons from the high energies at which they are released. A good moderator has high scattering cross section and low atomic weight. Graphite, beryllium, and water are widely used as moderators.

**moderator control** Control of a nuclear reactor by adjusting the position or quantity of the moderator, so neutrons are slowed to speeds at which they can excite fission in the fuel used.

**mode separation** The frequency difference between resonator modes of oscillation in a microwave oscillator.

**mode shift** A change in the mode of magnetron operation during the interval of a pulse.

**mode skip** Failure of a magnetron to fire on successive pulses.

**mode transducer** A device for transforming an electromagnetic wave from one mode of propagation to another. Also called mode changer and mode transformer.

**mode transformer** *Mode transducer.*

**modifier** A quantity used to alter the address of an operand in a computer, such as the cycle index.

**modify** 1. To alter the address of the operand in a computer instruction. 2. To alter a computer subroutine according to a defined parameter.

**modular construction** Construction involving the use of integral multiples of a given length for the dimensions of electronic components and electronic equipment, as well as for spacings of holes in a chassis or printed-wiring board.

**modulate** To vary the amplitude, frequency, or phase of a wave or vary the velocity of the electrons in an electron beam in some characteristic manner.

**modulated amplifier** The amplifier stage in a transmitter at which the modulating signal is introduced to modulate the carrier.

**modulated-beam photoelectric system** A photoelectric intrusion-detector system in which reliable beam ranges of several thousand feet are obtained by interrupting the light beam at the source with a rotating punched or slotted disk, so the output of the phototube is an a-c signal that can readily be amplified.

**modulated carrier** An r-f carrier whose amplitude or frequency has been varied in accordance with the intelligence to be conveyed.

**modulated continuous wave** [abbreviated mcw] A form of emission in which the carrier is modulated by a constant a-f tone. In telegraphic service, the carrier is keyed.

**modulated light** Light that has been made to vary in intensity in accordance with variations in an audio, facsimile, or code signal.

**modulated stage** The r-f stage to which the modulator is coupled and in which the carrier wave is modulated by an audio, video, code, or other intelligence signal.

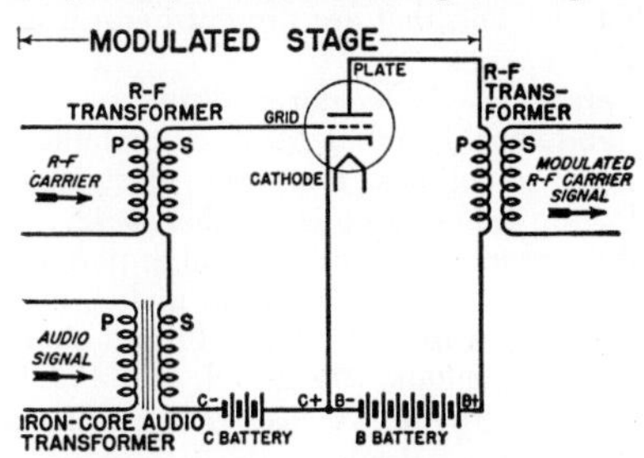


Modulated stage for radio transmitter.

**modulated wave** A carrier wave whose amplitude, frequency, or phase varies in accordance with the value of the intelligence signal being transmitted.

**modulating-anode klystron** A klystron having between the cathode and the drift tube section an electrode that can be used to turn the electron beam on and off for pulse generation.

**modulating electrode** An electrode to which a potential is applied to control the magnitude of the beam current in a cathode-ray tube.

**modulating signal** A signal that causes a variation of some characteristic of a carrier. Also called modulating wave.

**modulating wave** *Modulating signal.*

**modulation** The process by which some characteristic of one wave is varied in accordance with another wave. In radio broadcasting, some stations use amplitude modulation while others use frequency modulation. In television, the picture portion of the program uses amplitude modulation, while the sound portion uses frequency modulation.

**modulation capability** The maximum percentage modulation that is possible without objectionable distortion.

**modulation envelope** A curve drawn through the peaks of a graph showing the waveform of a modulated signal. The modulation envelope represents the waveform of the intelligence carried by the signal.

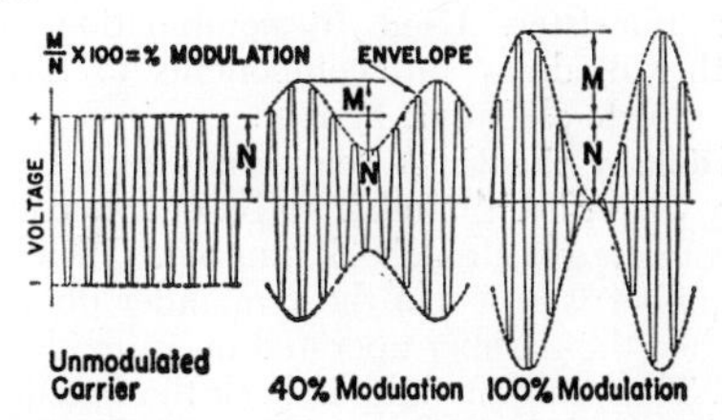


Modulation envelope and modulation percentage.

**modulation factor** The ratio of the peak variation in the modulation actually used in a transmitter to the maximum variation for which the transmitter was designed. Often expressed as a percentage. In amplitude modulation, the modulation factor is the ratio of half the difference between the maximum and minimum amplitudes of an amplitude-modulated wave to the average amplitude. For frequency modulation, it is the ratio of the actual frequency swing to the frequency swing required for 100% modulation, expressed in percentage.

**modulation index** The ratio of the frequency deviation to the frequency of the modulating wave in a frequency-modulation system when using a sinusoidal modulating wave.

**modulation meter** An instrument for measuring the modulation factor of a modulated wave train at a transmitter. The readings are usually expressed in per cent. Also called modulation monitor (deprecated).

**modulation monitor** Deprecated term for *modulation meter.*

**modulation noise** Noise that is caused by the modulating signal, making the noise level a function of the strength of the signal. Also called noise behind the signal.

**modulation percentage** The percentage value obtained by multiplying the modulation factor by 100.

**modulator** 1. The transmitter stage that supplies the modulating signal to the modulated amplifier stage, or triggers the modulated amplifier stage to produce pulses at desired instants as in radar. 2. A device that produces modulation by any means, such as by virtue of a nonlinear characteristic or by controlling some circuit quantity in accordance with the waveform of a modulating signal. 3. One of the electrodes of a spacistor.

**module** A packaged assembly of wired components, built in a standardized size and having standardized plug-in or solderable terminations. It may or may not include an amplifying element such as a tube or transistor. Used in combination with other modules and components to form a complete electronic system.

**modulo *n* check** A computer check system in which each number to be operated on is associated with a check number. This check number is equal to the remainder obtained when the number operated on is divided by *n*. Thus, in a modulo 4 check the remainder serving as the check number will be 0, 1, 2, or 3.

**moire** A spurious pattern in a reproduced television picture, resulting from interference beats between two sets of periodic structures in the image. Moires may be produced, for example, by interference between patterns in the original subject and that of the target grid in an image orthicon, or between subject patterns and the line and phosphor-dot patterns of a color picture tube.

**mold** The metal part derived from the master by electroforming in disk recording. A mold has grooves similar to those of a recording.

**molded capacitor** A capacitor that has been encased in a molded plastic insulating material to keep out dust and moisture.

**molecular beam** A unidirectional beam of neutral molecules. Measurement of the transmission of the beam through appropriate magnetic and electric fields in a vacuum can yield accurate values for such quantities as nuclear magnetic moments.

**molecular distillation** An isotope separation process in which molecules are evaporated from a surface at extremely low pressure and are condensed before undergoing collisions.

**molecular electronics** The branch of electronics that deals with the production of complex electronic circuits in microminiature form by producing semiconductor devices and circuit elements integrally while growing multizoned crystals in a furnace. Also called moletronics.

**molecular microwave amplifier** A solid-state amplifier in which the interaction is between uncharged molecular matter and the microwave field. Examples include the maser and the parametric amplifier.

**molecular pump** A vacuum pump in which the molecules of the gas to be exhausted are carried away by the friction between them and a rapidly revolving disk or drum.

**molecular stopping power** The energy loss per molecule of a compound, per unit area normal to the motion of the particle, when a charged particle passes through a medium. It is very nearly if not exactly equal to the sum of the atomic stopping powers of the constituent atoms.

**molecule** The smallest particle into which an element or compound may be divided and still retain the chemical properties of the element or compound in mass.

**mole fraction** The number of atoms of a certain isotope of an element, expressed as a fraction of the total number of atoms of that element, that are present in an isotopic mixture.

**moletronics** *Molecular electronics.*

**molybdenum** [symbol Mo] A metallic element, sometimes used for electrodes of electron tubes. Atomic number is 42.

**molybdenum permalloy** A high-permeability alloy consisting chiefly of nickel, molybdenum, and iron. A typical formulation has 4% molybdenum, 79% nickel, and 17% iron.

**monatomic layer** A coating consisting of a single layer of atoms.

**monaural recorded tape** Magnetic tape that has been recorded for use in single-channel audio systems, rather than for stereo systems.

**monaural recorder** A tape recorder having a single-channel audio system consisting of one microphone, one amplifier, and one recording head, used for producing monaural recorded tape.

**monaural sound system** A sound-reproducing system in which one or more microphones feed a single channel that terminates in an earphone or headphones worn by the listener.

**monitor** 1. An instrument used to measure continuously or at intervals a condition that must be kept within prescribed limits, such as radioactivity at some point in a nuclear reactor, the image picked up by a television camera, the sound picked up by a microphone at a radio or television studio, a variable quantity in an automatic process

Monitor console at television station.

control system, the transmissions in a communication channel or band, or the position of an aircraft in flight. 2. A person who watches a monitor.

**monitor head** An additional playback head provided on some tape recorders to permit playing back the recorded sounds off the tape while the recording is being made.

**monitoring** Using a monitor.

**monitoring amplifier** A power amplifier used primarily for evaluation and supervision of a program.

**monitoring antenna** An antenna used to pick up the r-f output of a transmitter at the transmitter site for over-all monitoring purposes.

**monitor ionization chamber** 1. An ionization chamber mounted in an x-ray beam and connected to a continuously reading instrument, to serve as an indicator of constancy of x-ray output. 2. An ionization chamber used to detect the presence of undesirable radiation in connection with health protection.

**monkey chatter** *Adjacent-channel interference.*

**mono-acceleration cathode-ray tube** An electrostatic cathode-ray tube in which all acceleration of the electron beam is produced before the beam passes through the deflection electrodes.

**monochromatic** Having only one color, corresponding to a negligibly small region of the spectrum.

**monochromatic radiation** Electromagnetic radiation having a single wavelength, or photons all having the same energy. Although no actual radiation is strictly monochromatic, it can have an extremely narrow band of wavelengths, as does sodium light and the spectrum of mercury-198.

**monochromatic sensitivity** The response of a device to light of a given color.

**monochrome** Having only one chromaticity. This is achromatic in television, involving only shades of gray between black and white.

**monochrome bandwidth** The video bandwidth of the monochrome channel or the monochrome signal in color television.

**monochrome channel** Any path intended to carry the monochrome signal in a color television system. This path may also carry other signals, such as the chrominance signal.

**monochrome signal** 1. A signal wave used for controlling luminance values in monochrome television. 2. The portion of a signal wave that has major control of the luminance values in a color television system, regardless of whether the picture is displayed in color or in monochrome.

**monochrome television** Television in which the final reproduced picture is monochrome, having only shades of gray between black and white. Also called black-and-white television.

**monochrome transmission** Transmission of a signal wave that controls the luminance values in a television picture but not the chromaticity values. The result is a monochrome picture.

**monoenergetic radiation** Radiation consisting of particles of a given type, all of which have the same energy.

**monogroove stereo record** *Stereo record.*

**monophonic sound system** A sound-reproducing system in which one or more microphones feed a single channel that terminates in one or more loudspeakers. A monaural sound system is similar except that it terminates in earphones. Examples include phonographs, radio, television, sound motion pictures, magnetic-tape reproducers, and public-address systems.

**monopulse radar** Radar in which directional information is obtained with high precision by using a receiving antenna system having two or more partially overlapping lobes in the radiation patterns. Sum and difference channels in the receiver compare the amplitudes of the antenna outputs (sum-and-difference monopulse radar) or compare the phases of the antenna outputs (phase-sensing monopulse radar). Also called monopulse tracking, simultaneous lobing, and static split tracking.

**monopulse tracking** *Monopulse radar.*

**monoscope** A signal-generating electron-beam tube in which a picture signal is produced by scanning an electrode that has a predetermined pattern of secondary-emission response over its surface. This fixed image is printed on the electrode during manufacture of the tube, to give a useful test pattern for testing and adjusting television equipment.

**monostable** Having only one stable state.

**monostable multivibrator** A multivibrator with one stable state and one unstable state. A trigger signal is required to drive the unit into the unstable state, where it

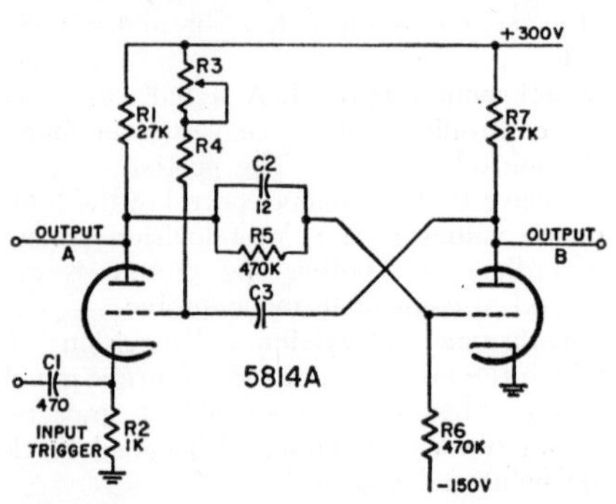


Monostable multivibrator, using preferred circuit design by National Bureau of Standards, gives positive or negative gate output pulse having 210-volt peak when triggered by a 25-volt positive input pulse. Sum of R3 and R4 must be between 0.5 and 3 megohms; when sum is 2 megohms, output pulse duration in microseconds is equal to value in micromicrofarads used for C3.

remains for a predetermined time before returning to the stable state. Also called one-shot multivibrator, single-shot multivibrator, and start-stop multivibrator.

**Monte Carlo method** A method of solving a group of physical problems by making a series of statistical experiments. Accuracy of results depends on the number of trials. Usually practical only when set up to be performed by a computer.

**Moore code** An equal-length message-transmitting code in which 70 possible characters have four marking intervals and four spacing intervals. If a received character has other than eight intervals, an error signal is printed or repetition is automatically requested. A 7-unit version having 35 characters uses three marking and four spacing intervals.

**mopa** Abbreviation for *master oscillator–power amplifier.*

**moptar** [Multi-Object Phase Tracking And Ranging] A single-site dme-cotar system, time-multiplexed to provide complete trajectory information on the flight of several missiles in flight simultaneously. Coding modulation is used to interrogate the transponder of the desired missile during each sampling period.

**Morse code** A message-transmitting code consisting of dot and dash signals. The international Morse code is universally used for radiotelegraphy. The American Morse code is used only for wire telegraphy.

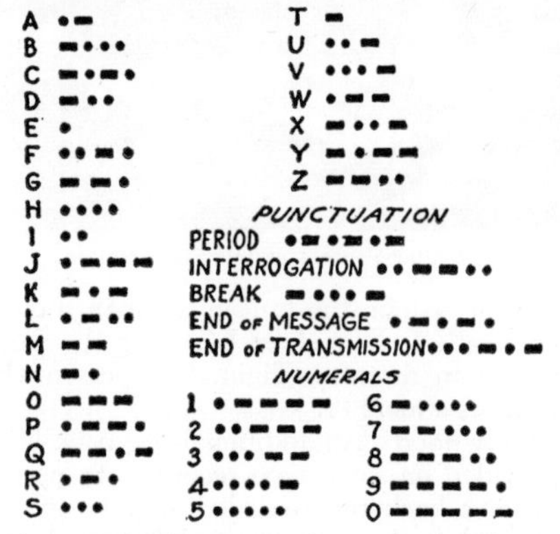


Morse code (international) used in radiotelegraphy.

**mosaic** A light-sensitive surface used in television camera tubes, consisting of a thin mica sheet coated on one side with a large number of tiny photosensitive silver-cesium globules, insulated from each other. The picture is optically focused on the mosaic, and the resulting charges on the globules are scanned by the electron beam of the camera tube.

**mother** A mold derived by electroforming from a master. Used to produce the stampers from which disk records are molded in large quantities. Also called metal positive.

**mother aircraft** An aircraft that carries the electronic equipment needed to direct a drone.

**mother crystal** A raw piezoelectric crystal as found in nature or grown artificially.

**motional impedance** The complex remainder after the blocked impedance of a transducer has been subtracted from the loaded impedance. Also called loaded motional impedance.

**motion-picture pickup** Use of a television

camera to pick up scenes directly from motion-picture film.

**motor** A machine that converts electric energy into mechanical energy by utilizing forces exerted by magnetic fields produced by current flow through conductors. Also called electric motor.

Motor connections, showing three ways in which commutator of d-c motor can be connected to field: left—shunt motor; center—series motor; right—compound motor.

**motor board** The platform on which the motor, reels, heads, and controls of a tape recorder are mounted.

**motorboating** Oscillation in a system or component, usually manifested by a succession of pulses occurring at a very low audio-frequency rate. Generally due to excessive positive feedback, such as through a common power supply. When a loudspeaker is connected to the system, as in a radio, the pulses produce a put-put sound like that of a motorboat.

**motor control** *Electronic motor control.*

**motor effect** The repulsion force exerted between adjacent conductors carrying currents in opposite directions.

**motor-field induction heater** An induction heater in which the induction winding typifies that of an induction motor of rotary or linear design.

**motor-generator set** A motor and generator that are coupled mechanically for use in changing one power-source voltage to other desired voltages or frequencies.

**Mott scattering formula** The formula giving the differential cross section for the scattering of identical particles due to a coulomb force.

**mount** 1. A shock or vibration isolator, usually consisting of one elastic member and one or more relatively inelastic members, used as a support for the equipment to be isolated. 2. The flange or other means by which a switching tube, or tube and cavity, is connected to a waveguide.

**mountain effect** The effect of rough terrain on radio-wave propagation, causing reflections that produce errors in radio direction-finder indications.

**mouth** The end of a horn that has the larger cross-sectional area.

**movable contact** The relay contact that is mechanically displaced to engage or disengage one or more stationary contacts. Also called armature contact.

**movement** Deprecated term for *moving element.*

**moving-armature loudspeaker** *Magnetic-armature loudspeaker.*

**moving-coil galvanometer** A galvanometer in which the current to be measured is sent through a coil suspended or pivoted in a fixed magnetic field.

**moving-coil hydrophone** A moving-coil microphone that responds to waterborne sound waves.

**moving-coil loudspeaker** *Dynamic loudspeaker.*

**moving-coil meter** A meter in which a pivoted coil is the moving element.

**moving-coil microphone** *Dynamic microphone.*

**moving-coil pickup** A pickup in which the electric output is due to motion of a coil or conductor in a constant magnetic field. In a moving-coil phono pickup the coil is moved by the needle as it follows the grooves of a record. Also called dynamic pickup and dynamic reproducer.

**moving-conductor hydrophone** A moving-conductor microphone that responds to waterborne sound waves.

**moving-conductor loudspeaker** A loudspeaker in which the mechanical forces result from reactions between a steady magnetic field and the magnetic field produced by current flow through a moving conductor.

**moving-conductor microphone** A microphone whose electric output results from motion of a conductor or coil in a magnetic field. Examples include dynamic microphone and velocity microphone.

**moving element** The instrument element that moves as a direct result of a variation in the quantity that the instrument is measuring. Also called movement (deprecated).

**moving-iron instrument** An instrument that depends on current in one or more fixed coils acting on one or more pieces of soft iron, at least one of which is movable.

**moving-magnet instrument** An instrument that depends on the action of a movable permanent magnet in aligning itself in the resultant field produced either by a fixed permanent magnet and an adjacent coil or coils carrying current, or by two or more current-carrying coils whose axes are displaced by a fixed angle.

**moving-magnet magnetometer** A magnetometer that depends for its operation on the torques acting on a system of one or more permanent magnets that can turn in the field to be measured.

**moving-target indicator** [abbreviated mti] A device that limits the display of radar information primarily to moving targets. Signals due to reflections from stationary objects are canceled by a memory circuit.

**MS** Designation for Military Standard.

**M scan** *M display.*

**M scope** A radarscope that produces an M display.

**M shell** The third layer of electrons about the nucleus of an atom, having electrons characterized by the principal quantum number 3.

**mti** Abbreviation for *moving-target indicator.*

**mti subclutter visibility** The gain in signal-to-clutter power ratio produced by a moving-target indicator.

**mu** [Greek letter μ] 1. Symbol for *amplification factor.* 2. Symbol for *permeability.* 3. Abbreviation for prefix *micro-*. 4. Abbreviation for *micron,* a unit of length equal to one-millionth of a meter.

**μa** Abbreviation for *microampere.*

**μc** Abbreviation for *microcurie.*

**muf** Abbreviation for *maximum usable frequency.*

**μf** Abbreviation for *microfarad.*

**μ-factor** The ratio of the magnitude of an infinitesimal change in the voltage at the $j$th electrode of an $n$-terminal electron tube to the magnitude of an infinitesimal change in the voltage at the $l$th electrode, with the current to the $m$th electrode remaining unchanged.

**μh** Abbreviation for *microhenry.*

**multi-anode tube** An electron tube having two or more main anodes and a single cathode. This term is used chiefly for pool-cathode tubes.

**multiar** An amplitude-comparison circuit that delivers a pulse when two amplifiers achieve the same amplitude.

**multiband antenna** An antenna that may be used satisfactorily on more than one frequency band.

**multicavity magnetron** A magnetron in which the circuit includes a plurality of cavities, generally cut into the solid cylindrical anode in such a way that the mouths of the cavities face the central cathode.

**multicellular horn** 1. A cluster of horn antennas having mouths that lie in a common surface. The horns are fed from openings spaced one wavelength apart in one face of a common waveguide, and serve to provide a desired directional radiation pattern

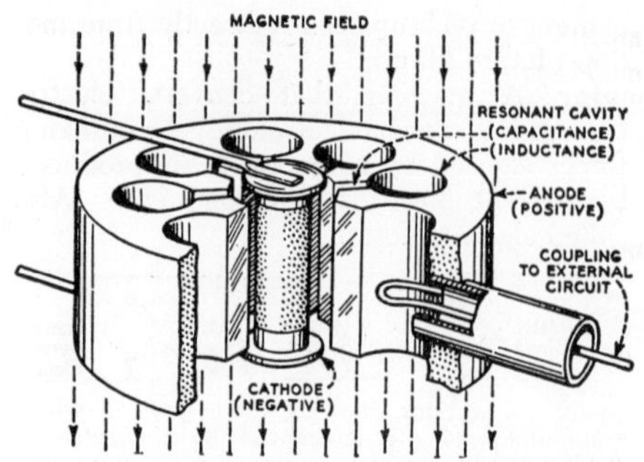

Multicavity magnetron, showing central cathode, coupling loop, and direction of magnetic field.

for the radiated energy. 2. A combination of individual horn loudspeakers having individual driver units or joined in groups to a common driver unit. Subdivision of a large horn into smaller horns makes it easier to control the pressure and phase of the acoustic waves across the mouths of the horns. Also called cellular horn.

**multichannel analyzer** *Pulse-height analyzer.*

**multichannel radio transmitter** A radio transmitter having two or more complete r-f portions capable of operating on different frequencies, either individually or simultaneously.

**multichannel stereo** A stereophonic sound system having more than two transmitting or recording channels.

**multielectrode tube** An electron tube containing more than three electrodes associated with a single electron stream.

**multifrequency transmitter** A radio transmitter capable of operating on two or more selectable frequencies, one at a time, using preset adjustments.

**multigrid tube** An electron tube having more than one grid electrode.

**multihop transmission** Transmission in which radio waves are reflected and refracted between the earth and ionosphere several times in their path of travel to a receiver far beyond the direct transmission range. Also called multiple-hop transmission.

**Multimeter** Trademark of Rawson Electrical Instrument Co. for their line of volt-ohm-milliammeters.

**multimeter** *Volt-ohm-milliammeter.*

**multipactor gap loading** The electronic gap admittance resulting from a sustained secondary-emission discharge produced in a gap by the motion of secondary electrons in synchronism with the electric field in the gap.

**multipath** *Multipath transmission.*

**multipath cancellation** Effectively complete cancellation of radio signals because of the relative amplitude and phase differences of components arriving over different paths.

**multipath transmission** The propagation phenomenon that results in signals reaching a radio receiving antenna by two or more paths, causing distortion in radio and ghost images in television. At least one of the paths involves reflection from some object. Also called multipath.

**multiple** *Parallel.*

**multiple-address code** A computer instruction code in which more than one address or storage location is specified. The instruction may give the locations of the operands, the destination of the result, and the location of the next instruction.

**multiple-break contacts** Contacts so arranged that a circuit is interrupted in two or more places when the contacts open.

**multiple circuit** A circuit in which two or more identical or different circuits are connected in parallel to a common signal or power source.

**multiple-contact switch** *Selector switch.*

**multiple course** One of a family of lines of position defined by a navigation system, any one of which may be selected as a course line.

**multiple decay** *Branching.*

**multiple disintegration** *Branching.*

**multiple grating** *Ultrasonic cross grating.*

**multiple-hop transmission** *Multihop transmission.*

**multiple modulation** Modulation in which the modulated wave from one process becomes the modulating wave for the next. Thus, in an amplitude-modulated pulse-position-modulation system one or more signals are used to position-modulate their respective pulse subcarriers, which are spaced in time and are used to amplitude-modulate a carrier.

**multiple-purpose tester** *Volt-ohm-milliammeter.*

**multiple scattering** Scattering in which the final displacement is the vector sum of many small displacements. Multiple scattering is greater than single scattering and plural scattering.

**multiple sound track** A group of sound tracks printed adjacently on a common base, independent in character but having a common time relationship, such as for stereophonic sound recording.

**multiple-speed floating action** Floating action in which a final control element is moved at two or more speeds, each corresponding to a definite range of values of deviation.

**multiple-spot scanning** Scanning by two or more scanning spots simultaneously, each analyzing its fraction of the total scanned area of the subject copy.

**multiple-track range** A range system using two closely spaced synchronized pulse stations, similar to gee. The indicator in the aircraft has several predetermined time difference settings by which a number of approximately parallel tracks may be flown.

**multiple-trip echo** An echo returned from a target so distant that the time required for the radar pulse to go out to the target and return to the set is longer than the interval between two successive pulses. These echoes show up at false ranges, indentifiable because they change when the pulse repetition rate is changed. Called a second-trip echo when the echo arrives between the second and third pulses.

**multiple tube counts** Spurious counts induced by previous tube counts in a radiation-counter tube.

**multiple-tuned antenna** A low-frequency antenna having a horizontal section connected to a multiplicity of tuned vertical sections.

**multiple-unit tube** An electron tube containing within one envelope two or more groups of electrodes associated with independent electron streams, such as a duodiode, duotriode, diode-pentode, duodiode-triode, duodiode-pentode, and triode-pentode.

**multiplex** The simultaneous transmission of two or more programs or signals over a single r-f channel, such as by time division, frequency division, or phase division. Also called multiplexing and multiplex transmission.

**multiplex channel** A carrier channel that provides two or more services simultaneously in the same or opposite directions.

**multiplexer** A device for combining two or more signals, as for multiplex or for creating the composite color video signal from its components in color television.

**multiplexing** *Multiplex.*

**multiplex operation** Simultaneous transmission of two or more messages in either or both directions over a multiplex channel.

**multiplex radio transmission** The simultaneous transmission of two or more signals by radio, using a common carrier wave.

**multiplex transmission** *Multiplex.*

**multiplication** The ratio of neutron flux in a subcritical reactor to that supplied by a neutron source. It is the factor by which, in effect, the reactor multiplies the source strength.

**multiplication constant** *Reactivity.*

**multiplication point** A mixing point whose output is obtained by multiplication of its inputs in a feedback control system.

**multiplier** 1. A device that has two or more inputs and an output that is a representation of the product of the quantities represented by the input signals. Voltages are the quantities commonly multiplied. 2. A resistor used in series with a voltmeter to increase the voltage range. Scale readings are then multiplied by a factor equal to the ratio of total resistance (meter resistance plus multiplier resistor value) to meter resistance. Also called multiplier resistor. 3. *Electron multiplier.* 4. *Frequency multiplier.*

**multiplier phototube** A phototube with one or more dynodes between its photocathode and the output electrode. The electron stream from the photocathode is reflected

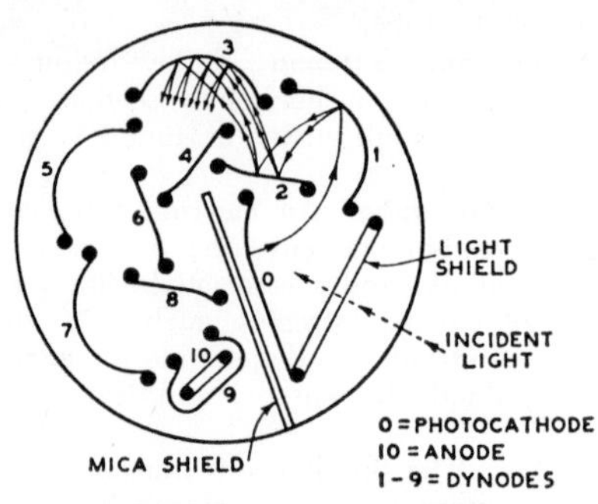


Multiplier phototube operation.

off each dynode in turn, with secondary emission adding electrons to the stream at each reflection. Also called electron-multiplier phototube and photoelectric multiplier. A deprecated synonym is photomultiplier tube.

**multiplier resistor** *Multiplier.*

**multiplying factor** The number by which the reading of a given meter must be multiplied to obtain the true value.

**multipolar** Having more than one pair of magnetic poles.

**multipole moment** A measure of the charge, current, and magnet distributions of a system of particles.

**multiport network** A network having more than one port.

**multiposition action** Automatic control action in which a final control element is moved to one of three or more predetermined positions, each corresponding to a definite value or range of values of the controlled variable.

**multisegment magnetron** A magnetron with an anode divided into more than two segments, usually by slots parallel to its axis.

**multitrack recording system** A recording system that provides two or more recording paths on a medium, carrying related or unrelated recordings having a common time relationship.

**multivibrator** A relaxation oscillator using two tubes, transistors, or other electron devices, with the output of each coupled to the input of the other through resistance-capacitance elements or other elements to obtain in-phase feedback voltage. The fundamental frequency is determined by the time constants of the coupling elements and may be further controlled by an external voltage. When such circuits are normally in a nonoscillating state and a trigger signal is required to start a single cycle of operation, the circuit is commonly called a one-shot multivibrator, a flip-flop circuit, or a start-stop multivibrator.

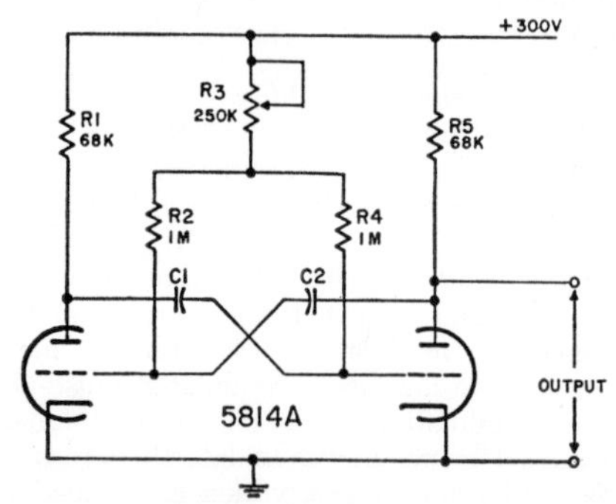


Multivibrator, using preferred circuit design by National Bureau of Standards, gives square-wave 260-volt output waveform having pulse-repetition frequency *f* determined by values used for capacitors; *f* in cps is 790,000 divided by the value of C1 or C2 in micromicrofarads.

**mu meson** A meson having a mean life of $2.1 \times 10^{-6}$ second and a mass 210 times that of an electron. Also called muon. Formerly called mesotron.

**Mumetal** Trademark of Allegheny Ludlum Steel Corp. for a magnetic alloy having high permeability and low hysteresis, usually consisting of iron, nickel, and small quantities of other metals. Used to shield components, tubes, and equipment from stray magnetic fields.

**μμ** Abbreviation for prefix *micromicro-*.

**μμf** Abbreviation for *micromicrofarad.*

**Munsell chroma** *Chroma.*

**Munsell value** The dimension, in the Munsell system of object-color specification, that indicates the apparent luminous transmittance or reflectance of the object on a scale having approximately equal perceptual steps under the usual conditions of observation.

**muon** *Mu meson.*

**muonium** A bound system consisting of a positive mu meson and an electron.

**Murray loop test** A method of localizing a fault in a cable by replacing two arms of a Wheatstone bridge with a loop formed by the cable under test and a good cable connected to the far end of the defective cable.

**musa** [Multiple-Unit Steerable Antenna] An electrically steerable receiving antenna whose directional pattern can be rotated by varying the phases of the contributions of the individual units. These units are usually stationary rhombic antennas.

**μsec** Abbreviation for *microsecond.*

**musical echo** A flutter echo that is periodic and has a flutter whose frequency is in the audio range.

**muting** Silencing, or reducing in volume.

**muting switch** 1. A switch used in connection with automatic tuning systems to silence the receiver while tuning from one station to another. 2. A switch used to ground the output of a phonograph pickup automatically while a changer is in its change cycle.

**mutual conductance** *Transconductance.*

**mutual-conductance meter** *Transconductance meter.*

**mutual inductance** [symbol $M$] A measure of the amount of inductive coupling that exists between two coils. It is related to the flux linkages produced in one coil by current in the other coil. Measured in henrys, millihenrys, and microhenrys the same as for inductance.

**μv** Abbreviation for *microvolt.*

**μw** Abbreviation for *microwatt.*

**mv** Abbreviation for *millivolt.*

**mw** Abbreviation for *milliwatt.*

**Mycalex** Trademark of Mycalex Corp. of America for their molded insulating material consisting chiefly of ground mica and lead borate glass fused together under high pressure at high temperature.

**Mylar** Trademark of Du Pont for their polyester film that is widely used for insulating purposes and as a backing for magnetic tape.

**Mylar capacitor** A capacitor that uses Mylar film as a dielectric between rolled strips of metal foil.

# N

**N** Symbol for *nitrogen*.

***N*** 1. Symbol for *neutron number*. 2. Symbol for number of turns.

**NAB** Abbreviation for National Association of Broadcasters, known before January 1, 1958, as NARTB.

**NAB curve** The standard playback equalization curve adopted by the National Association of Broadcasters for disk recordings.

**nancy** A system of visual blinker-signal communication using invisible infrared rays, used for ship-to-ship signaling at night.

**nancy ray** U. S. Navy term for *infrared radiation*.

**nancy receiver** U. S. Navy term for *infrared receiver*.

**nano-** A prefix representing $10^{-9}$.

**nanofarad** [abbreviated nf] One-billionth of a farad, equal to $10^{-9}$ farad, 0.001 microfarad, and 1,000 micromicrofarads. Used chiefly in Europe.

**nanosecond** [abbreviated nsec] One-billionth of a second, or $10^{-9}$ second.

**napier** *Neper*.

**Napierian logarithm** *Natural logarithm*.

**narrow-band axis** The direction of the phasor representing the coarse chrominance primary in color television, having a bandwidth extending from 0 to 0.5 mc.

**narrow-band frequency modulation** An f-m broadcasting system used primarily for two-way voice communication. For police, fire, taxicab, and other mobile communications, 15 kc is the maximum permissible deviation. For amateur radio, the deviation limit is 3 kc.

**narrow-band frequency-modulation adapter** An adapter that converts an amplitude-modulated communication receiver for f-m reception.

**NARTB** Abbreviation for National Association of Radio and Television Broadcasters, a title used for a period of several years by NAB. On January 1, 1958, the organization voted to return to its shorter original name.

**National Electrical Code** A set of regulations governing construction and installation of electric wiring and apparatus in the United States, established by the American National Board of Fire Underwriters for safety purposes.

**National Electrical Manufacturers Association** [abbreviated NEMA] An organization of manufacturers of electric products.

**National Electronics Conference** [abbreviated NEC] An annual technical conference held in Chicago, generally in October.

**National Television System Committee** [abbreviated NTSC] A committee organized in 1940 by representatives of United States companies and organizations interested in television. It formulated television standards for black-and-white television in 1940–1941 and for color television in 1950–1953 that were approved by the Federal Communications Commission.

**natural frequency** 1. A frequency at which a body or system will oscillate freely. 2. The lowest resonant frequency of an antenna, circuit, or component.

**natural logarithm** [abbreviated ln] A logarithm using the base 2.71828. Also called hyperbolic logarithm and Napierian logarithm.

**natural period** The period of a free oscillation of a body or system.

**natural radioactivity** Radioactivity exhibited by naturally occurring radionuclides.

**natural-uranium reactor** A nuclear reactor in which natural unenriched uranium is the principal fissionable material.

**natural wavelength** The wavelength corresponding to the natural frequency of an antenna or circuit.

**nautical mile** A measure of distance equal to 6,076.105 feet or approximately 1.15 miles. The nautical mile in navigation practice is one minute of a great circle. Because the earth is not a perfect sphere, different distances have been assigned to this measure. An Air Force regulation in 1954 specified its value as 6,076.10333 feet.

**navaglide** An instrument low approach system for aircraft that uses a single frequency on a time-sharing basis for both the localizer and glide-path beams, and also provides a distance indication.

**Navaglobe** A long-distance continuous-wave low-frequency navigation system of the amplitude-comparison type, providing bearing information automatically in an aircraft with respect to an omnidirectional radio range.

**Navaho** A surface-to-surface Air Force missile having a range of 5,000 miles, using celestial or inertial guidance.

**navar** [NAVigation And Ranging] A coordinated series of radar air navigation and traffic-control aids utilizing transmissions at wavelengths of 10 cm and 60 cm to provide in an aircraft the distance and bearing from a given point, along with a display of other aircraft in the vicinity and commands from the ground. The system also provides on the ground a display of all aircraft in the vicinity, with their altitudes, identities, and means for transmitting certain commands.

**navarho** [NAVigation Aid RHO] A long-distance continuous-wave low-frequency navigation system providing simultaneous bearing and distance information.

**navigation** The process of directing a vehicle to reach the intended destination.

**navigation aid** A device or system that provides a navigator with some or all of such navigation data as present position, heading, speed, location of fixed objects and other craft, right-left steering directions or automatic steering control, and altitude. Also called navigation instrument.

**navigation beacon** A light, radio beacon, or radar beacon that provides navigation aid to aircraft and ships.

**navigation computer** A computer that uses electronic or electric circuits to compute two or more navigation factors such as altitude, direction, and velocity, or to receive such data and compute course information.

**navigation coordinate** A quantity whose measurement serves to define a surface of position containing the vehicle, or a line of position if one surface is already known.

**navigation instrument** *Navigation aid.*

**navigation parameter** A visual or aural output of a navigation aid, having a specific relation to navigation coordinates.

**NC** 1. Symbol for *no connection.* Used on tube base diagrams. 2. Symbol for *normally closed,* used with reference to relay contacts.

**N display** A modified radarscope K display in which the target appears as a pair of vertical deflections from the horizontal time base. Direction is indicated by the relative amplitudes of the vertical deflections. Target distance is determined by moving an adjustable pedestal signal along the baseline until it coincides with the horizontal position of the vertical deflections. The pedestal control is calibrated in distance. Also called N scan.

**near-end crosstalk** A type of interference that may occur at carrier telephone repeater stations when output signals of one repeater leak into the same end of the other repeater.

**near field** 1. The acoustic radiation field that is close to the loudspeaker or other acoustic source. 2. The electromagnetic field that exists in the near region, within a distance of 1 wavelength from a transmitting antenna.

**near region** The region immediately surrounding a transmitting antenna, extending out to a distance of 1 wavelength, in which the strength of the induction field varies inversely with the square of the distance. Also called near zone.

**near zone** *Near region.*

**NEC** Abbreviation for *National Electronics Conference.*

**neck** The small tubular part of the envelope of a cathode-ray tube, extending from the funnel to the base and housing the electron gun.

**needle** *Stylus.*

**needle chatter** *Needle talk.*

**needle drag** *Stylus drag.*

**needle force** *Stylus force.*

**needle pressure** Deprecated term for *stylus force.*

**needle scratch** *Surface noise.*

**needle talk** Sounds produced directly as acoustic output by vibrations of a phonograph needle and associated parts of a phonograph pickup. Also called needle chatter.

**negative** 1. A terminal or electrode having more electrons than normal. Electrons flow out of the negative terminal of a voltage source. 2. A designation used to describe an opposite character to positive, as in negative feedback, negative image, negative resistance, and negative transmission.

**negative bias** A grid-bias voltage that makes the control grid of an electron tube negative with respect to its cathode.

**negative charge** An electric charge in which the object in question has more electrons than protons.

**negative electricity** A negative charge, such as that produced in a resin object by rubbing with wool.

**negative electron** An electron, as distinguished from a positive electron or positron. Also called negatron.

**negative feedback** Feedback in which a portion of the output of a circuit, device, or machine is fed back 180° out of phase with the input signal. Negative feedback decreases amplification in such a way as to stabilize the amplification with respect to time or frequency, and also serves to reduce distortion and noise. Also called degeneration, inverse feedback, and stabilized feedback.

**negative-feedback amplifier** An amplifier that uses negative feedback to improve stability and/or widen the frequency response.

**negative ghost image** A ghost image in which white is black and vice versa, due to particular phase and amplitude relations between the direct-path and reflection-path signals.

**negative glow** The luminous glow in a glow-discharge cold-cathode tube, occurring between the cathode dark space and the Faraday dark space.

**negative image** An image in which dark areas are bright and bright areas are dark. Also called reversed image.

**negative impedance** An impedance such that when the current through it increases, the voltage drop across the impedance decreases.

**negative ion** An atom having more electrons than normal, and therefore having a negative charge.

**negative modulation** 1. Modulation in which an increase in brightness corresponds to a decrease in amplitude-modulated transmitter power. Used in U. S. television transmitters and in some facsimile systems. 2. Modulation in which an increase in brightness corresponds to a decrease in the frequency of a frequency-modulated facsimile transmitter. Also called negative transmission.

**negative picture phase** The video signal phase in which the signal voltage swings in a negative direction for an increase in brilliance.

**negative plate** The internal plate structure that is connected to the negative terminal of a storage battery. Electrons flow from the negative terminal through the external load circuit to the positive terminal.

**negative proton** *Antiproton.*

**negative resistance** A resistance such that when the current through it increases, the voltage drop across the resistance decreases. This characteristic is possessed by an electric arc and by some electron-tube circuits.

**negative-resistance magnetron** A magnetron operated in such a way that it acts as a negative resistance.

**negative-resistance oscillator** An oscillator in which a parallel-tuned resonant circuit is connected to a vacuum tube in such a way that the combination acts as the negative resistance needed for continuous oscillation. Dynatron and transitron oscillators are examples.

**negative temperature coefficient** The condition wherein the resistance, length, or some other characteristic of a material decreases when temperature increases.

**negative terminal** The terminal of a battery or other voltage source that has more electrons than normal. Electrons flow from the negative terminal through the external circuit to the positive terminal.

**negative thermion** *Thermoelectron.*

**negative-transconductance oscillator** An electron-tube oscillator in which the output of the tube is coupled back to the input without phase shift, the phase condition for oscillation being satisfied by the negative transconductance of the tube.

**negative transmission** *Negative modulation.*

**negatron** 1. A four-electrode vacuum tube having the characteristics of a negative resistance. 2. *Electron.* 3. *Negative electron.*

**N electron** An electron having an orbit in the N shell, which is the fourth shell of electrons surrounding the atomic nucleus, counting out from the nucleus.

**NEMA** Abbreviation for *National Electrical Manufacturers Association.*

**nemo** [Not Emanating from Main Office] A radio or television broadcast originating outside the studio. Also called remote.

**neodymium** [symbol Nd] A rare-earth element. Atomic number is 60.

**neon** [symbol Ne] An inert gaseous element used in neon signs and in some electron tubes. It produces a characteristic bright red glow when ionized. Atomic number is 10.

**neon glow lamp** A glow lamp containing neon gas. It produces a characteristic red glow. Wattages of standard sizes range from 1/25 watt to 3 watts. Also called neon lamp.

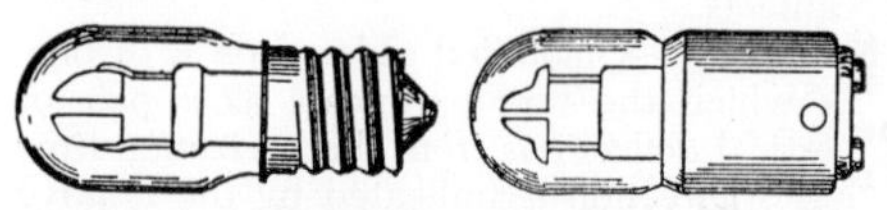

Neon glow lamp with screw base (left) and bayonet base (right).

**neon lamp** *Neon glow lamp.*

**neon oscillator** An oscillator circuit consisting of a neon glow lamp and a capacitor, sometimes also with a resistor.

**neon tubing** A glow lamp in which neon gas is ionized by the flow of electric current through long lengths of gas tubing, to produce a luminous red glow discharge. Used chiefly in outdoor advertising signs.

**neper** A unit used to express the ratio of two voltages, two currents, or two power values in a logarithmic manner. The number of nepers is the natural (Napierian) logarithm of the square root of the ratio of the two values being compared. The neper thus uses the base of 2.71828, whereas the decibel uses the common-logarithm base of 10. One neper is equal to 8.686 decibels. Also called napier.

**neptunium** [symbol Np] A transuranic radioactive element formed by radioactive decay of uranium-239, which emits a beta particle (high-energy electron) to become neptunium-239. Atomic number is 93.

**neptunium series** The series of nuclides resulting from the decay of the long-lived synthetic nuclide $Np^{237}$. Mass numbers of all members are given by $4n + 1$, where $n$ is an integer, so the sequence is also known as the $4n + 1$ series.

**Nernst bridge** A four-arm bridge containing capacitors instead of resistors, used for measuring capacitance values at high frequencies.

**Nernst effect** If heat flows through a strip of metal whose surface is perpendicular to a magnetic field, a voltage is developed between opposite edges of the strip.

**Nernst lamp** An electric lamp consisting of a short slender rod of zirconium oxide in open air, heated to brilliant white incandescence by current.

**Nesa glass** Trademark of Pittsburgh Plate Glass Co. for glass having a transparent conductive coating. Electric current can be sent through the coating to heat the surface to prevent condensation of moisture or to melt ice.

**nesistor** A negative-resistance semiconductor device that is basically a bipolar field-effect transistor.

**net** A number of communication stations equipped for communication with each other, often on a definite time schedule and in a definite sequence.

**network** 1. A combination of electric elements, such as interconnected resistors, coils, and capacitors. An active network also contains a source of energy, whereas a passive network does not. 2. A number of radio or television broadcast stations connected by coaxial cable, radio, or wire lines, so all stations can broadcast the same program simultaneously.

**network analyzer** An analog computer in which networks are used to simulate power-line systems or physical systems and obtain solutions to various problems before the systems are actually built.

**network constant** One of the values of resistance, inductance, mutual inductance, or capacitance that make up a network.

**neuroelectricity** A current or voltage generated in the nervous system.

**neutral** Having the same number of electrons as protons, hence in a normal condition wherein there is no electric charge.

**neutral atom** An atom in which the number of positive charges in the nucleus is equal to the number of electrons that surround the nucleus.

**neutral-density faceplate** A television picture-tube faceplate in which a neutral-density filter has been incorporated to increase picture contrast by attenuating external light reflected from the screen. Reflected light must pass through the filter twice and be doubly attenuated, whereas the desired light from the screen passes through only once.

**neutral-density filter** An optical filter that reduces the intensity of light without appreciably changing its color. Also called gray filter.

**neutralization** A method of nullifying oscillation-producing voltage feedback from the output to the input of an amplifier through tube interelectrode impedances. An external feedback path is used to produce at the input a voltage that is equal in magnitude but opposite in phase to that fed back through the interelectrode capacitance.

**neutralize** To stop regeneration in an amplifier stage by means of neutralization.

**netutralizing capacitor** A capacitor, usually variable, used in an anode-to-grid feedback path in an amplifier circuit for neutralization purposes.

**neutralizing indicator** A neon lamp or other indicator used to show when an amplifier has been neutralized. It is usually coupled to the anode tank circuit.

**neutralizing tool** A small screwdriver or socket wrench, partly or entirely nonmetallic, used to adjust a neutralizing capacitor.

**neutralizing voltage** The a-c voltage that is fed from the anode circuit to the grid circuit of an amplifier, or vice versa, 180° out of phase with and equal in amplitude to the a-c voltage that is transferred between these circuits over an undesired path, usually that through the grid-to-anode tube capacitance.

**neutral relay** A relay in which the movement of the armature is independent of the direction of current flow through the relay coil. Also called nonpolarized relay.

**neutral zone** *Dead band.*

**neutrino** A hypothetical particle presumed to be emitted during radioactive decay. It can be differentiated from an antineutrino only by its momentum characteristics.

**neutrodyne** An amplifier circuit used in early tuned-radio-frequency receivers, in which a capacitor was connected between the anode and grid circuits of a triode stage for neutralization purposes.

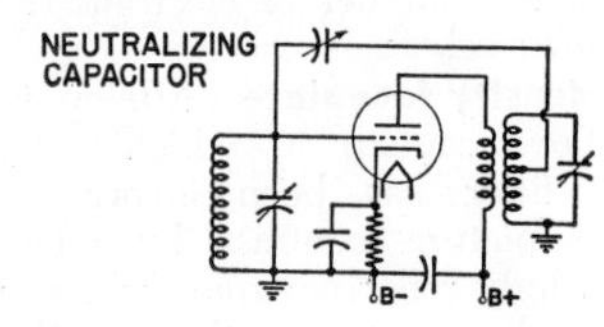


Neutrodyne circuit used as tuned r-f amplifier stage.

**neutron** An elementary nuclear particle having zero charge and mass number 1, making its mass approximately the same as that of a proton. Ionization is produced by the products of neutron collisions.

**neutron binding energy** The energy required to remove a single neutron from a nucleus. Most neutron binding energies are in the range from 5 to 12 million electron-volts.

**neutron bombardment** Bombardment of atomic nuclei with neutrons, generally to produce nuclear fission.

**neutron collision radius** The nuclear radius determined by fast-neutron transmission experiments.

**neutron crystallography** The study of crystals by neutron diffraction.

**neutron current density** The number of neutrons passing through a unit area of surface per unit time.

**neutron cycle** The life history of the neutrons in a nuclear reactor, extending from the initial fission process until all the neutrons have been absorbed or have leaked out.

**neutron density** The number of neutrons per unit volume in a nuclear reactor.

**neutron diffractometer** A diffractometer in which a beam of neutrons is used for diffraction analysis. The intensities of the diffracted beams at different angles are measured with an ionization chamber or radiation counter.

**neutron economy** The degree to which neutrons are used in desired ways in a nuclear reactor, instead of being lost by leakage or useless absorption.

**neutron excess** The number of neutrons in a nucleus in excess of the number of protons. Also called difference number and isotopic number.

**neutron flux** [symbol $nv$] The intensity of neutron radiation, expressed as the number of neutrons passing through a unit area per unit time.

**neutron hardening** The effect caused by the diffusion of thermal neutrons through a medium whose absorption cross section decreases as energy increases. Because the slower neutrons are preferentially absorbed, the average energy of the diffusing neutrons becomes greater.

**neutron inventory** The total number of neutrons in a reactor at any one time.

**neutron leakage** Leakage of neutrons through the boundaries of a nuclear reactor.

**neutron lifetime** The average time a neutron spends in slowing down and diffusion before capture.

**neutron number** [symbol $N$] The number of neutrons in the nucleus of an atom. It is equal to the difference between the mass number and the atomic number. In the symbol of a nuclide, the neutron number is added as a subscript following the element symbol; thus, in ${}_{26}Fe_{33}^{59}$, the neutron number for iron is 33, the mass number is 59, and the atomic number is 26.

**neutron producer** A nuclear reactor designed as a source of neutrons for isotope production.

**neutron radiography** Radiography in which a beam of neutrons is used in place of x-rays.

**neutron reflector** An external layer used on a nuclear reactor to scatter back into the reactor the neutrons that would otherwise escape. Typical reflector materials are carbon and beryllium.

**neutron source** 1. Any material that emits neutrons, such as a mixture of radium and beryllium. A neutron source may be introduced into a nuclear reactor as part of the start-up procedure. 2. A source that uses a nuclear reaction to generate neutrons.

**neutron spectrometer** A spectrometer used for determining the wavelengths of neutrons and the relative intensities of different wavelengths in a neutron beam.

**neutrons per absorption** The total number of neutrons emitted per neutron absorbed in fuel in a nuclear reactor.

**neutrons per fission** The total number of neutrons emitted per fission, usually including delayed neutrons.

**neutron therapy** Medical therapy involving irradiation with neutrons.

**neutron velocity selector** An instrument that isolates and detects neutrons having a particular range of velocities.

**new candle** *Candle.*

**nf** Abbreviation for *nanofarad.*

**Nichrome** Trademark of Driver-Harris Co. for an alloy of nickel, chromium, and sometimes also iron, having high electric resistance and the ability to withstand high temperatures for long periods of time. Used in wirewound resistors and electric heating elements.

**nickel** [symbol Ni] A metallic element. Atomic number is 28.

**Nicol prism** A prism made by cementing together two pieces of transparent crystalline Iceland spar with Canada balsam. It produces plane-polarized light from ordinary unpolarized light by eliminating the ordinary component of the original light by total reflection at the cementing layer. Only the extraordinary component passes.

**night effect** Deprecated term for *polarization error.*

**n-i junction** A semiconductor junction between n-type material and intrinsic material.

**Nike** [pronounced ny-kee] An Army surface-to-air guided missile used in place of antiaircraft guns. Named after the Greek goddess of victory. The first version, Nike-Ajax, has a speed of about mach 2, a range of about 30 miles, and is guided by radio command. The second version, Nike-Hercules, has a speed of about mach 3, a range of about 70 miles, is guided either by radar or radio command, and can carry an atomic warhead. The third version, Nike-Zeus, has still greater speed and range.

**niobium** [symbol Nb] A metallic element. Atomic number is 41. Formerly called columbium.

**Nipkow disk** A disk having one or more spirals of holes around the outer edge, with successive openings positioned so rotation of the disk provides mechanical scanning of a scene.

**nit** A unit of luminance equal to 1 candle per square meter.

**nitrogen** [symbol N] A gaseous element. Atomic number is 7. Often used in coaxial transmission lines and other enclosed volumes to keep out moisture, because it is chemically relatively inert and does not support combustion.

**n-n junction** A region of transition between two regions having different properties in n-type semiconducting material.

**NO** Symbol for *normally open.*

**nobelium** [symbol No] A transuranic radioactive element produced artificially by bombarding curium-244 with carbon-13 nuclei. Atomic number is 102.

**no connection** [symbol NC] An electron-tube pin for which there is no connection internally to electrodes. This generally permits use of the corresponding tube socket terminal as an insulated standoff terminal at which two or more leads may be connected together independently of the tube.

**nodal diagram** A diagram showing the order and mode of waves propagated in a waveguide.

**nodal-point keying** Keying of an arc transmitter at a point in the antenna circuit that is essentially at ground potential at all times.

**node** 1. A point, line, or surface in a standing-wave system where some characteristic of the wave has essentially zero amplitude. 2. A junction point in a network.

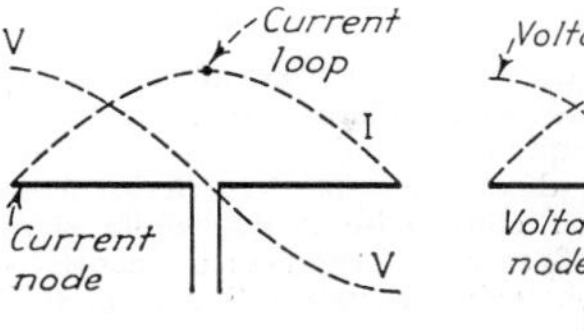

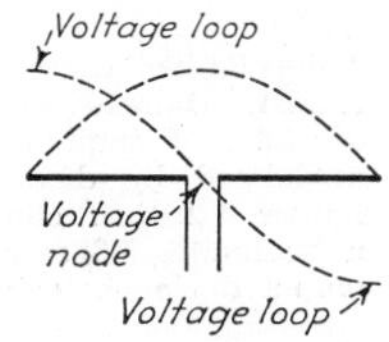


Nodes of current and voltage for a half-wave dipole.

**noise** An undesired electric disturbance or sound that tends to interfere with the normal reception or processing of a desired signal. In facsimile and television, noise voltages produce small black or white spots over the entire image area.

**noise behind the signal** *Modulation noise.*

**noise-canceling microphone** *Close-talking microphone.*

**noise-current generator** A current generator whose output is a random function of time.

**noise diode** A diode designed for operation at saturation in a noise-generating circuit.

**noise factor** The ratio of the total noise power per unit bandwidth at the output of a system to the portion of the noise power that is due to the input termination at the standard noise temperature of 290°K. Also called noise figure.

**noise figure** *Noise factor.*

**noise filter** A filter that is inserted in an a-c power line to block noise interference that would otherwise travel through the power line in either direction and affect the operation of receivers.

**noise level** The level of electric or acoustic noise at a particular location. Usually expressed in decibels with respect to a specified reference level. The value of the noise is integrated over a specified frequency range.

**noise limiter** A limiter circuit that cuts off all noise peaks that are stronger than the highest peak in the desired signal being

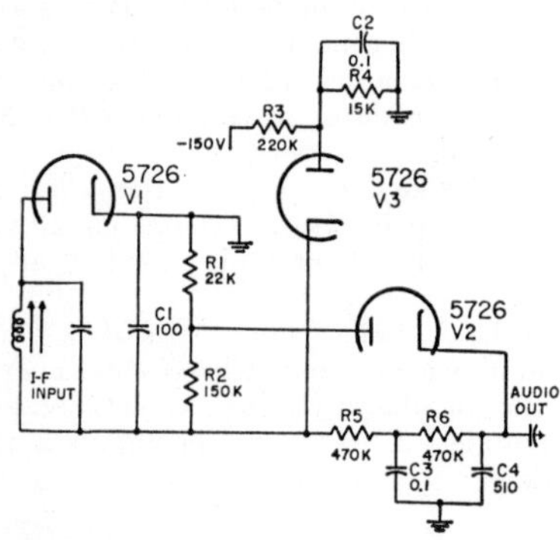


Noise limiter using preferred circuit design by National Bureau of Standards. Signal from i-f amplifier of amplitude-modulated receiver is demodulated by diode detector V1. Series noise limiter V2 then clips noise peaks, while shunt noise limiter V3 serves as short-circuit across i-f output during noise peaks, preventing them from operating the automatic gain control circuit.

received, thereby reducing the effects of atmospheric or manmade interference. Also called noise silencer and noise suppressor.

**noise-measuring set** *Circuit-noise meter.*

**noise quieting** The ability of a receiver to reduce noise background in the presence of a desired signal. Usually expressed in decibels.

**noise-reducing antenna system** A receiving antenna system so designed that only the antenna proper can pick up signals. The antenna is placed high enough to be out of the noise-interference zone, and is connected to the receiver with a shielded cable or twisted transmission line that is incapable of picking up signals.

**noise reduction** A process whereby the average transmission of the sound track of a sound motion-picture print, averaged across the track, is decreased for signals of

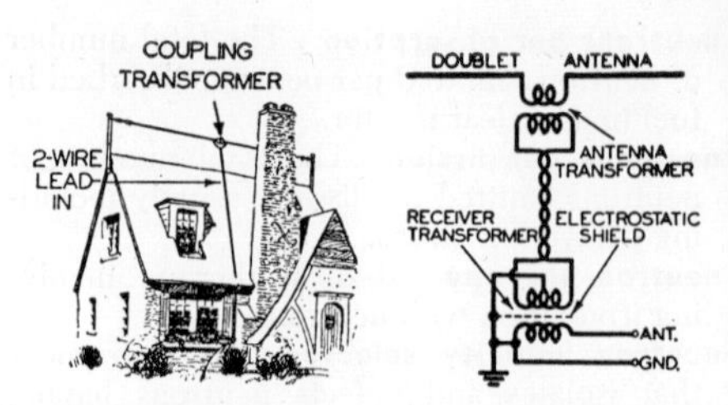


Noise-reducing antenna installation on home, for long-distance radio reception in locality having severe electric interference, such as near a power station.

low level and increased for signals of high level. Since background noise introduced by the sound track is less at low transmission, this process reduces film noise during soft passages.

**noise silencer** *Noise limiter.*

**noise source** A device for generating a random noise signal for test purposes. Common noise sources are a temperature-limited diode operated at cathode saturation, an electron multiplier, a crystal diode with positive bias, a nonoscillating reflex klystron, and a nonoscillating magnetron.

**noise suppressor** 1. A circuit that blocks the a-f amplifier of a radio receiver automatically when no carrier is being received, to eliminate background noise. Also called

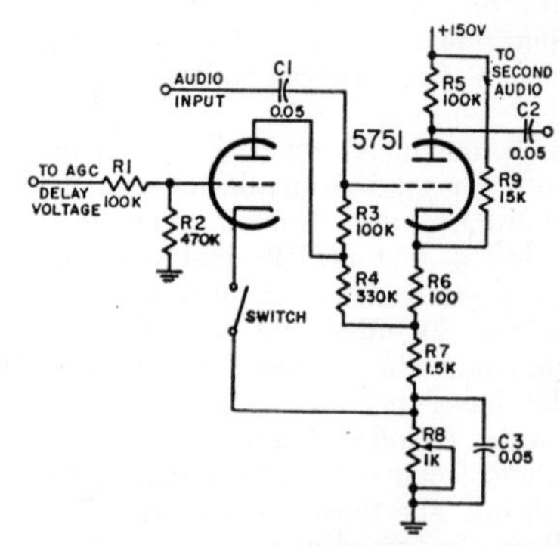


Noise suppressor using preferred circuit design by National Bureau of Standards.

intercarrier noise suppressor, interstation noise suppressor, and squelch circuit. 2. A circuit that reduces surface noise when playing phonograph records, generally by means of a filter that blocks out the higher frequencies where such noise predominates. 3. *Noise limiter.*

**noise temperature** The temperature at which the thermal noise power of a passive system per unit bandwidth is equal to the noise at the actual terminals. The standard reference temperature for noise measurements is 290°K.

**noise-voltage generator** A voltage generator whose output is a random function of time.

**no-load loss** The power loss of a device that is operated at rated voltage and frequency but is not supplying power to a load.

**nolo flight** The flight of a drone aircraft without a human safety pilot aboard.

**nominal band** A frequency band equal in width to that between zero frequency and the maximum modulating frequency in facsimile.

**nominal frequency** The specified frequency value of a crystal unit, as distinguished from the actual frequency measured during operation.

**nominal line width** The average separation between centers of adjacent scanning or recording lines.

**nomogram** *Nomograph.*

**nomograph** A chart having three or more scales, on which equations can be solved graphically by placing a straightedge on two known values and reading the answer where the straightedge crosses the scale for the unknown value. Also called abac, alignment chart, and nomogram.

**nonbridging** Switching action in which the movable contact leaves one fixed contact before touching the next.

**noncentral force** A nuclear force whose direction depends in part on the spin orientation of the nucleus. Also called tensor force.

**noncoherent mti** A moving-target indicator system in which ground clutter is used in place of a coherent reference oscillator as the reference signal. A moving target can be detected only when ground clutter exists at the same range and azimuth as the target.

**noncoherent radiation** Radiation in which there are no definite phase relations between different points in a cross-section of the beam.

**nonconductor** An insulating material.

**noncontacting piston** *Choke piston.*

**noncontacting plunger** *Choke piston.*

**noncooperative system** An instrumentation system in which data is transmitted from airborne missile equipment to a ground station for recording and processing, with no control of the transmission by the ground station.

**noncorrosive flux** Flux that is free from acid and other substances that might cause corrosion when used in soldering.

**nondegenerate gas** A gas formed by a system of particles whose concentration is sufficiently weak so the Maxwell-Boltzmann law applies. Examples are molecules or atoms of a body in the gaseous state, electrons emitted by a hot cathode, electrons and ions in a cloud or a plasma, electrons supplied to a conduction band by donor levels in an n-type semiconductor, and holes resulting from the passage of electrons from the normal band to an impurity band of acceptor levels in a p-type semiconductor.

**nondestructive readout** Readout of the magnetic state of a core without changing its state.

**nondirectional** *Omnidirectional.*

**nondirectional antenna** *Omnidirectional antenna.*

**nondirectional microphone** *Omnidirectional microphone.*

**nonerasable storage** A computer storage medium that cannot be erased and reused, such as punched paper tapes and punched cards.

**nonferrous** Not containing iron.

**nonhoming** Not returning to the starting or home position, as when the wipers of a stepping relay remain at the last-used set of contacts instead of returning to their home position.

**nonhoming tuning system** A motor-driven automatic tuning system in which the motor starts in the direction of previous rotation. If this is incorrect for the new station, the motor reverses after tuning to the end of the dial, then proceeds to the desired station.

**noninduced current** A current that flows through a winding of a transformer but is not due to a voltage induced in the transformer. The filament current in a bifilar winding is an example.

**noninduced voltage** An alternating or direct voltage that is applied uniformly to an entire winding in such a manner that no appreciable potential difference is induced along the winding. The filament voltage supplied to a load through a bifilar winding is an example.

**noninductive** Having negligible or zero inductance.

**noninductive capacitor** A capacitor constructed so it has practically no inductance. Foil layers are staggered during winding, so an entire layer of foil projects at either end for contact-making purposes. All currents then flow laterally rather than spirally around the capacitor.

**noninductive circuit** A circuit having practically no inductance.

**noninductive load** A load having practically no inductance. It may consist entirely of resistance or may be capacitive.

**noninductive resistor** A wirewound resistor constructed to have practically no inductance, either by using a hairpin winding or by reversing connections to adjacent sections of the winding.

**noninductive winding** A winding constructed so the magnetic field of one turn or section cancels the field of the next adjacent turn or section.

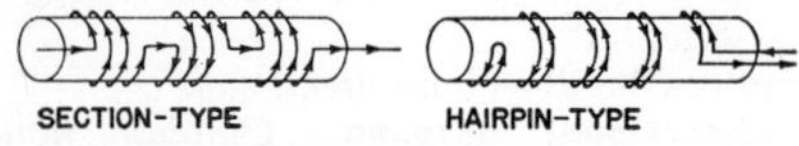


Noninductive windings.

**nonleakage probability** The probability that a fast neutron does not escape from a finite reactor before reaching a specified energy value.

**nonlinear** Not directly proportional.

**nonlinear amplifier** An amplifier in which the output is not directly related to the input.

**nonlinear circuit** A circuit not specifiable by linear differential equations in which time is the independent variable.

**nonlinear detection** Detection based on the curvature of a tube characteristic, such as square-law detection.

**nonlinear distortion** Distortion in which the output of a system or component does not have the desired linear relation to the input. Amplitude distortion, harmonic distortion, and intermodulation distortion are examples of nonlinear distortion.

**nonlinear element** An element in which an increase in applied voltage does not produce a proportional increase in current.

**nonlinear feedback control system** A feedback control system in which the relationships between the input and output signals are not linear.

**nonlinearity** The deviation of any functional relationship from direct proportionality. As an example, crowding of picture elements in one region of a television screen is due to nonlinearity of scanning action.

**nonlinear network** A network not specifiable by linear differential equations in which time is the independent variable.

**nonlinear taper** Nonuniform distribution of resistance throughout the element of a potentiometer or rheostat.

**nonmagnetic** Not magnetizable, and hence not affected by magnetic fields. Examples are air, glass, paper, and wood. All have a magnetic permeability of 1, the same as a vacuum.

**nonmagnetic shim** A nonmagnetic material attached to the armature or core of a relay to prevent iron-to-iron contact between armature and core when the relay is energized.

**nonmagnetic steel** A steel alloy containing about 12% manganese and sometimes a small quantity of nickel. It is practically nonmagnetic at ordinary temperatures.

**nonmicrophonic** *Antimicrophonic.*

**nonplanar network** A network that cannot be drawn on a plane without crossing of branches.

**nonpolarized electrolytic capacitor** An electrolytic capacitor in which the dielectric film is formed adjacent to both metal electrodes, so as to give the same construction in both directions of current flow.

**nonpolarized relay** *Neutral relay.*

**nonquantized system** A system of particles whose energies are assumed to be capable of varying in a continuous manner. The number of microscopic states of the system, defined by the positions and velocities of the particles at a given instant, is then infinite. Also called classical system.

**nonresonant line** A transmission line in which there are no standing waves.

**nonshorting switch** A selector switch in which the width of the movable contact is less than the distance between contacts, so the old circuit is broken before the new circuit is completed.

**nonsinusoidal wave** A wave whose form differs from that of a sine wave, and which therefore contains harmonics.

**nonspecular reflection** Reflection from a rough surface, producing diffraction and scattering of waves.

**nonstorage camera tube** A television camera tube in which the picture signal is at each instant proportional to the intensity of the illumination on the corresponding elemental area of the scene at that instant.

**nonsynchronous** Not related in phase, frequency, or speed to other quantities in a device or circuit.

**nonsynchronous vibrator** A vibrator that interrupts a direct-current circuit at a frequency unrelated to other circuit constants and does not rectify the resulting stepped-up alternating voltage.

**nonvolatile storage** A computer storage medium that retains information in the absence of power, such as a magnetic tape, drum, or core.

**NORAD** Abbreviation for North American Air Defense Command.

**normal** 1. The perpendicular to a line or surface at the point of contact. 2. The expected or regular value of a quantity.

**normal distribution** *Gaussian distribution.*

**normal induction** The limiting induction, either positive or negative, in a magnetic material that is under the influence of a magnetizing force varying between two specific limits.

**normalize** To multiply all quantities by a constant so they fall within the operating ranges of a computer or within the scale ranges of a graph.

**normalized admittance** The reciprocal of the normalized impedance.

**normalized impedance** An impedance divided by the characteristic impedance of a transmission line or waveguide.

**normalized plateau slope** A figure of merit for a radiation-counter tube, equal to the percentage change in counting rate divided by the percentage change in voltage. The midpoint of the plateau of the counting-rate–voltage characteristic of the tube is used as a reference.

**normally closed** [symbol NC] A term applied to relay contacts that are connected to complete a circuit when the relay is not energized.

**normally open** [symbol NO] A term applied to relay contacts that are connected to break a circuit when the relay is not energized.

**normal mode** A characteristic distribution of vibration amplitudes among the parts of a system, each part of which is vibrating freely at the same natural frequency and phase.

**normal permeability** The ratio of the normal induction $B$ of a magnetic material to the corresponding magnetic intensity $H$.

**normal position** The deenergized position of the contacts of a relay.

**normal state** *Ground state.*

**normal threshold of audibility** The minimum sound pressure level at the entrance to the ear, at a specified frequency, that produces an auditory sensation in a large percentage of normal persons in the age group from 18 to 30 years.

**north magnetic pole** The magnetic pole located approximately at 71° N latitude and 96° W longitude, about 1,140 nautical miles south of the North Pole.

**North Pole** A geographical point on the earth that is one end of the axis about which the earth revolves. It is almost directly beneath the star Polaris.

**north pole** The pole of a magnet at which magnetic lines of force are considered as leaving the magnet.

**north-stabilized ppi** *Azimuth-stabilized ppi.*

**nose cone** A hollow cone-shaped shield that fits over or serves as the nose of a guided missile. The cone is made from materials that will withstand high temperatures generated by friction with air particles during passage through the earth's atmosphere.

**nose whistler** A whistler whose frequency is determined primarily by the electron gyrofrequency at the top of the whistler path along a line of the earth's magnetic field.

**notch antenna** A microwave antenna in which the radiation pattern is determined by the size and shape of a notch or slot in a radiating surface.

**notch filter** A band-rejection filter that produces a sharp notch in the frequency response curve of a system. Used in television transmitters to provide attenuation at the low-frequency end of the channel, to prevent possible interference with the sound carrier of the next lower channel.

**note** A conventional sign used to indicate the pitch, the duration, or both, of a tone sensation. It is also the sensation itself or the vibration causing the sensation. The word serves when no distinction is desired between the symbol, the sensation, and the physical stimulus.

**not-gate** An electronic circuit whose output is energized only if its single input is not energized.

**noval base** An electron-tube base having positions for nine pins that extend directly through the glass envelope. The spacing between pins 1 and 9 is greater than the spacings between the other pins, for orienting purposes. Used on miniature tubes.

**nozzle** British term for an elementary antenna consisting of a waveguide in which neither transverse dimension is increased, but either or both may decrease toward the aperture. Also called spout.

**n-p-i-n transistor** An intrinsic-junction transistor in which the intrinsic region is sandwiched between the p-type base and n-type collector layers.

**n-p-i-p transistor** An intrinsic-junction transistor in which the intrinsic region is between p regions.

**n-p-n-p transistor** An n-p-n junction transistor having a transition or floating layer between p and n regions, to which no ohmic connection is made.

**n-p-n transistor** A junction transistor having a p-type base between an n-type emitter and an n-type collector. The emitter should then be negative with respect to the base, and the collector should be positive with respect to the base.

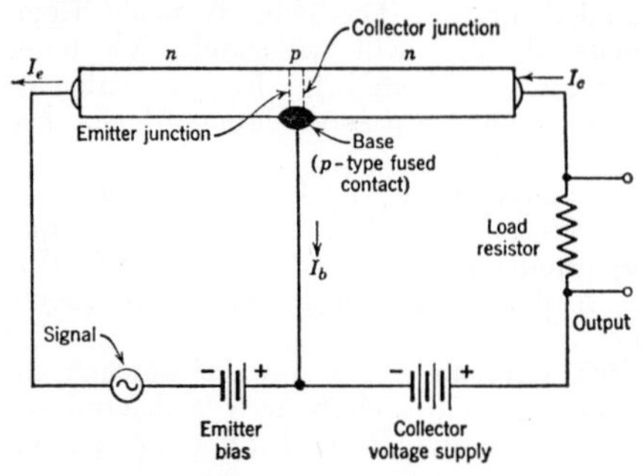


N-p-n transistor construction and circuit.

**N quadrant** One of the two quadrants in which the N signal of an A-N radio range is heard.

**n region** The region in a semiconductor where conduction electron density exceeds hole density. Also called n zone.

**N scan** *N display.*

**N scope** A radarscope that produces an N display.

**nsec** Abbreviation for *nanosecond.*

**N shell** The fourth layer of electrons about the nucleus of an atom, having electrons characterized by the principal quantum number 4.

**N signal** A dash-dot signal heard either in a bisignal zone or in an N quadrant of a radio range.

**NTSC** Abbreviation for *National Television System Committee.*

**n-type conductivity** The conductivity associated with conduction electrons in a semiconductor.

**n-type semiconductor** An extrinsic semiconductor in which the conduction electron density exceeds the hole density. The net ionized impurity concentration is donor-type.

**nuclear** Pertaining to the nucleus of an atom.

**nuclear absorption** Absorption of energy by the nucleus of an atom.

**nuclear battery** A primary battery in which the energy of radioactive material is converted into electric energy. It may have one or more nuclear cells, each using a radioisotope such as promethium-147. Also called atomic battery.

**nuclear binding energy** The energy that would be required to separate an atom of atomic number Z into Z hydrogen atoms.

**nuclear Bohr magneton** A unit of magnetic moment equal to 1/1,836th of the electronic Bohr magneton. Also called nuclear magneton.

**nuclear bomb** A bomb that releases explosive energy through nuclear fission, as in the atomic bomb, or through nuclear fusion, as in the hydrogen bomb.

**nuclear breeder** A nuclear reactor in which more fissionable material is formed in each generation than is used up in fission.

**nuclear cell** A single cell of a nuclear battery.

**nuclear charge** The sum of the charges of the protons in a nucleus.

**nuclear chemistry** The branch of chemistry concerned with substances produced or affected by nuclear reactions.

**nuclear cross section** The cross section of an atomic nucleus for a particular process.

**nuclear disintegration** Any transformation or change involving nuclei, including radiative capture, inelastic scattering, beta decay, and isomeric transition. If the disintegration is spontaneous, it is said to be radioactive; if it results from a collision, it is said to be induced.

**nuclear emulsion** A photographic emulsion specially designed to register individual tracks of ionizing particles.

**nuclear energy** Energy released by nuclear fission or nuclear fusion. Also called atomic energy, atomic power, and nuclear power.

**nuclear energy level** The energy of an atomic nucleus. Nuclei can assume only a discrete number of energy levels.

**nuclear engineering** Research and development associated with construction and practical use of nuclear reactors and their products.

**nuclear equation** An equation showing the changes in composition, usually in terms of mass number and charge, of an atomic nucleus during a nuclear reaction. The equation shows any particles captured or radiations absorbed by the nucleus, and also the particles or radiations emitted.

**nuclear fission** A nuclear reaction in which the nucleus of a heavy atom such as uranium or plutonium is split into two approximately equal parts by a neutron, charged particle, or photon. In an atomic bomb, nuclear fission results in heat, nuclear radiation, and blast. Also called atomic fission and fission.

**nuclear force** A nonelectromagnetic force peculiar to protons and neutrons in the atomic nucleus. Also called exchange force.

**nuclear fuel** The fissionable material used in a nuclear reactor.

**nuclear fusion** A thermonuclear reaction in which the nuclei of an element of low atomic weight unite under extremely high temperature and pressure to form a nucleus of a heavier atom. The associated loss in mass is released as energy. This reaction takes place in the hydrogen bomb, where hydrogen nuclei combine to form helium nuclei. Also called atomic fusion and fusion.

**nuclear gyromagnetic ratio** The ratio of the magnetic moment of the nucleus to the nuclear angular momentum quantum number.

**nuclear induction** Magnetic induction originating in the magnetic moments of nuclei. The effect depends on the unequal population of energy states available when the material is placed in a magnetic field.

**nuclear isobar** One of two or more nuclides having the same number of nucleons in their nuclei and therefore having about the same atomic mass.

**nuclear isomer** One of two or more nuclides having the same mass number and atomic number, but existing for measurable times in different quantum states with different energies and radioactive properties. The state of lowest energy is the ground state. Those of higher energies are metastable states.

**nuclear isomerism** The occurrence of nuclear isomers.

**nuclear magnetic moment** The magnetic moment of an electrically charged particle possessing angular momentum.

**nuclear magnetic resonance** Resonance encountered in energy transfers between an r-f magnetic field and a nucleus placed in a constant magnetic field that is sufficiently strong to decouple the nucleus from its orbital electrons. The amount of energy absorbed by the atoms at resonance is a clue to identification of the atoms involved.

**nuclear magneton** *Nuclear Bohr magneton.*

**nuclear medicine** 1. The branch of medicine concerned with the problems resulting from the use of nuclear energy and nuclear weapons. 2. Any medicine prepared by the use of fissionable material.

**nuclear model** *Independent-particle model.*

**nuclear number** *Mass number.*

**nuclear packing** The concentration of particles within the nucleus of an atom.

**nuclear paramagnetism** Paramagnetism associated with nuclear magnetic moments.

**nuclear particle** 1. A particle in the atomic nucleus. It may be either a proton or a neutron. Also called nucleon. 2. A particle emitted by an atomic nucleus, such as an alpha particle, beta particle, positron, neutrino, or meson.

**nuclear photodisintegration** *Nuclear reaction.*

**nuclear physics** The branch of physics concerned with atomic nuclei.

**nuclear pile** *Nuclear reactor.*

**nuclear polarization** Alignment of the spin magnetic moments of atomic nuclei in the same direction.

**nuclear potential** The potential energy of a nuclear particle as a function of its position in the field of a nucleus or of another nuclear particle.

**nuclear potential energy** The average total potential energy of all the protons and neutrons in a nucleus due to the nuclear forces between them, excluding the electrostatic potential energy.

**nuclear power** *Nuclear energy.*

**nuclear power plant** A power plant in which nuclear energy is converted into heat for use in producing steam for turbines. These in turn drive electric generators that produce electric power for commercial use.

**nuclear propulsion** Propulsion by means of nuclear energy.

**nuclear radiation** The radiation of neutrons, gamma particles, and other particles from an atomic nucleus as a result of nuclear fission or nuclear fusion. Also called radiation.

**nuclear radius** The radius of a spherical volume within which the density of protons and neutrons in a nucleus is effectively large. It is not a precisely determinable quantity.

**nuclear reaction** A process that involves transformation of an atomic nucleus by a photon, elementary particle, or another nucleus. Also called nuclear photodisintegration.

**nuclear reaction energy** The disintegration energy of a nuclear reaction. It is equal to the sum of the kinetic or radiant energies of the reactants minus the sum of the kinetic or radiant energies of the products.

**nuclear reactor** An apparatus in which nuclear fission may be sustained in a self-supporting chain reaction. The fissionable material used as fuel is uranium or plutonium. Most reactors use a moderating material such as carbon or heavy water to slow down the neutrons and thereby control the reaction. Also called atomic pile, atomic reactor, nuclear pile, pile, and reactor.

**nuclear rocket** A rocket propelled by the reaction to released nuclear energy.

**nuclear species** *Nuclide.*

**nuclear spin** The total angular momentum of the atomic nucleus when considered as a single particle.

**nuclear spontaneous reaction** *Radioactive decay.*

**nuclear structure** The internal structure of the atomic nucleus.

**nuclear warhead** A warhead that contains fissionable or fissionable-fusionable material.

**nuclear weapon** *Atomic weapon.*

**nucleogenesis** Large-scale formation of nuclei in nature.

**nucleon** *Nuclear particle.*

**nucleonics** A science that deals with the release and utilization of energy from the nuclei of atoms. Nucleonics combines nuclear physics and nuclear engineering.

**nucleon number** *Mass number.*

**nucleus** [plural nuclei] The central part of an atom, possessing a positive charge and containing nearly all the mass of the atom. The nucleus consists of protons and neutrons, together known as nucleons, except for the hydrogen nucleus, which consists only of one proton. Also called atomic nucleus.

**nuclide** A species of atom characterized by the number of protons, number of neutrons, and energy content in the nucleus, or alternatively by the atomic number, mass number, and atomic mass. To be regarded as a distinct nuclide, the atom must be capable of existing for a measurable lifetime, generally greater than $10^{-10}$ second. Also called nuclear species.

**nuclidic mass** *Atomic mass.*

**null** A position of minimum or zero indication or strength, such as that between adjacent lobes of an antenna radiation pattern.

**null astatic magnetometer** A magnetometer in which the magnetization of a small specimen is measured by balancing its magnetic moment against that of a small current-carrying coil.

**null detector** *Null indicator.*

**null indicator** A galvanometer or other device that indicates when voltage or current is zero. Used chiefly to determine when a bridge circuit is in balance. Also called null detector.

**nullity** The number of independent meshes that can be selected in a network.

**null method** A method of measurement in which the measuring circuit is balanced to bring the pointer of the indicating instrument to zero, as in a Wheatstone bridge, and the settings of the balancing controls are then read. Also called balance method and zero method.

**number** A figure, word, sequence of pulses, or other designation used to represent an arithmetic sum.

**number system** *Positional notation.*

**numerical positioning control** A control system for machine tools and some industrial processes, in which numerical values corresponding to desired positions of tools or controls are recorded on punched cards or magnetic tapes in such a way that they can be used to control the operation automatically.

**nupac monitoring set** A warning system that detects undesirable radiation emanating from a nuclear reactor and associated equipment. It may include such units as radiac computers, radiac indicators, radiac detectors, amplifiers, and aural or visual warning devices.

**nutating feed** An oscillating antenna feed for producing an oscillating deflection of a tracking radar beam, in which the plane of polarization remains fixed.

**nutation** A periodic variation in the inclination of the axis of a spinning gyroscope to the vertical, between certain limiting angles, or a corresponding motion of a radar dipole or other object.

**nutation field** The time-variant three-dimensional field pattern of a directional or beam-producing antenna having a nutating feed.

**nv** Symbol for *neutron flux.*

**nvt** The time integral of neutron flux, equal to neutron flux *nv* multiplied by time *t*.

**nylon** Generic name for a family of polyamide resins, widely used for molded coil forms and as a protective coating over insulation on wires.

**Nyquist criterion** A parameter used in servomechanism theory, corresponding to the open-loop harmonic response function.

**Nyquist diagram** A diagram from which stability of a control system may be determined. It is a closed polar plot of the loop transfer function of the system.

**n zone** *N region.*

# O

**obi** Abbreviation for *omnibearing indicator.*

**objective lens** The first lens through which rays pass in an optical or electronic lens system.

**oblique-incidence transmission** The transmission of a radio wave at a slant to the ionosphere and back to earth, as in long-distance radio communication.

**oboe** [Observer Bomber Over Enemy] A radar navigation system consisting of two ground stations that measure distance to an airborne responder beacon and relay the information to the aircraft.

**obs** Abbreviation for *omnibearing selector.*

**obsolescence-free** Designed to be as universal as possible, so the possibility of becoming outdated because of new developments is minimized. Used chiefly in connection with tube testers and other test instruments.

**occluded gas** Gas absorbed in a material, as in the electrodes, supports, leads, and insulation of a vacuum tube.

**octal base** An electron-tube base having a central aligning key and positions for eight equally spaced pins. Pins not needed for a particular tube are omitted without changing the positions of the remaining pins.

**octal digit** One of the symbols 0, 1, 2, 3, 4, 5, 6, and 7 when used as a digit in the scale of eight.

**octal notation** Notation using the scale of eight. Here the number 235 means 5 times 1, plus 3 times 8, plus 2 times 8 squared, which is equal to 157 in decimal notation.

**octal socket** A socket for an octal tube.

**octal tube** An electron tube having an octal base.

**octave** The interval between any two frequencies having a ratio of 2 to 1. Thus, going one octave higher means doubling the frequency. Going one octave lower means changing to one-half the original frequency.

**octave-band pressure level** The band pressure level for a frequency band corresponding to a specified octave of sound. The location of an octave-band pressure level on a frequency scale is usually specified as the geometric mean of the upper and lower frequencies of the octave. Also called octave pressure level.

**octave filter** A bandpass filter in which the upper cutoff frequency is twice the lower cutoff frequency.

**octave pressure level** *Octave-band pressure level.*

**octode** An eight-electrode electron tube containing an anode, a cathode, a control electrode, and five additional electrodes that are ordinarily grids.

**odd-even check** A forbidden-combination check in which an extra digit is carried along with each word to indicate whether the total number of ones in the word is odd or even.

**odd-even nuclei** Nuclei that have an odd number of protons and an even number of neutrons.

**odd-line interlace** Interlace in which each field contains an extra half-line. Thus, in the standard 525-line television picture, each field contains 262.5 lines.

**odd-odd nuclei** Nuclei that have an odd number of protons and an odd number of neutrons.

**odograph** An automatic electronic map-tracer that plots on a map or on cross-section paper the exact course taken by a jeep or other vehicle. Phototubes and thyratrons transfer the indications of a precision magnetic compass to a plotting unit actuated by the speedometer drive cable, causing a pen to trace the course.

**odoriferous homing** Homing on the ionized air produced by the exhaust gases of a snorkeling submarine. A hunter-killer aircraft equipped with sensitive electronic equipment approaches the suspected location from down-wind, to pick up the scent from miles away. Also called sniffer gear.

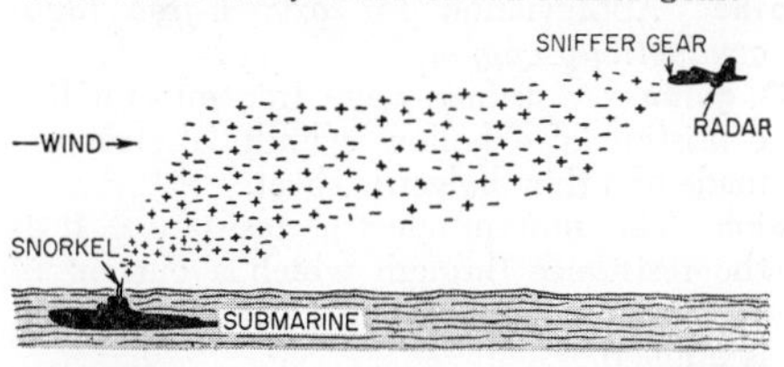


Odoriferous homing being used by hunter-killer aircraft.

**O electron** An electron having an orbit in the O shell, which is the fifth shell of electrons surrounding the atomic nucleus, counting out from the nucleus.

**oersted** The unit of magnetic field strength $H$ in the cgs electromagnetic system. The oersted replaced the gauss for this purpose by international agreement in 1930.

**off-center dipole** A rotating dipole mounted in a parabolic reflector at an angle to the axis of rotation, to give conical scanning.

**off-center ppi display** A plan-position-indicator display in which the zero position of the time base is not at the center of the display. The off-center arrangement permits enlargement of the display for a selected portion of the radar service area. Also called offset ppi.

**off period** The portion of an operating cycle in which an electron tube is nonconducting.

**off-scale** Beyond the normal indicating range, as when excessive current makes the pointer of a meter swing beyond the right-hand limit of the printed scale.

**offset** The steady-state difference between the desired control point and that actually obtained in a process control system. Offset is an inherent characteristic of positioning controller action.

**offset angle** The smaller of the two angles between the vibration axis of a phonograph pickup stylus and a line connecting the stylus point to the vertical pivot of the pickup arm. The angles are measured in the plane of the disk record.

**offset-course computer** *Course-line computer.*

**offset heads** *Staggered heads.*

**offset ppi** *Off-center ppi display.*

**offset stereophonic tape** *Staggered stereophonic tape.*

**off-target jamming** Use of a jammer at a location well removed from the main units of a force, to prevent utilization of the jamming signals by the enemy to his own advantage.

**ofhc** Abbreviation for *oxygen-free high-conductivity copper.*

**O guide** A surface-wave transmission line consisting of a hollow cylindrical structure made of a thin dielectric sheet.

**ohm** The unit of electric resistance. It is the resistance through which a current of 1 ampere will flow when a voltage of 1 volt is applied.

**ohmage** *Resistance.*

**ohm-centimeter** A unit of resistivity. The resistivity of a sample in ohm-centimeters is equal to its resistance $R$ in ohms multiplied by its cross-section area $A$ in square centimeters and the result divided by the sample length $L$ in centimeters (resistivity $= RA/L$).

**ohmic contact** A contact between two materials, possessing the property that the voltage drop across it is proportional to the current passing through.

**ohmic value** The resistance in ohms.

750,000 — 750 MΩ
750,000 ω — 750 Kω
750,000Ω — 750 KΩ
750 M — .75 MEG.
750 Mω — .75 MEG Ω

Ohmic values are specified in many different ways on circuit diagrams, as shown here for 750,000 ohms.

**ohmmeter** An instrument for measuring electric resistance. Its scale may be graduated in ohms or megohms.

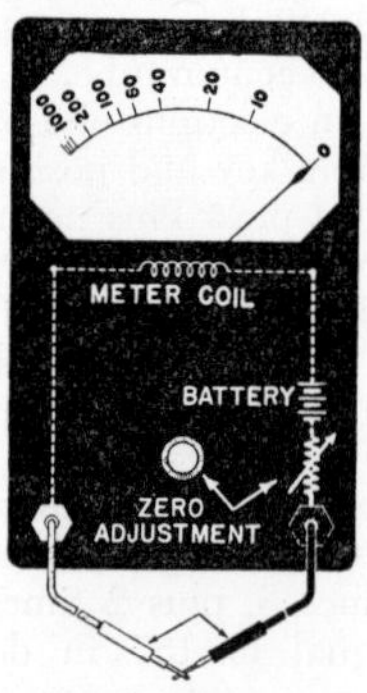


Ohmmeter circuit, showing zero adjustment and scale with zero value at right.

**Ohm's law** The current $I$ in a circuit is directly proportional to the total voltage $E$ in the circuit and inversely proportional to the total resistance $R$ of the circuit. The law may be expressed in three forms: $E = IR$; $I = E/R$; $R = E/I$.

**ohms per volt** A sensitivity rating for measuring instruments, obtained by dividing the resistance of the instrument in ohms at a particular range by the full-scale voltage value at that range. The higher the ohms-per-volt rating, the more sensitive is the meter and the less current it will draw from a circuit during a measurement.

**oil-cooled tube** An electron tube in which the heat produced is dissipated, directly or indirectly, by means of oil.

**oil diffusion pump** A diffusion pump that is similar to a mercury-vapor vacuum pump but uses oil instead of mercury vapor.

**omega** 1. A Greek letter, used in its capital form Ω to represent the word ohms. Used in its lower-case form $\omega$ as a letter symbol for a value equal to 6.28 times frequency

ω SMALL LETTER OMEGA MEANS $2\pi f$

Ω CAPITAL LETTER OMEGA MEANS OHMS

Omega uses in electronics.

($\omega = 2\pi f$). 2. A long-range navigation system originally developed for totally submerged submarines, giving worldwide coverage with six to ten ground stations. Operating principles are similar to those of delrac.

**omegatron** A miniature mass spectrograph, about the size of a receiving tube, that can be sealed to another tube and used to identify the residual gases left after evacuation. It can also be considered as a miniature cyclotron.

**omnibearing** A bearing indicated by a navigation receiver, using transmissions from an omnirange.

**omnibearing converter** An electromechanical device that combines an omnibearing signal with vehicle heading information to give the signals needed to drive the pointer of a radio magnetic indicator. The converter becomes an omnibearing indicator when a pointer and dial are added.

**omnibearing-distance facility** A radio facility consisting of an omnidirectional range in combination with distance-measuring equipment.

**omnibearing-distance navigation** Radio navigation that uses a system of polar coordinates as a reference in connection with omnibearing-distance facilities. It can furnish sufficiently accurate data to permit use of a computer that will provide arbitrary course lines anywhere within the coverage area of the system. Also called rho-theta navigation.

**omnibearing indicator** [abbreviated obi] An instrument that presents an automatic and continuous indication of an omnibearing.

**omnibearing line** One of an infinite number of straight lines radiating from the geographical location of a vhf omnirange.

**omnibearing selector** [abbreviated obs] An instrument that can be set manually to any desired omnibearing, to control a course-line deviation indicator.

**omnidirectional** Radiating or receiving equally well in all directions. Also called nondirectional.

**omnidirectional antenna** An antenna that has an essentially circular radiation pattern in azimuth and a directional pattern in elevation. Also called nondirectional antenna.

**omnidirectional beacon** A beacon that radiates radio signals equally well in all directions.

**omnidirectional hydrophone** A hydrophone that responds equally well in all directions to waterborne sound waves.

**omnidirectional microphone** A microphone whose response is essentially independent of the direction of sound incidence. Also called nondirectional microphone and astatic microphone.

**omnidirectional range** A radio facility that provides bearing information to or from its location at all azimuths within its service area, to serve as directional guidance for pilots. Also called omnirange.

**omnidistance** The distance between a vehicle and an omnibearing-distance facility.

**omnigraph** An instrument that converts Morse code signals on punched tape into corresponding buzzer-produced audio signals for training purposes.

**omnirange** *Omnidirectional range.*

**on-course curvature** The rate of change of the indicated navigation course with respect to distance along the course line or path.

**on-course signal** A signal indicating that the aircraft in which it is received is on course, following a guiding radio beam.

**ondograph** An instrument that draws the waveform of an a-c voltage step by step. A capacitor is charged momentarily to the amplitude of a point on the voltage wave, then discharged into a recording galvanometer, with the action being repeated a little further along on the waveform at intervals of about 100 cycles.

**ondoscope** A glow-discharge tube used to detect high-frequency radiation, as in the vicinity of a radar transmitter. The radiation ionizes the gas in the tube and produces a visible glow.

**one-address instruction** A digital computer programing instruction that explicitly describes one operation and one storage location. Also called single-address instruction.

**one-group model** A neutron-behavior model in which neutrons of all energies are treated as having the same characteristics.

**one-many function switch** A function switch in which only one input is excited at a time and each input produces a combination of outputs.

**one output** 1. The voltage response obtained from a magnetic cell in a one state by a reading or resetting process. 2. The integrated voltage response obtained from a magnetic cell in a one state by a reading or resetting process.

**one-port** A self-impedance, such as a choke.

**one-shot multivibrator** *Monostable multivibrator.*

**one state** A state of a magnetic cell wherein the magnetic flux through a specified cross-sectional area has a positive value, when determined from an arbitrarily specified direction for positive flux. A state wherein the magnetic flux has a negative value, when similarly determined, is a zero state.

**one-to-partial-select ratio** The ratio of a one output to a partial-select output from a magnetic cell.

**O network** A network composed of four impedance branches connected in series to form a closed circuit. Two adjacent points serve as input terminals, while the remaining two junction points serve as output terminals.

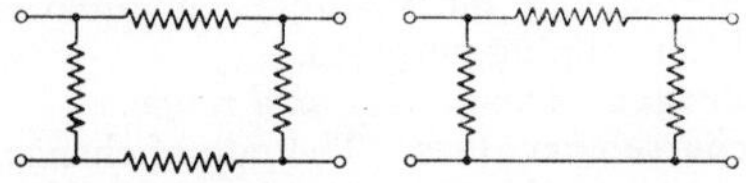

O network has four impedance branches, while pi network at right has only three branches.

**one-way repeater** A repeater that amplifies signals traveling in one direction over wire lines or through space as radio waves.

**on-line data reduction** The processing of information as rapidly as it is received by the computing system.

**on-line operation** Computer operation in which input data is fed into the computer directly from observing instruments or other input equipment, so the computer is used for the duration of the run.

**on-off control** A simple control system in which the device being controlled is either full on or full off, with no intermediate operating positions.

**on-off switch** A switch used to turn a receiver or other equipment on or off. Often combined with a volume control in radio and television receivers. Also called power switch.

**on period** The portion of an operating cycle in which an electron tube is conducting.

**on the air** Transmitting a radio signal.

**on the beam** Following a radio beam.

**opacimeter** *Turbidimeter.*

**opacity** The ability of a substance to block the transmission of radiant energy. Opacity is the reciprocal of transmission.

**opaque** Preventing the passage of radiation or particles.

**open** A break in a path for electric current.

**open-air ionization chamber** *Free-air ionization chamber.*

**open-center ppi** A plan-position-indicator display in which the initiation of the time base is shown before the transmission of the radar pulse.

**open circuit** An electric circuit that has been broken, so there is no complete path for current flow.

**open-circuit jack** A jack that has no circuit-shorting contacts. The circuit can be closed only by the connections made to the plug that is inserted in the jack.

**open-circuit parameter** One of a set of four transistor equivalent-circuit parameters that specify transistor performance when the input and output currents are chosen as the independent variables. When measuring these parameters, either the input or output circuit must be open for alternating current.

**open-circuit voltage** The voltage at the terminals of a source when no appreciable current is flowing.

**open-loop control system** A control system in which the controlled quantity is permitted to vary in accordance with the inherent characteristics of the control system, there being no self-correcting action for overshoots or undershoots.

**open subroutine** A computer subroutine that is inserted directly into the linear operational sequence, in contrast to a closed subroutine that must be entered by a jump. An open subroutine must be recopied at each point that it is needed in a routine.

**open-wire feeder** *Open-wire transmission line.*

**open-wire transmission line** A transmission line consisting of two spaced parallel wires supported by insulators at the proper distance to give a desired value of surge impedance. Such a line acts as a pure resistance when properly terminated. Also called open-wire feeder.

**operand** One of the quantities entering into or arising from a computer operation. An operand may be an argument, a parameter, a result, or the location code for the next instruction.

**operate time** The total elapsed time from the instant a relay coil is energized until the contacts have opened or firmly closed.

**operating angle** The electrical angle (portion of the grid voltage cycle) during which anode current flows in an amplifier or electron tube. Operating angles for three types of ampliers are: class A—360°; class B—180° to 360°; class C—less than 180°.

**operating frequency** The frequency at which a device is designed to operate.

**operating point** The point on the characteristic curve of an electron tube that corresponds to the direct voltage values being used for the grid and anode. Also called quiescent point.

**operating power** The power that is actually supplied to the antenna of a transmitter.

**operation** The action specified by a single computer instruction.

**operational** Immediately usable.

**operation code** 1. The list of operation parts occurring in an instruction code, together with the names of the corresponding operations. 2. *Operation part.*

**operation number** A number indicating the position of an operation or its equivalent subroutine in the sequence forming a problem routine for a computer.

**operation part** The part of a computer instruction that usually specifies the kind of operation to be performed, but not the location of the operands. Also called operation code.

**operation time** The time after simultaneous application of all electrode voltages for a current to reach a stated fraction of its final value.

**operator** 1. A person whose duties include operation, adjustment, and maintenance of a piece of equipment, such as a computer or transmitter. 2. A symbol that represents a mathematical operation to be performed.

**opposition** The condition in which the phase difference between two periodic quantities having the same frequency is 180°, corresponding to one half-cycle.

**optar** [OPTical Automatic Ranging] A guidance device for the blind.

**optical axis** 1. The straight line that passes through the centers of curvature of the surfaces of a lens. Light rays passing along this direction are neither refracted nor reflected. 2. In a quartz crystal, the *z* axis is the optical axis, and runs lengthwise through the mother crystal from apex to apex.

**optical density** *Photographic transmission density.*

**optical filter** A filter consisting of a pane of glass or other selectively transparent material that transmits only certain wavelength ranges in the visible, ultraviolet, and infrared spectrums.

**optical maser** A maser in which the pumping frequency is in the range of visible light or infrared radiation. The infrared maser is an example.

**optical pattern** A pattern observed when the surface of a laterally recorded phonograph record is illuminated by a light beam that is essentially parallel to the surface of the record. When bands of different constant frequencies are recorded, the width of the pattern is approximately related to the frequency response of the record. Also called Christmas-tree pattern.

**optical photon** A photon having energy corresponding to wavelengths in or near the visible spectrum, ranging from about 2,000 to 15,000 angstrom units.

**optical resonance** Luminescence in which the frequencies of the exciting and emitted radiation are essentially the same.

**optical sound recorder** *Photographic sound recorder.*

**optical sound reproducer** *Photographic sound reproducer.*

**optical sound track** A sound track consisting of a variable-width or variable-density photographic recording of sound signals, placed at one side of the frames of a motion-picture film.

**optical-track command guidance** Command guidance in which tracking of missile and target is done optically. The resulting data is converted into appropriate information for transmitting to the missile for correcting its flight path to the target.

**optical twinning** A defect occurring in natural quartz crystals. This generally results in small regions of unusable material that are discarded when cutting a crystal for piezoelectric use.

**optics** The branch of science that deals with the phenomena of light and vision, involving the portion of the electromagnetic spectrum between microwaves and x-rays. This range includes ultraviolet, visible, and infrared radiation.

**optimum bunching** The bunching condition

that produces maximum power at the desired frequency in an output gap of a microwave tube.

**optimum coupling** *Critical coupling.*

**optimum load** The load impedance value at which there is maximum transfer of power from source to load.

**optimum programing** Computer programing in which instructions and data are so stored that access time is a minimum.

**optimum working frequency** The most effective frequency at a specified time for ionospheric propagation of radio waves between two specified points.

**orbit** 1. The path described by a body in its revolution about another body, as by a manmade satellite revolving about the earth. 2. To revolve about another body.

**orbital** 1. Pertaining to electrons outside the nucleus of an atom. 2. An energy state or wave function from which the probability of finding an electron at a particular point can be calculated.

**orbital electron** An electron that is moving in an orbit around the nucleus of an atom.

**orbital-electron capture** Electron capture in which the electron usually comes from an orbit of the atom or molecule containing the transforming nucleus.

**orbital quantum number** A number equal to the angular momentum of an electron in its orbital motion around a nucleus. This number can have all whole values from zero to $n-1$, where $n$ designates the main quantum number.

**orbit-shift coil** One of the coils used to alter the orbit of beam particles in a betatron or synchrotron, to make the particles strike a target placed outside the stable orbit or enter a deflector to form an external beam.

**or-circuit** *Or-gate.*

**order** 1. The number of vibrations or half-period variations of a field along diameters of a circular waveguide or along the wider transverse axis of a rectangular waveguide. 2. A sequence of items or events. 3. *Instruction.*

**ordinary component** The component of light that is totally reflected at the cementing layer of a Nicol prism. Only the extraordinary component, which is plane-polarized, passes through the prism.

**ordinary-wave component** One of the two components into which a radio wave entering the ionosphere is divided under the influence of the earth's magnetic field. The ordinary-wave component has characteristics more nearly like those to be expected in the absence of a magnetic field. Also called O-wave component. The other component is the extraordinary-wave component.

**ordinate** The value that specifies distance in a vertical direction on a graph.

**ordir** [OmniRange DIgital Radar] A missile-detection radar having a range of about 2,500 miles, developed by Columbia University scientists. Operating frequency is about 3,000 mc. The transmitted signal is highly stable and frequency-modulated, permitting use of sustained coherence techniques to receive and analyze echo signals far below the noise level.

**organic-moderated reactor** An experimental nuclear reactor in which organic compounds are used as moderator and coolant.

**or-gate** A multiple-input gate circuit whose output is energized when any one or more of the inputs is in a prescribed state. Used in digital computers. An or-gate performs the function of the logical inclusive-or. Also called or-circuit.

**orientation** 1. The relationship between the length, width, and thickness directions of a quartz plate and the rectangular axes of the mother crystal. 2. The physical positioning of a directional antenna or other device having directional characteristics.

**orifice** An opening in a waveguide through which energy is transmitted.

**origin** The intersection of the reference axes on a graph.

**original master** The master produced by electroforming from the face of a wax or lacquer recording. Also called metal master and metal negative.

**origin distortion** An apparent loss of deflection sensitivity in the region of the undeflected position of the spot in a gas-focused electrostatic-deflection cathode-ray tube, due to a space-charge effect.

**orthicon** A camera tube in which a beam of low-velocity electrons scans a photoemissive mosaic that is capable of storing a pattern of electric charges. The orthicon has higher sensitivity than the iconoscope.

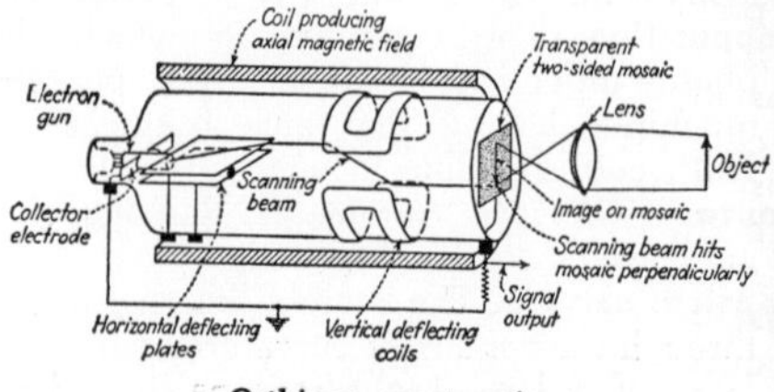


Orthicon construction.

**orthohydrogen** A hydrogen molecule in which the two nuclear spins are parallel, forming a triplet state.

**oscillating-crystal method** X-ray diffraction analysis in which the crystal is oscillated through an angle of a few degrees to simplify correlation between the diffracted beams and the crystal planes.

**oscillation** 1. A periodic change in a variable, as in the amplitude of an alternating current or the swing of a pendulum. 2. *Cycling.* 3. *Vibration.*

**oscillator** 1. A circuit that generates alternating current at a frequency determined by the values of its components. The oscillations are produced by using positive feedback with an electron tube, transistor, magnetic amplifier, or other amplifying device. Oscillations are maintained by drawing power from a battery or other source of power. For stable frequency characteristics the circuit may include a crystal, tuning fork, or other essentially unvarying source of vibrations. 2. The stage of a superheterodyne receiver that generates an r-f signal of the correct frequency to mix with the incoming signal and produce the i-f value of the receiver. 3. The stage of a transmitter that generates the carrier frequency of the station or some fraction of the carrier frequency.

**oscillator coil** The r-f transformer used in the oscillator circuit of a superheterodyne receiver or in other oscillator circuits to provide the feedback required for oscillation.

**oscillator harmonic interference** Interference occurring in a superheterodyne receiver when an undesired carrier signal beats with a harmonic of the local oscillator to produce the correct i-f value.

**oscillator-mixer-first detector** *Converter.*

**oscillator padder** An adjustable capacitor used in series with the oscillator tank circuit of a superheterodyne receiver to permit adjusting the tracking between the oscillator and preselector at the low-frequency end of the tuning dial.

**oscillator radiation** The field strength produced at a distance by the local oscillator of a television or radio receiver.

**oscillatory circuit** A circuit in which oscillations can be generated or sustained.

**oscillatory surge** A surge that includes both positive and negative polarity values. A unidirectional surge is a pulse.

**oscillogram** The permanent record produced by an oscillograph, or a photograph of the trace produced by an oscilloscope.

**oscillograph** 1. *Cathode-ray oscillograph.* 2. *Recorder.*

**oscilloscope** *Cathode-ray oscilloscope.*

**oscilloscope tube** A cathode-ray tube used to produce a visible pattern that is a graphic representation of electric signals.

**O shell** The fifth layer of electrons about the nucleus of an atom, having electrons characterized by the principal quantum number 5. It occurs with rubidium (atomic number 37) and all elements having higher atomic numbers.

**osmium** [symbol Os] A hard metallic element of the platinum group. Atomic number is 76. When alloyed with iridium, it is used in styli for disk recorders and phonographs.

**outdoor antenna** A receiving antenna erected outside a building, usually in an elevated location.

**outer-grid injection** Operation of a pentagrid mixer with the local oscillator signal applied to the third grid and the r-f signal applied to the first grid.

**outer marker** A marker located approximately 5 miles from the approach end of the runway in an instrument landing system for aircraft. It serves to provide a fix along the localizer course line.

**outer-shell electron** *Conduction electron.*

**outer space** The space beyond the atmosphere of the earth.

**outgassing** Heating an electron tube during evacuation, to remove residual gases occluded in the tube elements.

**outlet** A power-line termination from which electric power can be obtained by inserting the plug of a line cord. Also called convenience receptacle and receptacle.

**out of phase** Having waveforms that are of the same frequency but do not pass through corresponding values at the same instant.

**out-port** The exit for a network.

**output** 1. The useful energy delivered by a circuit or device. 2. The information that a computer feeds to an external device.

**output block** A portion of the internal storage of a computer that is reserved for receiving, processing, and transmitting data to be transferred out.

**output capacitance** The short-circuit transfer capacitance between the output terminal and all other terminals of an electron tube, except the input terminal, connected together.

**output gap** An interaction gap by means of which usable power can be abstracted from an electron stream in a microwave tube.

**output impedance** The impedance presented by a source to a load. For maximum power output, the output impedance should match the load impedance.

**output indicator** A meter or other device that is connected to a radio receiver to indicate variations in output signal strength for alignment and other purposes, without indicating the exact value of output.

**output meter** An a-c voltmeter connected to the output of a receiver or amplifier to measure output signal strength in volume units or decibels.

**output power** The power delivered by a system or component to its load.

**output resonator** The resonant cavity that is excited by density modulation of the electron beam in a klystron and delivers useful energy to an external circuit. Also called catcher and catcher resonator.

**output stage** The final stage in electronic equipment. In a radio receiver, it feeds the loudspeaker directly or through an output transformer. In an a-f amplifier, it feeds one or more loudspeakers, the cutting head of a sound recorder, a transmission line, or any other load. In a transmitter, it feeds the transmitting antenna.

**output transformer** The iron-core a-f transformer used to match the output stage of a radio receiver or a-f amplifier to its loudspeaker or other load.

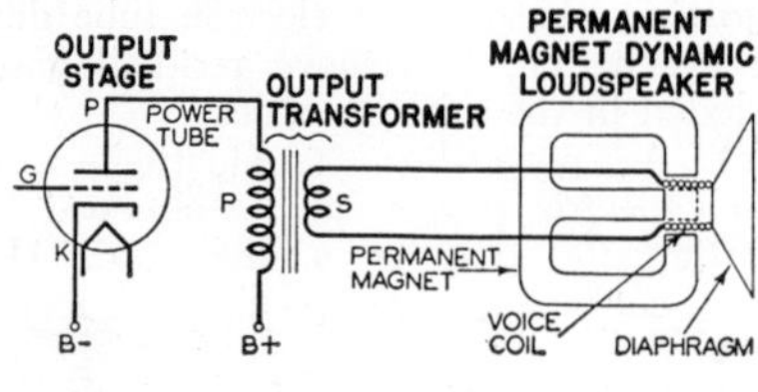


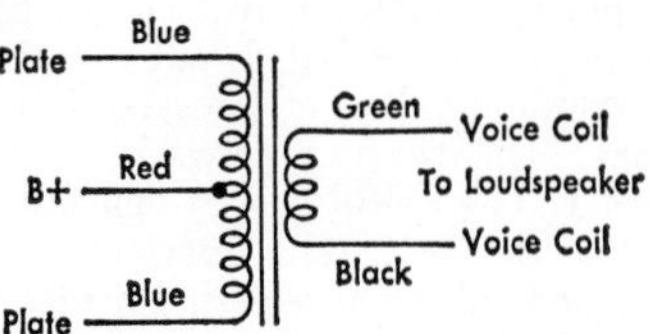


Output transformer connections and color code.

**output tube** A power-amplifier tube designed for use in an output stage.

**output winding** A winding other than the feedback winding of a saturable reactor, through which power is delivered to the load.

**over-all electric efficiency** The ratio of the power absorbed by the load material to the total power drawn from the supply lines in induction and dielectric heating.

**overbunching** The bunching condition produced by continuation of the bunching process beyond optimum bunching in a velocity-modulation tube.

**overcast bombing** Bombing of a target through an overcast above the target, using radar or other equipment to aid in sighting through the clouds.

**overcoupling** The condition in which two resonant circuits are tuned to the same frequency but coupled so closely that two response peaks are obtained. Used to obtain broadband response with substantially uniform impedance.

**overcutting** Recording on a disk at an excessively high signal level, so the stylus cuts through into adjacent grooves.

**overdamping** Damping greater than that required for critical damping.

**overdriven amplifier** An amplifier in which the input signal waveform is intentionally distorted by driving the grid past cutoff or into anode-current saturation.

**overflow** 1. The condition that arises when the result of an arithmetic operation exceeds the capacity of the number representation in a digital computer. 2. The carry digit arising from this condition.

**overlap** The amount by which the effective height of the scanning spot in a facsimile system exceeds the nominal width of the scanning line.

**overlap x** The amount by which the $x$ dimension of a recorded spot exceeds that necessary to form a constant-density line in facsimile equipment that produces a succession of discrete recorded spots in response to constant density in the subject copy.

**overlap y** The amount by which the $y$ dimension of a recorded spot exceeds the nominal line width in a facsimile system.

**overload** A load greater than that which a device is designed to handle. It may cause overheating of power-handling components and distortion in signal circuits.

**overload capacity** The current, voltage, or power level beyond which permanent damage occurs to the device in question.

**overload level** The level at which operation of a system ceases to be satisfactory because of signal distortion, overheating, or other effects.

**overload protection** Protection against excessive current by means of a device that automatically interrupts current flow when an overload occurs.

**overload relay** A relay that operates when current flow in a circuit exceeds the normal value for that circuit, to provide overload protection.

**overmodulation** Amplitude modulation greater than 100%, causing distortion because the carrier voltage is reduced to zero during portions of each cycle.

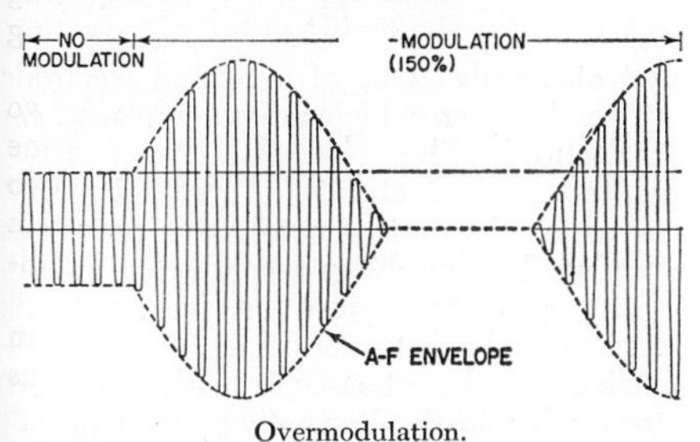


Overmodulation.

**over-radiation alarm** A radiation detector that trips an alarm when a predetermined level of radioactivity is reached.

**override** To cancel the influence of an automatic control by means of a manual control.

**overshoot** An excessive response to a change.

**overshoot distortion** *Overthrow distortion.*

**over-the-horizon communication** *Scatter propagation.*

**overthrow distortion** Distortion that occurs when the maximum amplitude of a signal wavefront exceeds the steady-state amplitude of the signal wave. Also called overshoot distortion.

**overtone** A component of a complex tone having a pitch higher than that of the fundamental pitch. The term overtone is sometimes used in place of harmonic, the $n$th harmonic being called the $(n-1)$st overtone.

**overtone crystal unit** A crystal unit in which the quartz plate is designed to operate at a higher order than the fundamental. Also called harmonic-mode crystal unit.

**overvoltage** The amount by which the applied voltage exceeds the Geiger threshold in a radiation-counter tube.

**overvoltage relay** A relay that is designed to operate when the voltage applied to its coil reaches a predetermined value.

**O-wave component** *Ordinary-wave component.*

**Owen bridge** A four-arm a-c bridge used to measure self-inductance in terms of capacitance and resistance. Bridge balance is independent of frequency.

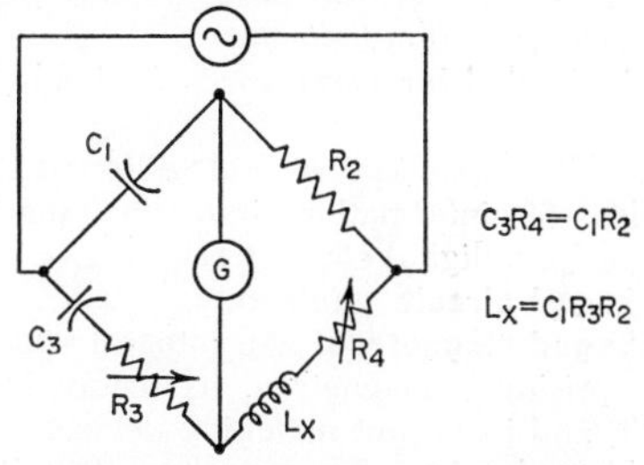


Owen bridge circuit and equations.

**oxide-coated cathode** A cathode that has been coated with oxides of alkaline-earth metals to improve electron emission at moderate temperatures. Also called Wehnelt cathode.

**oximeter** A photoelectric instrument for measuring continuously the oxygen content of the blood in a person, by photoelectric measurement of the intensity of a light beam passed through part of the ear.

**oxygen** [symbol O] A gaseous element. Atomic number is 8.

**oxygen-free high-conductivity copper** [abbreviated ofhc] Pure copper having 100% conductivity, used for the construction of high-power electron tubes because it does not release appreciable gas when hot.

# P

**P** 1. Symbol for *anode (plate)*. 2. Symbol for *primary winding,* used on circuit diagrams to identify the primary winding of a transformer.

**p$^3$i** Abbreviation for *precision plan position indicator.*

**p$^4$i** Abbreviation for *photographic projection plan position indicator.*

***P*** 1. Symbol for *permeance.* 2. Symbol for *power.*

**pack** To combine several different brief fields of information into one machine word for a digital computer.

**packaged circuit** *Rescap.*

**packaged magnetron** An integral structure comprising a magnetron, its magnetic circuit, and its output matching device.

**packing density** The number of units of desired information per unit length or unit area of a recording or storage medium for a digital computer, such as the number of binary digits of polarized spots per inch of magnetic-tape length.

**packing fraction** The mass defect of an atom divided by its mass number. It is negative for most atoms having mass numbers between 16 and 180, and positive for all atoms outside this range.

**pacor** [PAssive COrrelation Ranging] Passive detection combined with the use of a correlator to obtain useful range information from very weak signals reflected by an enemy target.

**pad** 1. An arrangement of fixed resistors used to reduce the strength of an r-f or a-f signal a desired fixed amount without introducing appreciable distortion. Also called fixed attenuator. The corresponding adjustable arrangement is called an attenuator. 2. *Terminal area.*

**padar** [PAssive Detection And Ranging] An electronic detection system that detects and tracks targets which are using radar, by picking up direct signals from the target or signals reflected from the target by equipment not at the tracking station.

**padder** A trimmer capacitor inserted in series with the oscillator tuning circuit of a superheterodyne receiver to control calibration at the low-frequency end of a tuning range.

**pad electrode** One of a pair of electrode plates between which a load is placed for dielectric heating.

**painted printed circuit** A printed circuit in which the desired conductive pattern is produced by applying a conductive liquid, using such techniques as spraying, silk screens, and offset printing.

**pairing** Faulty interlace in a television picture, wherein the lines of one field do not fall exactly between those of the preceding field. When this defect is serious, the lines of alternate fields tend to pair up and fall on one another, cutting the vertical resolution in half.

**pair production** The conversion of a photon into an electron and a positron when the photon traverses a strong electric field, such as that surrounding a nucleus or an electron.

**pair-production absorption** The absorption of gamma rays or other photons in the process of pair production.

**palladium** [symbol Pd] A metallic element. Atomic number is 46.

**Palmer scan** A combination of a circular or raster-type radar antenna scan with a conical scan. The beam is swung around the horizon concurrently with the conical scan.

**pam** Abbreviation for *pulse-amplitude modulation.*

**pam/f-m system** A carrier system in which several pulse-amplitude-modulated subcarriers are used to frequency-modulate a carrier.

**pan** To tilt or otherwise move a television camera vertically and horizontally to keep it trained on a moving object or secure a panoramic effect.

**pancake coil** A coil having a diameter appreciably greater than its length.

**panchromatic** Sensitive to all wavelengths within the visible spectrum, though not uniformly so.

**panel** 1. A metallic or nonmetallic sheet on which the operating controls of a receiver, transmitter, or other electronic unit are mounted. 2. A group of guests participating in a television or radio forum or quiz game.

**panning** Moving a television camera across a field of view.

**panoramic adapter** An adapter used with a search receiver to provide a visual presentation, on an oscilloscope screen, of a band of frequencies extending above and below the center frequency to which the search receiver is tuned.

**panoramic display** A display that simultaneously shows the relative amplitudes of all signals received at different frequencies.

**panoramic indicator** An indicator that is designed to be connected to a radio receiver, to show signals received over a band of frequencies centered about the specific frequency to which the receiver is tuned.

**panoramic radar** A nonscanning radar that transmits signals omnidirectionally over a wide beam in the direction of interest.

**panoramic receiver** A radio receiver that permits continuous observation, on a cathode-ray-tube screen, of the presence and relative strength of all signals in a wide frequency band through which the receiver is periodically tuned.

**paper capacitor** A fixed capacitor consisting of two strips of metal foil separated by oiled or waxed paper or other insulating material, and rolled together in compact

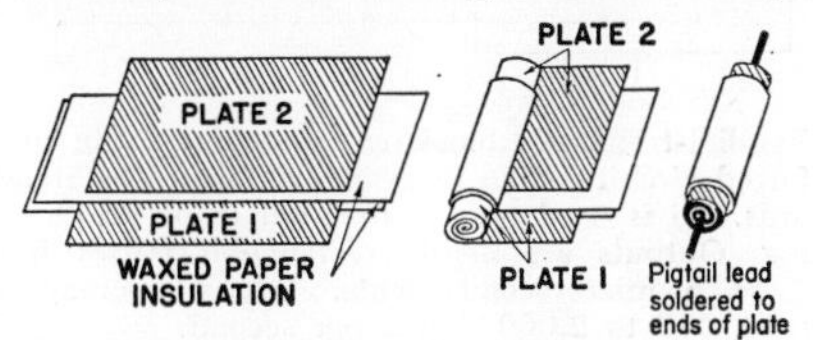


Paper capacitor construction.

tubular form. The foil strips are staggered so one projects from each end of the roll, and the connecting wires are attached to the projecting foil strips.

**par** Abbreviation for *precision approach radar.*

**parabolic antenna** A directional microwave antenna using some form of parabolic reflector to give directional characteristics.

**parabolic microphone** A microphone used at the focal point of a parabolic sound reflector to give improved sensitivity and directivity, as required for picking up a band marching down a football field.

**parabolic reflector** A reflector whose inner surface is shaped by rotating a parabola about its axis. When a microwave transmitting dipole, horn, or other antenna is placed at the focal point, the reflector concentrates the radiation into a parallel beam. For reception, incoming radiation is reflected to

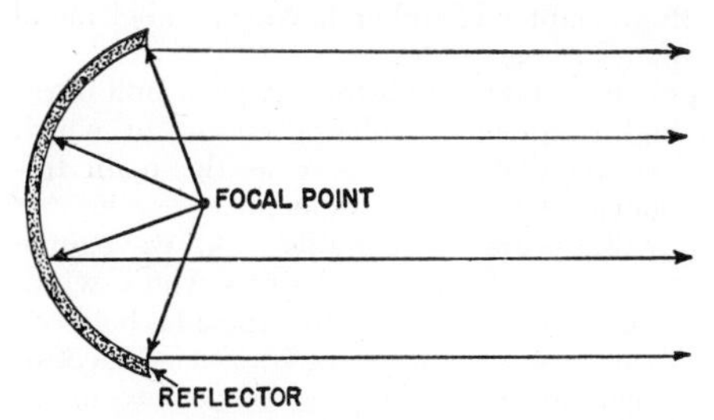


Parabolic reflector action.

the receiving antenna at the focal point. The reflector may be made of wire screen or sheet metal. Also called dish (slang).

**paraboloid** A reflecting surface formed by rotating a parabola about its axis of symmetry. Used as a reflector for sound waves and microwave radiation.

**parahydrogen** A hydrogen molecule in which the two nuclear spins are antiparallel, forming a singlet state.

**parallax** The apparent displacement of the position of an object that is due to a shift in the point of observation. Thus, the pointer of a meter will appear to be at different positions on the scale depending on the angle from which the meter is read.

**parallel** 1. Connected to the same pair of terminals. Also called multiple and shunt. 2. Simultaneous transmission of, storage of, or logical operations on the parts of a word, character, or other subdivision of a word in a computer, using separate facilities for the various parts. 3. Extending in the same direction and equally distant at all points, as of lines and surfaces.

**parallel circuit** A circuit in which the same voltage is applied to all components and the current divides among the components according to their resistances or impedances.

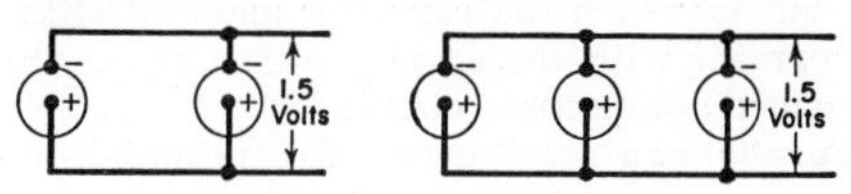


Parallel circuit for dry cells.

**parallel cut** A Y cut in a quartz crystal.

**parallel digital computer** A computer in which the digits are handled in parallel. The bits that comprise a digit may be handled either serially or in parallel.

**parallel feed** *Shunt feed.*

**parallel operation** The flow of information through a part or all of a computer over two or more lines or channels simultaneously.

**parallel-plate counter chamber** A radiation-counter chamber having parallel metal plates as electrodes.

**parallel-plate oscillator** A push-pull ultrahigh-frequency oscillator circuit in which two parallel plates serve as the main frequency-determining elements.

**parallel-plate waveguide** A waveguide consisting of two metal strips whose width is large compared to the spacing between them. For the dominant transverse electromagnetic wave, this waveguide has an infinite cutoff wavelength, and guide wavelength is equal to free-space wavelength.

**parallel-resistance formula** The combined resistance of resistors in parallel is less than that of the smallest resistor, and is equal to

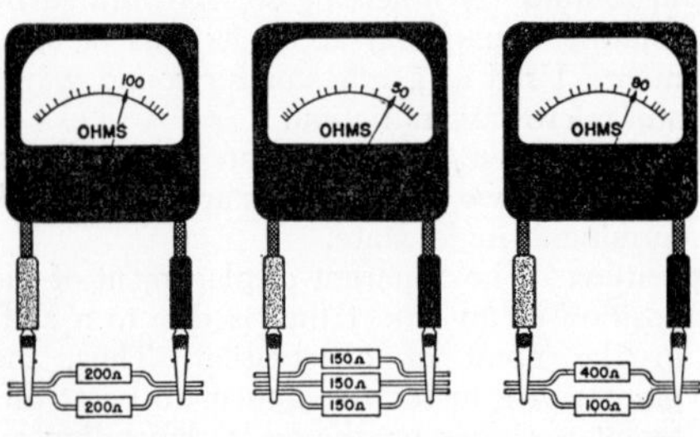


Parallel-resistance formula as illustrated by ohmmeter.

the product of the resistor values divided by their sum. For like values, divide the ohmic value of one resistor by the number of like resistors in parallel.

**parallel resonance** Resonance in a parallel resonant circuit, wherein the inductive and capacitive reactances are equal at the frequency of the applied voltage. The impedance of the parallel resonant circuit is then a maximum, so maximum signal voltage is developed across it. Also called antiresonance.

**parallel resonant circuit** A resonant circuit in which the capacitor and coil are in parallel with the applied a-c voltage. Also called antiresonant circuit.

**parallel-rod oscillator** An ultrahigh-frequency oscillator in which parallel rods form the tank circuits.

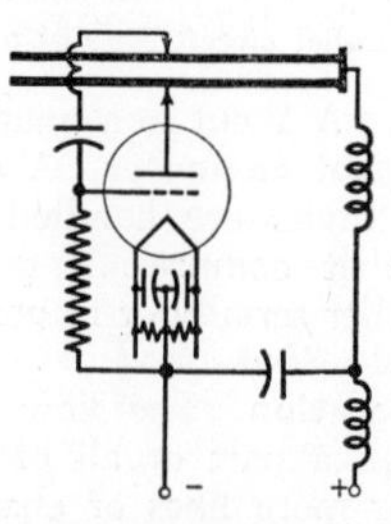


Parallel-rod oscillator in which the two rods are each a quarter-wavelength long.

**parallel storage** Computer storage in which all bits, characters, or words are essentially equally available.

**parallel transfer** Computer data transfer in which the characters of an element of information are transferred simultaneously over a set of parallel paths.

**parallel-triggered blocking oscillator** A triggered blocking oscillator in which the trigger pulse is applied to the anode of the blocking-oscillator tube rather than to the grid as in series triggering.

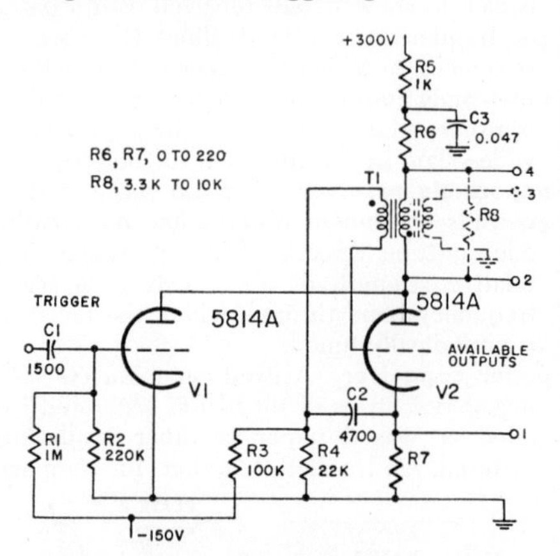


Parallel-triggered blocking oscillator using preferred circuit design by National Bureau of Standards. R8 is used only if required to prevent ringing. Outputs are nearly rectangular pulses from 1 to 5 microseconds wide. Triggering may be from 200 to 2,000 pulses per second.

**parallel two-terminal pair networks** Two-terminal pair networks having either their input or output terminals in parallel.

**parallel-wire line** A transmission line consisting of two parallel wires.

**parallel-wire resonator** A resonator consisting of a length of parallel-wire transmission line short-circuited at one end.

**paralysis** The overloading of an electron tube by a strong signal, causing the tube to charge the capacitances in the circuit to a point where they take too long to discharge and thus fail to amplify part of a succeeding signal. In radar, paralysis can obscure an echo from a target at short range because the capacitances do not have time to discharge between transmission of the pulse and reception of the echo.

**paramagnetic** Having a magnetic permeability greater than that of a vacuum and essentially independent of the magnetizing force. In ferromagnetic materials, the permeability varies with magnetizing force.

**paramagnetic amplifier** *Maser.*

**paramagnetic resonance** Resonance observable in a paramagnetic material as a peak in the energy absorption spectrum at a frequency related to the strength of the applied magnetic field and to the gyromagnetic ratio. Used in studying the energy states of nuclei, atoms, molecules, and crystal lattices.

**paramagnetism** Magnetism involving a permeability only slightly greater than unity.

**parameter** A quantity to which arbitrary values may be assigned, such as the value of a transistor or tube characteristic, or the value of a circuit component. A parameter is usually not changed during a given set of conditions.

**parametric amplifier** A microwave amplifier having as its basic element an electron tube or solid-state device whose reactance can be varied periodically by an a-c voltage

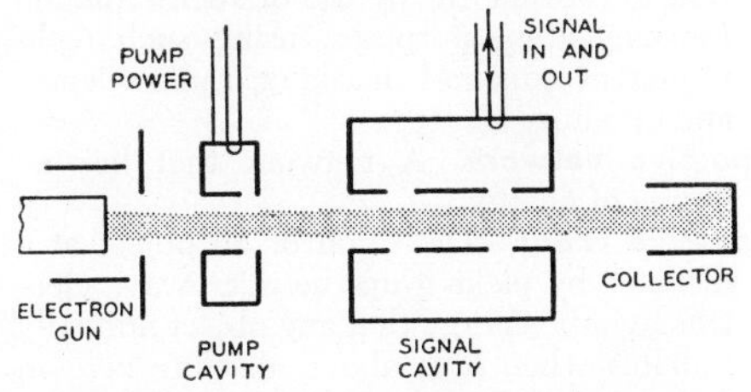

Parametric amplifier using electron-tube construction.

at a pumping frequency. Operation is at room temperature. The diode amplifier, ferromagnetic amplifier, and up-converter are examples. Also called mavar and reactance amplifier.

**parametric phase-locked oscillator** *Parametron.*

**parametron** A resonant circuit in which either the inductance or capacitance is

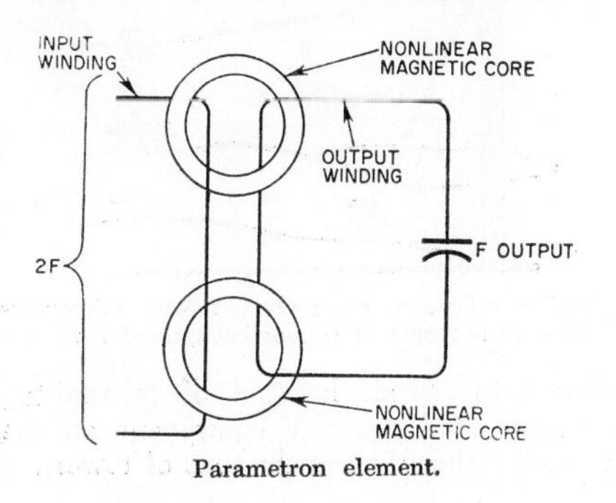

Parametron element.

made to vary periodically at one-half the driving frequency. Used as a digital computer element, in which the oscillation represents a binary digit. Also called parametric phase-locked oscillator, phase-locked oscillator, and phase-locked subharmonic oscillator.

**paraphase amplifier** An amplifier that uses the out-of-phase relation of the signal voltages at the anode and cathode of a tube to convert a single input signal into two out-of-phase output signals for driving a push-pull stage.

**parasitic** An undesired and energy-wasting signal current in an electronic circuit.

**parasitic array** An antenna array containing one or more parasitic elements.

**parasitic capture** Any absorption of a neutron that does not result in a fission or the production of a desired element.

**parasitic element** An antenna element that serves as part of a directional antenna array but has no direct connection to the receiver or transmitter. A parasitic element reflects or reradiates the energy that reaches it, in a phase relationship such as to give the desired radiation pattern. Also called passive element.

**parasitic oscillation** An undesired self-sustaining oscillation or a self-generated transient impulse in an oscillator or amplifier circuit, generally at a frequency above or below the correct operating frequency.

**parasitic suppressor** A suppressor, usually in the form of a coil and resistor in parallel, inserted in a circuit to suppress parasitic high-frequency oscillations.

**parent** A radionuclide that upon disintegration yields a specified nuclide, the daughter, either directly or as a later member of a radioactive series.

**parent peak** The component of a mass spectrum that results from the undissociated molecule.

**parity** A property of a wave function. The parity is 1 or even if the wave function is unchanged by an inversion of the coordinate system, and $-1$ or odd if the wave function is changed only in sign.

**parity check** A type of odd-even check that is used when the total number of ones or zeros in each permissible computer code expression is always made even or always made odd. A check may be made for even parity or odd parity.

**part** An article that is an element of a subassembly, is not normally useful by itself, and is not amenable to further disassembly for maintenance purposes. The term is used chiefly for structural members in electronic equipment, whereas tubes, transistors, resistors, capacitors, coils, switches, relays, transformers, and similar items that have distinct electrical characteristics are usually called components.

**partial** A sound sensation component that is distinguishable as a simple tone, cannot be further analyzed by the ear, and contributes to the character of the complex sound. The frequency of a partial may be higher or lower than the basic frequency and may or may not be an integral multiple or submultiple of the basic frequency.

**partial carry** The computer condition wherein a carry resulting from the addition of carries is not allowed to propagate.

**partial differential equation** A differential equation having more than one independent variable.

**partial node** The points, lines, or surfaces in a standing-wave system where some characteristic of the wave field has a minimum amplitude differing from zero.

**partial-read pulse** A current pulse that is applied to a magnetic memory to select a specific magnetic cell for reading.

**partial-select output** The voltage response produced by applying partial-read or partial-write pulses to an unselected magnetic cell.

**partial-write pulse** A current pulse that is applied to a magnetic memory to select a specific magnetic cell for writing.

**particle** Any very small part of matter, such as a molecule, atom, or electron.

**particle accelerator** *Accelerator.*

**particle velocity** The instantaneous velocity of a given infinitesimal part of a medium, with reference to the medium as a whole, due to the passage of a sound wave.

**partition noise** Noise that arises in an electron tube when the electron beam is divided between two or more electrodes, as between screen grid and anode in a pentode.

**party-line carrier system** A single-frequency carrier telephone system in which the carrier energy is transmitted directly to all other carrier terminals of the same channel.

**Paschen's law** The sparking potential between two parallel-plate electrodes in a gas is proportional to the product of gas pressure and electrode spacing.

**passband** A frequency band in which the attenuation of a filter is essentially zero.

**passive corner reflector** A corner reflector that is energized by a distant transmitting antenna. Used chiefly to improve the reflection of radar signals from objects that would not otherwise be good radar targets.

**passive detection** The detection of a target or other object by means that do not reveal the position of the detecting instrument.

**passive electronic countermeasures** Electronic countermeasures that are not detectable by the enemy.

**passive element** *Parasitic element.*

**passive homing** Homing that depends only on energy emanating naturally from the target, as in the form of infrared radiation, light, sound, electromagnetic radiation, ionization of air, or air pollution by exhaust gases.

**passive jamming** Use of confusion reflectors to return spurious and confusing signals to enemy radars. Also called mechanical jamming.

**passive navigation countermeasures** Control of transmissions from equipment capable of producing electromagnetic radiation to prevent enemy use of such radiation for navigation purposes, using such techniques as conelrad, masking, radio silence, and spoiling.

**passive network** A network that has no source of energy.

**passive radar** Detection of an object at a distance by picking up the microwave electromagnetic energy that any object normally radiates when it is above absolute zero in temperature. Passive radar requires an apparent temperature difference between the object and its surroundings, along with radio astronomy techniques in receivers to distinguish from noise a desired signal whose level may be less than one-millionth of a microwatt.

**passive reflector** A reflector used to change the direction of a microwave or radar beam. Often used on microwave relay towers to permit location of transmitter, repeater, and receiver equipment on the ground rather than at the tops of towers.

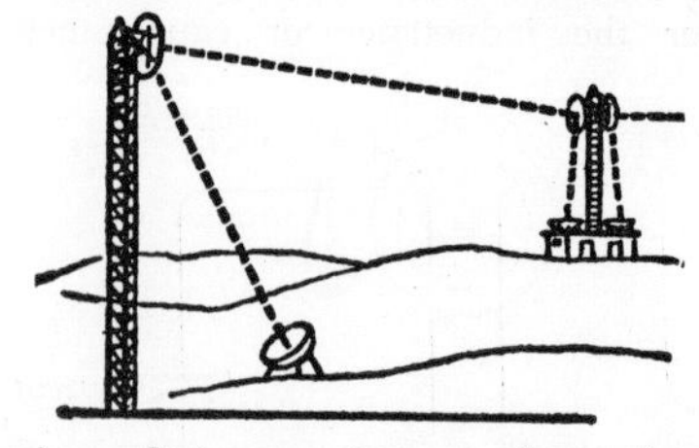

Passive reflectors on towers reflect microwave beams to antennas at convenient ground locations.

**passive sonar** Sonar that uses only underwater listening equipment, with no transmission of location-revealing pulses.

**passive transducer** A transducer containing no internal source of power.

**paste solder** Finely powdered solder metal combined with a flux.

**p-a system** Abbreviation for *public-address system.*

**patch** 1. A temporary connection between jacks or other terminations on a patchboard. 2. A section of coding inserted into a computer routine to correct a mistake or alter the routine.

**patchboard** A board or panel having a number of jacks at which circuits are terminated. Short cables called patchcords are plugged into the jacks to connect various circuits temporarily as required in broadcast, communication, and computer work. Patchboards for computers are often designed for quick removal without disturbing the patches made on the board, to permit plugging in a patchboard already set up for the next problem.

**patchcord** A cord equipped with plugs at each end, used to connect two jacks on a patchboard.

**patent** A document conferring on an inventor for a term of years the exclusive right to make, use, and sell his invention in practical form.

**path** A line connecting a series of points and constituting a proposed or traveled route.

**path length** The length of a magnetic flux line in a core.

**pattern generator** A signal generator used to generate a test signal that can be fed into a television receiver to produce on the screen a pattern of lines having usefulness for servicing purposes.

**Pauli exclusion principle** Any wave function involving several identical particles must change sign when the coordinates, including the spin coordinates, of any identical pair are interchanged. Only one particle of a given kind can occupy a particular quantum state. The principle applies to electrons, fermions, protons, and neutrons, but not to bosons. Also called exclusion principle.

**Pauli-Fermi principle** Each level of a quantized system can include one, two, or no electrons. If there are two electrons, they must have spins in opposite directions.

**pay television** *Subscription television.*

**P band** A band of radio frequencies extending from 225 to 390 mc.

**PbS** Symbol for *lead sulfide.*

**PbTe** Symbol for *lead telluride.*

**pcm** Abbreviation for *pulse-code modulation.*

**P display** *Plan position indicator.*

**pdm** Abbreviation for *pulse-duration modulation.*

**peak** The maximum instantaneous value of a quantity.

**peak alternating gap voltage** The negative of the line integral of the peak alternating electric field taken along a specified path across the gap in a microwave tube.

**peak amplitude** The maximum amplitude of an alternating quantity, measured from its zero value.

**peak anode current** The maximum instantaneous value of the anode current in an electron tube.

**peak cathode current** The maximum instantaneous value of a periodically recurring cathode current in an electron tube.

**peak clipper** *Limiter.*

**peak electrode current** The maximum instantaneous current that flows through an electrode.

**peak energy density** The maximum absolute value of the instantaneous energy density in a specified time interval.

**peaker** A small fixed or adjustable inductance used to resonate with stray and distributed capacitances in a broadband amplifier to increase the gain at the higher frequencies.

**peak field strength** *Peak magnetizing force.*

**peak flux density** The maximum flux density in a magnetic material in a specified cyclically magnetized condition.

**peak forward anode voltage** The maximum instantaneous anode voltage in the direction in which an electron tube is designed to pass current.

**peaking** Increasing the response of a circuit at a desired frequency or band of frequencies.

**peaking circuit** A circuit used to improve the high-frequency response of a broadband amplifier. In shunt peaking, a small coil is placed in series with the anode load. In series peaking, the coil is placed in series with the grid of the following stage. Used in video amplifiers, often with both types of peaking in the same stage.

**peaking coil** A small coil placed in a circuit to resonate with the distributed capacitance of the circuit at a frequency for which peak response is desired, as in a video amplifier near the cutoff frequency.

**peaking control** An adjustable resistor-capacitor circuit used to control the wave shape of the horizontal oscillator output pulses, as required to give a linear sweep.

**peaking transformer** A transformer in which the number of ampere-turns in the primary is high enough to produce many times the normal flux density values in the core. The flux changes rapidly from one direction of saturation to the other twice per cycle, inducing a highly peaked voltage pulse in a secondary winding. Used to fire ignitrons and thyratrons.

**peak inverse anode voltage** The maximum instantaneous anode voltage in the direction opposite to that in which an electron tube is designed to pass current.

**peak level** The maximum instantaneous level that occurs during a specified time interval, such as the peak sound pressure level in acoustics.

**peak limiter** *Limiter.*

**peak load** The maximum instantaneous load or the maximum average load over a designated interval of time.

**peak magnetizing force** The upper or lower limiting value of magnetizing force associated with a cyclically magnetized condition. Also called peak field strength.

**peak particle velocity** The maximum absolute value of the instantaneous particle velocity in a specified time interval.

**peak power** The maximum instantaneous power of a transmitted radar pulse. Since the resting time of a radar transmitter is long compared to its operating time, the average power is low compared to the peak power.

**peak power output** The output power of a radio transmitter as averaged over one carrier cycle at the maximum amplitude that can occur for any combination of transmitted signals.

**peak pulse amplitude** The maximum absolute peak value of a pulse, excluding spikes and other unwanted portions.

**peak pulse power** The power at the maximum of a pulse of power, excluding spikes.

**peak response** The maximum response of a device or system to an input stimulus.

**peaks** Momentary high volume levels during a radio program, making the volume indicator at the studio or transmitter swing upward.

**peak signal level** The maximum instantaneous signal power or voltage at a specified point in a facsimile system.

**peak sound pressure** The maximum absolute value of the instantaneous sound pressure in a specified time interval. The commonly used unit is the microbar.

**peak speech power** The maximum value of the instantaneous speech power within the time interval considered.

**peak-to-peak amplitude** The sum of the extreme swings of an alternating quantity in positive and negative directions from its zero value. For a sinusoidal waveform, the peak-to-peak amplitude is one-half of the peak amplitude in either direction.

**peak-to-peak voltmeter** A voltmeter that measures the voltage difference between the positive and negative peaks of a voltage. Two peak-reading voltmeters connected in series opposition can be used for this purpose.

**peak value** The maximum instantaneous value of a varying current, voltage, or power during the time interval under consideration. For a sine wave, it is equal to 1.414 times the effective value. Also called crest value.

**peak voltmeter** A voltmeter that reads peak values of an alternating voltage.

**peak volume velocity** The maximum absolute value of the instantaneous volume velocity in a specified time interval.

**pebble nuclear reactor** A nuclear reactor made by stacking pebbles of moderating and fissionable material. A coolant is circulated through the pores between the pebbles to cool the reactor.

**pedestal** 1. The structure that supports a radar antenna. 2. *Blanking level.*

**pedestal level** *Blanking level.*

**P electron** An electron having an orbit in the P shell, which is the sixth shell of electrons surrounding the atomic nucleus, counting out from the nucleus.

**pellet** A small piece of semiconductor material, used in crystal diodes and transistors.

**Peltier effect** The production or absorption of heat at the junction of two metals when a current is passed through the junction. Heat generated by current in one direction will be absorbed when the current is reversed.

**pencil beam** A narrow radar beam having an essentially circular cross-section.

**pencil-beam antenna** A unidirectional antenna so designed that cross-sections of its major lobe are approximately circular.

**pencil tube** A long, thin disk-seal tube designed for use as a uhf oscillator or amplifier.

**penetrating shower** A cosmic-ray shower in which some or all of the particles can penetrate over about 20 cm of lead. The particles are often pi mesons.

**penetration depth** 1. The nominal depth below the surface of a conductor within which current is concentrated by the skin effect during induction heating. The higher the frequency, the less is the penetration. 2. The depth to which an external magnetic field penetrates a superconductor.

**penetration frequency** *Critical frequency.*

**penetration probability** The probability of transmission of a particle through a potential barrier. Examples of barrier penetration are the passage of alpha particles through the barrier at the nuclear wall and the motions of electrons in a metal through the interatomic coulomb barriers. Also called transmission coefficient.

**penetrometer** An instrument that measures the penetrating power of a beam of x-rays or other penetrating radiation.

**pentagrid converter** A pentagrid tube used as a converter (combination oscillator, mixer, and first detector) in a superheterodyne receiver. The cathode and the first two grids are in the oscillator circuit, and the incoming r-f signal is applied to the third grid.

**pentagrid mixer** A pentagrid tube used to mix two signals. The first and third grids are control grids to which the signals are applied. The second and fourth grids are screen grids, and the fifth grid is a suppressor grid. When used in a superheterodyne receiver, the local oscillator signal may be applied either to the first or third grid.

**pentagrid tube** An electron tube having five grids.

**pentatron** A double-triode electron tube having a common cathode.

**pentode** A five-electrode electron tube containing an anode, a cathode, a control electrode, and two additional electrodes that are ordinarily grids.

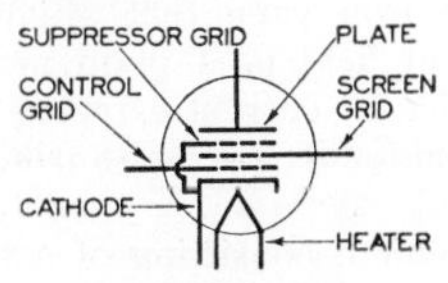

Pentode symbol.

**per cent beam modulation** An image orthicon rating equal to 100 times the ratio of the signal output current for highlight illumination to the dark current.

**per cent deafness** Per cent hearing loss.

**per cent depth dose** The amount of radiation delivered at a specified depth in tissue, expressed as a percentage of the amount delivered at the skin.

**per cent harmonic distortion** A measure of the harmonic distortion in a system or component. It is equal to 100 times the ratio of the square root of the sum of the squares of the root-mean-square voltages of each of the individual harmonic frequencies, to the root-mean-square voltage of the fundamental. Current values may be used in place of voltage values.

**per cent modulation** The modulation factor expressed as a percentage.

**per cent modulation meter** An instrument used to measure the per cent modulation of an amplitude-modulation transmitter.

**per cent ripple** The ratio of the effective value of the ripple voltage to the average value of the total voltage, expressed as a percentage.

**percussion welding** Welding in which a sudden discharge of electric current at the junction produces an arc that is extinguished by a percussive blow.

**perfect dielectric** A dielectric in which all the energy required to establish an electric field in the dielectric is returned to the electric system when the field is removed. A vacuum is the only known perfect dielectric.

**perhapsatron** A smaller version of British zeta apparatus for investigating controlled fusion of hydrogen atoms. The pinch effect is used to heat the gas to temperatures over 5,000,000°C, at which the gas dissociates into plasma.

**period** 1. The time required for one complete cycle of a regularly repeated series of events. 2. The time required for exponentially rising or falling neutron flux in a nuclear reactor to change by a factor of 2.71828.

**periodic** Having a repetition rate.

**periodic damping** Damping in which a pointer or other moving object oscillates about a new position before coming to rest. Also called underdamping.

**periodic duty** Intermittent duty in which the load conditions are regularly recurrent.

**periodic electromagnetic wave** A wave in which the electric field vector is repeated in detail at a fixed point after the lapse of a time known as the period.

**periodic law** Certain properties of the elements are periodic functions of their atomic numbers. When the elements are arranged in the order of their atomic numbers, these properties recur in regular cycles.

**periodic permanent magnet** An assembly of axially magnetized ring-shaped permanent magnets whose adjacent faces have like polarity. Used to produce a sinusoidal or otherwise nonuniform permanent magnetic field.

**periodic pulse train** A pulse train made up of identical groups of pulses, the groups being repeated at regular intervals.

**periodic quantity** An oscillating quantity whose values recur at equal increments of time, space, or some other independent variable.

**periodic rating** A rating defining the load that can be carried for specified alternate periods of load and rest.

**periodic table** A table in which the elements are arranged according to the periodic law, so elements with similar characteristics are logically grouped together.

**periodic wave** A wave that repeats itself at regular intervals, such as a sine wave.

**period meter** An instrument that indicates the period of a nuclear reactor in seconds. It may also operate interlocks and scram the reactor if the period is dangerously small.

**permalloy** A magnetic alloy having high permeability, usually consisting of iron, nickel, and small quantities of other metals. Used to shield components, tubes, and equipment from stray magnetic fields.

**permanent echo** A signal reflected from an object that is fixed with respect to a radar site.

**permanent magnet** [abbreviated p-m] A piece of hardened steel or other magnetic material that has been strongly magnetized and retains its magnetism indefinitely.

**permanent-magnet centering** Centering of the image on the screen of a television picture tube by means of magnetic fields produced by permanent magnets mounted around the neck of the tube.

**permanent-magnet dynamic loudspeaker** *Permanent-magnet loudspeaker.*

**permanent-magnet erasing head** An erasing head that uses the fields of one or more permanent magnets for erasing magnetic tape.

**permanent-magnet focusing** Focusing of the electron beam in a television picture tube by means of the magnetic field produced by one or more permanent magnets mounted around the neck of the tube.

**permanent-magnet loudspeaker** [abbreviated p-m loudspeaker] A moving-conductor loudspeaker in which the steady magnetic field is produced by a permanent magnet. Also called permanent-magnet dynamic loudspeaker.

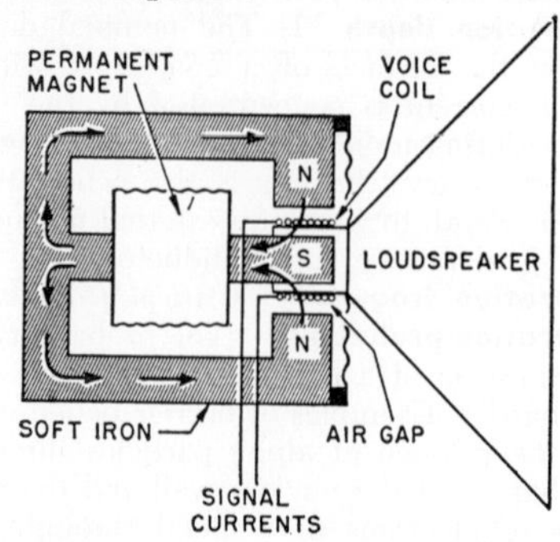


Permanent-magnet loudspeaker construction.

**permanent-magnet moving-coil instrument** A d-c meter movement consisting of a small coil of wire supported on jeweled bearings between the poles of a permanent magnet. Spiral springs serve as connections to the coil and keep the coil and its attached pointer at the zero position on the meter scale. When the direct current to be measured is sent through the coil, its magnetic field interacts with that of the permanent magnet to produce rotation of the coil. Also called d'Arsonval movement (deprecated).

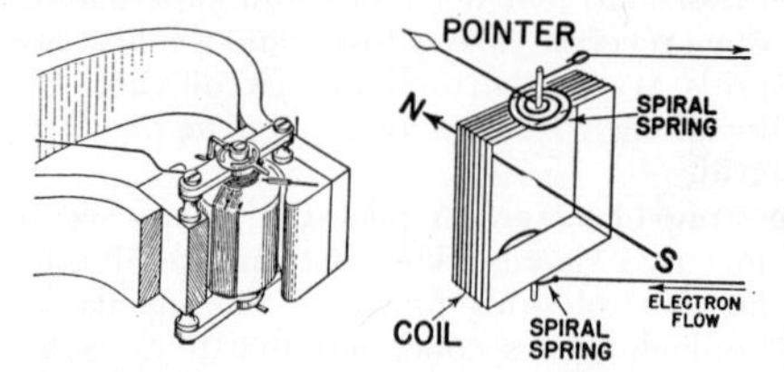


Permanent-magnet moving-coil instrument and diagram showing electron flow path through springs and coil.

**permanent-magnet moving-iron instrument** A meter that depends for its operation on a movable iron vane that aligns itself in the resultant field of a permanent magnet and an adjacent current-carrying coil.

**permanent-magnet second-harmonic self-synchronous system** A remote indicating arrangement consisting of a transmitter unit and one or more receiver units. All units have permanent-magnet rotors and toroidal stators using saturable ferromagnetic cores and excited with alternating current from a common external source. The coils are tapped at three or more equally spaced intervals, and the corresponding taps are connected together to transmit voltages that consist principally of the second harmonic of the excitation voltage. The rotors of the receiver units will assume the same angular position as that of the transmitter rotor.

**permanent memory** A computer memory that does not lose its stored data when computer power is turned off.

**permatron** A thermionic-cathode gas diode in which conduction is initiated by an external magnetic field instead of a grid. The resulting action is similar to that in a thyratron. Used chiefly as a controlled rectifier.

**permeability** [symbol $\mu$] A measure of how much better a given material is than air as a path for magnetic lines of force. The permeability of air is assumed as 1. Permeability is the magnetic induction $B$ in gauss divided by the magnetizing force $H$ in oersteds.

**permeability tuner** A television or radio tuner in which the tuning dial moves the powdered iron cores of coils in the tuning circuits.

**permeability tuning** Tuning of a resonant circuit by moving a ferrite core in or out of a coil, thereby changing the effective permeability of the core and the inductance of the circuit.

**permeameter** An instrument for measuring the magnetic flux or flux density produced in a test specimen of ferromagnetic material by a given magnetic intensity, to permit computation of the magnetic permeability of the material.

**permeance** [symbol $P$] A characteristic of a portion of a magnetic circuit, equal to magnetic flux divided by magnetomotive force. Permeance is the reciprocal of reluctance.

**permissible dose** The amount of radiation that may be safely received by an individual within a specified period. Formerly called tolerance dose.

**permittance** Obsolete term for *capacitance.*

**permittivity** *Dielectric constant.*

**perpendicular magnetization** In magnetic recording, magnetization of the recording medium in a direction perpendicular to the line of travel and parallel to the smallest cross-sectional dimension of the medium.

**Pershing** An Army surface-to-surface guided missile using a solid propellant.

**persistence** A measure of the length of time that the screen of a cathode-ray tube remains luminescent after excitation is removed. Long-persistence screens are used for ppi radar displays. Medium-persistence screens are used chiefly in television receivers. Short-persistence screens are used in cathode-ray oscilloscopes and some types of radar displays. The last number in the type designation of a cathode-ray tube indicates its persistence, on a scale ranging from 1 for short persistence to 7 for long persistence.

**persistence characteristic** The relation between luminance and time after excitation of a luminescent screen. Also called decay characteristic.

**persistence of vision** The ability of the eye to retain the impression of an image for a short time after the image has disappeared. This characteristic enables the eye to fill in the dark intervals between successive images in movies and television and give the illusion of motion.

**persistor** A miniature bimetallic printed circuit that is operated at temperatures near absolute zero. A critical current value in one metal loop changes that metal from a superconductive state to a normal resistive state, to give storage or readout for computers.

**persistron** A solid-state electroluminescent and photoconductive display panel that provides amplification of light.

**perspective representation** A radar ppi display in which the region ahead of the ship is produced practically in perspective, much as would be seen from the bridge when visibility is good.

**Perspex** Trademark of a British plastic similar to Plexiglas.

**perturbation** A change in a known system.

**perturbation theory** The study of the effect of small changes on the behavior of a system.

**perveance** The space-charge-limited cathode current of a diode divided by the three-halves power of the anode voltage.

**Petrel** A Navy air-to-surface and air-to-underwater guided missile using radar homing guidance.

**pf** Abbreviation for *picofarad.*

**pfm** Abbreviation for *pulse-frequency modulation.*

**phanotron** A hot-cathode gas diode used as a rectifier. The type 866 mercury-vapor rectifier tube is an example.

**phantastron** A monostable pentode circuit used to generate sharp pulses at an adjustable and accurately timed interval after receipt of a triggering signal.

**phantom** A volume of material having the radiation-absorbing characteristics of tissue, used to simulate a portion of the human body. Radiation measurements made at a point in a phantom permit determination of

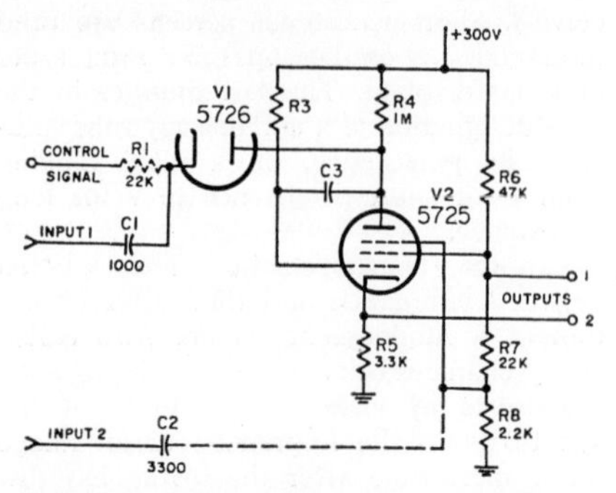


Phantastron, using preferred circuit design by National Bureau of Standards. Duration of output waveform is almost directly proportional to voltage of control signal. For durations under 1,000 microseconds, use 1.0 megohm for R3 and 100 to 1,000 micromicrofarads for C3.

the radiation dose delivered to a corresponding point within the body. Materials commonly used for x-rays are beeswax, Masonite (unit density), and water.

**phantom circuit** A communication circuit derived from two other communication circuits or from one other circuit and ground, with no additional wire lines.

**phantom target** *Echo box.*

**phase** The position of a point on the waveform of an alternating or other periodic quantity with respect to the start of the cycle. Usually expressed in degrees, with 360° representing one complete cycle.

**phase-angle meter** *Phasemeter.*

**phase constant** A rating for a line or medium through which a plane wave of a given frequency is being transmitted. It is the imaginary part of the propagation constant, and is the space rate of decrease of phase of a field component (or of the voltage or current) in the direction of propagation, in radians per unit length. The real part of the propagation constant is the attenuation constant.

**phase control** 1. A control that changes the phase angle at which the a-c line voltage fires a thyratron, ignitron, or other controllable gas tube. Also called phase-shift control. 2. *Hue control.*

**phase converter** A converter that changes the number of phases in an a-c power source without changing the frequency.

**phase delay** The very short time delay that occurs when a single-frequency wave is transferred from one point to another in a system. Phase delay is usually expressed as the ratio of the total phase shift in cps to the frequency in cps.

**phase detector** A circuit that provides a d-c output voltage which is related to the phase difference between an oscillator signal and a reference signal, for use in controlling the oscillator to keep it in synchronism with the reference signal.

**phase deviation** The peak difference between the instantaneous angle of a modulated wave and the angle of the sine-wave carrier.

**phase difference** The time in electrical degrees by which one wave leads or lags another.

**phase distortion** *Phase-frequency distortion.*

**phase equalizer** A network designed to compensate for phase-frequency distortion within a specified frequency band.

**phase focusing** An automatic action that helps to keep the electrons of a multicavity magnetron in phase with the rotating field. Lagging electrons receive energy from the radial component of the gap field to reduce the phase lag, while leading electrons give up energy to the gap field to reduce the phase lead.

**phase-frequency distortion** Distortion occurring because phase shift is not proportional to frequency over the frequency range required for transmission. Also called phase distortion.

**phase inverter** A circuit or device that changes the phase of a signal by 180°, as required for feeding a push-pull amplifier stage without using a coupling transformer, or for changing the polarity of a pulse. A triode is commonly used as a phase inverter.

**phase localizer** A localizer in which the on-course line is centered in an equiphase zone, and right-left deviations from this zone are detectable as reversals of phase of one of the two radiated signals.

**phase-locked oscillator** *Parametron.*

**phase-locked subharmonic oscillator** *Parametron.*

**phasemeter** An instrument for measuring the difference in phase between two alternating quantities having the same frequency. Also called phase-angle meter.

**phase-modulated transmitter** A transmitter that transmits a phase-modulated wave.

**phase modulation** [abbreviated p-m] Angle modulation in which the angle of a sine-wave carrier is caused to depart from the carrier angle by an amount proportional to the instantaneous value of the modulating wave. Combinations of phase and frequency modulation are commonly referred to as frequency modulation.

**phase modulator** A modulator that provides phase modulation of a carrier signal.

**phase-propagation ratio** The propagation ratio divided by its magnitude.

**phase quadrature** *Quadrature.*

**phaser** 1. A microwave ferrite phase shifter employing a longitudinal magnetic field along one or more rods of ferrite in a waveguide. 2. A device for adjusting facsimile equipment so the recorded elemental area bears the same relation to the record sheet as the corresponding transmitted elemental area bears to the subject copy in the direction of the scanning line.

**phase recovery time** The time required for a fired tr or pre-tr tube to deionize to such a level that a specified phase shift is produced in the low-level r-f signal transmitted through the tube.

**phase reversal** A change of 180° or one half-cycle in phase.

**phase-reversal switch** A switch used in a stereophonic sound system to reverse the connections to one loudspeaker, so its acoustic output is reversed 180° in phase.

**phase-sensing monopulse radar** Monopulse radar in which the receiving antenna has two or more apertures separated by several wavelengths, each with its own feed. The apertures give identical radiation patterns. Phase comparison of the arriving signals gives the desired directional information with high precision.

**phase-sensitive amplifier** A servoamplifier whose output signal polarity or phase is dependent on the phase relationship between an input voltage and a reference voltage.

**phase-shaped antenna** *Shaped-beam antenna.*

**phase shift** 1. A change in the phase relationship between two alternating quantities. 2. The phase relationship between a scattered wave and the incident wave associated with a particle or photon that undergoes scattering.

**phase-shift bridge** A mutual-inductance bridge that measures the ratio of two voltages in both magnitude and phase.

**phase-shift circuit** A network that provides a voltage component which is shifted in phase with respect to a reference voltage.

**phase-shift control** *Phase control.*

**phase-shift discriminator** A discriminator that uses two similarly connected diodes and requires a limiter in its input to remove amplitude variations from the frequency-modulated or phase-modulated input signal. The diodes are fed by a transformer that is tuned to the center frequency. When the frequency of the input signal swings away from this center frequency, one diode receives a stronger signal than the other. The net output of the diodes is then proportional to the frequency displacement. Also called Foster-Seeley discriminator.

**phase shifter** A device used to change the phase relation between two a-c values.

**phase-shift microphone** A microphone employing phase-shift networks to produce directional properties.

**phase-shift omnidirectional radio range** An omnidirectional radio range used to indicate the azimuthal position of an aircraft by means of two carrier waves, one of which is continuously changed in phase. The two waves are in phase only along a reference line that is usually north.

**phase-shift oscillator** An oscillator in which a network having a phase shift of 180° per stage is connected between the output and the input of an amplifier.

**phase space** Space having 6*N* dimensions, corresponding to the 3*N* coordinates of position and the 3*N* kinetic moments of the *N* particles considered, or to the generalized coordinates and to the Lagrange generalized kinetic moments.

**phase-splitter** A circuit that takes a single input signal voltage and produces two output signal voltages 180° apart in phase.

**phase-tuned tube** A fixed-tuned broadband tr tube in which the phase angle through the tube and the reflection introduced by the tube are controlled within limits.

**phase velocity** The velocity with which a point having a certain phase in an electromagnetic wave travels in the direction of propagation. In a waveguide the phase velocity may be greater than the wave velocity.

**phasing** *Framing.*

**phasing capacitor** A capacitor used in a crystal filter circuit to neutralize the capacitance of the crystal holder.

**phasing line** The portion of the length of a facsimile scanning line that is used for the phasing signal.

**phasing link** A delay line used to connect together the bays of a stacked antenna, so the signals from all bays are in phase at the transmission line.

**phasing signal** A signal used to adjust the picture position along the scanning line in a facsimile system.

**phasitron** An electron tube used to frequency-modulate an r-f carrier. Internal electrodes are designed to produce a rotating disk-shaped corrugated sheet of electrons. Audio input is applied to a coil surrounding the glass envelope of the tube, to produce a varying axial magnetic field that gives the desired phase or frequency modulation of the r-f carrier input to the tube.

**phasor** A quantity expressed in complex form, with or without time variation. A phasor may be used to represent a vector, but a vector does not involve a complex plane and hence is not a phasor.

**phenolic** A thermosetting plastic material available in many combinations of phenol and formaldehyde, often with added fillers to provide a broad range of physical, electrical, chemical, and molding properties.

**Phillips screw** A screw having in its head a recess in the shape of an indented cross. It is inserted or removed with a special Phillips screwdriver that automatically centers itself in the screw.

**pH indicator** An instrument that measures and indicates the hydrogen ion concentration of a solution on a scale of pH values from 0 to 14, where 7 indicates a neutral solution, lower numbers indicate acidity, and higher numbers indicate alkalinity.

**phi polarization** The state of an electromagnetic wave in which the E (electric) vector is tangent to the lines of latitude of some given spherical frame of reference. For theta polarization this vector is tangent to the meridian lines of the reference.

**phon** The unit of loudness level of a sound. It is numerically equal to the sound pressure level, in decibels relative to 0.0002 microbar, of a pure 1,000-cps tone that is judged by listeners to be equivalent in loudness to the sound under consideration.

**phone** *Headphone.*

**phone jack** A jack designed for standard ¼-inch-diameter phone plugs. Also called telephone jack.

**phone plug** A standard plug having a ¼-inch-diameter shank, used with headphones, microphones, and other audio equipment.

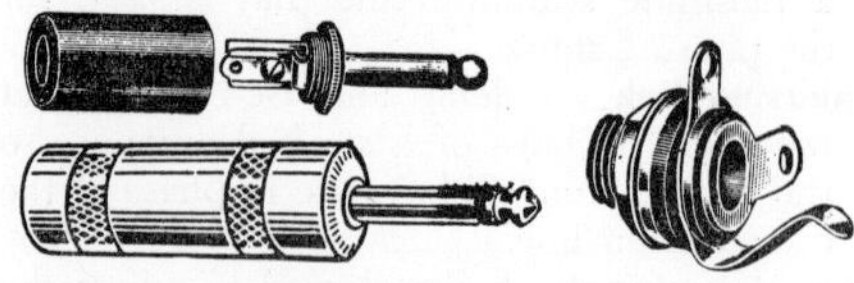

Phone plugs and jack.

Usually designed for use with either two or three conductors. Also called telephone plug.

**phono adapter** A device that slips under a tube or is otherwise connected to a radio receiver, and provides terminals to which a phonograph pickup can be connected so as to utilize the a-f system and loudspeaker of the receiver for reproduction of phonograph records.

**phonocardiogram** A graphic recording of the sounds of the heart.

**phono cartridge** *Phonograph pickup.*

**phonoelectrocardioscope** An electronic medical instrument employing a double-beam cathode-ray oscilloscope to show simultaneously the waveforms of two different quantities related to the heart.

**phonograph** An instrument for converting the sound groove variations of a phonograph record into sound waves. In an electric phonograph, the needle movements in the record grooves are converted into audio-frequency currents and amplified sufficiently for reproduction by a loudspeaker. In a mechanical phonograph, the needle actuates a sound-producing diaphragm directly. The turntable on which the record is placed may be driven by an electric motor or a spring motor. Also called gramophone (British).

**phonograph amplifier** An a-f amplifier designed to amplify the a-f output signal of a phonograph pickup.

**phonograph connection** Two terminals sometimes provided at the rear of a radio receiver, connected to the input of the first a-f amplifier stage. When a phonograph pickup is connected to these terminals, its output is amplified by the a-f amplifier and reproduced by the loudspeaker.

**phonograph oscillator** An r-f oscillator that can be modulated by a phonograph pickup. The resulting modulated r-f signal is fed through wires to the antenna and ground terminals of a radio receiver, so the entire radio receiver can serve for amplifying and reproducing phonograph records. In wireless phonograph oscillators, the output is fed to a small loop antenna and broadcast through space to the radio receiver, eliminating wire connections.

**phonograph pickup** A pickup that converts variations in the grooves of a phonograph record into corresponding electric signals. Also called cartridge, phono cartridge, and phono pickup. An acoustic pickup converts groove variations directly into sound waves.

**phonograph record** A shellac-composition or vinyl plastic disk, usually 7, 10, or 12 inches in diameter, on which sounds have been recorded as modulations in grooves. Common speeds used are 16⅔, 33⅓, 45, and

78 revolutions per minute. Also called disk, disk recording, and platter (slang).

**phono jack** A jack designed to accept a phono plug and provide a ground connection for the shield of the conductor connected to the plug.

**phonon** A unit of thermal energy in a crystal lattice, equal in value to the product of Planck's constant and the thermal vibration frequency.

**phono pickup** *Phonograph pickup.*

**phono plug** A plug designed for attaching to the end of a shielded conductor, for feeding a-f signals from a phonograph or other a-f source to a mating phono jack on a preamplifier or amplifier.

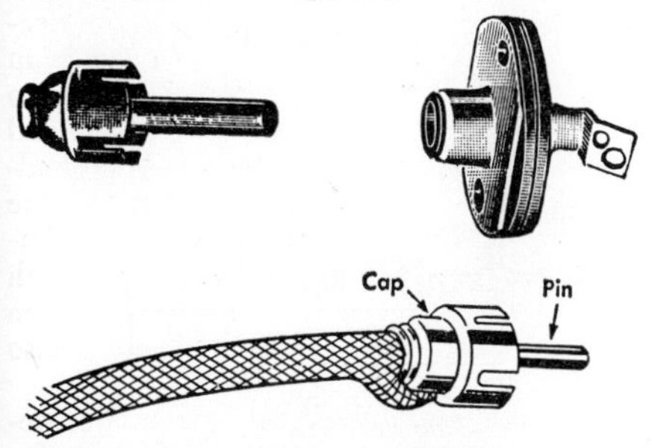


Phono jack and plug, and method of mounting plug on end of shielded conductor.

**phosphene** A visual sensation experienced by a human subject during the passage of current through the eye.

**phosphor** A substance capable of luminescence. Used for the screen of a cathode-ray tube, where the impact of the electron beam produces luminescence. Zinc oxide, zinc orthosilicate, zinc silicate, zinc sulfide, and cadmium tungstate are examples of phosphors. In color television, three different phosphors are used to obtain the three different receiver primary colors.

**phosphor bronze** An alloy of copper, tin, and phosphor, used for contact springs in switches and relays.

**phosphor dot** One of the tiny dots of phosphor material that are used in groups of three, one for each primary color, on the screen of a color television picture tube.

**phosphor-dot faceplate** The glass faceplate on which the trios of color phosphor dots are mounted in a shadow-mask three-gun color television picture tube.

**phosphorescence** A form of luminescence in which the emission of light continues more than $10^{-8}$ second after excitation by radiation having a shorter wavelength, such as by electrons, ultraviolet light, or x-rays. When emission of light occurs only during excitation, the result is fluorescence. Also called afterglow.

**phosphorogen** A substance that promotes phosphorescence in another substance, as manganese does in zinc sulfide.

**phosphorus** [symbol P] A nonmetallic element. Atomic number is 15.

**phot** A unit of illumination equal to 1 lumen per square centimeter.

**photocathode** A photosensitive surface that emits electrons when exposed to light or other suitable radiation. Used in phototubes, television camera tubes, and other light-sensitive devices.

**photocell** A solid-state photosensitive electron device whose current-voltage characteristic is a function of incident radiation. Examples include photoconductive cells, phototransistors, and photovoltaic cells. Also called electric eye (slang) and photoelectric cell. A phototube is not a photocell.

**photochemical activity** Chemical changes due to radiant energy, such as light.

**photoconductive cell** A photocell whose resistance varies with the illumination on the cell. The selenium cell is an example.

**photoconductive detector** A detector in which changes in radiant energy cause changes in electric resistance. When made from a semiconductor material such as lead sulfide, lead selenide, lead telluride, or germanium, it can give good response to infrared radiation. Cooling with liquid air or gas improves the infrared response of many photoconductive detectors.

**photoconductivity** Conductivity that varies with illumination.

**photoconductor** A semiconductor in which conductivity varies with illumination.

**photocurrent** An electric current that varies with illumination.

**photodiode** A semiconductor diode in which the reverse current varies with illumination. Examples include the alloy-junction photocell and the grown-junction photocell.

**photodisintegration** The disintegration of an atomic nucleus by radiant energy.

**photoelectric** Pertaining to the electrical effects of light, such as the emission of electrons, generation of a voltage, or a change in resistance when exposed to light.

**photoelectric abridged spectrophotometry** Analysis of color by means of from three to eight spectral filters used in a simplified spectrophotometer to isolate spectral bands that make up color. The process is approximate since the bands employed are considerably wider than in true spectrophotometry.

**photoelectric absorption** The absorption of photons in the photoelectric effect.

**photoelectric cathode** A cathode that functions primarily by photoelectric emission.

**photoelectric cell** *Photocell.*

**photoelectric character reader** *Character reader.*

**photoelectric color comparator** *Color comparator.*

**photoelectric colorimeter** A colorimeter that uses a phototube or photocell, a set of color filters, an amplifier, and an indicating meter for quantitative determination of color. Widely used to determine the constituents of a liquid in which the color varies with the constituents in a known manner.

**photoelectric color register control** A photoelectric control system used as a longitudinal position regulator for a moving material or web, to maintain a preset relationship between repetitive register marks when printing successive colors.

**photoelectric constant** A quantity that, when multiplied by the frequency of the radiation that is causing emission of electrons, gives the voltage absorbed by the escaping photoelectron. The constant is equal to $h/e$, where $h$ is Planck's constant and $e$ is the electronic charge.

**photoelectric control** Control of a circuit or piece of equipment in response to a change in incident light.

**photoelectric counter** A photoelectrically actuated device used to record the number of times a given light path is intercepted by an object.

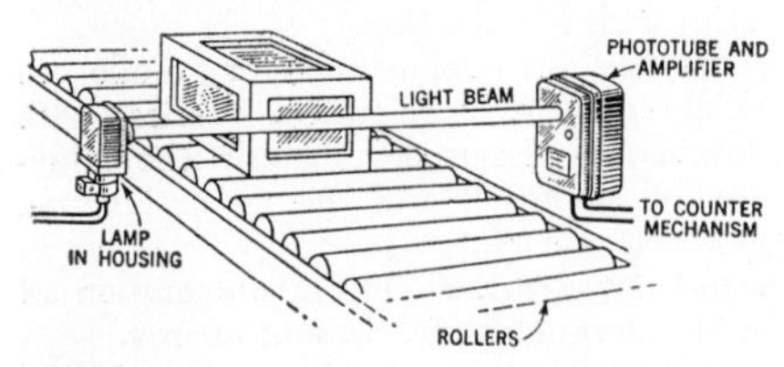


Photoelectric counter for cartons on roller conveyor.

**photoelectric current** A current of electrons emitted from the cathode of a phototube under the influence of light.

**photoelectric cutoff register controller** A photoelectric control system used as a longitudinal position regulator to maintain the position of the point of cutoff with respect to a repetitive pattern on a moving material.

**photoelectric densitometer** An electronic instrument used to measure the density or opacity of a film or other material. A beam of light is directed through the material, and the amount of light transmitted is measured with a photocell and meter.

**photoelectric directional counter** A photoelectrically actuated device used to record the number of times a given light path is intercepted by an object moving in a given direction.

**photoelectric door opener** A photoelectric control system used to open and close a power-operated door.

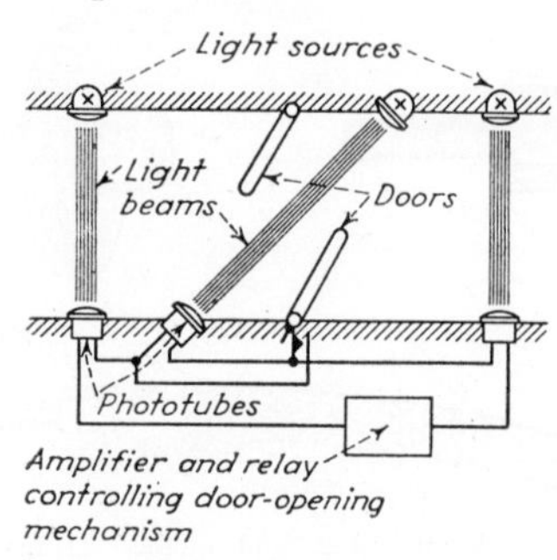


Photoelectric door opener, arranged so persons can pass in either direction and doors cannot close until person has passed through all three beams.

**photoelectric effect** The emission of electrons from a body due to visible, infrared, or ultraviolet radiant energy. The energy of a photon is absorbed for each electron emitted.

**photoelectric emission** The emission of electrons by certain materials upon exposure to radiation in and near the visible region of the spectrum.

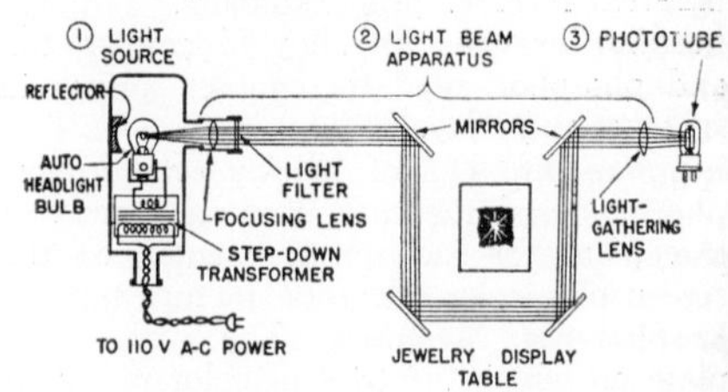


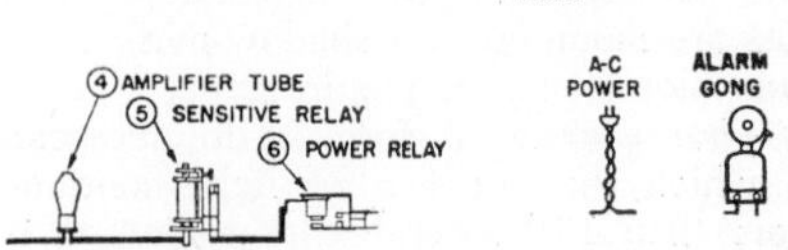


Photoelectric intrusion-detector setup for jewelry store or museum.

**photoelectric flame-failure detector** A photoelectric control that cuts off fuel flow when the fuel-consuming flame is extinguished.

**photoelectric intrusion-detector** A burglar-alarm system in which interruption of a light beam by an intruder reduces the illumination on a phototube and thereby closes an alarm circuit.

**photoelectric lighting controller** A photoelectric relay actuated by a change in illumination to control the illumination in a given area or at a given point.

**photoelectric loop control** A photoelectric control system used as a position regulator for a loop of material passing from one strip-processing line to another that may travel at a different speed. Also called loop control.

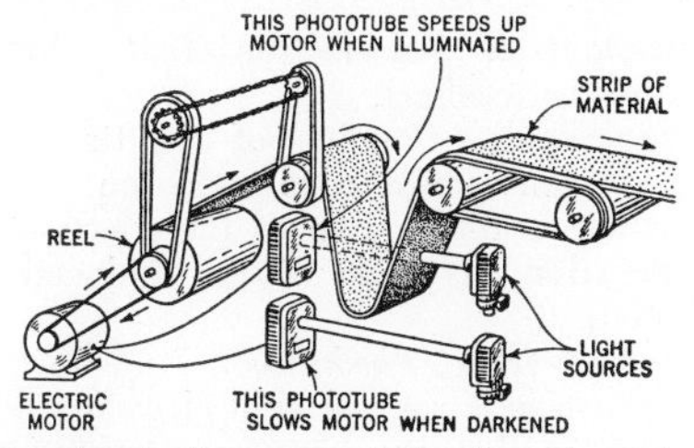

Photoelectric loop control for loop in strip of paper, metal, or fabric.

**photoelectric material** A material that will emit electrons when exposed to radiant energy in a vacuum. Examples are barium, cesium, lithium, potassium, rubidium, sodium, and strontium.

**photoelectric multiplier** *Multiplier phototube.*

**photoelectric opacimeter** *Photoelectric turbidimeter.*

**photoelectric phonograph pickup** A phonograph pickup in which a tiny mirror on the stylus varies the amount of light that reaches a photocell.

**photoelectric photometer** A photometer that uses a photocell, phototransistor, or phototube to measure the intensity of light. Also called electronic photometer.

**photoelectric pickup** A pickup that converts a change in light to an electric signal.

**photoelectric pinhole detector** A photoelectric control system that detects minute holes in an opaque material.

**photoelectric plethysmograph** A photoelectric medical instrument for measuring and recording ear opacity by means of a tiny phototube and lamp clipped to the ear, as a measure of the state of fullness of blood vessels. Also worn by aircraft pilots during high-altitude flights, as an alarm indicating the need for more oxygen.

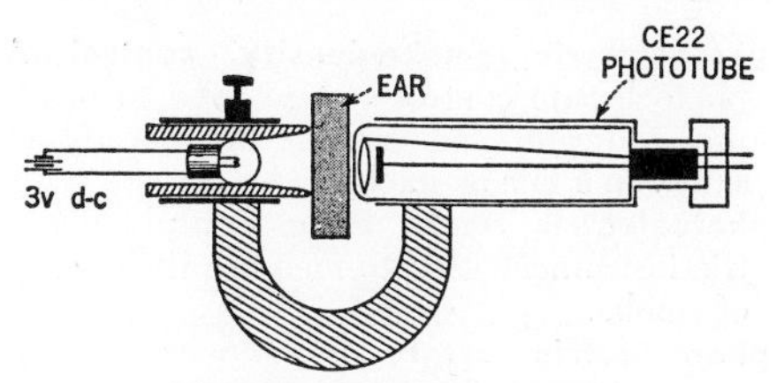

Photoelectric plethysmograph.

**photoelectric pyrometer** An instrument that measures high temperatures by using a photoelectric arrangement to measure the radiant energy given off by the heated object.

**photoelectric reflectometer** A reflectometer that uses a photocell or phototube to measure the diffuse reflection of surfaces, powders, pastes, and opaque liquids.

**photoelectric register control** A register control using a light source, one or more phototubes, a suitable optical system, an amplifier, and a relay to actuate control equipment when a change occurs in the amount of light reflected from a moving surface due to register marks, dark areas of a design, or surface defects. Also called photoelectric scanner.

**photoelectric relay** A relay combined with a phototube and amplifier, arranged so changes in incident light on the phototube make the relay contacts open or close. Also called light relay.

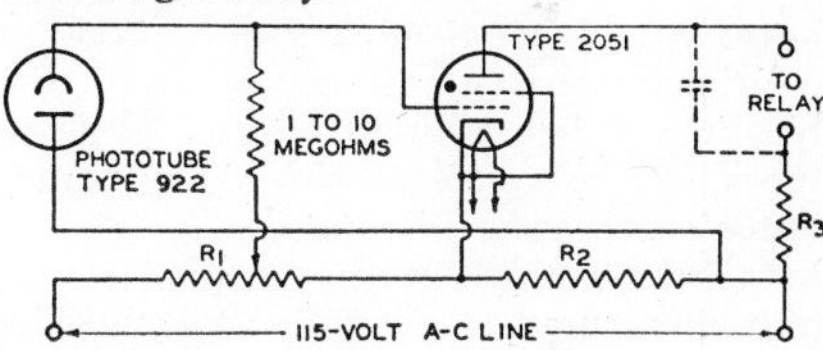

Photoelectric relay circuit.

**photoelectric scanner** *Photoelectric register control.*

**photoelectric scleroscope** A scleroscope that uses a phototube-light beam system to measure the rebound of a steel ball during hardness tests.

**photoelectric sensitivity** The ratio of photoelectric emission current to incident radiant energy. Also called photoelectric yield.

**photoelectric side-register control** A photoelectric control system used as a lateral position regulator to maintain the edge of a moving material or web at a fixed position.

**photoelectric smoke-density control** A photoelectric control system used to measure, indicate, and control the density of smoke in a flue or stack.

**photoelectric smoke meter** A photoelectric instrument used to measure the density of smoke.

**photoelectric sorter** A photoelectric control system used to sort objects according to color, size, shape, or other light-changing characteristics.

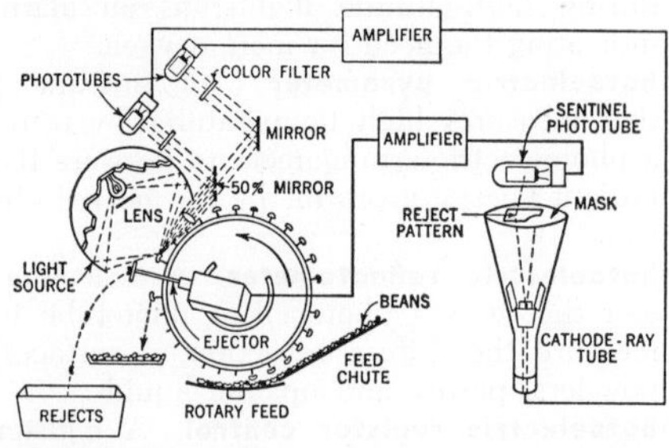


Photoelectric sorter as used to reject dark beans at high speed. Beans are held on sorting wheel by vacuum.

**photoelectric threshold** The quantum of energy just sufficient to release electrons from a given surface by the photoelectric effect.

**photoelectric timer** A timer that automatically turns off an x-ray machine when the film has received the correct exposure as determined by an integrating photoelectric measuring system that monitors a fluorescent screen placed behind the film.

**photoelectric tristimulus colorimeter** A colorimeter that uses three or more combinations of light sources, filters, and phototubes to measure colors with high accuracy.

**photoelectric tube** *Phototube.*

**photoelectric turbidimeter** A photoelectric instrument used to determine the turbidity of almost clear solutions. Also called photoelectric opacimeter.

**photoelectric work function** The energy required to transfer electrons from a given metal to a vacuum or other adjacent medium during photoelectric emission. Usually expressed in electron-volts.

**photoelectric yield** *Photoelectric sensitivity.*

**photoelectromagnetic effect** When light falls on a flat surface of an intermetallic semiconductor located in a magnetic field that is parallel to the surface, excess hole-electron pairs are created. These carriers diffuse in the direction of the light but are deflected by the magnetic field to give a current flow through the semiconductor that is at right angles to both the light rays and the magnetic field.

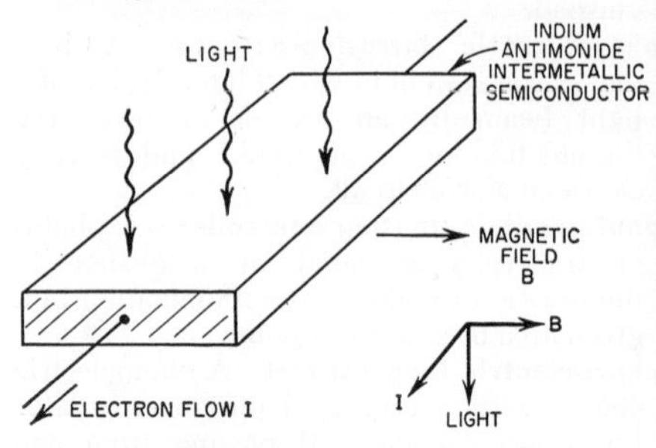


Photoelectromagnetic effect.

**photoelectron** An electron emitted by the photoelectric effect.

**photoemissive** Capable of emitting electrons upon exposure to radiation in and near the visible region of the spectrum.

**photofission** Nuclear fission induced by photons.

**photoflash tube** *Flash tube.*

**photoflash unit** A portable electronic light source for photographic use, consisting of a capacitor-discharge power source, a flash tube, a battery for charging the capacitor, and sometimes also a high-voltage pulse generator to trigger the flash.

**photoformer** A signal generator that delivers an output signal having a waveform corresponding to that of a mask cut to the shape of the desired function. The mask is placed on the face of a cathode-ray tube, and a phototube in front of the screen is connected to the deflection circuits in such a way as to make the trace follow the contour of the mask. The vertical deflection signal is then proportional to the desired function.

**photoglow tube** A gas-filled phototube used as a relay by making the operating voltage sufficiently high so ionization and a glow discharge occur, with considerable current flow, when a certain illumination is reached.

**photogoniometer** A goniometer that uses a phototube or photocell as a sensing device for studying x-ray spectra and x-ray diffraction effects in crystals.

**photographic projection plan position indicator** [abbreviated ppppi or p$^4$i] A plan position indicator used with an automatic camera, an automatic high-speed film-developing machine, and a film projector to project the ppi display on a large screen for plotting purposes. Pictures can be taken, developed, and projected at a rate of about six per minute.

**photographic recording** Facsimile recording in which a photosensitive surface is exposed by a signal-controlled light beam or spot.

**photographic sound recorder** A sound recorder having means for producing a modulated light beam and means for moving a light-sensitive medium relative to the beam to give a photographic recording of sound signals. Also called optical sound recorder.

**photographic sound reproducer** A sound reproducer in which an optical sound record on film is moved through a light beam directed at a light-sensitive device, to convert the recorded optical variations back into audio signals. Also called optical sound reproducer.

**photographic transmission density** The common logarithm of opacity. A film that transmits 100% of the light has a density of zero, while a film transmitting 10% has a density of 1. Also called optical density.

**photoionization** *Atomic photoelectric effect.*

**photo-island grid** The photosensitive surface of an image dissector tube for television cameras.

**photoluminescence** Luminescence stimulated by visible, infrared, or ultraviolet radiation.

**photomagnetic effect** The direct effect of light on the magnetic susceptibility of certain substances.

**photomeson** A meson, usually a pi meson, that is ejected from a nucleus by an impinging photon.

**photometer** An instrument for measuring the intensity of a light source or the amount of illumination on a surface.

**photometry** Measurement of luminous flux and related quantities, such as illuminance, luminance, luminosity, and luminous intensity.

**photomultiplier counter** A scintillation counter that has a built-in multiplier phototube.

**photomultiplier tube** Deprecated term for *multiplier phototube.*

**photon** A quantum of electromagnetic radiation, equal to Planck's constant multiplied by the frequency in cycles per second.

**photonegative** Having negative photoconductivity, hence decreasing in conductivity (increasing in resistance) under the action of light. Selenium sometimes exhibits this property.

**photon emission spectrum** The relative numbers of optical photons emitted by a scintillator material per unit wavelength as a function of wavelength. The emission spectrum may also be given in alternative units such as wave number, photon energy, or frequency.

**photon engine** A reaction engine in which thrust is obtained from a stream of light rays. Although the thrust is very small, it can be applied indefinitely in outer space to build up speeds approaching the speed of light.

**photoneutron** A neutron released from a nucleus in a photonuclear reaction.

**photonuclear reaction** A nuclear reaction induced by a photon.

**photopositive** Having positive photoconductivity, hence increasing in conductivity (decreasing in resistance) under the action of light. Selenium ordinarily has this property.

**photoproton** A proton released from a nucleus in a photonuclear reaction.

**photoscope reconnaissance** Photographic reconnaissance in which radarscope photography is used.

**photosensitive** *Light-sensitive.*

**photosensitive recording** Recording by the exposure of a photosensitive surface to a signal-controlled light beam or spot.

**phototelegraphy** *Facsimile.*

**phototransistor** A light-sensitive transistor.

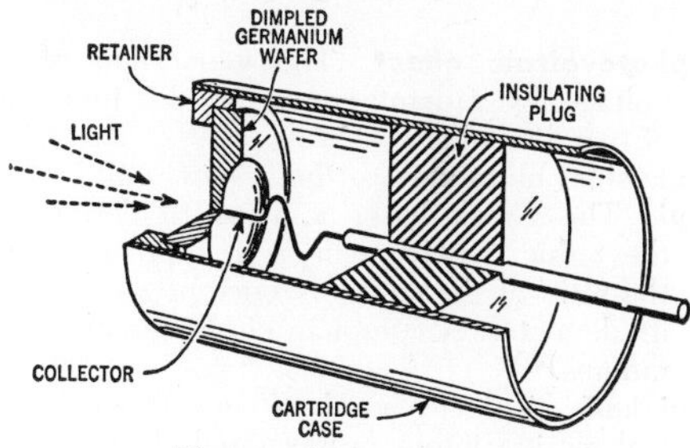

Phototransistor construction.

**phototube** An electron tube in which the output signal is related to the total radiation that is producing photoelectric emission

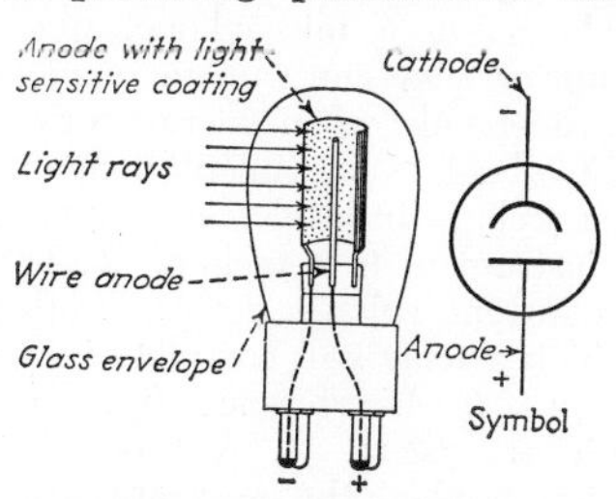

Phototube construction and symbol.

from the photocathode. The photocathode surface can be chosen for maximum response to a particular part of the visible, infrared, or ultraviolet spectrum. A phototube is not a photoelectric cell. Also called photoelectric tube.

**phototube relay** A photoelectric relay in which a phototube serves as the light-sensitive device.

**photovaristor** A varistor in which the current-voltage relation may be modified by illumination. Cadmium sulfide and lead telluride may be used in varistors for this purpose.

**photovoltaic** Capable of generating a voltage as a result of exposure to radiation.

**photovoltaic cell** A photocell in which radiant energy causes electrons to pass through the surface of contact between a conductor and semiconductor, thereby generating a voltage. Also called barrier-layer cell (deprecated).

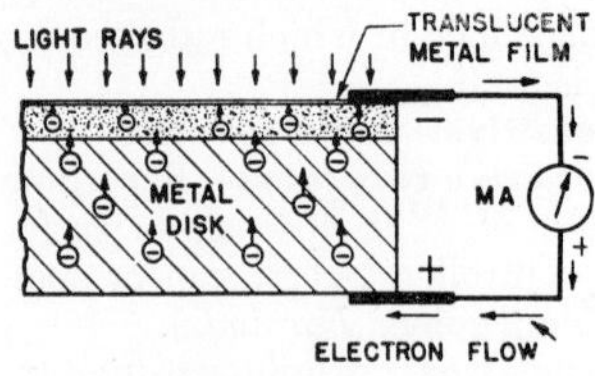

Photovoltaic cell.

**photovoltaic effect** The generation of a voltage by radiant energy at the junction of two dissimilar materials.

**physical mass unit** *Atomic mass unit.*

**pi** The Greek letter $\pi$, used to designate the value 3.1416, which is approximately the ratio of the circumference of a circle to its diameter. A complete circle contains $2\pi$ radians.

**pickoff** A device used to convert mechanical motion into a proportional electric signal.

**pickup** 1. A device that converts a sound, scene, measurable quantity, or other form of intelligence into corresponding electric signals, as in a microphone, phonograph pickup, or television camera. A pickup is a transducer only when energy conversion is also involved, as in a microphone or phonograph pickup. In a telemetering system the end instrument is a pickup. 2. The minimum current, voltage, power, or other value at which a relay will complete its intended function. 3. Interference from a nearby circuit or system. 4. A type of nuclear reaction in which the incident particle takes a nucleon from the target nucleus and proceeds with this nucleon bound to itself. Pickup is the inverse of stripping. 5. A potentiometer used in an automatic pilot to detect the motion of the airplane around the gyro and initiate corrective adjustments.

**pickup arm** A pivoted arm used to hold a phonograph pickup cartridge. Also called tone arm.

**pickup cartridge** A cartridge that contains the electromechanical translating elements and the reproducing stylus of a phonograph pickup.

**pickup current** The current at which a magnetically operated device starts to operate. Also called pull-in current.

**pickup spectral characteristic** The spectral response of a pickup that converts radiation to electric signals, as measured at the output terminals of the pickup tube.

**pickup tube** *Camera tube.*

**pickup value** The minimum voltage, current, or power at which the contacts of a previously deenergized relay will always assume their energized position. Also called pull-in value.

**pickup voltage** The voltage at which a magnetically operated device starts to operate.

**pico-** A prefix representing $10^{-12}$.

**picofarad** [abbreviated pf] European term for *micromicrofarad.*

**picowatt** *Micromicrowatt.*

**pictorial wiring diagram** A wiring diagram containing actual sketches of radio parts and showing clearly all connections between the parts. Used in service manuals.

**picture** The image on the screen of a television receiver.

**picture black** The signal produced at any point in a facsimile system by the scanning of a selected area of subject copy having maximum density.

**picture brightness** The brightness of the highlights of a television picture, usually expressed in foot-lamberts.

**picture detail** The total number of lines or elements that make up a picture on the screen of a television receiver.

**picture element** The smallest subdivision of a television or facsimile image. In a color television receiver it is one color phosphor dot. In a black-and-white television receiver it is a square segment of a scanning line whose dimension is equal to the nominal line width. Also called elemental area.

**picture frequency** 1. A frequency that results solely from scanning of subject copy in a facsimile system. 2. *Frame frequency.*

**picture inversion** Reversal of black and

white shades in the recorded copy in a facsimile system.

**picture line-amplifier output** The junction between the television studio facility and the line feeding either a relay transmitter, a visual transmitter, or a network.

**picture line standard** The number of horizontal lines in a complete television image. The U. S. standard is 525 lines.

**picture monitor** A cathode-ray tube and associated circuits, arranged to view a television picture or its signal characteristics at station facilities.

**picture signal** The signal resulting from the scanning process in television or facsimile.

**picture-signal amplitude** The difference between the white peak and the blanking level of a television signal.

**picture-signal polarity** The polarity of the signal voltage representing a dark area of a scene, with respect to the signal voltage representing a light area. Expressed as black negative or black positive.

**picture size** The useful viewing area on the screen of a television receiver, in square inches.

**picture synchronizing pulse** *Vertical synchronizing pulse.*

**picture transmission** The transmission, over wires or by radio, of a picture having a gradation of shade values.

**picture transmitter** *Visual transmitter.*

**picture tube** A cathode-ray tube used in television receivers to produce an image by varying the electron-beam intensity as the beam is deflected from side to side and up and down to scan a raster on the fluorescent screen at the large end of the tube. Also called kinescope and television picture tube.

**picture white** The signal produced at any point in a facsimile system by the scanning of copy having minimum density. Also called white.

**Pierce oscillator** An oscillator in which a piezoelectric crystal is connected between the anode and grid of a tube. It is basically a Colpitts oscillator with voltage division provided by the grid-to-cathode and anode-to-cathode capacitances of the circuit.

**pie winding** A winding having the shape of a washer, used in series-connected groups in some r-f chokes.

**piezodielectric** Having the ability to change in dielectric constant when a mechanical force is applied.

**piezoelectric** Having the ability to generate a voltage when mechanical force is applied, or to produce a mechanical force when a voltage is applied, as in a piezoelectric crystal.

**piezoelectric axis** One of the directions in a crystal in which either tension or compression will generate a voltage.

**piezoelectric crystal** A crystal having piezoelectric properties. Used in crystal loudspeakers, crystal microphones, and crystal cartridges for phono pickups.

**piezoelectric effect** Generation of a voltage between opposite faces of a piezoelectric crystal as a result of strain due to pressure or twisting, and the reverse effect in which application of a voltage to opposite faces causes deformation to occur at the frequency of the applied voltage.

**piezoelectric gage** A pressure-measuring gage that uses a piezoelectric material to develop a voltage when subjected to pressure. Used for measuring blast pressures resulting from explosions and pressures developed in guns.

**piezoelectricity** Electric energy resulting from the piezoelectric effect.

**piezoelectric loudspeaker** *Crystal loudspeaker.*

**piezoelectric microphone** *Crystal microphone.*

**piezoelectric oscillator** An oscillator circuit in which the frequency is controlled by a quartz crystal.

**piezoelectric pickup** *Crystal pickup.*

**piezoelectric pressure gage** A pressure gage in which deformation of a piezoelectric crystal by very high pressures, as in gun barrels during firing, produces a voltage related to pressure.

**piezoelectric vibrator** An element cut from piezoelectric material, usually in the form of a plate, bar, or ring, with electrodes attached to or supported near the element to excite one of its resonant frequencies.

**piezoid** *Finished crystal blank.*

**pigtail** A short, flexible wire, usually stranded or braided, used between a stationary terminal and a terminal having a limited range of motion, as in relay armatures.

**pigtail splice** A splice made by twisting together the bared ends of parallel conductors.

**pile** *Nuclear reactor.*

**pile oscillator** An arrangement for maintaining a neutron-absorbing body in periodic motion in a nuclear reactor. The resulting fluctuations of neutron flux can be used to measure the properties of the absorbing body of the reactor.

**pileup** A set of moving and fixed contacts,

insulated from each other, formed as a unit for incorporation in a relay or switch. Also called stack.

**pillbox antenna** A microwave antenna consisting of a cylindrical parabolic reflector enclosed by two plates perpendicular to the

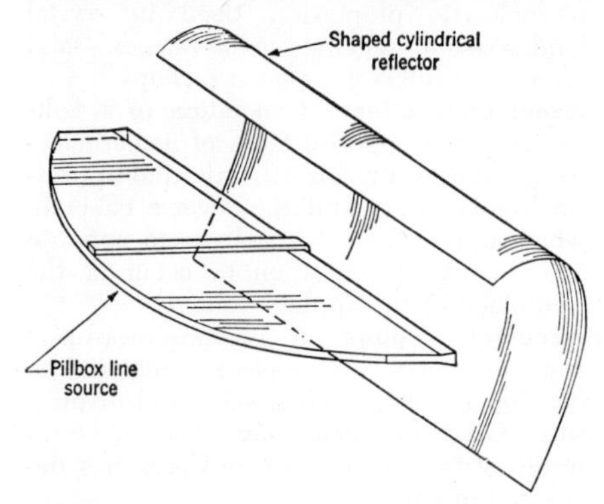


Pillbox antenna used in airborne navigation radar.

cylinder, so spaced as to permit the propagation of only one mode in the desired direction of polarization. It is fed on the focal line.

**pilotage** The process of directing the movement of a vehicle by reference to recognizable landmarks or soundings, without computing positions.

**pilot channel** A narrow channel over which a single frequency is transmitted to operate an alarm or automatic control.

**pilot circuit** The portion of a control circuit that carries the controlling signal from the master switch to the controller.

**pilot lamp** A small lamp used to indicate that a circuit is energized. Also called pilot light. When used to illuminate a dial, it is called a dial lamp.

**pilotless aircraft** An aircraft equipped to function without benefit of a human pilot. It may carry such functional equipment as cameras for photographic reconnaissance, bombs for bombing, guns for air defense, counters for radiation measurements, or radar for early warning. Target aircraft and drones are special examples of pilotless aircraft.

**pilot light** *Pilot lamp.*

**pilot regulator** A regulator that maintains a constant level at the receiving end of a carrier-derived circuit despite variations in the attenuation of the transmission line. The regulator usually monitors the resistance of a pilot wire that is exposed to essentially the same temperatures as the transmission circuit being regulated. Also called pilot-wire regulator.

**pilot spark** A weak spark used to ionize the air in preparation for a greater spark discharge. The keep-alive electrode in a tr tube serves this function.

**pilot wire** An auxiliary conductor used in connection with remote measuring devices or for operating apparatus at a distant point.

**pilot-wire regulator** *Pilot regulator.*

**pi meson** A meson having a mean life of $2.8 \times 10^{-8}$ second and a mass of about 270 times that of an electron. Also called pion.

**pi mode** The mode of operation of a magnetron for which the phases of the fields of successive anode openings facing the interaction space differ by $\pi$ radians.

**pin** A terminal on an electron-tube base, plug, or connector. Also called base pin and prong.

**pinch** A pressed glass stem used to support the internal leads of electron tubes. Also called press.

**pinch effect** 1. Constriction of ionized gas to a narrow thread in the center of a straight or doughnut-shaped electron tube through which a heavy current is passed. Also called rheostriction. 2. Constriction, and sometimes momentary rupture, of molten metal through which a heavy current is flowing. 3. Pinching of the reproducing stylus tip twice each cycle during reproduction of lateral disk recordings, due to a decrease of the groove angle cut by the recording stylus as it swings from a negative to a positive peak.

**pinchoff** The equivalent of collector cutoff in a field-effect transistor.

**pin connection** A connection made to a terminal pin at the base of an electron tube. The following abbreviations are used to identify pin connections: NC—no connection; IS—internal shield; IC—internal connection (but not an electrode connection); BS—base shield connection; P—anode; G—grid; SG—screen grid; K—cathode; H—heater; F—filament; RC—ray-control electrode; TA—target.

**pincushion distortion** Distortion in which all four sides of a received television picture are concave (curve inward).

**pine-tree array** An array of horizontal dipole antennas arranged to form a radiating curtain, with reflectors behind.

**pine-tree line** A chain of radar stations built by Canada and the United States along the Canadian-American border.

**pi network** A network having three impedance branches connected in series with each other to form a closed circuit, with the three junction points forming an input terminal, an output terminal, and a common input and output terminal.

**ping** A sonic or ultrasonic pulse sent out by an echo-ranging sonar.

**pinhole detector** A photoelectric device that detects extremely small holes and other defects in moving sheets of material.

**pin jack** A small jack designed for use with a plug whose thickness is comparable to that of an ordinary pin.

**pion** *Pi meson.*

**Pioneer** One of a series of lunar probes fired by the United States, the first on October 11, 1958.

**pip** The target echo indication on a radarscope. It may be a bright spot of light as on a ppi display, or a sharply peaked pulse as on an A display. Also called blip. The Navy uses pip while the Air Force, IRE Standards, and British texts use blip.

**piped program** A radio or television program sent over commercial transmission lines.

**pip-matching display** A navigation display in which the received signal appears as a pair of blips. Comparison of the characteristics of the blips provides a measure of the desired quantity. Used in the K, L, and N displays.

**pipology** The study and interpretation of radio target echoes. It includes determination of bearing, range, altitude, course, and speed, and analysis of the nature of the target.

**Pirani gage** A vacuum gage based on the fact that the temperature and resistance of a heated filament vary with the pressure of the surrounding gas. The less gas there is to conduct heat away from the filament, the hotter the filament becomes and the greater is its resistance.

**piston** A sliding metal cylinder used in waveguides and cavities for tuning purposes or for reflecting essentially all of the incident energy. Also called plunger and waveguide plunger.

**piston action** Movement of the entire diaphragm of a loudspeaker as a unit when driven at low audio frequencies.

**piston attenuator** A microwave attenuator inserted in a waveguide to introduce an amount of attenuation that can be varied by moving an output coupling device along its longitudinal axis.

**pistonphone** A small chamber equipped with a reciprocating piston having a measurable displacement. Used to establish a known sound pressure in the chamber, as for testing microphones.

**pitch** 1. The attribute of auditory sensation that depends primarily on the frequency of the sound stimulus, but also depends on the sound pressure and the waveform of the stimulus. Pitch determines the position of a sound on a musical scale, for which the standard pitch is 440 cps for the tone A. With this standard, middle C is 261.6 cps. 2. The rising and falling motion of the bow of a ship or the tail of an airplane as the structure oscillates about its transverse axis. 3. The distance between the peaks of two successive grooves on a disk recording or on a screw.

**pitch attitude** The angle between the longitudinal axis of the vehicle and a horizontal plane.

**pitchblende** An ore consisting largely of uranium oxides, used as a source of uranium and radium.

**place** A position corresponding to a given power of the base in positional notation. A digit located in any particular place is a coefficient of a corresponding power of the base. Places are usually numbered from right to left, zero being the place at the right if there is no decimal, binary, or other point, or the column immediately to the left of the point if there is one. Also called column.

**planar diode** A diode having planar electrodes in parallel planes.

**planar-electrode tube** An electron tube having parallel planar electrodes.

**planar network** A network that can be drawn on a plane without crossing of branches.

**Planckian locus** A line drawn through all points on a chromaticity diagram representing light radiation from a blackbody radiator at various temperatures up to 10,000°K.

**Planck's constant** [symbol $h$] A universal constant equal to $6.625 \times 10^{-27}$ erg-second. It is the proportionality factor that, when multiplied by the frequency of a photon, gives the energy of the photon.

**Planck's law** The fundamental law of quantum theory, stating that energy transfers associated with radiation are proportional to the frequency of the radiation. The proportionality factor is Planck's constant.

**plane-earth factor** The ratio of the electric field strength that would result from propagation over an imperfectly conducting plane earth to that which would result from propagation over a perfectly conducting plane.

**plane of polarization** The plane containing the electric field vector and the direction of propagation of a plane-polarized wave. In a horizontally polarized wave, this plane is horizontal.

**plane-polarized light** Light in which the electric vectors of all components of the radiation are in the same fixed plane.

**plane-polarized wave** An electromagnetic wave whose electric field vector at all times lies in a fixed plane that contains the direction of propagation through a homogeneous isotropic medium.

**plane wave** A wave whose equiphase surfaces form a family of parallel planes.

**planigraphy** *Laminography.*

**plan position indicator** [abbreviated ppi] A radarscope display in which targets appear as bright spots at the same locations as they would on a circular map of the area being scanned, the radar antenna being at the center of the map. The sweep moves

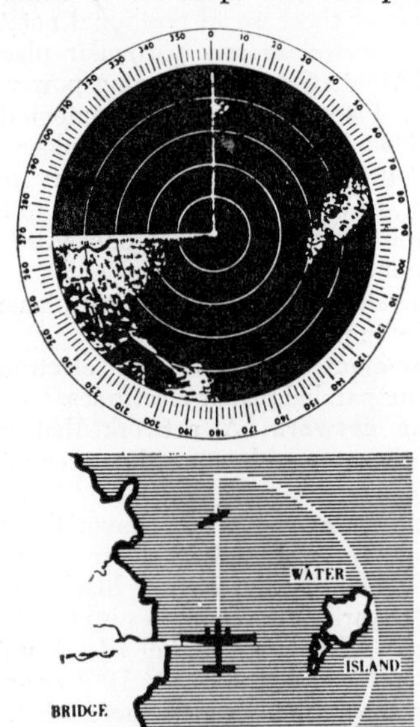


Plan position indicator presentation obtained with search radar in airplane, and map showing location of plane. Radar antenna is pointing at 270°.

radially outward from the center of the screen and the sweep line rotates synchronously with the radar antenna. The radial distance from the center at which an echo appears is an indication of range, and the angle measured clockwise from true north (usually at the top of the screen) is an indication of bearing. Used also in sonar. Also called P display.

**Plante cell** A type of lead-acid cell in which the active material is formed on the plates by electrochemical means during repeated charging and discharging, instead of being applied as a prepared paste.

**plasma** An ionized gaseous medium that contains approximately equal numbers of electrons and positive ions, making the resultant space charge essentially zero.

**plasma engine** An engine for space travel in which neutral plasma is accelerated and directed by external magnetic fields that interact with the magnetic field produced by current flow through the plasma. A nuclear power plant provides the necessary power.

**plasma pinch** Application of the pinch effect to plasma in attempts to produce controlled nuclear fusion. A large current is passed through the stream of plasma in an electron tube. A magnetic field constricts the plasma into a smaller diameter, keeping it away from the envelope and raising the temperature of the plasma.

**plasmatron** A gas-discharge tube in which independently generated plasma serves as a conductor between a hot cathode and an anode. The anode current is modulated by varying either the conductivity or the effective cross-section of the plasma.

**plastic** One of a large and varied group of organic materials that can be formed into shape by flow at some stage, usually by applying heat alone or with pressure.

**plastisol** A mixture of resins and plasticizers that can be molded, cast, or converted to continuous films by applying heat.

**plate** 1. One of the conducting surfaces in a capacitor. 2. One of the electrodes in a storage battery. 3. [symbol P] *Anode.*

**plateau** The portion of the plateau characteristic of a counter tube in which the counting rate is substantially independent of the applied voltage.

**plateau characteristic** The relation between counting rate and voltage for a counter tube when radiation is constant, showing a plateau after the rise from the starting voltage to the Geiger threshold. Also called counting-rate–voltage characteristic.

**plateau length** The range of applied voltage over which the plateau of a radiation-counter tube extends.

**plate bypass capacitor** *Anode bypass capacitor.*

**plate circuit** *Anode circuit.*

**plate current** *Anode current.*

**plated circuit** A printed circuit produced by electrodeposition of a conductive pattern on an insulating base. In one method, a negative of the required pattern is printed on a metal-clad or metal-coated sheet with a plating resist. Silver or solder is electroplated on the area not protected by the resist, to give the desired wiring pattern. The resist is then removed and the unwanted metal under it is etched away with a solution that does not attack the silver or solder plating. Also called plated printed circuit.

**plated crystal unit** A crystal unit in which the electrodes are metal films deposited directly on the quartz surfaces.

**plate detection** *Anode detection.*

**plated printed circuit** *Plated circuit.*

**plated-through interface connection** An interface connection formed by electrodeposition of metal on the sides of a hole in the base of a printed-wiring board.

**plate efficiency** *Anode efficiency.*

**plate input power** *Anode input power.*

**plate keying** *Anode keying.*

**plate load impedance** *Anode load impedance.*

**plate modulation** *Anode modulation.*

**plate neutralization** *Anode neutralization.*

**plate penetrameter** A plate of material similar to that of a specimen under radiographic examination, having a thickness of about 2% of the specimen thickness, and having holes of different diameters. Also called strip penetrameter.

**plate power input** *Anode power input.*

**plate pulse modulation** *Anode pulse modulation.*

**plate saturation** *Anode saturation.*

**plate supply** *Anode supply.*

**plate voltage** *Anode voltage.*

**platform stabilization** Radar antenna stabilization in which a platform is pivoted so it can be driven to a horizontal position regardless of the motion of the vehicle. Gyroscope signals actuate drive motors that maintain this horizontal position as the vehicle pitches and rolls.

**platinotron** A microwave tube that may be used as a high-power saturated amplifier or oscillator in pulsed radar applications. It requires a permanent magnet just as does a magnetron. The Raytheon trademark for this type of electron tube is Amplitron.

**platinum** [symbol Pt] A heavy, almost white metal that resists the action of practically all acids, is capable of withstanding high temperatures, and is little affected by sparks. Used for contact points in switches and relays. Atomic number is 78.

**platter** Slang term for *phonograph record.*

**playback** Reproduction of a recording.

**playback head** A head that converts a changing magnetic field on magnetic tape into corresponding electric signals. Also called reproduce head.

**playback loss** *Translation loss.*

**plethysmograph** An instrument for measuring changes in the size of a part of the body by measuring changes in the amount of blood in that part.

**Plexiglas** Trademark of Rohm & Haas Co. for their clear acrylic molding powders and clear acrylic sheets.

**pliotron** General term for any hot-cathode vacuum tube having one or more grids.

**plug** 1. The half of a connector that is normally movable and is generally attached to a cable or removable subassembly. A plug is inserted in a jack, outlet, receptacle, or socket. 2. Radiation-blocking material used to close an aperture in the shield of a nuclear reactor. 3. A mention of a commercial product in a radio or television program.

**plugboard** A removable computer panel containing an array of terminals that may be interconnected by short leads to provide a specific program. Prewired plugboards are easily interchanged and stored for future use.

**plug fuse** A fuse designed for use in a standard screw-base lamp socket.

**plugging** Braking an electric motor by reversing its connections, so it tends to turn in the opposite direction. The circuit is opened automatically when the motor stops, so the motor does not actually reverse.

**plug-in unit** A component or subassembly having plug-in terminals so all connections can be made simultaneously by pushing the unit into a suitable socket.

**plumbing** Slang term for the pipe-like waveguide circuit elements used in microwave radio and radar equipment.

**plunger** *Piston.*

**plunger relay** A relay in which a plunger moves through the center of the coil by solenoid action when the relay is energized. Contacts are mounted on one or both ends of the plunger that serves also as the core. Also called solenoid relay.

**plural scattering** Scattering in which the

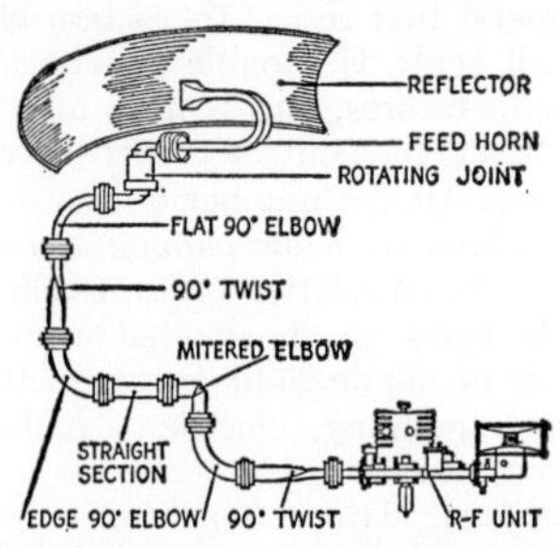


Plumbing setup used to feed microwave energy from radar transmitter to rotating antenna.

final displacement is the vector sum of a small number of displacements. Plural scattering is intermediate between single and multiple scattering.

**plutonium** [symbol Pu] A transuranic radioactive element formed by the decay of certain isotopes of neptunium. Plutonium does not occur in nature. Atomic number is 94.

**plutonium bomb** An atomic bomb using plutonium.

**plutonium reactor** A nuclear reactor in which plutonium is the principal fissionable material.

**p-m** 1. Abbreviation for *permanent magnet.* 2. Abbreviation for *phase modulation.*

**p-m loudspeaker** Abbreviation for *permanent-magnet loudspeaker.*

**p-n boundary** A surface in a p-n junction at which the donor and acceptor concentrations are equal.

**pneumatic detector** An infrared detector consisting of a small gas-filled chamber having an infrared transmitting window, a material inside the chamber to absorb radiation, and a means for converting pressure changes in the chamber into measurable signal output that is usually electric or optical.

**pneumatic loudspeaker** A loudspeaker in which the acoustic output results from controlled variation of an air stream.

**p-n hook** A thin region of p-type material sandwiched between regions of n-type base and collector material in a transistor to create a potential hook at which holes are trapped in such a way as to increase the current gain.

**p-n-i-n transistor** An intrinsic-junction transistor in which the intrinsic region is between n regions.

**p-n-i-p transistor** An intrinsic-junction transistor in which the intrinsic region is sandwiched between the n-type base and the p-type collector.

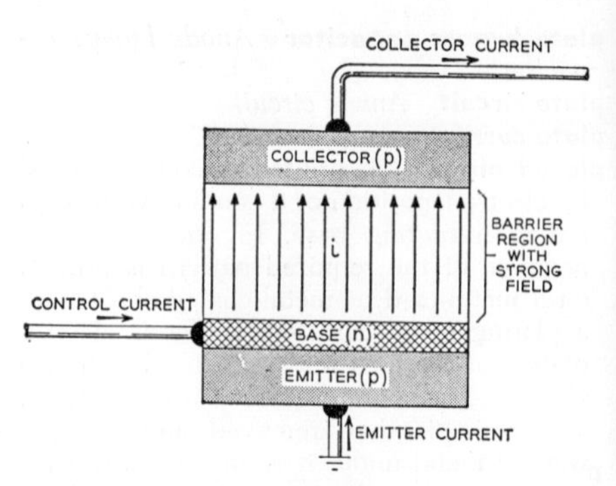


P-n-i-p transistor.

**p-n junction** A region of transition between p-type and n-type semiconducting material.

**p-n-p-n transistor** A p-n-p junction transistor having a transition or floating layer between p and n regions, to which no ohmic connection is made.

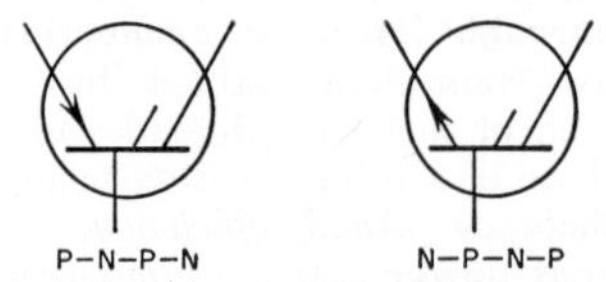


P-n-p-n and n-p-n-p transistor symbols, in which unconnected slant line represents transition or floating layer. Off-center base lead indicates remote base connection.

**p-n-p transistor** A junction transistor having an n-type base between a p-type emitter and a p-type collector. The emitter should then be positive with respect to the base, and the collector should be negative with respect to the base.

**point** The character or space that separates the integral and fractional parts of a numerical expression in positional notation, such as the binary point in binary notation, and the decimal point in decimal notation. Also called base point and radix point.

**point contact** Pressure contact between a semiconductor body and a metallic point.

**point-contact diode** A semiconductor diode that makes use of a point contact to provide rectifying action. Used as a detector in r-f and microwave circuits.

**point-contact transistor** A transistor having a base electrode and two or more point contacts located near each other on the surface of n-type germanium. Pressure of the points creates a small volume of p-type material under each point to produce the necessary junctions for a p-n-p transistor. Comparable n-p-n point-contact transistors have been made in the laboratory from

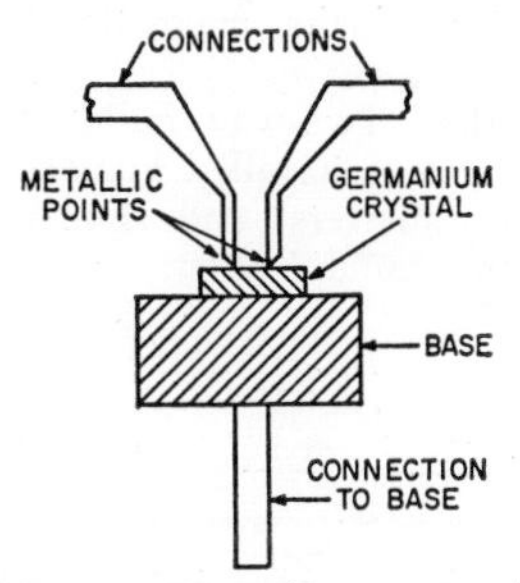

Point-contact transistor construction.

p-type germanium, but are not commercially available.

**point-contact transistor tetrode** A point-contact transistor having a collector and two emitters.

**point counter tube** A radiation-counter tube, using gas amplification, in which the central electrode is a point or a small sphere.

**pointer** The needle-shaped rod that moves over the scale of a meter.

**pointing element** The element in a fire-control system that points the missile-launching projector as directed by the other elements of the system.

**point-junction transistor** A transistor having a base electrode and a point contact along with junction electrodes.

**point source** A radiation source whose dimensions are small compared with the distance from which the radiation is used.

**point-to-point communication** Radio communication between two fixed stations.

**point-to-point wiring** Wiring in which the connections between components or parts are made directly, without intermediate terminals or supports, by using the leads of the components.

**poison** A contaminating substance that impairs the efficiency of a material. It may reduce phosphorescence in a luminescent material, reduce emission of a cathode surface, or reduce the reactivity of a nuclear reactor by absorbing electrons unproductively.

**polar** Pertaining to, measured from, or having a pole, such as the poles of the earth or of a magnet.

**polar coordinates** A system of coordinates in which the location of a point is specified by its distance from a fixed point and the angle that the line from this fixed point to the given point makes with a fixed reference line.

**polar diagram** A diagram employing polar coordinates to show the magnitude of a quantity in some or all directions from a point. Examples include directivity patterns and radiation patterns.

**polar grid** *Grid north.*

**polarimeter** An instrument for measuring the state of polarization of polarized light.

**Polaris** A Navy surface-to-surface intermediate-range ballistic missile designed to be launched from submarines and surface ships for accurate bombardment of small target areas with conventional or nuclear warheads at ranges up to 1,500 miles. It uses inertial guidance in conjunction with sins (Ship's Inertial Navigation System), so true north, the ship's position, and the ship's speed are accurately fed into a computer in the missile at the instant of firing.

**polarity** The characteristic wherein an object exhibits opposite properties within itself, such as opposite charges (positive and negative) or opposing magnetic poles (north and south).

**polarity of picture signal** The polarity of the black portion of a picture signal with respect to the white portion. In a black-negative picture, the potential corresponding to the black areas of the picture is negative with respect to the potential corresponding to the white areas of the picture. In a black-positive picture the potential corresponding to the black areas of the picture is positive.

**polarity-reversing switch** A switch that interchanges the connections to a device.

**polarization** 1. The direction of the electric field as radiated from a transmitting antenna. Horizontal polarization is standard for television in the United States, and vertical polarization is standard in Great Britain. 2. Vibration of the vectors of a beam of light or other electromagnetic radiation in a particular direction or manner, as in plane-polarized light. 3. A chemical change occurring in dry cells during use, increasing the internal resistance of the cell and shortening its useful life. 4. A displacement of bound charges in a dielectric when placed in an electric field.

**polarization-diversity reception** Diversity reception involving use of a horizontal dipole and a vertical dipole at the same location, with the individual receiver outputs being combined just as in space-diversity reception. The arrangement counteracts the changes in polarization of a received radio wave during fading.

**polarization error** An error in a radio direction-finder indication due to changes in the polarization of the received wave as atmospheric conditions change. The error is generally greatest at night. Also called night effect (deprecated).

**polarization receiving factor** The ratio of the power received by an antenna from a given plane wave of arbitrary polarization to the power received by the same antenna from a plane wave of the same power density and direction of propagation, whose state of polarization has been adjusted for maximum received power.

**polarized electrolytic capacitor** An electrolytic capacitor in which the dielectric film is formed adjacent to only one metal electrode. The impedance to the flow of current is then greater in one direction than in the other.

**polarized light** Light that vibrates in only one plane.

**polarized meter** A meter having a zero-center scale, with the direction of deflection of the pointer depending on the polarity of the voltage or the direction of the current being measured.

**polarized plug** A plug that can be inserted in its receptacle only when in a predetermined position.

**polarized receptacle** A receptacle designed for use with a polarized plug, to insure that the grounded side of an a-c line or the positive side of a d-c line is always connected to the same terminal on a piece of equipment.

**polarized relay** A relay in which the direction of movement of the armature depends on the direction of the energizing current in the relay coil.

**polarizer** A Nicol prism or other device for polarizing light.

**polarizing current** The direct current that is sent through the coil of an iron-core component to establish a reference value of magnetic flux.

**Polaroid** Trademark of Polaroid Corp. for a sheet material that produces plane-polarized light.

**pole** 1. A region in a magnet that has polarity, such as the north pole or the south pole. 2. A characteristic of a function used in circuit analysis. 3. An output terminal on a switch; a double-pole switch has two output terminals.

**pole face** The end of a magnetic core that faces the air gap in which the magnetic field performs useful work.

**pole piece** A piece of magnetic material forming one end of an electromagnet or permanent magnet, shaped to control the distribution of the magnetic flux in the adjacent air gap.

**police radio** Two-way radio communication equipment installed in police cars for car-to-headquarters and sometimes also car-to-car communication.

Police radio installation.

**polonium** [symbol Po] A radioactive element. Atomic number is 84.

**polychlorotrifluoroethylene resin** A fluorocarbon resin having high dielectric strength, widely used for electrical insulation. Marketed as Kel-F (M. W. Kellogg Co.).

**polydirectional microphone** A microphone having provisions for changing its directional characteristics.

**polyester** A resin used as a base for several types of plastic.

**polyester film** A plastic film made from a polyester such as Mylar. Used as a backing for magnetic tape to obtain high strength and resistance to humidity change.

**polyethylene** A tough, flexible plastic compound having excellent insulating properties at ultrahigh frequencies. Widely used as insulating material in coaxial cables and other transmission lines.

**polyphase** Having or utilizing two or more phases of an a-c power line.

**polyplexer** A radar unit that combines the functions of duplexing and lobe switching.

**polyrod antenna** A microwave antenna consisting of a parallel arrangement of rods made from polystyrene or other good dielectric material. When excited at one end, as from the end of a waveguide, the rods will radiate from their other ends.

**polystyrene** A thermoplastic material having excellent dielectric properties, crystal clarity, and good chemical resistance. Used for insulators, coil insulation, insulation between plates in capacitors, insulating beads in coaxial cables, and optical lenses.

**polystyrene capacitor** A capacitor that uses film polystyrene as a dielectric between rolled strips of metal foil.

**polytetrafluoroethylene resin** A fluorocarbon resin having high dielectric strength and a slippery feel, widely used for electrical insulation. The standard designation of the Society of the Plastics Industry is

TFE-fluorocarbon resin. Trademark names include Teflon (Du Pont) and Fluon (Imperial Chemical Industries, England).

**polyvinyl chloride** A thermoplastic synthetic resin used for electrical insulation, as for insulating coatings on wire.

**pool cathode** A cathode at which the principal source of electron emission is a cathode spot on a liquid metal electrode, usually mercury.

**pool-cathode tube** *Pool tube.*

**pool tube** A gas-discharge tube having a mercury-pool cathode. Also called mercury-pool tube and pool-cathode tube.

**poor geometry** A nuclear measuring arrangement such that the angular aperture between source and detector is large, thus introducing an uncertainty in particle energy measurements.

**popi** [Post Office Position Indicator] A British continuous-wave low-frequency navigation system of the phase-comparison type. The phase difference between sequential transmissions on a single frequency is measured to obtain bearing information.

**porcelain** A fired ceramic material used for insulators.

**porous reactor** A nuclear reactor composed of a porous material or an aggregate of small particles, with coolant or fluid fuel flowing through the pores.

**port** 1. An opening in a waveguide component, through which energy may be fed or withdrawn, or measurements made. 2. An opening in a base-reflex enclosure for a loudspeaker, designed and positioned to improve bass response. 3. An entrance or exit for a network.

**portable receiver** A completely self-contained radio receiver operating from batteries and having a built-in antenna.

**portable transmitter** A transmitter that can be readily carried by a person and may or may not be operated while in motion.

**porthole** A defect, in a properly aligned camera tube employing low-velocity scanning, that increases target cutoff voltage and decreases sensitivity toward the corners of the picture.

**port radar installation** A radar installation near a harbor, used to guide ships entering or leaving the harbor during fog or darkness.

**position** The location of a vehicle as determined by specific values of three navigation coordinates.

**positional crosstalk** The variation in the path followed by any one electron beam as the result of a change impressed on any other beam in a multibeam cathode-ray tube.

**positional notation** A system of notation in which the significance of each digit of a number depends on its position in the sequence. Successive digits are interpreted as coefficients of successive powers of an integer called the base. Also called number system.

**position-finding element** The element in a fire-control system that locates the position of the target in space.

**position fix** The intersection of two plotted bearing lines on a map.

**positioning action** Automatic control action in which there is a predetermined relation between the value of a controlled variable and the position of a final control element.

**position of effective short** The distance between a specified reference plane and the apparent location of the short-circuit produced by a fired switching tube in its mount.

**position telemeter** A telemeter that measures and transmits angular or linear position.

**position-type telemeter** Deprecated term for *ratio-type telemeter.*

**positive** 1. Having fewer electrons than normal, and hence having the ability to attract electrons. 2. Having the same rendition of light and shade as in the original scene.

**positive bias** A bias such that the control grid of an electron tube is positive with respect to the cathode.

**positive charge** The charge existing in a body that has fewer electrons than normal.

**positive column** The luminous glow, often striated, that occurs between the Faraday dark space and the anode in a glow-discharge tube. The positive column is a plasma, having equal numbers of electrons and positive ions.

**positive electricity** The positive charge that is produced in a glass object by rubbing it with silk.

**positive electrode** The electrode that serves as the anode in a primary cell when the cell is discharging. It is connected to the positive terminal of the cell. Electrons flow through the external circuit to the positive electrode.

**positive electron** *Positron.*

**positive feedback** Feedback in which a portion of the output of a circuit or device is fed back in phase with the input so as to increase the total amplification. Excessive positive feedback causes instability and distortion. When positive feedback is sufficiently high, oscillation occurs. In a sound system, excessive transfer of acoustic energy back from the loudspeaker to the microphone causes howling. Also called reaction (British), regeneration, and retroaction (British).

**positive ghost** A ghost image having the same tonal variations as the primary television image.

**positive-going** Increasing in a positive direction.

**positive-grid oscillator** *Retarding-field oscillator.*

**positive image** A picture as normally seen on a television picture tube, having the same rendition of light and shade as in the original scene being televised.

**positive ion** An atom having less electrons than normal, and therefore having a positive charge.

**positive-ion emission** Thermionic emission of positive particles that are ions of the metal used as the cathode of a vacuum tube or are due to some impurity in the cathode.

**positive magnetostriction** Magnetostriction in which the application of a magnetic field causes expansion of a material.

**positive modulation** Modulation in which an increase in brightness increases the transmitted power in an amplitude-modulation facsimile or television system, or increases the transmitted frequency in a frequency-modulation facsimile system. Used in British television systems. Also called positive transmission.

**positive ray** A stream of positively charged atoms or molecules, produced by a suitable combination of ionizing agents, accelerating fields, and limiting apertures.

**positive temperature coefficient** The condition wherein the resistance, length, or some other characteristic of a substance increases when temperature increases. All metals and most metallic alloys have a positive temperature coefficient of resistance, whereas that of carbon is negative.

**positive terminal** The terminal of a battery or other voltage source toward which electrons flow through the external circuit.

**positive transmission** *Positive modulation.*

**positron** [POSItive elecTRON] A nuclear particle having the mass of an electron and a positive charge that is exactly equal in magnitude to the negative charge of an electron. Positrons are formed in the beta decay of many radionuclides. Also called positive electron.

**positronium** A quasi-stable system consisting of a positron and an electron bound together. The set of energy levels is similar to that of a hydrogen atom.

**post** *Waveguide post.*

**post-accelerating electrode** *Intensifier electrode.*

**post-acceleration** Acceleration of beam electrons after deflection in an electron-beam tube. Also called post-deflection acceleration.

**post-acceleration cathode-ray tube** An electrostatic cathode-ray tube in which the electron beam is accelerated to its final high velocity after passing through the deflection electrodes.

**post-deflection accelerating electrode** *Intensifier electrode.*

**post-deflection acceleration** *Post-acceleration.*

**post-deflection focus** Focusing of beam electrons after deflection, as in the chromatron.

**post-emphasis** *Deemphasis.*

**post-equalization** *Deemphasis.*

**postforming** The forming, bending, or shaping of cured laminate sheets by applying heat rapidly and pressing the sheet over a mold.

**post mortem** A digital computer diagnostic routine that prints out automatically or on demand the contents of the registers and storage locations of the computer when a problem has become stalled, to assist in locating the error in coding the problem.

**post office bridge** A form of Wheatstone bridge in which desired resistance values are obtained by inserting or removing plugs that fit between special terminals.

**pot** Slang term for *potentiometer*.

**potassium** [symbol K] An alkali metal having photosensitive characteristics, used on the cathodes of phototubes when maximum response is desired to blue light. It has a moderate thermal neutron-absorption cross section. Atomic number is 19.

**potassium dihydrogen phosphate crystal** [abbreviated kdp crystal] A piezoelectric crystal used in sonar transducers.

**potential** The degree of electrification as referred to some standard, such as the earth. The potential at a point is the amount of work required to bring a unit quantity of electricity from infinity to that point. Potential and voltage are often used interchangeably.

**potential barrier** A semiconductor region

in which the voltage is such that moving electric charges attempting to pass through it encounter opposition and may be turned back. Also called barrier and potential hill.

**potential difference** The voltage existing between two points. More often called voltage.

**potential divider** *Voltage divider.*

**potential energy** Energy due to the position of a particle or piece of matter with respect to other particles.

**potential gradient** The difference in the values of the voltage per unit length along a conductor or through a dielectric.

**potential hill** *Potential barrier.*

**potential scattering** Scattering that has its origin in reflection from the nuclear surface, leaving the interior of the nucleus undisturbed, or scattering of an incident wave by reflection at a change or discontinuity.

**potential transformer** *Voltage transformer.*

**potential trough** The region of an energy-level diagram delimited by the sides of two neighboring potential barriers.

**potential well** The region adjacent to the minimum in the potential energy curve for nuclear potential.

**potentiometer** 1. A resistor having a continuously adjustable sliding contact that is generally mounted on a rotating shaft.

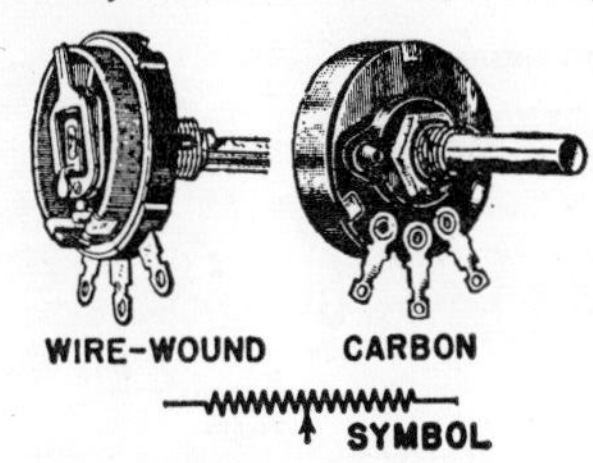


Potentiometers and potentiometer symbol.

Used chiefly as a voltage divider. Also called pot (slang). 2. An instrument for measuring a voltage by balancing it against a known voltage.

**potentiometer recorder** A recorder that gives a permanent record of a varying voltage by using a potentiometer to balance the unknown voltage against a known voltage.

**Potter-Bucky grid** An assembly of lead strips resembling an open Venetian blind, placed between a patient being x-rayed and the screen or film, to reduce the effects of scattered radiation. The grid is kept in motion during exposure to eliminate grid shadows. Also called Bucky diaphragm and grid.

**potting** A process for protecting a component or assembly by mounting it in a can and pouring in an insulating compound.

**powdered iron core** *Ferrite core.*

**powder metallurgy** The production of magnetic cores, permanent magnets, and other molded metal objects by compressing finely powdered metals in molding dies, then heating to a temperature below the fusion point of the metal.

**powder pattern** A diffraction pattern obtained from a sample consisting of many small crystals oriented at random. The developed x-ray film pattern consists of a central spot surrounded by a series of concentric rings from whose diameters the Bragg angles may be computed.

**power** 1. [symbol *P*] The rate at which electric energy is fed to or taken from a device, measured in watts. Power is a definite quantity, whereas level is a relative quantity. 2. The result obtained when a number is multiplied by itself a particular number of times. Thus, 125 is the third power of 5.

**power amplification** *Power gain.*

**power amplifier** An a-f or r-f amplifier designed to deliver maximum output power to a loudspeaker or other load, rather than maximum voltage gain, for a given per cent distortion.

**power-amplifier stage** 1. An a-f amplifier stage that is capable of handling considerable a-f power without distortion. 2. An r-f amplifier stage that serves primarily to increase the power of the carrier signal in a transmitter.

**power-amplifier tube** *Power tube.*

**power attenuation** *Power loss.*

**power breeder** A nuclear reactor designed to produce both useful power and fuel.

**power coefficient** The change of reactivity with increase in power in a nuclear reactor.

**power control rod** A control rod that produces only a small change in reactivity of a nuclear reactor, as required for controlling power level. The design is such that it is impossible to go from delayed to prompt critical by moving the power control rod through its entire range. A regulating rod sometimes serves as a power control rod.

**power converter** A converter used for changing d-c power to a-c power.

**power cord** *Line cord.*

**power density** The power generation per unit volume of a nuclear reactor core.

**power detector** A detector circuit that

will handle strong input signals without objectionable distortion.

**power divider** A device used to produce a desired distribution of power at a branch point in a waveguide system.

**power dump** The removal of all power from a computer, accidentally or intentionally.

**power factor** The ratio of active power to apparent power. As a percentage rating, it is equal to the resistance of a part or circuit divided by the impedance at the operating frequency, with the result multiplied by 100. A pure resistor has a power factor of 100%. A pure capacitor has a power factor of 0% leading, while a pure coil has a power factor of 0% lagging. Power factor is equal to the cosine of the phase angle between the current and voltage when both are sinusoidal.

**power-factor correction** Addition of capacitors to an inductive circuit to increase the power factor. The capacitors offset part or all of the inductive reactance, making the total circuit current more nearly in phase with the applied voltage.

**power-factor meter** A direct-reading instrument for measuring power factor.

**power gain** 1. The ratio of the power delivered by a transducer to the power absorbed by the input circuit of the transducer. The power gain in decibels is 10 times the logarithm of the ratio of the power values. Also called power amplification. 2. An antenna rating equal to 12.56 times the ratio of the radiation intensity in a given direction to the total power delivered to the antenna.

**power level** 1. The amount of power being transmitted past any point in an electric system. When expressed in decibels, it is equal to 10 times the logarithm to the base 10 of the ratio of the given power to a reference power value. Also expressed in volume units. 2. The power production of a nuclear reactor in watts, equal to the product of total number of nuclear fissions per second, the energy in ergs released per nuclear fission, and the number of watts per erg per second. There are $3.1 \times 10^{10}$ nuclear fissions per second for a power of 1 watt.

**power-level indicator** An a-c voltmeter calibrated to read a-f power levels directly in decibels or volume units.

**power line** Two or more wires conducting electric power from one location to another.

**power-line filter** *Line filter.*

**power loss** The ratio of the power absorbed by the input circuit of a transducer to the power delivered to a specified load. Usually expressed in decibels. Also called power attenuation.

**power output** The a-c power in watts delivered by an amplifier to a load.

**power output tube** *Power tube.*

**power pack** A power supply unit that converts the available power line or battery voltage to the voltage values required by a piece of electronic equipment.

**power rating** The power available at the output terminals of a component or piece of equipment that is operated according to the manufacturer's specifications.

**power reactor** A nuclear reactor designed to provide useful power, as for submarines, aircraft, ships, vehicles, and power plants.

**power relay** 1. A relay that functions at a predetermined value of input power. 2. The final relay in a sequence of relays controlling a load or a magnetic contactor.

**power spectrum level** The power level for the acoustic power in a band 1 cps wide, centered at a specified frequency.

**power supply** A power line, generator, battery, power pack, or other source of power for electronic equipment.

**power switch** *On-off switch.*

**power transformer** An iron-core transformer having a primary winding that is

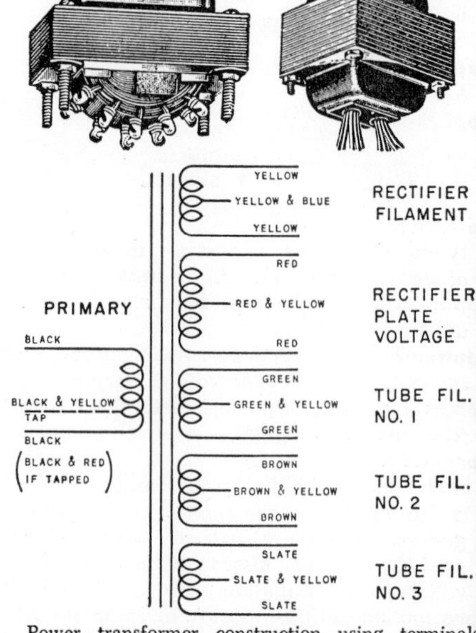

Power transformer construction using terminals and using insulated leads, with standard EIA color code for leads.

connected to an a-c power line and one or more secondary windings that provide different alternating voltage values.

**power transistor** A junction transistor designed to handle high current and power. Used chiefly in audio and switching circuits.

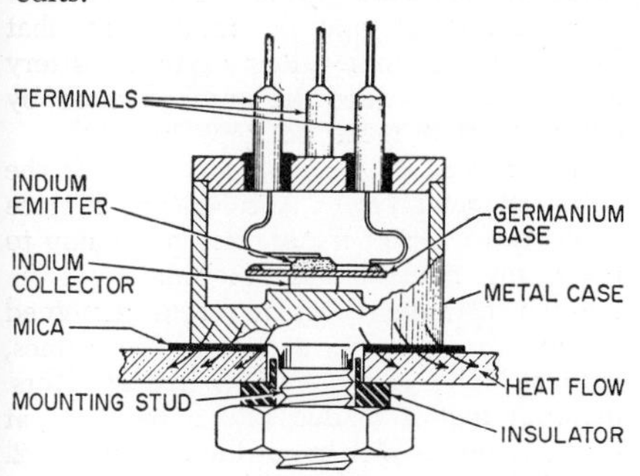


Power transistor construction.

**power tube** An electron tube capable of handling more current and power than an ordinary voltage-amplifier tube. It is used in the last stage of an a-f amplifier or in high-power stages of an r-f amplifier. Also called power-amplifier tube and power output tube.

**power winding** A saturable reactor winding that receives power from a local source.

**Poynting's vector** A vector that represents the direction and amount of energy flow at a point in a wave at a given instant of time.

**ppi** Abbreviation for *plan position indicator.*

**ppi repeater** A radarscope unit that duplicates at a remote location the plan-position-indicator display at the main radar installation. Also called remote ppi.

**p-p junction** A region of transition between two regions having different properties in p-type semiconducting material.

**ppm** Abbreviation for *pulse-position modulation.*

**pppi** Abbreviation for *precision plan position indicator.*

**ppppi** Abbreviation for *photographic projection plan position indicator.*

**practical system** A system of electrical units in which the units are convenient multiples or submultiples of centimeter-gram-second units. Practical units are the ampere, coulomb, farad, henry, joule, ohm, volt, watt, and watthour.

**praseodymium** [symbol Pr] A rare-earth element. Atomic number is 59.

**preamble** The portion of a commercial radio-telegraph message that is sent first, containing the message number, office of origin, date, and other numerical data not part of the original message.

**preamplifier** An amplifier whose primary function is boosting the output of a low-level a-f, r-f, or microwave source to an intermediate level so the signal may be further processed without appreciable degradation of the signal-to-noise ratio of the system. A preamplifier may include provisions for equalizing and mixing.

**precession** A change in the orientation of the axis of a gyroscope or other rotating body.

**precipitation static** Static interference due to the discharge of large charges built up on an aircraft or other object by rain, sleet, snow, or electrically charged clouds.

**precipitator** An electronic apparatus for removing smoke, dust, oil mist, or other small particles from air. A high direct voltage, of the order of 10,000 volts, is obtained from high-voltage rectifier tubes and

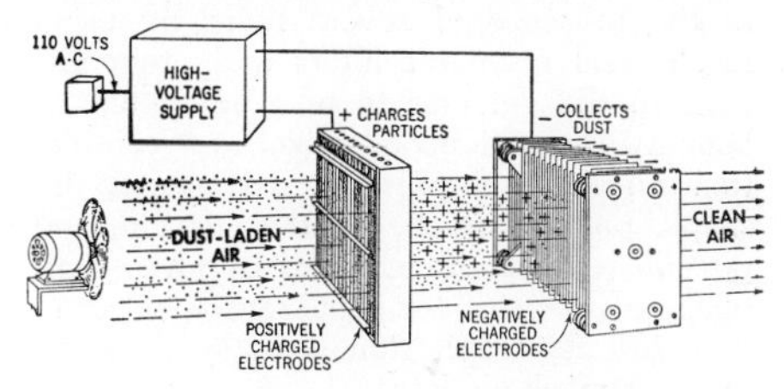


Precipitator operating principles.

applied to a fine wire mesh through which the air is drawn by a fan. Particles in the air are charged by this screen, and are then drawn through a system of parallel charged plates that attract the particles and remove them from the air. Also called electrostatic precipitator.

**Precipitron** Trademark of Westinghouse Electric Corp. for their line of precipitators.

**precision** The quality of being exactly or sharply defined or stated. A six-place table has greater precision than a four-place table. The accuracy of either table would be reduced by errors in compilation or printing, however.

**precision approach radar** [abbreviated par] A radar set located on an airport, used in ground-controlled approach to provide a display that shows the distance, azimuth, and elevation of an airplane moving along the final approach path. Precision approach radar and airport surveillance radar together constitute a ground-controlled approach system.

**precision plan position indicator** [abbreviated pppi or p³i] A plan position indicator combined with a B display radarscope for precise measurement of the coordinates of a target. A range strobe and a mechanical azimuth marker are adjusted roughly to select the desired target on the ppi display, then adjusted more accurately while watching the same target as it appears on the B display. Range and bearing can then be read directly on dials.

**precision sweep** A small portion of a normal radar sweep, usually 2,000 yards, that is expanded over the entire screen to permit precise range measurements.

**precision switch** A switch whose output is a precise function of the position of the actuating lever.

**preconduction current** The low value of anode current flowing in a thyratron or other grid-controlled gas tube prior to the start of conduction.

**predicted-wave signaling** A communication system in which detection is optimized in the presence of severe noise by using mechanical resonator filters and other circuits in the detector to take advantage of known information on arrival and completion times of each pulse, as well as on pulse shape, pulse frequency and spectrum, and possible data content. Used in phase-shift telegraph systems for synchronous or asynchronous telegraph signals or synchronous business machine data.

**predicting element** The element in a fire-control system that predicts, from data provided by the position-finding, tracking, and ballistic data elements, the position of the target at the instant that the missile is expected to reach it.

**predissociation** A process by which a molecule that has absorbed energy dissociates before it loses energy by radiation.

**predistortion** *Preemphasis.*

**preemphasis** The first part of a process for increasing the strength of some frequency components with respect to others, to help these components override noise or to reduce distortion. Used chiefly for emphasizing the higher audio frequencies in frequency-modulated and phase-modulated transmitters and in sound recording systems. Also called accentuation, emphasis, predistortion, and preequalization. The original relations are restored by the complementary process of deemphasis before reproduction of the sounds.

**preemphasis network** An RC filter inserted in a system to emphasize one range of frequencies with respect to another. Also called emphasizer.

**preequalization** *Preemphasis.*

**preferential recombination** Recombination that takes place immediately after an ion pair is formed.

**preferred numbers** A series of numbers adopted by EIA and the military services for use as nominal values of resistors and capacitors, to reduce the number of different sizes that must be kept in stock for replacements.

**preferred tube type** A tube type recommended to designers of electronic equipment for general use, to minimize the number of tube types required for stock supply.

**preform** The small slab of record stock material that is loaded into a press to be formed into a disk recording. Also called biscuit (deprecated).

**p region** The region in a semiconductor where hole density exceeds conduction electron density. Also called p zone.

**preoscillation current** *Starting current.*

**prerecorded tape** *Recorded tape.*

**preselector** A tuned r-f amplifier stage used ahead of the frequency converter in a superheterodyne receiver to increase the selectivity and sensitivity of the receiver.

**presence** The impression, as created by a recording or radio receiver, that the original program source is in the room.

**presentation** The form that radar echo signals are made to take on the screen of a cathode-ray tube.

**preset guidance** Guidance in which the path of a missile is determined by controls that are set before launching.

**preset parameter** A parameter whose value is not changed during the running of a given computer subroutine.

**press** 1. To mold a phonograph record from a stamper. 2. *Pinch.*

**pressed cathode** A dispenser cathode made by compacting and heating a mixture of tungsten-molybdenum alloy and barium-calcium aluminate, in a ratio such that evaporated electron-emitting surface barium is continuously replenished from within the cathode body.

**pressed-glass base** An electron-tube base in which heated powdered glass is pressed around the electrode leads and supports.

**pressed-powder printed circuit** A type of printed circuit formed from particles by application of pressure alone, or heat and pressure.

**pressing** A phonograph record produced

in a record-molding press from a master or stamper.

**press-to-talk switch** A switch mounted directly on a microphone to provide a convenient means for switching two-way radiotelephone equipment or electronic dictating equipment to the talk position.

**pressure** *Effective sound pressure.*

**pressure air-gap crystal unit** A crystal unit in which the electrodes are recessed metal plates held firmly against opposite faces of the quartz plate.

**pressure altimeter** An altimeter that measures and indicates altitude by means of differences in atmospheric pressure.

**pressure hydrophone** A pressure microphone that responds to waterborne sound waves.

**pressure microphone** A microphone whose output varies with the instantaneous pressure produced by a sound wave acting on a diaphragm. Examples are capacitor, carbon, crystal, and dynamic microphones.

**pressure pad** A felt pad mounted on a spring-brass arm, used to hold magnetic tape in close contact with the head on some tape recorders.

**pressure pickup** A device that converts changes in the pressure of a gas or liquid into corresponding changes in some more readily measurable quantity such as inductance or resistance.

**pressure potentiometer** A pressure transducer in which changes in pressure change the position of the movable contact on a potentiometer.

**pressure spectrum level** The effective sound pressure level for the sound energy contained within a band 1 cps wide, centered at a specified frequency.

**pressure switch** A switch that is actuated by a change in pressure of a gas or liquid.

**pressure-type capacitor** A fixed or variable capacitor mounted in a metal tank filled with nitrogen at a pressure that may be as great as 300 pounds per square inch. The high pressure permits a voltage rating several times that of the air rating. Used chiefly in transmitters.

**pressure welding** A welding process in which mechanical pressure is applied during welding. Examples are percussion welding, resistance welding, seam welding, and spot welding.

**pressurization** Use of an inert gas or dry air, at a pressure several pounds above atmospheric pressure, inside the components of a radar system or in a sealed coaxial line. Pressurization prevents corrosion by keeping out moisture, and minimizes high-voltage breakdown at high altitudes.

**pressurized water reactor** A nuclear reactor in which water is circulated under enough pressure to prevent it from boiling, while serving as moderator and coolant for the uranium fuel. The heated water is then used to produce steam for a power plant.

**prestore** To store a quantity in an available computer location before it is required in a routine.

**pre-tr tube** [pre-Transmit-Receive tube] A gas-filled r-f switching tube used to protect the tr tube from excessively high power and to protect the radar receiver from frequencies other than the fundamental.

**prf** Abbreviation for *pulse repetition frequency.*

**pri** Abbreviation for *primary winding.*

**primaries** The colors of constant chromaticity and variable amount which, when mixed in proper proportions, are used to produce or specify other colors.

**primary** *Primary winding.*

**primary battery** A battery consisting of one or more primary cells.

**primary carrier flow** The current flow that is responsible for the major properties of a semiconductor device. Also called primary flow.

**primary cell** A cell that delivers electric current as a result of an electrochemical reaction that is not efficiently reversible.

**primary color** A color that cannot be matched by any combination of other primary colors. In color television, the three primary colors emitted by the phosphors in the color picture tube are red, green, and blue.

**primary-color unit** The area within a color cell in a color picture tube that is occupied by one primary color.

**primary cosmic rays** *Cosmic rays.*

**primary current** The current flowing through the primary winding of a transformer.

**primary dark space** A narrow nonluminous region appearing between the cathode and the cathode glow of some gas-discharge tubes.

**primary detector** *Sensing element.*

**primary electron** An electron emitted directly by a material rather than as a result of a collision.

**primary element** *Sensing element.*

**primary emission** Electron emission due directly to the temperature of a surface, irradiation of a surface, or the application of an electric field to a surface.

**primary feedback** Feedback that is obtained from the controlled variable and is compared with the reference input to obtain the actuating signal for a feedback control system. Also called feedback signal.

**primary filter** A sheet of material, usually metal, that is placed in a beam of radiation to absorb the less penetrating components.

**primary flow** *Primary carrier flow.*

**primary frequency standard** The national standard of frequency as maintained by the National Bureau of Standards, Washington, D. C. The operating frequency of a radio station is determined by comparison with multiples of this standard frequency as broadcast by station WWV.

**primary grid emission** *Thermionic grid emission.*

**primary ionization** 1. The ionization produced by primary particles in a collision, as contrasted to the total ionization which includes the secondary ionization produced by delta rays. 2. The total ionization produced in a counter tube by incident radiation without gas amplification.

**primary ionizing event** *Initial ionizing event.*

**primary ion pair** An ion pair produced directly by the causative primary particle or photon. An ion cluster is a group of ion pairs produced at or near the site of a primary ionizing event; it includes the primary ion pair and any secondary ion pairs formed.

**primary radar** Radar in which the incident beam is reflected from the target to form the return signal.

**primary radiation** Radiation arriving directly from its source without interaction with matter.

**primary radiator** The portion of an antenna system from which energy leaves the transmission system. The distribution of the energy may be subsequently modified by other parts of the antenna system.

**primary radionuclide** A radionuclide that has a lifetime exceeding several hundred million years.

**primary service area** The area in which the ground wave of a broadcast station is not subject to objectionable interference or fading.

**primary skip zone** The area around a radio transmitter that is beyond the ground-wave range but not far enough out for good skip-distance reception. Radio reception is not reliable in the primary skip zone.

**primary specific ionization** The number of ion clusters produced per unit track length.

**primary spectrum** The first-order spectrum produced by a diffraction grating.

**primary standard** A unit directly defined and established by some authority, against which all secondary standards are calibrated.

**primary transit-angle gap loading** The electronic gap admittance that results from the traversal of the gap by an initially unmodulated electron stream.

**primary voltage** The voltage applied to the terminals of the primary winding of a transformer.

**primary winding** [abbreviated pri; symbol P] The transformer winding that receives signal energy or a-c power from a source. Also called primary.

**prime** To charge or discharge storage elements to a potential suitable for feeding data into a charge storage tube.

**prime contractor** A contractor having a direct contract for an entire project. A prime contractor may in turn assign portions of the work to subcontractors.

**priming speed** The rate of priming successive storage elements in a charge storage tube.

**principal axis** A reference direction for angular coordinates used in describing the directional characteristics of a transducer. It is usually an axis of structural symmetry or the direction of maximum response.

**principal E plane** A plane that contains the direction of maximum radiation and the electric vector.

**principal H plane** A plane that contains the direction of maximum radiation and the magnetic vector. The electric vector is everywhere perpendicular to the H plane.

**principal mode** *Fundamental mode.*

**printed circuit** A conductive pattern that may or may not include printed components, formed in a predetermined design on the surface of an insulating base in an accurately repeatable manner. The most common types of printed circuits are etched printed circuits and plated printed circuits.

**printed-circuit assembly** A printed-circuit board on which separable components, terminals, and hardware have been added.

**printed-circuit board** An insulating board serving as a base for a printed circuit. It may include printed components as well as printed wiring, completely processed as far as the printed portion is concerned.

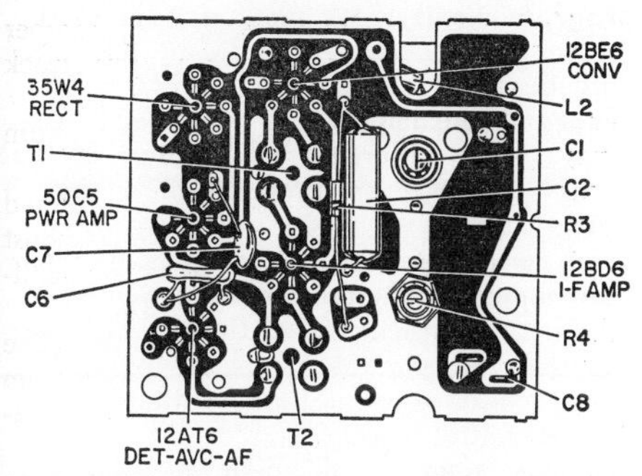


Printed-circuit assembly for table-model radio.

**printed component** A printed coil, resistor, capacitor, switch, transmission line, or other component that provides an electric or magnetic function other than point-to-point connections or shielding. Printed components are formed on printed-circuit boards along with connections.

**printed-component assembly** A printed-component board on which separable components have been added.

**printed-component board** An insulating board serving as a base for a pattern of printed components, completely processed as far as the printed portion is concerned. Printed wiring here serves principally for terminations.

**printed contact** The portion of a printed circuit that provides a contacting surface for some form of connector.

**printed wiring** A type of printed circuit intended primarily to provide point-to-point electric connections or shielding.

**printed-wiring assembly** A printed-wiring board on which separable components have been added.

**printed-wiring board** An insulating board serving as a base for printed wiring, completely processed as far as the printed portion is concerned, and consisting almost entirely of point-to-point connectors and shielding.

**printer** A computer output mechanism that prints characters one at a time or one line at a time.

**printer telegraph code** A five-unit or seven-unit code used to operate telegraph-type printers. The seven-unit code requires only three marking elements out of the seven available per character. If fewer than three are received because of fading, or if more than three impulses arrive because of static, the receiving machine prints a special error-indicator sign rather than an improper letter.

**print-through** Transfer of signals from one recorded layer of magnetic tape to the next on a reel.

**probability of collision** The probability that an electron will collide with an atom or molecule when moving through a distance of 1 cm.

**probability of ionization** The ratio of the number of collisions followed by ionization to the total number of collisions in a gas during a specified time.

**probable error** The amount of error that is most likely to occur during a measurement. Half the results will have a greater error than this value, and the other half will have less error.

**probe** 1. A metal rod that projects into but is insulated from a waveguide or resonant cavity. Used to provide coupling to an external circuit for injection or extraction of energy. When a probe is movable in a slot, it can be used to measure the standing-wave ratio. Also called waveguide probe. 2. A pointed metal tip on an instrument test lead, attached to an insulating handle, for use in making good temporary contact with a circuit terminal for measurement purposes.

**probe microphone** A small microphone used to measure the sound pressure at a point without significantly altering the sound field in the neighborhood of the point.

**process control** Automatic control of a complex industrial process.

**product demodulator** A demodulator whose output is the product of an amplitude-modulated carrier input voltage and a local oscillator signal voltage at the carrier frequency. With proper filtering, the output is proportional to the original modulation.

**production control** The procedure for planning, routing, scheduling, dispatching, and expediting the flow of materials, parts, subassemblies, and assemblies within a plant, from the raw state to the finished product, in an orderly and efficient manner.

**production model** A model in its final mechanical and electrical form, made by the production tools, jigs, fixtures, and methods to be used in turning out subsequent units.

**production reactor** A nuclear reactor designed primarily for large-scale production of transmutation products, such as plutonium.

**product modulator** A modulator whose output is proportional to the product of the carrier and modulating-signal voltages. The carrier is then normally suppressed.

**professional engineer** An engineer who is competent by virtue of fundamental education and training to apply the scientific method and outlook to the analysis and solution of engineering problems.

**profile chart** A vertical cross-section drawing of a microwave path between two stations, indicating terrain, obstructions, and antenna height requirements.

**Profilometer** Trademark of Micrometrical Manufacturing Co. for instruments used to measure surface roughness.

**program** 1. A precise sequence of coded instructions to be used by a digital computer in solving a problem. A complete program includes plans for transcription of data and effective use of results. 2. A sequence of audio signals alone, or audio and video signals, transmitted for entertainment or information.

**program amplifier** *Line amplifier.*

**program circuit** A telephone circuit that has been equalized to handle a wider range of frequencies than is required for ordinary speech signals, for use in transmitting musical programs to the stations of a radio network.

**program control** A controller in which the desired value is automatically changed from time to time in accordance with a predetermined program.

**programed check** 1. A computer check in which a sample problem with known answer, selected to have programing similar to that of the next problem to be run, is put through the computer. 2. A series of self-checking tests inserted in the computer program for a problem.

**programer** 1. A person who prepares sequences of instructions for a computer, without necessarily converting them into the detailed codes. 2. A device used to control the motion of a missile in accordance with a predetermined plan.

**program failure alarm** A signal-actuated electronic relay circuit that gives a visual and aural alarm when the program fails on the line being monitored. A time delay is provided to prevent the relay from giving a false alarm during intentional short periods of silence.

**programing** Preparing a detailed sequence of operating instructions for a particular problem to be run on a digital computer.

**program level** The level of the program signal in an audio system, expressed in volume units.

**program monitor** A monitor used to observe the quality of a radio or television broadcast.

**program parameter** A parameter that may have different values during the course of a given program in a digital computer.

**program register** The register in the control unit of a digital computer that stores the current instruction of the program and controls the operation of the computer during the execution of that instruction. Also called control register.

**program tape** A magnetic or punched-paper tape that contains the sequence of instructions required by a digital computer for solving a particular problem.

**progressive scanning** Scanning all lines in sequence, without interlace, so all picture elements are scanned during one vertical sweep of the scanning beam. Also called sequential scanning.

**progressive wave** A wave that is propagated freely in a medium.

**progressive-wave antenna** *Traveling-wave antenna.*

**projection cathode-ray tube** A television cathode-ray tube designed to produce an intensely bright but relatively small image that can be projected onto a large viewing screen by an optical system.

**projection optics** A system of mirrors and lenses used to project the image onto a screen in projection television. The Schmidt system is an example. Also called reflective optics.

**projection television receiver** A television receiver that uses a system of lenses and mirrors to project on a large-size screen an intensely bright image formed on the screen of a projection cathode-ray tube. Used chiefly for theater television.

**projector** 1. A horn designed to project sound chiefly in one direction from a loudspeaker. 2. A machine that projects film images onto a screen. 3. *Underwater sound projector.*

**projector efficiency** *Transmitting efficiency.*

**projector power response** *Transmitting power response.*

**promethium** [symbol Pm] A rare-earth element that has no known stable isotopes in nature. Atomic number is 61. Formerly called illinium.

**prompt critical** Capable of sustaining a chain reaction without the aid of delayed neutrons.

**prompt fission neutron** A fission neutron

emitted during nuclear fission, as contrasted to delayed fission neutrons emitted by fission products.

**prompt gamma** Gamma radiation emitted at the time of fission of a nucleus.

**prompt neutron** A neutron released coincident with the fission process, as opposed to neutrons subsequently released. About 99% of the neutrons released by an atomic explosion are prompt neutrons. The remainder, called delayed neutrons, appear in groups having half-lives up to about 56 seconds.

**prompt radiation** Radiation emitted within a time too short for measurement, including gamma rays, characteristic x-rays, conversion and Auger electrons, prompt neutrons, and annihilation radiation.

**prong** *Pin.*

**propagation** The travel of electromagnetic waves or sound waves through a medium, or the travel of a sudden electric disturbance along a transmission line. Also called wave propagation.

**propagation anomaly** A change in propagation characteristics due to a discontinuity in the medium of propagation.

**propagation constant** A rating for a line or medium through which a plane wave of a given frequency is being transmitted. It is a complex quantity; the real part is the attenuation constant in nepers per unit length, and the imaginary part is the phase constant in radians per unit length.

**propagation factor** *Propagation ratio.*

**propagation loss** The attenuation of signals passing between two points on a transmission path.

**propagation ratio** The ratio, for a wave propagating from one point to another, of the complex electric field strength at the second point to that at the first point. Also called propagation factor. The field strength is a vector, the magnitude of which is less than one and is the attenuation ratio.

**propagation time delay** The time required for a wave to travel between two points on a transmission path.

**propagation velocity** The velocity of propagation of radio waves, within the accuracy demanded of radar equipment, is usually taken as the velocity of light, equal to $2.998 \times 10^8$ meters per second or 299.8 meters per microsecond. This corresponds to 983.6 feet per microsecond, 0.1863 mile per microsecond, or 0.1618 nautical mile per microsecond.

**proportional band** The range of values of the controlled variable that will cause a controller to operate over its full range.

**proportional control** Control in which the amount of corrective action is proportional to the amount of error.

**proportional counter** A radiation counter consisting of a proportional counter tube and its associated circuits.

**proportional counter tube** A radiation-counter tube operated at voltages high enough to produce ionization by collision and adjusted so the total ionization per count is proportional to the ionization produced by the initial ionizing event.

**proportional ionization chamber** An ionization chamber in which the initial ionization current is amplified by electron multiplication in a region of high electric field strength, as it is in a proportional counter. Used for measuring ionization currents or charges over a period of time, rather than for counting.

**proportional navigation** A homing guidance technique in which the missile turn rate is directly proportional to the turn rate of the missile-target line-of-sight.

**proportional plus derivative control** *Error-rate damping.*

**proportional-position action** Control action in which there is a continuous linear relation between the value of the controlled variable and the position of a final control element.

**proportional region** The range of applied voltage for a radiation-counter tube in which the charge collected per isolated count is proportional to the charge liberated by the initial ionizing event.

**proportional response** *Rate control.*

**proportional-speed floating action** Floating action in which the final control element is moved at a speed proportional to the deviation.

**proportioning reactor** A saturable-core reactor used for regulation and control. Increasing the input control current from zero to rated value makes output current increase in proportion from cutoff up to full load value.

**protactinium** [symbol Pa] A radioactive element that yields actinium by disintegration. Atomic number is 91. Formerly spelled protoactinium.

**protection survey** Evaluation of the radiation hazards incidental to the production, use, or existence of radioactive materials or other sources of radiation under a specific set of conditions.

**protective resistance** A resistance used in

series with a gas tube or other device to limit current flow to a safe value.

**protector tube** A glow-discharge cold-cathode tube that becomes conductive at a predetermined voltage, to protect a circuit against overvoltage.

**protium** A name sometimes applied to the lightest hydrogen isotope, having a mass number of 1 and consisting of a single proton and electron. The other isotopes of hydrogen are deuterium and tritium.

**proton** An elementary particle having a positive charge equal in magnitude to the negative charge of the electron. The atomic number of an element indicates the number of protons in the nucleus of each atom of that element. The rest mass of a proton is $1.67 \times 10^{-24}$ gram.

**proton binding energy** The energy required to remove a single proton from a nucleus. Most proton binding energies are in the range of 5 to 12 million electron-volts.

**proton microscope** A microscope that is similar to the electron microscope but uses protons instead of electrons as the charged particles.

**proton-proton chain** A series of thermonuclear reactions initiated by a reaction between two protons.

**proton synchrotron** A synchrotron designed to accelerate protons. Also called bevatron and cosmotron.

**proximity detector** A sensing device that produces an electric signal when approached by an object or when approaching an object.

**proximity effect** The redistribution of current in a conductor due to the presence of another current-carrying conductor.

**proximity fuze** A fuze that detonates a warhead when the target is within some specified region near the fuze. Radio, radar, photoelectric, or other devices may be used as activating elements. Also called influence fuze and variable-time fuze.

**prr** Abbreviation for *pulse repetition rate.*

**pseudo-code** An arbitrary code, independent of the hardware of a computer, that must be translated into computer code before it can be used to direct the computer.

**P shell** The sixth layer of electrons about the nucleus of an atom, having electrons whose principal quantum number is 6.

**psychogalvanometer** An instrument for testing mental reaction by determining how skin resistance changes when a voltage is applied to electrodes in contact with the skin.

**psychointegroammeter** *Lie detector.*

**psychosomatograph** An instrument for recording muscular action currents or physical movements during tests of mental-physical coordination.

**ptm** Abbreviation for *pulse-time modulation.*

**p-type conductivity** The conductivity associated with holes in a semiconductor, which are equivalent to positive charges.

**p-type semiconductor** An extrinsic semiconductor in which the hole density exceeds the conduction electron density, so the majority carriers are holes. The net ionized impurity concentration is acceptor-type.

**public-address amplifier** An a-f amplifier that provides sufficient output power to loudspeakers for adequate sound coverage at public gatherings.

**public-address system** [abbreviated p-a system] A complete system for amplifying sounds and providing adequate volume for large public gatherings.

**pull-in current** *Pickup current.*

**pulling** An effect that forces the frequency of an oscillator to change from a desired value. Causes include undesired coupling to another frequency source or the influence of changes in the oscillator load impedance.

**pulling figure** The total frequency change of an oscillator when the phase angle of the reflection coefficient of the load impedance varies through 360°, the absolute value of this reflection coefficient being constant at 0.20.

**pull-in value** *Pickup value.*

**pulsatance** Angular velocity in radians, equal to $2\pi$ times frequency in cps.

**pulsating current** A direct current that increases and decreases in magnitude.

**pulsation welding** Resistance welding in which the current is applied in timed pulses rather than continuously, to improve the transfer of surface heat to the water-cooled welding electrodes and thereby increase electrode life.

**pulse** A momentary, sharp change in a current, voltage, or other quantity that is normally constant. A pulse is characterized by a rise and a decay, and has a finite duration. Also called impulse.

**pulse amplifier** An amplifier designed specifically to amplify electric pulses without appreciably changing their waveforms.

**pulse amplitude** The peak, average, effective, instantaneous, or other magnitude of a pulse, usually with respect to the normally constant value. The exact meaning should be specified when giving a numerical value.

**pulse-amplitude modulation** [abbreviated pam] Amplitude modulation of a pulse carrier.

**pulse analyzer** An instrument used to measure pulse widths and repetition rates, and to display on a cathode-ray screen the waveform of a pulse.

**pulse-averaging discriminator** A telemetering subcarrier discriminator that uses resistive and capacitive tuning components to give an output that is proportional to average pulse width.

**pulse bandwidth** The bandwidth outside of which the amplitude of a pulse is below a prescribed fraction of the peak amplitude.

**pulse carrier** A pulse train used as a carrier.

**pulse code** A code consisting of various combinations of pulses, such as the Morse code, Baudot code, and the binary code used in computers.

**pulse-code modulation** [abbreviated pcm] Modulation in which the peak-to-peak amplitude range of the signal to be transmitted is divided into a number of standard values each having its own three-place code. Each sample of the signal is then transmitted as the code for the nearest standard amplitude.

**pulse coder** *Coder.*

**pulse counter** A device that indicates or records the total number of pulses received during a time interval.

**pulsed altimeter** A radar altimeter that emits pulses of r-f energy.

**pulsed doppler system** A pulsed radar system that uses the doppler effect to obtain information about a target.

**pulse decay time** The interval of time required for the trailing edge of a pulse to decay from 90% to 10% of the peak pulse amplitude.

**pulse-delay network** A network consisting of two or more components such as resistors, coils, and capacitors, used to delay the passage of a pulse.

**pulse demoder** A circuit that responds only to pulse signals that have the specified spacing between pulses for which the device is adjusted. Also called constant-delay discriminator.

**pulse discriminator** A discriminator circuit that responds only to a pulse having a particular duration or amplitude.

**pulsed maser** *Two-level maser.*

**pulsed oscillator** An oscillator that generates a carrier-frequency pulse or a train of carrier-frequency pulses as the result of self-generated or externally applied pulses.

**pulsed plasma accelerator** A plasma engine for space travel in which a plasma current loop expands outward due to its self-induced magnetic field.

**pulse droop** A distortion of an otherwise essentially flat-topped rectangular pulse, characterized by a decline of the pulse top.

**pulse duration** The time interval between the first and last instants at which the instantaneous amplitude reaches a stated fraction of the peak pulse amplitude. Also called pulse length (deprecated) and pulse width (deprecated).

**pulse-duration coder** *Coder.*

**pulse-duration discriminator** A circuit in which the sense and magnitude of the output are a function of the deviation of the pulse duration from a reference.

**pulse-duration error** An error caused by pulse duration, which makes certain targets appear longer or thicker than they actually are in the direction of the radar beam.

**pulse-duration modulation** [abbreviated pdm] A form of pulse-time modulation in which the duration of a pulse is varied. Also called pulse-length modulation (deprecated) and pulse-width modulation (deprecated).

**pulse duty factor** The ratio of average pulse duration to average pulse spacing. This is equivalent to the product of average pulse duration and pulse repetition rate.

**pulse excitation** A method of producing oscillator current in which the duration of the impressed voltage in the circuit is relatively short compared with the duration of the current produced. Also called impulse excitation.

**pulse-forming line** A continuous line or ladder network having parameters that give a specified shape to the modulator pulse in a radar modulator.

**pulse-forming network** A network used to shape the leading and/or trailing edge of a pulse.

**pulse-frequency modulation** [abbreviated pfm] A form of pulse-time modulation in which the pulse repetition rate is the characteristic varied. A more precise term for pulse-frequency modulation would be pulse repetition-rate modulation.

**pulse-frequency spectrum** *Pulse spectrum.*

**pulse generator** A generator that produces repetitive pulses or single signal-initiated pulses. Also called impulse generator.

**pulse-height analyzer** An instrument capable of indicating the number of occurrences of pulses falling within each of one or more specified amplitude ranges. Used to obtain the energy spectrum of nuclear radiations. Also called kick-sorter (British) and multichannel analyzer.

**pulse-height discriminator** A circuit that produces a specified output pulse when and only when it receives an input pulse whose amplitude exceeds an assigned value.

**pulse-height selector** A circuit that produces a specified output pulse only when it receives an input pulse whose amplitude lies between two assigned values. Also called differential pulse-height discriminator.

**pulse interleaving** A process in which pulses from two or more sources are combined in time-division multiplex for transmission over a common path.

**pulse interrogation** The triggering of a transponder by a pulse or pulse mode.

**pulse interval** Deprecated term for *pulse spacing.*

**pulse-interval modulation** Deprecated term for *pulse-spacing modulation.*

**pulse ionization chamber** An ionization chamber used to detect individual ionizing events. Also called counting ionization chamber.

**pulse jitter** A relatively small variation of the pulse spacing in a pulse train. The jitter may be random or systematic, depending on its origin, and is generally not coherent with any pulse modulation imposed.

**pulse length** Deprecated term for *pulse duration.*

**pulse-length modulation** Deprecated term for *pulse-duration modulation.*

**pulse mode** A finite sequence of pulses in a prearranged pattern, used for selecting and isolating a communication channel.

**pulse-mode multiplex** A process or device for selecting channels by means of pulse modes. This process permits two or more channels to use the same carrier frequency. Also called pulse multiplex (deprecated).

**pulse moder** A device for producing a pulse mode.

**pulse-modulated waves** Recurrent wavetrains in which the duration of the trains is short compared with the interval between them. Used in radar.

**pulse modulation** Modulation of a carrier by a pulse train.

**pulse modulator** A device that applies pulses to the element in which modulation takes place.

**pulse multiplex** Deprecated term for *pulse-mode multiplex.*

**pulse navigation system** A navigation system that depends on the time required for a pulse of r-f energy to travel a given distance. Examples include gee, loran, radar, and shoran.

**pulse noise** Noise due to a succession of separated pulses. Also called impulse noise.

**pulse-numbers modulation** Modulation in which the pulse density per unit time of a pulse carrier is varied in accordance with a modulating wave by making systematic omissions without changing the phase or amplitude of the transmitted pulses. As an example, omission of every other pulse could correspond to zero modulation; reinserting some or all pulses then corresponds to positive modulation, and omission of more than every other pulse corresponds to negative modulation.

**pulse-phase modulation** *Pulse-position modulation.*

**pulse-position modulation** [abbreviated ppm] A form of pulse-time modulation in which the position in time of a pulse is varied. Also called pulse-phase modulation.

**pulser** A generator used to produce high-voltage, short-duration pulses, as required by a pulsed microwave oscillator or a radar transmitter. In a vacuum-tube pulser, the pulse is produced by discharging a capacitor through the load. In a line-type pulser, an unterminated transmission line is charged through a high impedance and discharged through the load.

**pulse radar** Radar in which the transmitter sends out high-power pulses that are spaced far apart in comparison with the duration of each pulse. The receiver is active for reception of echoes in the interval following each pulse.

**pulse recurrence interval** The time, usually expressed in microseconds, between pulses in radar and loran.

**pulse recurrence rate** *Pulse repetition rate.*

**pulse repetition frequency** [abbreviated prf] *Pulse repetition rate.*

**pulse repetition period** The reciprocal of the pulse repetition frequency.

**pulse repetition rate** [abbreviated prr] The number of times per second that a pulse is transmitted. In radar the pulse repetition rate is usually between 400 and 3,000 pulses per second. Also called pulse recurrence rate and pulse repetition frequency.

**pulse spacing** The interval between the

corresponding pulse times of two consecutive pulses. Also called pulse interval (deprecated).

**pulse-spacing modulation** A form of pulse-time modulation in which the pulse spacing is varied. Also called pulse-interval modulation (deprecated).

**pulse regeneration** The process of restoring pulses to their original relative timings, forms, and magnitudes.

**pulse repeater** A device used for receiving signal pulses from one circuit and transmitting corresponding pulses into another circuit. It may also change the frequency and waveform of the pulses and perform other functions.

**pulse reply** The transmission of a pulse or pulse mode by a transponder as the result of an interrogation.

**pulse rise time** The interval of time required for the leading edge of a pulse to rise from 10% to 90% of the peak pulse amplitude.

**pulse scaler** A scaler that produces an output signal when a prescribed number of input pulses has been received.

**pulse selector** A circuit or device for selecting the proper pulse from a sequence of telemetering pulses.

**pulse separation** The time interval between the trailing edge of one pulse and the leading edge of the succeeding pulse.

**pulse shaper** A transducer used for changing one or more characteristics of a pulse, such as a pulse regenerator or pulse stretcher.

**pulse shaping** Intentionally changing the shape of a pulse.

**pulse spectrum** The frequency distribution of the sinusoidal components of a pulse in relative amplitude and in relative phase. Also called pulse-frequency spectrum.

**pulse spike** An unwanted pulse of relatively short duration superimposed on a main pulse.

**pulse spike amplitude** The peak pulse amplitude of a pulse spike.

**pulse stretcher** A pulse shaper that produces an output pulse whose duration is greater than that of the input pulse and whose amplitude is proportional to the peak amplitude of the input pulse.

**pulse synthesizer** A circuit used to supply pulses that are missing from a sequence due to interference or other causes.

**pulse test** An insulation test in which the applied voltage is a pulse having a specified wave shape. Also called impulse test.

**pulse tilt** A distortion in an otherwise essentially flat-topped rectangular pulse, characterized by either a decline or a rise of the pulse top.

**pulse-time modulation** [abbreviated ptm] Modulation in which the time of occurrence of some characteristic of a pulse carrier is varied from the unmodulated value. Examples include pulse-duration, pulse-interval, and pulse-position modulation.

**pulse train** A sequence of pulses having similar characteristics.

**pulse-train spectrum** The frequency distribution of the sinusoidal components of a pulse train, in amplitude and in phase angle.

**pulse transformer** A transformer capable of operating over a wide range of frequencies, used to transfer nonsinusoidal pulses without materially changing their waveforms.

**pulse transmitter** A pulse-modulated transmitter whose peak power-output capabilities are usually large with respect to the average power-output rating.

**pulse-type altimeter** *Radar altimeter.*

**pulse-type scanning sonar** Scanning sonar in which a pulse of sound power is transmitted in all directions simultaneously. The volume of surrounding water is then scanned rapidly, for viewing of all echoes on a ppi screen.

**pulse-type telemeter** A telemeter that employs characteristics of intermittent electric signals other than their frequency as the means of conveying information.

**pulse valley** The portion of a pulse between two specified maxima.

**pulse width** Deprecated term for *pulse duration.*

**pulse-width modulation** Deprecated term for *pulse-duration modulation.*

**pulsing circuit** A circuit designed to provide abrupt changes of voltage or current in some characteristic pattern.

**pumped tube** An electron tube that is continuously connected to evacuating equipment during operation. Large pool-cathode tubes are often operated in this manner.

**pumping** The process of applying to a solid-state maser a microwave signal having a frequency such that the populations of the molecules in two different energy states in the solid-state material are made nearly equal so as to induce transitions to a third energy state.

**pumping frequency** The frequency at which pumping is provided in a maser, quadrupole amplifier, or other amplifier requiring high-frequency excitation.

**punched card** A card of constant size and shape, suitable for punching in a pattern that has meaning, and for being handled mechanically. The punched holes are sensed electrically by wire brushes, mechanically by metal fingers, or photoelectrically. Also called card.

**punched tape** Paper tape that is punched in coded patterns of holes to convey information. Also called tape.

**punch-through voltage** The transistor collector-to-base voltage at which the space-charge layer of the collector has widened until it touches the emitter junction.

**puncture** *Breakdown.*

**puncture voltage** The voltage at which a test specimen is electrically punctured.

**Pupin coil** *Loading coil.*

**pup jack** *Tip jack.*

**pure tone** *Simple tone.*

**purity** 1. The degree to which a primary color is pure and not mixed with the other two primary colors used in color television. 2. A ratio of distances on the CIE chromaticity diagram that serves to compare a sample color with a reference standard light. Also called excitation purity.

**purity coil** A coil mounted on the neck of a color picture tube, used to produce the magnetic field needed for adjusting color purity. The direct current through the coil is adjusted to a value that makes the magnetic field orient the three individual electron beams so each strikes only its assigned color of phosphor dots.

**purity control** A potentiometer or rheostat used to adjust the direct current through the purity coil.

**purity magnet** An adjustable arrangement of one or more permanent magnets used in place of a purity coil in a color television receiver.

**purple boundary** A straight line drawn between the ends of the spectrum locus on the CIE chromaticity diagram.

**pursuit-course guidance** Guidance in which the missile is always moving directly toward the target. The missile thereby follows a curved path and eventually collides with the target due to its greater speed.

**pushback hookup wire** Tinned copper wire covered with loosely braided insulation that can be pushed back with the fingers sufficiently to expose enough bare wire for making a connection.

**pushbutton control** Control of machines, missiles, and other complex equipment entirely by means of relays and automatic electric or electronic control circuits, once a human operator has operated a pushbutton switch.

**pushbutton station** A unit assembly of one or more pushbutton switches, with or without indicating lamps.

**pushbutton switch** A master switch that is operated by finger pressure on the end of an operating button.

**pushbutton tuner** A device that automatically tunes a radio receiver or other piece of equipment to a desired frequency when the button assigned to that frequency is pressed. The button may connect a set of preadjusted trimmer capacitors or adjustable coils into tuning circuits, may control a small motor that drives the regular gang tuning capacitor, or may apply force to a lever or cam system that rotates the gang tuning capacitor to the correct position.

**pushing figure** The amount of change in oscillator frequency produced by a specified change in oscillator electrode current, excluding thermal effects.

**push-pull amplifier** A balanced amplifier employing two similar electron tubes or equivalent amplifying devices working in phase opposition.

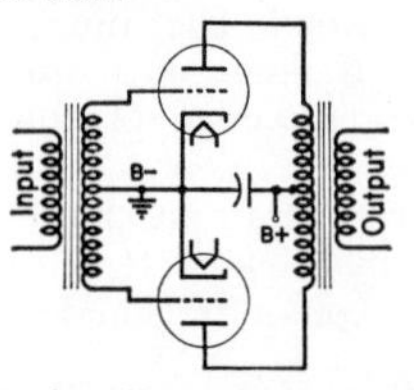


Push-pull amplifier circuit.

**push-pull currents** *Balanced currents.*

**push-pull microphone** A microphone that makes use of two identical microphone elements that are actuated 180° out of phase by a sound wave, as in a double-button carbon microphone.

**push-pull oscillator** A balanced oscillator employing two similar electron tubes or equivalent amplifying devices in phase opposition.

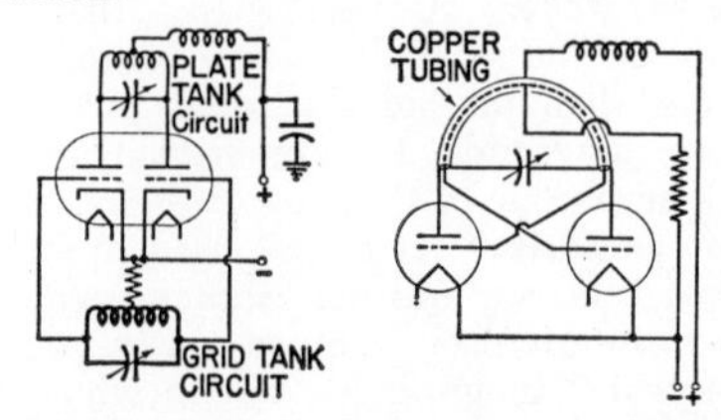


Push-pull oscillator circuit having tuned-grid and tuned-anode tank circuits, with uhf version at right.

**push-pull-parallel amplifier** A push-pull amplifier having two or more tubes or transistors in parallel in each half of the circuit, to obtain higher output power.

**push-pull track** A sound track having two recordings so arranged that the modulation in one is exactly 180° out of phase with that in the other.

**push-pull transformer** An a-f transformer having a center-tapped winding and designed for use in a push-pull amplifier.

**push-pull voltages** *Balanced voltages.*

**push-push amplifier** An amplifier employing two similar electron tubes with grids connected in phase opposition and with anodes connected in parallel to a common load. Usually used as a frequency multiplier to emphasize even-order harmonics. Transistors may be used in place of tubes.

**push-push currents** Currents flowing in the two conductors of a balanced line which, at every point along the line, are equal in magnitude and in the same direction. Also called totally unbalanced currents.

**push-push voltages** Voltages (relative to ground) on the two conductors of a balanced line which, at every point along the line, are equal in magnitude and have the same polarity.

**pylon antenna** A vertical antenna constructed of one or more cylinders of sheet metal, each having a lengthwise slot.

**pyramidal horn** An electromagnetic horn whose sides form a pyramid.

**pyranometer** An instrument that measures the intensity of the radiation received from any portion of the sky.

**pyrheliometer** An instrument for measuring the total intensity of solar radiation.

**pyroconductivity** Electric conductivity that develops in a material only at high temperature, chiefly at fusion, in solids that are practically nonconductive at atmospheric temperatures.

**pyroelectric effect** The development of charges in certain crystals when unequally heated or cooled.

**pyromagnetic** Pertaining to the interaction of heat and magnetism.

**pyrometer** An instrument for measuring temperatures by electric means, especially temperatures beyond the range of mercury thermometers. Examples include the radiation pyrometer, the resistance pyrometer, and the thermoelectric pyrometer.

**pyrotron** A machine that uses magnetic mirrors in a long straight tube to reflect charged particles and prevent end leaks. Used in controlled-fusion research. The tube is surrounded by current-carrying coils, with more coils or higher currents at the ends to produce the stronger magnetic fields that act as mirrors.

**Pythagorean scale** A musical scale such that the frequency intervals are represented by the ratios of integral powers of the numbers 2 and 3.

**p zone** *P region.*

# Q

**Q** A figure of merit for an energy-storing device, tuned circuit, or resonant system. It is equal to reactance divided by resistance. The *Q* of a capacitor, coil, circuit, or system thus determines the rate of decay of stored energy; the higher the *Q*, the longer it takes for the energy to be released. Also called *Q* factor and quality factor.

**Q antenna** A dipole that is matched to its transmission line by stub-matching.

**qcw signal** Abbreviation for *quadrature-phase subcarrier signal.*

**Q demodulator** The demodulator in which the chrominance signal and voltage from the color-burst oscillator signal are combined to recover the Q signal in a color television receiver.

**Q electron** An electron having an orbit in the Q shell, which is the seventh shell of electrons surrounding the atomic nucleus, counting out from the nucleus.

**Q factor** *Q.*

**Q meter** An instrument that measures the *Q* of a circuit or circuit element, by determining the ratio of reactance to resistance. Also called quality-factor meter.

**Q multiplier** A filter that gives a sharp response peak or a deep rejection notch at a particular frequency, equivalent to boosting the *Q* of a tuned circuit at that frequency.

**Q shell** The seventh layer of electrons about the nucleus of an atom, having electrons characterized by the principal quantum number 7.

**Q signal** 1. The quadrature component of the chrominance signal in color television, having a bandwidth of 0 to 0.5 mc. It consists of $+0.48(R-Y)$ and $+0.41(B-Y)$, where Y is the luminance signal, R is the red camera signal, and B is the blue camera signal. 2. A three-letter abbreviation starting with Q, used in the International List of Abbreviations for radiotelegraphy to represent complete sentences.

**QSL card** A card sent by one radio amateur to another to verify radio communication with each other.

**quad** An assembly of four separately insulated conductors.

**quadrant** 1. A 90° sector of a circle. 2. The fourth part of something, as one of the quadrants in a four-course radio range.

**quadrantal error** The angular error in a measured bearing that is due to the presence of metal in the vicinity of the direction-finding antenna, such as the metal structure and engines of an airplane or the hull of a ship.

**quadrantal heading** A heading to the northeast, southeast, southwest, or northwest.

**quadrant electrometer** An electrometer for measuring voltages and charges by means of electrostatic forces between a suspended metal plate and a surrounding metal cylinder that is divided into four insulated parts connected oppositely in pairs. The voltage to be measured is applied between the two pairs of quadrants.

**quadrature** Separated in phase by 90° or one quarter-cycle. Also called phase quadrature.

**quadrature amplifier** An amplifier that shifts the phase of a signal 90°. Used in a color television receiver to amplify the 3.58-mc chrominance subcarrier and shift its phase 90° for use in the Q demodulator.

**quadrature component** The reactive component of a current or voltage, due to inductive or capacitive reactance in a circuit.

**quadrature-phase subcarrier signal** [abbreviated qcw signal] The portion of the chrominance signal that leads or lags the in-phase portion by 90°.

**quadricorrelator** [QUADRature Information CORRELATOR] A circuit sometimes added to the automatic phase control loop in a color television receiver to obtain improved performance under severe interference conditions.

**quadripole** *Two-terminal pair network.*

**quadruplex circuit** A telegraph circuit designed to carry two messages in each direction at the same time.

**quadrupole amplifier** A low-noise parametric amplifier consisting of an electron-beam tube in which quadrupole fields act on the fast cyclotron wave of the electron beam to produce high amplification at frequencies in the range of 400 to 800 mc. The cyclotron frequency is approximately equal to the frequency of the signal to be amplified, and the pumping frequency is about twice this value.

**quadrupole moment** A term used in specifying mathematically the field due to a given distribution of electric or magnetic charges.

**quality factor** *Q*.

**quality-factor meter** *Q meter*.

**quantity of electricity** 1. The amount of electric charge stored in a capacitor, measured in coulombs or similar units. 2. The amount of current flowing through a circuit in a given time, measured in coulombs. One coulomb is one ampere flowing for one second.

**quantity of radiation** The total radiated energy passing through a unit area per unit of time. Expressed in ergs per square centimeter or watt-seconds per square centimeter.

**quantization** Division of the range of values of a wave into a finite number of smaller subranges, each of which is represented by an assigned or quantized value within the subrange.

**quantization distortion** Inherent distortion introduced in process of quantization. Also called quantization noise.

**quantization level** One of the subrange values obtained by quantization.

**quantization noise** *Quantization distortion*.

**quantize** To restrict a variable to a discrete number of possible values.

**quantized field theory** A theory in which electromagnetic fields and the fields of matter are represented by mathematical operators that describe the elementary processes of creation and annihilation of particles or photons.

**quantized pulse modulation** Pulse modulation that involves quantization, such as pulse-numbers modulation and pulse-code modulation.

**quantized system** A system of particles whose energies can have only discrete values.

**quantizer** A device that measures the magnitude of a time varying quantity in multiples of some fixed unit or quantum, at a specified instant or specified repetition rate, and delivers a proportional response that is usually in pulse code or digital form. The amplitude of the response signal is at each instant proportional to the number of quanta measured.

**quantizing** The process of representing any value between certain limits by the nearest of a limited number of values selected to cover the range. Used in pulse-code modulation.

**quantum** [plural quanta] The smallest quantity of energy that can be associated with a given phenomenon. The quantum of electromagnetic radiation is the photon.

**quantum efficiency** The average number of electrons photoelectrically emitted from a photocathode per incident photon of a given wavelength in a phototube.

**quantum mechanics** The study of atomic structure and related problems in terms of quantities that can actually be measured.

**quantum number** A number assigned to one of the various values of a quantized quantity in its discrete range. As an example, the principal quantum number of an electron determines the energy level with respect to the minimum-energy level or ground state that has a quantum number of 1.

**quantum state** One of the states in which an atom may exist permanently or momentarily.

**quantum statistics** Statistics giving the distribution of a particular type of particle among its possible quantized energy values. Also called statistics.

**quantum theory** A theory that atoms or molecules emit or absorb energy by a process that takes place in a series of steps, each step being the emission or absorption of an amount of energy called the quantum. For light or other radiation the quantum is the photon, the energy of which is equal to the frequency of the radiation in cps multiplied by Planck's constant, which is $6.625 \times 10^{-27}$ erg-second.

**quantum voltage** The voltage through which an electron must be accelerated to acquire the energy corresponding to a particular quantum.

**quantum yield** The number of photon-induced reactions of a specified type per photon absorbed.

**quarter-phase** *Two-phase*.

**quarter-wave** Having an electrical length of one quarter-wavelength.

**quarter-wave antenna** An antenna whose electrical length is equal to one quarter-wavelength of the signal to be transmitted or received.

**quarter-wave attenuator** An arrangement of two wire gratings spaced an odd number of quarter-wavelengths apart in a waveguide, used to attenuate waves traveling through in one direction. The wave reflected from the first grating is canceled by that reflected from the second, so all energy reaching the attenuator is either transmitted through the gratings or absorbed by them, with no resultant reflection.

**quarter-wavelength** The distance corresponding to an electrical length of a quarter of a wavelength at the operating frequency of a transmission line, antenna element, or other device.

**quarter-wave line** *Quarter-wave stub.*

**quarter-wave plate** A plate of mica or other doubly refracting crystal material of such thickness as to introduce a phase difference of one quarter-cycle between the ordinary and the extraordinary components of light passing through.

**quarter-wave stub** A section of transmission line that is one quarter-wavelength long at the fundamental frequency being transmitted. When shorted at the far end, it has a high impedance at the fundamental frequency and all odd harmonics, and a low impedance for all even harmonics. Also called quarter-wave line and quarter-wave transmission line.

**quarter-wave termination** A nonreflecting waveguide termination consisting of an energy-absorbing wire grating or semiconducting film stretched across the waveguide one quarter-wavelength from a metal-plate termination. The wave reflected by the grating is canceled by the wave reflected from the plate.

**quarter-wave transformer** A section of transmission line approximately one quarter-wavelength long, used for matching a transmission line to an antenna or load.

**quarter-wave transmission line** *Quarter-wave stub.*

**quartz** A natural or artificially grown piezoelectric crystal composed of silicon dioxide, from which thin slabs or plates are

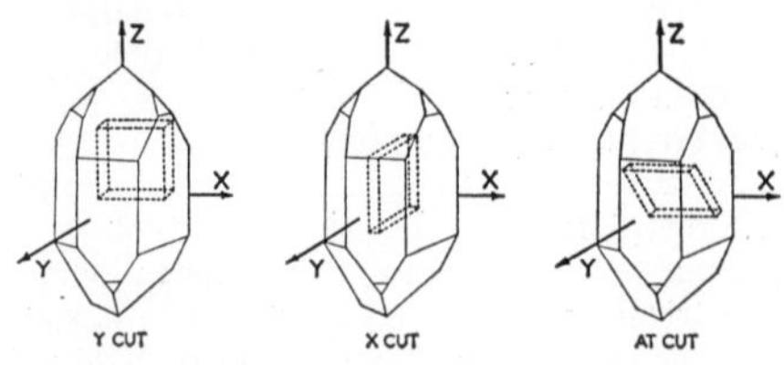


Quartz in natural form, as cut for three types of crystal plates.

carefully cut and ground to serve as a crystal plate for controlling the frequency of an oscillator.

**quartz delay line** An acoustic delay line in which quartz is used as the medium of sound transmission.

**quartz lamp** A mercury-vapor lamp having a transparent envelope made from quartz instead of glass. Quartz resists heat, permitting higher currents, and passes ultraviolet rays that are absorbed by ordinary glass.

**quartz plate** *Crystal plate.*

**quasi-active homing guidance** Homing guidance in which the missile contains only the transmitter that illuminates the target. The receiver of the reflected energy is at some point external to the missile. Control signals must then be transmitted to the missile.

**quasi-bistable circuit** An astable circuit that is triggered at a rate which is high compared to its own natural frequency.

**quasi-conductor** A conductor having a $Q$ much less than unity.

**quasi-dielectric** A dielectric having a $Q$ greater than unity.

**quasi-linear feedback control system** A feedback control system in which the relationships between the input and output signals are substantially linear despite the existence of nonlinear elements.

**quasi-monostable circuit** A monostable circuit that is triggered at a rate which is high compared to its own natural frequency.

**quasi-optical** Having properties similar to light waves, such as being limited to line-of-sight ranges.

**quasi-single-sideband** Transmitting parts of both sidebands to simulate single-sideband transmission.

**quench** 1. To stop abruptly. 2. To cool suddenly.

**quenched spark gap** A spark gap having provisions for rapid deionization. One form consists of many small gaps between electrodes that have relatively large mass and are good radiators of heat. The electrodes serve to cool the gaps rapidly and thereby stop conduction.

**quenched spark-gap converter** A spark-gap generator that uses the oscillatory discharge of a capacitor through a coil and a quenched spark gap as a source of r-f power. The spark gap usually consists of closely spaced gaps operating in series, to give good quenching action.

**quenching** 1. The process of terminating a discharge in a gas-filled radiation-counter tube by inhibiting reignition. 2. Cooling suddenly, as in heat-treating metals.

**quenching circuit** A circuit that diminishes, suppresses, or reverses the voltage applied to a counter tube in order to inhibit multiple discharges from an ionizing event.

**quenching frequency** The frequency at which the oscillations in a superregenerative receiver are suppressed or quenched.

**quick-break fuse** A fuse designed to draw out the arc and break the circuit rapidly when the fuse wire melts, generally by separating the broken ends with a spring.

**quick-break switch** A switch that breaks a circuit rapidly, independently of the rate at which the switch handle is moved, to minimize arcing.

**quiescent** Without an input signal.

**quiescent-carrier modulation** A system of modulation in which the carrier is suppressed during intervals when no modulation is applied.

**quiescent-carrier telephony** A radiotelephone system in which the carrier is suppressed when there are no voice signals.

**quiescent current** The electrode current corresponding to the electrode bias voltage.

**quiescent point** *Operating point.*

**quiescent push-pull amplifier** A push-pull amplifier in which the control grids are biased so negatively that little anode current flows when there is no signal. There is thus no noise when tuning between stations, but tone quality is poor for weak signals.

**quiescent value** The voltage or current value for an electron-tube electrode when no signals are present.

**quiet automatic volume control** *Delayed automatic gain control.*

**quieting sensitivity** The least signal input for which the output signal-to-noise ratio does not exceed a specified limit in f-m receivers.

**Q value** *Disintegration energy.*

# R

**R** 1. Symbol for *resistor.* 2. Abbreviation for *Rankine.*

**r** Abbreviation for *roentgen.*

***R*** Symbol for *resistance.*

**rabal** [RAdiosonde BALloon] 1. A system involving use of a radiosonde balloon to determine atmospheric conditions at various altitudes. 2. The report obtained by radio from a radiosonde balloon.

**rabbit** A small container that is propelled, usually pneumatically or hydraulically, through a tube in a nuclear reactor. Used to expose samples to the radiation and neutron flux, then remove them rapidly for measurements of atoms having short half-lives. Also called shuttle.

**races** Abbreviation for *radio amateur civil emergency service.*

**race track** An assembly of several calutron isotope separators in the shape of a race track, having a common magnetic field. Also called track.

**raceway** A channel used to hold and protect wires, cables, or busbars.

**rack** *Relay rack.*

**rack-and-panel construction** Construction of equipment in such a way that all chassis units and panels can be mounted on a standard relay rack.

**rack panel** A panel designed for mounting on a relay rack. The panel width is 19″, height is a multiple of 1¾″, and the mounting notches are standardized as to size and position.

**racon** [RAdar beaCON] *Radar beacon.*

**rad** A unit of absorbed dose, equal to 100 ergs per gram.

**radac** [RApid Digital Automatic Computation] An electronic system for analyzing complex data rapidly and accurately, as required for fire control against rockets and missiles.

**radan** [RAdar Doppler Automatic Navigator] A doppler radar navigation system for aircraft, operating independently of ground-based stations.

**radar** [RAdio Detecting And Ranging] 1. A system using beamed and reflected r-f energy for detecting and locating objects, measuring distance or altitude, navigating, homing, bombing, and other purposes. In detecting and ranging, the time interval between transmission of the energy and reception of the reflected energy establishes the range of an object in the beam's path. In primary radar, the return signal is produced by reflection of the transmitted energy from the target. In secondary radar, the transmitted energy triggers a responder beacon that sends an entirely new signal back to the radar set. Originally called radiolocation in Great Britain. 2. *Radar set.*

**radar aid** *Radar navigation aid.*

**radar altimeter** A high-altitude radio altimeter used in aircraft to give accurate absolute altitude indications at altitudes far above the common 5,000-foot limit of frequency-modulated radio altimeters. Simple pulse-type radar equipment is used to send a pulse straight down and measure its total time of travel to the surface and back to the aircraft. Lowest useful altitude is about 250 feet. Also called high-altitude radio altimeter and pulse-type altimeter.

**radar altitude** Absolute altitude as determined by a radar altimeter. Also called radio altitude.

**radar astronomy** The study of astronomical bodies and the earth's atmosphere by means of radar pulse techniques, including tracking of meteors and the reflection of radar pulses from the moon and the planets.

**radar band** One of the frequency ranges used in radar. Early designations for these bands were: P band—225–390 mc; L band —390–1,550 mc; S band (10-cm band)—1,550–5,200 mc; X band (3-cm band) —5,200–11,000 mc; K band (1-cm band)—11,000–33,000 mc.

**radar beacon** A radar receiver-transmitter that transmits a strong coded radar signal whenever its radar receiver is triggered by an interrogating radar on an aircraft or ship. The coded beacon reply can be used by the navigator to determine his own position in terms of bearing and range from the beacon. Also called beacon, racon, radar transponder, and transponder. The transmitter section of a radar beacon is called the responder.

**radar beam** The movable beam of r-f energy produced by a radar transmitting antenna. Its shape is commonly defined as the loci of all points at which the power has decreased to one-half of that at the center of the beam.

**radar bombardier** A person trained in radar bombing.

**radar bombing** Bombing in which radar is used to locate the target or aiming point, to aid in positioning the bombing aircraft at the proper release point for bombing, and/or to release bombs automatically, especially under conditions of poor visibility.

**radar bomb scoring** A scoring system in which ground-based radar and plotting equipment are used in determining the theoretical points of bomb impact during a simulated bombing mission. Information obtained from the ground radar is combined with ballistic data and information transmitted from the aircraft regarding the wind, true heading, and true airspeed at the time of release.

**radar bombsight** An airborne radar set used to sight the target, solve the bombing problem, and release the bombs.

**radar calibration** The process of determining the extent and accuracy of the radar coverage of a given aircraft-warning or tactical air control radar installation. Calibration includes computing the theoretical coverage, making flights to check coverage, and preparing calibration charts, diagrams, and overlays.

**radar camera** A special manual or automatic camera used to photograph images on a radarscope.

**radar camouflage** The use of special coverings or surfaces that reduce the reflection of r-f energy back to a radar set, to minimize chances for detection of the object by an enemy radar set.

**radar chart** A special map used in radar navigation. For radar-equipped ships, radar charts emphasize the coastline, hills, large buildings, and other objects that give prominent radar echoes to ships well off shore. For air navigation, radar charts similarly show outlines of cities, rivers, lakes, bridges, railroads, and other objects that appear on airborne radar screens.

**radar check point** A point on the surface of the earth that gives an outstanding return on an airborne radar screen for use in air navigation.

**radar chronograph** A radar system for measuring projectile velocities.

**radar clutter** *Clutter.*

**radar command guidance** A missile guidance system in which radar equipment at the launching site determines the positions of both target and missile continuously, computes the missile course corrections required, and transmits these by radio to the missile as commands.

**radar contact** Recognition and identification of an echo on a radar screen. An aircraft is said to be on radar contact when its radar echo can be seen and identified on a ppi display.

**radar control** Control of an aircraft, guided missile, or gun battery by means of radar.

**radar control area** The area or airspace in which radar control of aircraft or guided missiles is exercised.

**radar controller** 1. A person who exercises radar control over aircraft or guided missiles. 2. A radar device used in radar control.

**radar countermeasure** [abbreviated rcm] An electronic countermeasure used against enemy radar, such as jamming and the use of confusion reflectors.

**radar coverage indicator** A device that gives the maximum range at which a given aircraft can normally be tracked by a radar station, taking into account the aircraft db rating, the altitude of the aircraft, and the characteristics of the radar.

**radar cross-section** *Echo area.*

**radar data display board** A board used for displaying information derived from a radar set or associated equipment.

**radar display** The pattern representing the output data of a radar set, generally produced on the screen of a cathode-ray tube.

**radar drift** The drift of an aircraft as determined by a timed series of bearings taken on a fixed radar target.

**radar echo** *Echo.*

**radar equation** An equation that relates the transmitted and received powers and antenna gains of a primary radar system to the echo area and distance of the radar target.

**radar fence** A network of radar warning stations maintained as a barrier against surprise attack.

**radar fire control** Fire control by means of radar.

**radar fix** A determination of position by means of radar.

**radar gun-layer** A radar device that tracks a target and aims a gun automatically.

**radar homing** 1. Homing on the source of a radar beam. 2.Homing in which a missile-borne radar locks onto a target and guides the missile to that target.

**radar homing set** A complete radar set used in a missile or other vehicle to provide self-guidance to a target.

**radar horizon** The lowest elevation angle at which a radar can operate effectively at a particular location, taking into account the terrain in the vicinity and the curvature of the earth.

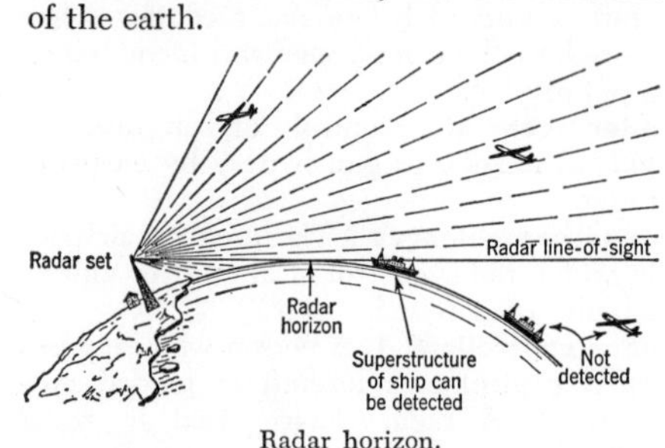

Radar horizon.

**radar illumination** Illumination of a target by a high-power radar external to a missile, to produce echo signals suitable for homing use by receiving equipment in the missile.

**radar indicator** A cathode-ray tube and associated equipment used to provide a visual indication of the echo signals picked up by a radar set.

**radarman** A person who operates a radar set and evaluates its indications.

**radar mapping central** A group of radar sets used for mapping, sometimes including facilities for automatic navigation of the mapping aircraft and retransmission of mapping data to a ground installation.

**radar missile-tracking central** A group of radar sets equipped with facilities for recording and/or indicating signals received from a transponder-equipped missile, to provide free-space position data of the missile.

**radar mission** An air mission that requires or uses radar.

**radar modulator** A modulator that varies the amplitude, phase, frequency, pulse repetition rate and/or pulse duration of a signal generated by a radar transmitter to which it is directly connected.

**radar nautical mile** The time interval of approximately 12.367 microseconds that is required for the r-f energy of a radar pulse to travel one nautical mile and return.

**radar navigation** Navigation by means of radar equipment and radar navigation aids.

**radar navigation aid** Any radar device designed as a navigation aid, such as a radar altimeter. Also called radar aid.

**radar navigator** A person trained in radar navigation.

**radar net** A network of radar installations set up to detect aircraft entering a defined airspace. Also called radar screen.

**radar observer** An aircraft observer trained in the use of airborne radar equipment for navigation, interception, search, or bombing.

**radar operator** 1. A person who operates a radar set. 2. An aircraft observer who operates radar equipment, such as a radar observer or radar bombardier.

**radar paint** A coating that absorbs radar waves.

**radar performance figure** The ratio of the pulse power of a radar transmitter to the power of the minimum signal detectable by the receiver.

**radar picket** A ship or aircraft equipped with early-warning radar and operating at a distance from the area being protected, to extend the range of radar detection.

**radar pilotage** Pilotage in which check points on the ground are viewed on a radarscope.

**radar pilotage equipment** Equipment utilizing primary radar techniques, carried on a vehicle to determine bearing and distance of recognizable landmarks and indicate the relative positions of other vehicles.

**radar plot** A plot of the positions of aircraft or ships, made from data obtained by means of radar.

**radar prediction** A graphic representation of what may be expected to show on a radar screen when an actual radar scan is made.

**radar prediction device** A relief map of enemy territory, on which a tiny lamp is mounted at the location of an enemy radar set. The shadows cast by mountains or other features of the terrain then indicate blind spots in the enemy's radar detection system.

**radar range** The maximum distance at which a radar set is ordinarily effective in detecting objects. Usually assumed to be the distance at which a radar set can detect a specified object at least 50% of the time. The range in free space varies directly with receiver power sensitivity and target echo area, but varies as the square of antenna gain and as the fourth power of transmitted power.

**radar range marker** A mark or line that is scribed or electronically formed on the face of a radarscope to indicate the range to the object detected. Also called range mark.

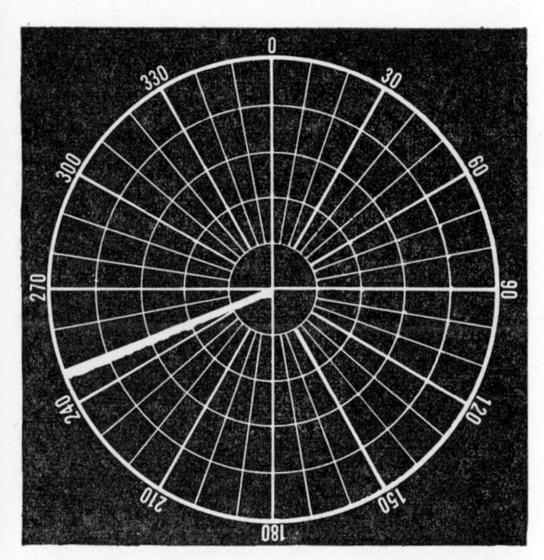

Radar range marker generates circles electronically on plan position indicator display, while bearing lines are usually scribed on transparent plastic sheet mounted over face of radarscope.

**radar receiver** A high-sensitivity radio receiver that is designed to amplify and demodulate radar echo signals and feed them to a radarscope or other indicator.
**radar reflector** A reflector that reflects or deflects radar waves.
**radar reflectoscope** An arrangement of mirrors used to produce a composite image of a chart and a radarscope presentation.
**radar relay** Equipment used for relaying radar video and synchronizing signals to a remote location. Also called relay radar (deprecated).
**radar repeat-back guidance** A missile guidance system in which a search radar in the missile transmits information to the point of control.
**radar resolution** The resolution of a radar set or system.
**radar safety beacon** An airborne radar beacon, used chiefly for identifying aircraft in connection with air traffic control at airports. Also called airborne beacon.
**radar scan** The circular, spiral, rectangular, or other motion of a radar beam as it searches for a target.
**radarscope** The cathode-ray oscilloscope or screen used as an indicator in a radar set. Also called scope.
**radarscope display** The visual presentation on a radar screen.
**radar screen** 1. A cathode-ray screen in a radar set. 2. *Radar net.*
**radar set** A complete assembly of radar equipment for detecting and ranging, consisting essentially of a transmitter, antenna, receiver, and indicator. Also called radar.
**radar shadow** A region shielded from radar illumination by an intervening reflecting or absorbing medium such as a hill.
**radar signal simulator** An instrument whose electric output can be applied directly to the indicator of a radar set to produce artificial radar echo indications.
**radar silence** A period of time during which radar transmission is stopped, generally for security reasons.
**radarsonde set** An electronic system for automatically measuring and transmitting high-altitude meteorological data from a balloon, kite, or rocket by pulse-modulated radio waves when triggered by a radar signal.
**radar station** The ground, air, or sea location at which a radar set transmits or receives signals.
**radar storm detection** The detection of certain storms or stormy conditions by means of radar. Liquid or frozen water drops within the storm reflect radar echoes.
**radar surveillance** Use of one or more radar sets to locate distant targets and provide approach information. Surveillance may include facilities for identifying a target as friend or foe and means for relaying data to appropriate information and/or control centers.
**radar synchronous bombing** *Synchronous radar bombing.*
**radar target** An object being tracked or watched on a radar screen.
**radar track command guidance** Command guidance that uses two radars external to a missile, one for tracking the target and the other for tracking the missile. The radar receiver outputs are fed to a computer, and the output of the computer is in turn fed into a data transmitter that transmits flight information to the missile for correcting its flight path to the target.
**radar tracking** Tracking a moving object by means of radar.
**radar trainer** A trainer for teaching radar techniques and operations by simulating various radar target displays.
**radar transmitter** The transmitter portion of a radar set.
**radar transponder** *Radar beacon.*
**radar warning net** A radar net used for warning.
**radar wave** A transmitted or reflected radio wave used in radar.
**radiac** [RAdioactivity Detection, Identification, And Computation] 1. Detection,

identification, and measurement of the intensity of nuclear radiation in an area. 2. *Radiac set.*

**radiac computer** A computer that scales, integrates, or counts information received from a radiac detector.

**radiac data transmitting set** The equipment required to detect radioactivity and transmit radioactivity data as modulation on a carrier.

**radiac detector** A detector that is sensitive to radioactivity or free nuclear particles, and produces a reaction that can be interpreted or measured by other components.

**radiac-detector charger** An electrostatic generator that provides an electrostatic charge for a radiac detector.

**radiac instrument** *Radiac set.*

**radiacmeter** *Radiac set.*

**radiac set** A complete radioactivity detecting, identifying, and measuring system. Also called radiac, radiac instrument, and radiacmeter.

**radial** One of a number of radial lines of position defined by an azimuthal radio navigation facility, and identified in terms of the bearing (usually magnetic) of all points on that line from the facility.

**radial-beam tube** A vacuum tube in which a radial beam of electrons is rotated past circumferentially arranged anodes by an external rotating magnetic field. Used chiefly as a high-speed switching tube or commutator.

**radial component** A component acting along a radius, as contrasted to a tangential component that acts at right angles to a radius.

**radial field** A field of force directed toward or away from a point in space.

**radial lead** A wire lead coming from the side of a component rather than axially from the end.

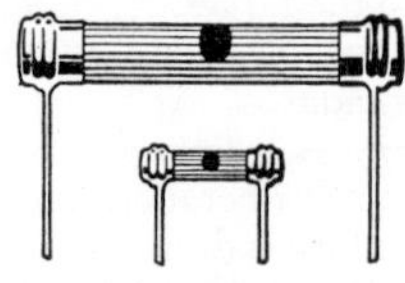

Radial leads on resistors.

**radial transmission line** A pair of parallel conducting planes used for propagating uniform circularly cylindrical waves having their axes normal to the planes.

**radian** The angle that intercepts an arc whose length is equal to its radius. A complete circle contains $2\pi$ radians. One radian is 57.2957795 degrees, and one degree is 0.01745 radian.

**radiance** The radiant flux per unit solid angle per unit of projected area of the source. The usual unit is the watt per steradian per square meter. This is the radiant analog of luminance.

**radianlength** The distance between points in a sinusoidal wave that differ in phase by an angle of one radian. One radianlength is equal to the wavelength divided by $2\pi$.

**radian per second** A unit of angular velocity.

**radiansphere** The boundary between the near and far fields of a small antenna. It is a spherical surface whose radius is equal to the wavelength divided by 6.28.

**radiant** Emitted or transmitted along radii, as from a point source.

**radiant energy** Energy transmitted in the form of electromagnetic radiation, such as radio waves, heat waves, and light waves.

**radiant flux** The time rate of flow of radiant energy.

**radiant flux density** The amount of radiant power per unit area that flows across or onto a surface. Also called irradiance.

**radiant intensity** The energy emitted per unit time, per unit solid angle about the direction considered. Usually expressed in watts per steradian.

**radiant sensitivity** The signal output current of a camera tube or phototube divided by the incident radiant flux at a given wavelength.

**radiate** To send out energy, such as electromagnetic waves, into space.

**radiated power** The total power emitted by a transmitting antenna.

**radiating circuit** A circuit capable of sending electromagnetic waves into space, such as the antenna circuit of a radio transmitter.

**radiating element** A basic subdivision of an antenna that in itself is capable of radiating or receiving r-f energy.

**radiating guide** A waveguide designed to radiate energy into free space through slots, gaps, or horns.

**radiation** 1. A stream of high-energy particles from a cyclotron or other accelerator. 2. Electromagnetic energy, such as light waves, sound waves, radio waves, x-rays, and heat rays, traveling through a material or through space. 3. *Nuclear radiation.*

**radiation absorber** An insulating material in sheet form and having a conductive backing, used as dielectric and reflecting elements for absorbing unwanted r-f energy.

**radiation belt** A radioactive belt surrounding the earth in space, believed due to nuclear particles from the sun that are held in position by the magnetic field of the earth. The belt is made up of two layers, one between 1,500 and 3,500 miles altitude, and the other at 8,000 to 12,000 miles altitude. Also called Van Allen layer.

**radiation burn** A burn caused by overexposure to radiant energy.

**radiation counter** An instrument used for detecting or measuring nuclear radiation by counting the resultant ionizing events. Examples include Geiger counters and scintillation counters. Also called counter.

**radiation-counter tube** *Counter tube.*

**radiation damage** The effect of gamma rays, fission fragments, and neutrons on substances.

**radiation danger zone** A zone within which the maximum permissible constant dose rate is exceeded.

**radiation detector** A device for converting radiant energy to a form more suitable for observation. Used in radiac sets.

**radiation efficiency** The ratio of the power radiated to the total power supplied to an antenna at a given frequency.

**radiation excitation** *Radiation ionization.*

**radiation field** The electromagnetic field that breaks away from a transmitting antenna and radiates outward into space as electromagnetic waves. The other type of electromagnetic field associated with an energized antenna is the induction field.

**radiation hazard** A health hazard arising from exposure to ionizing radiation.

**radiation intensity** The power radiated from an antenna per unit solid angle in a given direction.

**radiation ionization** Ionization of the atoms or molecules of a gas or vapor by the action of electromagnetic radiation. Also called radiation excitation.

**Radiation Laboratory** A government-financed electronic equipment development laboratory at the Massachusetts Institute of Technology, begun in November 1940 and discontinued in December 1945.

**radiation length** The mean path length required to reduce the energy of relativistic charged particles by the factor $1/e$ or 0.368 as they pass through matter. The radiation length for relativistic electrons in air is 0.5 cm in lead.

**radiation lobe** The portion of a radiation pattern that is bounded by one or two cones of nulls.

**radiation loss** The portion of the transmission loss that is due to radiation of r-f power from a transmission system.

**radiation pattern** A graphical representation of the radiation of an antenna as a function of direction. Also called field pattern.

**radiation potential** The voltage corresponding to the energy in electron-volts required to excite an atom or molecule and cause emission of one of its characteristic radiation frequencies.

**radiation pressure** The extremely small pressure exerted on a surface by electromagnetic radiation, or the larger pressure exerted on a surface or interface by a sound wave.

**radiation pyrometer** A pyrometer in which the radiant power from the object or source to be measured is focused on a thermocouple, thermopile, bolometer, or other suitable detector that provides electric output for an indicating instrument. Also called radiation thermometer.

**radiation resistance** The total radiated power of an antenna divided by the square of the effective antenna current measured at the point where power is supplied to the antenna.

**radiation shield** A shield or wall of lead or other material that effectively absorbs nuclear radiation.

**radiation sickness** The complex symptoms caused by overexposure to nuclear or other ionizing radiation, generally commencing a few hours after exposure, including nausea, vomiting, internal bleeding, and loss of white corpuscles.

**radiation survey** An investigation of nuclear installation or process factors that could give rise to a radiation hazard.

**radiation therapy** Treatment of disease with any type of radiation. Treatment of disease with ionizing radiation is called radiotherapy.

**radiation thermometer** *Radiation pyrometer.*

**radiation window** A window that is transparent to alpha, beta, gamma, and/or x-rays while protecting from foreign matter the item that it covers.

**radiative capture** A nuclear capture process whose prompt result is the emission of electromagnetic radiation only.

**radiator** 1. The part of an antenna or transmission line that radiates electromagnetic waves either directly into space or against a reflector for focusing or directing. 2. A body that emits radiant energy.

**radio** 1. The transmission of signals

through space by means of electromagnetic waves. Usually applied to the transmission of sound and code signals, although television and radar also depend on electromagnetic waves. 2. *Radio receiver.*

**radio-** 1. A prefix denoting radioactivity or a relationship to it, as in radiocarbon. 2. A prefix denoting the use of radiant energy, particularly radio waves.

**radioacoustic position-finding** A method of determining distances through water. A radio circuit is closed at the instant that a charge is exploded under water at one point. The times required for the radio signal to travel through air and the acoustic shock wave to travel through water to observing stations are measured, distances are computed from the time differences, and position is computed by triangulation.

**radioactinium** [symbol RdAc] A thorium isotope in the actinium series, produced naturally by beta decay of actinium-227. It emits alpha particles having a half-life of 18.8 days, and thereby changes to radium-223.

**radioactive** Pertaining to or exhibiting radioactivity. Also called active.

**radioactive chain** *Radioactive series.*

**radioactive contamination** A condition in which radioactive material has spread to places where it may harm persons, spoil experiments, or make products or equipment unsuitable or unsafe for some specific use.

**radioactive decay** The spontaneous transformation of a nuclide into one or more different nuclides. The process involves (a) the emission from the nucleus of alpha particles, electrons, positrons, and gamma rays, (b) the nuclear capture or ejection of orbital electrons, or (c) fission. The rate of radioactive decay is expressed in terms of the half-life of the nuclide. Also called decay, nuclear spontaneous reaction, radioactive disintegration, and radioactive transformation.

**radioactive decay constant** *Decay constant.*

**radioactive decay series** *Radioactive series.*

**radioactive disintegration** *Radioactive decay.*

**radioactive displacement law** A law governing radioactive transformations. When a nucleus emits an alpha particle, the new nucleus has an atomic number two less than the parent and a mass number of four less. When a nucleus emits a negative beta particle, the atomic number of the new nucleus is one greater than the parent and the mass number remains the same. The emission of a positron or the capture of an orbital electron decreases the atomic number by one without changing the mass number. Isomeric transition and gamma emission do not change atomic number or mass number. Also called displacement law.

**radioactive element** An element that disintegrates spontaneously, giving off various rays and particles. Examples include promethium, radium, thorium, and uranium.

**radioactive emanation** A radioactive gas given off by certain radioactive elements. Thus, radium gives off radon, thorium gives off thoron, and actinium gives off actinon. Also called emanation.

**radioactive equilibrium** The condition in which the rate of decay of the atoms of a radioactive parent is equal to the rate of formation of the atoms of the radioactive descendant. This condition can exist only when no activity longer-lived than that of the parent is interposed in the decay chain.

**radioactive half-life** The time required for a particular radioisotope to decrease to half its initial value.

**radioactive heat** *Radiogenic heat.*

**radioactive isotope** *Radioisotope.*

**radioactive nuclide** *Radionuclide.*

**radioactive period** The period of a nuclear reactor.

**radioactive series** A succession of nuclides, each of which transforms by radioactive disintegration into the next until a stable nuclide results. The first member is called the parent, the intermediate members are called daughters, and the final stable member is called the end product. Also called decay chain, decay family, decay series, disintegration chain, disintegration family, disintegration series, radioactive chain, radioactive decay series, series decay, and transformation series.

**radioactive standard** A sample of radioactive material, usually with a long half-life, in which the number and type of radioactive atoms at a definite reference time is known. Used for calibrating radiation-measuring equipment. Also called reference source.

**radioactive thickness gage** An instrument for measuring the thickness of the metal wall of a pipe or tank from one side, by directing a beam of gamma rays through the wall at an angle and measuring the amount of backscattering with a radiation detector. Thicker walls give greater scat-

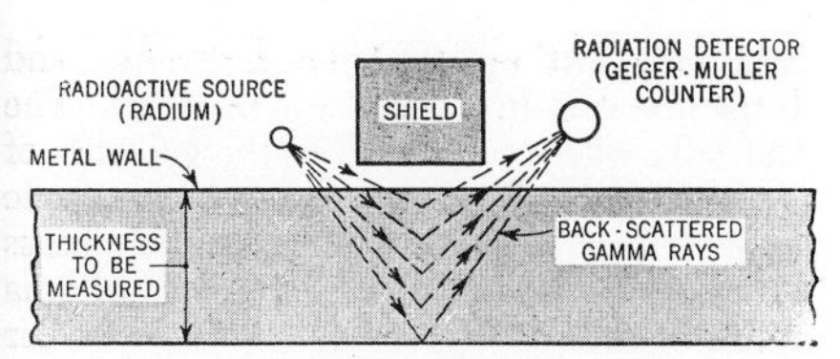

Radioactive thickness gage.

tering and a correspondingly higher meter reading.

**radioactive tracer** A small quantity of radioisotope used to trace the progress of a biological, chemical, or other process.

**radioactive transformation** *Radioactive decay.*

**radioactivity** Spontaneous nuclear disintegration, a property possessed by elements like radium, uranium, thorium, and their products. Alpha or beta particles and sometimes also gamma rays are emitted by disintegration of the nuclei of atoms. Also called activity.

**radio aid** *Radio navigation aid.*

**radio alert** A Department of Defense order to operate all radio stations in accordance with conelrad requirements for a period of time.

**radio all clear** A Department of Defense order to discontinue conelrad requirements.

**radio altimeter** An absolute altimeter that depends on the reflection of radio waves from the earth for the determination of altitude, as in a frequency-modulated radio altimeter and a radar altimeter. Also called electronic altimeter and reflection altimeter.

**radio altitude** *Radar altitude.*

**radio amateur civil emergency service** [abbreviated races] A temporary radio communication service carried on by licensed amateur radio stations while operating on specifically designated segments of amateur frequency bands under the direction of local, regional, or federal civil defense officials.

**radio astronomy** The study of radio waves emitted by astronomical bodies.

**radioautograph** Deprecated term for *autoradiograph.*

**radio-autopilot coupler** A coupler that permits direct control of an automatic pilot by a radio navigation aid in an aircraft, to give automatic flight.

**radio beacon** A nondirectional radio transmitting station in a fixed geographic location, emitting a characteristic signal from which bearing information can be obtained by a radio direction finder on a ship or aircraft. Some types operate continuously. Other types transmit only in response to an interrogation signal and may also provide range information. Also called aerophare, beacon, and radiophare.

**radio beam** A concentrated stream of r-f energy as used in radio ranges and microwave radio relays. A radar beam is a radio beam used for a particular application and in a particular manner.

**radio bearing** A bearing taken with respect to a radio transmitter, obtained with a radio direction finder.

**radiobiology** That branch of biology which deals with the effects of radiation on living tissue.

**radio blackout** *Radio fadeout.*

**radio bomb fuze** An electronic bomb fuze that is triggered by radio waves reflected from the target. It may be set to detonate at any desired distance from the target or ground by use of doppler effect.

**radio broadcast** A program broadcast from a radio transmitter for general reception.

**radio broadcasting** Radio transmission intended for general reception.

**radio broadcast station** A station that transmits radio programs in the broadcast band, intended to be received by the general public.

**radiocarbon** Carbon-14, a weak radioisotope used in biological and agricultural tracer studies. Half-life is 5,740 years.

**radiocarbon age** The age of a once-living material as calculated from the specific radioactivity of the carbon-14 remaining in it. Radiocarbon dating is possible because carbon-14 is produced in the atmosphere by cosmic rays and incorporated in all living objects. After death, the carbon-14 activity decays exponentially.

**radiocesium** Cesium-137, a radioisotope recovered from the waste of nuclear reactors. Useful for sterilizing food and as a substitute for radium in medical work. Half-life is 37 years.

**radio channel** A band of frequencies of a width sufficient to permit its use for radio communication. The width of a channel depends on the type of transmission and the tolerance for the frequency of emission.

**radiochemistry** Chemistry involving the use of radionuclides.

**radio circuit** 1. An arrangement of parts and connecting wires for radio purposes. 2. *Radio communication circuit.*

**radiocolloid** A grouping of radioactive atoms to form colloidal aggregates.

**radio command** A radio control signal to

which a guided missile or other remote-controlled vehicle or device responds.

**radio communication** Communication by means of radio waves, such as by radio facsimile, radiotelegraph, radiotelephone, and radioteletypewriter.

**radio communication circuit** A radio system for carrying out one communication at a time in either direction between two points. Also called radio circuit.

**radio compass** *Automatic direction finder.*

**radio control** The control of stationary or moving objects by means of signals transmitted through space by radio.

**radio deception** The use of radio to deceive the enemy, as by sending false dispatches or using enemy call signs.

**radio determination** The determination of position, or the obtaining of information relating to position, by means of the propagation properties of radio waves.

**radio direction finder** [abbreviated rdf] A radio aid to navigation that uses a rotatable loop or other highly directional antenna arrangement to determine the direction of arrival of a radio signal. Examples include aural-null direction finders and automatic direction finders. Also called direction finder.

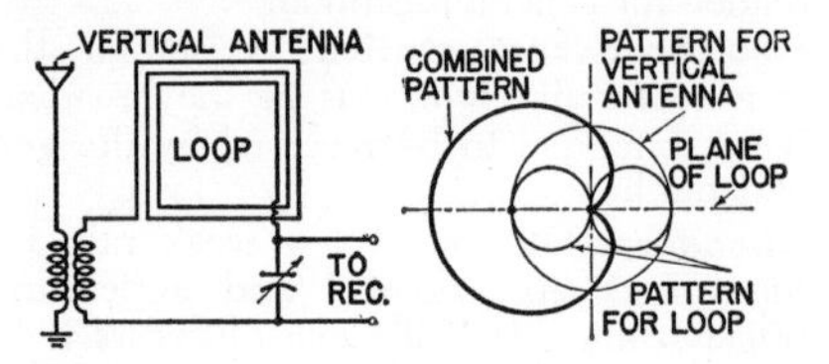


Radio direction finder antenna arrangement in which a unilateral radiation pattern is obtained by combining a vertical antenna with a loop antenna.

**radio direction-finding** [abbreviated rdf] Finding the direction to a radio station by using a radio direction finder.

**radioelement** An element tagged with one or more radioisotopes.

**radio engineering** The field of engineering that deals with the generation, transmission, and reception of radio waves and with the design, manufacture, and testing of associated equipment.

**radio facsimile** Facsimile communication by radio.

**radio fadeout** A sudden and abnormal increase in ionization in the lower layers of the ionosphere, causing increased absorption of radio waves passing through these regions. Signals at receivers then fade out or disappear. The fadeout occurs suddenly and may last up to 1 hour. Frequencies from about 3 to 10 mc are most affected, but only where part or all of the signal path is in daylight. Transmission on frequencies below about 100 kc is usually simultaneously improved. Also called blackout and radio blackout.

**radio fan marker beacon** *Fan marker beacon.*

**radio field strength** The effective value of the electric or magnetic field strength at a point due to the passage of radio waves of a specified frequency. Usually expressed as the electric field intensity in microvolts or millivolts per meter.

**radio field-to-noise ratio** The ratio of radio field strength to that of noise at a given location.

**radio fix** 1. Determination of the position of the source of radio signals by obtaining cross bearings on the transmitter with two or more radio direction finders in different locations, then computing the position by triangulation. 2. Determination of the position of a vessel or aircraft equipped with direction-finding equipment by obtaining radio bearings on two or more transmitting stations of known location and computing the position by triangulation. 3. Determination of position of an aircraft in flight by identification of a radio beacon or by locating the intersection of two radio beams.

**radio frequency** [abbreviated r-f] A frequency at which coherent electromagnetic radiation of energy is useful for communication purposes. Radio frequencies are designated by the FCC as follows: very low frequency, below 30 kc; low frequency, 30–300 kc; medium frequency, 300–3,000 kc; high frequency, 3–30 mc; very high frequency, 30–300 mc; ultrahigh frequency, 300–3,000 mc; superhigh frequency, 3,000–30,000 mc; extremely high frequency, 30,000–300,000 mc.

**radiogenic** Produced by radioactive transformation.

**radiogenic heat** Heat produced within the earth by the disintegration of radioactive nuclides. Also called radioactive heat.

**radiogoniometer** A goniometer used as part of a radio direction finder. In the Bellini-Tosi system, two loop antennas positioned at right angles to each other are connected to two field coils in the radiogoniometer. Bearings are obtained by a rotatable search coil that is inductively coupled to the field coils.

**radiogram** A message transmitted by radio.

**radiograph** A photographic image produced by a beam of penetrating ionizing radiation after passing through an object, showing the variations in density or absorption in the object. When produced by x-rays, a radiograph is called an x-ray photograph.

**adiographic putty** A blocking medium used in radiography to reduce the effect of scattered radiation and to shield portions of the x-ray film that would otherwise be overexposed.

**adiographic stereometry** The process of finding the position and dimensions of details within an object by measurements made on radiographs taken from different directions.

**adiography** Photography involving the use of x-rays, gamma rays, and other penetrating ionizing radiations to produce shadow images corresponding to differences in thickness, density, or absorption in the subject being examined. Widely used for medical and dental diagnosis and for nondestructive internal inspection of metal and other objects.

**adio-guided bomb** An aerial bomb guided by radio control from outside the missile.

**adio homing beacon** *Homing beacon.*

**adio horizon** The locus of points at which direct rays from a transmitter become tangential to the surface of the earth. The distance to the radio horizon is affected by atmospheric refraction.

**adio inertial guidance** A radio command guidance system having an auxiliary inertial guidance system in the missile for partial guidance in case of radio guidance failure or to provide more up-to-date data for correcting radar guidance information.

**adio interference** *Interference.*

**adioisotope** An artificially produced isotope that is radioactive. Many elements have as many as 10 radioisotopes, produced in a cyclotron or by neutron bombardment in a nuclear reactor. Widely used in industry, medicine, and other fields as radioactive tracers and as sources of ionizing radiation. Also called radioactive isotope.

**adio knife** A surgical knife that uses a high-frequency electric arc at its tip to cut tissue and simultaneously sterilize the edges of the wound.

**adio landing beam** A radio beam used for vertical guidance of aircraft during descent to a landing surface.

**adiolead** A radioisotope of lead.

**adio line of position** A line of position obtained with a radio direction finder.

**adio link** A radio system used to provide a communication or control channel between two specific points.

**radiolocation** 1. Determination of one or more navigation coordinates by means of measurements based on the known velocity of radio waves. 2. Obsolete British term for *radar.*

**radio log** A log of radio messages sent and received, together with other pertinent information, maintained by radio operators.

**radiological** Pertaining to nuclear radiation, radioactivity, and atomic weapons.

**radiological defense** Defense against the effects of radioactivity from atomic weapons, including detection and measurement of radioactivity, protection of persons from radioactivity, and decontamination of areas and equipment.

**radiological indicator** An indicator designed to display the occurrence of radioactivity above a predetermined value.

**radiological warfare** Warfare involving weapons that produce radioactivity, such as atomic bombs and shells.

**radiologist** A medical specialist skilled in the use of x-rays, gamma rays, and other penetrating ionizing radiations.

**radiology** The science that deals with the production and use of x-rays, gamma rays, and other penetrating ionizing radiations.

**radioluminescence** Luminescence produced by radiant energy, as by x-rays, radioactive emissions, alpha particles, or electrons.

**radio magnetic indicator** [abbreviated rmi] An indicating instrument that presents a display combining vehicle heading, relative bearing, magnetic bearing, and omnibearing of the radio station being utilized for navigation purposes.

**radioman** *Radio operator.*

**radio marker beacon** *Marker.*

**radio metal locator** *Metal detector.*

**radiometallography** Examination of the crystalline structure and other characteristics of metals and alloys with x-ray equipment.

**radiometeorograph** *Radiosonde.*

**radiometer** 1. An instrument for measuring radiant energy. Examples include the bolometer, microradiometer, and thermopile. 2. A receiver for detecting microwave thermal radiation and similar weak wideband signals that resemble noise and are obscured by receiver noise. No signal periodicity and no prior knowledge of signal characteristics are required. Examples include the Dicke radiometer, subtraction-type radiometer, and two-receiver radiometer. Also called microwave radiometer and radiometer-type receiver.

**radiometer-type receiver** *Radiometer.*

**radiometric analysis** Quantitative chemical analysis that is based on measurement of the absolute disintegration rate of a radioactive component having a known specific activity.

**radiometry** The measurement of radiation, as with a radiometer.

**radiomicrometer** Deprecated term for *microradiometer.*

**radio multiplexing** 1. Dividing a radio channel into a number of voice or code channels through frequency division or time division. 2. Connecting two or more transmitters or receivers to the same antenna through appropriate coupling networks.

**radio navigation** Navigation by means of radio signals, using such equipment as radio direction finders, radio ranges, radio beacons, and loran.

**radio navigation aid** A navigation aid that uses radio signals, as contrasted to a radar navigation aid. Also called radio aid.

**radio navigation guidance** The guidance or control of a guided missile along a course established by external radio transmitters.

**radio net** A net of radio stations established for communication purposes.

**radionuclide** A substance that exhibits radioactivity. Also called radioactive nuclide.

**radio operator** A person who operates radio transmitting and receiving equipment. Also called radioman.

**radiopaque** Not appreciably penetrable by x-rays or other forms of radiation.

**radiophare** *Radio beacon.*

**radiophone** *Radiotelephone.*

**radiophoto** 1. A photograph transmitted by radio to a facsimile receiver. 2. *Facsimile.*

**radiophotoluminescence** Luminescence exhibited by minerals such as fluorite and kunzite as a result of irradiation with beta and gamma rays followed by exposure to light.

**radio propagation prediction** A prediction of the quality or nature of radio propagation as influenced by such factors as sunspots and seasonal changes, published periodically by the National Bureau of Standards.

**radio prospecting** Use of radio and electronic equipment to locate mineral or oil deposits.

**radio proximity fuze** A proximity fuze that contains a miniature radio transmitter and uses radio echoes from a target to trigger the fuze within predetermined limits of distance from the target.

**radio range** A radio transmitting facility that radiates signals which can be used by aircraft to determine bearings from the transmitting site. The A-N radio range provides four courses. Also called radio-range beacon and range.

**radio-range beacon** *Radio range.*

**radio-range leg** One of the courses or beams in a four-course radio range. Also called leg.

**radio-range monitor** An instrument that automatically monitors the signal from a radio-range beacon, giving a warning to attendants when the transmitter deviates a specified amount from its correct bearings and transmitting a distinctive warning to approaching planes when trouble exists at the beacon.

**radio receiver** A receiver that converts radio waves into intelligible sounds or other perceptible signals. Also called radio, radio set, and receiving set.

**radio reception** Reception of messages, programs, or other intelligence by radio.

**radio relay system** *Radio repeater.*

**radio repeater** A repeater that acts as an intermediate station in transmitting radio communication signals or radio programs from one fixed station to another. It serves to extend the reliable range of the originating station. A microwave repeater is an example. Also called radio relay system and relay system.

**radioresistance** The relative resistance of cells, tissues, organs, organisms, or substances to the injurious action of radiation.

**radio scattering** *Scattering.*

**radiosensitive** Sensitive to damage by radiant energy.

**radio serviceman** A serviceman who is qualified to repair and maintain radio equipment. Also called radio technician.

**radio set** 1. A radio receiver and radio transmitter used together for two-way communication. 2. *Radio receiver.* 3. *Radio transmitter.*

**radio signal** A signal transmitted by radio.

**radio silence** A period during which transmissions by a radio station are stopped, such as to permit reception of signals from other stations or permit reception of weak distress signals.

**radiosonde** [pronounced radio sond] A meteorograph combined with a radio transmitter. When carried aloft by a balloon, it transmits radio signals that can be recorded at a ground station and interpreted in terms of the pressure, temperature, and humidity at regular intervals during the ascent. Also called radiometeorograph. The equipment is lowered by parachute when the balloon bursts.

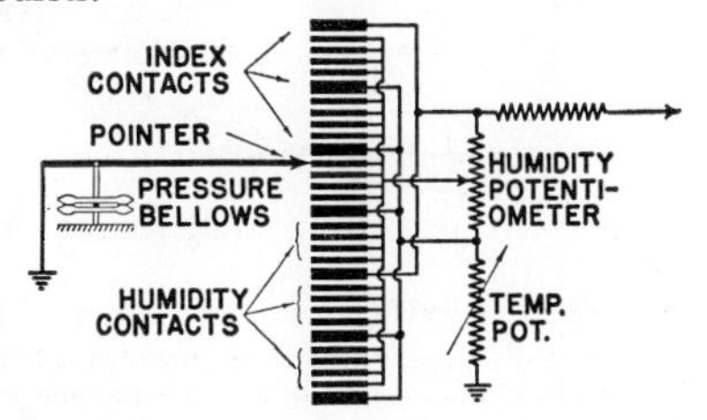


Radiosonde sensing and switching elements, used to feed amplifier connected to transmitting antenna.

**radiosonde balloon** A sounding balloon carrying a radiosonde.

**radio sonobuoy** *Sonobuoy.*

**radio spectrum** The entire range of frequencies in which useful radio waves can be produced, extending from the audio range to about 300,000 mc. The radio spectrum is divided into eight bands (see band or radio frequency). Also called r-f spectrum.

**radio star** A source of radio signals among the stars, within or beyond the galaxy. Some radio stars appear to contain gases in violent motion.

**radio station** A station equipped to engage in radio communication or radio broadcasting.

**radio-station interference** Interference caused by radio stations other than that from which reception is desired.

**radio strontium** *Strontium-90.*

**radio system** A complete radio equipment installation providing multichannel communication between two points.

**radio technician** *Radio serviceman.*

**radiotelegraph** Pertaining to telegraphy over radio channels.

**radiotelegraph transmitter** A radio transmitter that is capable of handling code signals (type A1 and B emissions).

**radiotelegraphy** Telegraphy involving the use of radio waves in place of wire lines. The international Morse code is generally used in radiotelegraphy.

**radiotelephone** 1. Pertaining to telephony over radio channels. 2. A radio transmitter and radio receiver used together for two-way telephone communication by radio. Also called radiophone.

**radiotelephone distress call** The word "mayday," corresponding to the French pronunciation of m'aider, spoken under the same conditions that the signal SOS would be transmitted in code by radiotelegraphy.

**radiotelephone transmitter** A radio transmitter capable of handling a-f modulation, such as voice and music.

**radiotelephony** Two-way voice communication (telephony) carried on by means of radio waves, without connecting wires between stations.

**radio telescope** A sensitive radio receiver used with a large and highly directional antenna to receive signals from radio stars.

**radioteletypewriter** A teletypewriter and the associated equipment needed for operation over a radio channel rather than over wires.

**radiotherapy** Treatment of disease with any ionizing radiation, including x-rays and gamma rays but not ultraviolet rays.

**radiothermoluminescence** Luminescence exhibited by certain vitreous and crystalline substances as a result of irradiation with beta and gamma rays followed by heating.

**radiothermy** *Diathermy.*

**radiothorium** [symbol RdTh] A thorium isotope having mass number 228 and a half-life of 1.90 years. It is produced naturally by beta decay of actinium-228, and emits gamma particles to give radium-224.

**radio transmission** The transmission of signals through space at radio frequencies by means of radiated electromagnetic waves.

**radio transmitter** A transmitter that produces r-f power for transmission through space in the form of radio waves. Also called radio set.

**radiotransparent** Permitting passage of x-rays or other forms of radiation.

**radiotropism** Turning or bending of a plant or other organism in response to some form of radiation.

**radio tube** *Electron tube.*

**radio watch** *Watch.*

**radio wave** An electromagnetic wave produced by reversal of current in a conductor at a frequency in the range from about 10 kc to 30,000,000 mc. Radio waves travel through space at approximately the speed of light. Also called Hertzian wave.

**radio wave propagation** The transfer of energy through space by electromagnetic radiation at radio frequencies.

**radio window** A band of frequencies extending from about 6 to 30,000 mc, in which radiation from the outer universe can enter and travel through the atmosphere of the earth.

**radium** [symbol Ra] A highly radioactive metallic element that gives off alpha, beta, and gamma rays. Atomic number is 88.

**radium age** The age of a mineral as calculated from the numbers of radium atoms present originally, now, and when equilibrium is established with ionium.

**radium cell** A sealed thin-wall tube containing radium.

**radium needle** A radium cell in the form of a needle, usually of platinum-iridium or gold alloy, designed primarily for insertion in tissue.

**radium parameter** The effective radius of a nucleus divided by the cube root of its mass number. The value is approximately the same for all nuclei and is about $1.4 \times 10^{-13}$ cm.

**radium plaque** A radium container in which the radium is distributed over a surface. The shielding is usually small in one direction so as to permit transmission of beta rays as well as gamma rays.

**radium therapy** Radiotherapy using the radiations from radium.

**radix** *Base.*

**radix notation** A positional notation in which the successive digits are interpreted as coefficients of successive integral powers of a number called the radix or base. The represented number is equal to the sum of this power series. Thus, 5,762 is the sum of the power series $5 \times 10^3 + 7 \times 10^2 + 6 \times 10^1 + 2 \times 10^0$, where 10 is the base.

**radix point** *Point.*

**radnos** [transposition of NO RADio plus S] A radio fadeout encountered chiefly in arctic regions, considered to be caused by solar explosions, sunspots, or the aurora borealis.

**radome** [RAdar DOME] A protective housing for a radar antenna, made from dielectric material that is transparent to r-f energy. When it protrudes beyond the normal skin contours, it is called a blister.

**radon** [symbol Rn] *Emanation.*

**radon seed** A small metal or glass tube containing radon.

**rad per unit time** A unit of absorbed dose rate.

**radux** A long-distance continuous-wave low-frequency navigation system of the phase comparison type, providing hyperbolic lines of position.

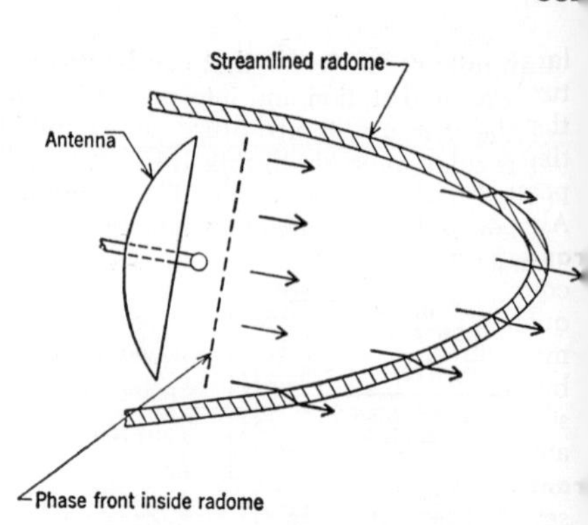


Radome and radar antenna in nose of aircraft, showing how waves passing through radome walls at an angle are refracted and distorted.

**railing** The jamming of radar transmissions by transmitters at a pulse rate of 50 to 150 kc, causing images resembling fence railings to appear on radar screens.

**railroad radio service** A radio communication service used in connection with the operation and maintenance of a railroad common carrier.

**rainbow generator** A signal generator that generates a signal which, when fed into a color television receiver, produces the entire color spectrum on the screen, with the colors merging together.

**rain return** Clutter due to rain.

**ramark** [RAdar MARKer] A fixed radar beacon that emits radar waves continuously to provide a bearing indication to radar-equipped ships and aircraft.

**Ramsauer effect** The low attenuation of slow-moving electrons by inert gases.

**Rand Corporation** [Research ANd Development] A nongovernmental nonprofit organization engaged in research for the welfare and national security of the United States.

**random access** Access to computer storage under conditions wherein the next register from which information is to be obtained is chosen at random.

**random-access programing** Computer programing without regard for the time required for access to stored information.

**random coincidence** *Accidental coincidence.*

**random error** An error that can be predicted only on a statistical basis.

**random interlace** Interlace based on less precise timing of sweep frequencies than is required for television broadcasts. Sometimes used in industrial television.

**random noise** Noise characterized by a

large number of overlapping transient disturbances occurring at random, such as thermal noise and shot noise. It is sometimes produced intentionally for test purposes or for jamming enemy transmissions. Also called fluctuation noise.

**random-noise testing** Testing in which a complex wave having randomly varying frequencies and amplitudes is applied to a mechanical shake table. The signal may be obtained experimentally, as from a missile telemetering record, or may be generated electronically.

**random number** A number formed by a set of digits in which each successive digit is equally likely to be any of *n* digits to the base *n*.

**random variable** A discrete or continuous variable that may assume any one of a number of values, each of which has a fixed probability.

**random walk** The path followed by a particle as it makes random scattering collisions in a medium.

**random winding** A coil winding in which the turns are positioned haphazardly rather than in layers.

**range** 1. The distance from a radar set or weapon to a target. 2. The distance capability of an aircraft, missile, gun, radar, or radio transmitter. 3. The maximum thickness of a given medium that can be penetrated by a given ionizing particle. 4. The difference between the maximums and minimums of a variable quantity. 5. A line defined by two fixed landmarks, used for missile, vehicle, and other test purposes. 6. A line of bearing defined by a radio range. 7. *Radio range.*

**range-amplitude display** A radar display in which a time base provides the range scale on which echoes appear as deflections normal to the base. The base is usually a straight line as in the A display, or a circle as in the J display.

**range-bearing display** *B display.*

**range circle** A radar range marker in the form of a circle.

**range control** Control exercised over the range of a guided missile.

**range discrimination** Deprecated term for *distance resolution.*

**range-energy relation** The graphical relation between range and energy of specified particles.

**range-height indicator display** [abbreviated rhi display] A radar display that presents visually the scalar distance between a reference point and a target, along with the vertical distance between a reference plane and the target.

**range mark** *Radar range marker.*

**range marker generator** A signal generator that generates the signal required for the production of radar range markers on a radarscope. Its action is initiated by the sync pulse that starts the time base.

**range resolution** Deprecated term for *distance resolution.*

**range selector** A control used to select the range scale on a radar indicator.

**range straggling** The variation in the range of particles having the same initial energy.

**ranging** Determining distance.

**rank** The number of independent cut-sets that can be selected in a network. The rank is equal to the number of nodes minus the number of separate parts.

**Rankine** [abbreviated R] An absolute temperature scale that uses Fahrenheit degrees but with the entire scale shifted so 0°R is at absolute zero. In this scale, water freezes at 459.6°R and boils at 639.6°R. Add 427.6 to a Fahrenheit value to get the corresponding Rankine value.

**rapcon** [Radar APproach CONtrol] Use of radar for direct control of aircraft in the vicinity of an airport and during the approach to the runway. Both surveillance radar and precision approach radar are required.

**Raphael bridge** A type of slide-wire Wheatstone bridge used for locating faults in transmission lines.

**rapid memory** The section of a computer from which stored information can be obtained most rapidly. Also called high-speed memory.

**rapid scanning** Scanning with a narrow radar beam at the rate of 10 sweeps per second or more, as required for tracking fast-moving targets.

**rare earth** An element having an atomic number in the range from 57 to 71 inclusive. The rare earths are cerium, dysprosium, erbium, europium, gadolinium, holmium, lanthanum, lutecium, neodymium, praseodymium, promethium, samarium, terbium, thulium, and ytterbium.

**Rascal** An Air Force air-to-surface guided missile capable of carrying a nuclear warhead. Designed for launching from B-47 Stratojet and other bombers. Range is over 75 miles. Command guidance is used.

**raster** A predetermined pattern of scanning lines that provides substantially uniform coverage of an area. In television the raster is seen as closely spaced parallel lines, most evident when there is no picture.

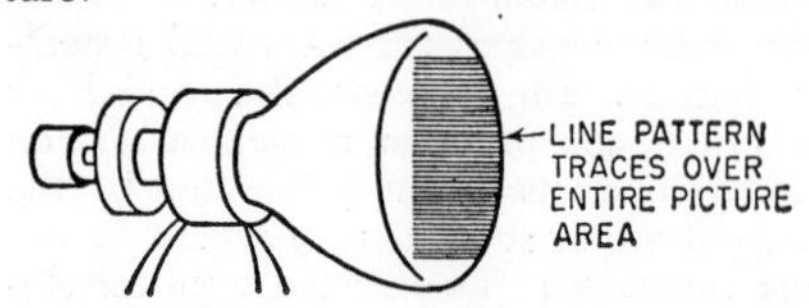


Raster is line pattern seen on screen of television picture tube when no video signal is present.

**raster burn** A change in the characteristics of the scanned area on the target of a camera tube, resulting in a spurious signal when a larger or tilted raster is scanned.

**rat** Abbreviation for *rocket-assisted torpedo.*

**ratchet relay** A stepping relay actuated by an armature-driven ratchet.

**rate action** *Derivative action.*

**rate control** Control of the rate of change of the independent variable in an automatic control system. Also called proportional response and throttling control.

**rated accuracy** The advertised accuracy of a manufactured instrument.

**rated coil current** The steady-state coil current at which a relay is designed to operate.

**rated coil voltage** The coil voltage at which a relay is designed to operate.

**rated contact current** The current that contacts are designed to carry for their rated life.

**rated output** The output power, voltage, current, or other value at which a machine, device, or apparatus is designed to operate under specified normal conditions.

**rated power output** The normal r-f power output capability of a transmitter under optimum conditions of adjustment and operation, as specified by its manufacturer.

**rate-grown junction** A grown junction produced by varying the rate of semiconductor crystal growth periodically and using a melt that contains both n-type and p-type impurities, so the two types of impurities alternately predominate. Also called graded junction.

**rate-grown transistor** A transistor having rate-grown junctions. Also called graded-junction transistor.

**rate gyroscope** A spinning or vibratory gyroscope that measures the rate of change of direction of an aircraft.

**rate meter** *Counting-rate meter.*

**rate-of-climb indicator** An instrument that indicates the rate of climb or descent of an aircraft, usually in feet per minute.

**rate of closure** The speed at which two airborne aircraft or other moving objects close the distance between them. With aircraft approaching each other, the rate of closure is the sum of their speeds.

**rate of decay** The time rate at which the sound pressure level, velocity level, or sound-energy density level is decreasing at a given point and at a given time. The practical unit is the decibel per second.

**rate-of-turn control** A gyroscopic instrument that furnishes a rate-of-turn signal to an automatic pilot system in an aircraft.

**rating** A designation of an operating limit for a machine, apparatus, or device used under specified conditions.

**ratio** The value obtained when one quantity is divided by another of the same kind, to indicate their relative proportions.

**ratio arms** Two adjacent arms of a Wheatstone bridge, designed so they can be set to provide a variety of indicated resistance ratios.

**ratio control** Control in which a predetermined ratio between two physical quantities is maintained.

**ratio detector** An f-m detector circuit that uses two diodes and requires no limiter at its input. The audio output is determined by the ratio of two developed i-f voltages whose relative amplitudes are a function of frequency.

**ratio meter** A meter that measures the quotient of two electrical quantities. The deflection of the meter pointer is proportional to the ratio of the currents flowing through two coils.

**rationalized unit** A unit in a system of measurement that is designed to minimize occurrence of the constant $4\pi$ in equations.

**ratio of transformation** The ratio of the secondary voltage of a transformer to the primary voltage under no-load conditions, or the corresponding ratio of currents in a current transformer.

**ratio-type telemeter** A telemeter that uses the phase or magnitude relations of two or more electrical quantities as the translating means. Also called position-type telemeter (deprecated).

**rat race** *Hybrid ring.*

**rawin** [RAdar WINd or RAdio WINd; pronounced ray win] 1. Determination of wind direction and velocity by radar or by radio direction-finding in conjunction with a radiosonde, radiosonde balloon, or a balloon carrying a radar reflector. 2. Wind information gathered by using radar tracking or radio direction-finding in connection with a specially equipped balloon.

**rawin balloon** A radiosonde balloon or other specially equipped balloon used in determining the movement and velocity of winds.

**rawinsonde** A radiosonde used in rawin.

**ray** *Beam.*

**ray-control electrode** [symbol RC] The electrode used to control the position of the electron beam on the screen of a cathode-ray tuning indicator.

**raydist** A navigation system in which a continuous-wave signal emitted from a vehicle is received at three or more ground stations. The received signals are compared in phase to determine the position of the vehicle.

**rayl** [named in honor of Lord Rayleigh] The magnitude of a specific acoustic resistance, reactance, or impedance for which a sound pressure of 1 microbar produces a linear velocity of 1 centimeter per second.

**Rayleigh cycle** A cycle of magnetization that does not extend beyond the initial portion of the magnetization curve, between zero and the upward bend. In this region the permeability is low and there is little hysteresis.

**Rayleigh disk** An acoustic radiometer used to measure particle velocity. A thin disk suspended by its edge from a fine fiber tends to take a position perpendicular to the horizontal component of sound particle velocity.

**Rayleigh distribution** A mathematical statement of a natural distribution of random variables.

**Rayleigh line** A spectrum line in scattered radiation that has the same frequency as the corresponding incident radiation. It arises from ordinary or Rayleigh scattering.

**Rayleigh scattering** Selective scattering of light by very small particles suspended in air, as by dust.

**Rayleigh wave** A type of wave that may be propagated near the surface of a solid and is characterized by elliptical motion of particles.

**razon** [Range AZON] A World War II radio-controlled missile similar to azon but having provisions for radio control of range also.

**rbe** Abbreviation for *relative biological effectiveness.*

**rbe dose** A radiation equal numerically to the product of the dose in rads and an agreed conventional value of rbe with respect to a particular form of radiation effect. The unit of rbe dose is the rem.

**RC** 1. Symbol for *ray-control electrode.* Used on tube-base diagrams. 2. Alternate abbreviation for *resistance-capacitance.*

**r-c** Abbreviation for *resistance-capacitance.*

**RCA-licensed** Manufactured under a licensing agreement that permits use of patents owned by Radio Corporation of America.

**r-c amplifier** *Resistance-coupled amplifier.*

**r-c circuit** Abbreviation for *resistance-capacitance circuit.*

**r-c coupling** *Resistance coupling.*

**r-c differentiator** Abbreviation for *resistance-capacitance differentiator.*

**r-c filter** Abbreviation for *resistance-capacitance filter.*

**rcg circuit** Abbreviation for *reverberation-controlled gain circuit.*

**rcm** Abbreviation for *radar countermeasure.*

**r-c oscillator** Abbreviation for *resistance-capacitance oscillator.*

**rd** Abbreviation for *rutherford.*

**RdAc** Symbol for *radioactinium.*

**rdf** 1. Abbreviation for *radio direction finder.* 2. Abbreviation for *radio direction-finding.*

**R display** A radarscope display that is essentially an expanded A display, in which an echo can be magnified for close examination.

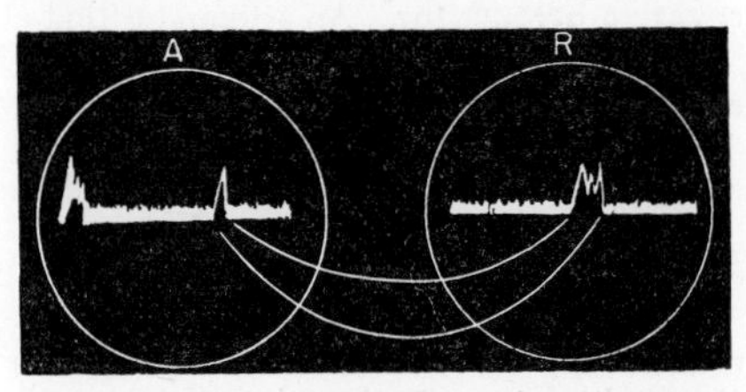


R display used as supplemental indicator on some late-model radars, with corresponding A display for same echo shown at left.

**RdTh** Symbol for *radiothorium.*

**reactance** [symbol $X$] The opposition offered to the flow of alternating current by pure inductance or capacitance in a circuit, expressed in ohms. It is the component of impedance that is not due to resistance. Inductive reactance is due to inductance, while capacitive reactance is due to capacitance.

**reactance amplifier** *Parametric amplifier.*

**reactance modulator** A modulator whose reactance may be varied in accordance with the instantaneous amplitude of the modulating voltage. This is normally an electron-tube circuit, used to produce phase or frequency modulation.

**reactance tube** An electron tube connected and operated in such a way that it acts as an inductive or capacitive reactance. The magnitude of the reactance can be changed by adjusting the grid bias voltage. Used in reactance modulators and in automatic frequency control of oscillators.

**reactatron** A low-noise microwave amplifier using a semiconductor diode.

**reaction** 1. An action wherein one or more substances are changed into one or more new substances, as in a nuclear reaction. 2. British term for *positive feedback*.

**reaction motor** A synchronous motor whose rotor contains salient poles but has no windings and no permanent magnets.

**reaction rate** The rate at which fission takes place in a nuclear reactor. This rate is fundamentally expressed as the number of nuclei undergoing fission per unit time. The reaction rate determines the reactivity of a nuclear reactor.

**reactivation** Application of a higher voltage than normal to the thoriated filament of an electron tube for a few seconds, to bring a fresh layer of thorium atoms to the filament surface and thereby improve electron emission.

**reactive** Pertaining to inductive or capacitive reactance.

**reactive attenuator** An attenuator that absorbs very little energy.

**reactive factor** The ratio of reactive power to apparent power.

**reactive-factor meter** A meter for measuring and indicating reactive factor.

**reactive load** A load having inductive or capacitive reactance.

**reactive power** The power value obtained by multiplying together the effective value of current in amperes, the effective value of voltage in volts, and the sine of the angular phase difference between current and voltage. Also called reactive volt-amperes and wattless power. The unit of reactive power is the var.

**reactive volt-ampere** *Volt-ampere reactive.*

**reactive volt-ampere meter** *Varmeter.*

**reactive volt-amperes** *Reactive power.*

**reactivity** A measure of the departure of a reactor from critical. Positive values of reactivity correspond to reactors above critical, while negative values correspond to reactors below critical. Reactivity is equal to the ratio of the number of neutrons present at a given time to the number present one generation or lifetime earlier. Also called multiplication constant.

**reactor** 1. A device that introduces either inductive or capacitive reactance into a circuit, such as a coil or capacitor. 2. *Nuclear reactor.*

**reactor period** The time required for the power level of a reactor to change by the factor $e = 2.71828$.

**reactor-rectifier amplifier** *Self-saturating circuit.*

**reactor-start motor** A split-phase motor designed for starting with a reactor in series with the main winding. The reactor is short-circuited or otherwise made ineffective and the auxiliary circuit opened when the motor has attained a predetermined speed.

**read** 1. To acquire information, usually from some form of storage in a computer. 2. To generate an output corresponding to the pattern stored in a charge storage tube. 3. To understand clearly, as in radio communication.

**read-around number** The number of times that priming, writing, reading, or erasing operations can be performed on storage elements adjacent to any given element in a charge storage tube without loss of information from that element. Also called read-around ratio (deprecated).

**read-around ratio** Deprecated term for *read-around number.*

**reading** 1. The indication shown by an instrument. 2. To observe the readings of one or more instruments.

**reading speed** The rate of reading successive storage elements in a charge storage tube.

**read number** The number of times that a storage element is read without rewriting in a charge storage tube.

**readout** The presentation of output information by means of lights, printed or punched tape or cards, or other methods.

**read pulse** A pulse that causes information to be acquired from a magnetic cell or cells.

**read time** The time interval between the instant at which information is called for from storage and the instant at which delivery is completed in a computer. Also called access time.

**ready-to-receive signal** A signal sent back to a facsimile transmitter to indicate that a facsimile receiver is ready to accept the transmission.

**real power** The component of apparent power that represents true work. Real power is expressed in watts, and is equal to volt-amperes multiplied by the power factor.

**real time** The performance of a computation during the time of a related physical process, so the results are available for guiding the physical process.

**rear projection** A projection television system in which the picture is projected on a ground-glass screen for viewing from the opposite side of the screen.

**rebecca** The airborne interrogator-responsor of a rebecca-eureka system.

**rebecca-eureka system** An aircraft radar homing system in which an airborne interrogator-responsor (rebecca) homes on a ground radar beacon (eureka) that has been dropped or set up in advance. The system can also give the distance from the rebecca radar to the eureka beacon.

**rebecca-H system** A British H system using a rebecca radar that has been modified to display two beacon responses simultaneously.

**rebroadcast** Repetition of a radio or television program at a later time.

**recalescent point** The temperature at which there is a sudden liberation of heat as a heated metal is cooled.

**receiver** The complete equipment required for receiving modulated radio waves and converting them into the original intelligence, such as into sounds or pictures, or converting to desired useful information as in a radar receiver.

**receiver bandwidth** The frequency range between the half-power points on the frequency-response curves of a receiver.

**receiver gating** The application of operating voltages to one or more stages of a receiver only during that part of a cycle of operation when reception is desired.

**receiver noise figure** The ratio of noise in a given receiver to that in a theoretically perfect receiver.

**receiver primaries** *Display primaries.*

**receiver pulse delay** *Transducer pulse delay.*

**receiver radiation** Radiation of interfering electromagnetic fields by the oscillator of a receiver.

**receiver synchro** *Synchro receiver.*

**receiving antenna** An antenna used to convert electromagnetic waves to modulated r-f currents.

**receiving current sensitivity** *Free-field current response.*

**receiving set** *Radio receiver.*

**receiving station** A station used for reception of radio signals or messages.

**receiving voltage sensitivity** *Free-field voltage response.*

**receptacle** *Outlet.*

**reciprocal-energy theorem** A theorem due to Rayleigh: If an electromotive force $E_1$ in one branch of a circuit produces a current $I_2$ in any other branch, and if an electromotive force $E_2$ inserted in this other branch produces a current $I_1$ in the first branch, then $I_1E_1 = I_2E_2$. This is closely related to the reciprocity theorem.

**reciprocal impedance** Two impedances $Z_1$ and $Z_2$ are said to be reciprocal impedances with respect to an impedance Z (invariably a resistance) if they satisfy the equation $Z_1Z_2 = Z^2$.

**reciprocal transducer** A transducer that satisfies the reciprocity principle.

**reciprocal velocity region** The energy region in which the capture cross section for neutrons by a given element is inversely proportional to neutron velocity.

**reciprocity constant** The sensitivity of a transducer used as a microphone divided by the sensitivity of the same transducer used as a source of sound.

**reciprocity principle** The relation between the sensitivity of a reversible electroacoustic transducer when used as a microphone and the sensitivity when used as a source of sound is independent of the type and construction of the transducer.

**reciprocity theorem** If a voltage located at one point in a network produces a current at any other point in the network, the same voltage acting at the second point will produce the same current at the first point.

**reclosing relay** A relay that functions to reclose a circuit automatically under certain conditions.

**recognition differential** The signal strength above noise level that gives a 50% probability of detection of an aural signal.

**recoil electron** An electron that has been set into motion by a collision.

**recoil nucleus** A nucleus that recoils when it collides with a nuclear particle or ejects a nuclear particle.

**recoil particle** A particle that has been set into motion by a collision or by a process involving the ejection of another particle.

**recoil radiation** Radiation emitted during nuclear disintegration in such a way that there is an observable recoil of the nucleus.

**recombination** The combination and resultant neutralization of particles or objects having unlike charges, such as a hole and an electron or a positive ion and a negative ion.

**recombination velocity** The normal component of the electron or hole current density on a semiconductor surface divided by the excess electron or hole charge density at the surface.

**recommutation** Failure of load current to be completely commutated from one ignitron to another within the required time, with the result that current is commutated back to the original tube.

**reconditioned-carrier receiver** *Exalted-carrier receiver.*

**reconnaissance satellite** An earth satellite designed to provide strategic information, as by television or photography.

**reconstituted conductive material** Conductive material formed by compressing finely divided particles.

**reconstituted mica** Mica sheets or shaped objects made by breaking up scrap natural mica, combining with a binder, and pressing into forms suitable for use as insulating material.

**recontrol time** *Deionization time.*

**record** 1. To preserve for later reproduction or reference. 2. *Recording.*

**record changer** A record player that plays a number of records automatically in succession.

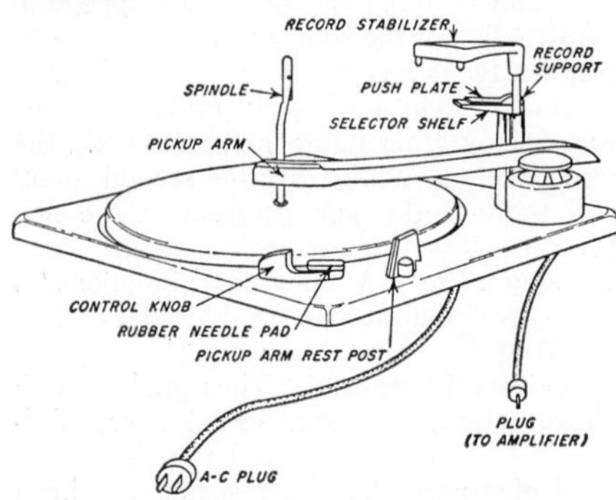

Record-changer construction.

**record compensator** An adjustable filter used in audio systems to compensate for the differing frequency-response curves used by phonograph-record manufacturers for accentuating treble frequencies and attenuating bass frequencies during the recording process. Also called record equalizer.

**recorded program** A radio program that uses phonograph records, electric transcriptions, magnetic tapes, or other means of reproduction.

**recorded spot** The image of the recording spot on the record sheet in a facsimile system.

**recorded tape** 1. A recording that is commercially available on magnetic tape. Also called prerecorded tape. 2. Any magnetic tape that has been recorded.

**record equalizer** *Record compensator.*

**recorder** An instrument that makes a permanent record of varying electrical quantities or signals. A common industrial version records one or more quantities as a function of another variable, usually time. Other types include the cathode-ray oscillograph, facsimile recorder, kinescope recorder, magnetic-tape recorder, and sound film recorder. Also called oscillograph and recording instrument.

**record head** *Recording head.*

**recording** 1. Any process for preserving signals, sounds, data, or other information for future reference or reproduction, such as disk recording, facsimile recording, ink-vapor recording, magnetic-tape or wire recording, and photographic recording. 2. The end product of a recording process, such as the recorded magnetic tape, disk, or record sheet. Also called record.

**recording blank** *Recording disk.*

**recording camera** A camera designed to photograph radarscope displays and instrument readings for record purposes.

**recording channel** One of a number of independent recorders in a recording system or independent recording tracks on a recording medium.

**recording characteristic** A graph showing the intentional attenuation of bass frequencies and accentuation of treble frequencies used in making a disk recording.

**recording disk** An unrecorded or blank disk designed for recording sounds by means of a stylus. Also called recording blank.

**recording head** 1. A magnetic head used only for recording. Also called record head. 2. *Cutter.*

**recording instrument** *Recorder.*

**recording lamp** A lamp whose intensity can be varied at an a-f rate, for exposing variable-density sound tracks on motion-picture film and for exposing paper on film in photographic facsimile recording.

**recording level** The amplifier output level required to drive a particular recorder.

**recording loss** The difference between the amplitude recorded on a mechanical recording medium and that executed by the recording stylus

**recording noise** Noise that is introduced during a recording process.

**recording-playback head** A magnetic head used for both recording and reproduction.

**recording spot** The image area formed at the record medium by a facsimile recorder.

**recording stylus** A tool that inscribes the grooves in a mechanical recording medium.

**record medium** The physical medium on which a facsimile recorder forms an image of the subject copy.

**record player** A motor-driven turntable used with a phonograph pickup to obtain a-f signals from a phonograph record. These signals must be fed into an a-f amplifier for additional amplification before they can be reproduced as sound waves by a loudspeaker. In an electric phonograph the amplifier and loudspeaker are combined with the record player.

**record sheet** The medium used to produce a visible image of the subject copy in a facsimile system.

**recovery package** An instrumentation package carried by a missile and designed for ejection before impact. The package usually also contains a transmitter that emits signals after ejection, to aid in recovery.

**recovery time** 1. The time required for the control electrode of a gas tube to regain control after anode-current interruption. 2. The time required for a fired tr or pre-tr tube to deionize to such a level that the attenuation of a low-level r-f signal transmitted through the tube is decreased to a specified value. 3. The time required for a fired atr tube to deionize to such a level that the normalized conductance and susceptance of the tube in its mount are within specified ranges. 4. The minimum time from the start of a counted pulse to the instant a succeeding pulse can attain a specific percentage of the maximum value of the counted pulse in a Geiger counter. 5. The interval required, after a sudden decrease in input signal amplitude to a system or component, to attain a specified percentage (usually 63%) of the ultimate change in amplification or attenuation due to this decrease. 6. The time required for a radar receiver to recover to half sensitivity after the end of the transmitted pulse, so it can receive a return echo.

**rectangular coordinate** A coordinate with respect to one of three mutually perpendicular axes, used for specifying the location of a point in space.

**rectangular horn antenna** A horn antenna having a rectangular cross-section, with one or both transverse dimensions increasing linearly from the small end or throat to the mouth, used at the end of a rectangular waveguide to radiate radio waves directly into space.

**rectangular picture tube** A television picture tube having an essentially rectangular faceplate and screen, to show with maximum economy of screen space the total scene area scanned by a television camera.

**rectangular scanning** A two-dimensional sector scan in which a slow sector scan in one direction is superimposed on a rapid sector scan in a perpendicular direction.

**rectangular waveguide** A waveguide having a rectangular cross-section.

**rectification** The process of converting an alternating current to a unidirectional current.

**rectification factor** The change in the average current of an electrode divided by the change in the amplitude of the alternating sinusoidal voltage applied to the same electrode, the direct voltages of this and other electrodes being maintained constant.

**rectified current** The direct current resulting from the process of rectification.

**rectifier** A device that converts alternating current into a current having a large unidirectional component, such as a gas tube, metallic rectifier, semiconductor diode, or vacuum tube.

**rectifier instrument** A d-c meter used with a rectifying device, such as a copper-oxide rectifier, to measure alternating currents or voltages.

**rectifier stack** A stacked assembly of semiconductor rectifier disks or wafers.

**rectify** To convert an alternating current into a unidirectional current.

**rectifying element** A circuit element that provides rectification. It may be a single cell or several cells in series, parallel, or series-parallel.

**rectilineal compliance** The mechanical compliance that opposes a change in the applied force, such as the springiness that opposes a force acting on the diaphragm of a loudspeaker or microphone.

**rectilinear** Following a straight line.

**rectilinear scanning** The process of scanning an area in a predetermined sequence of narrow, straight parallel strips.

**recurrence rate** *Repetition rate.*

**recycling** Returning to an original condition, as to zero or one in a counting circuit.

**recycling detector** A detector in which the capacitor across the detector output is discharged by a switching circuit just before each new carrier cycle, to give more complete elimination of the carrier and higher output.

**red gun** The electron gun whose beam strikes phosphor dots emitting the red primary color in a three-gun color television picture tube.

**rediffusion** Wired radio in which a radio program is picked up from a broadcast station and distributed to the loudspeakers of subscribers over telephone or other wire circuits. Used in some parts of Great Britain.

**redistribution** The alteration of charges on an area of a storage surface by secondary electrons from any other area of the surface in a charge storage tube or television camera tube.

**redox system** A chemical reduction-oxidation system, in which a potential is set up at an electrode made from an inert material such as platinum. This action is the basis of operation of the solion.

**red restorer** The d-c restorer for the red channel of a three-gun color television picture tube circuit.

**Redstone** An Army surface-to-surface guided missile having inertial guidance and a range of about 350 miles. It can carry either conventional or nuclear warheads. Used to extend and supplement the range and firepower of artillery. Developed at Redstone Arsenal, Huntsville, Alabama.

**redundancy check** A forbidden-combination check that uses redundant digits called check digits to detect errors made by a computer.

**redundant digit** A digit that is not necessary for an actual computation, but serves to reveal a malfunction in a digital computer.

**red video voltage** The signal voltage output from the red section of a color television camera, or the signal voltage between the matrix and the grid of the red gun in a three-gun color television picture tube.

**reed frequency meter** *Vibrating-reed frequency meter.*

**reed-type relay** A relay in which two flat magnetic strips mounted inside a coil are attracted to each other when the coil is energized. The relay contacts are mounted on the strips.

**reel** A container consisting of a core and flanged ends, used for magnetic tape.

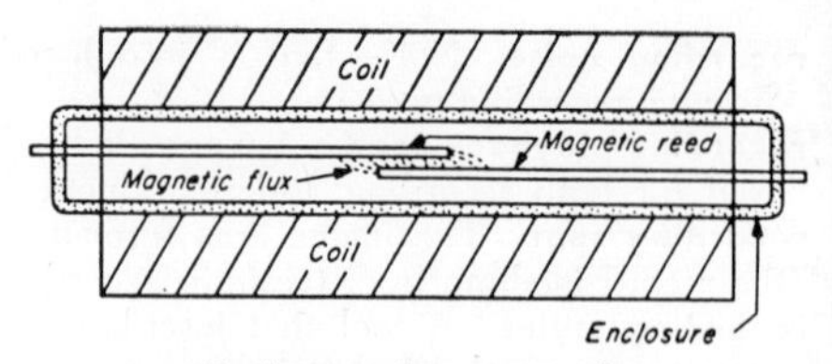

Reed-type relay construction.

**reentrant** Having one or more sections directed inward, as in certain types of cavity resonators.

**reentrant oscillator** An oscillator in which three coaxial-line resonators serve as tuning and feedback elements.

**reentry** The return of a missile or other object to the atmosphere of the earth after traveling through outer space.

**reference black level** The picture signal level corresponding to a specified maximum limit for black peaks.

**reference coupling** The coupling between two circuits that gives a reading of 0 dba (adjusted decibels) on a specified noise-measuring set connected to the disturbed circuit when a test tone of 90 dba is impressed on the disturbing circuit. Expressed in decibels above reference coupling, abbreviated dbx.

**reference dipole** A straight half-wave dipole tuned and matched for a given frequency and used as a unit of comparison in antenna measurement work.

**reference direction** The direction used as a reference for angular measurements.

**reference input** An independently established signal used as a standard of comparison in a feedback control system.

**reference input element** A feedback control system element that establishes the relationship between the reference input and the command.

**reference level** The level used as a basis of comparison when designating the level of an a-f signal in decibels or volume units. A common reference value in voltage, current, and power designations is 0.006 watt for 0 decibel. For sound loudness, the reference level is usually the threshold of hearing. For communication receivers, the commonly used level is 60 microwatts.

**reference line** A line from which angular measurements are made.

**reference noise** The power level used as a basis of comparison when designating noise power in dbrn. The reference usually used is $10^{-12}$ watt ($-90$ dbm) at 1,000 cps.

**reference phase** The phase of the color-burst signal voltage in a color television receiver, or the phase of the master-oscillator voltage in a color television transmitter.

**reference pressure** The pressure standard used in the sealed chamber of a differential pressure gage. It is usually 1 atmosphere.

**reference recording** Recording a radio program for future reference.

**reference source** *Radioactive standard.*

**reference stimulus** A reference quantity that is applied to a telemetering system, such as a reference phase, pressure, or voltage, for calibrating purposes.

**reference temperature** A standard temperature source used as a reference for calibration of high-temperature measuring devices.

**reference time** An instant near the beginning of switching that is chosen as a reference for time measurements in a digital computer. It may be the first instant at which the instantaneous value of the drive pulse, the voltage response of the magnetic cell, or the integrated voltage response reaches a specified fraction of its peak pulse amplitude.

**reference voltage** An a-c voltage used for comparison, usually to identify an in-phase or out-of-phase condition in an a-c circuit.

**reference volume** The audio volume level that gives a reading of 0 vu on a standard volume indicator. The sensitivity of the volume indicator is adjusted so reference volume or 0 vu is read when the instrument is connected across a 600-ohm resistance to which is delivered a power of 1 milliwatt at 1,000 cps.

**reference white** 1. The light from a nonselective diffuse reflector that receives the normal illumination of a scene. 2. The standard white color used as a reference for specifying all other colors. The reference white used in color television approximates direct sunlight or sky light having a color temperature of 6,500°K. Primary colors are specified in units such that one unit of each primary will combine to produce reference white.

**reference white level** The picture signal level corresponding to a specified maximum limit for white peaks.

**reflectance** *Reflection factor.*

**reflected-beam kinescope** A cathode-ray tube having approximately the diameter and thickness of an automobile wheel, with a transparent glass front face and a phosphor screen on the inside of the curved rear surface of the tube. The electron gun directs its beam forward to the front face in the conventional manner, but here the electron beam is reflected and goes back to the phosphor screen. A 180° deflection angle is thus readily obtained. Chief uses are for military and commercial radar.

**reflected impedance** The impedance value that appears to exist across the primary of a transformer when an impedance is connected as a load across the secondary.

**reflected-light scanning** The scanning of changes in the magnitude of the light reflected from the surface of an illuminated web.

**reflected resistance** The resistance value that appears to exist across the primary of a transformer when a resistive load is across the secondary.

**reflected wave** A wave reflected from a surface, discontinuity, or junction of two different media, such as the sky wave in radio, the echo wave from a target in radar, or the wave that travels back to the source end of a mismatched transmission line.

**reflecting curtain** A vertical array of half-wave reflecting antennas, generally used one quarter-wavelength behind a radiating curtain of dipoles to form a high-gain antenna.

**reflecting target** A target that reflects radar waves.

**reflection** The return or change in direction of light, sound, radar, or radio waves striking a surface or traveling from one medium into another. If the reflecting surface is smooth enough so each incident ray gives rise to a reflected ray in the same plane, the effect is known as direct, regular, specular, and mirror reflection. If the surface is so rough that reflected rays are distributed in all directions according to the cosine law, the effect is called diffuse reflection. Reflections of electromagnetic energy can occur at a mismatch in a transmission line, causing standing waves.

**reflection altimeter** *Radio altimeter.*

**reflection coefficient** The ratio of some quantity associated with a reflected wave to the corresponding quantity in the incident wave at a given point, for a given frequency and mode of transmission. Also called mismatch factor, reflection factor, and transition factor.

**reflection color tube** A color picture tube that produces an image by means of electron reflection techniques in the screen region.

**reflection error** An error in bearing indication due to wave energy that reaches a navigation receiver by undesired reflections.

**reflection factor** 1. The ratio of the load current that would be delivered by a generator to a particular load without matching to the load current obtained when generator and load impedances are matched. 2. The ratio of the total luminous flux reflected by a given surface to the incident flux. Also called reflectance. 3. The ratio of electrons reflected to electrons entering a reflector space, as in a reflex klystron. 4. *Reflection coefficient.*

**reflection grating** A wire grating placed in a waveguide to reflect one desired wave while allowing one or more other waves to pass freely.

**reflection interval** The time interval between the transmission of a radar pulse or wave and the reception of the reflected wave at the point of transmission. Multiplying this time interval in microseconds by 492 gives target distance in feet, while multiplying the time interval by 0.0931 gives target distance in standard miles.

**reflection law** The angle of incidence is equal to the angle of reflection.

**reflection loss** 1. The portion of the transition loss that is due to the reflection of power at the discontinuity. 2. The ratio in decibels of the power arriving at a discontinuity to the difference between the incident power and the reflected power.

**reflection plotter** An optical device used to superimpose on the face of a radar ppi tube a virtual image of a navigation chart. Used in radar piloting of ships near land. Also called virtual ppi reflectoscope.

**reflections** Radio waves that have been reflected from a building, hill, or other conductive or semiconductive surface during their travel to a television receiving antenna. The resulting longer travel time causes ghost images on the screen.

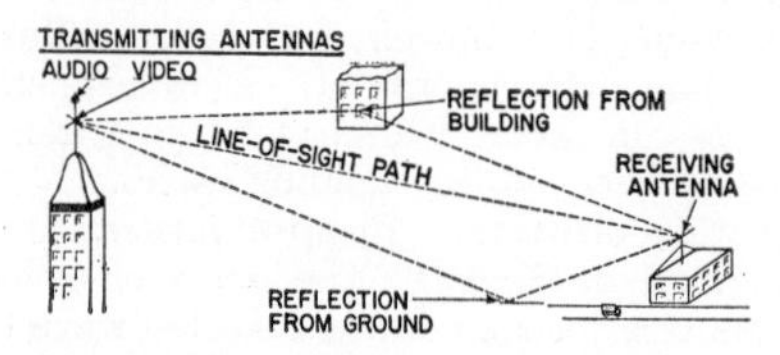


Reflections in a television system can cause ghost images on a receiver screen.

**reflection seismograph** A seismograph used in prospecting for underground oil deposits. A dynamite explosion near the surface produces sound waves that travel down to the oil strata and are there reflected back to the surface. Measurements of arrival times of the waves at seismographs give data for calculating the depth and extent of the underground oil pool.

**reflective code** *Gray code.*

**reflective optics** *Projection optics.*

**reflectivity** The fraction of incident radiant energy that is reflected from a uniformly irradiated surface.

**reflectometer** 1. A directional coupler designed to measure the power flowing in both directions through a waveguide, as a means of determining the reflection coefficient or standing-wave ratio. 2. A photoelectric instrument for measuring the optical reflectance of a reflecting surface.

**reflector** 1. A single rod, system of rods, metal screen, or metal sheet used behind an antenna to increase its directivity. 2. A metal sheet or screen used as a mirror to change the direction of a microwave radio beam. 3. A layer of water, graphite, beryllium, or other scattering material placed around the core of a nuclear reactor to reduce the loss of neutrons. Also called tamper. 4. *Repeller.*

**reflector electrode** *Repeller.*

**reflector element** A single rod or other parasitic element serving as a reflector in an antenna array.

**reflector saving** The amount by which the half-thickness of the active volume of a nuclear reactor is decreased by the use of a reflector.

**reflector space** The space in a reflex klystron that follows the buncher space and is terminated by the reflector.

**reflector voltage** The voltage between the reflector electrode and the cathode in a reflex klystron.

**reflex baffle** A loudspeaker baffle in which a portion of the radiation from the rear of the diaphragm is propagated forward after controlled shift of phase or other modification, to increase the over-all radiation in some portion of the a-f spectrum. Also called vented baffle.

**reflex bunching** The bunching that occurs in an electron stream which has been made to reverse its direction in the drift space.

**reflex circuit** A circuit in which the signal is amplified twice by the same amplifier tube or tubes, once as an i-f signal before detection and once as an a-f signal after detection.

**reflex klystron** A single-cavity klystron in which the electron beam is reflected back through the cavity resonator by a repelling electrode having a negative voltage. Used as a microwave oscillator.

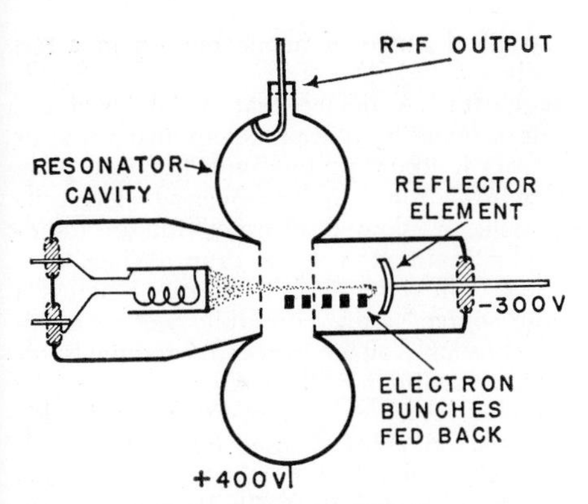

Reflex klystron construction.

**refracted wave** The portion of an incident wave that travels from one medium into a second medium. Also called transmitted wave.

**refraction** The bending of a heat, light, radar, radio, or sound wave as it passes obliquely from one medium to another in which the velocity of propagation is different.

**refraction error** A radio bearing error due to bending of one or more wave paths by undesired refraction.

**refraction loss** The portion of the transmission loss that is due to refraction resulting from nonuniformity of the medium.

**refractive index** The ratio of the phase velocity of a wave in free space to that in a given medium. The refractive index of air at sea level is 1.00029.

**refractive modulus** The excess over unity of the modified index of refraction in the troposphere, expressed in millionths and computed as $(n + h/a - 1)10^6$, where $n$ is the index of refraction at a height $h$ above sea level and $a$ is the radius of the earth.

**refractivity** The refractive index minus 1.

**refractometer** An instrument for measuring the refractive index of a liquid or solid, usually by measuring the critical angle at which total reflection occurs.

**refrangible** Capable of being refracted.

**regeneration** 1. Replacement or restoration of charges in a charge storage tube to overcome decay effects, including loss of charge by reading. 2. Restoration of contaminated nuclear fuel to a usable condition. 3. *Positive feedback.*

**regeneration control** A variable capacitor, variable inductor, potentiometer, or rheostat used in a regenerative receiver to control the amount of feedback and thereby keep regeneration within useful limits.

**regeneration period** The time interval in which the screen of a cathode-ray storage tube is scanned by the beam to regenerate the charge distribution that represents the stored information. Also called scan period.

**regenerative amplifier** An amplifier that uses positive feedback to give increased gain and selectivity.

**regenerative braking** *Dynamic braking.*

**regenerative detector** A vacuum-tube detector circuit in which r-f energy is fed back from the anode circuit to the grid circuit

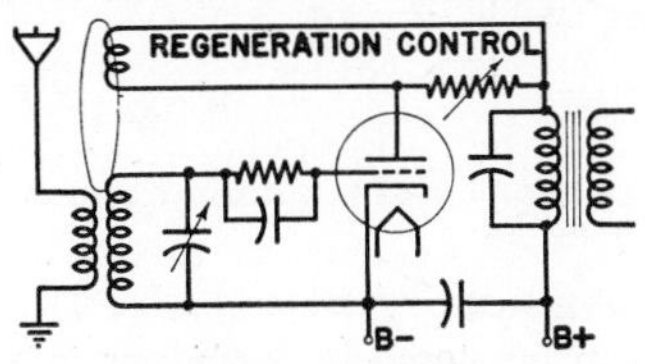

Regenerative detector circuit.

to give positive feedback at the carrier frequency, thereby increasing the amplification and sensitivity of the circuit.

**regenerative reactor** A nuclear reactor that produces fissionable material in addition to energy. When loaded with fissionable uranium-235 and nonfissionable uranium-238 or thorium, it converts the uranium-238 or thorium into fissionable materials which are then used as fuel in the core of the reactor. The reactor thus regenerates itself.

**regenerative receiver** A radio receiver that uses a regenerative detector.

**regenerative repeater** A repeater that performs pulse regeneration to restore the original shape of a pulse signal used in teletypewriter and other code circuits. Each code element is replaced by a new code element having specified timing, waveform, and magnitude.

**regional channel** A standard radio broadcast channel in which several stations may operate with powers not in excess of 5 kilowatts.

**region of limited proportionality** The range of applied voltage below the Geiger threshold, in which the gas amplification depends on the charge liberated by the initial ionizing event in a radiation-counter tube. Also called limited-proportionality region.

**register** 1. The computer hardware for storing one machine word. When in the main internal memory, it is called a storage register. 2. Accurate matching or superimposition of two or more patterns, such as the three images in color television or the patterns on opposite sides of a printed-circuit board. 3. To superimpose two or more designs or images with precise alignment. 4. *Registration.*

**register control** Automatic control of the position of a printed design with respect to reference marks or some other part of the design, as in photoelectric register control.

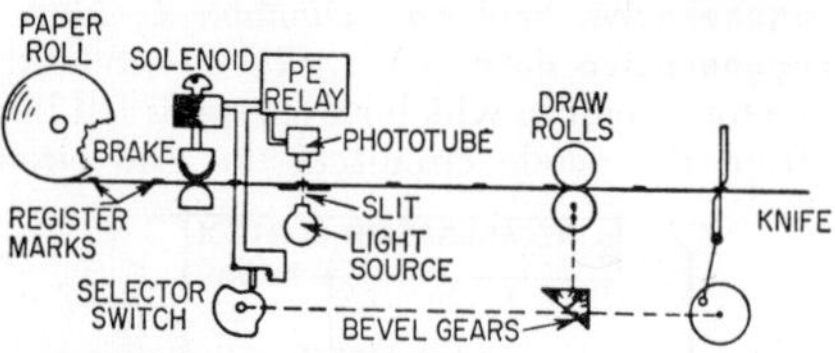


Register control used to cut paper strip at printed register marks.

**register length** The number of characters that can be stored by a register in a computer.

**register mark** A mark or line printed or otherwise impressed on a web of material for use as a reference to maintain register.

**registration** Exact superimposition of all three color images on the screen of a color television receiver, or superimposition of all colors of a design on a printed sheet. Also called register.

**regular reflection** *Direct reflection.*

**regulated power supply** A power supply containing means for maintaining essentially constant output voltage or output current under changing load conditions.

**regulating rod** A control rod intended to accomplish rapid, fine, and sometimes continuous adjustment of the reactivity of a nuclear reactor. It usually is capable of moving much more rapidly than a shim rod but makes a smaller change in reactivity. A regulating rod is generally moved automatically by a servosystem.

**regulating system** *Automatic control system.*

**regulation** 1. The process of holding constant a quantity such as speed, temperature, voltage, or position by means of an electronic or other system that automatically corrects errors by feeding back into the system the condition being regulated. Regulation thus is based on feedback, whereas control is not. 2. The change in output voltage that occurs between no load and full load in a transformer, generator, or other source. Dividing this change by the rated full-load value and multiplying the result by 100 gives per cent regulation. 3. The difference between the maximum and minimum tube voltage drops within a specified range of anode current in a gas tube.

**regulator** A device that maintains a desired quantity at a predetermined value or varies it according to a predetermined plan.

**regulator tube** A two-electrode gas tube having a glow discharge with an essentially constant voltage drop. When connected in series with a resistance across a d-c source, a regulator tube will maintain a constant voltage across its terminals for wide variations in the source voltage.

**Regulus** A Navy surface-to-surface guided missile. Regulus I used radio command guidance and had a range of about 500 miles. Canceled in 1959. Regulus II uses inertial guidance, has a speed of about mach 2, has a range of over 1,000 miles, and can carry either conventional or nuclear warheads.

**reignition** A process by which multiple counts are generated within a radiation-counter tube by atoms or molecules excited or ionized in the discharge accompanying a tube count.

**reignition voltage** The voltage that is just sufficient to reestablish conduction if applied to a gas tube during the deionization period. The value of this voltage varies inversely with time during the deionization period. Also called restriking voltage.

**Reinartz crystal oscillator** A crystal-controlled vacuum-tube oscillator in which the crystal current is kept low by placing in the cathode lead a resonant circuit tuned to half the crystal frequency. The resulting regeneration at the crystal frequency improves efficiency without the danger of uncontrollable oscillation at other frequencies.

**reinserter** *D-c restorer.*

**reinsertion of carrier** Combining a locally generated carrier signal with an incoming suppressed-carrier signal.

**rejection band** The band of frequencies below the cutoff frequency in a uniconductor waveguide.

**rejector** *Trap.*

**rejector circuit** *Band-elimination filter.*

**rel** A unit of reluctance, equal to 1 ampere-turn per magnetic line of force.

**relative abundance** The fraction or percentage of the atoms of an element in a given isotope.

**relative address** A designation for the position of a memory location in a computer routine or subroutine. Relative addresses are translated into absolute addresses by adding a reference address, such as that at which the first word of the routine is stored.

**relative aperture** The ratio of the minimum vertical or horizontal clearance for particle passage to the particle orbit radius in the accelerating chamber of an accelerator.

**relative bearing** A bearing in which the heading of the vehicle serves as the reference line.

**relative biological effectiveness** [abbreviated rbe] The effectiveness of an ionizing radiation in producing a specific biological damage, as compared to the effectiveness of a reference radiation such as 200-kilovolt x-rays.

**relative coding** Computer coding in which all addresses refer to an arbitrarily selected position, or in which all addresses are represented symbolically.

**relative damping** The ratio of damping torque at a given angular velocity of a moving element in an instrument to the damping torque that would produce critical damping at that same angular velocity.

**relative heading** Deprecated term for *heading*.

**relative humidity** The ratio of the amount of water vapor present in air to the amount that would saturate it at a given temperature.

**relative luminosity** The ratio of the value of the luminosity at a particular wavelength to the value at the wavelength of maximum luminosity.

**relative permeability** *Specific permeability.*

**relative plateau slope** The per cent change in counting rate per unit change of applied voltage near the midpoint of the plateau of a radiation-counter tube.

**relative refractive index** The ratio of the refractive indices of two media.

**relative response** The ratio, usually expressed in decibels, of the response under some particular conditions to the response under reference conditions.

**relative specific ionization** The specific ionization for a particle of a given medium, relative either to that for the same particle and energy in a standard medium or the same particle and medium at a specified energy.

**relative stopping power** The ratio of the stopping power of a given substance to that of a standard substance such as aluminum, oxygen, or air. Also called equivalent stopping power.

**relative target bearing** The bearing of a radar target expressed relative to the heading of a ship or aircraft.

**relative time delay** The difference in time delay encountered by the audio signal and the composite picture signal or between the components of the picture signal traveling over a television relay system.

**relative velocity** The time rate of change of a position vector of a point with respect to a reference frame.

**relativistic mass** The mass of a particle moving at a velocity exceeding about one-tenth the velocity of light. The relativistic mass is significantly larger than the rest mass.

**relativistic mass equation** The equation for the relativistic mass of a particle or body having a given rest mass and velocity.

**relativistic particle** A particle with a velocity so large that its relativistic mass exceeds its rest mass by a significant amount.

**relativistic velocity** A particle velocity sufficiently large that mass and other properties of the particle are significantly different from the at-rest values. Velocity is generally considered to be relativistic when it exceeds about one-tenth the velocity of light.

**relativity** A principle that postulates the interdependence of matter, space, and time in the universe, for various frames of reference.

**relaxation generator** *Relaxation oscillator.*

**relaxation inverter** An inverter that uses a relaxation oscillator circuit to convert d-c power to a-c power.

**relaxation oscillator** An oscillator whose fundamental frequency is determined by the time of charging or discharging a capacitor or coil through a resistor, producing waveforms that may be rectangular or sawtooth. Also called relaxation generator.

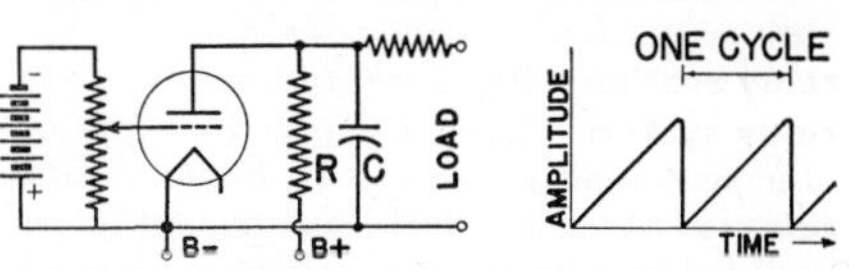


Relaxation oscillator circuit and its sawtooth output waveform.

**relaxation time** 1. The time constant required for an abrupt change of magnetizing force to make the magnetic induction reach a specified per cent of its new value. 2. The travel time of an electron in a metal before it is scattered and loses its momentum.

**relay** [symbol K] 1. A device that is operated by a variation in the conditions in one electric circuit and serves to make or break one or more connections in the same or another electric circuit. The commonest types are electromagnetic relays and thermal relays. 2. The process of passing on a signal from one microwave link to another.

**relay armature** The movable iron part of a relay.

**relay broadcast station** A station licensed to transmit, from points where wire facilities are not available, programs for broadcast by one or more broadcast stations.

**relay channel** The band of frequencies used in transmitting a single television relay signal, including the guard bands.

**relay computer** A computer consisting chiefly of electromagnetic relays or similar electromechanical devices.

**relay contact** One of the pair of contacts that are closed or opened by the movement of the armature of a relay.

**relay magnet** The electromagnet that attracts the armature of a relay when energized.

**relay rack** A standardized steel rack designed to hold 19-inch panels of various heights, on which are mounted radio receivers, amplifiers, and other units of electronic equipment. Mounting holes for the panels, usually drilled and threaded for 10-32 machine screws, are spaced apart in multiples of ½ inch and ⅝ inch, to accommodate notch spacings of correspondingly standardized panels. Originally designed to hold relay panels in telephone exchanges. Also called rack.

**relay radar** Deprecated term for *radar relay*.

**relay receiver** A receiver that accepts a television relay input signal and delivers a television relay output signal to the transmitter portion of a repeater station.

**relay station** *Repeater station.*

**relay system** 1. An assembly of relays used for switching purposes. 2. *Radio repeater.*

**release** A mechanical arrangement of parts for holding or freeing a device or mechanism as required.

**release factor** The ratio, expressed in per cent, of relay dropout current to rated current, or the corresponding voltage ratio.

**release time** The total elapsed time from the instant that relay coil current starts to drop until the make contacts have opened or the break contacts have closed.

**reliability** The probability that a device will perform its purpose adequately for the period of time intended under the operating conditions encountered.

**reliability engineering** A field of engineering that deals with the prevention and correction of malfunctions in equipment.

**reliability index** A quantitative figure of merit related to the reliability of a piece of equipment, such as the number of failures per 1,000 operations or the number of failures in a specified number of operating hours.

**reliability test** A test designed specifically to evaluate the level and uniformity of reliability of equipment under various environmental conditions.

**relieving anode** An auxiliary anode that provides an alternative conducting path for reducing the current to another electrode in a pool tube.

**reluctance** A measure of the opposition presented to magnetic flux in a magnetic circuit. Reluctance is the reciprocal of permeance, and is therefore equal to magnetomotive force divided by magnetic flux.

**reluctance microphone** *Variable-reluctance microphone.*

**reluctance motor** A synchronous motor, similar in construction to an induction motor, in which the member carrying the secondary circuit has salient poles but no d-c excitation. It starts as an induction motor but operates normally at synchronous speed.

**reluctance pickup** *Variable-reluctance pickup.*

**reluctivity** The ratio of the magnetic intensity in a region to the magnetic induction in the same region. Reluctivity is the reciprocal of magnetic permeability. Also called specific reluctance.

**rem** Abbreviation for *roentgen equivalent man.*

**remanence** The magnetic flux density that remains in a magnetic circuit after the removal of an applied magnetomotive force. If the magnetic circuit has an air gap, the remanence will be less than the residual flux density.

**remodulation** Transferring modulation from one carrier to another.

**remodulator** A circuit that converts amplitude modulation to audio frequency-shift modulation for transmission of facsimile signals over a voice-frequency radio channel. Also called converter.

**remote** *Nemo.*

**remote control** Control of equipment from a distance over wires or by means of light, radio, sound, or ultrasonic waves. Also called telecontrol.

**remote cutoff** The characteristic wherein a large negative bias is required for complete cutoff of output current in an electron tube or other amplifying device.

**remote-cutoff tube** *Variable-mu tube.*

**remote indicator** 1. An indicator located at a distance from the data-gathering sensing element, with data being transmitted to the indicator mechanically, electrically over wires, or by means of light, radio, or sound waves. 2. *Repeater.*

**remote line** A program transmission line running between a remote-pickup point and a broadcast studio or transmitter site.

**remote metering** *Telemetering.*

**remote pickup** Picking up a radio or television program at a remote location and transmitting it to the studio or transmitter over wire lines or a short-wave or microwave radio link.

**remote ppi** *Ppi repeater.*

**renormalization of mass** Adding to the mechanical mass of a particle its extra mass due to self-interaction, to give a sum equal to the measured mass. Used in quantized field theory.

**rep** 1. Abbreviation for *roentgen equivalent physical.* 2. Abbreviation for *representative.*

**repeatability** 1. A measure of the variation in the readings of an instrument when identical tests are made under fixed conditions. Also called reproducibility. 2. The ability of a voltage regulator or voltage reference tube to attain the same voltage drop at a stated time after the beginning of any conducting period.

**repeater** 1. An amplifier or other device that receives weak signals and delivers corresponding stronger signals with or without reshaping of waveforms. It may be either a one-way or two-way repeater. Carrier, telegraph, and telephone repeaters are used in wire lines, whereas radio repeaters act directly on radio waves. 2. An indicator that shows the same information as is shown on a master indicator. Synchros are widely used for this purpose. Also called remote indicator. 3. *Repeating coil.*

**repeater jammer** A jammer that intercepts an enemy radar signal and reradiates the signal after modifying it to incorporate erroneous data on azimuth, range, or number of targets.

**repeater station** A station containing one or more repeaters. Also called relay station.

**repeating coil** A transformer used to provide inductive coupling between two sections of a telephone line when a direct connection is undesirable. The transformer ratio is usually 1 to 1. Also called repeater.

**repeating timer** A timer that continues repeating its operating cycle until excitation is removed.

**repeat-point tuning** *Double-spot tuning.*

**repeller** An electrode whose primary function is to reverse the direction of an electron stream in an electron tube. Also called reflector and reflector electrode.

**repetition frequency** *Repetition rate.*

**repetition rate** The rate at which recurrent signals are produced or transmitted. Also called recurrence rate and repetition frequency.

**repetitive error** The maximum deviation of the controlled variable from the average value upon successive return to specified operating conditions following specified deviation therefrom in an automatic control system.

**replacement tube** An electron tube suitable for use in place of another tube having the same type number but not necessarily made by the same manufacturer.

**reply** An r-f signal or combination of signals transmitted by a transponder in response to an interrogation. Also called response.

**representative** [abbreviated rep] *Sales representative.*

**reprocessing** Processing of fuel after use in a nuclear reactor, to permit reuse.

**reproduce head** *Playback head.*

**reproducibility** *Repeatability.*

**reproducing stylus** *Stylus.*

**reproduction speed** The area of copy recorded per unit of time by a facsimile receiver.

**repulsion** A mechanical force tending to separate bodies having like electric charges or like magnetic polarity, or in the case of adjacent conductors, having currents flowing in opposite directions.

**repulsion-induction motor** A repulsion motor that has a squirrel-cage winding in the rotor in addition to the repulsion-motor winding.

**repulsion motor** An a-c motor having stator windings connected directly to the source of a-c power and rotor windings connected to a commutator. Brushes on the commutator are short-circuited and are positioned to produce the rotating magnetic field required for starting and running. This type of motor varies considerably in speed as load is changed.

**repulsion-start induction motor** An a-c motor that starts as a repulsion motor. At a predetermined speed the commutator bars are short-circuited to give the equivalent of a squirrel-cage winding for operation as an induction motor with constant-speed characteristics.

**reradiation** Undesirable radiation of signals generated locally in a radio receiver, causing interference or revealing the location of the receiver.

**rerecording** The process of making a recording by reproducing a recorded sound source and recording this reproduction.

**rerecording system** A system of reproducers, mixers, amplifiers, and recorders used to combine or modify various sound recordings to provide a final sound record.

**rerun** To run a program or a portion of it over on a computer. Also called rollback.

**rerun point** A point in a computer program at which all information is available for rerunning the last-run portion of a problem when an error is detected. Several such points are usually provided for in a program, to eliminate the need for rerunning the entire problem.

**rerun routine** A routine used after a computer malfunction, a coding error, or an operating mistake to reconstitute a routine from the last previous rerun point. Also called rollback routine.

**rescap** A capacitor and resistor assembly manufactured as a packaged encapsulated circuit. Also called capacitor-resistor unit, capristor, packaged circuit, and resistor-capacitor unit.

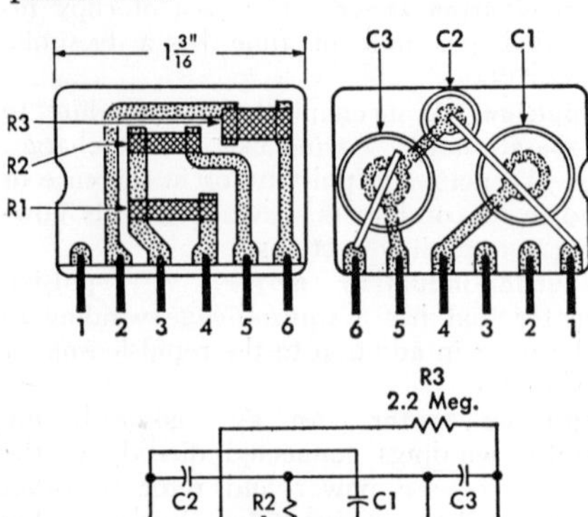


Rescap construction and circuit.

**research** Scientific investigation aimed at discovering and applying new facts, techniques, and natural laws.

**research reactor** A reactor designed primarily as a research tool. It may be used to supply neutrons, other particles, and gamma radiation for research purposes, expose samples to these radiations, and produce transmutation products.

**reserve battery** A battery having long shelf life in an unenergized state. Electrolyte is added when power is desired. When used in missiles and proximity fuzes, an acceleration-sensitive device breaks the electrolyte container and thereby activates the battery.

**reserve cell** A cell used in a reserve battery.

**reset** *Clear.*

**reset action** Floating action in which the final control element is moved at a speed proportional to the extent of proportional-position action. The speed of corrective action is thus determined by both the amount and duration of the deviation from the desired value.

**reset control circuit** The magnetic amplifier circuit that resets the flux in the core of the saturable reactor.

**reset flux level** The difference between the saturation and reset levels of the flux in the core of a saturable reactor.

**reset pulse** 1. A drive pulse that tends to reset a magnetic cell in the storage section of a digital computer. 2. A pulse used to reset an electronic counter to zero or to some predetermined position.

**reset rate** The number of corrections made per minute by a control system.

**reset switch** A machine-operated switch that restores normal operation to a control system after a corrective action.

**resetting half-cycle** The half-cycle in the a-c supply voltage of a magnetic amplifier in which the flux in the core of the saturable reactor is reset.

**resetting interval** The portion of the resetting half-cycle in which the flux in the core of the saturable-reactor core is actually changing from the saturation level to the reset flux level.

**reshaping circuit** A circuit that changes the waveform of a current, such as a limiter.

**residual activity** Radioactivity remaining in a substance or system at a specified time after a period of decay.

**residual charge** The charge remaining on the plates of a capacitor after an initial discharge.

**residual current** The current flowing through a thermionic diode when there is no anode voltage, due to the velocity of the electrons emitted by the heated cathode.

**residual discharge** A discharge of the residual charge remaining after the initial discharge of a capacitor.

**residual error** The sum of random errors and uncorrected systematic errors.

**residual field** The magnetic field left in an iron core after excitation has been removed.

**residual flux density** The magnetic flux density at which the magnetizing force is zero when the material is in a symmetrically and cyclically magnetized condition. Also called residual induction, residual magnetic induction, and residual magnetism.

**residual gap** The length of the magnetic air gap between the armature and the center of the core face of an energized relay.

**residual gas** The small amount of gas remaining in a vacuum tube after the best possible exhaustion by vacuum pumps. Much of this residual gas is removed by the getter.

**residual induction** *Residual flux density.*

**residual ionization** Ionization of air or other gas in a closed chamber, not accounted for by recognizable neighboring agencies. It is now attributed to cosmic rays.

**residual magnetic induction** *Residual flux density.*

**residual magnetism** *Residual flux density.*

**residual modulation** *Carrier noise level.*

**residual nucleus** The heavy nucleus that is the end product of a nuclear transformation.

**residual pin** A nonmagnetic pin or screw attached to the armature or core of a relay to prevent the armature from touching and sticking to the core.

**residual radiation** Nuclear radiation emitted by radioactive material deposited after an atomic burst, including fission products, unfissioned nuclear material, and material in which radioactivity may have been induced by neutron bombardment.

**residual range** The distance in which a particle can still produce ionization after having lost some of its energy in passing through matter.

**residual resistance** The portion of the electric resistance of a metal that is independent of temperature.

**resin** A natural or synthetic organic product used widely in plastics, adhesives, and surface coatings.

**resist** An acid-resistant nonconducting coating used to protect desired portions of a wiring pattern from the action of the etchant during manufacture of printed-wiring boards.

**resistance** [symbol *R*] The opposition that a device or material offers to the flow of direct current, measured in ohms, kilohms, or megohms. In a-c circuits, resistance is the real component of impedance. Also called d-c resistance and ohmage. A resistor is a component designed to provide a desired known value of resistance.

**resistance box** A box containing a number of precision resistors connected to panel terminals or contacts in such a way that a desired resistance value can be obtained by withdrawing plugs (as in a post office bridge) or by setting multicontact switches. Usually constructed as a decade box, wherein individual resistance values vary in submultiples and multiples of 10.

**resistance braking** *Dynamic braking.*

**resistance bridge** *Wheatstone bridge.*

**resistance-capacitance** [abbreviated r-c] Containing both resistance and capacitance, as provided by resistors and capacitors. An alternate abbreviation is RC.

**resistance-capacitance circuit** [abbreviated r-c circuit] A circuit containing resistors and capacitors that determine the time constant. The time constant in seconds is equal to the product of resistance in ohms and capacitance in farads.

**resistance-capacitance-coupled amplifier** An amplifier in which resistance coupling is used between stages.

**resistance-capacitance coupling** *Resistance coupling.*

**resistance-capacitance differentiator** [abbreviated r-c differentiator] A resistance-capacitance circuit used to produce an output voltage whose amplitude is proportional to the rate of change of the input voltage. A square-wave input thus produces sharp output voltage spikes.

**resistance-capacitance filter** [abbreviated r-c filter] A filter containing only resistance and capacitance elements (no inductance). Widely used in rectifier-type power supplies to reduce ripple.

**resistance-capacitance oscillator** [abbreviated r-c oscillator] An oscillator in which the frequency is determined by resistance-capacitance elements.

**resistance-coupled amplifier** Commonly used term for an amplifier in which a capacitor provides a path for signal currents from one stage to the next, with resistors connected from each side of the capacitor to the power supply or to ground. It can amplify a-c signals but cannot handle small changes in direct currents. Also called r-c amplifier.

**resistance coupling** Coupling in which resistors are used as the input and output impedances of the circuits being coupled.

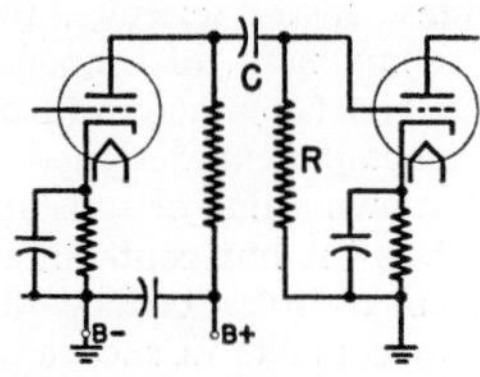

Resistance coupling between two tube stages.

A coupling capacitor is generally used between the resistors to transfer the signal from one stage to the next. Also called r-c coupling, resistance-capacitance coupling, and resistive coupling.

**resistance drop** 1. The voltage drop occurring between two points on a conductor due to the flow of current through the resistance of the conductor. Multiplying the resistance in ohms by the current in amperes gives the voltage drop in volts. 2. *IR drop.*

**resistance element** An element of resistive material in the form of a grid, ribbon, or wire, used singly or built into groups to form a resistor for heating purposes, as in an electric soldering iron.

**resistance furnace** An electric furnace in which the heat is developed by the passage of current through a suitable internal resistance that may be the charge itself, a resistor imbedded in the charge, or a resistor surrounding the charge.

**resistance hybrid** A hybrid junction consisting entirely of resistors. Operation is essentially independent of frequency up to several hundred megacycles, but attenuation along desired paths is greater than with a hybrid transformer. Also called resistance junction.

**resistance junction** *Resistance hybrid.*

**resistance loss** Power loss due to current flowing through resistance. Its value in watts is equal to the resistance in ohms multiplied by the square of the current in amperes.

**resistance magnetometer** A magnetometer that depends for its operation on variations in the electric resistance of a material immersed in the magnetic field to be measured.

**resistance material** A material having sufficiently high resistance per unit length or volume to permit its use in the construction of resistors.

**resistance pad** A pad employing only resistances. Used to provide attenuation without altering frequency response.

**resistance pyrometer** A pyrometer in which the heat-sensing element is a length of wire whose resistance varies greatly with temperature.

**resistance standard** *Standard resistor.*

**resistance-start motor** A split-phase motor having a resistance connected in series with the auxiliary winding. The auxiliary circuit is opened when the motor attains a predetermined speed.

**resistance strain gage** A strain gage consisting of a strip of material that is cemented to the part under test, and changes in resistance with elongation or compression.

**resistance thermometer** A thermometer in which the sensing element is a resistor whose resistance is an accurately known function of temperature.

**resistance welding** Welding in which the metals to be joined are heated to melting temperature at their points of contact by a localized electric current while pressure is applied. The heating current, which may be thousands of amperes, is usually controlled electronically as to amplitude and duration. Examples of resistance welding include percussion, seam, and spot welding.

**resistance wire** Wire made from a metal or alloy having high resistance per unit length, such as Nichrome. Used in wire-wound resistors and heating elements.

**resistive conductor** A conductor used primarily because it has high resistance per unit length.

**resistive coupling** *Resistance coupling.*

**resistive-wall amplifier** An electron-beam amplifier in which gain is obtained by interaction between the electron beam and a charge induced by the beam in an adjacent resistive wall. The resulting bunching action increases exponentially with distance, giving a gain comparable to that of other traveling-wave tubes.

**resistivity** The resistance in ohms that a unit volume of a material offers to the flow of current. Resistivity is the reciprocal of the conductivity of a material. Measured in ohm-centimeters. The resistivity of a wire sample in ohm-centimeters is equal to the resistance $R$ in ohms multiplied by the cross-sectional area $A$ in square centimeters and the result divided by the sample length

*L* in centimeters (resistivity = $RA/L$). Resistivity is also expressed in ohms per circular mil foot, which is the resistance of a sample that is 1 circular mil in cross-section and 1 foot long. Also called specific resistance.

**resistor** [symbol R] A device designed intentionally to have a definite amount of resistance. Used in circuits to limit current flow or to provide a voltage drop.

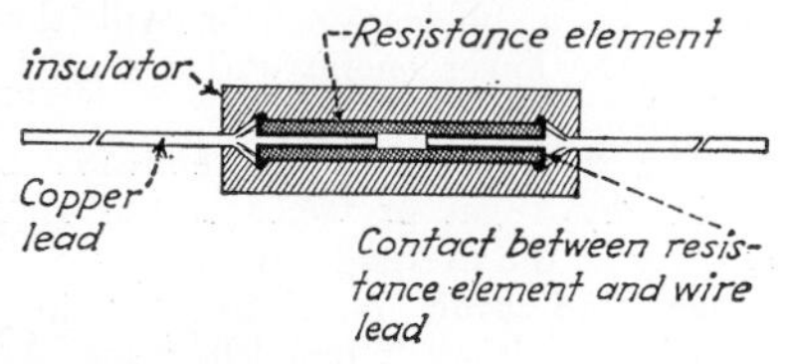


Resistor having carbon resistance element and axial leads.

**resistor-capacitor unit** *Rescap.*

**resistor color code** A method of marking the value in ohms on a resistor by means of dots or bands of colors as specified in the EIA color code.

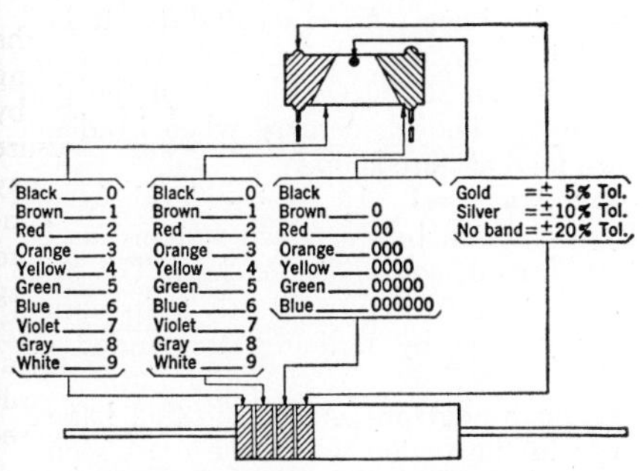


Resistor color code. On radial-lead resistors as at top, body color is read first, end color next, then dot; tolerance color on other end is easily recognized because it is always either gold or silver if present.

**resistor core** The insulating support on which a resistor element is wound or otherwise placed.

**resistor element** The portion of a resistor that provides resistance. It may be pure metal, an alloy, a metallic coating, a carbon-cement mixture, or a plastic containing finely powdered metal.

**resistor furnace** A resistance furnace in which the heat is developed in a resistor that is not a part of the charge.

**resnatron** [RESoNAtor-TRON] A microwave beam tetrode containing cavity resonators, used chiefly for generating large amounts of continuous power at high frequencies. Noise-modulated 50-kilowatt versions were used for jamming German airborne radars in World War II.

**resolution** A measure of ability to delineate detail or distinguish between nearly equal values of a quantity. In television, resolution is usually expressed as the maximum number of lines that can be discerned on the screen in a distance equal to tube height; this ranges from 350 to 400 lines for most receivers. In radar, resolution is the minimum separation between two targets in angle or range at which they can be distinguished on the radar screen. Also called resolving power.

**resolution chart** *Test pattern.*

**resolution in azimuth** The angle by which two targets at the same range must be separated in azimuth to be distinguishable on the screen of a particular radar set.

**resolution in range** The distance by which two targets at the same bearing must be separated in range to be distinguishable on the screen of a particular radar set.

**resolution sensitivity** The minimum change of measured variable that actuates an automatic control system.

**resolution time** The minimum time interval at which two successive voltage pulses can be registered by a counter.

**resolution wedge** A group of gradually converging lines on a test pattern, used to measure resolution in television.

**resolver** A synchro or other device whose rotor is mechanically driven to translate rotor angle into electric information corresponding to the sine and cosine of rotor angle. Used for interchanging rectangular and polar coordinates. Also called sine-cosine generator and synchro resolver.

**resolving power** 1. The ability of a mass spectrometer to separate adjacent mass spectrum lines. 2. The reciprocal of the beam width of a unidirectional antenna, measured in degrees. 3. The ability of a radar set to form distinguishable images. 4. *Resolution.*

**resolving time** The minimum time interval between two distinct events that will permit them to be counted or otherwise detected by a particular circuit or device.

**resonance** [noun; also used as adjective in place of resonant] 1. The condition existing in a circuit when the inductive reactance balances out the capacitive reactance. 2. The condition existing in a body when the frequency of an applied vibration equals the natural frequency of the body. A body vibrates most readily at its resonant frequency. Also called velocity resonance. 3.

The condition wherein a nuclear system in motion reacts to an external field or force applied at a natural vibration frequency, as in nuclear magnetic resonance.

**resonance absorption** The absorption of neutrons having a narrow range of energies corresponding to a nuclear resonance level of the absorber in a nuclear reactor.

**resonance bridge** A four-arm a-c bridge used to measure inductance, capacitance, or frequency. The inductor and the capacitor, which may be either in series or in

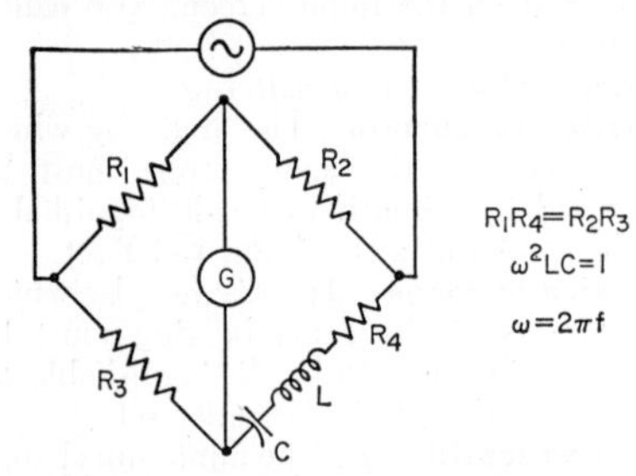


Resonance bridge circuit and equations.

parallel, are tuned to resonance at the frequency of the source before the bridge is balanced.

**resonance capture** The capture of an incident particle by a nucleus in such a way that the particle enters a resonance level of the resultant compound nucleus.

**resonance characteristic** *Resonance curve.*

**resonance current step-up** The ability of a parallel resonant circuit to circulate through its coil and capacitor a current that is many times greater than the current fed into the circuit.

**resonance curve** An amplitude-frequency response curve showing the current or voltage response of a tuned circuit to frequencies at and near the resonant frequency. Also called resonance characteristic.

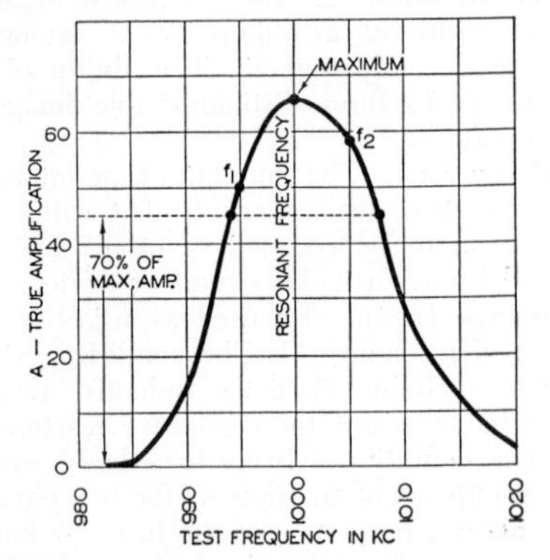


Resonance curve at 1,000 kc for a typical r-f amplifier.

**resonance energy** The kinetic energy of a particle that will be captured or scattered preferentially because of the presence of an appropriate resonance level in the compound nucleus formed by the incident particle and the target nucleus.

**resonance escape probability** The probability that a neutron at the higher of two energies in a nuclear reactor will slow down to the lower energy rather than be absorbed.

**resonance fluorescence** The emission of radiation by a gas or vapor at the same frequency as the exciting radiation.

**resonance indicator** A device that indicates when a circuit is tuned to resonance. It may be a meter, neon lamp, headphones, or a cathode-ray tuning indicator.

**resonance integral** The negative of the natural logarithm of the resonance escape probability multiplied by the slowing-down power per atom of absorber in a nuclear reactor.

**resonance lamp** An evacuated quartz bulb containing mercury, which acts as a source of radiation at the wavelength of the pure resonance line of mercury when irradiated by a mercury-arc lamp.

**resonance level** An energy level formed by a collision between two systems, as between a nucleon and a nucleus. The atom in question can return directly to its normal energy level by radiating the added energy.

**resonance neutron** A neutron that has energy in the region where the cross section of the nuclide or element is particularly large because of the occurrence of a resonance. Thus, cadmium resonance neutrons have energies between 0.05 and about 0.3 electron-volt.

**resonance penetration** The penetration of a nucleus by a charged particle whose energy corresponds to one of the energy levels in the nucleus.

**resonance radiation** The emission of radiation by a gas or vapor as a result of excitation of atoms to higher energy levels by incident photons at the resonance frequency of the gas or vapor. The radiation is characteristic of the particular gas or vapor atom but is not necessarily the same frequency as the absorbed radiation.

**resonance radiometer** A radiometer used for making relative measurements of weak radiation in infrared spectrometers.

**resonance scattering** Scattering arising

from the part of the incident wave that penetrates the surface and interacts with the interior of a nucleus.

**resonance spectral line** One of the spectral lines emitted when an electron undergoes transfer to a lower energy state in a given atom.

**resonance spectrum** The spectrum of light emitted by an excited atom during its return to the ground state.

**resonance transformer** A high-voltage transformer in which the secondary circuit is tuned to the frequency of the power supply.

**resonant** Pertaining to resonance. The term resonance is often used as an adjective in place of resonant.

**resonant capacitor** A tubular capacitor that is intentionally wound to have inductance in series with its capacitance. Often used as a bypass capacitor in i-f amplifiers, where it is made to be resonant at the intermediate-frequency value to give more effective bypassing of i-f signals.

**resonant cavity** *Cavity resonator.*

**resonant-cavity maser** A maser in which the paramagnetic active material is placed in a cavity resonator.

**resonant chamber** *Cavity resonator.*

**resonant-chamber switch** A waveguide switch in which a tuned cavity in each waveguide branch serves the functions of switch contacts. Detuning of a cavity blocks the flow of energy in the associated waveguide.

**resonant charging choke** A choke that resonates with the effective capacitance of a pulse-forming network in a modulator to produce oscillation at the resonant frequency.

**resonant circuit** A circuit that contains inductance and capacitance of such values as to give resonance at an operating frequency. The frequency in cps at which resonance occurs is $1/2\pi\sqrt{LC}$, where $L$ is in henrys and $C$ is in farads. When the capacitance and inductance are in series, the combination has a lower impedance at resonance than either alone. When in parallel, the combination has a higher impedance than either alone.

**resonant diaphragm** A diaphragm that has no reactance at a specified frequency.

**resonant element** *Cavity resonator.*

**resonant frequency** 1. The frequency at which the inductive reactance of a given resonant circuit is equal to the capacitive reactance, so resonance exists. 2. The frequency at which a quartz crystal, loudspeaker diaphragm, or other object will vibrate readily.

**resonant gap** The small internal gap in which the electric field of a tr tube is concentrated.

**resonant iris** A resonant window in a circular waveguide, so called because of its resemblance to an optical iris.

**resonant line** A transmission line having values of distributed inductance and distributed capacitance such as to make the line resonant at the frequency it is handling. Parallel resonance exists when the line is an odd number of quarter-wavelengths long and is short-circuited at the load, and series resonance exists for the same line when open at the load end.

**resonant-line oscillator** An oscillator in which one or more sections of transmission line serve as resonant circuits.

**resonant-line tuner** A television tuner in which resonant lines are used to tune the antenna, r-f amplifier, and r-f oscillator circuits. Tuning is achieved by moving shorting contacts that change the electrical lengths of the lines.

**resonant resistance** The resistance of a resonant circuit or resonant line at the frequency of resonance.

**resonant voltage step-up** The ability of a coil and a capacitor in a series resonant circuit to deliver a voltage several times greater than the input voltage of the circuit.

**resonant window** A parallel combination of inductive and capacitive diaphragms, used in a waveguide structure to provide transmission at the resonant frequency and reflection at other frequencies.

**resonant-window switch** A waveguide switch in which a resonant window in each waveguide branch serves the function of switch contacts. Detuning of the window blocks the flow of energy in the associated waveguide.

**resonate** To bring to resonance, as by tuning.

**resonating piezoid** A piezoid (finished crystal blank) used as a resonator or oscillator rather than as a transducer.

**resonator** A device that exhibits resonance at a particular frequency, such as an acoustic resonator or cavity resonator.

**resonator grid** A grid that is attached to a cavity resonator in a velocity-modulated tube to provide coupling between the resonator and the electron beam.

**resonator mode** The operating mode for which an electron stream introduces a negative conductance into the coupled circuit of an oscillator.

**resonator wavemeter** Any resonant circuit used to determine wavelength, such as a cavity-resonator frequency meter.

**responder** The transmitter section of a radar beacon.

**response** 1. A quantitative expression of the output of a device or system as a function of the input. 2. *Amplitude-frequency response.* 3. *Reply.*

**response characteristic** *Amplitude-frequency response.*

**responser** *Responsor.*

**response time** The time required for the output of a control system or element to reach a specified fraction of its new value after application of a step input or disturbance. Usually given in seconds, but for magnetic amplifiers the response time is often specified in cycles of the power-line frequency. For indicating instruments the time for the pointer to come to rest at its new position is used.

**responsor** The receiving section of an interrogator-responsor. Also called responser.

**resting frequency** *Carrier frequency.*

**rest mass** The mass of a particle at rest or when moving with a velocity low compared with that of light. At higher velocities the particle acquires an additional relativistic mass.

**restore** To return a computer word to its initial value.

**restorer** *D-c restorer.*

**restoring spring** The spring that moves the armature of a relay away from the core when the relay is deenergized.

**restricted radiation device** A device in which r-f energy is intentionally generated and conducted along wires or radiated, but the total electromagnetic field does not exceed 15 microvolts per meter at a distance in feet equal to 15,700 divided by the frequency in kc.

**restriking voltage** *Reignition voltage.*

**resultant** A force that combines the effects of two or more forces acting on an object.

**retained image** A partial image that remains on a camera-tube target for a large number of frames after the removal of a previously stationary light image, causing a spurious signal corresponding to that light image. Also called image burn.

**retainer** A clamp or other device specifically designed to restrain the movement of an electron tube, fuse, or other removable component mounted in a socket or holder.

**retardation coil** A high-inductance coil used in telephone circuits to permit passage of direct current or low-frequency ringing current while blocking the flow of a-f currents.

**retarded potential** The instantaneous electric potential produced at a point in space by a distribution of charge and current that existed earlier at a source and required a finite time for travel to the point.

**retarding field** An electric or magnetic field that slows up electrons traveling through an interelectrode space in an electron tube.

**retarding-field oscillator** An oscillator employing an electron tube in which the electrons oscillate back and forth through a grid that is maintained positive with respect to both the cathode and anode. The frequency depends on the electron transit time and may also be a function of associated circuit parameters. The field in the region of the grid exerts a retarding effect that draws electrons back after they pass through the grid in either direction. Also called positive-grid oscillator. Barkhausen-Kurz and Gill-Morell oscillators are examples.

**retarding-field tube** An electron tube designed for use in a retarding-field oscillator.

**retention** The percentage of radioactive atoms that cannot be separated from the target materials after production of the atoms by nuclear reaction or radioactive decay.

**retentivity** The property of a magnetic material that is measured by the residual flux density corresponding to the saturation induction for the material.

**RETMA** Abbreviation for Radio-Electronics-Television Manufacturers Association, now *Electronic Industries Association.*

**RETMA color code** Former name of *EIA color code.*

**retrace** *Flyback.*

**retrace blanking** Blanking a television picture tube during vertical retrace intervals to prevent retrace lines from showing on the screen. Voltage pulses for blanking are derived from vertical sweep oscillator or vertical deflection circuits and are applied to the control grid of the picture tube.

**retrace ghost** A ghost image produced on a television receiver screen during retrace periods. Generally due to insufficient blanking of the camera tube at the transmitter.

**retrace interval** The interval of time in which the blanked scanning beam of a television picture tube or camera tube returns to the starting point of a line or field.

It is about 7 microseconds for horizontal retrace and 500 to 750 microseconds for vertical retrace in U. S. television broadcasting. Also called retrace period, retrace time, return interval, return period, and return time.

**retrace line** The line traced by the electron beam in a cathode-ray tube in going from the end of one line or field to the start of the next line or field. Also called return line.

**retrace period** *Retrace interval.*

**retrace time** *Retrace interval.*

**retransmission unit** A control unit used at an intermediate station for feeding one radio receiver-transmitter unit automatically to another receiver-transmitter unit for two-way communication.

**retroaction** British term for *positive feedback.*

**return** 1. To go back to a planned point in a computer program and rerun a portion of the program, usually when an error is detected. Rerun points are usually not more than 5 minutes apart. 2. *Echo.*

**return interval** *Retrace interval.*

**return line** *Retrace line.*

**return loss** 1. The difference between the power incident upon a discontinuity in a transmission system and the power reflected from the discontinuity. 2. The ratio in decibels of the power incident upon a discontinuity to the power reflected from the discontinuity.

**return period** *Retrace interval.*

**return time** *Retrace interval.*

**return trace** *Flyback.*

**return transfer function** The transfer function that relates a feedback control loop return signal to the corresponding loop input signal.

**return wire** The ground wire, common wire, or negative wire of a direct-current power circuit.

**reverberation** The persistence of sound at a given point after direct reception from the source has stopped. In air it may be due to repeated reflections from a small number of boundaries or to free decay of normal modes of vibration that were excited by the sound source. In water it may be due to scattering from a large number of inhomogeneities in the medium or to reflections from bounding surfaces.

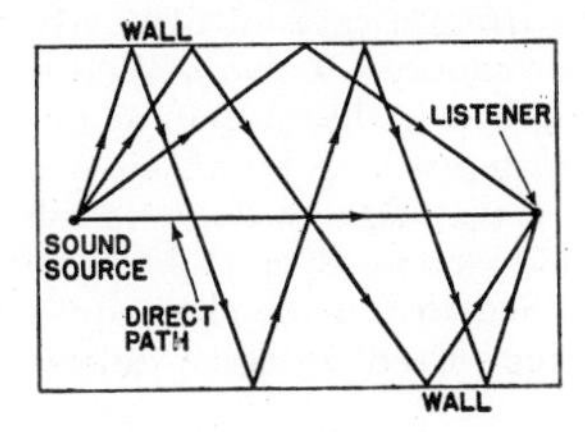

Reverberation paths for sound waves in room.

**reverberation absorption coefficient** The term used for the sound absorption coefficient when the distribution of incident sound is completely random.

**reverberation chamber** An enclosure in which all surfaces have been made as sound-reflective as possible. Used for some types of acoustic measurements. Also called reverberation room.

**reverberation-controlled gain circuit** [abbreviated rcg circuit] A circuit used in underwater sound equipment to vary the gain of the receiving amplifier in proportion to the strength of undesired reverberations associated with the desired echo.

**reverberation reflection coefficient** The term used for the sound reflection coefficient when the distribution of incident sound is completely random.

**reverberation room** *Reverberation chamber.*

**reverberation time** The time in seconds required for the average sound-energy density at a given frequency to reduce to one-millionth of its initial steady-state value after the sound source has been stopped. This corresponds to a decrease of 60 decibels.

**reverberation-time meter** An instrument for measuring the reverberation time of an enclosure.

**reverberation transmission coefficient** The term used for the sound transmission coefficient when the distribution of incident sound is completely random.

**reverse bias** A bias voltage applied to a diode or a semiconductor junction with polarity such that little or no current flows. It is the opposite of forward bias.

**reverse current** The small value of direct current that flows when a metallic rectifier or semiconductor diode has reverse bias.

**reverse-current relay** A relay that operates whenever current flows through the relay coil in the opposite of the normal direction.

**reversed image** 1. A mirror image, in which the right and left sides of the picture are interchanged. 2. *Negative image.*

**reverse direction** *Inverse direction.*

**reverse emission** The flow of electrons in the reverse direction (from anode to cathode) in a vacuum tube during that part of a cycle in which the anode is negative with respect to the cathode. The action is similar to arcback in gas tubes. Also called back emission.

**reverse resistance** The resistance of a metallic rectifier cell as measured at a specified value of reverse voltage or reverse current.

**reverse voltage** A voltage applied to a metallic rectifier with the opposite of normal polarity.

**reversible motor** A motor in which the direction of rotation can be reversed by means of a switch that changes motor connections when the motor is stopped.

**reversible permeability** The term used for incremental permeability when the change in magnetic induction is vanishingly small.

**reversible transducer** A transducer in which the transducer loss is independent of the direction of transmission.

**reversing motor** A motor for which the direction of rotation can be reversed by changing electric connections or by other means while the motor is running at full speed. The motor will then come to a stop, reverse, and attain full speed in the opposite direction.

**reversing switch** A switch intended to reverse the connections of one part of a circuit.

**rewind** 1. The components on a magnetic-tape recorder that serve to return the tape to the supply reel at high speed. 2. To return a magnetic tape to its starting position.

**rewrite** The process of restoring a storage device to its state prior to reading. Used when the information-storing state may be destroyed by reading.

**r-f** Abbreviation for *radio frequency.*

**r-f alternator** A rotating-type alternator designed to produce high power at frequencies above power-line values but generally lower than 100,000 cps. Used today chiefly for high-frequency heating.

**r-f amplification** Amplification of a radio signal at the same carrier frequency at which it travels through space.

**r-f amplifier** An amplifier that is used to increase the voltage or power of r-f signals at the carrier frequency. In a superheterodyne receiver, an r-f amplifier is sometimes used ahead of the converter.

**RFC** Symbol for *r-f choke.*

**r-f cable** A cable having electric conductors separated from each other by a continuous homogeneous dielectric or by touching or interlocking spacer beads, designed primarily to conduct r-f energy with low losses.

**r-f choke** [symbol RFC] An r-f coil designed and used specifically to block the flow of r-f current while passing lower frequencies or direct current.

**r-f coil** A coil having one continuous untapped winding, specifically designed to furnish inductive reactance for tuning purposes in a circuit carrying r-f current.

**r-f converter** A power source for producing electric power at a frequency above about 10,000 cps.

**r-f current** An alternating current having a frequency higher than about 10,000 cps.

**r-f energy** Alternating-current energy at any frequency in the radio spectrum between about 10,000 cps and 300,000,000 mc.

**r-f generator** A generator capable of supplying sufficient r-f energy at the required frequency for induction or dielectric heating.

**r-f harmonic** A harmonic of a carrier frequency. The frequency of the second r-f harmonic is twice the carrier frequency.

**r-f head** A radar transmitter and part of a radar receiver, contained in a package for ready installation and removal.

**r-f heating** *Electronic heating.*

**r-f indicator** An indicator that shows the presence of r-f energy at or near its own resonant frequency. It usually consists of a coil or parallel line connected to an incandescent lamp.

**r-f intermodulation distortion** Intermodulation distortion that originates in the r-f stages of a receiver.

**r-f oscillator** An oscillator that generates alternating current at radio frequencies.

**r-f pattern** A fine herringbone pattern occurring in a television picture due to high-frequency interference.

**r-f pentode** A pentode in which the screen grid is extended beyond the anode, to minimize electrostatic coupling between grid and anode and thereby reduce feedback at radio frequencies.

**r-f power probe** A probe used to extract r-f energy from a transmission system.

**r-f power supply** A high-voltage power supply in which the output of an r-f oscillator is stepped up by an air-core transformer to the high voltage required for the second anode of a cathode-ray tube, then rectified to provide the required high d-c voltage. Used in some television receivers.

**r-f pulse** An r-f carrier that is amplitude-modulated by a pulse. The amplitude of the modulated carrier is zero before and after the pulse.

**r-f resistance** *High-frequency resistance.*

**r-f response** The response of a receiver to radio frequencies within and outside the channel being received.

**r-f signal generator** A test instrument that generates the various radio frequencies required for alignment and servicing of radio, television, and electronic equipment. Also called service oscillator.

**r-f spectrum** *Radio spectrum.*

**r-f stage** A single stage of r-f amplification.

**r-f transformer** A transformer having a tapped winding or two or more windings designed to furnish inductive reactance and/or to transfer r-f energy from one circuit to another by means of a magnetic field. It may have an air core or some form of ferrite core.

**r-f transmission line** A transmission line designed primarily to conduct r-f energy, consisting of two or more conductors supported in a fixed spatial relationship along their own length.

**rhenium** [symbol Re] A metallic element. Atomic number is 75.

**rheo-** Prefix meaning a flow of current.

**rheobase** The intensity of the steady current just sufficient to excite a tissue when suddenly applied.

**rheostat** A resistor whose value may be changed readily by means of a control knob, to control the current in a circuit. A rheostat has one fixed terminal and one that is connected to the sliding or rolling contact. Also called variable resistor. A potentiometer is a rheostat having an additional fixed terminal at the other end of the resistance element.

**rheostriction** *Pinch effect.*

**rheotron** Former name for *betatron.*

**rhi display** Abbreviation for *range-height indicator display.*

**rhm** Abbreviation for *roentgen-per-hour-at-one-meter.*

**rhodium** [symbol Rh] A metallic element. Atomic number is 45.

**rhombic antenna** A horizontal antenna having four conductors forming a diamond or rhombus. Usually fed at one apex and terminated with a resistance or impedance at the opposite apex. Also called diamond antenna.

**rho meson** Former name for a meson that was stopped in a nuclear emulsion without any apparent decay event or nuclear interaction. Most of the rho mesons were mu mesons, the decay event being unobserved because of insufficient emulsion sensitivity.

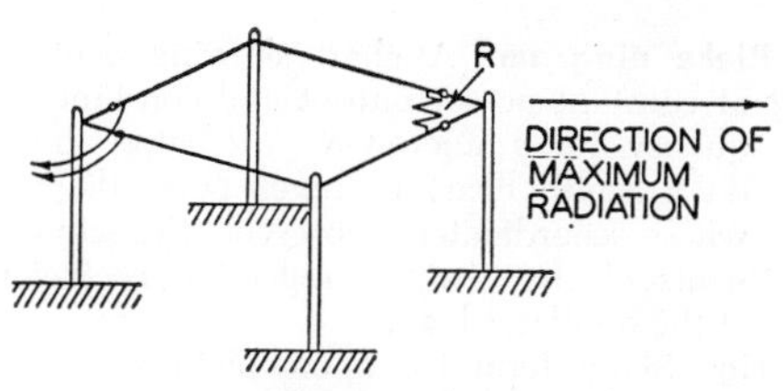

Rhombic antenna.

**rhometal** A high-resistivity magnetic alloy having an initial permeability of 250 to 2,000.

**rho-theta navigation** *Omnibearing-distance navigation.*

**rhumbatron** *Cavity resonator.*

**RIAA curve** 1. Recording Industry Association of America curve representing standard recording characteristics for long-play records. 2. The corresponding equalization curve for playback of microgroove records.

**ribbon microphone** A velocity microphone in which the moving element is a thin corrugated metal ribbon mounted between the poles of permanent magnets. The ribbon cuts magnetic lines of force as it is moved back and forth in proportion to the velocity of air particles in a sound wave, so an a-f output voltage is induced in the ribbon.

**Rice neutralizing circuit** An r-f amplifier circuit that neutralizes the grid-to-anode capacitance of the amplifier tube.

**Richardson effect** *Edison effect.*

**Richardson equation** An equation that gives the density of thermionic emission at saturation current in terms of the absolute temperature of the filament or cathode of an electron tube.

**ride gain** To control the volume range of an a-f circuit while watching a volume indicator.

**ridge waveguide** A circular or rectangular waveguide having one or more longitudinal internal ridges that serve primarily to increase transmisson bandwidth by lowering the cutoff frequency.

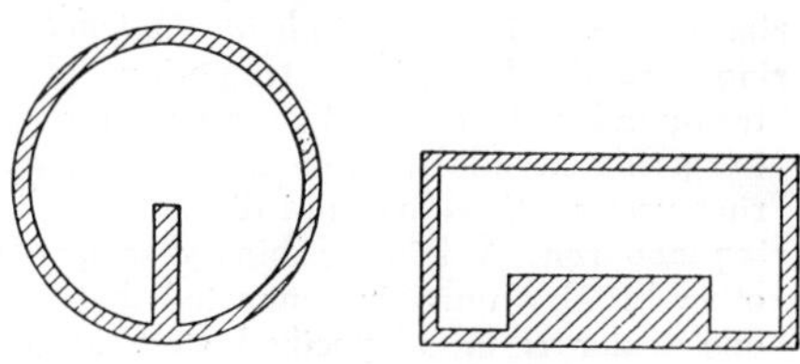
Ridge waveguides.

**Rieke diagram** A chart showing contours of constant power output and constant frequency for a microwave oscillator, drawn on a Smith chart or other polar diagram whose coordinates represent the components of the complex reflection coefficient at the oscillator load.

**rig** Slang term for a complete system of components, such as a complete amateur radio station.

**Righi-Leduc effect** Development of a difference in temperature between the two edges of a strip of metal in which heat is flowing longitudinally, when the plane of the strip is perpendicular to magnetic lines of force.

**right-hand polarized wave** An elliptically polarized transverse electromagnetic wave in which the rotation of the electric field vector is clockwise for an observer looking in the direction of propagation. Also called clockwise polarized wave.

**right-hand rule** 1. For a current-carrying wire: If the fingers of the right hand are placed around the wire in such a way that the thumb points in the direction of current flow, the fingers will be pointing in the direction of the magnetic field produced by the wire. For electron flow (the opposite of current flow), the left-hand rule is used. 2. For a movable current-carrying wire or an electron beam in a magnetic field: If the thumb, first, and second fingers of the right hand are extended at right angles to one another, with the first finger representing the direction of magnetic lines of force and the second finger representing the direction of current flow, the thumb will be pointing in the direction of motion of the wire or beam. Also called Fleming's rule.

**right-hand taper** Taper in which there is greater resistance in the clockwise half of the operating range of a potentiometer or rheostat (looking from the shaft end) than in the counterclockwise half.

**rim drive** A phonograph or sound recorder drive in which a rubber-covered drive wheel is in contact with the inside of the rim of the turntable.

**rim magnet** *Field-neutralizing magnet.*

**ring-around** Undesired triggering of a transponder by its own transmitter, or triggering at all bearings so as to give a ring presentation on a ppi display.

**ring counter** A loop of binary scalers or other bistable units so connected that only one scaler is in a specified state at any given time. As input signals are counted, the position of the one specified state moves in an ordered sequence around the loop.

**ring head** A magnetic head in which the magnetic material forms an enclosure having one or more air gaps. The magnetic recording medium bridges one of these gaps and is in contact with or in close proximity to the pole pieces on one side only.

**ringing** An oscillatory transient occurring in the output of a system as a result of a sudden change in input. In television it may produce a series of closely spaced images or may produce a black line immediately to the right of a white object.

**ringing time** 1. The time required for the output of an oscillatory circuit to decrease to a predetermined level after input power is removed. 2. The time between termination of a transmitted radar pulse and the instant at which the reradiated power from an echo box falls below the minimum required to produce an indication. Used as a measure of over-all radar performance.

**ring modulator** A modulator in which four diode elements are connected in series to form a ring around which current flows readily in one direction. Input and output connections are made to the four nodal points of the ring. Used as a balanced modulator, demodulator, or phase detector.

**ring oscillator** Two or more pairs of tubes operating as push-pull oscillators around a ring, usually with alternate successive pairs of grids and anodes connected to tank circuits. Adjacent tubes around the ring operate in phase opposition. The load is coupled to the anode circuits.

**ring scaler** A scaler in which the asymmetrical condition is passed along to the next tube in line, with the last tube feeding back to the first to complete the ring.

**ring-seal tube** An electron tube in which the grid and anode are radially symmetrical, with the grid connected to a metal ring sealed into the glass envelope. The construction permits insertion of the tube in a coaxial chamber with the grid connected to the outer cylinder and the anode to the inner conductor for operation as a grounded-grid amplifier.

**ring time** The time during which the output of a radar echo box remains above a specified level. The interval starts when a pulse is transmitted and is usually considered ended when the energy reradiated by the echo box falls below the minimum needed for an indication on the radar screen.

**ripple** The a-c component in the output of

a d-c power supply, arising within the power supply from incomplete filtering or from commutator action in a d-c generator.

**ripple factor** The ratio of the effective value of the a-c component of a pulsating d-c voltage to the average value.

**ripple filter** A low-pass filter designed to reduce ripple while freely passing the direct current obtained from a rectifier or d-c generator.

**ripple frequency** The frequency of the ripple present in the output of a d-c source.

**ripple voltage** The alternating component of the unidirectional voltage from a rectifier or generator used as a source of d-c power.

**rise time** The time required for a signal pulse to rise from 10% to 90% of its final steady value. It is a measure of the steepness of the wavefront.

**rising-sun magnetron** A multicavity magnetron in which resonators having two different resonant frequencies are arranged

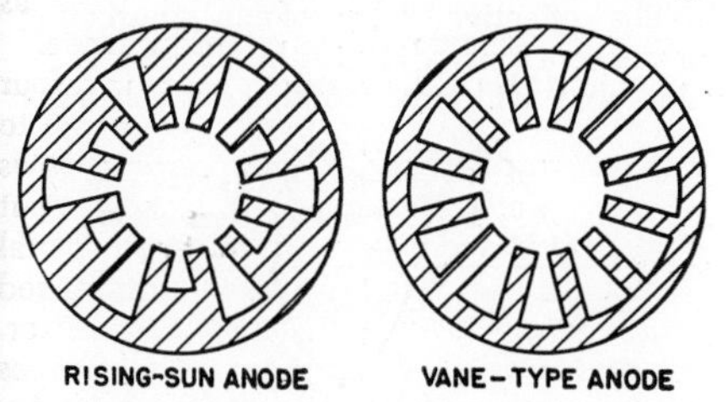


Rising-sun magnetron anode, with vane-type anode shown for comparison.

alternately for the purpose of mode separation. The cavities appear as alternating long and short radial slots around the perimeter of the anode structure, resembling the rays of the sun.

**RMA** Abbreviation for Radio Manufacturers Association, now *Electronic Industries Association.*

**RMA color code** Former name of *EIA color code.*

**r-meter** An ionization instrument calibrated to indicate the intensity of gamma rays, x-rays, and other ionizing radiation in roentgens.

**rmi** Abbreviation for *radio magnetic indicator.*

**rms current** Abbreviation for *root-mean-square current.*

**rms value** Abbreviation for *root-mean-square value.*

**robot** A completely self-controlled electronic, electric, or mechanical device.

**Rochelle-salt crystal** A crystal of sodium potassium tartrate, having a pronounced piezoelectric effect. Used in crystal microphones and crystal pickups.

**rocket** An unmanned self-propelled vehicle, with or without a warhead, designed to travel above the surface of the earth without control of its trajectory or course during flight.

**rocket-assisted torpedo** [abbreviated rat] A Navy surface-to-underwater guided missile system in which an acoustic homing torpedo having a rocket booster is launched into the air from a surface vessel and lowered into the water by a parachute when the target is within range of the homing system. When submerged, the torpedo releases the parachute, sheds its nose cap, and begins a spiraling search path. This changes to a homing path as soon as a submarine is picked up by the special sonar equipment in the torpedo.

**rocket missile** A missile using rocket propulsion.

**rocking** Back-and-forth rotation of the tuning control in a superheterodyne receiver while adjusting the oscillator padder near the low-frequency end of the tuning dial, to obtain more accurate alignment.

**rod** A relatively long and slender body of material used in or in conjunction with a nuclear reactor. It may contain fuel, absorber, or fertile material, or other material in which activation or transmutation is desired.

**roentgen** [abbreviated r] The international unit of quantity for x-rays and gamma rays. One roentgen of radiation will ionize dry air sufficiently to produce 1 electrostatic unit of electricity per 0.001293 gram of air.

**roentgen equivalent man** [abbreviated rem] A unit of ionizing radiation, equal to the amount that produces the same damage to man as 1 roentgen of high-voltage x-rays.

**roentgen equivalent physical** [abbreviated rep] A unit of ionizing radiation, equal to the amount that causes absorption of 93 ergs of energy per gram of soft tissue. Also called equivalent roentgen.

**roentgen meter** A meter for measuring the cumulative quantity of x-rays or gamma rays, without reference to time.

**roentgenogram** Deprecated term for *x-ray photograph.*

**roentgenography** Radiography by means of x-rays.

**roentgenology** That branch of radiology that deals with x-rays, especially their use for diagnosis or treatment in medicine and dentistry. Radiology is a slightly more general term covering also the use of gamma rays and other extremely high-frequency ionizing radiation.

**roentgenotherapy** Medical treatment by x-rays.

**roentgen-per-hour-at-one-meter** [abbreviated rhm] A unit of gamma-ray source strength, corresponding to a dose rate of 1 roentgen per hour at a distance of 1 meter in air.

**roentgen-rate meter** An electrically operated instrument that measures radioactivity and is calibrated in roentgens per unit time or any multiple thereof.

**roentgen ray** *X-ray.*

**roger** 1. A code word used in communications, meaning that a message has been received and understood. 2. An expression of agreement.

**Roget spiral** A helix of wire that contracts in length when a current is sent through, owing to mutual attraction between adjacent turns.

**roll** Slow upward or downward movement of the entire image on the screen of a television receiver, due to a lack of vertical synchronization.

**roll-and-pitch control** A control designed for use with automatic pilots and remote attitude indicators, consisting of a gyroscope that provides signals for controlling an aircraft about its lateral and longitudinal axes and/or signals for providing a visual presentation of attitude to the pilot.

**rollback** *Rerun.*

**rollback routine** *Rerun routine.*

**rolloff** Gradually increasing attenuation as frequency is changed in either direction beyond the flat portion of the amplitude-frequency response characteristic of a system or component.

**roll out** To read a computer register or counter by adding a one to each digit column simultaneously until all have returned to zero, with a signal being generated at the instant when a column returns to zero.

**romotar** A doppler ranging system similar to doran but having fewer modulating frequencies. Slant range is obtained directly by having the receiver and transmitter at the same ground station.

**roof filter** A low-pass filter used in carrier telephone systems to limit the frequency response of the equipment to frequencies needed for normal transmission, thereby blocking unwanted higher frequencies induced in the circuit by external sources. A roof filter improves runaround crosstalk suppression and minimizes high-frequency singing.

**room tone** A sound track recorded in a particular studio when unoccupied, for use between portions of dialog recorded previously in that studio to provide the quality of silence determined by the acoustics and temperature of the studio.

**rooter amplifier** A nonlinear amplifier in which negative feedback is used to make the output voltage vary as the square root or some other root of the input voltage. Used in video amplifiers of television transmitters for gamma correction to compensate for camera-tube characteristics.

**root-mean-square current** [abbreviated rms current] *Effective current.*

**root-mean-square particle velocity** *Effective particle velocity.*

**root-mean-square sound pressure** *Effective sound pressure.*

**root-mean-square value** [abbreviated rms value] 1. The square root of the average of the squares of a series of related values. 2. The effective value of an alternating current, corresponding to the direct-current value that will produce the same heating

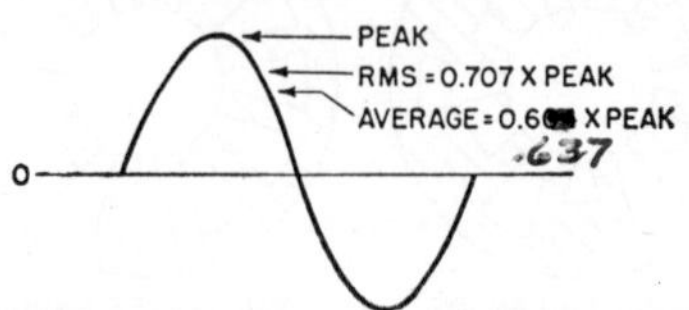


Root-mean-square (rms) value of sine wave, as compared with peak and average values.

effect. The rms value is computed as the square root of the average of the squares of the instantaneous amplitudes for one complete cycle. For a sine wave, the rms value is 0.707 times the peak value. Unless otherwise specified, alternating quantities are assumed to be rms values. Also called effective value.

**root-sum-square value** The square root of the sum of the squares of a series of related values. Commonly used to express total harmonic distortion.

**rope** A long form of confusion reflector, sometimes used with tiny parachutes to reduce the rate of fall of the long strips of metal foil.

**rosebud** An airborne radar beacon used in radar control and iff systems.

**rosin-core solder** Solder made up in tubular or other hollow form, with the inner space filled with rosin flux to serve as a noncorrosive flux for soldering joints.

**rosin joint** A soldered joint in which one of the wires is surrounded by an almost invisible film of insulating rosin, making the joint intermittently or continuously open even though it looks good. Insufficient heating of the parts before applying solder is the cause.

**rotary actuator** A device that converts electric energy into controlled rotary force. It usually consists of an electric motor, gear box, and limit switches.

**rotary amplifier** *Rotating magnetic amplifier.*

**rotary beam antenna** A highly directional short-wave antenna system mounted on a mast in such a manner that it can be rotated to any desired position either manually or by an electric motor drive.

**rotary converter** *Dynamotor.*

**rotary coupler** *Rotating joint.*

**rotary gap** *Rotary spark gap.*

**rotary generator** An a-c generator adapted to be rotated by a motor or other prime mover, as for generating power for induction heating.

**rotary joint** *Rotating joint.*

**rotary spark gap** A spark gap in which sparks occur between one or more fixed electrodes and a number of electrodes projecting outward from the circumference of a motor-driven metal disk. Also called rotary gap.

**rotary stepping relay** *Stepping relay.*

**rotary stepping switch** *Stepping relay.*

**rotary switch** A switch that is operated by rotating its shaft.

**rotary voltmeter** *Generating voltmeter.*

**rotatable loop antenna** A loop antenna that can be rotated in azimuth, for use in direction-finding.

**rotatable-loop radio compass** An automatic direction finder employing a loop antenna that is rotated manually to determine the relative bearing between an aircraft or ship and a transmitter.

**rotating amplifier** *Rotating magnetic amplifier.*

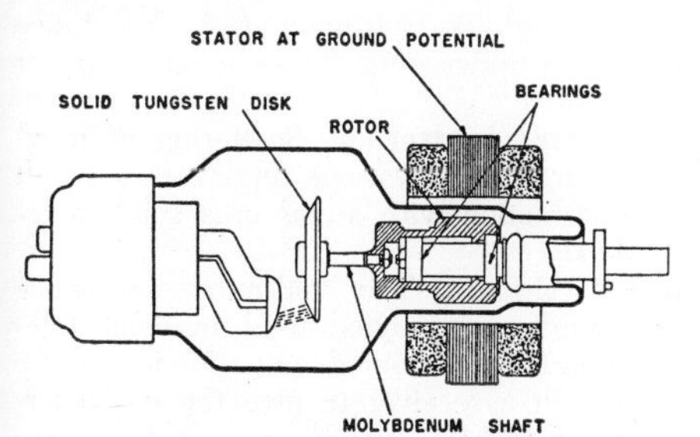


Rotating-anode tube, with rotor inside tube and stator surrounding glass neck outside tube. X-ray beam emerges downward after reflection from rotating tungsten anode disk.

**rotating-anode tube** An x-ray tube in which the anode rotates continually to bring a fresh area of its surface into the beam of electrons, allowing greater output without melting the target.

**rotating-crystal method** Rotation of a crystal while it is irradiated by monochromatic x-rays and recording of the reflected beams as spots on a photographic plate, for analysis of crystal structure.

**rotating joint** A joint that permits one section of a transmission line or waveguide to rotate continuously with respect to another while passing r-f energy. Also called rotary coupler and rotary joint.

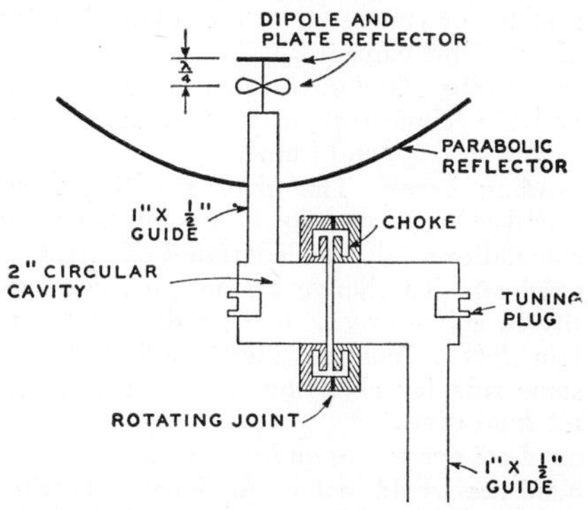


Rotating joint in waveguide feed to radar antenna

**rotating magnetic amplifier** A prime-mover-driven d-c generator whose power output can be controlled by small field input powers, to give power gain as high as 10,000. Used as power sources for the d-c drive motors of large radar antennas. as well as for other motor drives in automatic control systems. Examples include the amplidyne and metadyne. Also called rotary amplifier and rotating amplifier.

**rotating radio beacon** A radio transmitter arranged to radiate a concentrated beam that rotates in a horizontal plane at constant speed and transmits different signals in different directions so ships and aircraft can determine their bearings without directional receiving equipment.

**rotating-type scanning sonar** Scanning sonar using electric means to obtain a rotating receiving-beam pattern. Stationary transducer units arranged in a circle are connected in succession to the receiver by a commutator.

**rotation photograph** The photographic record of the diffracted beams produced when a slender beam of x-rays is directed on a rotating single crystal.

**rotation therapy** Radiation therapy in which either the patient or the source of radiation is rotated, to permit a larger dose at the center of rotation within the patient's body than on any area of the skin.

**rotation wave** *Shear wave.*

**rotator** A device that rotates the plane of polarization, such as a twist in a rectangular waveguide.

**rotoflector** An elliptically shaped rotating radar reflector used to reflect a vertically directed radar beam at right angles so it radiates horizontally.

**rotor** The rotating member of a machine or device, such as the rotating armature of a motor or generator, or the rotating plates of a variable capacitor.

**rotor plate** One of the rotating plates of a variable capacitor, usually directly connected to the metal frame.

**rounding error** The error resulting from round-off of a quantity in a computer or in calculations. Also called round-off error.

**round off** To change a more precise quantity to a less precise one by dropping certain less significant digits and applying some rule for changing the last significant retained digit.

**round-off error** *Rounding error.*

**round-the-world echo** A signal occurring every one-seventh second when a radio wave repeatedly encircles the earth at its speed of 186,000 miles per second.

**routine** A set of instructions arranged in proper sequence to cause a computer to perform a desired operation, such as the solution of a mathematical problem. Also called master routine.

**r parameter** A transistor parameter relating to resistivity.

**rpm** Abbreviation for revolutions per minute.

**rps** Abbreviation for revolutions per second.

**R scope** A radarscope that produces an R display.

**rt box** Deprecated term for *atr tube.*

**RTMA** Abbreviation for Radio-Television Manufacturers Association, now *Electronic Industries Association.*

**rt switch** Deprecated term for *atr tube.*

**rubber-covered wire** Wire insulated with rubber.

**rubidium** [symbol Rb] A photosensitive metallic element. Atomic number is 37.

**ruby maser** A maser that uses a ruby crystal in the cavity resonator.

**ruggedization** Making electronic equipment and components resistant to severe shock, temperature changes, high humidity, or other detrimental environmental influences.

**rumble** *Turntable rumble.*

**run** One performance of a program or routine on a computer.

**runaround crosstalk** Crosstalk resulting from coupling between the high-level end of one repeater and the low-level end of another repeater, as at a carrier telephone repeater station.

**running rabbits** Random spots drifting across a radar screen, due to interference from nearby radar sets.

**runway localizing beacon** A small radio range used to provide accurate directional guidance along the runway of an airport and for some distance beyond, for instrument landings.

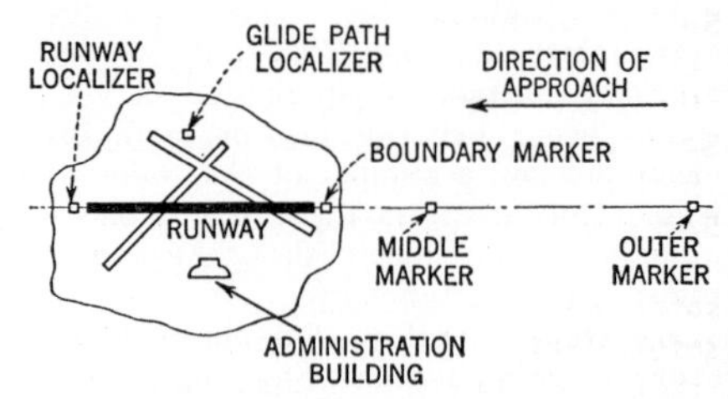


Runway localizing beacon for instrument landings, at opposite end of runway from marker beacons.

**rural radio service** A radio service used to provide public message communication service between a central office and subscribers located in rural areas to which it is impracticable to run wire lines.

**ruthenium** [symbol Ru] A metallic element. Atomic number is 44.

**rutherford** [abbreviated rd] 1. A unit used to express the decay rate of radioactive material, equal to $10^6$ disintegrating atoms per second. 2. That amount of a substance which is undergoing $10^6$ disintegrations per second.

**Rutherford scattering** Scattering of moving particles at various angles as a result of interaction with atoms of a solid material.

**R − Y signal** The red-minus-luminance color-difference signal used in color television. It is combined with the luminance signal in a receiver to give the red color-primary signal.

# S

**S** Symbol for *secondary winding,* used on circuit diagrams to identify the secondary winding of a transformer.

**sabin** A unit of sound absorption for a surface, equivalent to 1 square foot of perfectly absorbing surface. Also called square-foot unit of absorption.

**Sabine absorption** The sound absorption defined by the equation in which reverberation time in seconds is 0.049 $V/A$, where $V$ is the volume of the room in cubic feet and $A$ is the total Sabine absorption in sabins.

**Sabine coefficient** The Sabine absorption of a sound-absorptive surface divided by the area of the surface.

**SAC** [pronounced as a word] Abbreviation for Strategic Air Command.

**SAE** Abbreviation for Society of Automotive Engineers.

**safety base** *Acetate base.*

**safety factor** The amount of load, above the normal operating rating, that a device can handle without failure.

**safety rod** A control rod capable of shutting down a reactor quickly in case of failure of the ordinary control system using regulating rods and shim rods. A safety rod may be suspended above the core by a magnetic coupling and allowed to fall in when power reaches a predetermined level. Also called scram rod.

**sage** [SemiAutomatic Ground Environment] An air defense system in which air surveillance data is processed for transmission to computers at direction centers. Here the data is further processed, evaluated, and analyzed automatically to produce weapon assignment and guidance orders.

**Saint Elmo's fire** A visible electric discharge sometimes seen on the mast of a ship, on metal towers, and on projecting parts of aircraft, due to concentration of the atmospheric electric field at such projecting parts.

**sal ammoniac** Common name for ammonium chloride, a chemical compound used as an electrolyte in some types of dry cells.

**sal ammoniac cell** A cell in which the electrolyte consists primarily of a solution of ammonium chloride.

**sales representative** A salesman who sells the products of a number of different manufacturers in a given territory and represents those manufacturers in other ways. Also called representative.

**salient pole** A structure of magnetic material on which is mounted a field coil of a generator, motor, or similar device.

**samarium** [symbol Sm] A rare-earth element. Atomic number is 62.

**sampler** *Boxcar detector.*

**sample size** The number of units in a sample.

**sampling** Selecting a small statistically determined portion of the total group under consideration for tests used to infer the value of one or several characteristics of the entire group.

**sampling action** Control action in which the difference between the set point and the value of the controlled variable is measured and correction made only at intermittent intervals.

**sampling gate** A gate circuit that extracts information from the input waveform only when activated by a selector pulse.

**sampling plan** A plan that states sample sizes and the criteria for accepting, rejecting, or taking another sample during inspection of a group of items.

**sampling theorem** Equispaced data, in which there are two or more points per cycle of highest frequency, permits reconstruction of band-limited functions. Used in the study of information theory.

**sanaphant** A linear time-delay circuit similar to the sanatron, differing chiefly in the connections between the two pentodes.

**sanatron** A variable time-delay circuit having two pentodes and two diodes, used to produce very short gate waveforms having time durations that vary linearly with a reference voltage.

**sand load** An attenuator used as a power-dissipating terminating section for a coaxial line or waveguide. The dielectric space in the line is filled with a mixture of sand and graphite that acts as a matched-impedance load, preventing standing waves.

**sapphire** A pure variety of gem corundum occurring in nature and also produced artificially. Used for tips of phonograph needles because it has a hardness of 9 and takes a fine polish.

**sarah** [Search And Rescue And Homing] A radio system used to facilitate rescue when an aircraft goes down at sea. A small radio beacon transmitter attached to a lifebelt is used to send a coded pulse to a suitable receiver in search craft.

**Sargent curve** A curve in which decay constants of various radioactive materials are plotted against maximum particle energy.

**satellite** 1. A booster or translator operated by a television station to improve its signal strength in certain portions of its coverage area. 2. *Artificial satellite.*

**satelloid** A manned vehicle designed to orbit the earth as a satellite and then return to earth as an airplane.

**saturable-core magnetometer** A magnetometer that depends for its operation on the changes in permeability of a ferromagnetic core as a function of the magnetic field to be measured.

**saturable-core reactor** *Saturable reactor.*

**saturable reactor** An iron-core reactor having an additional control winding carrying direct current whose value is adjusted to change the degree of saturation of the core, thereby changing the reactance that the a-c winding offers to the flow of alternating current. With appropriate external circuits a saturable reactor can serve as a magnetic amplifier. Also called saturable-core reactor and transductor.

**saturable-reactor-controlled oscillator** An oscillator having a saturable reactor in its tuning circuit to control the output frequency.

**saturable transformer** A saturable reactor having additional windings to provide voltage transformation or isolation from the a-c supply.

**saturated activity** The maximum activity obtainable by activation in a definite flux in a nuclear reactor.

**saturated color** A pure color, not contaminated by white.

**saturated diode** A diode that is passing the maximum possible current, so further increases in applied voltage have no effect on current.

**saturating signal** A radar signal whose amplitude is greater than the dynamic range of the receiving system.

**saturation** 1. The condition in which a further increase in one variable produces no further increase in the resultant effect. 2. The condition in which the decay rate of a given radionuclide is equal to its rate of production in an induced nuclear reaction. 3. The condition in which the voltage applied to an ionization chamber is high enough to collect all the ions formed by radiation but not high enough to produce ionization by collision. 4. *Anode saturation.* 5. *Color saturation.* 6. *Magnetic saturation.* 7. *Temperature saturation.*

**saturation current** The maximum possible current that can be obtained as the voltage applied to a device is increased. In a gas tube it is the current at which the applied voltage is sufficient to attract all ions.

**saturation curve** A curve showing the manner in which a quantity such as current or magnetic flux reaches saturation.

**saturation flux density** *Saturation induction.*

**saturation induction** The maximum intrinsic induction possible in a material. Also called saturation flux density.

**saturation interval** The portion of the a-c power supply cycle in which the core of a saturable-core device is saturated.

**saturation magnetostriction** The value of magnetostriction that would be reached if the applied magnetizing force were increased indefinitely.

**saturation point** The point beyond which an increase in one quantity produces no further increase in another quantity.

**saturation reactance** The reactance of the gate winding of a magnetic amplifier during the saturation interval.

**saturation voltage** The minimum voltage needed to produce saturation current.

**sawtooth current** A current having a sawtooth waveform.

**sawtooth generator** A generator whose output voltage has a sawtooth waveform. Used to produce sweep voltages for cathode-ray tubes.

**sawtooth voltage** A voltage having a sawtooth waveform.

**sawtooth wave** A periodic wave whose amplitude varies linearly with time between two values, the interval required for one direction of progress being longer than that for the other, as in a sawtooth waveform.

**sawtooth waveform** A waveform characterized by a slow rise time and a sharp fall, resembling the tooth of a saw.

**saxophone** A linear antenna array that is fed at its vertex to give a cosecant-squared radiation pattern.

**sba** Abbreviation for *standard beam approach.*

**S band** A band of radio frequencies extending from 1,550 to 5,200 mc, corresponding to wavelengths of 19.37 to 5.77 cm.

**scalar** Having magnitude but not direction.

**scalar function** A scalar quantity that has a definite value for each value of some other scalar quantity. Thus, the resistance of a given conductor is a scalar function of the temperature of the conductor.

**scalar quantity** A quantity that has only magnitude, such as resistance, time, or temperature.

**scale** 1. A series of markings used for reading the value of a quantity or setting. 2. A series of musical notes arranged from low to high by a specified scheme of intervals suitable for musical purposes. 3. To change the magnitudes of the units in which a problem is expressed, so as to bring all magnitudes within the capacity of a computer.

**scale division** The portion of a scale between two adjacent scale marks.

**scale factor** The factor by which the reading of an instrument or the solution of a problem should be multiplied to give the true final value when a corresponding scale factor is used initially to bring the magnitude within the range of the instrument or computer.

**scale-of-eight circuit** A counting circuit that recycles at every eighth pulse.

**scale-of-ten circuit** *Decade scaler.*

**scale-of-two circuit** *Binary scaler.*

**scaler** A circuit that produces an output pulse when a prescribed number of input pulses is received. A single binary scaler stage delivers an output pulse for every two input pulses, while a decade scaler delivers an output pulse for every ten input pulses. Also called counter and scaling circuit.

**scaling** Counting pulses with a scaler when the pulses occur too fast for direct counting by conventional means.

**scaling circuit** *Scaler.*

**scaling factor** The number of input pulses per output pulse of a scaling circuit. Also called scaling ratio.

**scaling ratio** *Scaling factor.*

**scan** 1. To examine an area or a region in space point by point in an ordered sequence, as when converting a scene or image to an electric signal or when using radar to monitor an airspace for detection, navigation, or traffic control purposes. 2. One complete circular, up-and-down, or left-to-right sweep of the radar, light, or other beam or device used in making a scan.

**scan axis** The axis used as a reference for specifying target displacement in radar or sonar.

**scandium** [symbol Sc] A metallic element. Atomic number is 21.

**scanner** 1. A radar antenna and reflector assembly designed to oscillate back and forth about a center position during search, then move in any required plane during tracking of a target. 2. That part of a facsimile transmitter which systematically translates the densities of the elemental areas of the subject copy into corresponding electric signals. 3. A device that automatically samples, measures, or checks a number of quantities or conditions in sequence, as in process control.

**scanning** The process of examining an area, a region in space, or a portion of the radio spectrum point by point in an ordered sequence.

**scanning antenna** A directional antenna used in radar for scanning a region.

**scanning antenna mount** A radar antenna support that provides mechanical means for moving the antenna to give a desired scanning pattern, along with means to take off information for indication and control.

**scanning beam** A beam of light, an electron beam, or a radar beam used in scanning.

**scanning disk** A rotating metal disk having one or more spirals of holes near the circumference, used in early mechanical television systems to break up a scene into elemental areas at a television camera or to reconstruct a scene in a television receiver.

**scanning head** A light source and phototube combined as a single unit for scanning a moving strip of paper, cloth, or metal in photoelectric side-register control systems.

**scanning line** The single continuous narrow strip covered during one scan of a television image or one scan of the subject copy or record sheet in facsimile.

**scanning linearity** The uniformity of speed with which a scanning line is traversed.

**scanning-line frequency** The number of times per minute that a fixed line perpendicular to the direction of scanning is crossed in one direction by a scanning or recording spot in a facsimile system. In most mechanical systems this is equal to drum speed. Also called stroke speed.

**scanning-line length** The scanning-spot speed divided by the scanning-line frequency in a facsimile system.

**scanning loss** The reduction in sensitivity when scanning across a radar target as compared with the sensitivity obtained when the radar beam is directed constantly at the target. Scanning loss is due to the change in antenna position during the interval in which the signal travels from the antenna to the target and back.

**scanning sonar** Sonar in which all targets of interest are shown simultaneously, as on a radar ppi display or sector display. The sound pulse may be transmitted in all directions simultaneously and picked up by a rotating receiving transducer, or transmitted and received in only one direction at a time by a scanning transducer.

**scanning speed** *Spot speed.*

**scanning spot** The area that is viewed instantaneously by the pickup system of a facsimile scanner or a television camera.

**scanning transducer** A multielement sonar transducer in which the elements are arranged in a circle and electronically switched in sequence to give the equivalent of scanning without mechanical movement.

**scan period** *Regeneration period.*

**scan rate** The angular velocity of a radar beam. It is generally in the range of 4 to 12 revolutions per minute.

**scatterband** The total bandwidth occupied by the frequency spread of numerous interrogations having the same nominal radio frequency in a pulse system.

**scattered radiation** Radiation that has been changed in direction during its passage through a substance. It may also be increased in wavelength.

**scattering** 1. The change in direction of a particle or photon owing to a collision with another particle or a system. 2. Diffusion of acoustic waves because of nonuniformity of the transmitting medium. 3. Diffusion of electromagnetic waves in a random manner by air masses in the upper atmosphere, permitting long-range reception as in scatter propagation. Also called radio scattering.

**scattering amplitude** A quantity closely related to the intensity of scattering of a wave by a central force field such as that of a nucleus.

**scattering angle** The angle between the initial and final lines of motion of a scattered particle.

**scattering cross section** The cross section for elastic scattering with kinetic energy conserved plus the cross section for elastic scattering with absorption followed by emission of a neutron of lower energy. It is equal to 0.657 barn per electron. Also called classical scattering cross section and Thomson cross section.

**scattering loss** The portion of the transmission loss that is due to scattering within the medium or roughness of the reflecting surface.

**scattering matrix** A square array of complex numbers consisting of the transmission and reflection coefficients of a waveguide component.

**scatter propagation** Transmission of radio waves far beyond line-of-sight distances by using high power and a large transmitting antenna to beam the signal upward into the atmosphere and using a similar large receiving antenna to pick up the small portion of the signal that is scattered by the atmosphere. In ionospheric scatter the phenomenon occurs in the lower E layer of the ionosphere. In tropospheric scatter the phenomenon is entirely in the earth's lower atmosphere, from ground level to about 30,000 feet. In meteoric scatter the trail of a passing meteor scatters the radio waves back to earth. Also called beyond-the-horizon communication, forward-scatter propagation, and over-the-horizon communication.

**scc** Abbreviation for single-cotton-covered wire.

**sce wire** Abbreviation for single-cotton-over-enamel insulation on a wire.

**schematic circuit diagram** A circuit diagram in which component parts are represented by simple, easily drawn symbols. Also called circuit diagram.

**Schering bridge** A four-arm a-c bridge used to measure capacitance and dissipation factor. Bridge balance is independent of frequency.

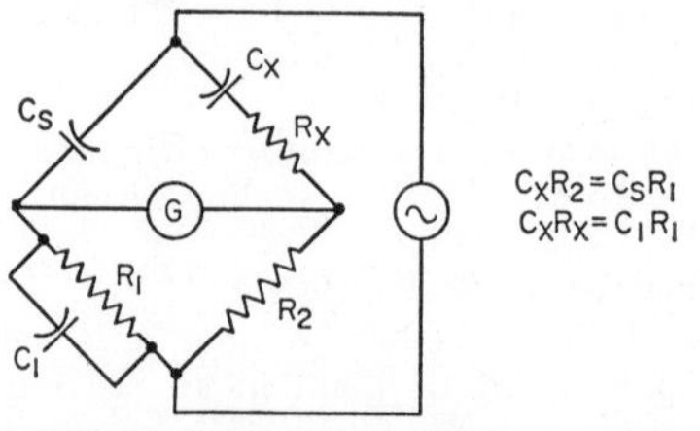


Schering bridge circuit and equations.

**schlieren photography** An optical system used to photograph changes in gas density due to sound waves, shock waves, or turbulence in wind tunnels. A knife edge at the focal point serves to transmit or cut off a light beam when the refraction of the intervening gas varies with density.

**Schmidt line** One of the two lines, in the plot of nuclear magnetic moment versus nuclear spin, that show the relation of these quantities in the independent-particle model of the nucleus.

**Schmidt system** A projection optics system consisting of a spherical mirror, a curved image plate at the focal plane of the mirror, and a corrector plate for correcting the spherical aberration of the mirror. Used to project a television image from a projection cathode-ray tube onto a screen.

**Schottky effect** An increase in anode current of a thermionic tube beyond that predicted by the Richardson equation, due to lowering of the work function of the cathode when an electric field is produced at the surface of the cathode by the anode.

**Schottky noise** *Shot noise.*

**Schroedinger equation** A wave equation that gives the relation between wave function, particle mass, total energy, potential energy, and Planck's constant.

**Schroedinger wave function** A wave function that determines the state of a system and satisfies the Schroedinger equation. For a particle or photon, the square of the wave function is proportional to the probability that the particle will be at a particular point at a particular time.

**scientist** A person having the training, ability, and desire to seek new knowledge, new principles, and new materials in some field of science.

**scintillation** 1. A flash of light (optical photons) produced in a phosphor by an ionizing particle or photon. 2. A rapid apparent displacement of a target indication from its mean position on a radar display. One cause is shifting of the effective reflection point on the target. Also called target glint, target scintillation, and wander.

**scintillation counter** A counter in which the scintillations produced in a fluorescent material by an ionizing radiation are detected and counted by a multiplier phototube and associated circuits. Widely used in medical research, nuclear research, and prospecting for radioactive ores.

**scintillation-counter cesium resolution** The scintillation-counter energy resolution for the gamma ray or conversion electron from cesium-137.

**scintillation-counter head** The combination of scintillators and phototubes or photocells used to produce electric pulses in response to ionizing radiation.

**scintillation-counter time discrimination** The smallest interval of time that can separate two events individually discernible by a given scintillation counter.

**scintillation decay time** The time required for the rate of emission of optical photons of a scintillation to decrease from 90% to 10% of its maximum value.

**scintillation duration** The time interval from the emission of the first optical photon of a scintillation until 90% of the optical photons of the scintillation have been emitted.

**scintillation rise time** The time required for the rate of emission of optical photons of a scintillation to increase from 10% to 90% of its maximum value.

**scintillation spectrometer** A scintillation counter adapted to the study of energy distributions.

**scintillator** The body of scintillator material together with its container.

**scintillator conversion efficiency** The ratio of the optical photon energy emitted by a scintillator to the incident energy of a particle or photon of ionizing radiation.

**scintillator material** A material that emits optical photons in response to ionizing radiation. The five major classes of scintillator materials are: (a) inorganic crystals such as sodium iodide, thallium single crystals, and zinc sulfide-silver screens; (b) organic crystals such as anthracene and transstilbene; (c) liquid, plastic, or glass solution scintillators; (d) gaseous scintillators; (e) Cerenkov scintillators.

**scintillator photon distribution** The curve giving the number of optical photons produced by a scintillator when it totally absorbs the energy of a monoenergetic particle or photon of ionizing radiation.

**scintillator total conversion efficiency** The ratio of the optical photon energy produced to the energy of a particle or photon of ionizing radiation that is totally absorbed in the scintillator material.

**scope** 1. *Cathode-ray oscilloscope.* 2. *Radarscope.*

**Scophony television system** An early mechanical television system using a rotating mirror and Kerr cell, developed in Great Britain.

**scram** A sudden shutting down of a nuclear reactor, usually by dropping safety rods, when a predetermined neutron flux or other dangerous condition occurs.

**scrambled speech** Speech that has been made unintelligible by inversion, such as for privacy of transoceanic radiotelephone calls. At the receiving end, the signals are inverted again to restore the original sounds. Also called inverted speech.

**scrambler** A circuit that divides speech frequencies into several ranges by means of filters, then inverts and displaces the frequencies in each range so the resulting reproduced sounds are unintelligible. In the simplest system the entire speech frequency range is combined with the output of a fixed-frequency oscillator, and difference frequencies are used as the inverted signal. The process is reversed at the receiving apparatus to restore intelligible speech. Also called secrecy system, speech inverter, and speech scrambler.

**scram rod** *Safety rod.*

**scratch filter** A low-pass filter circuit inserted in the circuit of a phonograph pickup to suppress higher audio frequencies and thereby minimize needle-scratch noise.

**screen** 1. The surface on which a television, radar, x-ray, or cathode-ray oscilloscope image is made visible for viewing. It may be a fluorescent screen having a phosphor layer that converts the energy of an electron beam to visible light, or a translucent or opaque screen on which the optical image is projected. Also called viewing screen. 2. *Screen grid.*

**screen factor** The ratio of the actual area of a grid structure in an electron tube to the total area of the surface containing the grid.

**screen grid** [symbol SG] A grid placed between a control grid and an anode of an electron tube and usually maintained at a fixed positive potential, to reduce the electrostatic influence of the anode in the space between the screen grid and the cathode. Also called screen.

**screen-grid modulation** Modulation produced by introducing the modulating signal into the screen-grid circuit of a multigrid tube in which the carrier is present.

**screen-grid tube** An electron tube having a screen grid.

**screen-grid voltage** The d-c voltage value applied between the screen grid and the cathode of an electron tube.

**screening** 1. The reduction of the electric field about a nucleus by the space charge of the surrounding electrons. 2. *Shield.*

**screening constant** The atomic number of an element minus the apparent atomic number that is effective for a given process.

**screening elevation** The elevation angle from a radar antenna to the crest that screens the radar.

**Scriptron** Trademark of Machlett Laboratories for their cathode-ray tube that produces a display in the form of letters or numbers.

**S distortion** *Spiral distortion.*

**sea clutter** Clutter on an airborne radar due to reflection of signals from the sea. Also called sea return and wave clutter.

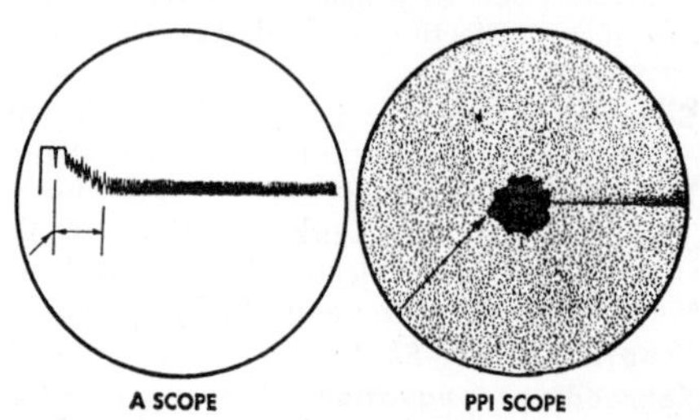


Sea clutter as seen on two types of radarscope displays.

**seal** A joint between two pieces of glass, two pieces of metal, or between glass and metal. For electron tubes the joint must be hermetically tight.

**sealed circuit** A printed circuit coated or encapsulated with an insulating material that keeps out moisture and protects against mechanical damage.

**sealed crystal unit** A crystal unit in which the quartz plate is sealed in its holder, usually by means of a gasket under pressure, for protection against humidity or a contaminated atmosphere.

**sealed tube** An electron tube that is hermetically sealed. This term is used chiefly for pool tubes.

**sealing compound** A compound used in dry batteries, capacitor blocks, transformers, and other components to keep out air and moisture.

**sealing off** Final closing of the envelope of an electron tube after evacuation.

**seam welding** 1. Spot welding in which the lapped metal sheets are passed between roller contacts and overlapping spot welds are made. 2. Use of dielectric heating for uniting two pieces of thermoplastic material along a prescribed line.

**search** To explore a region in space with radar.

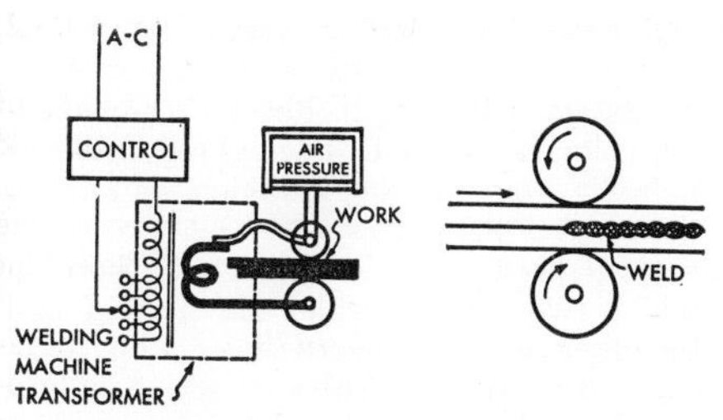


Seam-welding machine and nature of weld.

**search antenna** A radar antenna or antenna system designed especially for search.

**search coil** *Exploring coil.*

**searching gate** A gate pulse that is made to search back and forth over a certain range.

**searchlight-control radar** Ground radar used to direct searchlights at aircraft.

**searchlighting** Projecting a radar beam continuously at a target, instead of scanning the area containing the target.

**searchlight sonar** A sonar system in which a directional transducer is used to concentrate the outgoing pulse of sound energy into a narrow beam and receive the echo reflected from an underwater target. The bearing of a target is determined by aiming the transducer for maximum echo strength.

**search radar** A radar intended primarily to cover a large region of space and display targets as soon as possible after they enter the region. Used for early warning, and in connection with ground-controlled approach, ground-controlled interception, and air traffic control.

**search receiver** A radio receiver that can be tuned over a wide frequency range for use in detecting and measuring r-f signals transmitted by the enemy.

**sea return** *Sea clutter.*

**sec** 1. Abbreviation for *second.* 2. Abbreviation for *secondary winding* as applied to a transformer winding.

**seco** [SEquential COntrol] A teletype control system that permits a primary station to control automatically the sequential transmission of data stored on perforated tape at secondary stations on the teletype circuit. Also called sequential control.

**second** [abbreviated sec] 1. A unit of time equal to 1/86,400th part of a mean solar day. 2. A unit of angle, equal to 1/3,600th part of a degree.

**second anode** *Ultor.*

**secondary** *Secondary winding.*

**secondary battery** *Storage battery.*

**secondary cell** *Storage cell.*

**secondary cosmic rays** Radiation produced when primary cosmic rays enter the atmosphere and collide with atomic nuclei. Each such collision splits one or both particles into smaller nuclear fragments such as mesons, protons, neutrons, electrons, and photons, and these in turn collide with other nuclei to produce additional high-speed particles. All of these results of collisions are known as secondary cosmic rays.

**secondary electron** 1. An electron emitted as a result of bombardment of a material by an incident electron. 2. An electron whose motion is due to a transfer of momentum from primary radiation.

**secondary-electron gap loading** The electronic gap admittance that results from the traversal of a gap by secondary electrons originating in the gap.

**secondary-electron multiplier** *Electron multiplier.*

**secondary emission** The emission of electrons from a solid or liquid as a result of bombardment by electrons or other charged particles.

**secondary-emission ratio** The average number of electrons emitted from a surface per incident primary electron.

**secondary filter** A sheet of material having a low atomic number relative to that of a primary filter used in radiology, placed in the filtered beam of radiation to remove characteristic radiation produced in the primary filter.

**secondary grid emission** Electron emission from a grid as a direct result of bombardment of the grid surface by electrons or other charged particles.

**secondary lobe** *Minor lobe.*

**secondary parameter** An additional rating or characteristic needed to evaluate the operation of a part beyond its normal limits, such as its temperature coefficient.

**secondary radar** Radar involving transmission of a second signal when the incident signal triggers a responder beacon.

**secondary radiation** Particles or photons produced by the action of primary radiation on matter, such as Compton recoil electrons, delta rays, secondary cosmic rays, and secondary electrons.

**secondary radionuclide** A radionuclide that has a geologically short lifetime and is a decay product of natural primary radionuclides. All presently known members of this class belong to the elements from thallium to uranium.

**secondary service area** The area served by the sky wave of a broadcast station and not subject to objectionable interference. The signal is subject to intermittent variations in intensity.

**secondary standard** 1. A unit, as of length, capacitance, or weight, used as a standard of comparison in individual countries or localities, but checked against the one primary standard in existence somewhere. 2. A unit defined as a specified multiple or submultiple of a primary standard, such as the centimeter.

**secondary storage** Storage that is not an integral part of the computer but is directly linked to and controlled by the computer.

**secondary voltage** The voltage across the secondary winding of a transformer.

**secondary winding** [abbreviated sec; symbol S] A transformer winding that receives energy by electromagnetic induction from the primary winding. A transformer may have several secondary windings, and they may provide a-c voltages that are higher, lower, or the same as that applied to the primary winding. Also called secondary.

**secondary x-ray** An x-ray given off by a material when irradiated by x-rays. The frequency of the secondary rays is characteristic of the material.

**second-channel attenuation** *Selectance.*

**second-channel interference** Interference in which the extraneous power originates from a signal of assigned type in a channel two channels removed from the desired channel.

**second detector** The detector that separates the intelligence signal from the i-f signal in a superheterodyne receiver.

**second-harmonic magnetic modulator** A magnetic modulator in which the output frequency is twice the power supply frequency.

**second-time-around echo** A radar or sonar echo received after an interval exceeding the pulse recurrence interval.

**secor** [SEquential COrrelation Ranging] An instrumentation system designed to provide distance and position information during missile tests. A missile-borne transponder, when interrogated, sends signals to several ground stations to provide distance data. Bearing data is obtained by phase comparison techniques.

**secrecy system** *Scrambler.*

**section** Each individual transmission span in a radio relay system. A system has one more section than it has repeaters.

**sectionalized vertical antenna** A vertical antenna that is insulated at one or more points along its length. Reactances or driving voltages are applied across the insulated points to modify the radiation pattern in the vertical plane.

**sectoral horn** An electromagnetic horn in which two opposite sides are parallel and the other two sides diverge.

**sector display** A display in which only a sector of the total service area of a radar system is shown. Usually the sector is selectable.

**sector scan** A radar scan through a limited angle, as distinguished from complete rotation.

**secular equilibrium** Radioactive equilibrium in which the parent has such a small decay constant that there has been no appreciable change in the quantity of parent present by the time the decay products have reached radioactive equilibrium. Equal numbers of atoms of all members of the series then disintegrate in unit time.

**secular variation** The slow variation in the strength of the magnetic field of the earth, requiring many years for a complete cycle.

**security classification** The classification assigned to defense information or material to denote the degree of danger to the nation that would result from unauthorized disclosure. The usual classifications are confidential, secret, and top secret.

**security clearance** A clearance that permits a person to have access to classified material or information up to and including a given security classification, provided he can establish a need-to-know.

**Seebeck effect** Development of a voltage due to differences in temperature between two junctions of dissimilar metals in the same circuit. Discovered by J. T. Seebeck, German physicist, in 1821. Also called thermoelectric effect.

**seed** A small single crystal of semiconductor material, used to start the growth of a large single crystal for use in cutting semiconductor wafers.

**seeker** A missile or other device that finds its target by means of heat, light, radio waves, sound, or other radiation emitted by the target.

**segment** A portion of a digital-computer routine short enough to be stored entirely in the internal storage yet containing the coding necessary to call in and jump automatically to other segments of the routine.

**seismic detector** A microphone used to detect acoustic waves transmitted through the earth.

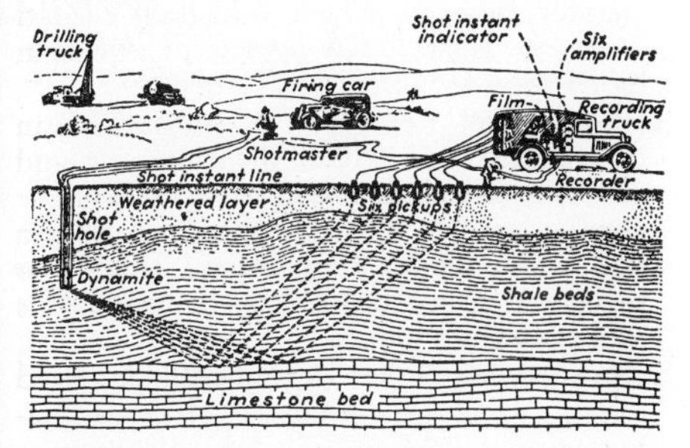


Seismic detectors being used in prospecting for petroleum.

**seismograph** An instrument for recording the time, direction, and intensity of earthquakes or of earth shocks produced by explosions during geophysical prospecting.

**selectance** The reciprocal of the ratio of the sensitivity of a receiver tuned to a specified channel to its sensitivity at another channel separated by a specified number of channels from the one to which the receiver is tuned. Generally expressed as a voltage or field-strength ratio. Also called adjacent-channel attenuation and second-channel attenuation.

**selection check** A check made by a digital computer to verify that the correct register or other device is selected for performance of the next instruction.

**selection ratio** The least ratio of a magnetomotive force used to select a magnetic cell to the maximum magnetomotive force used that is not intended to select a cell, in the storage section of a digital computer.

**selection rules** Statements that classify transitions in terms of quantum numbers of the initial and final states of the systems involved, in such a way that transitions of a given order of inherent probability are grouped together. Transitions having highest probability of taking place per unit time are called allowed transitions; all others are called forbidden transitions.

**selective absorption** Absorption of radiation at some function of frequency.

**selective calling system** A radio communication system in which the central station transmits a coded call that activates only the receiver to which that code is assigned.

**selective fading** Fading that is different at different frequencies in a frequency band occupied by a modulated wave, causing distortion that varies in nature from instant to instant.

**selective jamming** Jamming in which only a single radio channel is jammed.

**selectivity** The characteristic of a receiver that determines its ability to separate a desired signal frequency from all other signal frequencies.

**selectivity control** A control that adjusts the selectivity of a radio receiver.

**selector** An automatic or other device for making connections to any one of a number of circuits, such as a selector relay or selector switch.

**selector pulse** A pulse used to identify for selection one event in a series of events.

**selector relay** A relay capable of selecting one circuit automatically from a number of circuits.

**selector switch** A manually operated multiposition switch. Also called multiple-contact switch.

**selectron** An electron tube that can serve as a computer memory. One type stores 256 binary digits.

**selenium** [symbol Se] A nonmetallic element having photosensitive properties. Its resistance varies inversely with illumination. Used also as a rectifying layer in metallic rectifiers. Atomic number is 34.

**selenium cell** A photoconductive cell in which a thin film of selenium is used between suitable electrodes. The resistance of the cell decreases when illumination is increased.

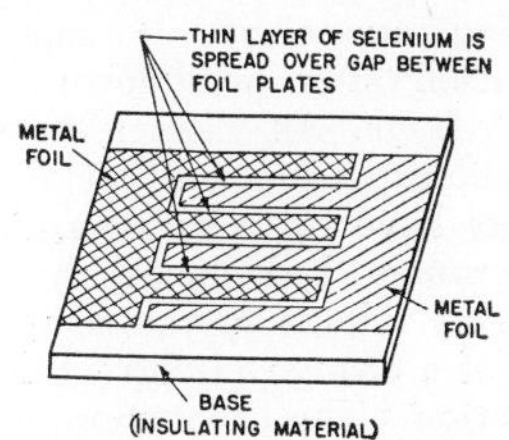


Selenium cell construction.

**selenium rectifier** A metallic rectifier in which a thin layer of selenium is deposited on one side of an aluminum plate and a conductive metal coating is deposited on

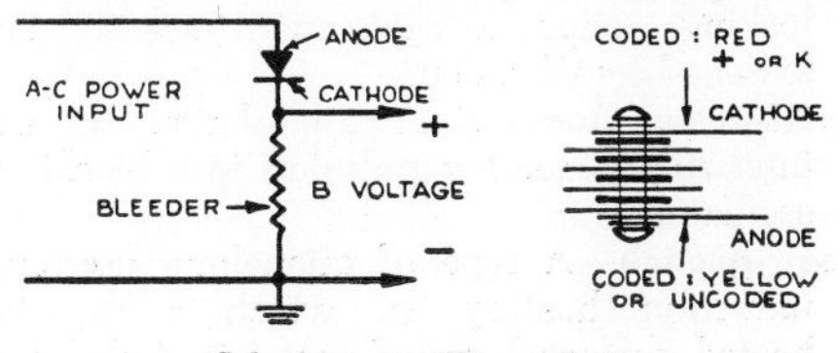


Selenium rectifier polarity.

the selenium. Electrons flow more freely in the direction from the metallic coating to the selenium than in the opposite direction, thus giving rectifying action.

**self-absorption** Absorption of radiation by the material that emits the radiation, reducing the radiation level against which further shielding must be provided.

**self-bias** Grid bias provided automatically by the flow of electrode currents through a resistor in the cathode or grid circuit of an electron tube. The resulting voltage drop across the resistor serves as the grid bias. Also called automatic bias, automatic C bias, and automatic grid bias.

**self-capacitance** *Distributed capacitance.*

**self-checking code** A computer code in which errors in a code expression produce a forbidden combination. Also called error-detecting code.

**self-cleaning contact** *Wiping contact.*

**self-energy** The energy equivalent of the rest mass of a particle. The self-energy of an electron is 0.511 million electron-volts.

**self-excited** Operating without an external source of a-c power.

**self-excited oscillator** An oscillator that depends on its own resonant circuits for initiation of oscillation and frequency determination.

**self-focused picture tube** A television picture tube having automatic electrostatic focus incorporated into the design of the electron gun.

**self-guided** Directed only by built-in self-reacting devices, as in a homing missile.

**self-healing capacitor** A capacitor that repairs itself after breakdown due to excessive voltage. Air capacitors, some wet electrolytic capacitors, and metallized paper capacitors have this characteristic.

**self-inductance** Inductance that produces an induced voltage in the same circuit as a result of a change in current flow.

**self-induction** The production of a voltage in a circuit by a varying current in that same circuit.

**self-instructed carry** A carry in which information goes to succeeding locations automatically as soon as it is generated.

**self-locking nut** A nut having an inherent locking action, so it cannot readily be loosened by vibration.

**self-pulse modulation** Modulation by an internally generated pulse, as in a blocking oscillator.

**self-quench** A type of quenching used in induction heating, in which a rapidly heated surface layer of metal is cooled by the rapid conduction of heat into the cold interior.

**self-quenched counter tube** A radiation-counter tube in which reignition of the discharge is inhibited by gas or other internal means.

**self-quenched detector** A superregenerative detector in which the time constant of the grid leak and grid capacitor is sufficiently large to cause intermittent oscillation above audio frequencies. This stops regeneration just before it spills over into a squealing condition.

**self-rectifying tube** A hot-cathode x-ray tube in which an a-c anode voltage is used but current flows in only one direction as long as the anode is kept cool.

**self-regulation** Inherent stability of a nuclear reactor, wherein deviation from a certain power level affects reactivity in such a way as to tend to restore power to the previous level without intervention of the control system.

**self-resetting** Automatically returning to an original position when current is interrupted or normal conditions are restored. Applied chiefly to relays and circuit breakers.

**self-saturating circuit** A magnetic-amplifier circuit in which a rectifier is used in series with the output winding to give higher gain and faster response. The rectifier conducts and saturates the core during positive half-cycles of the applied a-c voltage, and blocks load current during negative half-cycles to permit resetting of the core by the control current. Also called reactor-rectifier amplifier and self-saturating magnetic amplifier.

**self-saturating magnetic amplifier** *Self-saturating circuit.*

**self-scattering** Scattering of radiation by the material that emits the radiation, increasing the measured activity over that expected for a weightless sample.

**self-shielding** Shielding of the inner portion of the fuel in a nuclear reactor by the outer portion of the fuel.

**self-starting synchronous motor** A synchronous motor provided with the equivalent of a squirrel-cage winding, to permit starting as an induction motor.

**self-supporting antenna tower** An antenna tower that requires no guy wires.

**self-synchronous device** *Synchro.*

**self-wiping contact** *Wiping contact.*

**selsyn** [SELf SYNchronous] *Synchro.* Not a trademark.

**selsyn generator** *Synchro transmitter.*

**selsyn motor** *Synchro receiver.*
**selsyn receiver** *Synchro receiver.*
**selsyn system** *Synchro system.*
**selsyn transmitter** *Synchro transmitter.*
**semiactive homing** Homing in which the transmitter that illuminates the target is not on the missile. The missile contains only the receiver for energy reflected from the target.
**semiconductor** A material whose resistivity is between that of insulators and conductors. The resistivity is often changed by light, heat, an electric field, or a magnetic field. Current flow is often achieved by transfer of positive holes as well as by movement of electrons. Examples include germanium, lead sulfide, lead telluride, selenium, silicon, and silicon carbide. Widely used in diodes, photocells, thermistors, and transistors.
**semiconductor device** An electron device in which conduction takes place within a semiconductor.
**semiconductor diode** *Crystal diode.*
**semiconductor rectifier** *Metallic rectifier.*
**semiduplex operation** Operation of a communication circuit with one end duplex and the other end simplex. When used in mobile systems, the base station is usually duplex and the mobile stations are simplex.
**semiremote control** Remote control of a radio transmitter by devices connected to but not an integral part of the transmitter.
**semitone** The interval between two sounds whose basic frequency ratio is approximately equal to the twelfth root of two. The interval, in equally tempered semitones, between any two frequencies is 12 times the logarithm to the base 2 (or 39.86 times the logarithm to the base 10) of the frequency ratio. Also called half-step.
**semitransparent photocathode** A camera tube or phototube photocathode in which radiant flux on one side produces photoelectric emission from the opposite side.
**sending** Transmitting, as Morse code.
**sending-end impedance** The input impedance of a transmission line.
**sensation level** *Level above threshold.*
**sense** 1. The relation of a change in the indication of a radio navigation facility to the change in the navigation parameter being indicated. 2. To resolve a 180° ambiguity in a reading. 3. To determine the arrangement or position of a device or the value of a quantity. 4. To read punched holes in tape or cards.
**sense antenna** An auxiliary antenna used with a directional receiving antenna to resolve a 180° ambiguity in the directional indication. Also called sensing antenna.
**sense indicator** A flight instrument used to determine whether an aircraft is flying toward or away from a vhf omnirange.
**sensing** The process of determining the sense of an indication.
**sensing antenna** *Sense antenna.*
**sensing element** An element that detects a change in a selective physical quantity and converts that change into a useful input signal for a measuring, recording, or control system. Also called primary detector, primary element, and sensor.
**Sensistor** Trademark of Texas Instruments for their silicon resistors.
**sensitive altimeter** An altimeter sufficiently sensitive for use in instrument flying.
**sensitive relay** A relay that will operate at small currents, usually below 10 milliamperes.
**sensitive switch** *Snap-action switch.*
**sensitive time** The duration of supersaturation required for track formation following expansion in a cloud chamber.
**sensitive volume** The portion of a radiation-counter tube that responds to a specific radiation.
**sensitivity** A figure of merit that expresses the ability of a circuit or device to respond to an input quantity. Expressed as divisions per volt or ohms per volt for a measuring instrument, as spot displacement per volt of deflection voltage or ampere of deflection current for a cathode-ray tube, as output current per unit incident radiation density for a camera tube or other photoelectric device, and as microvolts of input signal when specifying minimum signal strength to which a receiver will respond.
**sensitivity control** A control that adjusts the amplification of r-f amplifier stages in a receiver.
**sensitivity-time control** [abbreviated stc] An automatic control circuit that changes the gain of a receiver at regular intervals, to obtain desired relative output levels from two or more sequential and unequal input signals. In a loran receiver it keeps output signal amplitude essentially constant as the receiver is tuned between input signals of different strength. In a radar receiver it reduces the gain after transmission of a pulse so nearby echo signals do not overload the system, then gradually restores the gain to the maximum value required for more distant targets. Also called amplitude balance control, anti-

clutter gain control, differential gain control, gain-time control, swept gain control, temporal gain control, and time-varied gain control.

**sensitization** *Activation.*

**sensitizer** *Activator.*

**sensitometer** An instrument for measuring the sensitivity of light-sensitive materials.

**sensitometry** The measurement of the light-response characteristics of photographic film under specified conditions of exposure and development.

**sensor** *Sensing element.*

**sentinel** A symbol marking the beginning or end of an element of computer information. Also called tag.

**separately excited** Obtaining excitation from a source other than the machine or device itself.

**separately instructed carry** A carry in which information goes to the next location only on receipt of a specific signal.

**separation circuit** A circuit that sorts signals according to amplitude, frequency, or some other characteristic.

**separation energy** *Binding energy.*

**separation factor** The abundance ratio of two isotopes after processing, divided by their abundance ratio before processing.

**separation filter** A filter used to separate one band of frequencies from another, as in carrier systems.

**separative element** An isotope-separating unit, such as a centrifuge, fractionating tower, calutron, or diffuser, that may be considered as the basic element of a system containing many similar units.

**separative power** A measure of the useful amount of separation accomplished in unit time by a separative element.

**separator** 1. A circuit that separates one type of signal from another by clipping, differentiating, or integrating action. 2. A porous insulating sheet used between the plates of a storage battery.

**septate waveguide** A waveguide containing one or more septa placed across the waveguide to control the transmission of microwave power.

**septum** [plural septa] A metal plate placed across a waveguide and attached to the walls by highly conducting joints. The plate usually has one or more windows or irises designed to give inductive, capacitive, or resistive characteristics.

**sequence control** The automatic control of a series of operations in a predetermined order.

**sequence relay** A relay that opens or closes two or more sets of contacts in a predetermined sequence.

**sequence weld timer** A timer that controls the sequence and duration of each portion of a complete resistance-welding cycle.

**sequential color television** A color television system in which the primary color components of a picture are transmitted one after the other. The three basic types are the line-sequential, dot-sequential, and field-sequential color television systems. Also called sequential system.

**sequential control** 1. Control of a digital computer in such a way that instructions are fed into the computer in a given sequence during the running of a problem. 2. *Seco.*

**sequential-control line** One of the wire lines that interconnect the radio stations in a conelrad system, to permit turning the stations on and off in a given sequence from a central control point.

**sequential interlace** Interlace in which the lines of one field fall directly under the corresponding lines of the preceding field.

**sequential scanning** *Progressive scanning.*

**sequential system** *Sequential color television.*

**Sergeant** An Army surface-to-surface guided missile, reported to be using inertial guidance. Range is about 75 miles. It can carry either conventional or nuclear warheads.

**serial** Pertaining to time-sequential transmission of, storage of, or logical operations on the parts of a word in a digital computer, using the same facilities for successive parts.

**serial by bit** Digital-computer storage in which the individual bits that make up a computer word appear in time sequence.

**serial by character** Digital-computer storage in which the characters for coded-decimal or other nonbinary numbers appear in time sequence.

**serial by word** Digital-computer storage in which the words within a given group appear one after the other in time sequence.

**serial digital computer** A digital computer in which the digits are handled serially, although the bits that comprise a digit may be handled either serially or in parallel.

**serial operation** The flow of information through a computer in time sequence, using only one digit, word, line, or channel at a time.

**serial radiography** Making a number of radiographs of the same subject in succession.

**serial storage** Computer storage in which time is one of the coordinates used to locate any given bit, character, or word. Access time therefore includes a variable waiting time ranging from zero to many word times.

**serial transfer** Transfer of the characters of an element of information in sequence over a single path in a digital computer.

**series** 1. An arrangement of circuit components end to end to form a single path for current. 2. The indicated sum of a set of terms in a mathematical expression, as in an alternating series or an arithmetic series.

**series circuit** A circuit in which all parts are connected end to end to provide a single path for current.

**series coil** The coil that carries the main current in a rotating machine or other device. The shunt coil is connected across the line and usually carries only a small current.

**series connection** A connection that forms a series circuit.

**series decay** *Radioactive series.*

**series disintegration** The successive radioactive transformations in a radioactive series. Also called chain decay and chain disintegration.

**series element** A two-terminal element connected to complete the only path existing between two nodes of a network. Any mesh including one series element must include all the other series elements of the mesh.

**series excitation** Obtaining field excitation in a motor or generator by allowing the armature current to flow through the field winding.

**series-fed vertical antenna** A vertical antenna that is insulated from the ground and energized at its base. Also called end-fed vertical antenna.

**series feed** Application of direct voltage to the anode of a vacuum tube through the load that is carrying the output signal current. In shunt feed, the direct voltage is applied to the anode through a choke, and only signal current flows through the load.

**series loading** Loading in which reactances are inserted in series with the conductors.

**series modulation** Modulation in which the modulating tube, the modulated amplifier tube, and the anode voltage supply are all in series.

**series motor** A commutator-type motor having armature and field windings in series. Characteristics are high starting torque, variation of speed with load, and dangerously high speed on no-load. Also called series-wound motor.

**series-parallel switch** A switch used to change the connections of lamps or other devices from series to parallel, or vice versa.

**series peaking** Use of a peaking coil and resistor in series as the load for a video amplifier to produce peaking at some desired frequency in the passband, such as to compensate for previous loss of gain at the high-frequency end of the passband.

**series resonance** Resonance in a series resonant circuit, wherein the inductive and capacitive reactances are equal at the frequency of the applied voltage. The reactances then cancel each other, reducing the impedance of the circuit to a minimum purely resistive value. Signal current is then a maximum, and the signal voltage developed across either the coil or capacitor may be several times the voltage applied to the combination.

**series resonant circuit** A resonant circuit in which the capacitor and coil are in series with the applied a-c voltage.

**series tee junction** A tee junction having an equivalent circuit in which the impedance of the branch waveguide is predominantly in series with the impedance of the main waveguide at the junction.

**series-triggered blocking oscillator** A triggered blocking oscillator in which the

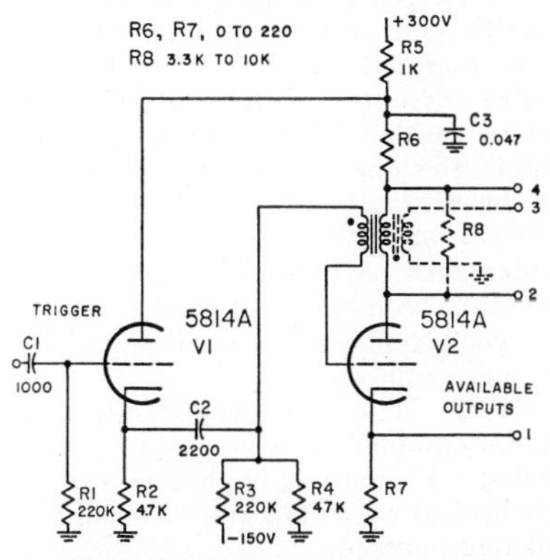


Series-triggered blocking oscillator using preferred circuit design of National Bureau of Standards. Input trigger may be 30 to 50 volts, at 200 to 2,000 pulses per second.

trigger pulse is applied to the grid of the blocking oscillator tube rather than to the anode.

**series two-terminal pair networks** Two-terminal pair networks having either their input or output terminals in series.

**series-wound motor** *Series motor.*

**serrasoid modulator** A modulator that uses a crystal-controlled sawtooth oscillator in conjunction with a low-frequency modulating voltage to produce variable triggering time for a blocking oscillator.

**serrated pulse** A pulse having notches or sawtooth indentations in its waveform.

**serrated rotor plate** *Slotted rotor plate.*

**serrated vertical pulse** A vertical synchronizing pulse that is broken up by five notches extending down to the black level of a television signal, to give six component pulses, each lasting about 0.4 line and serving to keep the horizontal sweep circuits in step during the vertical sync pulse interval.

**serrodyne** A phase modulator using transit-time modulation of a traveling-wave tube or klystron.

**service area** The area that is effectively served by a given radio or television transmitter, navigation aid, or other type of transmitter. Also called coverage.

**service band** A band of frequencies allocated to a given class of radio service.

**service engineer** An engineer who is qualified to maintain and repair complex electronic equipment after it has been installed.

**service life** The length of time that a battery or other device will provide specified performance under specified conditions of use.

**serviceman** A person engaged in the maintenance and repair of equipment, such as a radio serviceman.

**service oscillator** *R-f signal generator.*

**service routine** A routine designed to assist in the operation of a computer, such as block location, correction, and tape comparison routines.

**service test** A test made under simulated or actual conditions of use to determine the characteristics, capabilities, and limitations of a product.

**servicing** Preventive maintenance and repair of equipment.

**serving** A wrapping applied to a cable for mechanical protection rather than for insulating purposes.

**servo** *Servomotor.*

**servoamplifier** An amplifier used in a servomechanism.

**servomechanism** A feedback control system in which one or more of the system signals represent mechanical motion. Usually used to control an output position mechanically in response to input signal changes. A servomechanism may also be a regulator under certain conditions. Also called servosystem.

**servomotor** The electric, hydraulic, or other type of motor that serves as the final control element in a servomechanism. It receives power from the amplifier element and drives the load with a linear or rotary motion. Also called servo.

**servosystem** *Servomechanism.*

**sesqui-sideband transmission** Transmission of a carrier modulated by one full sideband and half of the other sideband.

**set** 1. A radio or television receiver. 2. A combination of units, assemblies, and parts connected or otherwise used together to perform an operational function, such as a radar set. 3. To place a storage device in a prescribed state, such as to place a binary cell in the one state.

**set analyzer** A test instrument used to measure voltages and currents in electronic equipment by using a cable and plugs to transfer the test points to the instrument. Also called analyzer.

**set point** The value selected to be maintained by an automatic controller.

**set pulse** A drive pulse that tends to set a magnetic cell in a computer.

**setscrew** A small headless machine screw, usually having a point at one end and a recessed hexagonal socket or a slot at the other end, used for such purposes as holding a knob or gear on a shaft or holding a playing needle in a phonograph pickup.

**settling time** *Correction time.*

**setup** The ratio between reference black level and reference white level in television, both measured from blanking level. Usually expressed as a percentage.

**sexadecimal number system** A digital system based on powers of 16, as compared to the use of powers of 10 in the decimal number system.

**sferics** [coined from atmospherics] *Atmospheric interference.*

**sferics set** An electronic system used to detect, analyze, and determine the position of electromagnetic disturbances generated by any atmospheric phenomena.

**SG** Symbol for *screen grid.*

**shaded-pole motor** A single-phase induction motor having one or more auxiliary short-circuited windings acting on only a portion of the magnetic circuit. Generally the winding is a closed copper ring imbedded in the face of a pole. The shaded pole provides the required rotating field for starting purposes.

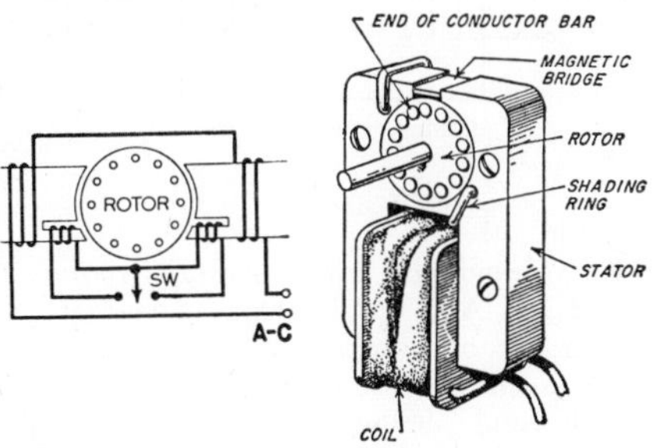


Shaded-pole motor with switch for reversing direction of rotation, and single-direction shaded-pole motor used in phonographs.

**shading** A variation in brightness over the area of a reproduced television picture, due to spurious signals generated in a television camera tube during the retrace intervals. These spurious signals are generally due to redistribution of secondary electrons over the mosaic in a storage-type camera tube, and vary from scene to scene as background illumination changes.

**shading coil** *Shading ring.*

**shading generator** One of the signal generators used at a television transmitter to generate waveforms that are 180° out of phase with the undesired shading signals produced by a television camera. An operator watches the picture on a monitor and adjusts the controls of the shading generators as required to give essentially uniform scene brightness.

**shading ring** 1. A heavy copper ring sometimes placed around the central pole piece of an electrodynamic loudspeaker to serve as a shorted turn that suppresses the hum

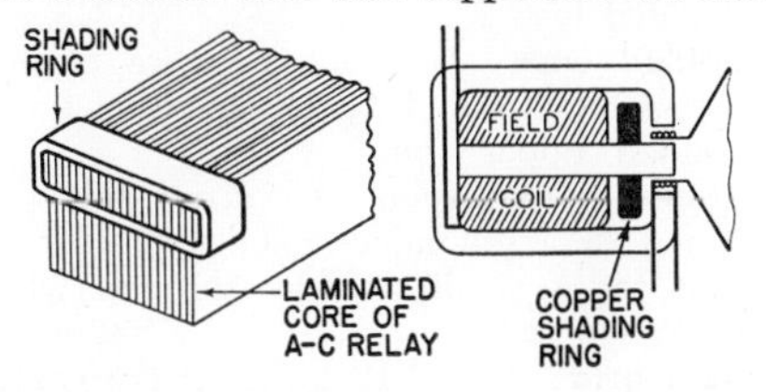


Shading ring as used on relay and on excited-field loudspeaker.

voltage produced by the field coil. 2. The copper ring used in a shaded-pole motor to produce a rotating magnetic field for starting purposes. Also used around part of the core of an a-c relay to prevent contact chatter. Also called shading coil.

**shadow factor** The ratio of the electric field strength that would result from propagation of waves over a sphere to that which would result from propagation over a plane under comparable conditions.

**shadow mask** A thin perforated metal mask mounted just back of the phosphor-dot faceplate in a three-gun color picture

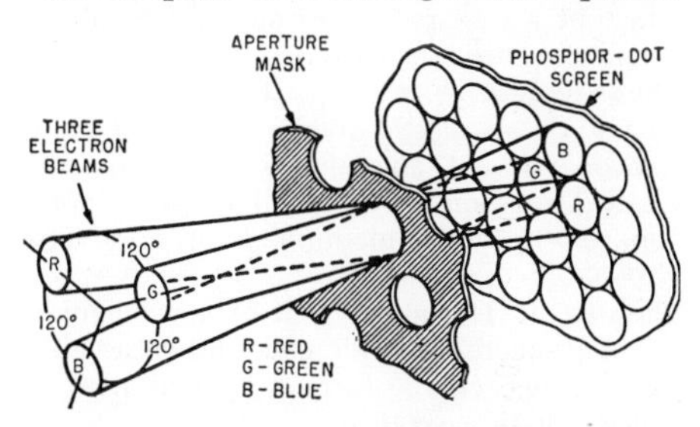


Shadow-mask details, showing positions of color phosphor dots with respect to apertures in mask.

tube. The holes in the mask are positioned to insure that each of the three electron beams strikes only its intended color phosphor dot. Also called aperture mask.

**shadow-mask color picture tube** A three-gun color picture tube that uses a shadow mask.

**shadow region** The region in which received field strength is so reduced by some obstruction that effective reception of signals or radar detection of objects is normally improbable.

**shadow tuning indicator** A tuning indicator that shows by means of a moving shadow the accuracy with which a receiver is tuned. Older forms used the shadow of a meter pointer. Newer electron-tube types use an electron stream to produce a shadow on a small fluorescent screen, as in a cathode-ray tuning indicator.

**shank** The portion of a phonograph needle that is clamped or otherwise anchored in a phonograph pickup.

**shaped-beam antenna** A unidirectional antenna whose major lobe differs materially from that provided by an aperture that gives uniform phase. Also called phase-shaped antenna.

**shaped-beam tube** A character-writing tube in which a character is formed all at once by an electron beam whose cross-section corresponds to the shape of the character.

**shape factor** *Form factor.*

**shaping network** *Corrective network.*

**sharp-cutoff tube** An electron tube in which the control-grid openings are uniformly spaced. Anode current then decreases linearly as grid voltage is made more negative, and cuts off sharply at a particular grid voltage.

**sharp tuning** Having high selectivity, and therefore responding only to a desired narrow range of frequencies.

**shaving** Removing material from the surface of a disk recording medium to obtain a new recording surface.

**shear wave** A wave that causes an element of an elastic medium to change its shape without changing its volume. Also called rotation wave. A shear plane wave in an isotropic medium is a transverse wave.

**sheath** 1. The metal wall of a waveguide. 2. A protective outside covering on a cable. 3. A space charge formed by ions near an electrode in a gas tube.

**sheath-reshaping converter** A wave converter in which the change of wave pattern is achieved by gradual reshaping of the sheath of the waveguide and use of conducting metal sheets mounted longitudinally in the waveguide.

**sheet grating** A grating consisting of thin longitudinal metal sheets extending inside a waveguide a distance of about one wavelength. Used to suppress undesired modes of propagation.

**shelf life** The time that elapses before an unused battery or other device becomes inoperative due to age or deterioration.

**shell** A group of electrons that form part of the outer structure of an atom and have a common energy level.

**shellac** A purified lac resin, once widely used in insulating materials and commercial phonograph records.

**shell model** *Independent-particle model.*

**shell structure** The arrangement of the quantum states of nucleons (protons or neutrons) of a given kind in a nucleus in groups of approximately the same energy. Each such group is called a shell. The number of nucleons in each shell is limited by the Pauli exclusion principle. A closed shell is one containing the maximum number. A nucleus having all of its protons and/or neutrons in closed shells has greater than average stability.

**shell-type transformer** A power transformer in which the primary and secondary coils are wound over each other on the center leg of the iron core structure. There are no coils on the outer two legs, which serve as return paths for the magnetic circuit.

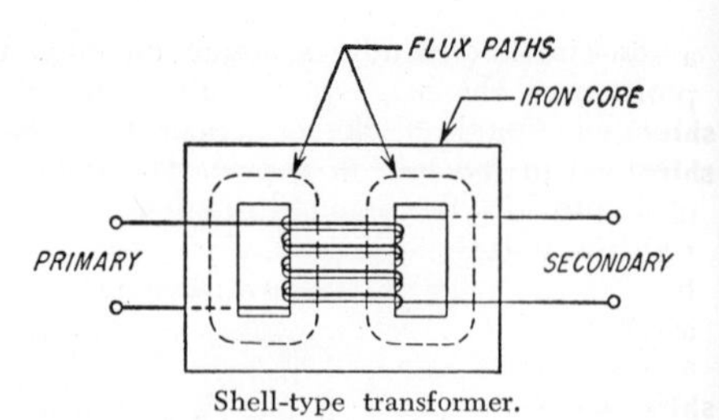

Shell-type transformer.

**shf** Abbreviation for *superhigh frequency.*

**shield** 1. A metal housing placed around a circuit or component to suppress the effect of an electric or magnetic field within or beyond definite regions. Also called screening and shielding. 2. Material placed around a nuclear reactor or other source of radiation to reduce escaping radiation or particles to a permissible level. Also called shielding.

**shielded-arc welding** Arc welding in which the metal electrode is coated with a flux that produces an envelope of protective inert gas.

**shielded cable** A cable having a conducting envelope around its insulated conductors. Braided wire is most often used for this purpose.

**shielded line** *Shielded transmission line.*

**shielded nuclide** A nuclide having a charge that is one unit larger than that of a stable nuclide having the same mass number. A shielded nuclide is generally a primary fission product.

**shielded pair** A two-wire transmission line surrounded by a metallic sheath.

**shielded transmission line** A transmission line whose elements essentially confine propagated electric energy to a finite space inside a conducting sheath, to prevent the line from radiating radio waves. Also called shielded line.

**shielded wire** An insulated wire covered with a metal shield, usually made of tinned braided copper wire.

**shielded x-ray tube** An x-ray tube enclosed in a grounded metal container except for a small window through which x-rays emerge.

**shield grid** A grid that shields the control grid of a gas tube from electrostatic fields, thermal radiation, and deposition of thermionic emissive material. The shield grid may also be used as an additional control electrode.

**shield-grid thyratron** A thyratron having

a shield grid, usually operated at cathode potential.

**shielding** *Shield.*

**shift** Displacement of an ordered set of characters one or more places to the left or right in a digital computer. If the characters are the digits of a numerical expression, a shift may be equivalent to multiplying by a power of the base.

**shift pulse** A drive pulse that initiates shifting of characters in the register of a digital computer.

**shift register** A computer circuit that converts a sequence of input signals into a parallel binary number or vice versa.

**shimming** Adjustment of the strength of a magnetic field by means of thin spacers, shims of soft iron, or compensating coils.

**shim rod** A control rod used for making occasional coarse adjustments in the reactivity of a nuclear reactor.

**ship error** A radio direction finder error due to reradiation of radio waves by the metal structure of a ship.

**shock** 1. The sudden pain, convulsion, unconsciousness, or death produced by the passage of electric current through the body. Also called electric shock. 2. *Shock motion.*

**shock excitation** Excitation produced by a voltage or current variation of relatively short duration. Used to initiate oscillation in the resonant circuit of an oscillator.

**shock-excited oscillation** *Free oscillation.*

**shock motion** Transient mechanical motion characterized by suddenness and by significant relative displacements. Also called shock.

**shock mount** A mount used with sensitive equipment to reduce or prevent transmission of shock motion to the equipment.

**shock therapy** *Electroshock therapy.*

**shock wave** A sound wave produced by a sudden change in pressure and particle velocity. It may travel faster than the velocity of sound.

**shoran** [SHOrt-RAnge Navigation] A precision position-fixing system using a pulse transmitter and receiver in an aircraft or other vehicle and two transponder beacons at fixed points. A receiver in the aircraft measures the round-trip times of the signals and converts these into distances to the fixed ground stations. Ordinary triangulation on a map then gives position.

**shoran bombing** Bombing done after positioning the aircraft to the bomb-release point by shoran.

**shore effect** Bending of waves toward the shoreline when traveling over water near a shoreline, due to the slightly greater velocity of radio waves over water than over land. This effect causes errors in radio direction finder indications.

**shore radar station** A shore radio navigation station using radar to determine the direction and distance of ships from the station at night or during fog.

**shore-to-ship communication** Radio communication between a shore station and a ship at sea.

**short** *Short-circuit.*

**short-circuit** A low-resistance connection across a voltage source or between both sides of a circuit or line, usually accidental and usually resulting in excessive current flow that may cause damage. Also called short.

**short-circuit driving-point admittance** The driving-point admittance between the *j*th terminal of an *n*-terminal network and the reference terminal when all other terminals have zero alternating components of voltage with respect to the reference point.

**short-circuit feedback admittance** The short-circuit transadmittance from the physically available output terminals of an electron-tube transducer to the physically available input terminals of a specified socket, associated filters, and tube.

**short-circuit forward admittance** The short-circuit transadmittance from the physically available input terminals of an electron-tube transducer to the physically available output terminals of a specified socket, associated filters, and tube.

**short-circuit input admittance** The short-circuit driving-point admittance of an electron-tube transducer at the physically available input terminals of a specified socket, associated filters, and tube.

**short-circuit input capacitance** The effective capacitance of an *n*-terminal electron tube as determined from the short-circuit input admittance.

**short-circuit output admittance** The short-circuit driving-point admittance of an electron-tube transducer at the physically available output terminals of a specified socket, associated filters, and tube.

**short-circuit output capacitance** The effective capacitance of an *n*-terminal electron tube as determined from the short-circuit output admittance.

**short-circuit parameter** One of a set of four transistor equivalent-circuit parameters that specify transistor performance when the input and output voltages are chosen as the independent variables. Two of the four measurements require short-circuiting of the input for alternating current.

**short-circuit transadmittance** The transfer admittance obtained when all terminals except that under consideration have zero complex alternating components of voltage with respect to the reference point.

**short-circuit transfer capacitance** The effective capacitance as determined from the short-circuit transadmittance of an electron tube.

**short-distance navigation aid** A navigation aid that is useful primarily at distances within radio line of sight.

**shorted out** Made inactive by connecting a heavy wire or other low-resistance path around a device or portion of a circuit.

**shorting-contact switch** A selector switch in which the width of the movable contact is greater than the distance between fixed contacts, so the new circuit is contacted before the old one is broken. This avoids noise during switching.

**short-range radar** Radar whose maximum line-of-sight range, for a reflecting target having 1 square meter of area perpendicular to the beam, is between 50 and 150 miles.

**short-time rating** A rating defining the load that a machine, apparatus, or device can carry for a specified short time.

**short wave** [abbreviated s-w] A general term applied to a wavelength shorter than 200 meters, corresponding to frequencies higher than the highest broadcast-band frequency.

**short-wave antenna** An antenna designed for reception of frequencies above the broadcast band, in the range from about 1,600 to 30,000 kc.

**short-wave converter** A converter used between a receiver and its antenna system to convert incoming high-frequency signals to a lower carrier frequency to which the receiver can be tuned. A converter usually contains a local oscillator and a mixer, as in a superheterodyne receiver.

**short-wave receiver** A radio receiver designed to tune in stations in the range from about 1,600 to 30,000 kc or some portion of that range.

**short-wave transmitter** A radio transmitter that radiates short waves, generally for communication purposes or for international broadcasting.

**shot effect** *Shot noise.*

**shot noise** Noise voltage developed in a thermionic tube because of the random variations in the number and velocity of electrons emitted by the heated cathode. The effect causes sputtering or popping sounds in radio receivers and snow effects in television pictures. Also called Schottky noise and shot effect.

**shower** *Cosmic-ray shower.*

**shower unit** The mean path length required to reduce the energy of relativistic charged particles to half value as they pass through matter. One shower unit is equal to 0.693 radiation length.

**Shrike** An Air Force surface-to-air guided missile.

**shunt** 1. A precision low-value resistor placed across the terminals of an ammeter to increase its range by allowing a known fraction of the circuit current to go around

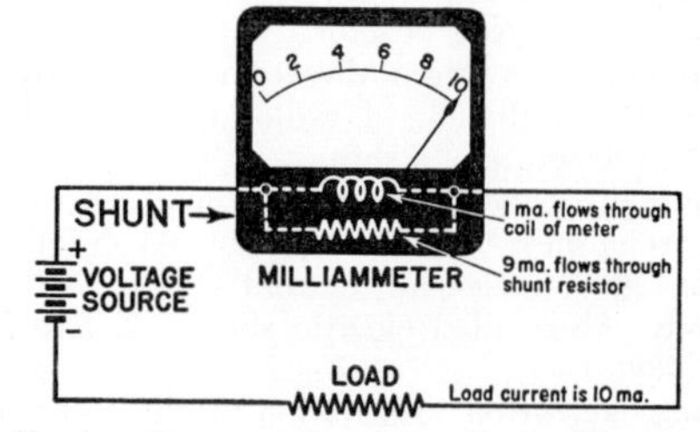


Shunt resistor across moving coil of milliammeter.

the meter. 2. A piece of iron that provides a parallel path for magnetic flux around an air gap in a magnetic circuit. 3. To place one part in parallel with another. 4. *Parallel.*

**shunted monochrome** A color television technique in which the luminance or monochrome signal is shunted around the chrominance modulator or chrominance demodulator. Also called bypassed monochrome (deprecated).

**shunt-excited** Having field windings connected across the armature terminals, as in a d-c generator.

**shunt-fed vertical antenna** A vertical antenna connected to ground at the base and energized at a point suitably positioned above the grounding point.

**shunt feed** The feed of direct operating voltages to the electrodes of an electron tube through a choke coil that is parallel to and therefore separated from the signal circuit. Also called parallel feed.

**shunt loading** Loading in which reactances are connected between the conductors of a transmission line.

**shunt neutralization** *Inductive neutralization.*

**shunt peaking** Use of a peaking coil in a parallel circuit branch connecting the output load of one stage to the input load of the following stage, to compensate for high-frequency loss due to the distributed capacitances of the two stages.

**shunt tee junction** A waveguide tee junction having an equivalent circuit in which the impedance of the branch guide is predominantly in parallel with the impedance of the main waveguide at the junction.

**shunt-wound** Having armature and field windings in parallel, as in a d-c generator or motor.

**shutdown** Stopping a chain reaction by bringing a nuclear reactor to a subcritical condition.

**shutter** 1. A movable plate of absorbing material used to cover a window or beam hole in a reactor when radiation is not desired, or used to shut off a flow of slow neutrons. 2. A device that prevents light from reaching the light-sensitive surface of a television camera except during the desired period of exposure.

**shuttle** *Rabbit.*

**SiC** Symbol for *silicon carbide.*

**sideband** A band of frequencies on each side of the carrier frequency of a modulated radio signal, produced by the process of modulation. The upper sideband contains the frequencies that are the sums of the carrier and modulation frequencies, while the lower sideband contains the difference frequencies.

**sideband attenuation** Attenuation in which the transmitted relative amplitude of some sideband component of a modulated signal is smaller than that produced by the modulation process.

**sideband interference** *Adjacent-channel interference.*

**sideband power** The power contained in the sidebands. A receiver responds to sideband power rather than carrier power when receiving a modulated wave.

**sideband splash** *Adjacent-channel interference.*

**side frequency** One of the frequencies of a sideband.

**side lobe** *Minor lobe.*

**side-lobe echo** A radar echo due to a side lobe of the radar beam.

**side thrust** The radial component of the force on a pickup arm due to stylus drag.

**Sidewinder** A Navy air-to-air guided missile having infrared homing guidance and a speed of over mach 2. When launched from an airplane, the missile seeks and hits an enemy bomber or fighter target by homing on the heat emitted by the target. Developed by Naval Ordnance Test Station, China Lake, California. Named after the sidewinder, a desert rattlesnake that is also a heat-seeker.

**siemen** A name proposed at one time for a unit of energy equal to 1 watthour, and later for the unit of conductance that is now called the mho.

**Siemens** Trademark of Siemens and Halske for their line of synchros.

**sigma meson** A term used for a short time to designate a meson that produces a star. These mesons were soon identified as pi mesons.

**sigma particle** A hyperon having an extremely short life (about $10^{-10}$ second), a mass between that of neutrons and deuterons, and either a positive or negative charge.

**sigma pile** An assembly of moderating material containing a neutron source, used to study the absorption cross sections and other neutron properties of the material.

**sigmatron** A cyclotron and betatron operating in tandem to produce billion-volt x-rays.

**signal** 1. A visual, aural, or other indication used to convey information. 2. The intelligence, message, or effect to be conveyed over a communication system. 3. *Signal wave.*

**signal comparator** A circuit that correlates information from two or more signals.

**signal conditioner** A circuit used to shape or adapt a signal to the requirements of a data transmission line.

**signal contrast** The ratio in decibels between the white signal and the black signal in facsimile.

**signal data converter** A circuit that converts a data-modulated signal from one form to another.

**signal electrode** The electrode from which the signal output is obtained in a television camera tube.

**signal frequency shift** The numerical difference between the frequencies corresponding to white signal and black signal at any point in a frequency-shift facsimile system.

**signal generator** A test instrument that can be set to generate an unmodulated or tone-modulated r-f signal voltage at a known frequency, as needed for aligning or servicing receivers and amplifiers.

**signaling** A procedure for indicating to

the receiving end of a communication circuit that intelligence is to be transmitted.

**signaling channel** A tone channel used for signaling purposes.

**signaling communication** One-way communication from a base station to a mobile receiver for the purpose of actuating a signaling device in the mobile unit or for communicating information to the desired mobile unit.

**signaling key** *Key.*

**signal level** The difference between the level of a signal at a point in a transmission system and the level of an arbitrarily specified reference signal. For audio signals the difference is conveniently expressed as a ratio in decibels since the reference level is then zero decibels.

**signal light** A light specifically designed for the transmission of code messages by means of visible light rays that are interrupted or deflected by electric or mechanical means.

**signal output current** The absolute value of the difference between output current and dark current in a camera tube or phototube.

**signal plate** The metal plate that backs up the mica sheet containing the mosaic in one type of cathode-ray television camera tube. The capacitance existing between this plate and each globule of the mosaic is acted on by the electron beam to produce the television signal.

**signal-shaping network** A network inserted in a telegraph circuit, usually at the receiving end, to improve the waveform of the code signals.

**signal strength** The strength of the signal produced by a radio transmitter at a particular location, usually expressed as microvolts or millivolts per meter of effective receiving antenna height.

**signal-strength meter** A meter that is connected to the automatic volume control circuit of a communication receiver and calibrated in decibels or arbitrary S units to read the strength of a received signal. Also called S meter and S-unit meter.

**signal-to-noise ratio** The ratio of the amplitude of a desired signal at any point to the amplitude of noise signals at that same point. Often expressed in decibels. The peak value is usually used for pulse noise, while the rms value is used for random noise.

**signal tracer** A test instrument designed to facilitate signal-tracing.

**signal-tracing** A servicing technique that involves tracing the progress of a signal through each stage of a receiver to locate the faulty stage.

**signal voltage** The effective (root-mean-square) voltage value of a signal.

**signal wave** A wave whose characteristics permit some intelligence, message, or effect to be conveyed. Also called signal.

**signal-wave envelope** The contour of a signal wave that consists of a series of r-f cycles.

**signal winding** The control winding to which the control signals are applied in a saturable reactor.

**sign digit** A character used to designate the algebraic sign of a number in a digital computer.

**significant digit** A digit appearing in the coefficient of a number when the number is written as a coefficient between 1.000 ...... and 9.999...... times a power of 10. Thus, 0.009407, which is equal to 9.407 times $10^{-3}$, has four significant digits.

**silent discharge** A gradual, nondisruptive discharge of electricity from a conductor into the atmosphere. Sometimes accompanied by the production of ozone.

**silent period** A three-minute period twice each hour, starting at 15 minutes and 45 minutes after the hour, during which International Radio Regulations require that normal transmissions must cease on all frequencies within a designated frequency band centered on 500 kc, to permit listening for weak distress calls.

**silent speed** The speed at which silent motion pictures are fed through a projector, equal to 16 frames per second.

**silent zone** *Skip zone.*

**silica gel** A chemically inert and highly hygroscopic form of hydrated silica, used for absorbing water and vapors of solvents in electronic equipment cabinets.

**silicon** [symbol Si] A metallic element used in pure form as a semiconductor, and mixed with iron or steel during smelting to improve their magnetic properties for use as transformer core material. Atomic number is 14.

**silicon capacitor** A capacitor in which a pure silicon crystal slab serves as the dielectric. A silicon capacitor can have high *Q* at frequencies up to 5,000 mc in low-voltage circuits. When the crystal is grown to have a p zone, a depletion zone, and an n zone, the capacitance varies with the externally applied bias voltage. The higher the bias voltage of this voltage-controlled capacitor, the lower is its capacitance.

**silicon carbide** [symbol SiC] A semiconductor having a forbidden band gap of 3.3 electron-volts and a maximum operating temperature of 1,300°C when used in a transistor.

**silicon-carbide rectifier** A semiconductor rectifier capable of withstanding temperatures up to 1,200°F.

**silicon controlled rectifier** A three-junction semiconductor device that normally represents an open circuit, but switches rapidly to the conducting state of a single-junction silicon rectifier when an appropriate gate signal is applied to the gate terminal. It is equivalent to a thyratron.

**silicon detector** *Silicon diode.*

**silicon diode** A crystal diode that uses silicon as a semiconductor. Widely used as a detector in uhf and shf circuits. Also called silicon detector.

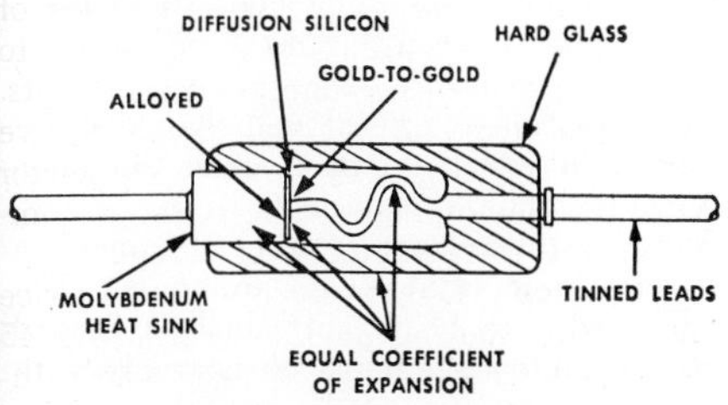


Silicon diode construction.

**silicone** A polymeric organosilicon compound having excellent insulating, lubricating, water-resisting, and heat-resisting properties.

**silicon rectifier** A metallic rectifier in which rectifying action is provided by an alloy junction formed in a high-purity silicon slab.

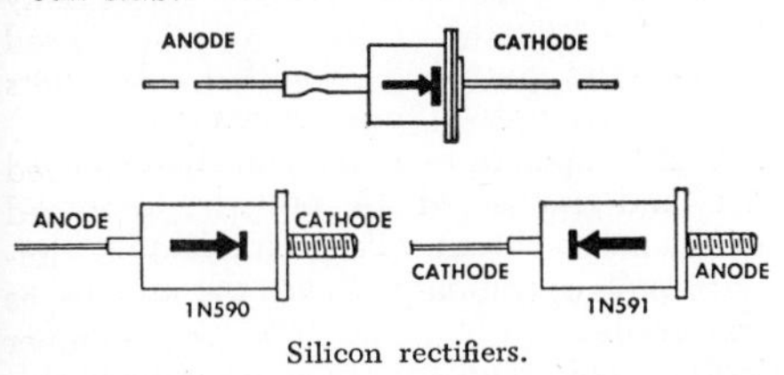


Silicon rectifiers.

**silicon resistor** A resistor using silicon semiconductor material as a resistance element, to obtain a positive temperature coefficient of resistance that does not appreciably change with temperature. Used as a temperature-sensing element.

**silicon solar cell** A solar cell consisting of p and n silicon layers placed one above the other to form a p-n junction at which radiant energy is converted into electricity. Theoretical maximum efficiency is 22%, and actual efficiencies better than 11% have been achieved. Used as a power source for satellite instrumentation and for portable radios.

**silicon steel** An alloy steel containing 3 to 5% silicon, having desirable magnetic qualities for iron cores of transformers and other a-c devices.

**silver** [symbol Ag] A precious-metal element having better electric conductivity than copper, used for contact points of relays and switches and as a plating on electronic components because it does not readily corrode. Atomic number is 47.

**silvered mica capacitor** A mica capacitor in which a coating of silver is deposited directly on the mica sheets to serve in place of conducting metal foil.

**silver oxide cell** A primary cell in which depolarization is accomplished by an oxide of silver.

**silver sensitization** A process of depositing a thin layer of silver on photosensitive surfaces during formation, to increase the sensitivity.

**silver solder** A solder composed of silver, copper, and zinc, having a melting point lower than silver but higher than lead-tin solder.

**silver storage battery** An alkaline storage battery in which the positive active material is silver oxide and the negative active material contains zinc.

**simple harmonic wave** A wave whose amplitude at any point is a simple harmonic function of time.

**simple scanning** Scanning that involves the use of only one scanning spot at a time in a facsimile system.

**simple sound source** A source that radiates sound uniformly in all directions under free-field conditions.

**simple target** A radar target having a reflecting surface such that the amplitude of the reflected signal does not vary with the aspect of the target. A metal sphere is an example.

**simple tone** 1. A sound wave whose instantaneous sound pressure is a simple sinusoidal function of time. 2. A sound sensation characterized by singleness of pitch. Also called pure tone.

**simplex** *Simplex operation.*

**simplex operation** A method of radio operation in which communication between two stations takes place in only one direction at a time. This includes ordinary transmit-receive operation, press-to-talk operation, voice-operated carrier, and other forms of manual or automatic switching from transmit to receive. Also called simplex.

**simulation** The representation of physical systems and phenomena by computers, models, or other equipment.

**simulator** A computer or other piece of equipment that simulates a desired system or condition and shows the effects of various applied changes, such as a flight simulator.

**simulcast** A program broadcast simultaneously by two different types of stations, as by radio and television stations or by f-m and a-m broadcast stations.

**simultaneous color television** A color television system in which the phosphors for the three primary colors are excited at the same time, not one after another. The shadow-mask color picture tube gives a simultaneous display.

**simultaneous lobing** *Monopulse radar.*

**sine-cosine generator** *Resolver.*

**sine potentiometer** A potentiometer whose d-c output voltage is proportional to the sine of the shaft angle. Used as a resolver in computer and radar systems.

**sine wave** A wave whose amplitude varies as the sine of a linear function of time.

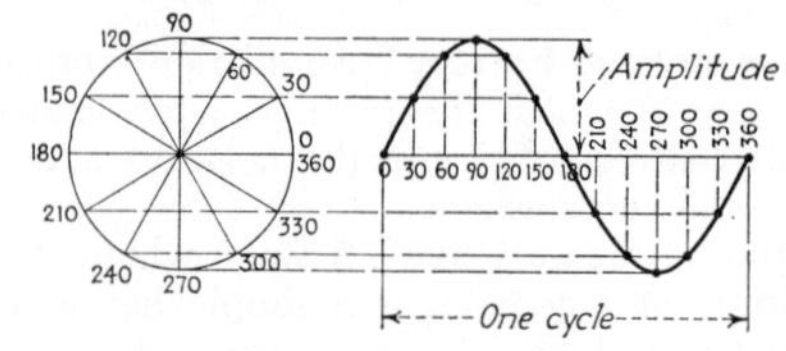


Sine-wave construction procedure:

1. Draw circle with radius equal to amplitude of desired sine wave.
2. Divide circle into 12 equal parts with a 30°–60° triangle, and label circle intersections from 0° to 360°.
3. At right of center of circle, draw a horizontal line with length corresponding to desired time distance for one cycle.
4. Divide the line into 12 equal parts and label them from 0° to 360°.
5. Project vertically and horizontally from each pair of corresponding labeled points, and place a dot at the point of intersection.
6. Connect dots together with smooth curve to get sine wave.

**sine-wave clipper** A clipper circuit that cuts off the top of a sine wave so it resembles a square wave.

**sine-wave modulation** Modulation in which the envelope of the modulated carrier signal has the waveform of a sine wave.

**sine-wave response** *Amplitude-frequency response.*

**singing** An undesired self-sustained oscillation in a system or component, at a frequency in or above the passband of the system or component. Generally due to excessive positive feedback.

**singing margin** The difference in level, usually expressed in decibels, between the singing point and the operating gain of a system or component.

**singing point** The minimum value of gain of a system or component that will result in singing.

**single-address instruction** *One-address instruction.*

**single-anode tube** An electron tube having a single main anode. This term is used chiefly for pool tubes.

**single-button carbon microphone** A carbon microphone having a carbon-filled button-like container on only one side of its flexible diaphragm.

**single-channel simplex** Simplex operation that provides nonsimultaneous radio communication between stations using the same frequency channel.

**single crystal** A crystal, usually artificially grown, in which all parts have the same crystallographic orientation.

**single-crystal camera** An x-ray camera in which single crystals are examined. The diffracted x-ray beams are usually recorded on a cylindrical film placed coaxially with the axis of rotation of the crystal.

**single-dial control** Control of a number of different devices or circuits by means of a single control knob, as when using a gang capacitor.

**single-ended** Unbalanced, as when one side of a transmission line or circuit is grounded.

**single-ended amplifier** An amplifier in which each stage normally employs only one tube, so operation is asymmetric with respect to ground.

**single-ended push-pull amplifier** An amplifier having two transmission paths designed to operate in a complementary manner and connected to provide a single unbalanced output. This circuit provides push-pull operation without the use of a transformer.

**single-ended tube** A tube in which all electrode connections, including the control grid, are made to base pins, so there is no top cap. The letter S after the first numerals in a receiving-tube designation indicates a single-ended tube, as in 12SK7.

**single-frequency duplex** Duplex carrier communication that provides communication in opposite directions, but not simultaneously, over a single-frequency carrier channel, the transfer between transmitting

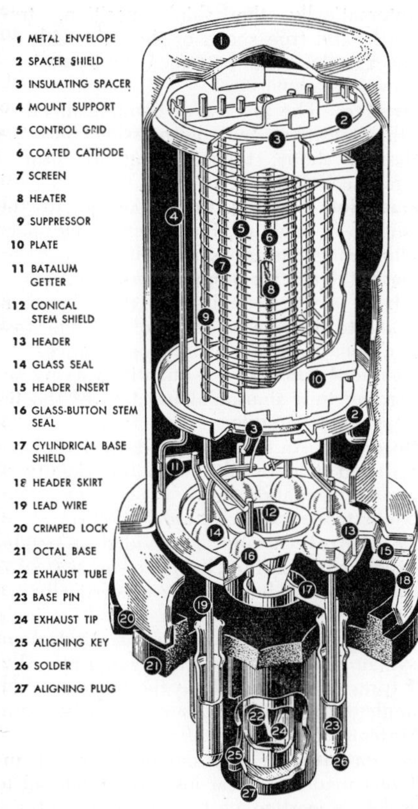


Single-ended tube, with portions cut away to show internal construction. A cylindrical base shield (17) inside the aligning plug permits bringing the control grid lead out through a base pin without feedback trouble.

and receiving conditions being automatically controlled by the voices of the communicating parties.

**single-frequency simplex** Single-frequency carrier communication in which manual rather than automatic switching is used to change over from transmission to reception.

**single-groove stereo record** *Stereo record.*

**single-gun color tube** A color television picture tube having only one electron gun and one electron beam. The beam is sequentially deflected across phosphors for the three primary colors to form each color picture element, as in the chromatron.

**single hop** The range of a radio wave that is radiated at a small angle to the horizontal, so it penetrates the ionosphere only a small amount before being reflected back to the surface of the earth. The maximum range that can be spanned by single-hop transmission is about 1,500 miles for E-layer transmissions, and is obtained when the radio wave is radiated horizontally.

**single-particle model** *Independent-particle model.*

**single-phase** Energized by a single alternating voltage.

**single-phase circuit** A circuit energized by a single alternating voltage, applied through two wires.

**single-polarity pulse** Deprecated term for *unidirectional pulse.*

**single-polarity pulse train** Deprecated term for *unidirectional pulse train.*

**single-pole double-throw** [abbreviated spdt] A three-terminal switch or relay contact arrangement that connects one terminal to either of two other terminals.

**single-pole-piece magnetic head** A magnetic head having a single pole piece on one side of the recording medium.

**single-pole single-throw** [abbrevated spst] A two-terminal switch or relay contact arrangement that opens or closes one circuit.

Single-pole single-throw and single-pole double-throw knife switches.

**single scattering** The deflection of a particle from its original path owing to one encounter with a single scattering center in the material traversed. Plural scattering involves several successive encounters, while multiple scattering involves many successive encounters.

**single-shot multivibrator** *Monostable multivibrator.*

**single-sideband communication** A communication system in which the r-f carrier and one sideband are suppressed. Less power is then required at the transmitter for the same effective signal at the receiver, a narrower frequency band can be used, and the signal is less affected by selective fading or manmade interference.

**single-sideband converter** A converter that connects to the i-f amplifier output of an amplitude-modulation radio receiver and converts the receiver into a single-sideband receiver.

**single-sideband modulation** [abbreviated ssb modulation] Modulation resulting from elimination of all components of one sideband from an amplitude-modulated wave.

**single-sideband receiver** A radio receiver

designed for the reception of single-sideband modulation, having provisions for restoring the carrier.

**single-sideband transmission** Transmission of a carrier and substantially only one sideband of modulation frequencies, as in television where only the upper sideband is transmitted completely for the picture signal. The carrier wave may be either transmitted or suppressed.

**single-sideband transmitter** A transmitter in which one sideband is transmitted and the other is effectively eliminated.

**single-signal receiver** A highly selective superheterodyne receiver for code reception, having a crystal filter in the i-f amplifier.

**single-speed floating action** Floating action in which a final control element is moved at a single speed.

**single-station assembly machine** An assembly machine in which all insertion of components is done at a single station to which boards are fed or through which boards are circulated.

**single-stub transformer** A shorted section of coaxial line that is connected to a main coaxial line near a discontinuity to provide impedance matching at the discontinuity.

**single-stub tuner** A section of transmission line terminated by a movable short-circuiting plunger or bar, attached to a main transmission line for impedance-matching purposes.

**single-tone keying** Keying in which the modulating wave causes the carrier to be modulated with a single tone for one condition, which may be either marking or spacing. The carrier is unmodulated for the other condition.

**single track** A variable-density or variable-area sound track in which both positive and negative halves of the signal are linearly recorded. Also called standard track.

**single-track recorder** A magnetic-tape recorder that records only one track on the tape.

**sink** 1. A power-consuming device, such as the load in a circuit. 2. The region of a Rieke diagram where the rate of change of frequency with respect to phase of the reflection coefficient is maximum for an oscillator. Operation in this region may lead to unsatisfactory performance by reason of cessation or instability of oscillations.

**sins** [Ship's Inertial Navigation System] A navigation system that uses a gyroscope-stabilized platform and associated inertial guidance and sonar equipment to provide automatically the ship's position, true north, and true speed of the ship with respect to the ocean bottom. Used in nuclear-powered submarines.

**sintering** The process of bonding metal or other powders by cold-pressing into the desired shape, then heating to form a strong cohesive body.

**sinusoidal** Varying in proportion to the sine of an angle or time function. Ordinary alternating current is sinusoidal.

**sinusoidal electromagnetic wave** A wave whose electric field strength is proportional to the sine or cosine of an angle that is a linear function of time or distance.

**sinusoidal quantity** A quantity varying according to a sinusoidal function of the independent variable.

**site error** An error due to distortion of the radiated field by objects in the vicinity of a radio navigation aid.

**six-electrode tube** *Hexode.*

**size control** A control provided on a television receiver for changing the size of a picture either horizontally or vertically.

**skew** The deviation of a received facsimile frame from rectangularity due to lack of synchronism between scanner and recorder. Expressed numerically as the tangent of the angle of this deviation.

**skiatron** *Dark-trace tube.*

**skin antenna** A flush-mounted aircraft antenna made by using insulating material to isolate a portion of the metal skin of the aircraft.

**skin depth** The depth below the surface of a conductor at which the current density has decreased one neper below the current density at the surface due to the action of the electromagnetic waves associated with the high-frequency current flowing through the conductor.

**skin dose** The dose at the center of the irradiation field on skin. It is the sum of the air dose and backscattering, plus the exit dose from other parts if this is significant.

**skin effect** The tendency of alternating current to concentrate in the surface layer of a conductor. The effect increases with frequency, and serves to increase the effective resistance of the conductor.

**skip** A digital-computer instruction to proceed to the next instruction.

**skip distance** The minimum distance that radio waves can be transmitted between two points on the earth by reflection from the ionosphere, at a specified time and frequency. The skip distance thus includes the maximum ground-wave range and the width of the skip zone. In multihop transmission, the distance of each succeeding hop from earth to ionosphere and back is also the skip distance.

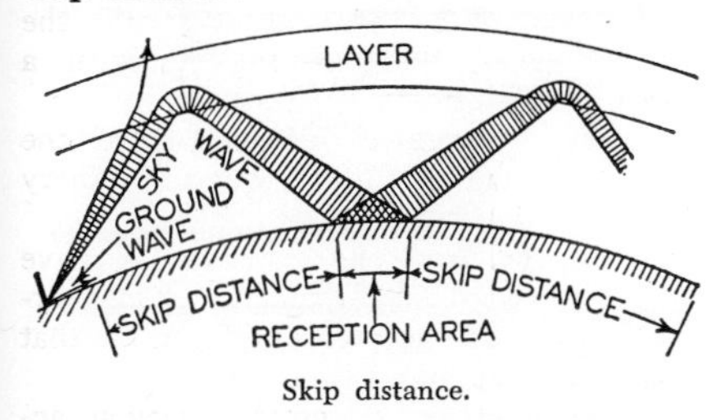


Skip distance.

**skip zone** A ring-shaped area around a radio transmitter, surrounding the ground-wave reception region, within which no radio signals are received. The outer edge of the skip zone is at the minimum distance for reception of sky-wave signals, while the inner edge is at the maximum limit for reception of ground-wave signals. Also called silent zone and zone of silence.

**sky error** Deprecated term for *ionospheric error.*

**sky wave** A radio wave that travels upward into space and may or may not be returned to earth by reflection from the ionosphere. Also called ionospheric wave.

**sky-wave correction** A correction for sky-wave propagation errors, applied to measured position data. The amount of the correction is based on an assumed position and an assumed ionosphere height.

**sky-wave error** Deprecated term for *ionospheric error.*

**sky-wave station error** The error in station synchronization in sky-wave-synchronized loran that is due to the effect of ionospheric variations on the time of transmission of the synchronizing signal from one station to the other.

**sky-wave-synchronized loran** [abbreviated ss loran] A type of loran in which the transmitting stations are synchronized by signals reflected from the ionosphere. Used to obtain greater range and more accurate nighttime navigation.

**slab** A relatively thick crystal cut from which blanks are obtained by subsequent transverse cutting.

**slant distance** The distance between two points not at the same elevation. Also called slant range (deprecated).

**slant range** Deprecated term for *slant distance.*

**slaved gyromagnetic compass** A compass in which the gyroscope is synchronized to a magnetic force.

**slave station** A navigation station in which some characteristic of its emission is controlled by a master station, such as the B station in loran.

**slave sweep** A time base that is triggered by an external waveform. Used in navigation systems to display the same information at different locations.

**slc** Abbreviation for *straight-line capacitance.*

**sleeve** 1. The cylindrical contact that is farthest from the tip of a phone plug. 2. Insulating tubing used over wires or components. Also called sleeving.

**sleeve antenna** A single vertical half-wave radiator, the lower half of which is a metallic sleeve through which the concentric feed line runs. The upper radiating portion, one quarter-wavelength long, connects to the center of the line.

**sleeve-dipole antenna** A dipole antenna surrounded in its central portion by a coaxial sleeve.

**sleeve stub** An antenna consisting of half of a sleeve-dipole antenna projecting from a large metal surface.

**sleeving** *Sleeve.*

**slewing** Moving a radar antenna or sonar transducer rapidly in a horizontal or vertical direction, or both.

**slewing motor** A motor used to drive a radar antenna at high speed for slewing to pick up or track a target.

**slf** Abbreviation for *straight-line frequency.*

**slide-back voltmeter** An electronic voltmeter in which an unknown voltage is measured indirectly by adjusting a calibrated voltage source until its voltage equals the unknown voltage.

**slider** A sliding type of movable contact.

**slide-rule dial** A dial in which a pointer moves in a straight line over long straight scales resembling the scales of a slide rule.

**slide switch** A switch that is actuated by sliding a button, bar, or knob.

**slide-wire bridge** A bridge circuit in which the resistance in one or more branches is controlled by the position of a sliding contact on a length of resistance wire stretched along a linear scale.

**slide-wire rheostat** A rheostat in which a sliding contact rides over a long single-layer coil of resistance wire.

**sliding contact** *Wiping contact.*

**slip** The difference between synchronous and operating speeds of an induction machine.

**slip ring** A conducting ring mounted on but insulated from a rotating shaft, used with a stationary brush to join fixed and moving parts of a circuit.

**slope** 1. The projection of a flight path in the vertical plane. 2. The degree of deviation of an essentially straight portion of a characteristic curve from the horizontal or vertical.

**slope angle** The angle in the vertical plane between the horizontal and the flight path of an aircraft or missile. Also called glide-slope angle.

**slope detector** A discriminator that uses a single tuned circuit and single diode to react to differences in frequency. Operation is on one of the slopes of the response curve for the tuned circuit. Seldom used in f-m receivers because the linear portion of the curve is too short for large-signal operation.

**slope deviation** The difference between the projection in the vertical plane of the actual path of movement of an aircraft and the planned slope for the aircraft. Expressed either in terms of angular or linear measurement.

**slope equalizer** An equalizer used with an amplifier to make the attenuation of a section of transmission line constant over the frequency band being transmitted.

**slope filter** A filter having a response that rises or falls with frequency over a given frequency range.

**slot antenna** An antenna formed by cutting one or more narrow slots in a large metal surface fed by a coaxial line or waveguide. For unidirectional radiation the rear of the metal sheet is boxed in or the slot is energized directly by a waveguide. In another version, diagonal slots are cut into a length of waveguide at precisely spaced intervals.

**slot coupling** Coupling between a coaxial cable and a waveguide by means of two coincident narrow slots, one in a waveguide wall and the other in the sheath of the coaxial cable.

**slotted line** *Slotted section.*

**slotted rotor plate** A rotor plate having radial slots to permit bending different sections of the plate either inward or outward to adjust the total capacitance of a variable capacitor section during alignment. Also called serrated rotor plate and split rotor plate.

**slotted section** A section of waveguide or shielded transmission line in which the shield is slotted to permit the use of a traveling probe for examination of standing waves. Also called slotted line and slotted waveguide.

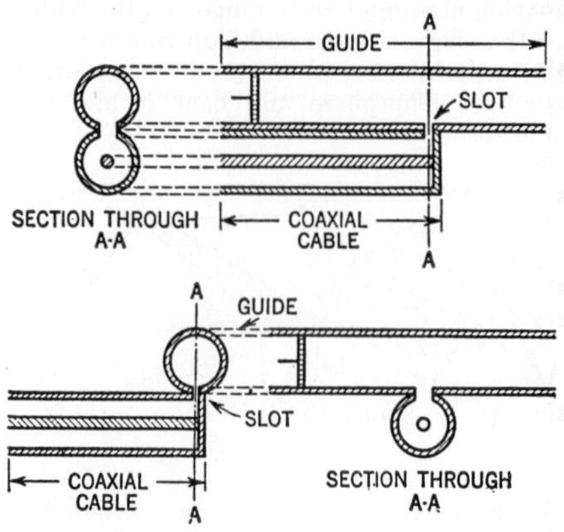


Slot coupling as used between coaxial cable and circular waveguide.

**slotted waveguide** *Slotted section.*

**slow-acting relay** A time-delay relay in which an interval of several seconds may exist between energizing of the coil and pulling up of the armature. The delay can be obtained electrically by placing a solid copper ring on the core of the relay. Also called slow-operate relay.

**slow-down video** A method of transmitting radar data over narrow-bandwidth circuits by storing the radar video signal over the time required for the antenna to move through one beam width. The stored signal is then sampled at a periodic rate such that each range interval of interest is sampled at least once per beam width or per azimuth quantum, and the resulting data is quantized for transmission.

**slowing-down** A decrease in the energy of a particle as a result of collisions with nuclei.

**slowing-down area** An area defined as one-sixth the mean square distance between a neutron source and the point where the neutron reaches a given energy in an infinite homogeneous medium.

**slowing-down density** The number of neutrons slowing down per unit volume per unit time.

**slowing-down kernel** The probability that a neutron will go from one position to another while slowing down through a specified energy range.

**slowing-down length** The square root of the slowing-down area.

**slowing-down power** The average decrease in the value of the natural logarithm of energy of a neutron per unit distance traveled by the neutron in a material.

**slow memory** A section of a computer memory from which information can be obtained automatically but not at the fastest rate.

**slow neutron** 1. A neutron having low kinetic energy, up to about 100 electron-volts. 2. *Thermal neutron.*

**slow-operate relay** *Slow-acting relay.*

**slow reactor** A nuclear reactor in which fission is induced primarily by slow neutrons, as in a thermal reactor.

**slow-release relay** A time-delay relay in which there is an appreciable delay between deenergizing of the coil and release of the armature.

**slow wave** A wave having a phase velocity less than the velocity of light, as in a ridge waveguide.

**slug** 1. A heavy copper ring placed on the core of a relay to delay operation of the relay. 2. A movable iron core for a coil. 3. A movable piece of metal or dielectric material used in a waveguide for tuning or impedance-matching purposes. 4. Nuclear fuel in the form of a short round bar or cylinder, inserted in a hole or channel in the active lattice of a nuclear reactor.

**slug tuner** A waveguide tuner containing one or more longitudinally adjustable pieces of metal or dielectric.

**slug tuning** A means for varying the frequency of a resonant circuit by introducing a slug of material into either the electric or magnetic fields or both, as in permeability tuning.

**slurry reactor** A nuclear reactor having the fissionable material in the form of particles suspended in a liquid.

**slw** Abbreviation for *straight-line wavelength.*

**small-signal analysis** Circuit analysis based on such small excursions of current and voltage from their quiescent operating points that linear operation may be assumed.

**small-signal forward transadmittance** The value of the forward transadmittance obtained when the input voltage is small compared to the beam voltage.

**smear** A television picture defect in which objects appear to be extended horizontally beyond their normal boundaries in a blurred or smeared manner. One cause is excessive attenuation of high video frequencies in the television receiver.

**S meter** *Signal-strength meter.*

**Smith chart** A special polar diagram containing constant-resistance circles, constant-reactance circles, circles of constant standing-wave ratio, and radius lines representing constant line-angle loci. Used in solving transmission-line and waveguide problems.

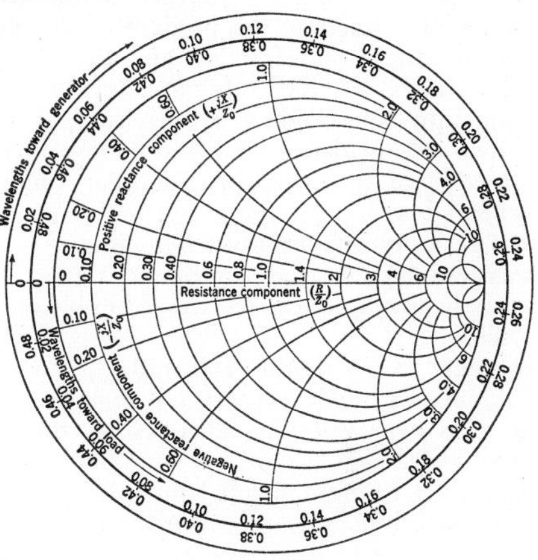


Smith chart.

**smoke detector** A photoelectric system for actuating an alarm when smoke in a chimney or other location exceeds a predetermined density.

**smoke-puff decoy** A countermeasure against enemy infrared devices.

**smoothing choke** An iron-core choke used in a power-supply filter circuit to remove ripple.

**smoothing factor** A factor expressing the effectiveness of a filter in smoothing ripple-voltage variations.

**SMPTE** Abbreviation for *Society of Motion Picture and Television Engineers.*

**snap-action switch** A switch that responds to very small movements of its actuating button or lever and changes rapidly and positively from one contact position to the other. The trademark of one version is Micro Switch. Also called sensitive switch.

**snap-on ammeter** An a-c ammeter having a magnetic core in the form of hinged jaws that can be snapped around the current-carrying wire. Also called clamp-on ammeter.

**snap switch** A switch in which the contacts are separated or brought together suddenly by the action of a spring placed under tension or compression by the operating knob or lever.

**Snark** An Air Force surface-to-surface guided missile having a range of about 5,000 miles and capable of carrying nuclear weapons. It is powered by an air-breathing turbojet engine, resembles an airplane, and has a self-contained all-weather guidance system.

**Snell's law** When a wave travels from one medium into another having a different index of refraction, the product of the sine of the angle of refraction and the refractive index of the refracting medium is equal to the product of the sine of the angle of incidence and the index of refraction of the medium containing the incident beam.

**sniffer gear** *Odoriferous homing.*

**sniperscope** A snooperscope designed for use in place of a telescopic sight on a rifle.

**snipper** An airborne f-m radar used for automatic bomb release on semi-isolated targets over water at altitudes of from 40 to 400 feet.

**snivet** A straight, jagged, or broken vertical black line appearing near the right-hand edge of a television receiver screen due to a horizontal amplifier tube fault that produces a discontinuity in the zero-bias anode current curve near its knee. It can be eliminated by changing the setting of almost any picture control or by changing horizontal amplifier tubes.

**snooperscope** An infrared source, an infrared image converter, and a battery-operated high-voltage d-c source constructed in portable form to permit a foot soldier or other user to see objects in total darkness. Infrared radiation sent out by the infrared source is reflected back to the snooperscope and converted into a visible image on the fluorescent screen of the image tube.

**snow** Small, random white spots produced on a television or radar screen by inherent noise signals originating in the receiver. Visible on television screens only when received signal strength is inadequate, because strong incoming signals will override the noise signals.

**snow static** Precipitation static caused by falling snow.

**Society of Motion Picture and Television Engineers** [abbreviated SMPTE] A nonprofit organization established for the advancement of theory and practice related to the production and utilization of motion pictures and television programs.

**socket** A device designed to provide electric connections and mechanical support for an electronic or electric component requiring convenient replacement.

**socket adapter** An adapter used between a tube socket and a tube to permit use of the tube in a socket designed for some other type of tube or to permit making tube current or voltage measurements while the tube is in use.

**sodar** [SOund Detecting And Ranging; SOund raDAR] A system for obtaining local weather data by projecting sound waves directly up and analyzing the echoes as shown on a cathode-ray oscilloscope.

**sodium** [symbol Na] A metallic alkali element having a melting point of 97.5°C and moderate thermal neutron cross section. Used in liquid form as a coolant for some types of nuclear reactors. Used on cathodes of phototubes when maximum response is desired at the violet end of the visible spectrum. Atomic number is 11.

**sodium-graphite reactor** A nuclear reactor that uses slightly enriched uranium as fuel, graphite as moderator, and liquid sodium as the coolant.

**sodium-vapor lamp** A discharge lamp containing sodium vapor, used chiefly for highway illumination.

**sofar** [SOund Fixing And Ranging] A system for fixing a position at sea by exploding a charge under water, measuring the times for the shock waves to travel through water to three widely separated shore stations, and calculating the position of the explosion by triangulation. The explosive can be dropped from a lifeboat by survivors of air or sea disasters.

**soft magnetic material** Magnetic material that is easily demagnetized.

**soft radiation** Less penetrating radiation.

**soft tube** 1. An x-ray tube having a vacuum of about 0.000002 atmosphere, the remaining gas being left in intentionally to give less penetrating rays than those of a more completely evacuated tube. 2. *Gassy tube.*

**soft x-ray** An x-ray having a comparatively long wavelength and poor penetrating power.

**solar cell** A photoelectric cell used to convert the radiant energy of the sun into electric power. The silicon solar cell is an example.

**solarimeter** An instrument for making direct readings of solar radiation intensity from sun and sky.

**solar radio noise** Radio noise originating at the sun, and increasing greatly in intensity during sunspots and flares. Heard as a hissing noise on short-wave radio receivers.

**solar satellite** A satellite that orbits the sun.

**solder** 1. An alloy that can be melted at a fairly low temperature, for use in joining metals having much higher melting points. An alloy of lead and tin in approximately equal proportions is the solder most often used for making permanent joints in circuits. 2. To join two metals with solder.

**solderability** The ability of a printed circuit or other metal surface to be wet by molten solder.

**soldering** The process of joining metals by fusion and solidification of an adherent alloy having a melting point below about 800°F.

**soldering flux** A material that dissolves oxides from surfaces being soldered. Rosin is widely used for this purpose when soldering electronic circuits, because of its noncorrosive qualities.

**soldering gun** A soldering tool having as its tip a fast-heating resistance element that serves as a short-circuit across a high-current, low-voltage secondary winding of a step-down transformer built into the unit. A trigger-type switch in a pistol-grip handle permits intermittent use. Also called gun.

**soldering iron** A tool used to apply heat to a joint preparatory to soldering. The source of heat is either a gas torch or an internal electric heating element.

**soldering lug** *Lug.*

**soldering paste** A soldering flux prepared in the form of a paste.

**soldering pliers** A soldering tool designed to send electric heating current directly through the terminal, joint, or other part being heated for soldering. The tool resembles a pair of pliers having carbon-electrode jaws, each connected to one terminal of a low-voltage, high-current secondary winding on a transformer.

**soldering tool** A tool used to produce the heat needed in soldering, such as a soldering gun, soldering iron, or soldering pliers.

**solderless connector** A connector that holds wire ends firmly together to provide a good connection without solder. A common form is a cap with tapered internal threads, twisted over the exposed ends of the wires.

**solderless wrapped connection** *Wire-wrap connection.*

**solenoid** A coil that surrounds a movable iron core. When the coil is energized by sending alternating or direct current through it, the core is pulled to a central position with respect to the coil. Used to convert electric energy into mechanical energy. Other forms of solenoids produce rotary rather than axial movement of the core. The core may be stationary and the coil movable.

**solenoid relay** *Plunger relay.*

**solenoid valve** A valve actuated by a solenoid, for controlling the flow of gases or liquids in pipes.

**solid conductor** A conductor consisting of a single wire rather than strands.

**solid-state maser** A maser in which the electromagnetic radiation interacts with the atoms or molecules of a solid paramagnetic material such as a ferrite. Examples include the two-level maser and three-level maser.

**solid-state physics** The branch of physics that deals with the structure and properties of solids, including semiconductors.

**solion** [SOLution ION] An electrochemical sensing and control device in which ions in solution carry electric charges to give amplification corresponding to that of vacuum tubes and transistors. In one form, a plastic housing holds platinum electrodes immersed in a solution of potassium iodide and iodine. Alternating mechanical pressure on diaphragms set into the housing is converted into a proportional electric output.

**solution ceramic** A nonbrittle inorganic ceramic insulating coating that can be applied to wires at a low temperature. Examples include ceria, chromia, titania, and zirconia.

**Sommerfeld's equation** An equation for ground-wave propagation that relates field strength at the surface of the earth at any distance from a transmitting antenna to the field strength at unit distance for given ground losses.

**sonar** [SOund Navigation And Ranging] Equipment used for sonic and ultrasonic underwater detection, ranging, sounding, and communication. The commonest type is echo-ranging sonar, an active sonar in which a sonic or ultrasonic pulse is transmitted, reflected from an object, and received back at the transmitter location. The elapsed time for the pulse gives target distance, and the directional characteristics of the transmitting-receiving transducer give target range. For underwater communication the pulses may be spaced according to international Morse code. Other versions are passive sonar, scanning sonar, and searchlight sonar. British term is asdic. Also called sonar set.

**sonaramic indicator** An indicator connected to a sonar receiver to present visually the signals received over a band of frequencies centered about the specific frequency to which the sonar receiver is tuned.

**sonar attack plotter** A system that coordinates information from a sonar installation, a ship's gyrocompass, and related devices, and presents graphically the information needed to plan an antisubmarine attack.

**sonar beacon** An underwater beacon that transmits sonic or ultrasonic signals for the purpose of providing bearing information. It may have receiving facilities that permit triggering by an external source.

**sonar data computer** A computer that calculates two or more factors of sonar data, such as range, bearing, depth, and sound velocity.

**sonar dome** A streamlined watertight enclosure that provides protection for a sonar transducer, sonar projector, or hydrophone and associated equipment, while offering minimum interference to sound transmission and reception.

**sonar modulator** A modulator designed to vary the frequency of a signal generated by a sonar transmitter.

**sonar projector** An electromechanical device used under water to convert electric energy to sound energy. A crystal or magnetostriction transducer is usually used for this purpose.

**sonar receiver** A receiver designed to intercept and amplify the sound signals reflected by an underwater target and display the accompanying intelligence in useful form. It may also pick up other underwater sounds.

**sonar receiver-transmitter** A single piece of equipment that combines the functions of generating energy for sonic or ultrasonic underwater transmission and receiving the resulting echo signals. Used primarily for detection and ranging. Secondary functions may include underwater communication.

**sonar resolver** A resolver used with echo-ranging and depth-determining sonar to calculate and record the horizontal range of a sonar target, as required for depth-bombing.

**sonar set** *Sonar.*

**sonar signal simulator** A signal generator that feeds synthetic signals to a sonar receiver or directly to its indicator to give a presentation similar to that observed under actual operating conditions.

**sonar sounding set** A sonar set used to determine depth of water from the point of installation on a vessel to the ocean bottom or to the surface of the ocean.

**sonar trainer** A trainer designed as a teaching aid to develop skills in sonar techniques.

**sonar train mechanism** A mechanism that rotates a sonar projector or transducer. It may also have facilities for tilting the transducer or projector.

**sonar transducer** A transducer used under water to convert electric energy to sound energy and sound energy to electric energy.

**sonar transducer scanner** A circuit or switch that permits sampling the output signals of individual elements or groups of elements in certain types of sonar transducers, to determine the direction of arriving signals.

**sonar transmitter** A transmitter that generates electric signals of the proper frequency and form for application to a sonar transducer or sonar projector, to produce sound waves of the same frequency in water. The sound waves may carry intelligence.

**sonar window** The portion of a sonar dome or sonar transducer that passes sound waves at sonar frequencies with little attenuation while providing mechanical protection for the transducer.

**sonde** An instrument used to obtain weather data during ascent and descent through the atmosphere, in a form suitable for telemetering to a ground station by radio as in a radiosonde.

**sone** A unit of loudness. By definition, a simple 1,000-cps tone that is 40 db above a listener's threshold of hearing produces a loudness of 1 sone. A sound that is judged by the listener to be $n$ times as loud as that of the 1-sone tone has a loudness of $n$ sones. A millisone is equal to 0.001 sone.

**sonic** Pertaining to sound waves and the speed of sound.

**sonic altimeter** An instrument for determining the height of an aircraft above the earth by measuring the time taken for sound waves to travel from the aircraft to the surface of the earth and back to the aircraft again. The method is based on sound having a velocity of 1,080 feet per second through dry air at 32°F.

**sonic barrier** The turbulence encountered by an aircraft as its speed approaches the speed of sound. Also called sound barrier and transonic barrier.

**sonic delay line** *Acoustic delay line.*

**sonic depth finder** *Fathometer.*

**sonic frequency** *Audio frequency.*
**sonic mine** *Acoustic mine.*
**sonic speed** *Speed of sound.*
**sonne** A radio navigation aid that provides a number of characteristic signal zones which rotate in a time sequence. A bearing may be determined by observation of the instant at which transition occurs from one zone to the following zone. Also called consol.
**sonobuoy** An acoustic receiver and radio transmitter mounted in a buoy that can be dropped from an aircraft by parachute to pick up underwater sounds of a submarine and transmit them to the aircraft. To track a submarine, several buoys are dropped in a pattern that includes the known or suspected location of the submarine, with each buoy transmitting an identifiable signal. An electronic computer then determines the location of the submarine by comparison of the received signals and triangulation of the resulting time-delay data. Also called radio sonobuoy.
**sonobuoy trainer** A trainer designed to give aircraft crews practical experience in locating submarines by simulating the tactics necessary to obtain a fix.
**sort** To arrange items of information in a computer according to specified rules.
**sorter** A machine that sorts punched cards according to the punches in a specified column of the card.
**SOS** The distress signal in radiotelegraphy, consisting of the letters S, O, and S of the international Morse code run together.
**sound** 1. The sensation of hearing, produced when sound waves act on the brain through the nerves of the ears. The extreme frequency limits for human hearing are from about 15 to 20,000 cps, but animals can hear much higher frequencies. Also called audio (slang) and sound sensation. 2. A vibration in an elastic medium at any frequency that produces the sensation of hearing. This vibration is propagated by the medium as a sound wave. Frequencies that produce sound are called sound or audio frequencies. Frequencies below the audio range are infrasonic, and frequencies above the audio range are ultrasonic.
**sound absorber** A material that absorbs a high percentage of incident sound energy.
**sound absorption** The process by which sound energy is diminished in passing through a medium or in striking a surface.
**sound absorption coefficient** The ratio of sound energy absorbed to that arriving at a surface or medium. Also called acoustic absorption coefficient and acoustic absorptivity.
**sound analyzer** A device for measuring the levels of the components of a complex sound as a function of frequency.
**sound articulation** The per cent articulation obtained when speech units are fundamental sounds, usually combined into meaningless syllables.
**sound bar** One of the two or more alternate dark and bright horizontal bars that appear in a television picture when a-f voltage reaches the video input circuit of the picture tube.
**sound barrier** *Sonic barrier.*
**sound box** *Acoustic pickup.*
**sound carrier** The television carrier that is frequency-modulated by the sound portion of a television program. The unmodulated center frequency of the sound carrier is 4.5 mc higher than the video carrier frequency for the same channel.
**sound channel** The series of stages that handles only the sound signal in a television receiver.
**sound effects** Mechanical devices or recordings used to provide lifelike imitations of various sounds.
**sound-effects filter** A filter, usually adjustable, designed to reduce the passband of a system at low and/or high audio frequencies to produce special effects.
**sound energy** Energy existing in a medium due to sound waves.
**sound-energy density** Sound energy per unit volume, generally expressed in ergs per cubic centimeter.
**sound-energy flux** The average rate of flow of sound energy for one period through any specified area, generally expressed in ergs per second.
**sound-energy flux density** *Sound intensity.*
**sound-energy flux density level** *Intensity level.*
**sounder** *Telegraph sounder.*
**sound field** A region containing sound waves.
**sound film** Motion-picture film having a sound track along one side for reproduction of the sounds that are to accompany the film.
**sound frequency** *Audio frequency.*
**sound gate** The gate through which film passes in a sound-film projector for conversion of the sound track into a-f signals that can be amplified and reproduced.
**sounding** Determining depth of water or

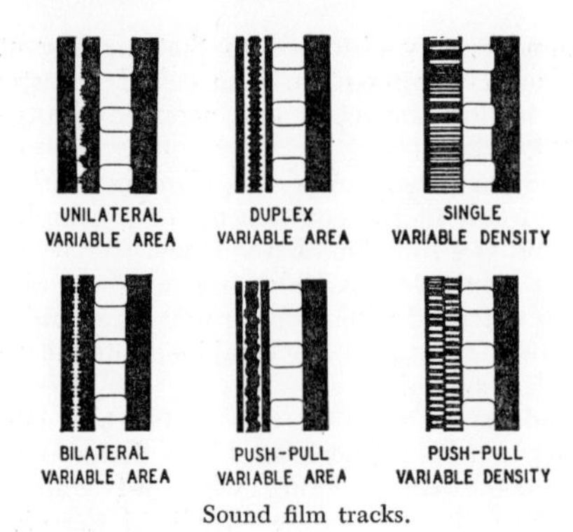

Sound film tracks.

altitude above land or sea by any of several methods.

**sounding balloon** A small free balloon used for carrying radiosonde equipment aloft.

**sounding electrode** A probe used to make measurements in a gas discharge.

**sounding rocket** A research rocket used to obtain upper-atmosphere data.

**sound intensity** The sound energy transmitted per unit of time through a unit area, expressed in ergs per second per square centimeter or in watts per square centimeter. Also called sound-energy flux density and specific sound-energy flux.

| dB | |
|---|---|
| 150 | RAM JET |
| | TURBO-JET WITH AFTERBURNER |
| | TURBO-JET, 7,000-LB THRUST |
| 140 | |
| | 50-HP VICTORY SIREN AT 100' |
| | F-84 TAKING OFF, 80' FROM TAIL |
| 130 | HYDRAULIC PRESS AT 3' |
| | BOILER SHOP (MAXIMUM LEVEL) |
| | PNEUMATIC CHIPPER AT 5' |
| 120 | |
| | SUBMARINE ENGINE ROOM |
| | TRUMPET AUTO HORN AT 3' |
| 110 | |
| | WOODWORKING SHOP |
| | INSIDE DC-6 AIRLINER |
| 100 | |
| | INSIDE CHICAGO SUBWAY CAR |
| | TRAIN WHISTLE AT 500' |
| 90 | 10-HP OUTBOARD AT 50' |
| | INSIDE SEDAN IN CITY TRAFFIC |
| 80 | OFFICE WITH TABULATING MACHINES |
| | HEAVY TRAFFIC AT 50' |
| 70 | |
| | AVERAGE TRAFFIC AT 100' |
| | CONVERSATIONAL SPEECH AT 3' |
| 60 | |
| 50 | PRIVATE BUSINESS OFFICE |
| | AVERAGE RESIDENCE |
| 40 | |
| | RESIDENTIAL AREA AT NIGHT |
| 30 | SPEECH BROADCASTING STUDIO |
| | MUSIC BROADCASTING STUDIO |
| 20 | WHISPER AT 5' |
| 10 | RUSTLE OF LEAVES |
| 0 | THRESHOLD OF HEARING, YOUNG EARS, AT 1,000 CPS |

Sound levels in decibels with respect to threshold of hearing.

**sound level** The weighted sound pressure level at a point in a sound field, determined in the manner specified by the American Standards Association. The meter reading in decibels corresponds to a value of the sound pressure integrated over the audible frequency range with a specified frequency weighting and integration time.

**sound-level meter** An instrument used to measure noise and sound levels in a specified manner. The meter may be calibrated in decibels or volume units. It includes a microphone, an amplifier, an output meter, and frequency-weighting networks. A volume-unit meter is an example.

**sound power** The total sound energy radiated by a source per unit time, generally expressed in ergs per second or watts.

**sound-powered telephone** A telephone operating entirely on current generated by the speaker's voice, with no external power supply. Sound waves cause a diaphragm

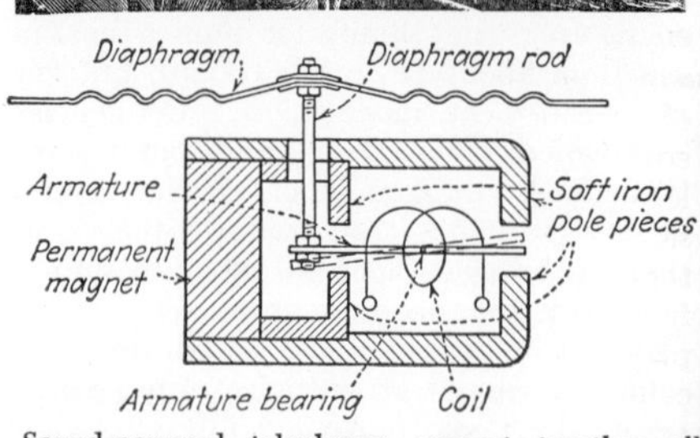

Sound-powered telephones connect together all battle stations on combat vessels, and operate even though the ship's entire electric power system is put out of commission. Diagram shows operating principle.

to move a coil back and forth between the poles of a powerful but small permanent magnet, generating the required a-f voltage in the coil.

**sound power level** A value in decibels

equal to 10 times the logarithm to the base 10 of the ratio of radiated sound power to a reference power.

**sound pressure** *Effective sound pressure.*

**sound pressure level** A value in decibels equal to 20 times the logarithm to the base 10 of the ratio of the pressure of this sound to a reference pressure. Reference pressures in common use are 0.0002 microbar and 1 microbar.

**sound probe** A probe used to explore a sound field without significantly disturbing the field in the region being explored. It may be a small microphone or a small tubular attachment added to a conventional microphone.

**soundproofing** *Damping.*

**sound ranging** Determining the location of a gun or other sound source by measuring travel times of the sound wave to microphones at three or more different known positions.

**sound recorder** A recorder that provides a permanent record of sounds, as on phonograph records, optical sound tracks, magnetic tape, or magnetic wire.

**sound recording system** A combination of transducing devices and associated equipment suitable for storing sound in a form capable of subsequent reproduction.

**sound reflection coefficient** The ratio of sound energy reflected from a surface to that reaching the surface. Also called acoustic reflection coefficient and acoustic reflectivity.

**sound-reinforcing system** A public-address system used in a theater or auditorium to make the sound-energy density as nearly uniform as possible throughout the audience.

**sound-reproducing system** A combination of transducing devices and associated equipment for picking up sound at one location and reproducing it either at the same location or some other location, at the same time or at some later time. The four types are monaural, binaural, monophonic, and stereo sound systems. Also called audio system and sound system.

**sound sensation** *Sound.*

**sound spectrum** A representation of the amplitudes of the components of a complex sound, arranged as a function of frequency.

**sound speed** The speed of sound motion-picture film, standardized at 24 frames per second.

**sound system** *Sound-reproducing system.*

**sound takeoff** The point at which the sound signal is separated from the video signal in a television receiver for separate i-f amplification, demodulation, and a-f amplification. Also called takeoff.

**sound track** A narrow band, usually along the margin of a sound film, that carries the sound record. It may be a variable-width or variable-density optical track or a magnetic track.

**sound transmission coefficient** The ratio of transmitted to incident sound energy at an interface in a sound medium. The value depends on the angle of incidence of the sound. Also called acoustic transmission coefficient and acoustic transmittivity.

**sound wave** The traveling wave produced in an elastic medium by vibrations in the frequency range of sound. Approximate velocities of sound waves are 1,080 feet per second in air at 32°F, 4,800 feet per second in water, and 16,400 feet per second in steel.

**source** The circuit or device that supplies signal power or electric energy to a transducer or load circuit.

**source impedance** The impedance presented by a source to a transducer or load circuit.

**south magnetic pole** The magnetic pole located approximately at 73°S latitude and 156°E longitude, about 1,020 nautical miles north of the South Pole.

**South Pole** A geographical point on the earth that is one end of the axis about which the earth revolves.

**south pole** The pole of a magnet at which magnetic lines of force are assumed to enter.

**space** 1. The open-circuit condition or the signal causing the open-circuit condition in telegraphic communication. The closed-circuit condition is the mark. 2. The universe, extending from the earth without limit.

**space attenuation** The power loss of a signal traveling in free space, due to such factors as absorption, reflection, and scattering. Usually expressed in decibels.

**space charge** The net electric charge distributed throughout a volume or space, such as the cloud of electrons in the space near the cathode of a thermionic vacuum tube or phototube.

**space-charge debunching** A process in which the mutual interactions between electrons in a stream spread out the electrons of a bunch.

**space-charge density** The net electric charge per unit volume.

**space-charge effect** The repulsion between the electrons emitted by the cathode of a thermionic vacuum tube and the electrons accumulated in the space charge near the cathode, resulting in a reduction in anode current.

**space-charge grid** A grid, usually positive, that controls the position, area, and magnitude of a potential minimum or of a virtual cathode in a region adjacent to the grid of an electron tube.

**space-charge layer** *Depletion layer.*

**space-charge-limited current** The current passing through an interelectrode space in a vacuum tube when a virtual cathode exists due to a space charge.

**space-charge-limited operation** Operation of an electromagnetic isotope separator with insufficient focusing of the ion beam due to repulsion of an excess of like charges in the beam.

**space-charge region** A semiconductor region in which the net charge density is significantly different from zero.

**spacecraft** A vehicle designed to travel beyond the atmosphere of the earth, through outer space, when given lift and propulsion by reaction to a jet stream or by other means.

**space current** The total current flowing between the cathode and all the other electrodes in a tube, including the anode current and the currents to all other electrodes.

**spaced antennas** Two or more antennas spaced several wavelengths apart and used to pick up the same signal for diversity reception.

**space-diversity reception** Radio reception involving the use of two or more antennas located several wavelengths apart, feeding individual receivers whose outputs are combined. The system gives an essentially constant output signal despite fading due to variable propagation characteristics, because fading affects the spaced-out antennas at different instants of time.

**space factor** 1. The ratio of the space occupied by the conductors in a winding to the total cubic content or volume of the winding, or the similar ratio of cross-sections. 2. The ratio of the space occupied by iron to the total cubic content of an iron core.

**space pattern** A geometric pattern on a test chart, used to measure geometric distortion in television equipment.

**space permeability** The factor that expresses the ratio of magnetic induction to magnetizing force in a vacuum. In the cgs electromagnetic system of units, the permeability of a vacuum is arbitrarily taken as unity.

**space phase** Reaching corresponding peak values at the same point in space.

**space quadrature** A difference of a quarter-wavelength in the position of corresponding points of a wave in space.

**space service** A radio communication service between space stations.

**space station** A radio communication station located on an object that is beyond, or intended to go beyond, the major portion of the earth's atmosphere.

**space wave** The component of a ground wave that travels more or less directly through space from the transmitting antenna to the receiving antenna. One part of the space wave goes directly from one antenna to the other; another part is reflected off the earth between the antennas.

**spacing wave** The signal emitted by a radiotelegraph transmitter during spacing portions of the code characters. Also called back wave.

**spacistor** A multiple-terminal solid-state device, similar to a transistor, that achieves high-frequency operation up to about 10,000 mc by injecting electrons or holes into a space-charge layer which forces these carriers rapidly to a collecting electrode. Input and output impedances can be as high as 30 megohms.

**spade bolt** A bolt having a spade-shaped flattened head with a transverse hole, used to fasten shielded coils, capacitors, and other components to a chassis.

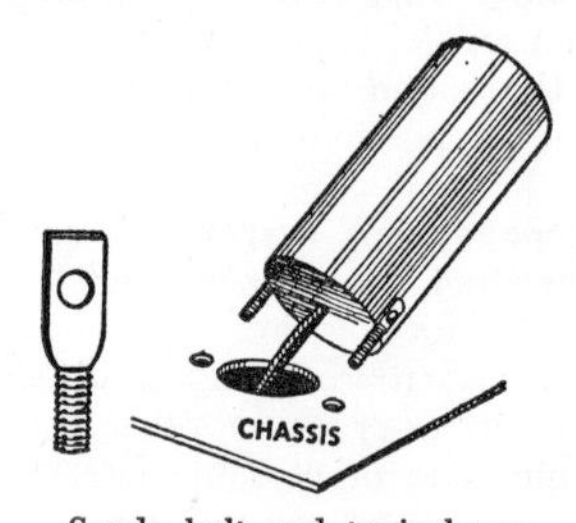


Spade bolt and typical use.

**spade lug** An open-ended flat termination for a wire lead, easily slipped under a terminal nut.

**spaghetti** Insulating tubing used over bare wires or as a sleeve for holding two or more insulated wires together. The tubing is usually made of varnished cloth or a plastic.

**spallation** A nuclear reaction in which the energy of each incident particle is so high that more than two or three particles are ejected from the target nucleus and both its mass number and atomic number are changed.

**spark** A short-duration electric discharge due to a sudden breakdown of air or some other dielectric material separating two terminals, accompanied by a momentary flash of light. Also called sparkover.

**spark coil** An induction coil used to produce spark discharges.

**spark gap** An arrangement of two electrodes between which a spark may occur. The insulation (usually air) between the electrodes is self-restoring after passage of the spark. Used as a switching device, as for protecting equipment against lightning or for switching a radar antenna from receiver to transmitter and vice versa.

**spark-gap generator** A high-frequency generator in which a capacitor is repeatedly charged to a high voltage and allowed to

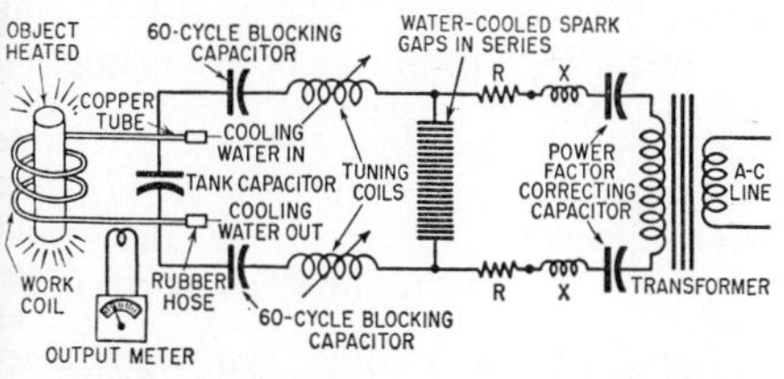

Spark-gap generator for induction heating.

discharge through a spark gap into an oscillatory circuit, generating successive trains of damped high-frequency oscillations.

**spark-gap modulation** A modulation process that produces one or more pulses of energy by means of a controlled spark-gap breakdown, for application to the element in which modulation takes place.

**spark-gap modulator** A modulator that uses a controlled spark gap to modulate a carrier.

**sparking** Intentional or accidental spark discharges, as between the brushes and commutator of a rotating machine, between contacts of a relay or switch, or at any other point at which a current-carrying circuit is broken.

**sparking voltage** The minimum voltage at which a spark discharge occurs between electrodes of given shape at a given distance apart under given conditions.

**sparkover** *Spark.*

**spark plate** A metal plate insulated from the chassis of an auto radio by a thin sheet of mica, and connected to the battery lead to bypass noise signals picked up by battery wiring in the engine compartment.

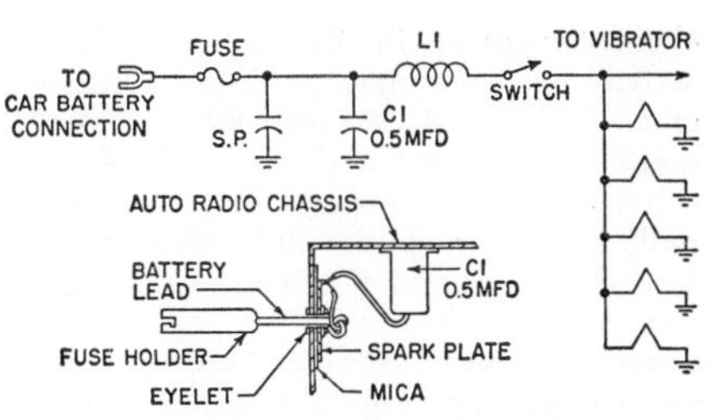

Spark plate installation and circuit.

**spark spectrum** The spectrum produced by a spark discharging through a gas or vapor. With metal electrodes, a spectrum of the metallic vapor is obtained.

**spark transmitter** A radio transmitter that utilizes the oscillatory discharge of a capacitor through an inductor and a spark gap as the source of r-f power.

**Sparrow** A Navy air-to-air guided missile having a speed of over 1,500 miles per hour, guided to its target by a radar beam

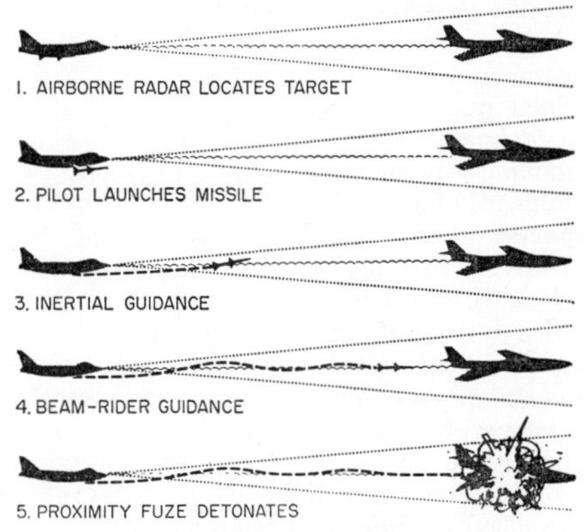

Sparrow I firing, using combination of simple inertial guidance and beam-rider guidance.

transmitted by the launching airplane. Several models exist, designated Sparrow I, Sparrow II, and Sparrow III.

**spdt** Abbreviation for *single-pole double-throw.*

**speaker** *Loudspeaker.*

**specific acoustic impedance** The complex ratio of sound pressure to particle velocity at a point in a medium. Also called unit area acoustic impedance.

**specific acoustic reactance** The imaginary component of the specific acoustic impedance.

**specific acoustic resistance** The real component of the specific acoustic impedance.

**specific activity** 1. The activity of a radioisotope of an element per unit weight of element present in the sample. 2. The activity per unit mass of a pure radionuclide. 3. The activity per unit weight of a sample of radioactive material. Specific activity can be expressed in such units as millicuries per gram, disintegrations per second per milligram, or counts per minute per milligram.

**specific address** *Absolute address.*

**specific coding** Digital-computer coding in which all addresses refer to specific registers and locations.

**specific conductivity** The conducting power of a material in mhos per cubic centimeter. It is the reciprocal of resistivity.

**specific dielectric strength** The dielectric strength per millimeter of thickness of an insulating material.

**specific electronic charge** The ratio of the electronic charge to the rest mass of the electron.

**specific emission** The rate of emission per unit area.

**specific gamma-ray emission** The exposure dose rate produced by the unfiltered gamma rays from a point source of a defined quantity of a radioactive nuclide at a defined distance. The unit of specific gamma-ray emission is the roentgen per millicurie hour at 1 cm.

**specific ionization** The number of ion pairs formed per unit distance along the track of an ion passing through matter. Also called total specific ionization.

**specific ionization coefficient** The average number of pairs of ions with opposite charges that are produced by electrons having a specified kinetic energy when traveling a unit distance in a gas at a specified pressure and temperature.

**specific permeability** The permeability of a substance divided by the permeability of a vacuum. Also called relative permeability.

**specific power** The power produced per unit mass of fuel present in a nuclear reactor.

**specific reluctance** *Reluctivity.*

**specific repetition rate** One of a set of closely spaced repetition rates derived from the basic rate and associated with a specific set of synchronized loran stations.

**specific resistance** *Resistivity.*

**specific routine** A routine expressed in computer coding designed to solve a particular problem, with addresses of registers and locations specifically stated.

**specific sound-energy flux** *Sound intensity.*

**specific sound-energy flux level** *Intensity level.*

**spectral characteristic** The relation between wavelength and some other variable, such as between wavelength and emitted radiant power of a luminescent screen per unit wavelength interval.

**spectral color** A color that appears in the spectrum of white light. The basic spectral colors are violet, blue, green, yellow, orange, and red.

**spectral purity** Having a single wavelength.

**spectral quantum yield** The average number of electrons photoelectrically emitted from a photocathode per incident photon of a given wavelength.

**spectral radiant intensity** The radiant intensity per unit wavelength interval, such as watts per steradian per micron.

**spectral response** *Spectral sensitivity characteristic.*

**spectral sensitivity characteristic** The relation between the radiant sensitivity and the wavelength of the incident radiation of a camera tube or phototube, under specified conditions of irradiation. Also called spectral response.

**spectrograph** A spectrometer that provides a permanent record of a spectrum of radiation.

**spectrometer** An instrument that disperses radiation into its component wavelengths and measures the magnitude of each component.

**spectrophotometer** An instrument that measures transmission or apparent reflectance of visible light as a function of wavelength, permitting accurate analysis of color or accurate comparison of luminous intensities of two sources at specific wavelengths.

**spectrophotometric analysis** A method of quantitative analysis based on spectral energy distribution in the absorption spectrum of a substance in solution.

**spectroradiometer** An instrument that measures the spectral energy distribution of any type of radiation, such as infrared radiation.

**spectroscope** An instrument that spreads individual wavelengths in a radiation to permit observation of the resulting spectrum.

**spectroscopy** The branch of physical science that deals with the measurement and analysis of spectra.

**spectrum** [plural is spectra] 1. All the frequencies used for a particular purpose. Thus, the radio spectrum extends from about 10,000 cps to well over 30,000 mc. 2. The result of dispersing an emission

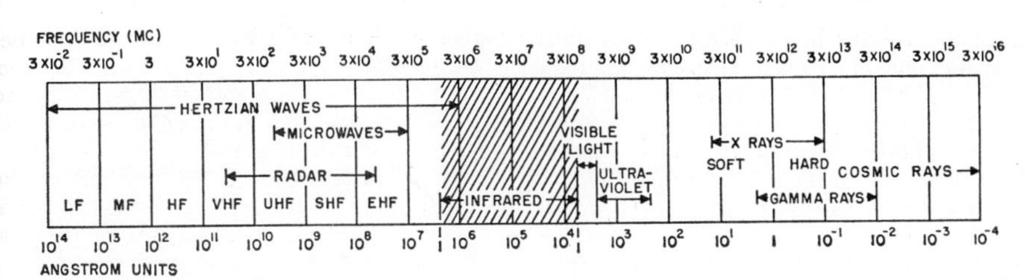


Spectrum of electromagnetic radiation, extending from 30,000 cps at left to cosmic rays at right. Radio or Hertzian waves overlap longer infrared wavelengths.

(such as light) in accordance with some progressive property, usually its frequency. 3. *Electromagnetic spectrum.*

**spectrum analyzer** An instrument that measures the amplitudes of the components of a complex waveform throughout the frequency range of the waveform. One version gives the energy at each frequency in the output of a pulsed magnetron.

**spectrum line** A line recorded by a spectrograph to represent a specific wavelength, atomic mass, or other spectral quantity.

**spectrum locus** The locus of points representing the chromaticities of spectrally pure stimuli in a chromaticity diagram.

**specular reflection** *Direct reflection.*

**specular transmission** Transmission in which only the emergent radiation parallel to the entrant beam is observed.

**specular transmission density** The value of the photographic density obtained when the light flux impinges normally on the sample and only the normal component of the transmitted flux is collected and measured.

**speech amplifier** An a-f amplifier designed specifically for amplification of speech frequencies, as for public-address equipment and radiotelephone systems.

**speech clipper** A clipper used to limit the peaks of speech-frequency signals, as required for increasing the average modulation percentage of a radiotelephone or amateur radio transmitter.

**speech coil** British term for *voice coil.*

**speech frequency** *Voice frequency.*

**speech inverter** *Scrambler.*

**speech power** The rate at which sound energy is being radiated by a speech source at a given instant, or the average of the instantaneous values over a given time interval.

**speech scrambler** *Scrambler.*

**speed** 1. A scalar quantity equal to the magnitude of velocity. Speed thus specifies no direction. 2. The angular velocity of a rotating shaft or device, generally expressed in revolutions per minute. 3. The rate of performance of an act. 4. The aperture of a lens. 5. The exposure time of a shutter. 6. The frequency of a relaxation oscillator.

**speed control** 1. A control that changes the speed of a motor or other drive mechanism, as for a phonograph or magnetic-tape recorder. 2. *Hold control.*

**speed of light** A physical constant equal to $2.99779 \times 10^{10}$ centimeters per second. This corresponds to 186,280 statute miles per second, 161,750 nautical miles per second, and 328 yards per microsecond. All electromagnetic radiation travels at this same speed in free space. The velocity of light is usually considered to be a vector quantity representing the speed of light in a particular direction.

**speed of sound** The speed at which sound waves travel through a given medium. In air under standard sea-level conditions, sound travels at approximately 1,080 feet per second. In water it is about 4,800 feet per second. In steel it is about 16,400 feet per second. The velocity of sound is usually considered to be a vector quantity representing the speed of sound in a particular direction. Also called sonic speed.

**speed regulator** A device that maintains the speed of a motor or other device at a predetermined value or varies it in accordance with a predetermined plan.

**speed variation** Any change in motor speed resulting from causes independent of the control system adjustment, such as line-voltage changes, temperature changes, or load changes.

**spheredop** A modified dovap missile-tracking system in which a stable oscillator serves in place of a transponder in the missile.

**sphere gap** A spark gap between two equal-diameter spherical electrodes.

**sphere-gap voltmeter** A high-voltage voltmeter consisting of an adjustable sphere gap. The electrodes are moved together until the voltage being measured produces a spark, and the voltage is calculated from gap spacing and electrode diameter.

**spherical aberration** An image defect due to the spherical form of an optical or electron lens or mirror, resulting in blurred focus and image distortion.

**spherical antenna** An antenna having the shape of a sphere, used chiefly in theoretical studies.

**spherical-earth factor** The ratio of the electric field strength that would result from propagation over an imperfectly conducting spherical earth to that which would result from propagation over a perfectly conducting plane.

**spherical faceplate** A television picture tube faceplate that is a portion of a spherical surface.

**spherical hyperbola** The locus of the points on the surface of a sphere having a specified constant difference in great-circle distances from two fixed points on the sphere.

**spherical wave** A wave whose equiphase surfaces form a family of concentric spheres. The direction of travel is always perpendicular to the surfaces of the spheres.

**sphygmogram** A graphic recording of human pulse waves.

**sphygmograph** An instrument for recording the waveforms of the pulse of a patient.

**sphygmophone** A microphone attached to the wrist to pick up the sounds of the pulse.

**spider** A highly flexible perforated or corrugated disk used to center the voice coil of a dynamic loudspeaker with respect to the pole piece without appreciably hindering in-and-out motion of the voice coil and its attached diaphragm.

**spike** A short-duration transient whose amplitude considerably exceeds the average amplitude of the associated pulse or signal.

**spike leakage energy** The r-f energy per pulse that is transmitted through a tr or pre-tr tube before and during establishment of the steady-state r-f discharge.

**spike train** A regular succession of pulses, used in electrobiology.

**spill** The loss of information from a storage element of a charge storage tube by redistribution.

**spin** The total angular momentum of a nucleus when considered as a single particle.

**spindle** A shaft, such as the upward-projecting shaft on a phonograph turntable, used for positioning the record.

**spinner** A rotating radar antenna and reflector assembly.

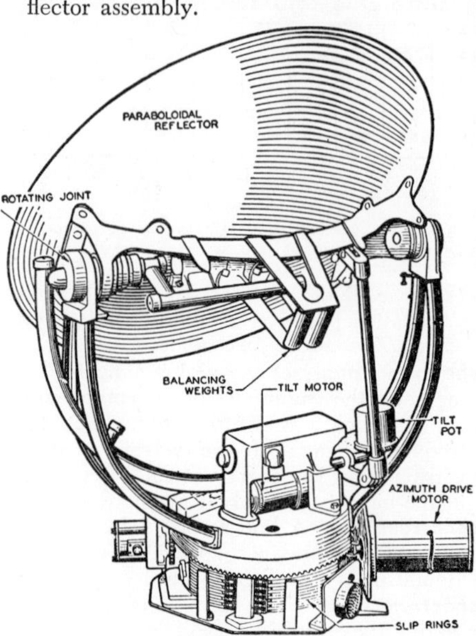


Spinner assembly for ground radar.

**spin-orbit coupling** The interaction between intrinsic and orbital angular momentum of a particle.

**spin quantum number** A number that gives the angular momentum of the electron considered as a small charged sphere rotating around an axis.

**spinthariscope** An instrument for viewing the scintillations of alpha particles on a luminescent screen, usually with the aid of a microscope.

**spiral distortion** A distortion in which image rotation varies with distance from the axis of symmetry of the electron optical system of a camera or image tube using magnetic focusing. Also called S distortion.

**spiral-four cable** *Spiral quad.*

**spiral quad** Four separately insulated telephone or telegraph conductors wound spirally around a supporting core. Also called spiral-four cable.

**spiral scanning** Scanning in which the direction of maximum radiation describes a portion of a spiral. The rotation is always in one direction. Used with some types of radar antennas.

**spiral tuner** A tuner having spiral coils and a tuning mechanism that slides a contact along the spiral of each coil. The sliding contacts change the inductance of the coils and thus change the frequency. Also called continuous tuner and inductuner.

**spkr** Abbreviation for *loudspeaker.*

**splash baffle** *Arc baffle.*

**splash ring** A metal shield used in an ignitron or other pool tube to prevent mercury from splashing on other electrodes.

**splatter** Distortion due to overmodulation of a transmitter by peak signals of short duration, particularly sounds containing high-frequency harmonics. It is a form of adjacent-channel interference.

**splice** A joint used to connect two lengths of conductor with good mechanical strength and good conductivity.

**splicing tape** A pressure-sensitive nonmagnetic tape used for splicing magnetic tape. It has a hard adhesive that will not ooze and gum up the recording head or cause adjacent layers of tape on the reel to stick together.

**split-anode magnetron** A magnetron having an anode divided into two equal segments, usually by slots parallel to its axis.

**split-flow reactor** A nuclear reactor in which the coolant enters at the center section and flows outward at both ends or vice versa.

**split hydrophone** A directional hydrophone in which each transducer or group of transducers produces a separate output voltage.

**split integrator** A digital-differential-analyzer integrator that can multiply its output or input by a constant.

**split-phase current** One of two different phases of current obtained from a single-phase a-c circuit by means of reactances.

**split-phase motor** A single-phase induction motor having an auxiliary winding connected in parallel with the main winding, but displaced in magnetic position from the main winding so as to produce the required rotating magnetic field for starting. The auxiliary circuit is generally opened when the motor has attained a predetermined speed.

**split projector** A directional projector in which electroacoustic transducing elements are so divided and arranged that each division may be energized separately through its own electric terminals.

**split rotor plate** *Slotted rotor plate.*

**split-series motor** A d-c series-connected motor having one series field winding for each direction of rotation. Used in servomechanisms having on-off control.

**split-sound system** A television receiver i-f system in which the audio and video i-f signals are separated after the mixer stage and amplified in separate i-f stages. Now largely replaced by the intercarrier sound system.

**split-stator variable capacitor** A variable capacitor having a rotor section that is common to two separate stator sections. Used in grid and anode tank circuits of transmitters for balancing purposes.

**spoiler** A rod grating mounted on a parabolic reflector to change the pencil-beam pattern of the reflector to a cosecant-squared pattern. Rotating the reflector and grating 90° with respect to the feed antenna changes from one pattern to the other.

**spoiling** A passive navigation countermeasure used to prevent use of a radio or radar transmission by the enemy for navigation purposes.

**spoking** A radar malfunction in which luminous spots continue on the screen for an abnormal length of time and form radial lines, interfering with presentations.

**sponsor** The advertiser who pays part or all of the cost of a television or radio program.

**spontaneous fission** Nuclear fission in which no particles or photons enter the nucleus from the outside. It can occur only in the heaviest elements, and gives very long half-lives for decay.

**spoofing** Deceiving or misleading the enemy in electronic operations, as by continuing transmission on a frequency after it has been effectively jammed by the enemy, using decoy radar transmitters to lead the enemy into useless jamming effort, or transmitting radio messages containing false information for intentional interception by the enemy.

**sporadic E layer** A layer of intense ionization that occurs sporadically within the E layer. It is variable in time of occurrence, height, geographical distribution, penetration frequency, and ionization density.

**sporadic reflection** *Abnormal reflection.*

**spot** 1. The luminous area produced on the viewing screen of a cathode-ray tube by the electron beam. 2. A commercial announcement of short duration, inserted in programs or broadcast between programs.

**spot gluing** Applying heat to a glued assembly by dielectric heating to make the glue set in spots that are more or less regularly distributed.

**spot jamming** Jamming of a specific channel or frequency.

**spot-noise factor** The noise factor that

exists when noise is a point function of input frequency.

**spot projection** An optical method of scanning or recording in which the scanning or recording spot is defined in the path of the reflected or transmitted light.

**spot size** The cross-section of an electron beam at the screen of a cathode-ray tube.

**spot-size error** An error in interpreting a radarscope presentation, occurring when the spot on the screen is so large that two or more objects appear as one.

**spot speed** The speed of a scanning or recording spot within the available line. Also called scanning speed.

**spot welding** 1. Resistance welding in which the fusion is limited to a small area. 2. Use of dielectric heating to join together sheets of thermoplastic material at a number of spots.

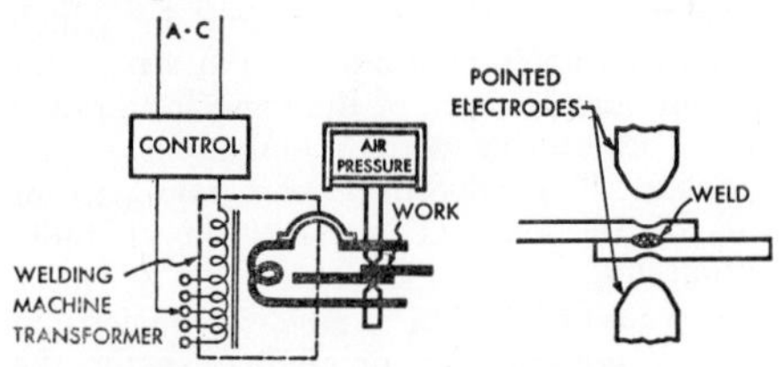

Spot-welding machine and nature of weld.

**spot wobble** An oscillating movement of an electron beam and its resultant spot on the screen, produced intentionally to suppress the pattern of horizontal lines across the picture.

**spout** *Nozzle.*

**sprayed printed circuit** A printed circuit formed by spraying particles of metal in molten or gaseous form onto an insulating base.

**spray point** One of the sharp points arranged in a row and charged to a high d-c potential, used to charge and discharge the conveyor belt in a Van de Graaff generator.

**spread** The range within which the values of a variable quantity occur.

**spreader** An insulating crossarm used to hold apart the wires of a transmission line or multiple-wire antenna.

**spreading resistance** The portion of the resistance of a point-contact rectifier that is due to the semiconducting material alone, not including the barrier-layer resistance.

**spring contact** A relay or switch contact mounted on a flat spring, usually of phosphor bronze.

**spring-return switch** A switch in which the contacts return to their original positions when the operating lever is released.

**spst** Abbreviation for *single-pole single-throw.*

**spurious count** A count from a radiation counter other than background counts and those due directly to ionizing radiation. Also called spurious tube count.

**spurious emission** *Spurious radiation.*

**spurious pulse** A pulse in a radiation counter other than one purposely generated or due directly to ionizing radiation.

**spurious pulse mode** An unwanted pulse mode formed by the chance combination of two or more pulse modes, and indistinguishable from a pulse interrogation or a pulse reply by a transponder.

**spurious radiation** Any emission from a radio transmitter at frequencies outside its frequency band. Also called spurious emission.

**spurious response** Response of a radio receiver to a frequency different from that to which the receiver is tuned.

**spurious response ratio** The ratio of the field strength at the frequency that produces a spurious response to the field strength at the desired frequency, each field being applied in turn to the receiver under specified conditions to produce equal outputs. Image ratio and i-f response ratio are special forms of spurious response ratio.

**spurious signal** An unwanted signal generated in the equipment itself, such as spurious radiation or undesired shading signals generated in a television camera tube.

**spurious tube count** *Spurious count.*

**sputnik** Russian term for an artificial earth satellite. Their first one was placed in orbit October 4, 1957.

**sputtering** A method of depositing a thin layer of metal on a glass, plastic, metal, or other surface in a vacuum. The object to be coated is placed in a large demountable vacuum tube having a cathode made of the metal to be sputtered. The tube is operated under conditions that promote cathode

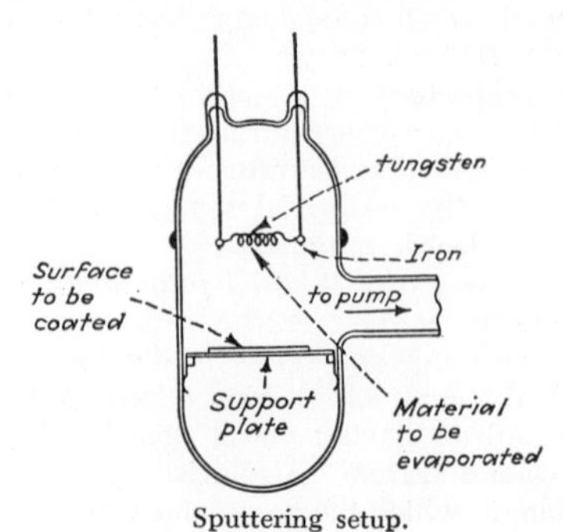

Sputtering setup.

bombardment by positive ions. As a result, extremely small particles of molten metal fall uniformly on the object and produce on it a thin, conductive metal coating. The action occurs to some extent in ordinary electron tubes but is undesirable there because the positive ions knock small particles of the coating off the cathode. Also called cathode sputtering.

**square-foot unit of absorption** *Sabin.*

**square-law demodulator** *Square-law detector.*

**square-law detection** Detection in which sinusoidal input gives an output proportional to the square of the input.

**square-law detector** A demodulator whose output voltage is proportional to the square of the amplitude-modulated input voltage. Also called square-law demodulator.

**square-loop ferrite** A ferrite that has an approximately rectangular hysteresis loop.

**squareness ratio** The ratio of the flux density at zero magnetizing force to the maximum flux density for a material in a symmetrically cyclically magnetized condition. The ratio may alternatively be based on a magnetizing force halfway between zero and its negative limiting value.

**square wave** A wave that alternately assumes two different fixed values for equal lengths of time, the time of transition being negligible in comparison with the duration of each fixed value.

**square-wave generator** A signal generator that generates a square-wave output voltage.

**square-wave modulator** A modulator that delivers a square-wave a-f output voltage, generally at a frequency of 1,000 cps, for modulating r-f signal sources such as klystrons.

**square-wave response** The ratio of the peak-to-peak camera-tube signal amplitude given by a test pattern consisting of alternate equal-width black and white bars to the difference in signal between large-area blacks and large-area whites having the same illumination as the black and white bars in the test pattern. Horizontal square-wave response is measured if the bars run perpendicular to the direction of horizontal scan. Vertical square-wave response is measured if the bars run parallel to the direction of horizontal scan.

**square-wave response characteristic** The relation between the square-wave response of a camera tube and the ratio of raster height to bar width in the square-wave response test pattern.

**square-wave testing** Use of a series of related step functions to determine the performance characteristics of a device or system.

**square well** A potential well that has a constant negative value.

**squaring circuit** 1. A circuit that reshapes a sine or other wave into a square wave. 2. A circuit that contains nonlinear elements and produces an output voltage proportional to the square of the input voltage.

**squealing** A condition in which a radio receiver produces a high-pitched note or squeal along with the desired radio program, due to interference between stations or to oscillation in some receiver circuit.

**squeezable waveguide** A section of waveguide whose width can be altered periodically to shift the phase of the r-f wave traveling through it. Also called squeeze box.

**squeeze box** *Squeezable waveguide.*

**squeeze time** The time interval between the initial application of electrode force to the work and the first application of current in spot welding.

**squeeze track** A variable-density sound track whose width is varied by the recording operator to provide an overriding control on the amplitude of the reproduced signal for the purpose of improving the signal-to-noise ratio.

**squegging** The condition wherein the start and stop of oscillation is determined by the charging and discharging action of a capacitor-resistor combination in the grid circuit of a vacuum-tube oscillator.

**squegging oscillator** *Blocking oscillator.*

**squelch** To quiet a receiver automatically by reducing its gain in response to a specified characteristic of the input, as by reducing gain to suppress background noise when there is no input signal.

**squelch circuit** *Noise suppressor.*

**squint** 1. The angle between the two major lobe axes in a radar lobe-switching antenna. 2. The angular difference between the axis of radar antenna radiation and a selected geometric axis, such as the axis of the reflector. 3. The angle between the full-right and full-left positions of the beam of a conical-scan radar antenna.

**squirrel-cage antenna** A four-bay stacked array of vertical dipoles mounted on a vertical column. Each bay is balun-fed at two points to obtain omnidirectional radiation in the horizontal plane.

**squirrel-cage induction motor** An induction motor in which the secondary circuit consists of a squirrel-cage winding arranged in slots in the iron core.

**squirrel-cage magnetron** A magnetron in which the anode consists of spaced bars concentric with and parallel to the axis of the cathode.

**squirrel-cage winding** A permanently short-circuited winding, usually uninsulated, having its conductors uniformly distributed around the periphery of the rotor and joined by continuous end rings.

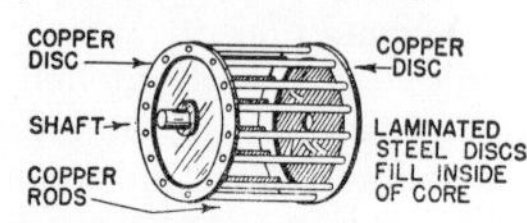


Squirrel-cage winding used as rotor of small induction motor.

**squitter** Random intentional or unintentional triggering of a transponder transmitter in the absence of interrogation, generally by noise signals.

**sre** Abbreviation for *surveillance radar element.*

**ssb modulation** Abbreviation for *single-sideband modulation.*

**ss loran** Abbreviation for *sky-wave-synchronized loran.*

**stability** 1. Freedom from undesired variations. 2. Ability to develop restoring forces that are equal to or greater than the disturbing forces in a control system, so equilibrium is restored. 3. Freedom from undesired oscillation.

**stabilization** 1. Maintenance of a desired orientation independent of the roll and pitch of a ship or aircraft. 2. Treatment of a magnetic material to improve the stability of its magnetic properties.

**stabilized feedback** *Negative feedback.*

**stabilized flight** Flight in which control information is obtained from inertia-stabilized references such as gyroscopes.

**stabilized platform** A platform whose attitude is maintained by two or more gyroscopes and associated servosystems, despite the pitch and roll of a vehicle in space, on water, or in water. Used to support radar antennas, sonar transducers, and inertial guidance systems.

**stable** Incapable of spontaneous changes in atomic or nuclear systems. A stable nuclide is one that is not radioactive.

**stable element** A navigation instrument or device that maintains a desired orientation independently of the motion of the vehicle.

**stable isotope** 1. A nonradioactive isotope of an element that also has radioactive isotopes. 2. A mixture of isotopic nonradioactive nuclides different in composition from any natural mixture. 3. Any stable nuclide.

**stable orbit** The constant-radius circular path of accelerated particles in a betatron or synchrotron. Also called equilibrium orbit.

**stack** 1. To assign different altitudes by radio to aircraft awaiting their turns to land at an airport. 2. A sonar equipment assembly in the sound room of a ship. 3. *Pileup.*

**stacked array** An array in which the antenna elements are stacked one above the other and connected in phase to increase the gain.

**stacked dipoles** Two or more dipole antennas arranged above each other on a vertical supporting structure and connected in phase to increase the gain.

**stacked heads** *In-line heads.*

**stacked stereophonic tape** *In-line stereophonic tape.*

**stage** A circuit containing a single section of an electron tube or equivalent device, or two or more similar sections connected in parallel, push-pull, or push-push. It includes all parts connected between the control-grid input terminal of the device and the input terminal of the next adjacent stage.

**stage-by-stage elimination method** A method of locating trouble in receivers by checking performance of one stage after another with a test signal introduced by a signal generator.

**stage efficiency** The ratio of useful a-c load power of a stage to d-c power input.

**stagger** A periodic error in the position of the recorded spot along a recorded facsimile line.

**staggered circuits** Adjacent circuits that are alternately tuned to two different frequencies to obtain broadband response, as in a video i-f amplifier.

**staggered heads** Magnetic-tape heads that are staggered or displaced 1 7/32 inches along the length of the tape. Used in early stereophonic tape players designed for staggered stereophonic tape. Now superseded by in-line heads. Also called offset heads.

**staggered stereophonic tape** A stereophonic tape recorded with the head gaps spaced 1 7/32 inches apart. Also called offset stereophonic tape.

**staggered tuning** Alignment of successive tuned circuits to slightly different frequencies in order to widen the over-all amplitude-frequency response curve.

**staggering** Adjusting tuned circuits to give staggered tuning.

**stagger-tuned amplifier** An amplifier that uses staggered tuning to give a wide bandwidth.

**staircase generator** A signal generator whose output voltage increases in steps, making its output waveform on an oscilloscope screen have the appearance of a staircase.

**stalo** [STAble Local Oscillator] A highly stable local r-f oscillator used for heterodyning signals to produce an intermediate frequency in radar moving-target indicators. Only echoes that have changed slightly in frequency due to reflection from a moving target produce an output signal.

**stalo cavity** A cavity resonator used with a klystron oscillator to stabilize the output frequency.

**stamped printed circuit** A type of printed circuit formed by die-stamping a foil or film to embed the conductive pattern in an insulating base.

**stamper** A negative, generally made of metal by electroforming, used for molding phonograph records.

**stamping** A transformer lamination that has been cut out of a strip or sheet of metal by a punch press.

**standard** 1. A reference used as a basis for comparison or calibration. 2. A concept that has been established by authority, custom, or agreement, to serve as a model or rule in the measurement of a quantity or the establishment of a practice or procedure.

**standard atmosphere** An arbitrary atmosphere used in comparing performance of aircraft. For official U. S. government use, the standard atmosphere is based on the assumptions that air is a dry perfect gas, that the ground temperature is 59°F, that the temperature gradient in the troposphere is 0.003566°F per foot, and that the temperature in the stratosphere is −67°F. The atmospheric pressure at sea level is then 29.92 inches of mercury.

**standard beam approach** [abbreviated sba] A vhf 40-mc continuous-wave low-approach system using a localizer and markers. The two main signal lobes are tone-modulated with the Morse code letters E and T, respectively. These modulations interlock to form a continuous on-course signal for visual or aural indications in an aircraft.

**standard broadcast band** *Broadcast band.*

**standard broadcast channel** The band of frequencies occupied by the carrier and two sidebands of a broadcast signal, with the carrier frequency at the center. Carrier frequencies are spaced 10 kc apart between 540 and 1,600 kc.

**standard cable** A cable of particular size and construction, used as a reference for specifying transmission line losses. The standard cable has a conductor weighing 20 pounds per mile, a loop resistance of 88 ohms per mile, a capacitance of 0.054 microfarad per mile, an inductance of 0.001 henry per mile, and an attenuation constant of 0.103.

**standard candle** The unit of candlepower, equal to a specified fraction of the visible light radiated by a group of 45 carbon-filament lamps preserved at the National Bureau of Standards, when the lamps are operated at a specified voltage. The standard candle was originally the amount of light radiated by a tallow candle of specified composition and shape.

**standard capacitor** A capacitor constructed in such a manner that its capacitance value is not likely to vary with temperature and is known to a high degree of accuracy. Also called capacitance standard.

**standard cell** A primary cell whose voltage is accurately known and remains sufficiently constant for instrument-calibration purposes. The Weston standard cell has a voltage of 1.018636 volts at 20°C.

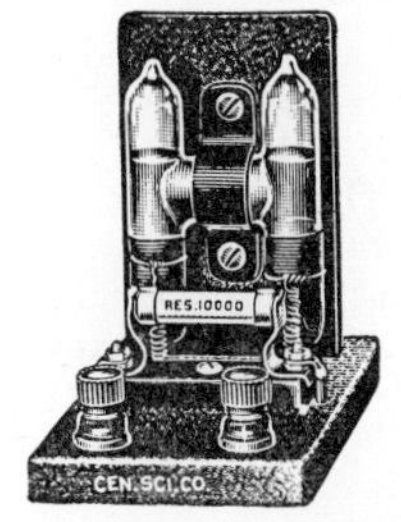

Standard cells.

**standard deviation** The root-mean-square value of the deviations of a series of like quantities from their mean.

**standard-frequency service** A radio communication service involving the transmission of standard and specified frequencies of known high accuracy, intended for general reception.

**standard-frequency signal** One of the highly accurate signals broadcast by the National Bureau of Standards radio station WWV on 2.5, 5, 10, 15, 20, 25, 30, and 35 mc at various scheduled times. Used for testing and calibrating radio equipment throughout the world.

**standard inductor** An inductor (coil) having high stability of inductance value, with little variation of inductance with current or frequency and with a low temperature coefficient. It may have an air core or an iron core. Used as a primary standard in laboratories and as a precise working standard for impedance measurements. Also called inductance standard.

**standardize** To adjust the exponent and coefficient of a floating-point result in a digital computer so the coefficient lies in the normal range of the computer.

**standard loran** Basic loran, in which medium-frequency pulses from transmitting stations are synchronized by means of ground waves rather than by sky waves as in sky-wave-synchronized loran.

**standard microphone** A microphone whose response is accurately known for the condition under which it is to be used.

**standard noise temperature** The standard reference temperature for noise measurements, equal to 290°K.

**standard observer** A hypothetical observer who requires standard amounts of primaries in a color mixture to match every color. The present standard primaries and the standard amounts required to match various wavelengths of the spectrum were established in 1931 by the International Commission on Illumination.

**standard pitch** A musical pitch based on 440 cps for tone A. With this standard the frequency of middle C is 261.6 cps.

**standard propagation** The propagation of radio waves over a smooth spherical earth of uniform dielectric constant and conductivity, under conditions of standard refraction in the atmosphere.

**standard refraction** The refraction that would occur in an idealized atmosphere in which the index of refraction decreases uniformly with height at the rate of $39 \times 10^{-6}$ per kilometer.

**standard resistor** A resistor that is adjusted with high accuracy to a specified value and is but slightly affected by variations in temperature. Also called resistance standard.

**standard sea-water conditions** Sea water at a static pressure of 1 atmosphere, a temperature of 15°C, and a salinity such that the velocity of propagation is exactly 1,500 meters per second. Density is then 1.02338 grams per cubic centimeter, characteristic acoustic impedance is 153.507 cgs units, and pressure spectrum level of thermal noise is 82.17 db below 1 microbar. The velocity of sound increases 0.018 meter per second per meter of depth.

**standard television signal** A signal that conforms to television transmission standards.

**standard time** Mean solar time, based on the transit of the sun over a specified meridian called the time meridian, and adopted for use over an area called a time zone.

**standard track** *Single track.*

**standard volume indicator** A volume indicator having the characteristics specified by the American Standards Association.

**stand by** A request to wait for additional messages to be transmitted a short time later.

**standby battery** A storage battery held in reserve as an emergency power source in event of failure of regular power facilities at a radio station or other location.

**standby transmitter** A transmitter installed and maintained for use during periods when a main transmitter is out of service for maintenance or repair.

**standing-on-nines carry** *High-speed carry.*

**standing wave** A wave in which the ratio of an instantaneous value at one point to that at any other point does not vary with time. A standing wave is produced by two waves traveling in opposite directions and having the same frequency, such as a wave and its reflection from a discontinuity.

**standing-wave antenna** An antenna or antenna system in which the current distributions are produced by standing waves of charges on the conductors.

**standing-wave detector** *Standing-wave-ratio indicator.*

**standing-wave indicator** *Standing-wave-ratio indicator.*

**standing-wave loss factor** The ratio of the transmission loss in an unmatched waveguide to that in the same waveguide when matched.

**standing-wave meter** *Standing-wave-ratio indicator.*

**standing-wave ratio** [abbreviated swr] The ratio of the maximum to the minimum amplitudes of corresponding components of a field, voltage, or current along a transmission line or waveguide in the direction of propagation and at a given frequency; or, alternatively, the reciprocal of this ratio.

**standing-wave-ratio bridge** A bridge used to measure the standing-wave ratio in a transmission line, generally to check the impedance match.

**standing-wave-ratio indicator** An instrument for measuring the standing-wave ratio in a transmission line or waveguide. It may include means for finding the locations of nodes and antinodes. The detecting device is generally a bolometer, crystal diode, or thermocouple. Also called standing-wave detector, standing-wave indicator, standing-wave meter, and standing-wave-ratio meter.

**standing-wave-ratio meter** *Standing-wave-ratio indicator.*

**standing-wave system** *Stationary-wave system.*

**standing-wave voltage ratio** [abbreviated swvr] The ratio of the maximum to the minimum voltage values along a transmission line.

**standoff insulator** An insulator used to support a conductor at a distance from the surface on which the insulator is mounted.

**star** A star-shaped group of tracks made by ionizing particles originating at a common point, either in a nuclear emulsion or in a cloud chamber. Some stars are produced by successive disintegrations of an atom in a radioactive series, and others by nuclear reactions of the spallation type such as those due to cosmic-ray particles.

**star chain** A radio navigation transmitting system comprising a master station about which three or more slave stations are symmetrically located.

**star connection** *Star network.*

**star network** A set of three or more branches with one terminal of each connected at a common node to give the form of a star. Also called star connection.

**starter** An auxiliary control electrode used in a gas tube to establish sufficient ionization to reduce the anode breakdown voltage. Also called trigger electrode.

**starter breakdown voltage** The starter voltage required to cause conduction across the starter gap of a glow-discharge tube when all other tube elements are at cathode potential before breakdown.

**starter gap** The conduction path between a starter and the other electrode to which starting voltage is applied in a gas tube.

**starter voltage drop** The starter-gap voltage drop after conduction is established in the starter gap of a glow-discharge tube.

**starting anode** An anode used to establish the initial arc in a mercury-arc rectifier.

**starting current** The value of electron-stream current through an oscillator at which self-sustaining oscillations will start under specified conditions of loading. Also called preoscillation current.

**starting electrode** An electrode used to establish a cathode spot in a pool tube.

**starting rheostat** A rheostat used to control the current taken by a motor during starting and acceleration.

**starting voltage** The minimum voltage that must be applied to a radiation-counter tube to obtain counts with a particular circuit.

**start-record signal** A signal used for starting the process of converting the electric signal to an image on the record sheet in a facsimile system.

**start signal** A signal used to transfer facsimile equipment from standby to an active condition.

**start-stop multivibrator** *Monostable multivibrator.*

**start-up procedure** A specific procedure for bringing a given nuclear reactor into operation at a desired power level. This may include establishing flow of coolant, starting reaction in the counter range, increasing reaction in the counter range, increasing control, leveling off, approaching criticality, and shifting from manual to automatic required power.

**starved amplifier** A direct-coupled pentode amplifier operated at unusually low screen-grid voltage and having a high anode circuit resistance to give high stage gain with fewer components and low current drain.

**stat-** A prefix used to identify cgs electrostatic units, as in statampere, statcoulomb, statfarad, stathenry, statmho, statohm, and statvolt.

**statampere** The cgs electrostatic unit of current. One statampere is equal to $3.3356 \times 10^{-10}$ ampere.

**statcoulomb** The cgs electrostatic unit of charge. One statcoulomb is equal to $3.3356 \times 10^{-10}$ coulomb.

**statfarad** The cgs electrostatic unit of charge. One statfarad is equal to $1.1126 \times 10^{-12}$ farad.

**stathenry** The cgs electrostatic unit of inductance. One stathenry is equal to $8.9876 \times 10^{11}$ henrys.

**static** 1. A hissing, crackling, or other sudden sharp sound that tends to interfere with the reception, utilization, or enjoyment of desired signals or sounds. When heard in an ordinary radio receiver it may be due to natural electric storms or to improperly operating electric devices in the vicinity. Crackling sounds heard when listening to long-playing plastic phonograph records are due to dust particles attracted to the record by surface electric charges built up by friction on dry days. Static appears as small white specks or flashes on a television picture, called snow. 2. Without motion or change.

**static breeze** *Convective discharge.*

**static characteristic** A relation between a pair of variables, such as electrode voltage and electrode current, with all other operating voltages for an electron tube, transistor, or other amplifying device maintained constant.

**static charge** An electric charge accumulated on an object.

**static convergence** Convergence of the three electron beams at an opening in the center of the shadow mask in a color picture tube. This is called static convergence because the beams must meet at this point when there are no scanning forces. Also called d-c convergence.

**static discharger** A rubber-covered cloth wick about 6 inches long, sometimes attached to the trailing edges of the surfaces of an aircraft to discharge static electricity in flight.

**static electricity** The transfer of a static charge from one object to another by actual contact or by means of a spark that bridges an air gap between the objects.

**static eliminator** A device intended to reduce the effect of atmospheric static interference in a radio receiver.

**static focus** The focus of the undeflected electron beam in a cathode-ray tube.

**static frequency converter** A frequency converter that has no moving parts.

**staticizer** A storage device for converting time-sequential information into static parallel information.

**static luminous sensitivity** The direct anode current of a photoelectric device divided by the incident luminous flux.

**static machine** A machine for generating electric charges, usually by electric induction, to build up high voltages for research purposes.

**static pressure** The pressure that would exist at a point in a medium with no sound waves present. In acoustics, the commonly used unit is the microbar. Also called hydrostatic pressure.

**static split tracking** *Monopulse radar.*

**static storage** Computer storage such that information is fixed in space and available at any time, as in flip-flop circuits, electrostatic memories, and coincident-current magnetic-core storage.

**static stylus force** *Stylus force.*

**static subroutine** A computer subroutine that involves no parameters other than the addresses of the operands.

**static switching** Switching of circuits by means of magnetic amplifiers, semiconductors, and other devices that have no moving parts.

**static transconductance test** A method of testing an electron tube in which anode current is measured for each step increment of grid bias while other operating voltages are held constant, and the transconductance value is computed from the change in anode current values.

**station** 1. An assembly-line or assembly-machine location at which a wiring board or chassis is stopped for insertion of one or more parts. 2. A location at which radio, television, radar, or other electronic equipment is installed. 3. *Broadcast station.*

**stationary-anode tube** An x-ray tube having a stationary anode.

**stationary contact** A contact that is rigidly fastened to the frame of a switch or relay and does not move during operation.

**stationary cpa axis** A fixed reference phase with respect to which a constant chrominance signal has equal and opposite phase angles for successive color television fields. This reference phase is the same for all chrominances.

**stationary state** A discrete energy state in which an electron, atom, or other quantized particle or system may exist.

**stationary wave** A standing wave in which amplitudes of the wave components are equal, so the energy flux is zero at all points.

**stationary-wave system** An interference pattern characterized by stationary nodes and antinodes. Also called standing-wave system.

**statistical error** An error in counting due to random time-distributions of the disintegrations of nuclear particles.

**statistical straggling** A variation in range, ionization, or direction of particles that is due to fluctuations in the distance between collisions in a medium and in the energy loss and deflection angle per collision.

**statistical test** A procedure used to determine whether observed values or quantities fit a hypothesis well enough so the hypothesis can be accepted.

**statistics** *Quantum statistics.*

**statitron** *Van de Graaff generator.*

**statmho** The cgs electrostatic unit of conductance. One statmho is equal to 1.1126 $\times 10^{-12}$ mho.

**statohm** The cgs electrostatic unit of resistance. One statohm is equal to 8.9876 $\times 10^{11}$ ohms.

**stator** 1. The portion of a rotating machine that contains the stationary parts of the magnetic circuit and their associated windings. 2. The stationary set of plates in a variable capacitor.

**stator plate** One of the fixed plates in a variable capacitor. Stator plates are generally insulated from the frame of the capacitor.

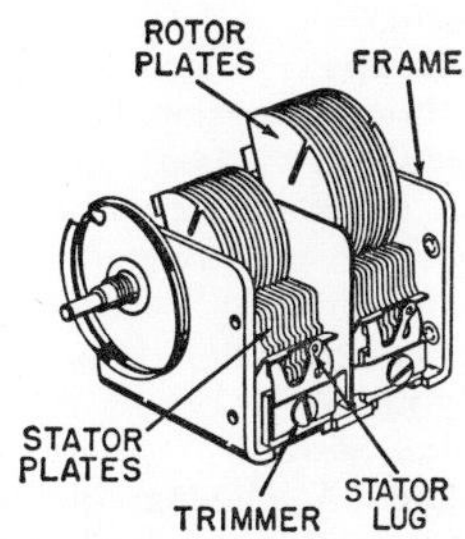


Stator plates in gang capacitor.

**statute mile** A unit of distance equal to 5,280 feet.

**statvolt** The cgs electrostatic unit of voltage. One statvolt is equal to 299.796 volts.

**stc** Abbreviation for *sensitivity-time control.*

**steady state** The condition in which circuit values remain essentially constant, after initial transients or fluctuating conditions have disappeared.

**steady-state error** The error that remains after transient conditions have disappeared in a control system.

**steady-state oscillation** Oscillation in which the motion at each point in the system is a periodic quantity. Also called steady-state vibration and undamped oscillation.

**steady-state vibration** *Steady-state oscillation.*

**steatite** A dense, nonporous heat-resisting ceramic consisting chiefly of a silicate of magnesium, having excellent insulating properties even at high frequencies. It is molded and fired in various shapes for tube sockets and insulators.

**steel-tank rectifier** A mercury-arc rectifier having a steel tank as an envelope.

**steerable antenna** A directional antenna whose major lobe can be readily shifted in direction.

**steering-wheel antenna** A nondirectional vertical dipole used with the iff interrogator of a surface search radar, so named because of the four-spoke wheel structure that surrounds the lower half of the dipole.

**Stefan-Boltzmann law** The total emitted radiant energy of a blackbody is proportional to the fourth power of its absolute temperature.

**Steinmetz formula** An empirical formula for the magnetic hysteresis loss per unit volume of material per magnetization cycle, specifying that the energy loss in ergs is proportional to the 1.6th power of the maximum flux density.

**stellarator** [STELLAR generATOR] An instrument used for studying controlled thermonuclear reactions, in which ionized gases in a glass tube surrounded by magnetizing coils are heated to stellar temperatures of several million degrees by means of plasma pinch.

**stellar guidance** *Celestial guidance.*

**stem** The inward-projecting portion of the glass envelope of an electron tube, through which the heavy leads pass that support and make connections to the electrodes.

**stem radiation** X-rays given off from parts of the anode other than the target in an x-ray tube.

**stenode circuit** An i-f amplifier circuit in which a crystal filter passes only signals at the exact i-f value, giving high selectivity.

**step** An offset in the side of a hole through a shield or in the corresponding plug or other closure for a nuclear reactor, used to give a zigzag joint that minimizes leakage of radiation.

**step attenuator** An attenuator in which the attenuation can be varied in precisely known steps by means of switches.

**step-by-step excitation** The successive transitions of an atom to higher levels of excitation.

**step-by-step system** 1. A control system in which the drive motor moves in discrete steps when the input element is moved continuously. 2. *Strowger system.*

**step-down transformer** A transformer in which the a-c voltages of the secondary windings are lower than that applied to the primary winding.

**step function** A signal having zero value before a certain instant of time and a constant nonzero value immediately after that instant.

**step-function response** The time variation of an output signal when a specified step-function input signal or disturbance is applied.

**stepping** *Zoning.*

**stepping relay** A relay whose contact arm

may rotate through 360° but not in one operation. Also called rotary stepping relay, rotary stepping switch, and stepping switch.

Stepping relay.

**stepping switch** *Stepping relay.*

**step up** To increase the value of electrical quantity.

**step-up** 1. An increase in the value of an electrical quantity. 2. Pertaining to an increase in the value of an electrical quantity.

**step-up transformer** A transformer in which the a-c voltages of the secondary windings are higher than that applied to the primary winding.

**step wedge** 1. A step-shaped block of material that absorbs x-rays, used to compare the radiographic effects of progressively weaker x-rays. 2. An optical negative whose density varies in steps, used for test purposes.

**step-wedge penetrameter** A penetrameter made from material similar to the specimen under radiographic examination, having steps ranging usually from 1 to 5% of the specimen thickness. Each step may contain one or more drilled holes for assessment of definition.

**steradian** A solid spherical angle, subtending a spherical surface whose area is equal to the square of the radius. The total solid angle about a point in space is $4\pi$ steradians.

**Sterba-curtain array** A stacked array with a curtain reflector, suspended from messenger cables running between two steel towers. The curtain may be parasitic or excited. Used with a high-power transmitter for highly directional long-range communications.

**stereo** 1. Pertaining to three-dimensional pickup or reproduction of sound, as achieved by using two or more separate audio channels. Also called stereophonic. 2. *Stereo sound system.*

**stereo-** A prefix used to designate a three-dimensional characteristic.

**stereo amplifier** An audio-frequency amplifier having two or more channels, as required for use in a stereo sound system.

**stereo broadcasting** Broadcasting two sound channels, for reproduction by a stereo sound system having a stereo tuner at its input. Also called stereocasting.

**stereocasting** *Stereo broadcasting.*

**stereocephaloid microphone** A microphone arrangement designed to simulate normal human hearing.

**stereo effect** Reproduction of sound in such a manner that the listener receives the sensation that individual sounds are coming from different locations, just as did the original sounds reaching the stereo microphone system.

**stereofluoroscopy** A fluoroscopic technique that gives three-dimensional images.

**stereo microphone system** An arrangement of two or more microphones spaced far enough apart to give two different output signals, as required for making a stereo recording or feeding a stereo sound system directly or by radio.

**stereophonic** *Stereo.*

**stereo pickup** A phonograph pickup designed for use with standard single-groove two-channel stereo records. The pickup cartridge has a single stylus that actuates two elements, one responding to stylus motion at 45° to the right of vertical and the other responding to stylus motion at 45° to the left of vertical. Each cartridge element feeds one of the channels in a stereo preamplifier or stereo amplifier. Also called forty-five/forty-five pickup and Westrex stereo pickup.

**stereo preamplifier** An audio-frequency preamplifier having two channels, for use in a stereo sound system.

**stereo record** A single-groove disk record having V-shaped grooves at 45° to the vertical. Each groove wall has one of the two recorded channels. Originally developed by Western Electric Co. Also called monogroove stereo record, single-groove stereo record, and stereo recording.

**stereo recorded tape** Recorded magnetic tape having two separate recordings, one for each channel of a stereo sound system. Also called stereo recording.

**stereo recording** 1. *Stereo record.* 2. *Stereo recorded tape.*

**stereoscopic** Pertaining to a three-dimensional visual image.

**stereoscopic television** Television in which the viewed images have a three-dimensional appearance.

**stereo sound system** A sound-reproducing system in which a stereo pickup, stereo tape recorder, stereo tuner, or stereo microphone system feeds two independent audio channels, each of which terminates in one or more loudspeakers arranged to give listeners the same audio perspective as they would get at the original sound source. Also called stereo.

**stereo tape recorder** A magnetic-tape recorder having two stacked playback heads, for reproduction of stereo recorded tape.

**stereo tuner** A tuner having provisions for receiving both channels of a stereo broadcast.

**sticking voltage** The highest voltage that can exist between the cathode and the fluorescent screen of a cathode-ray tube in which the phosphor backing is nonconducting. At higher voltages the screen acquires a negative charge that repels beam electrons.

**stiction** [STatic frICTION] Friction that tends to prevent relative motion between two movable parts at their null position.

**stilb** A unit of luminance equal to one candle per square centimeter.

**stimulator** A neurosurgical device that supplies a controlled a-c voltage to two electrodes that are applied to a patient.

**stimulus** A signal that affects the controlled variable in a control system.

**stirring effect** The circulation produced in a molten charge of metal due to the combined forces of motor and pinch effects.

**stochastic process** A random process.

**stochastic variable** A variable that is dependent on the random variable and is usually measured experimentally.

**stoichiometric impurity** A crystalline imperfection in a semiconductor.

**Stokes' law** The wavelength of luminescence excited by radiation is always greater than that of the exciting radiation.

**stop** The aperture or useful opening of a lens, usually adjustable by means of a diaphragm.

**stopping** The decrease in kinetic energy of an ionizing particle as a result of energy losses along its path through matter.

**stopping capacitor** *Coupling capacitor.*

**stopping cross section** *Atomic stopping power.*

**stopping equivalent** The thickness of a standard substance that is capable of producing the same energy loss as does a given thickness of the substance under consideration, when a charged particle passes through. Also called equivalent stopping power.

**stopping off** The application of a resist to any part of a cathode or other surface prior to electroplating or etching.

**stopping power** The effect of a substance on the kinetic energy of a charged particle passing through it.

**stopping voltage** The voltage required to stop an electron emitted by photoelectric or thermionic action.

**stop-record signal** A signal used for stopping the process of converting the electric signal to an image on the record sheet in a facsimile system.

**stop signal** A signal that initiates the transfer of facsimile equipment from active to standby conditions.

**storage** A computer device used primarily for storing information. The physical means of storing information may be electrostatic, ferroelectric, magnetic, acoustic, optical, chemical, electronic, electrical, or mechanical. Also called memory.

**storage battery** A connected group of two or more storage cells. Common usage permits application of this term to a single cell used independently. Also called accumulator (British) and secondary battery.

**storage capacity** The information-storing capacity of a storage device, expressed in words of a specified length or in bits. Also called memory capacity.

**storage cell** An electrolytic cell for generating electric energy, in which the cell after being discharged may be restored to a charged condition by sending an electric current through it in a direction opposite that of the discharging current. Also called secondary cell.

**storage effect** Temporary storage of injected excess minority carriers in the higher-resistivity side of a semiconductor junction.

**storage element** An area of a storage surface that retains information distinguishable from that of adjacent areas in a charge storage tube.

**storage location** A digital-computer storage position holding one machine word and usually having a specific address.

**storage register** A register in the main internal memory of a digital computer. storing one computer word. Also called location.

**storage time** Deprecated term for *decay time* and *maximum retention time* in charge storage tubes.

**storage tube** An electron tube employing cathode-ray beam scanning and charge storage for the introduction, storage, and removal of information. Also called electrostatic storage tube, graphechon, and memory tube (deprecated).

**storage-type camera tube** *Iconoscope.*

**store** To transfer an element of information to storage in a computer, for later extraction.

**stored energy** The potential energy of atomic displacements that is retained by a solid after irradiation.

**stored-energy welding** Welding in which electric energy is accumulated electrostatically, electromagnetically, or electrochemically at a relatively slow rate and released at the required rate for welding.

**stovepipe antenna** A nondirectional vertical dipole used with the iff interrogator of a surface search radar, so named because it approximates the diameter of a stovepipe.

**straggling** Random variations of some property of particles as a result of their passage through matter.

**straight dipole** A dipole consisting of a single straight conductor, usually broken at its center for connection to a transmission line.

**straight-line capacitance** [abbreviated slc] A variable-capacitor characteristic obtained when the rotor plates are shaped so capacitance varies directly with the angle of rotation.

**straight-line frequency** [abbreviated slf] A variable-capacitor characteristic obtained when the rotor plates are shaped so the resonant frequency of the tuned circuit containing the capacitor varies directly with the angle of rotation.

**straight-line path** The axis of the Fresnel-zone family of paths between two microwave antennas.

**straight-line wavelength** [abbreviated slw] A variable-capacitor characteristic obtained when the rotor plates are shaped so the wavelength for resonance in the tuned circuit containing the capacitor varies directly with the angle of rotation.

**strain gage** A strain-sensitive element designed to be attached to a member in which strain is to be measured. It is usually connected into a bridge circuit that feeds a recorder directly or through an amplifier. The commonest type is the resistance strain gage.

**strain insulator** An insulator used between sections of a stretched wire or antenna to break up the wire into insulated sections while withstanding the total pull of the wire.

**strand** One of the wires or groups of wires in a stranded wire.

**stranded conductor** *Stranded wire.*

**stranded wire** A conductor composed of a group of wires or a combination of groups of wires, usually twisted together. Also called stranded conductor.

**strap** A conductive link used between alternate resonator segments of a magnetron.

**strapping** Connecting together resonator segments having the same polarity in a multicavity magnetron, to suppress undesired modes of oscillation.

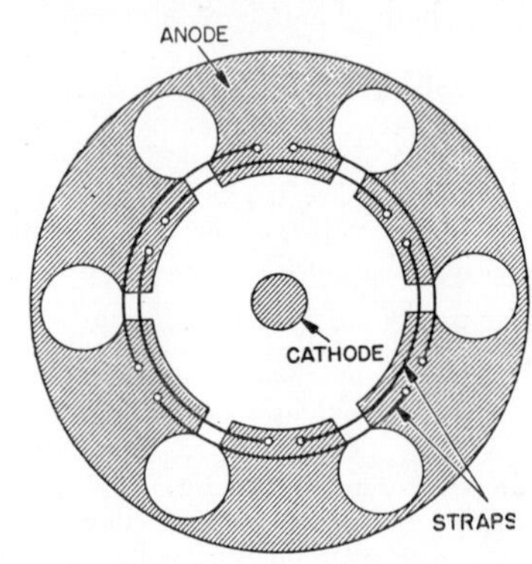


Strapping in multicavity magnetron.

**strategic missile** A guided missile designed for long-range action against an enemy's war-making capacity.

**stratosphere** A stratum of the earth's atmosphere above the troposphere, extending from about 7 miles up to about 50 miles above the earth. The temperature is essentially constant in the stratosphere.

**stratovision** A proposed system for increasing the range of television coverage by transmitting from an airplane circling in the stratosphere.

**stray capacitance** Undesirable capacitance between circuit wires, between wires and the chassis, or between components and the chassis of electronic equipment. Also called strays.

**stray field** Leakage magnetic flux that spreads outward from a coil and does no useful work.

**stray radiation** Radiation that serves no useful purpose.

**strays** *Stray capacitance.*

**streaking** A television picture condition in which white or black horizontal streaks or smudges appear to follow images across the screen. The effect is more apparent at vertical edges of objects where there is an abrupt transition from black to white or white to black, and may be due to excessive low-frequency response.

**streamer** An indefinite wavy band occurring when the gas pressure in a discharge tube is reduced below the value required for a glow discharge through the tube.

**streaming** The production of unidirectional flow currents in a medium, arising from the presence of sound waves.

**strength** The maximum instantaneous rate of volume displacement produced by a sound source when emitting a wave with sinusoidal time variation.

**striated discharge** An electric discharge characterized by alternate light and dark bands in the positive column adjacent to the anode of a glow-discharge tube.

**striation technique** A technique for making sound waves visible by using their individual abilities to refract light waves.

**striking** 1. Starting an electric arc by touching the electrodes together momentarily. 2. Electrodeposition of a thin initial film of metal, usually at a high current density.

**striking voltage** The grid-cathode voltage required to start the flow of anode current in a gas tube.

**string electrometer** An electrometer in which a conducting fiber is stretched midway between two oppositely charged metal plates. The electrostatic field between the plates displaces the fiber laterally in proportion to the voltage between the plates.

**stringer** A long plug used to close a hole in the shield of a nuclear reactor. The stringer is removed for insertion of experimental materials in the core.

**strip penetrameter** *Plate penetrameter.*

**stripper** A hand or motorized tool used to remove insulation from wires.

**stripping** 1. Removal of a metal coating, as when etching away undesired portions of a printed circuit. 2. An effect observed in bombardment with deuterons or heavier nuclei, whereby only part of the incident particle merges with the target nucleus. The remainder proceeds with most of its original momentum in practically its original direction.

**strip transmission line** A microwave transmission line consisting of a thin, narrow rectangular strip that is separated from a wide ground-plane conductor or mounted between two wide ground-plane conductors. Separation is usually achieved with a low-loss dielectric material on which the conductors are formed by printed-circuit techniques. Also called flat coaxial transmission line and microstrip.

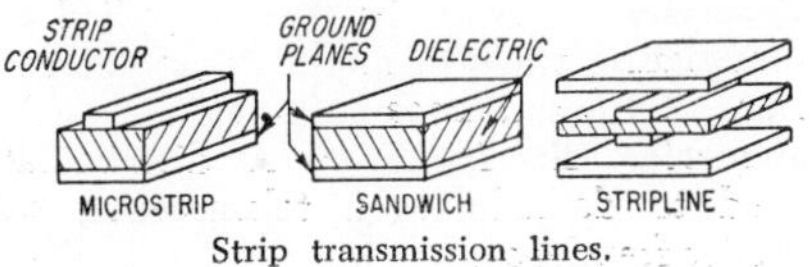


Strip transmission lines.

**strobe pulse** A pulse superimposed on the trace for the echo on the screen of a radar cathode-ray tube, to serve as a marker from which the range of the target can be measured.

**stroboscope** A controllable intermittent source of intense light, used to create the illusion of slowing down or stopping vibrating or rotating objects. The flashing frequency is adjusted until it corresponds to some multiple of the speed of vibration or rotation of the object under study.

**stroboscopic disk** A printed disk having a number of concentric rings each containing a different number of dark and light segments. When the disk is placed on a phonograph turntable or rotating shaft and illuminated at a known frequency by a flashing discharge tube, speed can be determined by noting which pattern appears to stand still or rotate slowly.

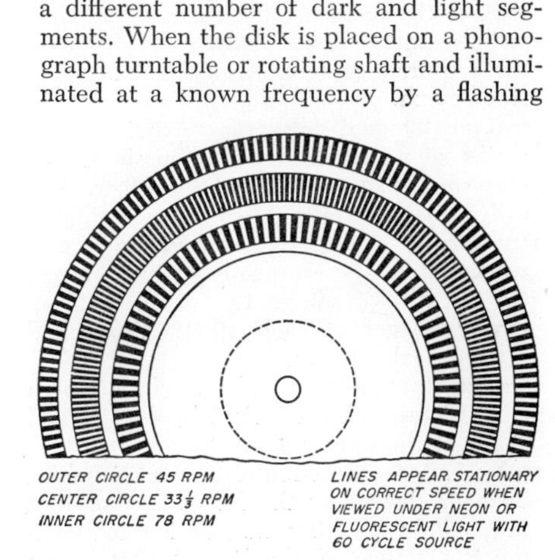


Stroboscopic disk for checking speed of phonographs.

**stroboscopic tachometer** A stroboscope having a scale that reads in flashes per minute or in revolutions per minute. The speed of a rotating device is measured by directing the stroboscopic lamp on the device, adjusting the flashing rate until the device appears to be stationary, then reading the speed directly on the scale of the instrument.

**stroboscopic tube** *Strobotron.*

**strobotron** [STROBOscope thyraTRON] A cold-cathode gas-filled arc-discharge tube having one or more internal or external grids to initiate current flow and produce intensely bright flashes of light for a stroboscope. Also called stroboscopic tube.

**stroke speed** *Scanning-line frequency.*

**strontium** [symbol Sr] A metallic element sometimes used on cathodes of phototubes to obtain maximum response to ultraviolet radiation. Atomic number is 38.

**strontium-90** A radioisotope having a half-life of about 25 years. Also called radio strontium.

**strophotron** A multitransit microwave tube developed in Sweden.

**Strowger system** An automatic telephone switching system that uses successive step-by-step selector switches actuated by current pulses produced by rotation of a telephone dial. The selectors are electromagnetically operated and contain a number of tiers of fixed contacts, each arranged in a semicircle. A moving contact arm first rises to the height of the desired tier, then swings around horizontally and stops over the required contact. Also called step-by-step system.

**structurally dual networks** A pair of networks so arranged that any mesh of one corresponds to a cut-set of the other. Also called dual networks.

**structurally symmetrical network** A network that can be arranged so a cut through the network produces two parts that are mirror images of each other. Also called symmetrical network.

**structural resolution** The resolution of a color picture tube as limited by the size and shape of the screen elements.

**stub** 1. A short section of transmission line, open or shorted at the far end, connected in parallel with a transmission line to match the impedance of the line to that of an

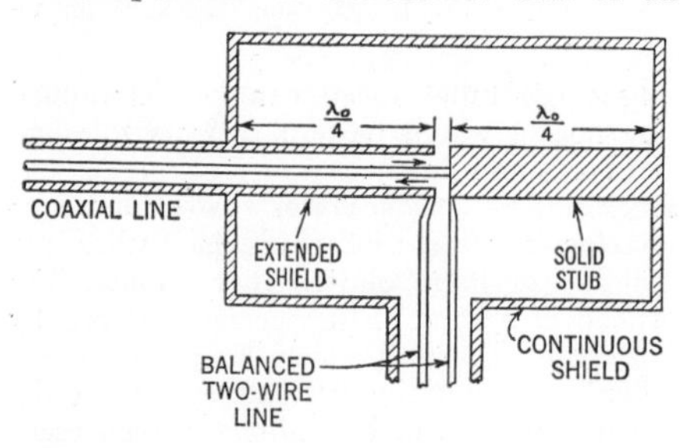


Stub used in cavity to support conductor during conversion from coaxial line to balanced two-wire line.

antenna or transmitter. 2. A solid projection one quarter-wavelength long, used as an insulating support in a waveguide or cavity.

**stub-matching** Use of a stub to match a transmission line to an antenna or load. Matching depends on the spacing between the two wires of the stub, the position of the shorting bar, and the point at which the transmission line is connected to the stub.

**stub-supported line** A transmission line that is supported by short-circuited quarter-wave sections of coaxial line. A stub exactly a quarter-wavelength long acts as an insulator because it has infinite reactance.

**stub tuner** An adjustable shorted stub used to adjust a transmission line for maximum power transfer.

**studio** A room in which television or radio programs are produced.

**stylus** [plural is styli] The portion of a phonograph pickup that follows the modulations of a record groove and transmits the

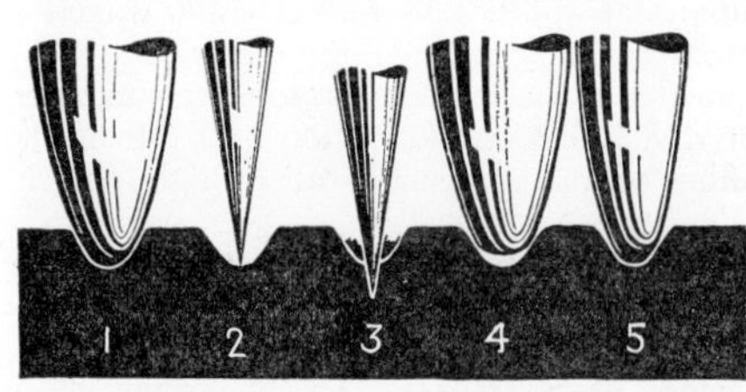


Stylus points: (1) theoretically ideal but impractical because needle angle changes; (2) too sharp, allowing free movement; (3) too sharp, causing gouging; (4) too blunt, scoring groove walls; (5) best practical shape.

resulting mechanical motions to the transducer element of the pickup for conversion to corresponding a-f signals. Also called needle and reproducing stylus.

**stylus drag** The force resulting from friction between the surface of a recording medium and the reproducing stylus. Also called needle drag.

**stylus force** The vertical force exerted on a stationary recording medium by the stylus when in its operating position. Also called needle force, needle pressure (deprecated), static stylus force, stylus pressure (deprecated), and vertical stylus force.

**stylus pressure** Deprecated term for *stylus force.*

**subassembly** Two or more components combined into a unit for convenience in assembling or servicing equipment. An i-f strip for a receiver is an example.

**subatomic** Pertaining to particles smaller than atoms, such as electrons, protons, and neutrons.

**subcarrier** 1. A carrier that is applied as a modulating wave to modulate another carrier. 2. *Chrominance subcarrier.*

**subcarrier discriminator** A discriminator used to demodulate a telemetering subcarrier frequency.

**subcarrier oscillator** 1. The crystal oscillator that operates at the chrominance subcarrier or burst frequency of 3.579545 mc in a color television receiver. This oscillator, synchronized in frequency and phase with the transmitter master oscillator, furnishes the continuous subcarrier frequency required for demodulators in the receiver. 2. An oscillator used in a telemetering system to translate variations in an electrical quantity into variations of a frequency-modulated signal at a subcarrier frequency.

**subcontractor** A manufacturer or other organization that receives a contract from a prime contractor for a portion of the work on a project.

**subcritical** Having an effective multiplication constant less than 1, so that a self-supporting chain reaction cannot be maintained in a nuclear reactor.

**subharmonic** A sinusoidal quantity having a frequency that is an integral submultiple of the frequency of some other sinusoidal quantity to which it is referred. A third subharmonic would be one-third the fundamental or reference frequency.

**subject copy** Material that is to be transmitted for facsimile reproduction. Also called copy.

**submarine detecting set** A complete airborne electronic detection system designed primarily to indicate the presence of submerged submarines and mines.

**subminiature tube** An extremely small electron tube designed for use in hearing aids and other miniaturized equipment. A typical subminiature tube is about 1½ inches long and 0.4 inch in diameter, with the pins emerging through the glass base.

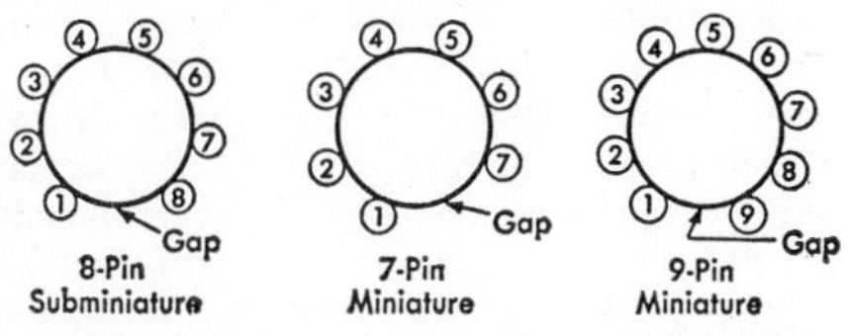


Subminiature-tube pin connections, with miniature tube connections shown for comparison. Diagrams represent bottom views of tubes and sockets. No. 1 pin is always next to gap in clockwise direction. Pins are never omitted.

**subminiaturization** Reduction of size and weight of electronic equipment, generally achieved through use of printed circuits, modules, and special heat-dissipating features.

**subprogram** A portion of a computer program.

**subrefraction** Refraction in which radar waves are bent upward by the atmosphere due to an excessive vertical temperature gradient or humidity gradient, reducing the range of a radar set.

**subroc** [SUBmarine ROCket] A Navy missile that is fired underwater from a conventional torpedo tube, but is rocket-powered and most of its course is through the air. Range is about 50 miles.

**subroutine** A portion of a routine that causes a computer to carry out a well-defined mathematical or logical operation. At its conclusion, control reverts to the master routine.

**subscription television** A television system in which programs are broadcast in coded or scrambled form, for reception only by subscribers who make individual or monthly payments for use of the decoding or unscrambling devices required to obtain a clear program. Either the picture or the sound, or both, may be scrambled. Also called pay television and toll television.

**subsonic** 1. Less than the velocity of sound in air, hence less than mach 1, which is about 738 miles per hour. Used chiefly in connection with airplanes and missiles. 2. Deprecated term for *infrasonic*.

**subsynchronous** Operating at a frequency or speed that is related to a submultiple of the source frequency.

**subtraction-type radiometer** A radio receiver in which a d-c voltage equal to the smoothed second detector output, when only noise is present at the receiver input, is subtracted from the smoothed second detector output. The resulting difference in voltage is applied to a galvanometer as an indication of an input signal.

**sudden ionospheric disturbance** A sudden increase of ionization density in low parts of the ionosphere due to a bright solar eruption. It usually lasts a few minutes, causing a sudden increase in absorption of radio waves and sometimes also simultaneous disturbances in earth currents and the earth's magnetic field.

**Suhl effect** When a strong transverse magnetic field is applied to an n-type semiconducting filament, holes injected into the filament are deflected to the surface, where they may recombine rapidly with electrons or be withdrawn by a probe. The over-all effect is an increase in conductance.

**sulfating** Formation of lead sulfate on the plates of lead-acid storage batteries, reducing the energy-storing ability of the battery and eventually causing failure.

**sulfur** [symbol S] An element. Atomic number is 16.

**sulfuric acid** A compound of sulfur, hydrogen, and oxygen, having the chemical formula $H_2SO_4$. Used as the electrolyte in lead-acid storage batteries.

**sum-and-difference monopulse radar** Monopulse radar in which the receiving antenna has one aperture and two or more closely spaced feeds, each of which produces a radiation pattern that is displaced from the antenna boresight axis. Signals arriving off the axis give unequal amplitudes in the two channels, while signals on the axis give equal amplitudes that produce a sharp null in the difference channel and a peak in the sum channel.

**sum channel** A combination of two stereophonic channels to give a program that can be recorded or transmitted by a single channel.

**summation bridge** A bridge that measures such values as temperature, frequency, speed of rotation, time, resistance, and capacitance by adding the original bridge current to the current needed for balance and presenting the result on an indicator or scale.

**summation check** A redundancy check in which groups of digits are summed, usually without regard for overflow, and that sum checked against a previously computed sum to verify the accuracy of a digital computer.

**summation network** *Summing network.*

**summing amplifier** An amplifier that delivers an output voltage which is proportional to the sum of two or more input voltages or currents.

**summing network** A passive electric network whose output voltage is proportional to the sum of two or more input voltages. Also called summation network.

**summing point** A mixing point whose output is obtained by addition, with prescribed signs, of its inputs in a feedback control system.

**sun follower** A photoelectric pickup and an associated servomechanism used to maintain a sun-facing orientation, as for a space vehicle. Also called sunseeker.

**S unit** An arbitrary unit of signal strength, sometimes used along with decibels on the scale of a signal-strength meter.

**S-unit meter** *Signal-strength meter.*

**sunlight recorder** A recorder that uses a phototube, capacitor-charging circuit, and thyratron-operated counter to record the integrated value of solar radiation over a period of time.

**sunseeker** *Sun follower.*

**sunspot** A dark spot on the sun, usually associated with magnetic storms on the earth that affect radio communication at the lower frequencies.

**sunspot cycle** A period of about 11 years in which the number and duration of sunspots pass through a cycle of buildup to a maximum value and then drop back to a minimum value.

**sunspot number** The predicted number of sunspots for a given month.

**sun strobe** The signal display seen on a radar ppi screen when the radar antenna is aimed at the sun. The pattern resembles that produced by continuous-wave interference, and is due to r-f energy radiated by the sun. Used for checking radar antenna azimuth and elevation calibration, by comparing with the known position of the sun.

**SUP** Symbol for *suppressor grid.* Used on circuit diagrams.

**super** *Superheterodyne receiver.*

**superconducting transition** A transition from a normal state to a superconducting state, occurring at a temperature that depends on the magnetic field as well as on the nature of the material.

**superconductivity** A low-temperature phenomenon in which the resistance of certain metals, alloys, and compounds drops essentially to zero at a critical temperature near absolute zero.

**superconductor** A material that exhibits superconductivity.

**supercontrol tube** *Variable-mu tube.*

**supercritical** Having an effective multiplication constant greater than 1, so that the rate of reaction rises rapidly in a nuclear reactor.

**superemitron camera** British term for *image iconoscope.*

**superhet** *Superheterodyne receiver.*

**superheterodyne receiver** A receiver in which all incoming modulated r-f carrier signals are converted to a common i-f carrier value for additional amplification and selectivity prior to demodulation, using

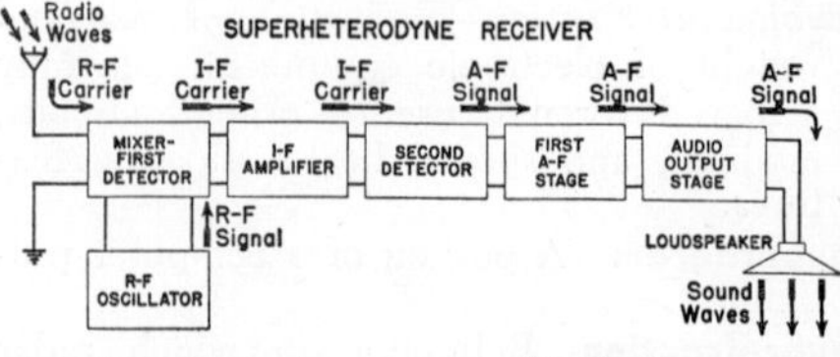


Superheterodyne receiver block diagram.

heterodyne action. The output of the i-f amplifier is then demodulated in the second detector to give the desired a-f signal. Also called super and superhet.

**superhigh frequency** [abbreviated shf] A Federal Communications Commission designation for the band from 3,000 to 30,000 mc in the radio spectrum.

**superlattice** A lattice in which the atoms of one element occupy regular positions in the lattice of another element without forming a compound.

**Supermalloy** Trademark of Arnold Engineering Co. for a magnetic alloy having a maximum permeability of over 1,000,000.

**superposed circuit** An additional channel obtained from one or more circuits normally provided for other channels, in such a manner that all the channels can be used simultaneously without mutual interference.

**superposition theorem** The current that flows at a point in a linear network during simultaneous application of a number of voltages throughout the network is the sum of the component currents at the point that would be produced by the individual voltages acting separately. Similarly, the voltage between any two points under such conditions is the sum of the voltages that would be produced between these two points by the individual voltages acting separately.

**superpower station** A broadcast station employing extremely high power, generally over 1,000,000 watts.

**superrefraction** Refraction in which radar waves are bent downward by the earth's atmosphere, generally over sea water that is at least 5°F colder than the air. The effect is comparable to that of an atmospheric duct. The range of a radar set is greatly increased under this condition.

**superregeneration** Regeneration in which the oscillation is broken up or quenched at a frequency slightly above the upper audibility limit of the human ear by a separate oscillator circuit connected between the grid and anode of the amplifier tube, to prevent the regeneration from exceeding the maximum useful amount.

**superregenerative detector** A detector in which superregeneration is used to obtain extremely high sensitivity with a minimum number of amplifier stages.

**superregenerative receiver** A tuned-radio-frequency receiver employing a superregenerative detector.

**supersensitive relay** A relay that operates on extremely small currents, generally below 250 microamperes.

**supersonic** 1. Faster than the velocity of sound in air, hence faster than mach 1, which is about 738 miles per hour. Used chiefly in connection with airplanes and missiles. 2. Deprecated term for *ultrasonic.*

**supersonics** The study of phenomena associated with aircraft and missile speeds higher than the speed of sound.

**supersync signal** A combination horizontal and vertical sync signal transmitted at the end of each television scanning line to synchronize the operation of a television receiver with that of the transmitter.

**superturnstile antenna** A modified turnstile antenna having wing-shaped dipole elements used in pairs mounted at right angles about a common vertical axis. The dipole pairs are fed in quadrature to give substantially omnidirectional radiation over a wide band for f-m and television transmitters. Also called batwing antenna.

**supervisory control system** A control system that provides both indication and control for remotely located equipment by electrical means, as by carrier-current channels on power lines.

**supervoltage** A voltage in the range of 500 to 2,000 kilovolts, used for some x-ray tubes.

**supply voltage** The voltage obtained from a power supply for operation of a circuit or device.

**suppressed carrier** A carrier that is suppressed at the transmitter. The chrominance subcarrier in a color television transmitter is an example.

**suppressed-carrier transmission** Transmission in which the carrier component of the modulated wave is suppressed, leaving only the sidebands to be transmitted.

**suppressed time delay** *Code delay.*

**suppressed-zero instrument** An indicating or recording instrument in which the zero position is below the lower end of the scale markings.

**suppression** Elimination of any component of an emission, as a particular frequency or group of frequencies in an a-f or r-f signal.

**suppressor** 1. A resistor used in series with a spark plug or distributor of an automobile engine or other internal-combustion engine to suppress spark interference that might otherwise interfere with radio reception. 2. *Suppressor grid.*

**suppressor grid** [symbol SUP] A grid placed between two positive electrodes in an electron tube primarily to reduce the

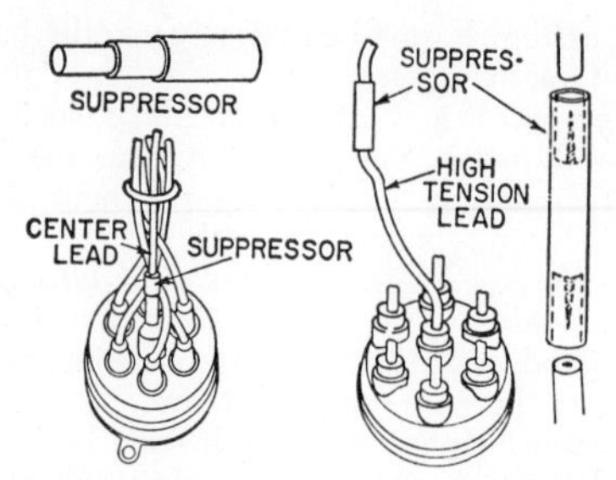


Suppressor installation in high-voltage center lead of distributor in automobile.

flow of secondary electrons from one electrode to the other. It is usually used between the screen grid and the anode. Also called suppressor.

**suppressor-grid modulation** Amplitude modulation in which the modulating signal is applied to the suppressor grid of a pentode that is amplifying the carrier signal.

**surface analyzer** An instrument that measures or records irregularities in a surface by moving the stylus of a crystal pickup or similar device over the surface, amplifying the resulting voltage, and feeding the output voltage to an indicator or recorder that shows the surface irregularities magnified as much as 50,000 times.

**surface barrier** A potential barrier formed at a surface of a semiconductor by the trapping of carriers at the surface. The effective area of the barrier is appreciably larger than for a point-contact transistor.

**surface-barrier transistor** A transistor triode in which surface barriers are formed on opposite sides of a thin wafer of n-type germanium by etching depressions, then electroplating the collector and emitter dots that serve as rectifying contacts.

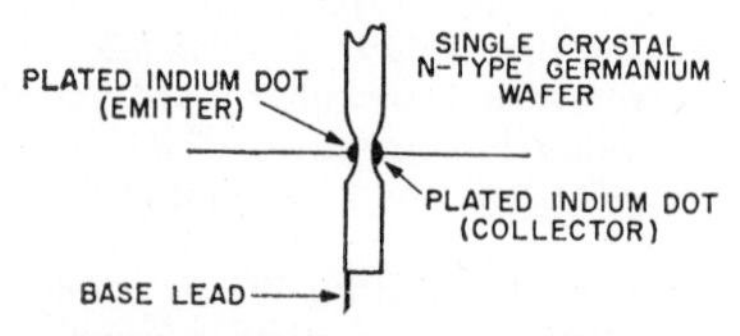


Surface-barrier transistor construction.

**surface-contact rectifier** A rectifier in which a surface barrier serves as the rectifying contact.

**surface density** The quantity per unit area of anything distributed over a surface. In nuclear physics, mass per unit area is used to indicate absorber thickness, source thickness, target thickness, and support thickness for thin radioactive sources.

**surface duct** An atmospheric duct for which the lower boundary is the surface of the earth.

**surface hardening** Hardening of a metallic surface by rapid induction heating and rapid quenching.

**surface leakage** The passage of current over the surface of an insulator.

**surface noise** The noise component in the electric output of a phonograph pickup due to irregularities in the contact surface of the groove. Also called needle scratch.

**surface of position** A surface defined by a constant value of some navigation coordinate.

**surface photoelectric effect** Ejection of an electron from the surface of a solid or liquid by an incident photon whose total energy is absorbed by the material.

**surface recombination rate** The time rate at which free electrons and holes recombine at the surface of a semiconductor.

**surface recombination velocity** The velocity with which electrons and holes drift to the surface of a semiconductor and recombine.

**surface resistivity** The electric resistance of the surface of an insulator, measured between the opposite sides of a square on the surface. The value in ohms is independent of the size of the square and the thickness of the surface film.

**surface-to-air guided missile** A guided missile designed to be fired at an airborne target from the ground or from the deck of a surface ship. Examples include Bomarc, Hawk, Nike, Talos, Tartar, Terrier, and Wizard.

**surface-to-surface guided missile** A guided missile designed to be fired at a surface target from a surface position on land or water. Examples include Atlas, Corporal, Dart, Jupiter, Lacrosse, Mace, Matador, Navaho, Pershing, Polaris, Redstone, Regulus, Sergeant, Snark, Thor, and Titan.

**surface transfer impedance** The ratio of the tangential component of the electric field at the surface of a conductor to the current in the conductor, when an electromagnetic wave is guided along the surface.

**surface wave** 1. A wave that can travel along an interface between two different media without radiation. The interface must be essentially straight in the direction of propagation. The commonest interface used is that between air and the surface of a circular wire. 2. *Ground wave.*

**surface-wave transmission line** A single-conductor transmission line energized in such a way that a surface wave is propagated along the line with satisfactorily low attenuation.

**surge** A momentary large increase in the current or voltage in an electric circuit.

**surge admittance** The reciprocal of characteristic impedance.

**surge-crest ammeter** A magnetometer used with magnetizable links to measure the peak value of transient electric currents.

**surge generator** A device for producing high-voltage pulses, usually by charging capacitors in parallel and discharging them in series.

**surge impedance** *Characteristic impedance.*

**surge suppressor** A suppressor that responds to the rate of change of a current or voltage to prevent a rise above a predetermined value. It may include resistors, capacitors, coils, gas tubes, and semiconducting disks.

**surveillance radar** A radio navigation aid employing primary radar to display on a map-like circular screen the positions of all aircraft within its range. Examples are airport surveillance radar and ground surveillance radar.

**surveillance radar element** [abbreviated sre] *Airport surveillance radar.*

**survey** Measurement of radiation in the vicinity of a nuclear reactor or other source.

**survey instrument** A portable instrument used to detect and measure radiation.

**survival curve** A curve showing the number or percentage of organisms or individuals surviving at specified intervals after a particular dose of radiation.

**susceptance** The imaginary component of admittance. Susceptance is measured in mhos and is the reciprocal of reactance.

**suspension** A fine wire or coil spring that supports the moving element of a meter.

**sustained oscillation** Continuous oscillation at a frequency essentially equal to the resonant frequency of the system.

**sustaining program** A program that has no commercial sponsor.

**SW** Symbol for *switch.* Used on circuit diagrams.

**s-w** Abbreviation for *short-wave.*

**swamping resistor** A resistor placed in the emitter lead of a transistor circuit to minimize the effects of temperature on the emitter-base junction resistance.

**sweep** 1. The steady movement of the electron beam across the screen of a cathode-ray tube, producing a steady bright line when no signal is present. The line is straight for a linear sweep and circular for a circular sweep. 2. The steady change in the output frequency of a signal generator from one limit of its range to the other.

**sweep amplifier** An amplifier used in a television receiver to amplify the sawtooth output voltage of the sweep oscillator and shape the waveform as required for the deflection circuits.

**sweep circuit** The sweep oscillator, sweep amplifier, and any other stages used to produce the deflection voltage or current for a cathode-ray tube.

**sweep frequency** The rate at which an electron beam is swept back and forth across the screen of a cathode-ray tube.

**sweep generator** A test instrument that generates an r-f voltage whose frequency varies back and forth through a given frequency range at a rapid constant rate. Used to produce an input signal for circuits or devices whose frequency response is to be observed on an oscilloscope. Also called sweep oscillator.

**sweep jamming** Jamming an enemy radarscope by sweeping the region of radar beam coverage with electromagnetic waves having the same frequency as those received by the radarscope.

**sweep oscillator** 1. An oscillator used to generate a sawtooth voltage that can be amplified to serve for deflecting the electron beam of a cathode-ray tube. Also called time-base generator and timing-axis oscillator. 2. *Sweep generator.*

**sweep-through jamming** Jamming in which a transmitter is swept through an r-f band in short steps to jam each frequency briefly, producing a sound like that of an aircraft engine.

**sweep voltage** The periodically varying voltage applied to the deflection plates of a cathode-ray tube to give a beam displacement that is a function of time. Also called time-base voltage.

**swept gain control** *Sensitivity-time control.*

**swimming-pool reactor** A low-power nuclear reactor in which the fuel elements are plates of enriched $U^{235}$ encased in aluminum alloy and suspended in a pool of water. The water serves as moderator, coolant, and radiation shield. Used for research.

**swing** The total variation in the frequency or amplitude of a quantity.

**swinging** Momentary variation in the frequency of a received radio wave.

**swinging choke** An iron-core choke having a core that can be operated almost at magnetic saturation. The inductance is then a maximum for small currents, and swings to a lower value as current increases. Used as the input choke in a power supply filter to provide improved voltage regulation.

**switch** [symbol SW] A manual or mechanically actuated device for making, breaking, or changing the connections in an electric circuit.

**switchboard** A single large panel or assembly of panels on which are mounted switches, circuit breakers, meters, fuses, and terminals essential to the operation of electric equipment.

**switch detent** A detent used on a switch to establish predetermined switching positions.

**switching coefficient** The derivative of applied magnetizing force with respect to the reciprocal of the resultant switching time. It is usually determined as the reciprocal of the slope of a curve in which the reciprocal of switching time is plotted against applied magnetizing force. The magnetizing force is applied in steps.

**switching control console** A console designed to control individual elements of a switching system in accordance with a predetermined plan.

**switching diode** A crystal diode that provides essentially the same function as a switch. Below a specified applied voltage it has high resistance corresponding to an open switch, while above that voltage it suddenly changes to the low resistance of a closed switch.

**switching key** *Key.*

**switching reactor** A saturable-core reactor that has several input control windings and one or more output windings that essentially duplicate the functions of a relay.

**switching time** 1. The time interval between the reference time and the last instant at which the instantaneous voltage response of a magnetic cell reaches a stated fraction of its peak value. 2. The time interval between the reference time and the first instant at which the instantaneous integrated voltage response of a magnetic cell reaches a stated fraction of its peak value.

**switching tube** A gas tube used for switching high-power r-f energy in the antenna circuits of radar and other pulsed r-f systems. Examples are atr tube, pre-tr tube, and tr tube.

**swr** Abbreviation for *standing-wave ratio.*

**swvr** Abbreviation for *standing-wave voltage ratio.*

**syllabic companding** Companding in which the effective gain variations are made at speeds allowing response to the syllables of speech but not to individual cycles of the signal wave.

**syllable articulation** The per cent articulation obtained when speech units are meaningless syllables.

**symbol** 1. A design used on diagrams to represent a component or identify specific characteristics, quantities, or objects. 2. A letter or abbreviation used on diagrams or in equations to represent a quantity or to identify an object.

**symbolic address** A label chosen to identify a particular word, function, or other information in a digital-computer programing routine, independent of the location of the information within the routine. Also called floating address.

**symbolic logic** Logic in which symbols suitable for calculation are used to express nonnumerical relations for a computer. Boolean algebra is an example.

**symmetrical cyclically magnetized condition** A condition of a magnetic material when it is in a cyclically magnetized condition and the limits of the applied magnetizing forces are equal and of opposite sign, so the limits of flux density are equal and of opposite sign.

**symmetrical network** *Structurally symmetrical network.*

**symmetrical transducer** A transducer in which all possible pairs of specified terminations may be interchanged without affecting transmission, because the input and output image impedances are all equal.

**symmetrical transistor** A junction transistor in which the emitter and collector electrodes are identical and their terminals interchangeable.

**sync** 1. *Synchronization.* 2. *Synchronize.* 3. *Synchronizing.*

**sync compression** The reduction in gain applied to the sync signal over any part of its amplitude range with respect to the gain at a specified reference level.

**sync generator** An electronic generator that supplies synchronizing pulses to television studio and transmitter equipment. Also called sync-signal generator.

**synchro** [SYNCHROnous] Any of several devices used for transmitting and receiving angular position or angular motion over wires, such as a synchro transmitter or synchro receiver. Also called mag-slip (British), self-synchronous device, and selsyn. Common trademarks are Autosyn, Magnesyn, and Siemens.

**synchro angle** The angular displacement

of a synchro rotor from its electrical zero position.

**synchro control transformer** A transformer having its secondary winding on a rotor. When its three input leads are excited by angle-defining voltages, the two output leads deliver an a-c voltage that is proportional to the sine of the difference between the electrical input angle and the mechanical rotor angle. The output voltage thus varies sinusoidally with rotor position, being essentially zero when mechanical and electrical angles are the same, and can be used for control purposes.

**synchro control transmitter** A high-accuracy synchro transmitter, having high-impedance windings.

**synchrocyclotron** A cyclotron in which the radio frequency of the electric field is frequency-modulated. Also called f-m cyclotron and frequency-modulated cyclotron.

**synchro differential receiver** A synchro receiver that subtracts one electrical angle from another and delivers the difference as a mechanical angle. One set of three input leads is excited by one set of angle-defining voltages. The other set of three input leads is excited by the other set of angle-defining voltages. The rotor rotates to the difference angle, with a torque proportional to the sine of the difference between the angles.

**synchro differential transmitter** A synchro transmitter that adds a mechanical angle to an electrical angle and delivers the sum as an electrical angle. When its three input leads are excited by the electrical angle-defining voltages, the three output leads deliver voltages that define an angle which is the sum of the electrical input angle and the mechanical rotor angle.

**synchro generator** *Synchro transmitter.*

**synchro motor** *Synchro receiver.*

**synchronism** The condition in which two or more varying quantities have the same speed or reach their peaks at the same instant of time.

**synchronization** The maintenance of one operation in step with another, as in keeping the electron beam of a television picture tube in step with the electron beam of the television camera tube at the transmitter. Also called sync.

**synchronization error** The error due to imperfect timing of two operations in a navigation system.

**synchronization indicator** An indicator that presents visually the relationship between two varying quantities or moving objects.

**synchronize** To produce synchronization. Also called sync.

**synchronized sweep** A sweep voltage that is controlled by an a-c voltage in such a way that the forward and return traces on a cathode-ray oscilloscope are exactly superimposed and appear as a single trace.

**synchronizing** Maintaining a fixed speed or phase relationship between two varying quantities or moving objects, as between two scanning processes. Also called sync.

**synchronizing signal** *Sync signal.*

**synchronometer** An instrument that counts the number of cycles produced by a signal source in a given time interval. It can serve as a master clock when driven by a frequency standard.

**synchronous** In step or in phase, as applied to two or more circuits, devices, or machines.

**synchronous capacitor** A synchronous motor running without mechanical load and drawing a large leading current like a capacitor. Used to improve the power factor and voltage regulation of an a-c power system.

**synchronous clock** An electric clock driven by a synchronous motor, for operation from an a-c power system in which the frequency is accurately controlled.

**synchronous computer** A digital computer in which all operations are controlled by signals from a master clock.

**synchronous converter** A converter in which motor and generator windings are combined on one armature and excited by one magnetic field. Normally used to change a-c power to d-c power. Also called converter.

**synchronous coupling** An electric coupling in which torque is transmitted by attraction between magnetic poles on both rotating members.

**synchronous demodulator** *Synchronous detector.*

**synchronous detector** 1. A detector that inserts a missing carrier signal in exact synchronism with the original carrier at the transmitter. When the input to the detector consists of two suppressed-carrier signals in phase quadrature as in the chrominance signal of a color television receiver, the phase of the reinserted carrier can be adjusted to recover either one of the signals. Two synchronous detectors, using carriers differing in phase by 90°, can thus extract the I and Q signals separately from the chrominance signal. Also called synchronous demodulator. 2. *Crosscorrelator.*

**synchronous gate** A time gate in which the output intervals are synchronized with an incoming signal.

**synchronous generator** *Alternator.*

**synchronous inverter** *Dynamotor.*

**synchronous machine** An a-c machine whose average speed is proportional to the frequency of the applied or generated voltage.

**synchronous motor** A synchronous machine that transforms a-c electric power into mechanical power, using field magnets excited with direct current.

**synchronous radar bombing** Radar bombing in which special airborne radar equipment containing rate and steering mechanisms is used to control the direction of flight of the bombing aircraft, solve the bombing problem, and automatically drop bombs at the proper release point. Also called radar synchronous bombing.

**synchronous rectifier** A rectifier in which contacts are opened and closed at correct instants of time for rectification by a synchronous vibrator or by a commutator driven by a synchronous motor.

**synchronous speed** A speed value related to the frequency of an a-c power line and the number of poles in a synchronous machine. Synchronous speed in revolutions per minute is equal to the frequency in cps divided by the number of poles, with the result multiplied by 120.

**synchronous switch** A thyratron circuit used to control the operation of ignitrons in such applications as resistance welding.

**synchronous vibrator** An electromagnetic vibrator that simultaneously converts a low direct voltage to a low alternating voltage

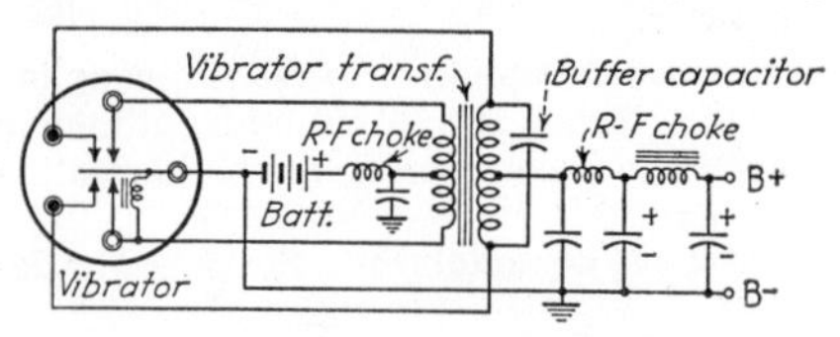


Synchronous vibrator circuit.

and rectifies a high alternating voltage obtained from a power transformer to which the low alternating voltage is applied. In power packs, it eliminates the need for a rectifier tube.

**synchronous voltage** The voltage required to accelerate electrons from rest to a velocity equal to the phase velocity of a wave in the absence of electron flow in a traveling-wave tube.

**synchro receiver** A synchro that provides an angular position related to the applied angle-defining voltages. When two of its input leads are excited by an a-c voltage and the other three input leads are excited by the angle-defining voltages, the rotor rotates to the corresponding angular position. The torque of rotation is proportional to the sine of the difference between the mechanical and electrical angles. Also called receiver synchro, selsyn motor, selsyn receiver, and synchro motor.

**synchro resolver** *Resolver.*

**synchroscope** 1. An instrument for indicating whether two periodic quantities are synchronous. The indicator may be a rotating-pointer device or a cathode-ray oscilloscope providing a rotating pattern. The position of the rotating pointer is a measure of the instantaneous phase difference between the quantities. 2. A cathode-ray oscilloscope designed to show a short-duration pulse by using a fast sweep that is synchronized with the pulse signal to be observed.

**synchro system** An electric system for transmitting angular position or motion. In the simplest form it consists of a synchro transmitter connected by wires to a synchro receiver. More complex systems include synchro control transformers and synchro differential transmitters and receivers. Also called selsyn system.

**synchro transmitter** A synchro that provides voltages related to the angular position of its rotor. When its two input leads are excited by an a-c voltage, the magnitudes and polarities of the voltages at the three output leads define the rotor position.

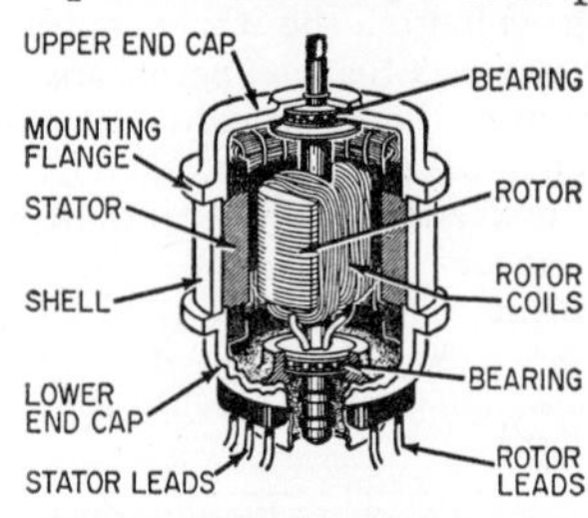


Synchro transmitter construction.

Also called selsyn generator, selsyn transmitter, synchro generator, transmitter, and transmitter synchro.

**synchrotron** An accelerator similar to a betatron but having in addition a higher-frequency magnetic field that is applied in synchronism with the orbiting and rapidly accelerating charged particles, to give much higher beam energy than in a betatron. The two types are the electron synchrotron and the proton synchrotron.

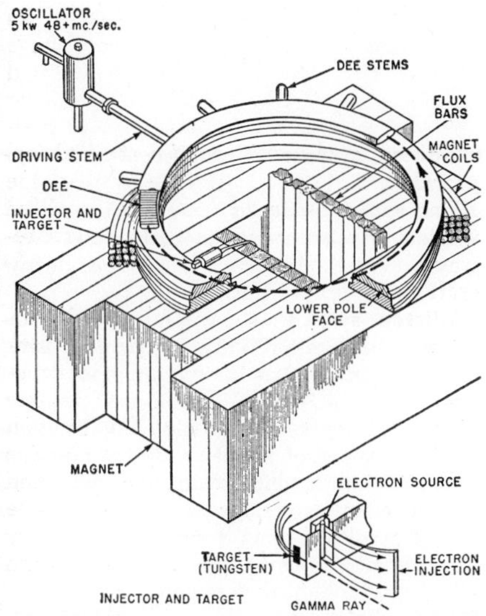


Synchrotron construction, with top half of magnet removed.

**synchro zeroing** Lining up the zero positions of a synchro system with the zero position of the associated indicator or mechanism being controlled.

**sync level** The level of the peaks of the synchronizing signal in a television system.

**sync limiter** A limiter circuit used in television to prevent sync pulses from exceeding a predetermined amplitude.

**sync pulse** One of the pulses that make up a sync signal.

**sync separator** A circuit that separates synchronizing pulses from the video signal in a television receiver. The signal for the sync separator is usually taken from the anode circuit of the video amplifier.

**sync signal** A signal transmitted after each line and field to synchronize the scanning process in a television or facsimile receiver with that of the transmitter. The picture signal, blanking signal, and sync signal together make up the composite picture signal in a television system. Also called synchronizing signal.

**sync-signal generator** *Sync generator.*

**synthetic mica** A fluor-phlogopite mica made artificially by heating a large batch of raw material in an electric resistance furnace and allowing the mica to crystallize from the melt during controlled slow cooling.

**system** A combination of several pieces of equipment integrated to perform a specific function. Thus, a fire control system may include a tracking radar, computer, and gun.

**systematic errors** Errors that have an orderly character and can be corrected by calibration.

**system deviation** The value of the ultimately controlled variable minus the ideal value in a feedback control system.

**system engineering** An engineering approach that takes into consideration all of the elements related in any way to the equipment under development, including utilization of manpower as well as the characteristics of each component of the system.

**system error** The ideal value minus the value of the ultimately controlled variable in a feedback control system.

**Szilard-Chalmers method** A chemical method of separating a radionuclide from isotopic target material following nuclear reactions that do not result in a change of atomic number.

# T

**T** 1. Symbol for *tritium.* 2. Symbol for *transformer.* Used on circuit diagrams.

**TA** Symbol for *target.* Used on tube-base diagrams.

**table-model receiver** A radio or television receiver having a cabinet of suitably small size for use on a table.

**tacan** [TACtical Air Navigation] An air navigation system in which a single uhf transmitter sends out signals that actuate airborne equipment to provide range and bearing indications with respect to the transmitter location, when interrogated by a transmitter in the aircraft. Each tacan station broadcasts a location-identifying Morse code signal at regular intervals. Also called tactical air navigation.

**tachometer** An instrument for measuring angular speed in revolutions per minute. An electric tachometer delivers an output voltage that is proportional to speed.

**tacitron** A thyratron in which anode current can be stopped by grid action because the grid limits ion generation to the grid-anode region.

**tactical air navigation** *Tacan.*

**tactical missile** A guided missile used in tactical operations against or in the presence of a hostile force.

**tactical radar** A radar set used in operations against or in the presence of a hostile force.

**tag** *Sentinel.*

**tagged atom** An atom of an isotopic tracer.

**tail** 1. A small pulse that follows the main pulse of a radar set and rises in the same direction. 2. The trailing edge of a pulse.

**tailing** Excessive prolongation of the decay of a signal. Also called hangover.

**tail warning radar** Radar installed in the tail of an aircraft to warn the pilot that an aircraft is approaching from the rear.

**takeoff** *Sound takeoff.*

**takeup reel** The reel that accumulates magnetic tape after it is recorded or played by a tape recorder.

**talbot** The unit of luminous energy in the mksa system. One talbot is equal to $10^7$ lumergs.

**talk-back circuit** *Interphone.*

**talk-down system** *Ground-controlled approach.*

**talking rebecca-eureka system** A modified version of rebecca-eureka that allows two-way voice communication between an aircraft and a ground station.

**talk-listen switch** A switch provided on intercommunication units to permit using the loudspeaker as a microphone when desired.

**Talos** A Navy surface-to-air guided missile having a speed of about mach 3, a range of about 25 miles, and beam-rider guidance. One version can carry nuclear warheads, and can also be used against ships and shore bombardment targets. Named after a Greek mythological demigod who guarded the island of Crete.

**tamper** 1. A substance that resists movement for a split microsecond in a thermonuclear bomb, so the active materials can build up greater pressure. 2. *Reflector.*

**tandem connection** *Cascade connection.*

**tandem transistor** Two transistors mounted in a single housing and connected in series internally.

**tangent galvanometer** A galvanometer in which a small compass is mounted horizontally in the center of a large vertical coil of wire. The current through the coil is proportional to the tangent of the angle of deflection of the compass needle from its normal position parallel to the magnetic field of the earth.

**tangential component** A component acting at right angles to a radius.

**tangential wave path** The path of propagation of a direct wave that is tangential to the surface of the earth. The path is curved by atmospheric refraction.

**tank** 1. A unit of acoustic delay-line storage containing a set of channels each forming a separate recirculation path. 2. The heavy metal envelope of a large mercury-arc rectifier or other gas tube having a mercury-pool cathode. 3. *Tank circuit.*

**tank circuit** A circuit capable of storing electric energy over a band of frequencies continuously distributed about the resonant frequency, such as a coil and capacitor in parallel. The selectivity of the circuit is proportional to the $Q$ factor, which is the ratio of the energy stored in the circuit to the energy dissipated. Also called tank.

**tantalum** [symbol Ta] A metallic element used for the electrodes of electron tubes because of its high absorption rate for residual gases. Also used for the plates of electrolytic capacitors. Atomic number is 73.

**tantalum detector** A detector consisting of the tip of a fine tantalum wire just touching the surface of a pool of mercury.

**tantalum electrolytic capacitor** An electrolytic capacitor that uses tantalum in the form of foil or a sintered slug as the anode, in an acid electrolyte. Both weight and volume are less than for comparable aluminum electrolytic capacitors. An insulating oxide formed on the tantalum serves as the dielectric.

**tantalum-foil electrolytic capacitor** An electrolytic capacitor that uses plain or etched tantalum foil for both electrodes, with a weak acid electrolyte.

**tantalum-slug electrolytic capacitor** An electrolytic capacitor that uses a sintered slug of tantalum as the anode, in a highly conductive acid electrolyte. Some types may be used at operating temperatures as high as 200°C.

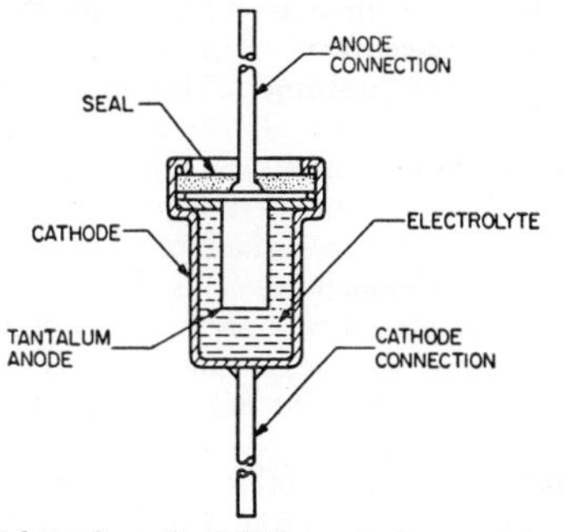


Tantalum-slug electrolytic capacitor construction.

**T antenna** An antenna consisting of one or more horizontal wires, with a lead-in connection being made at the approximate center of each wire.

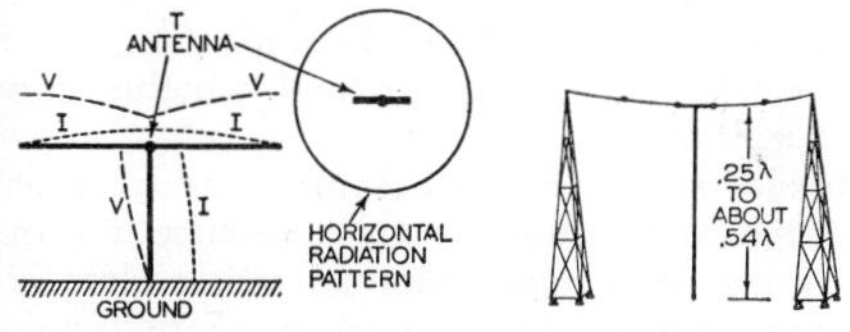


T antenna voltage and current distribution curves, radiation pattern, and construction when supported by two towers.

**tap** A connection made at some point other than the ends of a resistor or coil.

**tape** 1. *Magnetic tape.* 2. *Punched tape.*

**tape cartridge** A cartridge that holds a length of magnetic tape in such a way that the cartridge can be slipped into a tape recorder and played without threading the tape. The tape may be an endless loop feeding out from the center of a roll and back onto the roll on the outside, or it may be an ordinary length of tape that runs back and forth between two reels inside the cartridge. Also called cartridge.

**tape deck** A tape-recorder mechanism mounted on a motor board, including the tape transport and the bias and erase oscillators but no preamplifier, power amplifier, loudspeaker, or cabinet.

**tape guides** Grooved pins of nonmagnetic material mounted at either side of a magnetic recording head assembly to position the magnetic tape on the heads as it is being recorded or played.

**tape loop** A length of magnetic tape having the ends spliced together to form an endless loop. Used in message repeater units and in some types of tape cartridges, to eliminate the need for rewinding the tape.

**tape player** A machine designed only for playback of recorded magnetic tapes. Also called tape reproducer.

**taper** The manner in which resistance is distributed throughout the element of a potentiometer or rheostat. Uniform distribution, having the same resistance per unit length throughout the element, is called linear taper. Nonuniform distribution

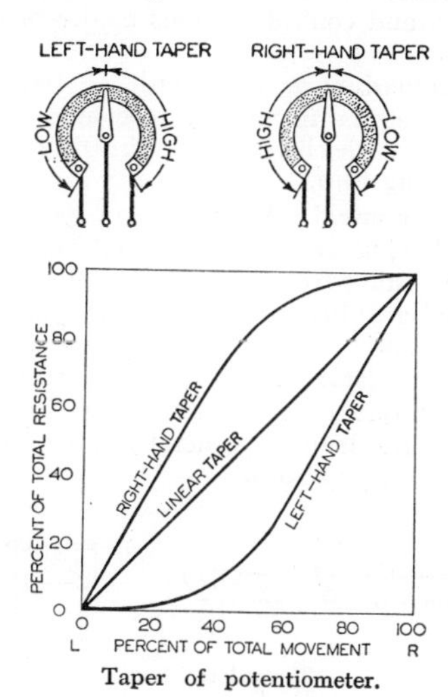


Taper of potentiometer.

is called nonlinear taper, left-hand taper, or right-hand taper.

**tape recorder** A recorder that records audio signals and other information on magnetic tape by selective magnetization of iron oxide particles that form a thin film on the tape. A recorder usually also includes provisions for playing back the recorded material.

**tapered transmission line** *Tapered waveguide.*

**tapered waveguide** A waveguide in which a physical or electrical characteristic changes continuously with distance along the axis of the waveguide. Also called tapered transmission line.

**tape reproducer** *Tape player.*

**tape speed** The speed at which magnetic tape moves past the recording head in a tape recorder. Standard speeds are 15/16, 1 7/8, 3 3/4, 7 1/2, 15, and 30 inches per second. Faster speeds give improved high-frequency response under given conditions.

**tape splicer** A device for splicing magnetic tape automatically or semiautomatically by using splicing tape or by fusing with heat.

**tape transmitter** 1. A code-transmitting machine actuated by previously punched paper tape. Used for high-speed sending because the tape can be fed through the machine much faster than it was originally punched. 2. A facsimile transmitter designed to transmit subject copy printed on narrow tape.

**tape transport** The mechanism on the deck of a tape recorder that holds the tape reels, drives the tape past the recording heads, and controls various modes of operation.

**tape-wound core** A length of ferromagnetic material in tape form, wound in such a way that each turn falls directly over the preceding turn.

**tapped control** A rheostat or potentiometer having one or more fixed taps along the resistance element, usually to provide a fixed grid bias or for automatic bass compensation.

**tapped resistor** A wirewound fixed resistor having one or more additional terminals along its length, generally for voltage-divider applications.

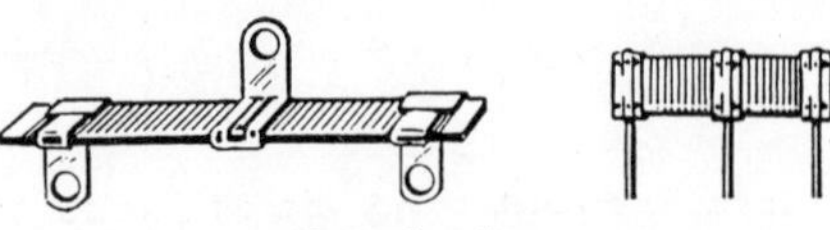

Tapped resistors.

**tap switch** A multicontact switch used chiefly for connecting one circuit lead to any one of a number of taps on a resistor or coil.

**target** [symbol TA] A substance or object exposed to bombardment or irradiation by nuclear particles, electrons, or electromagnetic radiation. In an x-ray tube the target is the anode or anticathode, from which x-rays are emitted as a result of electron bombardment. In a nuclear reaction it is the initially stationary atom or nucleus. In radar and sonar it is any object capable of reflecting the transmitted beam. In a television camera tube it is the storage surface that is scanned by an electron beam to generate an output signal current corresponding to the charge-density pattern stored there. In a cathode-ray tuning indicator tube it is one of the electrodes that is coated with a material that fluoresces under electron bombardment.

**target acquisition** The first appearance of a recognizable and useful echo signal from a new target in radar and sonar.

**target capacitance** The capacitance between the scanned area of the target and the backplate in a television camera tube.

**target cross-section** *Echo area.*

**target cutoff voltage** The lowest target voltage at which a detectable electric signal can be obtained in a television camera tube when an optical image is present on the photosensitive surface of the tube.

**target discrimination** The ability of a detection or guidance system to distinguish a target from its background or to discriminate between two or more targets that are close together.

**target drone** A pilotless aircraft controlled by radio from the ground or from a mother ship and used exclusively as a target for antiaircraft weapons.

**target fade** A momentary reduction in the strength of an echo signal from a radar or sonar target, due to interference or other phenomena. Tracking radar usually includes memory circuits that maintain tracking during this period, to prevent loss of the target.

**target glint** *Scintillation.*

**target identification** Identification of a target as to its nature and whether it is friend or foe.

**target noise** Statistical variations in a radar echo signal due to the presence on the target of a number of reflecting elements randomly oriented in space. Target noise can cause scintillation.

**target scintillation** *Scintillation.*

**target seeker** A missile having a self-contained system that provides homing guidance to the target. Also called homer.

**target timing** A radar technique for determining wind velocity by correlating the distance that a radar target travels across a radar screen with other known speed data. Used particularly in arctic regions.

**target voltage** The voltage between the thermionic cathode and the backplate of a television camera tube employing low-velocity scanning.

**Tartar** A Navy surface-to-air guided missile intended primarily for use on destroyers. It is a smaller version of Terrier, having about the same range.

**tau meson** A meson having a mean life of about $10^{-9}$ second and a mass 975 times that of an electron.

**taxi radar** *Airport surface detection equipment.*

**Taylor connection** A transformer connection for converting three-phase power to two-phase power, or vice versa.

**T circulator** A circulator in which three identical rectangular waveguides are joined asymmetrically to form a T-shaped structure, with a ferrite post or wedge at its center. Each port must be separately matched. Power entering any waveguide will emerge from only one adjacent waveguide.

**tearing** A television picture defect in which groups of horizontal lines are displaced in an irregular manner, due to inadequate horizontal synchronization.

**technetium** [symbol Tc] An element produced by bombarding molybdenum with deuterons and neutrons. It also occurs among the fission products of uranium. Atomic number is 43. Formerly called masurium.

**technical representative** [abbreviated tech rep] A person who represents one or more manufacturers in an area and gives technical advice on the application, installation, operation, and maintenance of their products in addition to selling the products.

**technician** An engineering assistant.

**tech rep** Abbreviation for *technical representative.*

**tecnetron** A semiconductor device similar to a triode in that it has cathode and anode connections at opposite ends of a small rod of germanium. In between is a deep groove containing an indium ring that serves as the grid connection. The electric field produced by the grid serves to control current flow.

**tee junction** A waveguide junction in which the longitudinal waveguide axes form a tee. Also called T junction and waveguide tee.

**Teflon** Trademark of Du Pont for their polytetrafluoroethylene resin.

**telautograph** A writing telegraph instrument in which manual movement of a pen at the transmitting position varies the current in two circuits in such a way as to cause corresponding movements of a pen at the remote receiving instrument. Ordinary handwriting can thus be transmitted over wires.

**tele-** Prefix meaning from a distance.

**teleammeter** A telemeter that measures and transmits current values to a remote point.

**telecamera** *Television camera.*

**telecast** [TELEvision broadCAST] The transmission of a television program intended for reception by the general public.

**telecasting** Broadcasting a television program.

**telecine** [TELEvision CINEma] Referring to motion-picture film used on a television program.

**telecine projector** A motion-picture projector adapted for use with a television camera tube to televise 24-frame-per-second motion-picture film at the 30-frame-per-second rate required for television. Also called television film scanner.

**telecommunication** Any transmission, emission, or reception of signals, writing, images, sounds, or intelligence of any nature by wire, radio, visual, or other electromagnetic systems. The terms telecommunication and communication are often used interchangeably.

**teleconference** A conference in which the participants are some distance apart but are able to talk to each other and sometimes also see each other by means of telecommunication.

**telecontrol** *Remote control.*

**telegenic** Suitable for televising.

**telegraph channel** A path suitable for the transmission of telegraph signals between two telegraph stations, either over wires or by radio. It may be one of several channels on a single radio or wire circuit, providing simultaneous transmission in the same frequency range, simultaneous transmission in different frequency ranges, or successive transmission.

**telegraph circuit** The complete wire or radio circuit over which signal currents flow between transmitting and receiving apparatus in a telegraph system.

**telegraph distributor** A device that effectively associates one direct-current or carrier telegraph channel in rapid succession with the elements of one or more sending or receiving devices.

**telegraph key** A hand-operated telegraph transmitter, used to form telegraph signals.

**telegraph level** The signal power at a specified point in a telegraph circuit when one telegraph channel is in the marking or continuous-tone condition and all other channels are silenced.

**telegraph modem** The complete equipment for modulating and demodulating one or more separate telegraph circuits, each containing one or more telegraph channels.

**telegraph-modulated wave** A wave obtained by varying the amplitude or frequency of a continuous wave by means of telegraphic keying.

**telegraph repeater** A repeater inserted at intervals in long telegraph lines to amplify weak code signals, with or without reshaping of pulses, and retransmit them automatically over the next section of the line.

**telegraph sounder** A telegraph receiving instrument in which an electromagnet attracts an armature each time a pulse arrives. The armature makes an audible sound as it hits against its stops at the beginning and end of each current pulse. The intervals between these sounds are varied in accordance with the telegraph code. Also called sounder.

**telegraph transmitter** A device for controlling a source for electric power so as to form telegraph signals for radio or wire transmission. It may be a telegraph key or a motor-driven sender using previously punched tape.

**telegraphy** Communication at a distance by means of code signals consisting of current pulses sent over wires or by radio.

**telemeter** 1. The complete measuring, transmitting, and receiving apparatus for indicating or recording the value of a quantity at a distance. Also called telemetering system. 2. To transmit the value of a measured quantity to a remote point over wires or by radio.

**telemeter band** One of up to 18 subcarrier bands used to modulate a carrier in the standard f-m/f-m telemetering system.

**telemeter channel** A complete circuit for transmission of one telemetered function, including pickup, commutator, modulator, transmitter, receiver, detector, decoder, and recorder.

**telemetering** Transmitting the readings of instruments to a remote location by means of wires, radio waves, or other means. Also called electric telemetering, remote metering, and telemetry.

**telemetering system** *Telemeter.*

**telemeter pickup** A device used to measure and convert data to be telemetered into a form suitable for modulation of a telemetering channel.

**telemetric data analyzer** An analyzer that amplifies telemetered data and separates the individual values.

**telemetric data monitor** A monitor that provides a continuous visual display of telemetered signals.

**telemetric data receiving set** A complete electronic system designed to intercept, demodulate, display, and record data propagated by the transmitter of a telemeter.

**telemetry** *Telemetering.*

**telephone** A system for converting sound waves into electric variations that can be sent over wires and reproduced at a distant point. Used primarily for voice communication. It consists essentially of a telephone transmitter and receiver at each station, interconnecting wires, signaling devices, a central power supply, and switching facilities.

**telephone capacitor** A small fixed capacitor connected in parallel with a telephone receiver to bypass higher audio frequencies and thereby reduce noise.

**telephone carrier current** A carrier current used for telephone communication over power lines or to obtain more than one channel on a single pair of wires.

**telephone channel** A one-way or two-way path suitable for the transmission of audio signals between two stations.

**telephone circuit** The complete circuit over which audio and signaling currents travel in a telephone system between the two telephone subscribers in communication with each other. The circuit usually consists of insulated conductors, as ground returns are now rarely used in telephony.

**telephone current** An electric current produced or controlled by the operation of a telephone transmitter.

**telephone dial** A switch operated by a finger wheel, used to make and break a pair of contacts the required number of times for setting up a telephone circuit to the party being called.

**telephone induction coil** A coil used in a

telephone circuit to match the impedance of the line to that of a telephone transmitter or receiver.

**telephone jack** *Phone jack.*

**telephone line** The conductors extending between telephone subscriber stations and central offices.

**telephone loading coil** *Loading coil.*

**telephone modem** A piece of equipment that modulates and demodulates one or more separate telephone circuits, each containing one or more telephone channels. It may include multiplexing and demultiplexing circuits, individual amplifiers, and carrier-frequency sources.

**telephone pickup** A large flat coil placed under a telephone set to pick up both voices during a telephone conversation for recording purposes.

**telephone plug** *Phone plug.*

**telephone receiver** The portion of a telephone set that converts the a-f current variations of a telephone line into sound waves.

**telephone relay** A relay having a multiplicity of contacts on long spring strips mounted parallel to the coil, actuated by a lever arm or other projection of the hinged armature. Used chiefly for switching in telephone circuits.

Telephone relay construction.

**telephone repeater** A repeater inserted at one or more intermediate points in a long telephone line to amplify telephone signals so as to maintain the required current strength.

**telephone repeating coil** A coil used in a telephone circuit for inductively coupling two sections of a line when a direct connection is undesirable.

**telephone retardation coil** A coil used in a telephone circuit to pass direct current while offering appreciable impedance to alternating current.

**telephone ringer** An electromagnetic device that actuates a clapper which strikes one or more gongs to produce a ringing sound. Used with a telephone set to signal a called party.

**telephone set** An assembly including a telephone transmitter, a telephone receiver, and associated switching and signaling devices.

**telephone switchboard** A switchboard for interconnecting telephone lines and associated circuits.

**telephone transmitter** The microphone used in a telephone set to convert speech into a-f signals.

**telephone-type relay** A relay in which a pileup of long contact springs is mounted parallel to the long axis of the relay coil. Widely used in telephone systems.

**telephony** The transmission of speech and sounds to a distant point for communication purposes.

**telephoto** *Facsimile.*

**telephotography** *Facsimile.*

**telephoto lens** A lens having a long focal length, used in television cameras to secure large images of distant objects.

**teleprinter** A device that responds to teletype signals and prints the corresponding characters on paper tape.

**teleran** [TELEvision Radar Air Navigation] A radar navigation system in which the positions of aircraft are determined by ground radar. The resulting ppi display is then superimposed on a map and transmitted to the aircraft by television, so each pilot can see the position of his aircraft in relation to others in the vicinity.

**Teletype** Trademark of Teletype Corp. for their teletypewriters.

**Teletypesetter** Trademark of Fairchild Camera & Instruments Corp. for apparatus used to make a linotype machine set and cast type automatically in response to punched paper tape or equivalent electric signals.

**teletypewriter** A special electric typewriter that produces coded electric signals corresponding to manually typed characters, and automatically types messages when fed with similarly coded signals produced by another machine. The signals may be transmitted directly over wires connecting the machines or used to drive a teletypewriter perforator that produces punched paper tape for later transmission. Also called twx machine.

**teletypewriter perforator** An electromechanical perforator that punches teletypewriter code combinations on a paper tape when connected to a manually operated teletypewriter.

**televiewer** A person who watches a program on a television receiver.

**televise** To pick up a scene with a television camera and convert it into corresponding electric signals for transmission by a television station.

**television** [abbreviated tv] A system for converting a succession of visual images into corresponding electric signals and transmitting these signals by radio or over wires to distant receivers at which the signals can be used to reproduce the original images.

**television broadcast band** The band extending from 54 to 890 mc in which are the 6-mc channels assignable to television broadcast stations in the United States. The frequencies are 54 to 72 mc (channels 2 through 4), 76 to 88 mc (channels 5 and 6), 174 to 216 mc (channels 7 through 13), and 470 to 890 mc (channels 14 through 83).

**television broadcast station** A station in the television broadcast band transmitting simultaneous visual and aural signals intended to be received by the general public.

**television cable** Coaxial cable capable of transmitting the bandwidth required for television signals, which is about 4.5 mc.

**television camera** The pickup unit used to convert a scene into corresponding electric signals. Optical lenses focus the scene to be televised on the photosensitive surface of a camera tube. This tube breaks down the visual image into small picture elements and converts the light intensity of each element in turn into a corresponding electric signal. Also called camera and telecamera.

**television camera tube** *Camera tube.*

**television channel** A band of frequencies 6 mc wide in the television broadcast band, available for assignment to a television broadcast station.

**television chart** A test chart used for checking the resolution of a television system.

**television engineering** The field of engineering that deals with the design, manufacture, and testing of equipment required for the transmission and reception of television programs.

**television film scanner** *Telecine projector.*

**television interference** [abbreviated tvi] Interference produced in television receivers by amateur radio and other transmitters.

**television picture photography** To obtain photographs of the screen of a television receiver, use a speed of 1/30th second (1/25 is closest on most cameras) to get one complete frame (two fields). Set the aperture according to the reading of an exposure meter, using the daylight exposure index. Very fast pan film, such as Tri-X with an ASA daylight index of 200, is needed. A tripod is essential. Turn out room lights, and use the maximum brilliance that still gives a clear picture.

**television picture tube** *Picture tube.*

**television receiver** A receiver that converts incoming television signals into the original scenes along with the associated sounds. Also called television set.

**television reconnaissance** Reconnaissance in which television is used to transmit a scene from the reconnoitering point to another location on the surface or in the air.

**television recording** *Kinescope recording.*

**television relay system** *Television repeater.*

**television repeat-back guidance** Command guidance in which a television camera and transmitter are mounted in a guided missile or pilotless vehicle to provide a view ahead for the operator at the remote-control location.

**television repeater** A repeater that transmits television signals from point to point by using radio waves in free space as a medium, such transmission not being intended for direct reception by the public. Also called television relay system.

**television screen** The fluorescent screen of the picture tube in a television receiver.

**television set** *Television receiver.*

**television signal** General term for the aural signal and visual signal that are broadcast together to provide the sound and picture portions of a television program.

**television studio-transmitter link** A fixed station used to transmit television program material and related communications from a studio to the transmitter of a television broadcast station.

**television transmission standards** The standards that specify the characteristics of a U. S. television signal. Channel width is 6 mc, the visual carrier is 4.5 mc lower than the aural center frequency, the aural center frequency is 0.25 mc lower than the upper frequency limit of the channel, there are 525 scanning lines per frame period and they are interlaced two to one, the frame frequency is 30 per second and the field frequency 60 per second, the aspect ratio of the transmitted television picture is four units horizontally to three units vertically, the scene is scanned from left to right horizontally and from top to bottom vertically at uniform velocities, and a decrease in initial light intensity increases the radiated power.

**television transmitter** A visual transmitter and an aural transmitter used together for transmitting a complete television signal.

**televoltmeter** A telemeter that measures voltage.

**telewattmeter** A telemeter that measures power.

**tellurium** [symbol Te] An element. Atomic number is 52.

**$TE_{m,n}$ mode** A mode in which a particular transverse electric wave is propagated in a waveguide. Also called $H_{m,n}$ mode (British).

**$TE_{m,n,p}$ mode** A mode of wave propagation in a cavity consisting of a hollow metal cylinder closed at its ends, for which the transverse field pattern is similar to that of the $TE_{m,n}$ mode in a corresponding cylindrical waveguide and for which $p$ is the number of half-period field variations along the axis. Also applicable to closed rectangular cavities.

**$TE_{m,n}$ wave** 1. In a circular waveguide, the transverse electric wave for which $m$ is the number of axial planes along which the normal component of the electric vector vanishes, and $n$ is the number of coaxial cylinders (including the boundary of the waveguide) along which the tangential component of the electric vector vanishes. The $TE_{0,1}$ wave is the circular electric wave having the lowest cutoff frequency, while the $TE_{1,1}$ wave is the dominant wave and has electric lines of force approximately parallel to a diameter of the waveguide. 2. In a rectangular waveguide, the transverse electric wave for which $m$ is the number of half-period variations of the electric field along the longer transverse dimension, and $n$ is the number of half-period variations of the electric field along the shorter transverse dimension. Also called $H_{m,n}$ wave (British), for both circular and rectangular waveguides.

**temperature characteristic** The performance of a device as temperature is varied through specified limits.

**temperature coefficient** The amount of change in the value of a performance characteristic per degree change in temperature.

**temperature coefficient of frequency** The rate of change of frequency with temperature.

**temperature coefficient of resistance** The rate of change in resistance value per degree change in temperature, usually expressed as ohms per ohm per degree centigrade.

**temperature coefficient of voltage drop** The change in the voltage drop of a glow-discharge tube divided by the change in ambient temperature or the change in the temperature of the envelope.

**temperature-compensating capacitor** A capacitor whose capacitance varies with temperature in a known and predictable manner. Used in resonant circuits to compensate for changes in the values of other parts with temperature.

**temperature-compensating network** A network whose components are so chosen that the network characteristics change with temperature in a predetermined manner.

**temperature compensation** The process of making some characteristic of a circuit or device independent of changes in ambient temperature.

**temperature control** A control used to maintain the temperature of an oven, furnace, or other enclosed space within desired limits.

**temperature-controlled crystal unit** A crystal unit containing, in addition to the quartz plate or plates, a heater device designed to maintain the temperature of the quartz plate within specified limits.

**temperature correction** A correction applied to a measured value to compensate for changes that are due to a temperature that is higher or lower than some standard temperature value.

**temperature derating** Lowering the rating of a device when it is to be used at elevated temperatures.

**temperature element** The sensing element of a temperature-measuring device or direct-reading temperature indicator.

**temperature inversion** A region in the troposphere at which temperature increases rather than decreases with altitude.

**temperature range** The total variation in ambient temperature for a given application, expressed in degrees centigrade.

**temperature saturation** The condition in which the anode current of a thermionic vacuum tube cannot be further increased by increasing the cathode temperature at a given value of anode voltage. The effect is due to the space charge formed near the cathode. Also called filament saturation and saturation.

**temporal gain control** *Sensitivity-time control.*

**temporary storage** An internal computer

storage location reserved for intermediate and partial results in a digital computer.

**TEM wave** Abbreviation for *transverse electromagnetic wave.*

**tensor force** *Noncentral force.*

**T-equivalent circuit** A transistor equivalent circuit in which the resistances of the electrodes are connected together at a common point in the form of a T.

**tera-** A prefix representing $10^{12}$.

**teracycle** *Megamegacycle.*

**teraohm** One million megohms, equal to $10^{12}$ ohms.

**terbium** [symbol Tb] A rare-earth element. Atomic number is 65.

**terminal** 1. A screw, soldering lug, or other point to which electric connections can be made. 2. The equipment at the end of a microwave relay system or other communication channel. 3. One of the electric input or output points of a circuit or component.

**terminal area** The enlarged portion of conductor material surrounding a hole for a lead on a printed circuit. Also called land and pad.

**terminal board** An insulating mounting for terminal connections. Also called terminal strip.

**terminal equipment** The equipment at a terminal of a communication channel.

**terminal guidance** Navigation control of a missile as it approaches its target.

**terminal impedance** The impedance of a circuit or device as measured between its input or output terminals.

**terminal lug** A soldering lug placed on a terminal board or at the end of a wire.

**terminal pair** An associated pair of accessible terminals, such as the input terminals or output terminals of a device or network.

**terminal phase** The path of a missile as it approaches its target. For a ballistic missile, the terminal phase is that part of the trajectory between reentry and impact.

**terminal strip** *Terminal board.*

**terminal vor** [abbreviated tvor] A vhf omnirange located at an airport to provide omnirange service, often as an approach aid.

**terminated line** A transmission line terminated in a resistance equal to the characteristic impedance of the line, so there are no reflections and no standing waves.

**termination** The load connected to the output end of a circuit, device, or transmission line.

**ternary fission** The splitting of a nucleus into three nuclear fragments.

**ternary notation** A system of notation using the base of 3 and the characters 0, 1, and 2.

**terrain-clearance indicator** *Frequency-modulated radio altimeter.*

**terrain-clearance warning indicator** A terrain-clearance indicator that gives a warning signal when the clearance between the aircraft and the earth immediately below reaches a predetermined minimum value.

**terrain echoes** *Ground clutter.*

**terrain error** The navigation error due to distortion of a radiated field by the nonhomogeneous characteristic of the terrain over which the radiation has propagated.

**terrestrial guidance** *Terrestrial-reference guidance.*

**terrestrial-magnetic guidance** Terrestrial-reference guidance in which the control system of the missile reacts to the magnetic field of the earth.

**terrestrial magnetism** Magnetism produced by the earth.

**terrestrial-reference flight** Stabilized flight in which control information is obtained from terrestrial phenomena such as the earth's magnetic field, atmospheric pressure, or gravity.

**terrestrial-reference guidance** Long-range missile guidance in which the control system of the missile reacts to magnetic, gravitational, or other properties of the earth. Also called terrestrial guidance.

**Terrier** A Navy surface-to-air guided missile using beam-rider guidance and having a range of about 10 miles.

**tesla** The unit of magnetic induction in the mksa system, equal to 1 weber per square meter.

**Tesla coil** An air-core transformer used with a spark gap and capacitor to produce a high voltage at a high frequency.

**test clip** A spring clip used at the end of an insulated wire lead to make temporary connections quickly for test purposes.

**test lead** A flexible insulated lead, usually with a test prod at one end, used for making tests, connecting instruments to a circuit temporarily, or making other temporary connections.

**test pattern** A chart having various combinations of lines, squares, circles, and graduated shading, transmitted from time to time by a television station to check definition, linearity, and contrast for the complete system from camera to receiver. Also called resolution chart.

**test prod** A metal point attached to an insulating handle and connected to a test lead for convenience in making a temporary connection to a terminal while tests are being made.

**test program** *Check routine.*

**test record** A phonograph record having recorded frequencies suitable for checking and adjusting audio systems.

**test routine** *Check routine.*

**test set** A combination of instruments needed for servicing a particular type of equipment.

**tetrad** A group of four pulses used to express a digit in the scale of 10 or 16.

**tetrafluoroethylene resin** A fluorocarbon used as a base for polytetrafluoroethylene resin, marketed as Teflon.

**tetrode** A four-electrode electron tube containing an anode, a cathode, a control electrode, and one additional electrode that is ordinarily a grid.

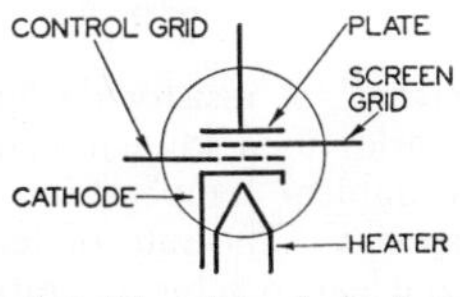


Tetrode symbol. The plate electrode is more properly called the anode.

**tetrode junction transistor** *Double-base junction transistor.*

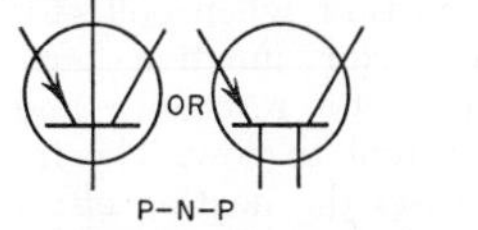


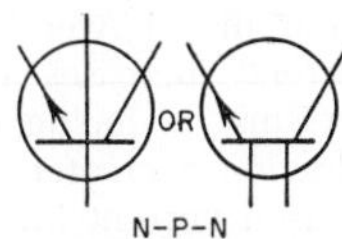


Tetrode junction transistor symbols, showing alternative methods of indicating two base connections.

**tetrode point-contact transistor** A point-contact transistor having three point contacts and one base connection.

**tetrode transistor** A four-electrode transistor, such as a tetrode point-contact transistor or double-base junction transistor.

**TE wave** Abbreviation for *transverse electric wave.*

**Texas tower** A radar tower built in the sea offshore, to serve as part of an early-warning radar network. It resembles offshore oil derricks in the Gulf of Mexico.

**TFE-fluorocarbon resin** Standard term of Society of the Plastics Industry for polytetrafluoroethylene resin, marketed as Teflon by Du Pont and as Fluon by Imperial Chemical Industries in Great Britain.

**thallium** [symbol Tl] An element. Atomic number is 81. Several thallium isotopes are members of the uranium, actinium, thorium, and neptunium radioactive series.

**thallium oxysulfide** A compound of thallium, oxygen, and sulfur having photoconductive properties.

**thalofide cell** A photoconductive cell in which the active light-sensitive material is thallium oxysulfide in a vacuum. It has maximum response at the red end of the visible spectrum and in the near infrared.

**theater television** A large projection-type television receiver used in theaters, generally for closed-circuit showing of important sport events.

**theoretical cutoff frequency** A frequency at which, disregarding the effects of dissipation, the image attenuation constant of a transducer changes from zero to a positive value or vice versa.

**therapy tube** An x-ray tube designed for use in x-ray therapy.

**thermal agitation** Random movements of the free electrons in a conductor, producing noise signals that may become noticeable when they occur at the input of a high-gain amplifier. Also called thermal effect.

**thermal-agitation voltage** The voltage produced in a circuit by thermal agitation.

**thermal ammeter** *Hot-wire ammeter.*

**thermal base** The operating temperature of a nuclear reactor.

**thermal battery** 1. A combination of thermal cells. 2. A voltage source consisting of a number of bimetallic junctions connected to produce a voltage when heated by a flame.

**thermal cell** A reserve cell that is activated by applying heat to melt a solidified electrolyte.

**thermal column** A column of moderating material extending through the shield into the reflector of a nuclear reactor, used to provide a source of thermal neutrons for research purposes.

**thermal conductivity** The quantity of heat that passes through a unit volume of a material in unit time when the difference in temperature of the two faces is 1°C.

**thermal converter** A converter consisting of one or more thermojunctions in thermal contact with an electric heater. The voltage developed at the output terminals by thermoelectric action is then a measure of the input current to its heater. Also called thermocouple converter, thermoelectric generator, and thermoelement (deprecated).

**thermal cross section** The cross section as measured with thermal neutrons.

**thermal cutout** A heat-sensitive switch that automatically opens the circuit of an electric motor or other device when the operating temperature exceeds a safe value.

**thermal detector** *Bolometer.*

**thermal diffusion method** A method of separating isotopes based on the difference in the rates of diffusion of gases or liquids across a temperature gradient.

**thermal effect** *Thermal agitation.*

**thermal flasher** An electric device that opens and closes a circuit automatically at regular intervals owing to alternate heating and cooling of a bimetallic strip that is heated by a resistance element in series with the circuit being controlled.

**thermal imager** A camera or other infrared mapping system that gives an infrared image of a scene in which various objects or areas differ in temperature. In arctic regions, ice crevasses concealed by snow bridges are revealed by a temperature differential of as little as 0.1°C over a crevasse.

**thermal inertia** The reciprocal of thermal response.

**thermal instability** A positive temperature coefficient in a nuclear reactor, especially in a component having low heat capacity. A thermally unstable reactor tends to run away unless well controlled.

**thermal instrument** An instrument that depends on the heating effect of an electric current, such as a thermocouple or hot-wire instrument.

**thermal ionization** The ionization of atoms or molecules by heat, as in a flame.

**thermalization** Slowing down of neutrons, as by repeated collisions with other particles, until they have approximately the same kinetic energy as thermal neutrons.

**thermal neutron** A neutron that is in thermal equilibrium with the substance in which it exists. At 15°C, a thermal neutron has a kinetic energy of about 0.025 electron-volt. Also called slow neutron.

**thermal noise** Electric noise produced by thermal agitation of electrons in conductors and semiconductors. This random motion of free electrons increases with temperature. Also called Johnson noise.

**thermal noise generator** A generator that uses the inherent thermal agitation of an electron tube to provide a calibrated noise source.

**thermal photograph** A photograph made of an image tube or similar device that shows objects on the earth differentiated by their radiations of heat or infrared waves.

**thermal photography** *Thermography.*

**thermal radiation** Radiation in the form of heat, emitted by all bodies that are not at absolute zero in temperature. The wavelength range extends from the shortest ultraviolet through visible light to the longest infrared wavelengths. Also called heat.

**thermal reactor** A nuclear reactor in which fission is induced primarily by neutrons of such low energy that they are in substantial thermal equilibrium with the material of the core.

**thermal relay** A relay operated by the heat produced by current flow.

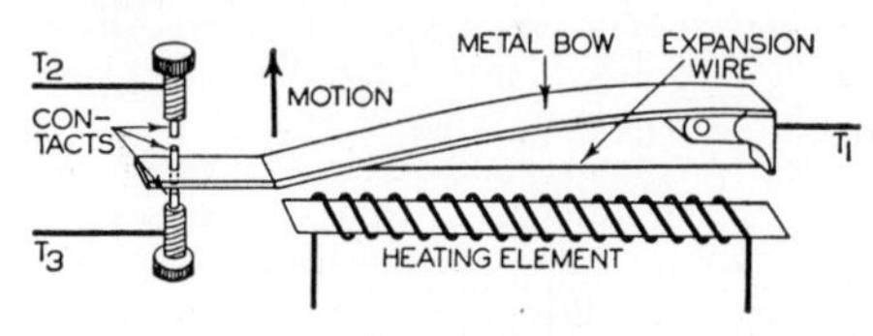


Thermal relay.

**thermal resistor** A resistor designed so its resistance varies in a known manner with changes in ambient temperature.

**thermal response** The rate of temperature rise in a nuclear reactor operating at its rated power if no heat is withdrawn by cooling. The reciprocal of this rate is called thermal inertia.

**thermal runaway** A condition that may occur in a power transistor when collector current increases collector junction temperature, reducing collector resistance and allowing a greater current to flow. The increased current increases the heating effect still more. The action may continue until the transistor is destroyed, particularly when the ambient temperature is high.

**thermal shield** A high-density heat-conducting portion of a shield placed close to the reflector of a nuclear reactor to absorb thermal neutrons, gamma rays, beta rays, and x-rays, whose absorption in the outer shield would generate excessive heat.

**thermal switch** A temperature-controlled switch.

**thermal tuner** A microwave tuner that uses thermal tuning of a cavity resonator.

**thermal tuning** The process of changing the operating frequency of a system by using controlled thermal expansion to alter the geometry of the system.

**thermal tuning rate** The initial time rate of change in frequency that occurs when the input power to a microwave tuner is instantaneously changed by a specified amount.

**thermal tuning sensitivity** The rate of change of the equilibrium frequency of a cavity resonator with respect to applied thermal tuner power.

**thermal tuning time** The time required to tune through a specified frequency range when thermal tuner power is instantaneously changed from zero to the specified maximum or vice versa in a microwave tube.

**thermal-tuning time constant** The time required for the frequency of a thermal tuner to change by a specified fraction $(1 - 1/2.718)$ of the change in equilibrium frequency after an incremental change of applied thermal tuner power.

**thermal utilization** The probability that a thermal neutron which is absorbed is absorbed usefully, as in a fissionable material.

**thermel** A thermoelectric device used to measure temperature, such as a thermocouple or thermopile.

**thermion** A charged particle emitted by a heated body, as by the hot cathode of a thermionic tube.

**thermionic** Pertaining to the emission of electrons as a result of heat.

**thermionic arc** An electric arc in which the cathode is heated by the arc current itself.

**thermionic cathode** *Hot cathode.*

**thermionic converter** A converter that converts heat energy directly into electric energy. In one version, two metal electrodes are separated by a gas at low pressure. When one electrode is heated to about 2,000°F, electrons boiled out of it travel through the gas to the other electrode to give an electric current. Also called thermionic generator and thermoelectron engine.

**thermionic current** A current due to directed movements of electrons or other thermions, such as the flow of emitted electrons from the cathode to the anode in a thermionic tube.

**thermionic detector** A detector using a hot-cathode tube.

**thermionic diode** A diode electron tube having a heated cathode.

**thermionic emission** The liberation of electrons or ions from a solid or liquid as a result of heat.

**thermionic generator** *Thermionic converter.*

**thermionic grid emission** The current produced by electrons thermionically emitted from a grid. Also called primary grid emission.

**thermionic tube** *Hot-cathode tube.*

**thermionic work function** The energy required to transfer electrons from a given metal to an adjacent medium during thermionic emission, as from a heated filament to a vacuum.

**thermistor** [THERMal resISTOR] A bolometer that makes use of the change in resistivity of a semiconductor with change in temperature. A thermistor has a high negative temperature coefficient of resistance, so its resistance decreases as temperature rises. Used in critical circuits to compensate for opposite temperature variations in other components. Used as a bolometer to measure temperatures and microwave energy. Used also as a nonlinear circuit element.

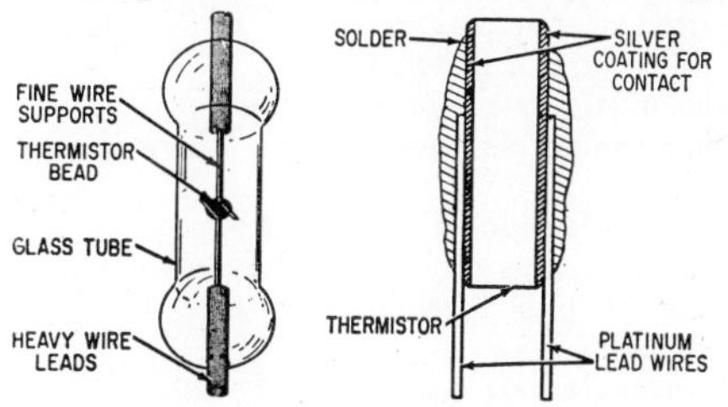


Thermistor construction details.

**thermistor mount** A waveguide mount in which a thermistor can be inserted to measure electromagnetic power.

**thermoammeter** An ammeter that is actuated by the voltage generated in a thermocouple through which is sent the current to be measured. Used chiefly for measuring r-f currents. Also called thermocouple ammeter, thermocouple instrument, and thermocouple meter.

**thermocouple** A device consisting of two dissimilar conductors welded together at their ends to form a junction. When this junction is heated, the voltage developed across it is proportional to the temperature rise. Used for measuring temperatures, as in a thermoelectric pyrometer, or for converting radiant energy into electric energy.

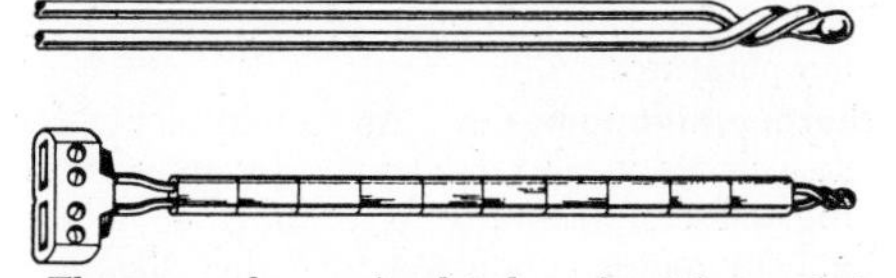

Thermocouples, uninsulated and with two-hole ceramic-bead insulators.

**thermocouple ammeter** *Thermoammeter.*

**thermocouple converter** *Thermal converter.*

**thermocouple instrument** *Thermoammeter.*

**thermocouple meter** *Thermoammeter.*

**thermocouple vacuum gage** A vacuum gage that depends for its operation on the thermal conduction of the gas present. Pressure is measured as a function of the voltage of a thermocouple whose measuring junction is in thermal contact with a heater that carries a constant current. Ordinarily used over a pressure range of $10^{-1}$ to $10^{-3}$ millimeter of mercury.

**thermoelectric effect** *Seebeck effect.*

**thermoelectric generator** *Thermal converter.*

**thermoelectricity** Electricity produced by direct action of heat, as by unequal heating of two thermojunctions in the same circuit.

**thermoelectric junction** *Thermojunction.*

**thermoelectric microrefrigerator** A refrigeration device that uses the Peltier effect for cooling small electronic components such as infrared detectors.

**thermoelectric module** A device that utilizes the Peltier effect to provide spot cooling of transistors, infrared detectors, and other components when energized by direct current. Also used for precise temperature control of liquids, solids, and gases.

**thermoelectric pyrometer** A pyrometer in which the sensing element is a thermocouple.

**thermoelectric series** A series of metals arranged in the order of their thermoelectric voltage-generating ratings with respect to some reference metal such as lead.

**thermoelectron** An electron liberated by heat, as from a heated filament. Also called negative thermion.

**thermoelectron engine** *Thermionic converter.*

**thermoelement** Deprecated term for *thermal converter.*

**thermofission** The fission of nuclear material under the influence of extremely high temperatures.

**thermogalvanometer** An instrument for measuring weak r-f currents by their heating effect, generally consisting of a d-c galvanometer connected to a thermocouple that is heated by a filament carrying the current to be measured.

**thermograph** 1. A far-infrared image-forming device that provides a thermal photograph by scanning a far-infrared image of an object or scene. 2. An instrument that senses, measures, and records the temperature of the atmosphere.

**thermography** Photography that uses radiation in the long-wavelength far-infrared region, emitted by objects at temperatures ranging from −170°F to over 300°F. Also called thermal photography.

**thermojunction** One of the surfaces of contact between the two conductors of a thermocouple. Also called thermoelectric junction.

**thermoluminescence** Luminescence produced in a material by moderate heat.

**thermomagnetic** Pertaining to the effect of temperature on the magnetic properties of a substance, or to the effect of a magnetic field on the temperature distribution in a conductor.

**thermometer** An instrument for measuring and indicating temperature.

**thermomilliammeter** A low-range thermoammeter.

**thermonuclear** Pertaining to nuclear reactions caused by intense heat.

**thermonuclear reaction** A nuclear reaction occurring at temperatures of several million degrees, in which light atoms are converted to heavier atoms and energy is released. Experimental devices being developed to control such a reaction include the astron, perhapsatron, pyrotron, stellarator, and zeta machines.

**thermophone** An electroacoustic transducer in which sound waves having an accurately known strength are produced by the expansion and contraction of the air adjacent to a conductor whose temperature varies in response to a current input. Used chiefly for calibrating microphones.

**thermopile** A group of thermocouples connected in series to give higher voltage output or in parallel to give higher current output, for measuring temperature, measuring radiant energy, or converting radiant energy into electric power.

**thermoplastic** A plastic that can be softened by heat and rehardened into a solid state by cooling. It may be remelted and remolded many times. Examples are cellulose acetate, cellulose nitrate, methyl methacrylate, polyethylene, polystyrene, and vinyls.

**thermoplastic recording** A recording process in which information is stored on a plastic tape by the action of a modulated electron beam, to give high storage density along with playback within a few milliseconds. The tape consists of a thermoplastic film on a transparent conducting film supported by an ordinary plastic tape base. The electron beam deposits charges on the thermoplastic film in accordance with beam modulation. Application of heat by r-f heating electrodes softens the film enough to produce deformation that is proportional to the stored electrostatic charges. Hardening preserves the deformation. An optical system is used for playback.

**thermoregulator** A high-accuracy or high-sensitivity thermostat. One type consists of a mercury-in-glass thermometer with sealed-in electrodes, in which the rising and falling column of mercury makes and breaks an electric circuit.

**thermorelay** *Thermostat.*

**thermosetting material** A plastic that solidifies when first heated under pressure and cannot be remelted or remolded without destroying its original characteristics. Examples are epoxies, melamines, phenolics, and ureas.

**thermostat** A device that opens or closes a circuit when the temperature deviates from a preset value or range of values, to actuate the controls of a heating element and thereby produce the required corrective action. Also called thermorelay.

**thermostatic switch** A temperature-operated switch that receives its operating energy by thermal conduction or convection from the device being controlled or operated.

**theta polarization** Wave polarization in which the E (electric) vector is tangent to the meridian lines of some given spherical frame of reference. For phi polarization this vector is tangent to the lines of latitude of the reference.

**Thevenin's theorem** If an impedance is connected between two points at which there exist a voltage and an impedance, the current through the added impedance will be equal to the voltage value divided by the sum of the impedance values.

**thickness vibration** Vibration of a piezoelectric crystal in the direction of its thickness.

**thimble ionization chamber** A small cylindrical or spherical ionization chamber, usually with walls of organic material or with air-filled walls.

**thin-film memory** A computer memory consisting of a thin film of magnetic material evaporated on a heated glass base in the presence of a d-c magnetic field parallel to the surface of the base. Large magnetic memory arrays with thousands of elements can be made in one operation.

**thin-wall counter tube** *Thin-window counter tube.*

**thin-window counter tube** A counter tube in which a portion of the enclosure has low absorption to permit the entry of short-range radiation. Also called thin-wall counter tube.

**third harmonic** A sine-wave component having three times the fundamental frequency of a complex wave.

**Thomson bridge** *Kelvin bridge.*

**Thomson coefficient** The ratio of the voltage existing between two points on a metallic conductor to the difference in temperature of those points.

**Thomson cross section** *Scattering cross section.*

**Thomson effect** When a current flows from a warmer to a cooler portion of a conductor, or vice versa, heat is liberated or absorbed depending on the material of which the conductor is made.

**Thomson scattering** Scattering of electromagnetic radiation by electrons. The scattering cross section for an electron is 0.657 barn.

**Thomson voltage** The voltage that exists between two points that are at different temperatures in a conductor.

**Thor** An Air Force surface-to-surface intermediate-range ballistic missile using inertial guidance and having a range of about 2,000 miles. It can carry nuclear warheads and can be adapted for launching a satellite.

**Thoraeus filter** A primary radiological filter of tin, combined with a secondary filter of copper to absorb the characteristic radiation of the tin and a third filter of aluminum to absorb the characteristic radiation of the copper. In the range of 200 to 400 kilovolts such a filter hardens x-rays more efficiently than the usual combination of copper and aluminum.

**thorianite** An ore of thorium and uranium, consisting largely of thorium oxides, oxides of the cerium metals, and uranium.

**thoriated emitter** *Thoriated tungsten filament.*

**thoriated tungsten filament** A vacuum-tube filament consisting of tungsten mixed with a small quantity of thorium oxide to give improved electron emission. Also called thoriated emitter.

**thorides** 1. A name proposed for elements in the last row of the periodic system, having an oxidation state of +4. 2. A name proposed for the series of elements immediately following thorium (atomic number 90), more commonly called actinides.

**thorium** [symbol Th] A metal that emits electrons liberally when heated. Sometimes incorporated in tungsten filaments of vacuum tubes. Important as a fertile material because thorium-232 may absorb neutrons and, decaying through the intermediate element protactinium, yield fissionable uranium-233. Atomic number is 90.

**thorium reactor** A nuclear reactor in which thorium surrounds the central enriched uranium core to give breeder operation.

**thorium series** The series of nuclides resulting from the decay of thorium-232.

**thoron** [symbol Tn] The common name for one of the gaseous radioactive members of the thorium series. It is an isotope of radon.

**thread** *Chip.*

**three-gun color picture tube** A color television picture tube in which three electron guns emit three electron beams, one for each primary color. Each beam is directed onto phosphor dots that emit only the corresponding primary color. Each gun is controlled by its appropriate primary color signal. The shadow-mask color picture tube is an example. Also called tri-gun color picture tube.

**three-junction transistor** A p-n-p-n transistor having three junctions and four regions of alternating conductivity. The emitter connection may be made to the p region at the left, the base connection to the adjacent n region, and the collector connection to the n region at the right. The remaining p region is allowed to float.

**three-level maser** A solid-state maser in which three energy levels are used. Successful operation has been obtained with crystals of gadolinium ethyl sulfate and crystals of potassium chromecyanide at the temperature of liquid helium.

**three-phase circuit** A circuit energized by a-c voltages that differ in phase by one-third of a cycle or 120°.

**three-pole switch** An arrangement of three single-pole single-throw switches coupled together to make or break three circuits simultaneously.

**three-way system** A three-unit loudspeaker system, consisting of a woofer to handle the lowest frequencies, a mid-range unit, and a tweeter for the high frequencies.

**threshold** The least value of a current, voltage, or other quantity that produces the minimum detectable response. Also called limen.

**threshold current** The minimum current value at which a nonself-sustained gas discharge changes to a self-sustained discharge.

**threshold dose** The minimum radiation dose that will produce a detectable specified effect.

**threshold effect** The inherent suppression of noise in a phase-modulated or frequency-modulated receiver by a carrier whose peak value is only slightly greater than that of the noise.

**threshold energy** The energy limit, for an incident particle or photon, below which a particular endothermic reaction will not occur or a particular nuclear reaction cannot be observed.

**threshold field** The least magnetizing force, in a direction that tends to decrease the remanence, that will cause a stated fractional change of remanence when applied either as a steady field of long duration or as a pulsed field appearing many times.

**threshold frequency** The frequency of incident radiant energy below which there is no photoemissive effect.

**threshold of audibility** The minimum effective sound pressure of a specified signal that is capable of evoking an auditory sensation in a specified fraction of the trials.

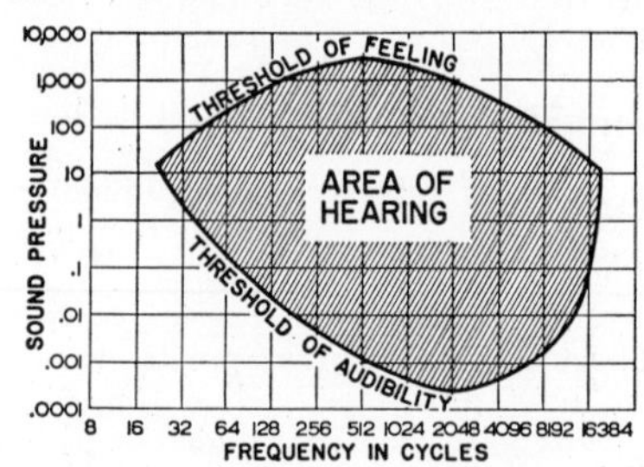


Threshold of audibility, shown as sound pressure in microbars plotted against frequency. At 1,000 cps, threshold of hearing is 0.0002 microbar, equal to 0 decibel when 0.0002 microbar is the reference level.

The threshold may be expressed in decibels relative to 0.0002 microbar or 1 microbar. Also called threshold of detectability and threshold of hearing.

**threshold of detectability** *Threshold of audibility.*

**threshold of feeling** The minimum effective sound pressure of a specified signal that, in a specified fraction of trials, will stimulate the ear to a point at which there is the sensation of feeling, discomfort, tickle, or pain. Customarily expressed in decibels relative to 0.0002 microbar or 1 microbar.

**threshold of hearing** *Threshold of audibility.*

**threshold of luminescence** *Luminescence threshold.*

**threshold sensitivity** The smallest amount of a quantity that can be detected by a measuring instrument or automatic control system.

**threshold signal** The smallest signal that gives a recognizable change in positional information in a navigation system.

**threshold value** The minimum input that produces a corrective action in an automatic control system.

**threshold voltage** The lowest voltage at which all pulses produced in a Geiger counter by any ionizing event are of the same size, regardless of the size of the initial ionizing event.

**threshold wavelength** The wavelength of the incident radiant energy above which there is no photoemissive effect.

**throat** The smaller end of a horn or tapered waveguide.

**throat acoustic impedance** The acoustic impedance at the throat end of a horn.

**throat microphone** A contact microphone that is strapped to the throat of a speaker and reacts to throat vibrations directly rather than to the sound waves they produce.

**throttling** Control by means of intermediate steps between full on and full off.

**throttling control** *Rate control.*

**through path** The transmission path from the loop input signal to the loop output signal in a feedback control loop.

**through transfer function** The transfer function of the through path in a feedback control loop.

**throwing power** The ability of an electroplating solution to deposit metal uniformly on an irregularly shaped cathode.

**throwout spiral** *Leadout groove.*

**thulium** [symbol Tm] A rare-earth element. Atomic number is 69.

**thump** A low-frequency transient disturbance in an audio system.

**thyratron** A hot-cathode gas tube in which one or more control electrodes initiate but do not limit the anode current except under certain operating conditions. Also called hot-cathode gas-filled tube. A thyratron with one control electrode is also called gas triode and hot-cathode gas triode. A thyratron with two control electrodes is also called a gas tetrode.

**thyratron inverter** An inverter circuit that uses thyratrons to convert d-c power to a-c power.

**thyristor** A transistor having a thyratron-like characteristic. As collector current is increased to a critical value, the alpha of the unit rises above unity to give high-speed triggering action.

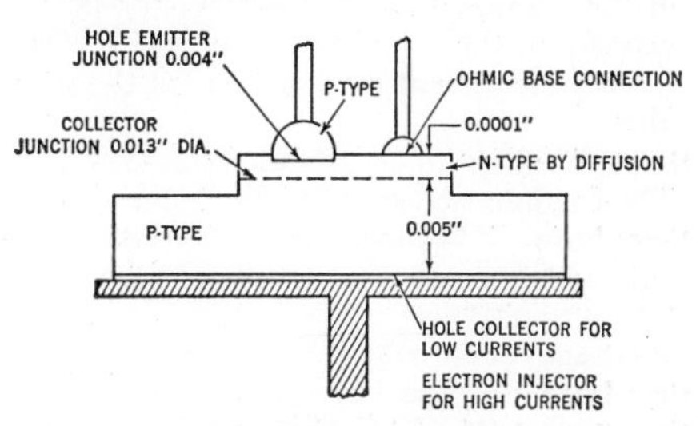


Thyristor construction.

**tickler** A small coil connected in series with the anode circuit of an electron tube and inductively coupled to a grid-circuit coil to provide feedback. Used chiefly in regenerative detector and oscillator circuits.

**tie-down point** One of the frequencies at which a radio receiver is aligned. For the broadcast band, the tie-down points are usually 600 and 1,400 kc.

**tie point** An insulated terminal to which two or more wires are connected.

**tie wire** A wire used to connect a number of terminals together.

**tight coupling** Inductive coupling in which practically all the magnetic flux of one coil links another coil.

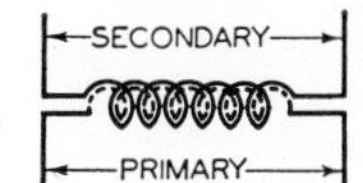


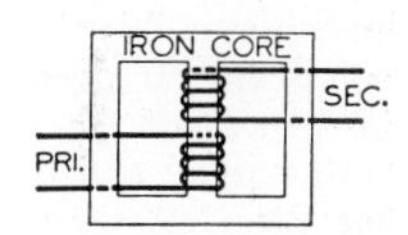


Tight coupling in air-core and iron-core transformers.

**tilt** The angle that an antenna axis forms with the horizontal.

**tilt angle** The angle between the axis of radiation of a radar beam in the vertical plane and a reference axis (normally the horizontal).

**tilt error** The component of ionospheric height error in navigation that is due to nonuniform height of the ionospheric layer.

**tilting** Forward inclination of the wavefront of radio waves traveling along the ground. The amount depends on the electrical constants of the ground.

**tilt stabilization** Stabilization of a radar antenna by using an additional servomotor to tilt the antenna up or down during scanning, as required to correct for pitch and roll of the ship or aircraft.

**timbre** That attribute of auditory sensation

in terms of which a listener can judge that two sounds similarly presented and having the same loudness and pitch are dissimilar. Timbre depends primarily on the spectrum of the stimulus, but it also depends on the waveform, the sound pressure, and the frequency location of the spectrum of the stimulus.

**time** A measure of duration of an event. The fundamental unit of time is the second.

**time base** The line formed by sweep-circuit action on the screen of a cathode-ray tube.

**time-base generator** *Sweep oscillator.*

**time-base voltage** *Sweep voltage.*

**time constant** The time required for a voltage or current in a circuit to rise to approximately 63% of its steady final value or to fall to approximately 37% of its initial value. The time constant of a coil having an inductance $L$ in henrys and resistance $R$ in ohms is $L/R$. The time constant of a capacitor having a capacitance $C$ in farads in series with a resistance $R$ in ohms is $RC$.

**time delay** The time required for a signal to travel between two points in a circuit or for a wave to travel between two points in space.

**time-delay circuit** A circuit that delays a signal or action a definite desired period of time.

**time-delay relay** A relay in which there is an appreciable interval of time between energizing or deenergizing of the coil and movement of the armature, such as a slow-acting relay and a slow-release relay.

**time discriminator** A circuit in which the sense and magnitude of the output is a function of the time difference between two pulses and their relative time sequence.

**time-distribution analyzer** An instrument that indicates the number or rate of occurrence of time intervals falling within one or more specified time-interval ranges. The time interval is delineated by the separation between pulses of a pulse pair.

**time-division multiplex** A device or process for transmitting two or more signals over a common path by using successive time intervals for different signals.

**time flutter** *Time jitter.*

**time gate** A circuit that gives an output only during chosen time intervals.

**time-interval counter** An electronic counter used to measure a time interval by counting the number of pulses received from an r-f signal generator in that time interval.

**time-interval selector** A circuit that produces a specified output pulse when and only when the time interval between two pulses lies between specified limits.

**time jitter** Variations in the synchronization of the components of a radar system, causing variations in the position of the observed pulse along the time base and reducing the accuracy with which the time of arrival of a pulse can be determined. Also called time flutter.

**time lag** The time between an event and a resultant effect, as between occurrence of a primary ionizing event and its count by a counter.

**time modulation** Modulation in which the time of occurrence of a definite portion of a waveform is varied in accordance with a modulating signal.

**time pattern** A television picture-tube presentation of horizontal and vertical lines or dot rows generated by two stable frequency sources operating at multiples of the line and field frequencies.

**time phase** Reaching corresponding peak values at the same instants of time though not necessarily at the same points in space.

**time quadrature** Differing by a time interval corresponding to one-fourth the time of one cycle of the frequency in question.

**timer** 1. A circuit used in radar and in electronic navigation systems to start pulse transmission and synchronize it with other actions such as the start of a cathode-ray sweep. 2. *Interval timer.*

**time response** The time required for the output of a control system to show the effect of application of a prescribed input signal.

**time-shared amplifier** An amplifier used with a synchronous switch to amplify signals from different sources one after another.

**time signal** A radio signal broadcast at accurately known times each day on a number of different frequencies by WWV and other stations, for use in setting clocks.

**time switch** A clock-controlled switch used to open or close a circuit at one or more predetermined times.

**time tick** An accurately controlled pulsed radio signal used for setting timepieces.

**time-varied gain control** *Sensitivity-time control.*

**timing-axis oscillator** *Sweep oscillator.*

**timing signal** Any signal recorded simultaneously with data on magnetic tape for use in identifying the exact time of each recorded event.

**tin** [symbol Sn] A metallic element. Atomic number is 50.

**tinned wire** Copper wire that has been coated during manufacture with a layer of tin or solder to prevent corrosion and simplify soldering of connections.

**tinsel** A type of confusion reflector.

**tinsel cord** A highly flexible cord used for headphone leads and test leads, in which the conductors are strips of thin metal foil or tinsel wound around a strong but flexible central cord.

**tip** 1. The contacting part at the end of a phone plug. 2. A small protuberance on the envelope of an electron tube, resulting from the closing of the envelope after evacuation.

**tip jack** A small single-hole jack for a single-pin contact plug. Also called pup jack.

**tissue dose** The dose received by a tissue in the region of interest, expressed in roentgens for x-rays and gamma rays.

**tissue-equivalent material** Material having the same elements in the same proportions as they occur in some particular biological tissue.

**Titan** An Air Force surface-to-surface intercontinental ballistic missile having a range of over 6,000 miles and a greater pay load of nuclear or conventional weapons than Atlas. It uses inertial guidance alone or combined with radar guidance.

**titanium** [symbol Ti] A metallic element having high strength and corrosion resistance. Atomic number is 22.

**titration control** An electronic control used in chemical processes to regulate acidity or alkalinity.

**T junction** *Tee junction.*

**$TM_{m,n}$ mode** A mode in which a particular transverse magnetic wave is propagated in a waveguide. Also called $E_{m,n}$ mode (British).

**$TM_{m,n,p}$ mode** A mode of wave propagation in a cavity consisting of a hollow metal cylinder closed at its ends, for which the transverse field pattern is similar to that of the $TM_{m,n}$ mode in a corresponding cylindrical waveguide and for which $p$ is the number of half-period field variations along the axis. Also applicable to closed rectangular cavities.

**$TM_{m,n}$ wave** 1. In a circular waveguide, the transverse magnetic wave for which $m$ is the number of axial planes along which the perpendicular component of the magnetic vector vanishes, and $n$ is the number of coaxial cylinders to which the electric vector is perpendicular. The $TM_{0,1}$ wave is the circular magnetic wave having the lowest cutoff frequency. 2. In a rectangular waveguide, the transverse magnetic wave for which $m$ is the number of half-period variations of the magnetic field along the longer transverse dimension, and $n$ is the number of half-period variations of magnetic field along the shorter transverse dimension. Also called $E_{m,n}$ wave (British) for both circular and rectangular waveguides.

**TM wave** Abbreviation for *transverse magnetic wave.*

**Tn** Symbol for *thoron.*

**T network** A network composed of three branches, with one end of each branch connected to a common junction point, and with the three remaining ends connected to an input terminal, an output terminal, and a common input and output terminal, respectively.

**toe and shoulder** The nonlinear portions of the H and D curve, located below and above the straight portion of this curve.

**to-from indicator** A sensing device used in an aircraft to show whether the numerical reading of an omnibearing selector represents a bearing toward or away from an omnidirectional range.

**toggle switch** A small switch that is operated by manipulation of a projecting lever that is combined with a spring to provide a snap action for opening or closing a circuit quickly.

**tolerance** A permissible deviation from a specified value, expressed in actual values or more often as a percentage of the nominal value.

**tolerance dose** Former term for *permissible dose.*

**toll television** *Subscription television.*

**tomography** *Laminography.*

**tone** 1. A sound wave capable of exciting an auditory sensation having pitch, or a sound sensation having pitch. 2. The quality of reproduction of a sound program.

**tone arm** *Pickup arm.*

**tone control** A control used in an a-f amplifier to change the frequency response so as to secure the most pleasing proportion of bass to treble. Individual bass and treble controls are provided in some amplifiers.

**tone generator** A signal generator used to generate an a-f signal suitable for signaling purposes or for testing a-f equipment.

**tone localizer** *Equisignal localizer.*

**tone-modulated wave** A continuous wave that is modulated by a single audio frequency.

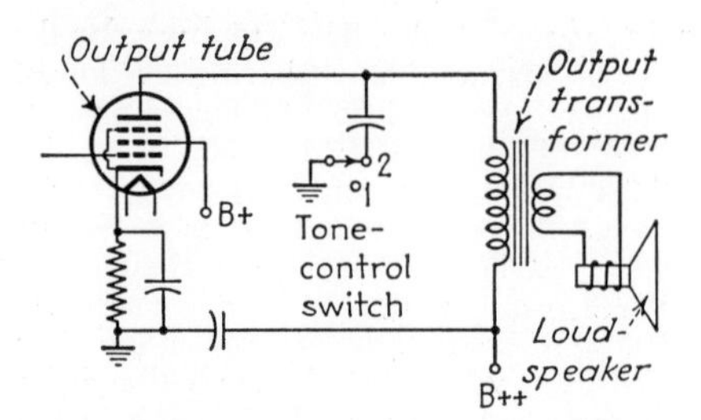


Tone control circuit bypasses higher audio frequencies to ground when switch is in position 2, to emphasize bass notes.

**tonlar** [Tone-Operated Net-Loss AdjusteR] A system for stabilizing the net loss of a telephone circuit by means of a tone transmitted between conversations.

**tonometer** An electronic instrument that measures hydrostatic pressure within the eye. When placed in position, a tiny movable plate is pressed against the eye, flattening a circular section of the cornea. No eyeball anesthesia is required. A current is then sent through a small electromagnet, of such value that it will just pull the plate away from the eye. The value of the current is then proportional to eye pressure. Used in diagnosis of glaucoma. A measurement can be made in about 1 second. Also called electronic tonometer.

**top cap** A metal cap positioned at the top of an electron tube and connected to one of the electrodes, usually the control grid.

**top-loaded vertical antenna** A vertical antenna that is wider at the top, to modify the current distribution and give a more desirable radiation pattern in the vertical plane. A coil may be connected between the enlarged portion of the antenna and the remaining structure.

**toroid** 1. A coil or transformer wound on a doughnut-shaped core. The toroidal core gives a maximum magnetic field within itself, with minimum magnetic flux leakage externally. 2. *Doughnut.*

**torpedo nose assembly** An assembly located in front of a torpedo warhead, containing the hydrophones and associated amplifier and control circuits required for acoustic homing on sounds made by the propellers of a ship.

**torque amplifier** An analog computer device having input and output shafts and supplying work to rotate the output shaft in positional correspondence with the input shaft without imposing any significant torque on the input shaft.

**torque-coil magnetometer** A magnetometer that depends for its operation on the torque developed by a known current in a coil that can turn in the field to be measured.

**torque motor** A motor designed primarily to exert torque while stalled or rotating slowly.

**torsiometer** An instrument for measuring the amount of power transmitted by a rotating shaft, as by measuring the twisting of the shaft under load or measuring the twisting of components mounted on a coupling device inserted between sections of the shaft.

**torsion galvanometer** A galvanometer in which the force between the fixed and moving systems is measured by the angle through which the supporting head of the moving system must be rotated to bring the moving system back to its zero position.

**torsion-string galvanometer** A sensitive galvanometer in which the moving system is suspended by two parallel fibers that tend to twist around each other.

**total cross section** The sum of the cross sections for all mutually exclusive processes for a given medium in a nuclear reactor. For removal of an incident particle from a beam, it is the sum of the separate cross sections for all processes by which the particle can be removed from the beam.

**total electrode capacitance** The capacitance between one electrode and all other electrodes connected together.

**total electron binding energy** The energy required to remove all of the electrons of an atom to infinite distance from the nucleus and from each other, leaving only the bare nucleus.

**total ionization** 1. The total electric charge on the ions of one sign when the energetic particle that has produced these ions has lost all of its kinetic energy. 2. The total number of ion pairs produced by the ionizing particle along its entire path.

**totally unbalanced currents** *Push-push currents.*

**total nuclear binding energy** The energy required to break up a nucleus into its constituent nucleons.

**total specific ionization** *Specific ionization.*

**touch control** A circuit that closes a relay when two metal areas are bridged by a finger or hand.

**tourmaline** A strongly piezoelectric natural crystal.

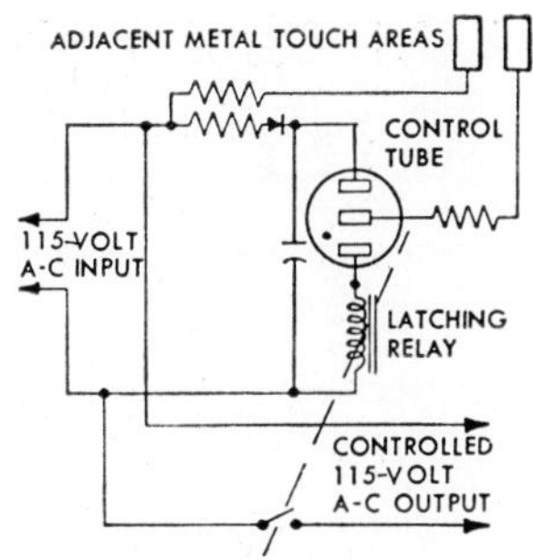

Touch control using neon gas triode.

**tower** A tall metal structure used as a transmitting antenna, or used with another such structure to support a transmitting antenna wire.

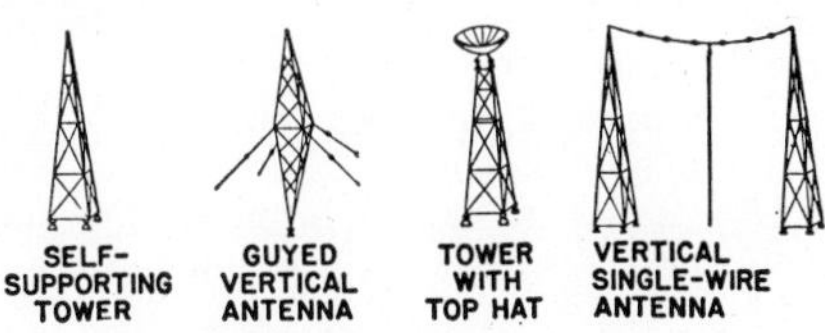

Tower designs for broadcast stations.

**tower radiator** A tall metal structure used as a transmitting antenna.

**Townsend avalanche** *Avalanche.*

**Townsend characteristic** The current-voltage characteristic curve for a phototube at constant illumination and at voltages below that at which a glow discharge occurs.

**Townsend coefficient** The number of ionizing collisions by an electron per centimeter of path length in the direction of the applied electric field in a radiation counter.

**Townsend discharge** A discharge in a gas at moderate pressure (above about 0.1 millimeter of mercury), corresponding to corona. It is free of space charges.

**Townsend ionization** *Avalanche.*

**T pad** A pad made up of resistors arranged as a T network.

**tracals** [air TRAffic Control, navigation, Approach, and Landing System] An Air Force system that is intended to provide and maintain the air traffic control, navigation, approach, and landing assistance required by Air Force aircraft anywhere in the world, through use of such subsystems as ground-controlled approach, instrument landing systems, tacan, and vhf omnirange.

**trace** 1. The visible path of a moving spot on the screen of a cathode-ray tube. Also called line. 2. An extremely small quantity of a substance.

**trace concentration** A concentration of a substance below the usual limits of chemical detection. Radionuclides are often observable in trace concentration by means of their radioactivity.

**trace interval** The time interval in which a sweep traces a desired pattern on the screen of a cathode-ray tube.

**tracer** 1. A foreign substance, usually radioactive, that is mixed with or attached to a given substance so the distribution or location of the latter can later be determined. 2. A thread of contrasting color woven into the insulation of a wire for identification purposes.

**tracing distortion** The nonlinear distortion introduced in the reproduction of a mechanical recording because the curve traced by the motion of the reproducing stylus is not an exact replica of the modulated groove.

**track** 1. A path for recording one channel of information on a magnetic tape, drum, or other magnetic recording medium. The location of the track is determined by the recording equipment rather than by the medium. 2. The horizontal component of the path actually followed by a vehicle, or (marine usage) the intended path. 3. The visible path of an ionizing particle in a cloud chamber or nuclear photographic emulsion. 4. The trace of a moving target on a ppi radar screen or an equivalent plot. 5. To follow the progress of a missile, aircraft, hurricane, or other moving object or action, generally by means of radar, radio direction finders, infrared, or optical equipment. 6. *Race track.*

**track-command guidance** Missile guidance in which the target and missile are tracked by separate radars, and corrective commands are sent to the missile by radio.

**track homing** The process of following a line of position known to pass through an objective.

**tracking** 1. The condition in which all tuned circuits in a receiver accurately follow the frequency indicated by the tuning dial over the entire tuning range. 2. A motion given to the major lobe of a radar or radio antenna such that some preassigned moving target in space is always within the major lobe. 3. The following of a groove by a phonograph needle. 4. Maintaining the same ratio of loudness in the two channels of a stereophonic sound system at all settings of the ganged volume control.

**tracking element** The element in a fire-control system that receives data from the position-finding element and computes the speed and direction of movement of the target, and sometimes also the rates of change in speed and direction.

**tracking error** Deviation of the vibration axis of a phonograph pickup from tangency with a groove. True tangency is possible for only one groove when the pickup arm is pivoted. The longer the pickup arm, the less is the tracking error.

**track in range** To adjust the gate of a radar set so it opens at the correct instant to accept the signal from a target that is changing in range.

**track made good** The resultant track of an aircraft, represented as a straight line between the departure point and the last point of fix on the surface.

**track while scan** An electronic system used to detect a radar target, compute its velocity, and predict its future position without interfering with continuous radar scanning.

**traffic** The messages transmitted and received over a communication channel.

**trailer** A bright streak at the right of a dark area or dark line in a television picture, or a dark area or streak at the right of a bright part. Usually due to insufficient gain at low video frequencies.

**trailing antenna** An aircraft radio antenna having one end weighted and trailing free from the aircraft when in flight.

**trailing edge** The major portion of the decay of a pulse.

**trailing-edge pulse time** The time at which the instantaneous amplitude of a pulse last reaches a stated fraction of the peak pulse amplitude.

**train** To aim or direct a radar antenna in azimuth.

**trainer** A piece of equipment used for training operators of radar, sonar, and other electronic equipment by simulating signals received under operating conditions in the field.

**trajectory** The path traced through space in a vertical plane by a missile.

**trajectory-controlled** Guided or directed so its trajectory will follow a predetermined curve, as for a missile.

**transadmittance** A specific measure of transfer admittance under a given set of conditions, as in forward transadmittance, interelectrode transadmittance, short-circuit transadmittance, small-signal forward transadmittance, and transadmittance compression ratio.

**transadmittance compression ratio** The ratio of the magnitude of the small-signal forward transadmittance of a klystron to the magnitude of the forward transadmittance at a given input signal level.

**transceiver** A radio transmitter and receiver combined in one unit and having switching arrangements such as to permit use of one or more tubes for both transmitting and receiving.

**transconductance** [symbol $G_m$] An electron-tube rating, equal to the change in anode current divided by the change in control-grid voltage. The unit of transconductance is the mho. Less strictly, transconductance is the amplification factor of the tube divided by its anode resistance. Also called grid-anode transconductance and mutual conductance.

**transconductance meter** An instrument for indicating the transconductance of a grid-controlled electron tube. Also called mutual-conductance meter.

**transcontinental ballistic missile** A ballistic missile having a range of at least 12,500 miles, so it can be fired from any point on the earth's surface and reach any surface target.

**transcribe** 1. To record, as to record a radio program by means of electric transcriptions or magnetic tape for future rebroadcasting. 2. To copy, with or without translating, from one external computer storage medium to another.

**transcriber** The equipment used to convert information from one form to another, as for converting computer input data to the medium and language used by the computer.

**transcription** A 16-inch-diameter, 33⅓-rpm disk recording of a complete program, made especially for broadcast purposes. Also called electric transcription.

**transducer** General term for any device that converts energy from one form to another, as from acoustic energy to electric or mechanical energy. Loudspeakers, microphones, phonograph pickups, and strain gages are examples of transducers. A television camera is not a transducer because a scene has no energy.

**transducer gain** The power that a transducer delivers to a specified load, divided by the available power of the specified source. Usually expressed in decibels.

**transducer insertion loss** The ratio of power delivered to a transmission system before insertion of a transducer to the power delivered after insertion of a transducer. Usually expressed in decibels.

**transducer loss** The ratio of the power available to a transducer from a specified source to the power that the transducer delivers to a specified load. Usually expressed in decibels.

**transducer pulse delay** The interval of time between a specified point on the input pulse and a specified point on the related output pulse of a transducer, such as a transmitter, receiver, amplifier, or oscillator. Also called receiver pulse delay and transmitter pulse delay.

**transducer scanner** A device that provides a means of sampling directional signals from individual magnetostriction transducers in a sonar transmitter array. Capacitor plates arranged radially on a disk rotate with respect to a stationary circular disk containing matching plates that are connected to the transducer elements, to give scanning of all elements once per revolution of the rotor disk.

**transducing piezoid** A piezoid used in a transducer.

**transductor** 1. *Magnetic amplifier.* 2. *Saturable reactor.*

**transfer** 1. To transmit or copy information from one computer device to another without changing its form. 2. *Jump.*

**transfer admittance** An admittance rating for electron tubes and other transducers or networks. It is equal to the complex alternating component of current flowing to one terminal from its external termination, divided by the complex alternating component of the voltage applied to the adjacent terminal on the cathode or reference side. All other terminals have arbitrary external terminations.

**transfer characteristic** 1. The relation between the voltage of one electrode in an electron tube and the current to another electrode. 2. The relation between the illumination on a camera tube and the corresponding signal output current.

**transfer check** A check on the accuracy of transfer of a word in a digital computer, usually made automatically.

**transfer constant** A transducer rating, equal to one-half the natural logarithm of the complex ratio of the product of the voltage and current entering a transducer to that leaving the transducer when the latter is terminated in its image impedance. Alternatively, the product may be that of force and velocity or pressure and volume velocity. The real part of the transfer constant is the image attenuation constant, and the imaginary part is the image phase constant.

**transfer current** The current to one electrode required to initiate breakdown to another electrode of a gas tube.

**transfer function** The mathematical relationship between the output of a control system and its input.

**transfer impedance** The ratio of the voltage applied at one pair of terminals of a network to the resultant current at another pair of terminals, all terminals being terminated in a specified manner.

**transfer instruction** A digital-computer instruction or signal that specifies the location of the next operation to be performed.

**transfer of control** *Jump.*

**transfer ratio** The transfer function from one system variable to another in a linear system, expressed as the ratio of the Laplace transform of the second variable to the Laplace transform of the first variable, assuming zero initial conditions.

**transferred printed circuit** A printed circuit in which the pattern is formed on a temporary base and transferred to a permanent base.

**transfer switch** A switch for transferring one or more conductor connections from one circuit to another.

**transfer time** The total elapsed time between the breaking of one set of contacts on a relay and the making of another set of contacts, after all contact bounce has ceased.

**Transfluxor** Trademark of RCA for a magnetic core having two or more apertures and three or more legs for flux. Used as a computer memory element, crossbar switch, channel commutator, or control element.

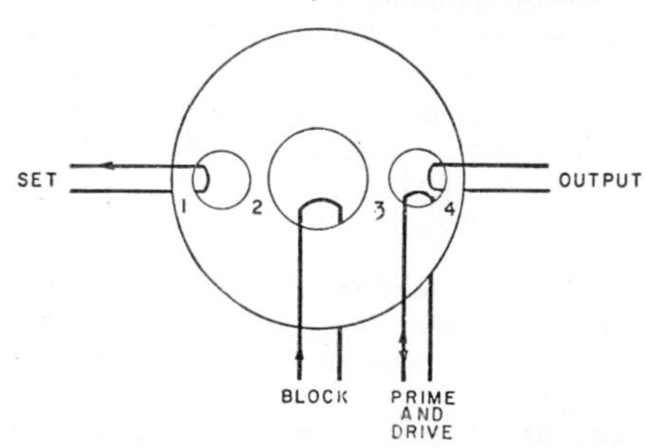


Transfluxor element having three apertures.

**transform** To change the form of digital-computer information without significantly altering its meaning.

**transformation constant** *Decay constant.*

**transformation series** *Radioactive series.*

**transformer** [symbol T] A component consisting of two or more coils that are coupled together by magnetic induction. Used to transfer electric energy from one or more circuits to one or more other circuits without change in frequency but usually with changed values of voltage and current.

**transformer bridge** A network consisting of a transformer and two impedances, in which the input signal is applied to the transformer primary and the output is taken

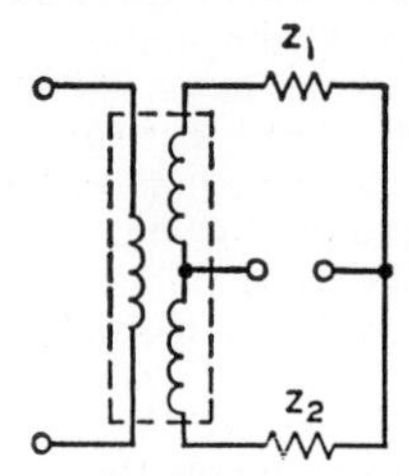


Transformer bridge, with input terminals at left.

between the secondary center-tap and the junction of the impedances that connect to the outer leads of the secondary. When used as a crystal filter, a capacitor is used as one impedance to balance the static capacitance of the crystal that serves as the other impedance, so there is no transmission except in the vicinity of crystal resonance.

**transformer-coupled amplifier** An a-f amplifier that uses untuned iron-core transformers to provide coupling between stages.

**transformer coupling** *Inductive coupling.*

**transformer hybrid** *Hybrid set.*

**transformer loss** The ratio of the power delivered by an ideal transformer to the power delivered by an actual transformer under specified conditions.

**transformer oil** A high-quality insulating oil in which windings of large power transformers are sometimes immersed to provide high dielectric strength, high insulation resistance, high flash point, freedom from moisture, and freedom from oxidization.

**transforming section** A length of waveguide or transmission line having a varying cross-section, used for impedance transformation.

**transient** A pulse, damped oscillation, or other temporary phenomenon occurring in a system prior to reaching a steady-state condition.

**transient analyzer** An analyzer that generates transients in the form of a succession of equal electric surges of small amplitude and adjustable waveform, applies these transients to a circuit or device under test, and shows the resulting output waveforms on the screen of an oscilloscope.

**transient distortion** Distortion due to inability to amplify transients linearly.

**transient equilibrium** Radioactive equilibrium in which the parent has such a large decay constant that the quantity of parent present decreases before radioactive equilibrium is reached.

**transient motion** An oscillatory or other irregular motion occurring while a quantity is changing to a new steady-state value.

**transient oscillation** A momentary oscillation occurring in a circuit during switching.

**transient overshoot** The maximum value of the overshoot of a quantity as a result of a sudden change in conditions.

**transient phenomena** Rapidly changing actions occurring in a circuit during the interval between closing of a switch and settling to steady-state conditions, or any other temporary actions occurring after some change in a circuit.

**transient radioactive equilibrium** Radioactive equilibrium that occurs if the lifetime of the parent is sufficiently short so the quantity of atoms present decreases appreciably during the period under consideration.

**transient response** The response of a circuit to a sudden change in an input quantity.

**transistor** [TRANSfer resISTOR] An active semiconductor device having three or more electrodes. The three main electrodes used are the emitter, collector, and base.

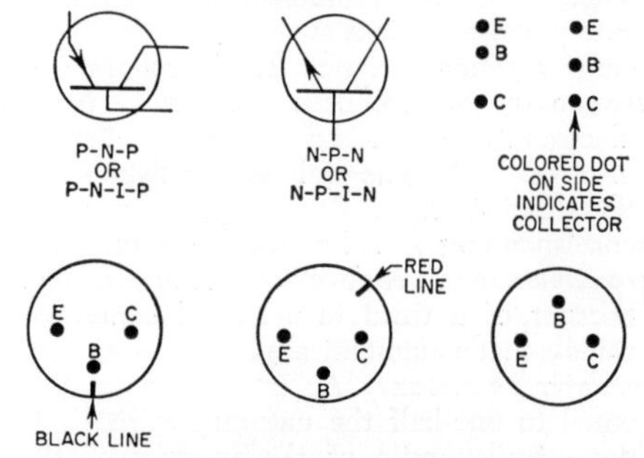


Transistor symbols and bottom views of bases. Leads may go straight out through circle as shown for n-p-n transistor, or may be bent for convenience in drawing associated circuit.

Conduction is by means of electrons and carriers or holes. Germanium and silicon are the materials most often used as the semiconductor material.

**transistor amplifier** An amplifier in which one or more transistors provide amplification comparable to that of electron tubes.

In a class A transistor amplifier, operation is in the linear region of the collector characteristic. For class B, amplification occurs only during half of each input signal cycle. For class AB, the collector current or voltage is zero for less than half of each input cycle. For class C, collector current or voltage is zero for more than half of each input cycle.

**transistorized** Constructed with transistors being used in place of electron tubes.

**transistor oscillator** An oscillator in which a transistor is used in place of an electron tube.

**transistor radio** A radio receiver in which transistors are used in place of electron tubes.

**transistor symbol** A schematic symbol used to represent a transistor in circuit diagrams. The base is represented by a straight line at right angles to its lead. The collector line intersects the base at an angle and has no arrow. The emitter line has an arrow, pointing toward the base for a p-n-p transistor and pointing away from the base for an n-p-n transistor.

**transit angle** The product of angular frequency and the time taken for an electron to traverse a given path.

**transitional coupling** The amount of inductive coupling between two coils that gives the widest passband and flattest response curve without double peaks.

**transition effect** A change in the intensity of the secondary radiation associated with a beam of primary radiation as the latter passes from a vacuum into a material medium or from one medium into another.

**transition element** An element used to couple one type of transmission system to another, as for coupling a coaxial line to a waveguide.

**transition factor** *Reflection coefficient.*

**transition frequency** The frequency corresponding to the intersection of the asymptotes to the constant-amplitude and constant-velocity portions of the frequency response curve for a disk recording. This curve is plotted with output voltage ratio in decibels as the ordinate, and the logarithm of the frequency as the abscissa. Below the transition frequency, the level is progressively reduced when cutting a record to prevent loud bass notes from overcutting the groove walls. One standard transition frequency value is 500 cps. Also called crossover frequency and turnover frequency.

**transition loss** 1. The difference between the power incident upon a transition or discontinuity between two media in a wave propagation system and the power transmitted beyond the discontinuity that would be observed if the medium beyond the discontinuity were match-terminated. 2. The ratio in decibels of the power incident upon a discontinuity to the power transmitted beyond the discontinuity that would be observed if the medium beyond the discontinuity were match-terminated.

**transition multipole moment** A multipole moment that determines radiative transitions between two states and therefore depends on both states.

**transition point** A point at which the constants of a circuit change in such a way as to cause reflection of a wave being propagated along the circuit.

**transition probability** The probability per unit time that a system in one state will undergo a transition to another state.

**transition region** The region between two homogeneous semiconductors in which the impurity concentration changes.

**transitron oscillator** A negative-resistance oscillator in which the screen grid is more positive than the anode, and a capacitor is connected between the screen grid and the suppressor grid. The suppressor grid periodically divides the current between the screen grid and the anode, thereby producing oscillation.

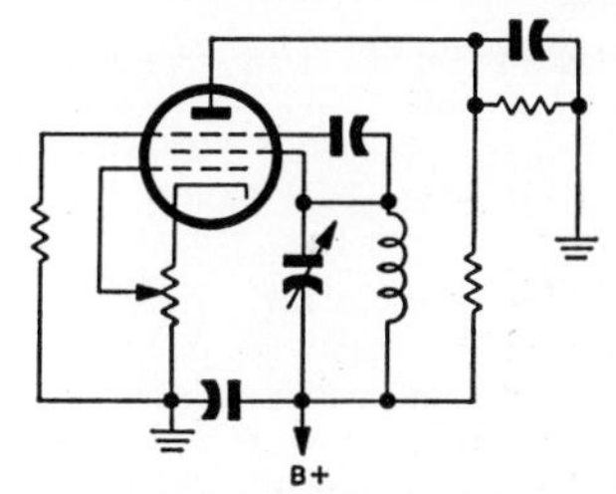

Transitron oscillator circuit.

**transit time** The time required for an electron or other charge carrier to travel between two electrodes in an electron tube or transistor.

**transit-time mode** A mode of oscillator operation corresponding to a limited range of drift-space transit angle for which the electron stream introduces a negative conductance into the coupled circuit.

**translate** To change computer information from one language to another without significantly affecting the meaning.

**translation loss** The amount by which the amplitude of motion of a stylus differs from the recorded amplitude in a disk record. Also called playback loss.

**translator** 1. A computer network or system having a number of inputs and outputs, so connected that when signals representing information expressed in a certain code are applied to the inputs, the output signals will represent the same information in a different code. Also called matrix. 2. A combination television receiver and low-power television transmitter, used to pick up television signals on one frequency and retransmit them on another frequency to provide reception in areas not served directly by television stations. A translator usually broadcasts on a uhf channel from No. 70 to No. 83.

**transmission** 1. The process of transferring a signal, message, picture, or other form of intelligence from one location to one or more other locations by means of wire lines, radio, light beams, infrared beams, or other communication systems. 2. A message, signal, or other form of intelligence that is being transmitted. 3. The ratio of the light flux transmitted by a medium to the light flux incident upon it. Transmission may be either diffuse or specular. Also called transmittance.

**transmission band** The frequency range above the cutoff frequency in a waveguide, or the comparable useful frequency range for any other transmission line, system, or device.

**transmission coefficient** 1. The ratio of transmitted to incident energy or some other quantity at a discontinuity in a transmission medium. For sound waves, it is called the sound transmission coefficient. 2. *Penetration probability.*

**transmission curve** *Absorption curve.*

**transmission gain** *Gain.*

**transmission grating** A diffraction grating produced on a transparent base so radiation is transmitted through the grating instead of being reflected from it.

**transmission level** The ratio of the signal power at any point in a transmission system to the signal power at some point in the system chosen as a reference point. Usually expressed in decibels.

**transmission limit** A limiting wavelength or frequency above or below which a given type of radiation is not appreciably transmitted by a given medium.

**transmission line** A waveguide, coaxial line, or other system of conductors used to transfer signal energy efficiently from one location to another.

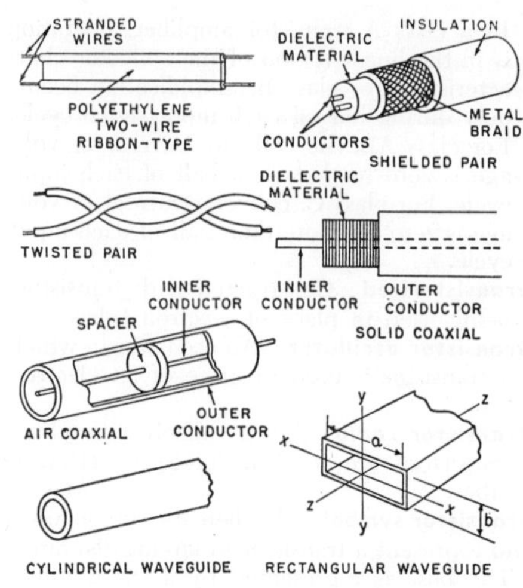

Transmission lines.

**transmission-line coupler** A coupler that permits the passage of electric energy in either direction between balanced and unbalanced transmission lines.

**transmission loss** 1. The ratio of the power at one point in a transmission system to the power at a point farther along the line. Usually expressed in decibels. 2. The actual power that is lost in transmitting a signal from one point to another through a medium or along a line. Also called loss.

**transmission measuring set** A measuring instrument consisting of a signal source and receiver having known impedances, used to measure the insertion loss or gain of a network or transmission path connected between those impedances.

**transmission mode** *Mode.*

**transmission plane** The plane of vibration of polarized light that will pass through a Nicol prism or other polarizer.

**transmission primaries** The set of three color primaries that correspond to the three independent signals contained in the color television picture signal. The three receiver primaries in the color picture tube form one set. The luminance primary and the two chrominance primaries, known as the Y, I, and Q primaries, form another possible set of transmission primaries.

**transmission security** The aspect of communication security that is concerned with the transmission of messages over wires or by radio.

**transmission target** An x-ray target in which the useful x-ray beam emerges from the surface remote from that on which the electron stream is incident.

**transmission time** The absolute time interval from transmission to reception of a signal.

**transmission unit** An early signal-level unit now known as the decibel.

**transmissivity** *Transmittivity.*

**transmissometer** A photoelectric instrument used to measure the visibility of the atmosphere.

**transmit** 1. To send a message, program, or other information to a person or place by wire, radio, or other means. 2. To reproduce information in a new location in a digital computer, replacing whatever was previously stored and clearing or erasing the source of the information.

**transmit negative** The transmission of facsimile signals intended for reception as a negative.

**transmit positive** The transmission of facsimile signals intended for reception as a positive.

**transmit-receive tube** *Tr tube.*

**transmittance** *Transmission.*

**transmitted-light scanning** The scanning of changes in the magnitude of light transmitted through a web.

**transmitted wave** *Refracted wave.*

**transmitter** 1. The equipment used for generating and amplifying an r-f carrier signal, modulating the carrier signal with intelligence, and feeding the modulated carrier

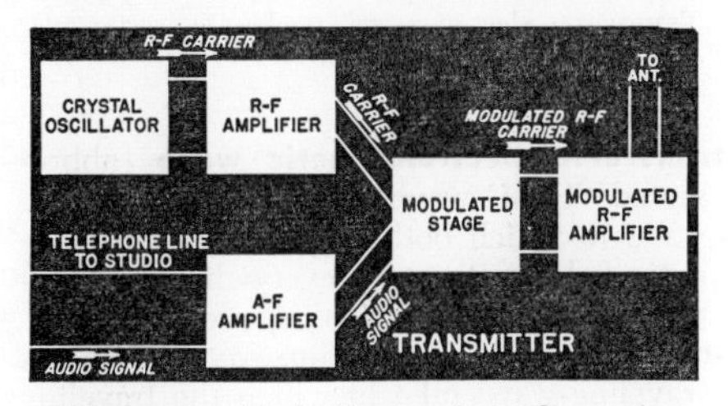


Transmitter block diagram for radio station.

to an antenna for radiation into space as electromagnetic waves. 2. In telephony, the carbon microphone that converts sound waves into a-f signals. 3. *Synchro transmitter.*

**transmitter input polarity** The polarity of the portion of a television picture signal representing a dark area of a scene, relative to the potential of a portion of the signal representing a light area.

**transmitter pulse delay** *Transducer pulse delay.*

**transmitter synchro** *Synchro transmitter.*

**transmitting current response** The ratio of the sound pressure at 1 meter to the current flowing at the electric input terminals of a loudspeaker or other electroacoustic transducer. Usually expressed in decibels above a reference current response of 1 microbar per ampere.

**transmitting efficiency** The ratio of total acoustic power output to electric power input for an electroacoustic transducer. Also called projector efficiency.

**transmitting power response** The ratio of the effective sound pressure at 1 meter to the electric power input of a loudspeaker or other electroacoustic transducer used for sound emission. Usually expressed in decibels above a reference response of 1 microbar squared per watt of electric power input. Also called projector power response.

**transmitting station** The location at which a transmitter, transmitting antenna, and associated transmitting equipment of a radio or television station are grouped.

**transmitting voltage response** The ratio of the sound pressure at 1 meter to the signal voltage applied at the electric input terminals of a loudspeaker or other acoustic transducer. Usually expressed in decibels above a reference voltage response of 1 microbar per volt.

**transmittivity** The ratio of the transmitted radiation to radiation arriving perpendicular to the boundary between two media. Also called transmissivity.

**trans-mu factor** The ratio of the magnitude of an infinitesimal change in the voltage at the control grid of any one beam of a multibeam electron tube to the magnitude of an infinitesimal change in the voltage at the control grid of a second beam. The current in the second beam and the voltage of all other electrodes are maintained constant.

**transmutation** A nuclear process in which a nuclide is transformed into the nuclide of a different element.

**transonic** Pertaining to transonic speed.

**transonic barrier** *Sonic barrier.*

**transonic speed** A speed in the range of about mach 0.8 to mach 1.2, corresponding to 600 to 900 miles per hour, at which one or more local points on the body of an aircraft or missile are moving at subsonic speed at the same time that one or more other points move at sonic or supersonic speed.

**transparent** Permitting passage of radiation or particles.

**transponder** 1. A radio device that receives

an interrogating or challenging radio signal and automatically transmits a response on the same or a different frequency. Used in conjunction with an interrogator to determine bearing or range, or both. 2. *Radar beacon.*

**transponder dead time** The time interval between the start of a pulse and the earliest instant at which a new pulse can be received or produced by a transponder.

**transponder reply efficiency** The ratio of the number of replies emitted by a transponder to the number of interrogations that the transponder recognizes as valid.

**transport** To convey as a whole from one storage device to another in a digital computer.

**transportable transmitter** A transmitter designed to be readily carried or transported from place to place, but not normally operated while in motion.

**transport cross section** The reciprocal of the transport mean free path.

**transport mean free path** A path length equal to three times the diffusion coefficient of neutron flux in a nuclear reactor when Fick's law is applicable.

**transport theory** A theory based on an approximation to Boltzmann's equation for conditions in which Fick's law is not applicable.

**transposition** Interchanging the relative positions of conductors at regular intervals along a transmission line to reduce crosstalk.

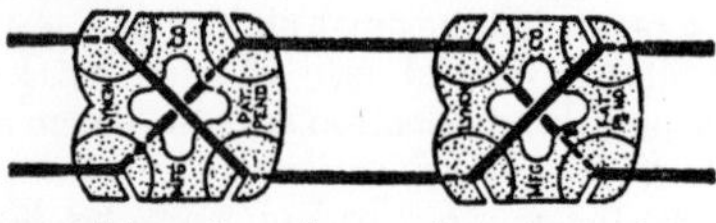

Transposition of conductors by means of ceramic insulators.

**transrectification** Rectification that occurs in one circuit when an alternating voltage is applied to another circuit.

**transrectification factor** The change in average current of an electrode divided by the change in amplitude of the alternating sinusoidal voltage applied to another electrode when all direct voltages of an electron tube or transistor are maintained constant.

**transtrictor** *Unipolar transistor.*

**transuranic element** An element having an atomic number greater than uranium (which is 92), such as neptunium, plutonium, americium, curium, berkelium, californium, einsteinium, fermium, mendelevium, and nobelium.

**transversal filter** A filter whose frequency transmission properties exhibit a periodic symmetry.

**transverse-beam traveling-wave tube** A traveling-wave tube in which the direction of motion of the electron beam is transverse to the average direction in which the signal wave moves.

**transverse electric wave** [abbreviated TE wave] An electromagnetic wave in which the electric field vector is everywhere perpendicular to the direction of propagation. Also called H wave (British).

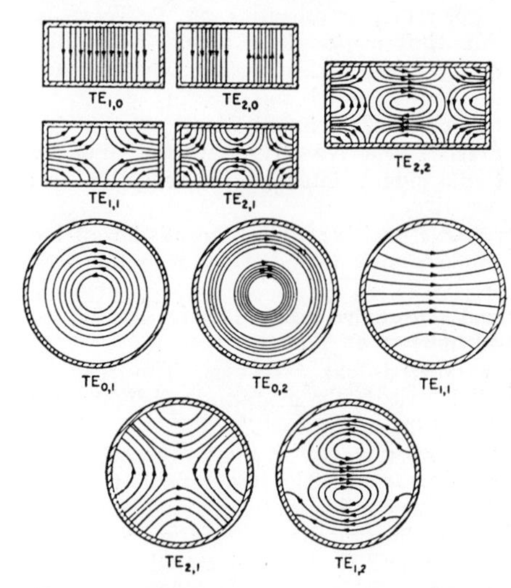


Transverse electric wave modes in rectangular and circular waveguides, with electric field configuration for each.

**transverse electromagnetic wave** [abbreviated TEM wave] An electromagnetic wave in which both the electric and magnetic field vectors are everywhere perpendicular to the direction of propagation.

**transverse-field traveling-wave tube** A traveling-wave tube in which the traveling electric fields that interact with the electrons are essentially perpendicular to the average motion of the electrons.

**transverse-film attenuator** An attenuator in which a conducting film is placed across a waveguide.

**transverse fuze** An electronic fuze in which the antenna is set at right angles to the longitudinal axis of the missile, to give sensitivity directly ahead of the nose.

**transverse heating** Dielectric heating in which the electric field is perpendicular to the layers of a laminated material being heated.

**transverse magnetic wave** [abbreviated TM wave] An electromagnetic wave in which the magnetic field vector is everywhere perpendicular to the direction of propagation. Also called E wave (British).

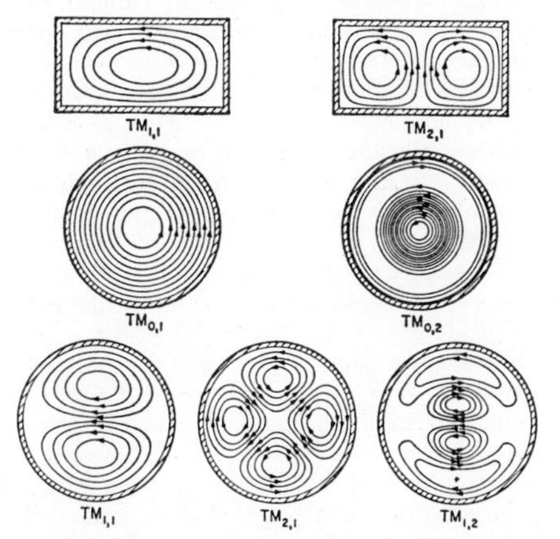


Transverse magnetic wave modes for rectangular and circular waveguides, with magnetic field configuration for each.

**transverse magnetization** Magnetization of a magnetic recording medium in a direction perpendicular to the line of travel and parallel to the greatest cross-sectional dimension.

**transverse plate** A plate of metal or highly resistant material used to close the end of a waveguide or used as an adjustable piston inside the waveguide. Also called transverse septum.

**transverse septum** *Transverse plate.*

**transverse wave** A wave in which the direction of displacement at each point of the medium is parallel to the wavefront.

**trap** 1. A tuned circuit used in the r-f or i-f section of a receiver to reject undesired frequencies. Traps in television receiver video circuits keep the sound signal out of the picture channel. Also called rejector. 2. A semiconductor imperfection that prevents carriers from moving through the material. 3. *Wave trap.*

**trapped mode** Propagation in which the energy radiated in the troposphere is almost entirely confined within a duct.

**trapping** A process wherein electrons are held at an irregularity in the crystal lattice of a semiconductor until released by thermal agitation.

**trapping spot** *Capture spot.*

**traveling detector** An r-f probe mounted in a slotted-line section of waveguide and used with a detector to measure standing-wave ratios.

**traveling plane wave** A plane wave in which each frequency component varies exponentially in amplitude and linearly in phase in the direction of propagation. Also called traveling wave.

**traveling wave** *Traveling plane wave.*

**traveling-wave accelerator** A plasma engine for space travel in which plasma is accelerated through a tube by a series of coils spaced along the tube and excited by polyphase r-f energy.

**traveling-wave amplifier** An amplifier that uses one or more traveling-wave tubes to provide useful amplification of signals at frequencies of the order of thousands of megacycles.

**traveling-wave antenna** An antenna in which the current distributions are produced by waves of charges propagated in only one direction in the conductors. Also called progressive-wave antenna.

**traveling-wave interaction** The interaction between an electron stream and a slow wave moving through a circuit in approximate synchronism with the velocity of the electrons.

**traveling-wave magnetron** A traveling-wave tube in which the electrons move in crossed static electric and magnetic fields that are substantially normal to the direction of wave propagation, as in practically all modern magnetrons.

**traveling-wave magnetron oscillation** Oscillation sustained by the interaction between the space-charge cloud of a magnetron and a traveling electromagnetic field whose phase velocity is approximately the same as the mean velocity of the cloud.

**traveling-wave maser** A maser in which the paramagnetic active material is placed in a nonresonant traveling-wave structure.

**traveling-wave tube** [abbreviated twt] An electron tube in which a stream of electrons interacts continuously or repeatedly with a guided electromagnetic wave moving substantially in synchronism with it, in such a way that there is a net transfer of energy

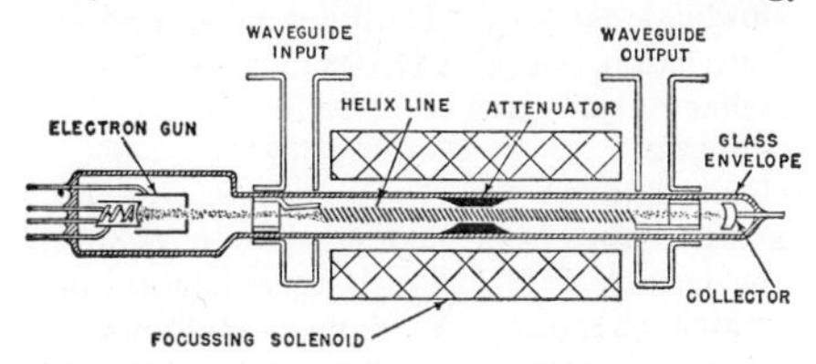


Traveling-wave tube uses solenoid to provide required longitudinal magnetic field.

from the stream to the wave. The tube is used as an amplifier or oscillator at frequencies in the microwave region.

**traveling-wave-tube interaction circuit** An extended electrode arrangement in a traveling-wave tube, designed to propagate an electromagnetic wave in such a manner that the traveling electromagnetic fields are retarded and extend into the space occupied by the electron stream.

**tr box** *Tr tube.*

**tr cavity** The resonant portion of a radar tr tube.

**tr cell** British term for *tr tube.*

**treble** High audio frequencies, such as those handled by a tweeter in a sound system.

**treble boost** Adjustment of the amplitude-frequency response of a system or component to accentuate the higher audio frequencies.

**tree** A set of connected circuit branches that includes no meshes. It responds uniquely to each of the possible combinations of a number of simultaneous inputs. Also called decoder.

**trf** Abbreviation for *tuned radio frequency.*

**triad** 1. A triangular group of three small phosphor dots, each emitting one of the three primary colors, on the screen of a three-gun color picture tube. 2. *Triplet.*

**triangulation** Determination of the position of a ship or aircraft by obtaining bearings of the moving object with reference to two fixed radio stations a known distance apart. This gives the values of one side and all angles of a triangle, from which the position can be computed.

**tribo-** A prefix meaning pertaining to or resulting from friction.

**triboelectricity** Electric charges generated by friction.

**triboelectric series** A list of materials that produce an electrostatic charge when rubbed together, arranged in such an order that a material has a positive charge when rubbed with a material substance below it in the list, and has a negative charge when rubbed with a material above it in the list.

**triboluminescence** Luminescence produced by friction between two materials.

**trichromatic coefficient** *Chromaticity coordinate.*

**trickle charge** A continuous charge of a storage battery at a low rate to maintain the battery in a fully charged condition.

**trickle charger** A device designed to charge a storage battery at a low rate continuously to keep it fully charged.

**triclinic** Pertaining to a crystal structure having three unequal axes intersecting at angles, not more than two of which are equal and not more than one of which is 90°.

**tricolor camera** A television camera that separates the light from a scene into three frequency groups and converts the light energies of these groups into corresponding color signals.

**tricolor picture tube** *Color picture tube.*

**tricon** A radio navigation system in which the airborne receiver accepts pulses from a triplet or chain of three stations pulsed in variable time sequence. The time sequences vary so pulses arrive at the same time along paths of various lengths.

**tridipole antenna** A horizontally polarized antenna having three curved dipoles mounted in a horizontal plane to form a circle.

**tridop** A doppler missile-tracking system consisting of a master station and three additional receiving stations on the ground. The master station triggers a continuous-wave transponder in the missile, and this in turn radiates signals to the ground station for comparison with a timing signal from the master station.

**triductor** An arrangement of iron-core transformers and capacitors used to triple a power-line frequency. A d-c voltage is applied to some of the windings to provide enough premagnetization so the applied a-c voltage can saturate the cores and thereby generate the desired third-harmonic output.

**trifluorochloroethylene resin** A fluorocarbon used as a base for polychlorotrifluoroethylene resin, marketed as Kel-F.

**trigatron** An electronic switch in which conduction is initiated by the breakdown of an auxiliary gap in a gas-filled envelope. The gap between the two main electrodes is normally nonconducting, but breaks down when a pulse is applied to a trigger electrode. Used in some radar modulators.

**trigger** 1. To initiate a sudden action, as by applying a pulse to a trigger circuit. 2. The pulse used to initiate the action of a trigger circuit. 3. *Trigger circuit.*

**trigger action** Use of a weak input pulse to initiate main current flow suddenly in a circuit or device.

**trigger circuit** 1. A circuit or network in which the output changes abruptly with an infinitesimal change in input at a predetermined operating point. Also called trigger. 2. A circuit in which an action is initiated by an input pulse, as in a radar modulator. 3. *Flip-flop circuit.*

**trigger control** Control of thyratrons, ignitrons, and other gas tubes in such a way that current flow may be started or stopped, but not regulated as to rate.

**triggered blocking oscillator** A blocking oscillator that can be reset to its starting condition by a trigger voltage. A parallel-triggered blocking oscillator has less effect on the trigger source than a series-triggered blocking oscillator, but the series-triggered type has less delay.

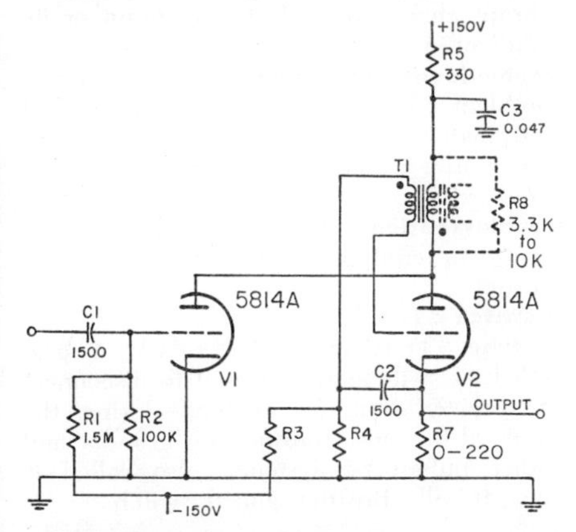


Triggered blocking oscillator using preferred circuit design by National Bureau of Standards. For pulse trigger spacings up to 60 microseconds, R3 is 47,000 ohms and R4 is 4,700 ohms; for spacings from 60 to 500 microseconds, R3 is 100,000 ohms and R4 is 10,000 ohms.

**triggered spark gap** A fixed spark gap in which the discharge passes between two electrodes but is initiated by an auxiliary trigger electrode to which low-power pulses are applied at regular intervals by a pulse amplifier.

**trigger electrode** *Starter.*

**triggering** Initiation of an action in a circuit, which then functions for a predetermined time, as for the duration of one sweep in a cathode-ray tube.

**trigger level** The minimum input to the receiver of a transponder that is capable of causing a transmitter to emit a reply.

**trigger pulse** A pulse that starts a cycle of operation. Also called tripping pulse.

**trigger switch** A switch that is actuated by pulling a trigger, and is usually mounted in a pistol-grip handle.

**trigger winding** A winding added to a pulse transformer to supply a low-voltage pulse to an external load, usually for synchronizing purposes.

**tri-gun color picture tube** *Three-gun color picture tube.*

**trihedral reflector** A corner reflector having three square or triangular sides meeting at a point. Used as an artificial radar reflector and for other applications where signals must be reflected back toward the transmitter over a greater angle than is practical with a plane sheet reflector.

**trimmer** A small variable or semiadjustable capacitor or variable inductance used in tuning circuits to adjust capacitance values for alignment purposes so all circuits can be tuned accurately by a single control.

**trimmer capacitor** A variable or semiadjustable capacitor used as a trimmer.

**trimming** Fine adjustment of capacitance or inductance.

**Trinity bomb** *Alamogordo bomb.*

**trinoscope** An arrangement of three picture tubes with color filters and projection lenses, used in theater television to project the superimposed red, blue, and green images required for full-color pictures.

**triode** A three-electrode electron tube containing an anode, a cathode, and a control electrode.

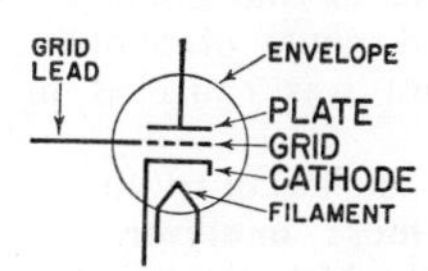


Triode symbol. The plate electrode is more properly called the anode.

**triode amplifier** An amplifier that uses only triodes.

**triode-hexode converter** A triode oscillator and a multigrid mixer in the same tube envelope.

**trip action** Instability occurring in a magnetic amplifier due to excessive feedback.

**trip coil** A coil used to open a circuit breaker or other protective device when coil current exceeds a predetermined value.

**triple-conversion receiver** A communication receiver having three different intermediate frequencies, to give higher adjacent-channel selectivity and greater image-frequency suppression.

**triple-diode triode** An electron tube having three diodes and one triode served by a common cathode.

**triple-stub transformer** A transformer in which three stubs are placed a quarter-wavelength apart on a coaxial line and adjusted in length to compensate for impedance mismatch.

**triplet** Three radio navigation stations

operated as a group for the determination of positions. Also called triad.

**tripping pulse** *Trigger pulse.*

**trip value** The voltage, current, or power at which a polarized relay will transfer from one contact to another.

**tristimulus colorimeter** A colorimeter that measures a color stimulus in terms of tristimulus values.

**tristimulus values** The amounts of each of the three primary colors that must be combined to establish a match with a given sample color. Also called color-mixture data (deprecated).

**tri-tet oscillator** A crystal-controlled tetrode oscillator in which the crystal circuit is isolated from the output circuit by using the screen grid as the oscillator anode. Used for multiband operation because it generates strong harmonics of the crystal frequency.

**tritium** [symbol T or $H^3$] The hydrogen isotope having mass number 3. It is one form of heavy hydrogen, the other being deuterium.

**Triton** A Navy surface-to-surface guided missile using inertial guidance and having an intended range of about 1,500 miles. Development was canceled in September 1957.

**triton** The nucleus of tritium.

**trochoidal mass analyzer** A mass spectrometer in which the ion beams traverse trochoidal paths within mutually perpendicular electric and magnetic fields.

**trochotron** [TROCHOidal magneTRON] *Beam-switching tube.*

**trombone** A U-shaped length of waveguide that is adjustable in length.

**tropicalize** To prepare for use in a tropical climate by applying a coating that resists moisture and fungi.

**tropopause** The discontinuity that separates the stratosphere from the troposphere.

**troposphere** The portion of the earth's atmosphere extending from the surface up to about 6 miles, in which temperature generally decreases with altitude, clouds form, and convection exists.

**tropospheric bending** Refraction of radio waves by adjacent layers of air masses having different temperature and humidity characteristics in the troposphere, making possible long-distance transmission of vhf radio waves.

**tropospheric duct** *Duct.*

**tropospheric scatter** A form of scatter propagation in which radio waves are scattered by the troposphere. The phenomenon is essentially independent of frequency, hence is useful for communication at distances of several hundred miles over the entire r-f spectrum.

**tropospheric superrefraction** A condition in the troposphere whereby radio waves are bent sufficiently to be returned to the earth.

**tropospheric wave** A radio wave that is propagated by reflection from a region of abrupt change in dielectric constant or its gradient in the troposphere.

**trouble-locating problem** A computer test problem whose incorrect solution supplies information as to the location of a fault. Used after a check problem shows that a fault exists.

**troubleshooting** Locating and repairing faults in equipment after they have occurred.

**tr switch** *Tr tube.*

**tr tube** [Transmit-Receive tube] A gas-filled r-f switching tube used to disconnect a receiver from its antenna during the interval for pulse transmission in radar and other pulsed r-f systems. Also called tr box, tr cell (British), and tr switch.

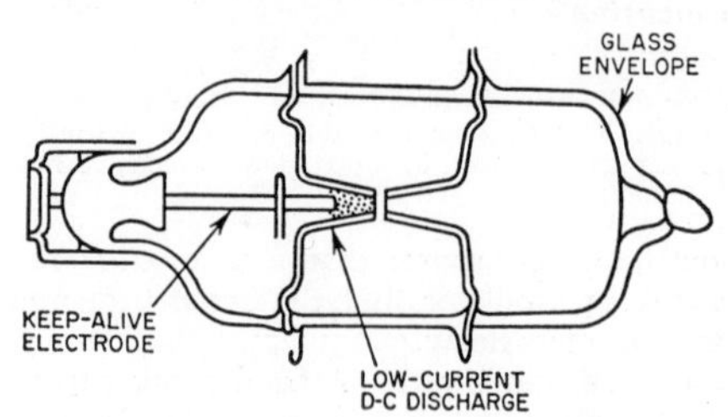


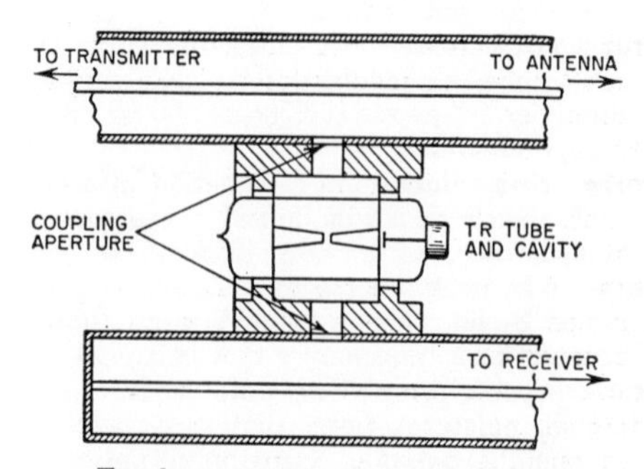


Tr tube construction and method of use.

**true altitude** The altitude above mean sea level.

**true bearing** A bearing given in relation to true geographic north. A magnetic bearing is given in relation to magnetic north, while a relative bearing is given in relation to the lubber line or other axis of an aircraft or vessel.

**true-bearing rate** The rate of change of true bearing.

**true-bearing unit** A unit added to a radar set to rotate the ppi display so true north is always at the top of the screen.

**true coincidence** Coincidence of counts in two or more counters due to the incidence of a single particle or several genetically related particles.

**true course** A course indicated by an angle measured clockwise from true north.

**true heading** A heading measured with respect to true north.

**true north** The direction of the North Pole from the observer, or a line showing this direction.

**truncate** To drop digits at the end of a numerical value. The number 3.14159265 is truncated to five figures in 3.1415, whereas it would be 3.1416 if rounded off to five figures.

**truncated paraboloid** A radar parabolic reflector in which a portion of the top and bottom have been cut away to broaden the radar beam in the vertical plane.

**truncation error** The computation error resulting from use of only a finite number of terms of an infinite series.

**trunk** 1. A path over which information is transferred in a computer. 2. A telephone line connecting two central offices.

**tuba** A high-power radar jamming transmitter operating in the range of 480 to 500 mc, using a tunable resnatron, that served during World War II to jam German nightfighter radar.

**tube** *Electron tube.*

**tube coefficients** The constants that describe the characteristics of an electron tube, such as amplification factor and transconductance.

**tube complement** The number of electron tubes required in a piece of electronic equipment.

**tube count** A terminated discharge produced by an ionizing event in a radiation-counter tube.

**tube heating time** The time required for the temperature of the condensed mercury to reach a specified value in a mercury-vapor tube.

**tube noise** Noise originating in an electron tube, such as that due to shot noise and thermal agitation.

**tube shield** A shield designed to be placed around an electron tube.

**tube socket** A socket designed to accommodate electrically and mechanically the terminals of an electron tube.

**tube tester** A test instrument designed to measure and indicate the condition of electron tubes used in electronic equipment.

**tube voltage drop** The anode voltage existing during the conduction period in an electron tube.

**tube voltmeter** *Vacuum-tube voltmeter.*

**tubular capacitor** A paper or electrolytic capacitor having the form of a cylinder, with leads usually projecting axially from the ends. The capacitor plates are long strips of metal foil separated by insulating strips, rolled into a compact tubular shape.

**tumbling** Loss of control in a two-frame free gyroscope, occurring when both frames of reference become coplanar.

**tunable echo box** An echo box consisting of an adjustable cavity operating in a single mode. It can be calibrated so the setting of the plunger at resonance indicates wavelength.

**tunable magnetron** A magnetron that can be tuned over a range of frequencies by electronic or mechanical means. Generally the capacitance or inductance of the resonant structure is varied mechanically to achieve tuning.

**tune** To adjust for resonance at a desired frequency.

**tuned amplifier** An amplifier in which the load is a tuned circuit. Load impedance and amplifier gain then vary with frequency.

**tuned-anode oscillator** A vacuum-tube oscillator whose frequency is determined by a tank circuit in the anode circuit, coupled to the grid to provide the required feedback.

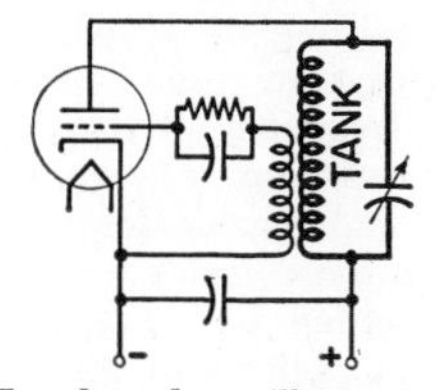

Tuned-anode oscillator circuit.

**tuned antenna** An antenna so designed that its inductance and capacitance values provide resonance at the desired operating frequency.

**tuned-base oscillator** A transistor oscillator in which the frequency-determining resonant circuit is located in the base circuit. This is comparable to a tuned-grid electron-tube oscillator.

**tuned cavity** *Cavity resonator.*

**tuned circuit** A circuit whose components can be adjusted to make the circuit responsive to a particular frequency in a tuning range. Also called tuning circuit.

**tuned-circuit oven** An electrically heated compartment designed to accommodate and maintain tuned-circuit elements at an essentially constant temperature to prevent drifting in frequency with changes in temperature.

**tuned-collector oscillator** A transistor oscillator in which the frequency-determining resonant circuit is located in the collector circuit. This is comparable to a tuned-anode electron-tube oscillator.

**tuned dipole** A dipole that provides resonance at its operating frequency.

**tuned filter** A filter that uses one or more tuned circuits to attenuate or to pass signals at the resonant frequency.

**tuned-grid oscillator** A vacuum-tube oscillator whose frequency is determined by a tank circuit in the grid circuit, coupled to the anode to provide the required feedback.

**tuned-grid tuned-anode oscillator** A vacuum-tube oscillator having parallel-tuned tanks in both anode and grid circuits, with feedback being obtained through the anode-to-grid interelectrode capacitance of the tube.

**tuned radio frequency** [abbreviated trf] Pertaining to a receiver in which all r-f amplification is carried out at the frequency of the transmitted carrier signal.

**tuned radio-frequency amplifier** An r-f amplifier in which all tuned circuits are adjusted to the frequency of the desired transmitted carrier signal.

**tuned radio-frequency receiver** A radio receiver consisting of a number of amplifier stages that are tuned to resonance at the carrier frequency of the desired signal by a gang capacitor. The amplified signals at the original carrier frequency are fed directly into the detector for demodulation, and the resulting a-f signals are amplified by an a-f amplifier and reproduced by a loudspeaker. Now largely replaced by superheterodyne receivers.

**tuned radio-frequency stage** A stage of amplification that is tunable to the carrier frequency of the signal being received.

**tuned-reed frequency meter** *Vibrating-reed frequency meter.*

**tuned relay** A relay having mechanical or other resonating arrangements that limit response to currents at one particular frequency.

**tuned rope** Long lengths of chaff cut to the various lengths required for tuning to a number of different frequencies.

**tuned transformer** A transformer whose associated circuit elements are adjusted to be resonant at the frequency of the alternating current supplied to the primary, thereby causing the secondary voltage to build up to higher values than would otherwise be obtained.

**tuner** The portion of a receiver that contains circuits which can be tuned to accept the carrier frequency of a desired transmitter while rejecting the carrier frequencies of all other stations on the air at that time. A television tuner commonly contains only the r-f amplifier, local oscillator and mixer stages, whereas a radio tuner also contains the i-f amplifier and second-detector stages.

**tungar tube** A gas tube having a heated thoriated tungsten filament serving as cathode and a graphite disk serving as anode in an argon-filled bulb at a low pressure. Used chiefly as a rectifier in battery chargers.

**tungsten** [symbol W] A hard metallic element having a melting point of 3,370°C. Sometimes used for filaments and other electrodes of electron tubes. Atomic number is 74. Usually called wolfram outside the United States.

**tuning** Adjusting circuits for optimum performance at a desired frequency.

**tuning capacitor** A variable capacitor used for tuning purposes.

**tuning circuit** *Tuned circuit.*

**tuning coil** A variable inductance used for tuning purposes.

**tuning control** The control knob that adjusts all tuning circuits simultaneously to a desired frequency.

**tuning core** A ferrite core that is designed to be moved in and out of a coil or transformer to vary the inductance.

**tuning fork** A U-shaped bar of hard steel, fused quartz, or other elastic material that vibrates at a definite natural frequency when struck or when set in motion by electromagnetic means. Used as a frequency standard.

**tuning-fork drive** Use of a tuning fork to control the frequency of an oscillator. A high harmonic of the fork frequency is picked up by a coil and amplified to control the frequency of the main oscillator in a transmitter or other piece of equipment.

**tuning-fork resonator** A tuning fork and associated coils that generate an a-c voltage related to the natural vibrating frequency of the fork.

**tuning indicator** A device that indicates when a radio receiver is tuned accurately to a radio station, such as a meter or a cathode-ray tuning indicator. It is connected to a circuit having a d-c voltage that varies with the strength of the incoming carrier signal.

**tuning meter** A d-c voltmeter or ammeter used as a tuning indicator.

**tuning probe** An essentially lossless probe that can be extended through the wall of a waveguide or cavity resonator to project an adjustable distance inside.

**tuning range** The frequency range over which a receiver or other piece of equipment can be adjusted by means of a tuning control.

**tuning screw** A screw that is inserted into the top or bottom wall of a waveguide and adjusted as to depth of penetration inside for tuning or impedance-matching purposes.

**tuning stub** A short length of transmission line, usually shorted at its free end, that is connected to a transmission line for impedance-matching purposes.

**tuning susceptance** The normalized susceptance of an atr tube in its mount due to the deviation of its resonant frequency from the desired resonant frequency.

**tuning wand** A rod of insulating material having a brass plug at one end and a ferrite core at the other end. Used for checking receiver alignment.

**tunnel diode** A heavily doped junction diode that has negative resistance in the forward direction over a portion of its operating range, due to quantum mechanical tunneling. It can be made from a variety of semiconductor materials, including germanium, silicon, gallium arsenide, and indium antimonide, for use as an oscillator or amplifier operating well up into microwave frequencies. Also called Esaki diode (after L. Esaki of Japan, who proposed the design).

**tunnel effect** The piercing of a rectangular potential barrier in a semiconductor by a particle that does not have sufficient energy to go over the barrier. The wave associated with the particle is almost totally reflected on the first slope of the barrier, but a small fraction passes through the barrier.

**turbidimeter** An instrument for measuring the turbidity of a liquid. In a photoelectric turbidimeter, the amount of light that passes through the liquid is measured. Also called opacimeter.

**turn** One complete loop of wire.

**turnover cartridge** A phonograph pickup having two styli and a pivoted mounting that places in playing position the correct stylus for a particular record. There is usually a stylus with 3-mil radius for 78-rpm records and a stylus with 1-mil radius for 45-rpm and 33⅓-rpm records.

**turnover frequency** *Transition frequency.*

**turns ratio** The ratio of the number of turns in a secondary winding of a transformer to the number of turns in the primary winding.

**turnstile antenna** An antenna consisting of one or more layers of crossed horizontal dipoles on a mast, usually energized so the currents in the two dipoles of a pair are equal and in quadrature. Used with television, f-m, and other vhf or uhf transmitters to obtain an essentially omnidirectional radiation pattern. The superturnstile antenna is a more elaborate version in which the dipole elements are wing-shaped.

**turntable** The rotating platform on which a disk record is placed for recording or playback.

**turntable rumble** Low-frequency vibration that is mechanically transmitted to a recording or reproducing turntable and superimposed on the reproduction. Also called rumble.

**turret tuner** A television tuner having one set of pretuned circuits for each channel, mounted on a drum that is rotated by the channel selector. Rotation of the drum connects each set of tuned circuits in turn to the receiver antenna circuit, r-f amplifier, and r-f oscillator.

**tv** Abbreviation for *television.*

**tvi** Abbreviation for *television interference.*

**tvor** Abbreviation for *terminal vor.*

**tweeter** A loudspeaker designed to handle only the higher audio frequencies, usually those well above 3,000 cps. Generally used in conjunction with a crossover network and a woofer.

**twin check** A continuous check of computer operation, achieved by duplication of equipment and automatic comparison of results.

**twin-lead** *Twin-line.*

**twin-line** A transmission line having two parallel conductors separated by insulating material. Line impedance is determined by the diameter and spacing of the conductors and the insulating material, and is usually 300 ohms for television receiving antennas. Also called balanced transmission line and twin-lead.

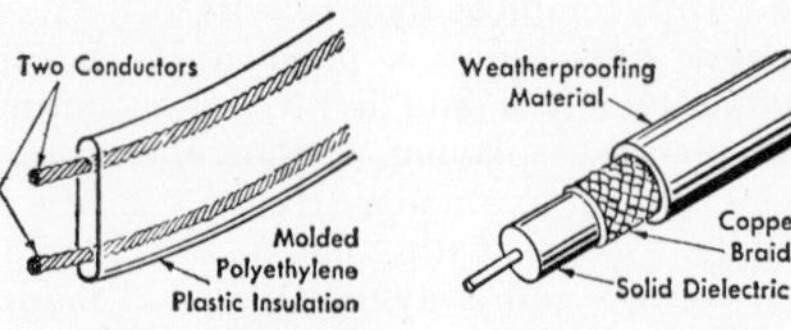

Twin-line (300-ohm impedance) and coaxial cable (72-ohm impedance) used as transmission lines between television receiving antennas and television receivers.

**twinning** A defect occurring in quartz crystals, resulting from structural misgrowth of otherwise perfect crystals. The two forms of twinning are electric twinning and optical twinning.

**twin-triode** Two triode vacuum tubes in a single envelope.

**twist** A waveguide section in which there is a progressive rotation of the cross-section about the longitudinal axis of the waveguide.

**twisted pair** A cable composed of two small insulated conductors twisted together without a common covering.

**twister** A piezoelectric crystal that generates a voltage when twisted.

**twistor** A computer memory element consisting of a helix of magnetic wire wound under tension at an angle of 45° on a short piece of nonmagnetic wire, with a fine-wire solenoid wound over the helix. Signal currents through the straight wire and the solenoid combine to produce a magnetic

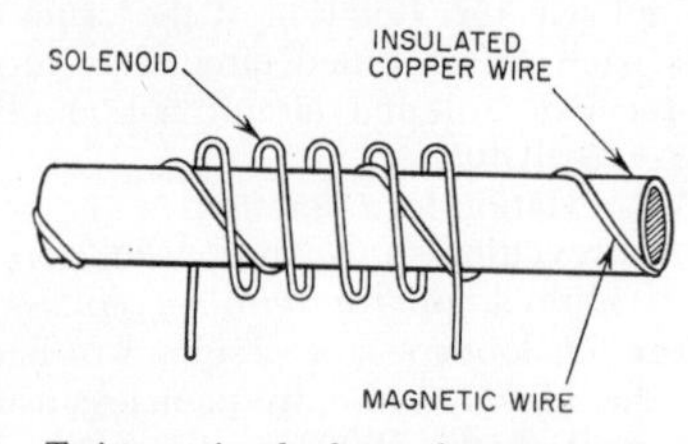

Twistor using barber-pole construction.

flux parallel to the helical wire. The direction of this magnetic flux is easily reversed by reversal of coil current. Readout is obtained by sending through a line of twistors a current strong enough to switch over all of them, and noting which ones deliver readout signals.

**two-address programing** Digital-computer programing in which each complete instruction includes an operation and specifies the location of two registers, usually one containing an operand and the other containing the result of the operation.

**two-course radio range** A radio range that beams on-course signals in opposite directions.

**two-frequency duplex** Carrier communication that provides simultaneous communication in both directions between two stations by using different carrier frequencies in the two directions.

**two-hole directional coupler** A directional coupler consisting of two parallel coaxial lines in contact, with holes or slots through their contacting walls at two points a quarter-wave apart. These permit extraction of a portion of the r-f energy traveling in one direction through the main line while rejecting energy traveling in the opposite direction. One end of the secondary line must be terminated in its characteristic impedance.

**two-level maser** A solid-state maser in which all of the molecules are excited simultaneously, permitting only pulsed operation. Each useful interval of oscillation or amplification must be followed by an interval of excitation. A paramagnetic solid material such as cerium ethyl sulfate is used. Also called pulsed maser.

**two-phase** Having a phase difference of one quarter-cycle or 90°. Also called quarter-phase.

**two-port** A transducer having an in-port and an out-port.

**two-position action** Automatic control action in which a final control element is moved from one of two fixed positions to the other, with no intermediate positions.

**two-position differential gap action** Automatic control action in which a final control element is moved from one of the two fixed positions to the other at a predetermined value of a controlled variable, and can go back to the first position only after the controlled variable has reached a second predetermined value.

**two-position single-point action** Automatic control action in which a final control element is moved from one of two fixed positions to the other at a single value of a controlled variable.

**two-receiver radiometer** A radiometer-type receiver using two independent receiver channels, the i-f outputs of which are multiplied and the product smoothed in a low-pass filter. A high signal-to-noise ratio is obtained at the output, as required for detection of extremely weak signals.

**two-source frequency keying** Keying in which the modulating wave switches the output frequency between predetermined values corresponding to the frequencies of independent sources.

**two-terminal network** A network that is connected to an external system by only two terminals.

**two-terminal pair network** A network with four accessible terminals grouped in pairs. One terminal of each pair may coincide with a network node. Also called four-pole network and quadripole.

**two-tone keying** Keying in which the modulating wave causes the carrier to be modulated with one frequency for the marking condition and with a different frequency for the spacing condition.

**two-value capacitor motor** A capacitor motor using different values of effective capacitance for starting and running.

**two-way communication** Communication between radio stations having both transmitting and receiving equipment.

**two-way correction** Register control that produces a correction in register in either direction.

**two-way repeater** A repeater used to amplify signals coming from either direction.

**two-wire repeater** A telephone repeater that provides for transmission in both directions over a two-wire telephone circuit.

**twt** Abbreviation for *traveling-wave tube.*

**twx** A message sent or received over a teletypewriter circuit.

**twx machine** *Teletypewriter.*

**Typotron** Trademark of Hughes Aircraft Co. for their character-writing storage tube.

# U

**uf** Alternate abbreviation for *microfarad.*

**ufo** Abbreviation for *unidentified flying object.*

**uhf** Abbreviation for *ultrahigh frequency.*

**ultor** The electrode or set of electrodes to which the highest d-c voltage is applied in a picture tube, to accelerate the electron beam before deflection. Also called second anode.

**ultra-audion oscillator** A Colpitts oscillator in which the two voltage-dividing capacitances of the tank circuit are the anode-to-cathode and grid-to-cathode capacitances of the tube.

**Ultrafax** Trademark of RCA for a system that combines radio, television, facsimile, and film recording techniques for transmitting printed information at high speed.

**ultrahigh frequency** [abbreviated uhf] A Federal Communications Commission designation for the band from 300 to 3,000 mc in the radio spectrum.

**ultrahigh-frequency converter** An electronic circuit that converts uhf signals to a lower frequency to permit reception on a vhf receiver. Used to convert uhf television signals to vhf signals for reception on vhf television receivers.

**ultrahigh-frequency generator** An oscillator capable of generating frequencies in the uhf range. The main types include conventional vacuum-tube oscillators, Barkhausen-Kurz oscillators, magnetron oscillators, and velocity-modulated oscillators such as the klystron.

**ultramicrometer** An instrument for measuring very small displacements by electrical means, such as by the variation in capacitance resulting from the movement being measured.

**ultrashort waves** Radio waves shorter than about 10 meters in wavelength (above 30 mc in frequency). Waves shorter than 30 cm (above 1,000 mc) are usually called microwaves.

**ultrasonic** Pertaining to signals, equipment, or phenomena involving frequencies just above the range of human hearing, hence above about 20,000 cps. Also called supersonic (deprecated).

**ultrasonic coagulation** The bonding of small particles into large aggregates by the action of ultrasonic waves.

**ultrasonic communication** Communication through water by keying the sound output of echo-ranging sonar on ships or submarines.

**ultrasonic cross grating** An ultrasonic space grating produced by crossing beams of ultrasonic waves. Also called multiple grating.

**ultrasonic delay line** A delay line in which use is made of the propagation time of sound through a medium such as fused quartz, barium titanate, or mercury to obtain a time delay of a signal. Also called ultrasonic storage cell.

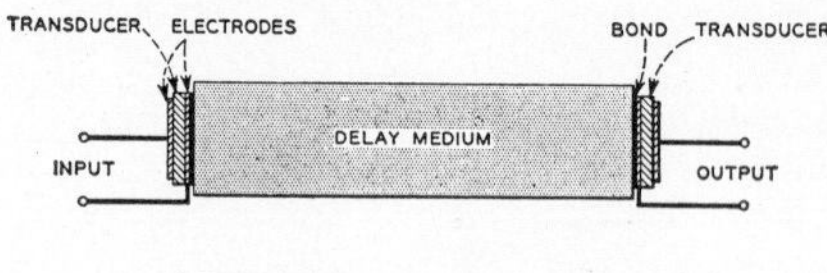


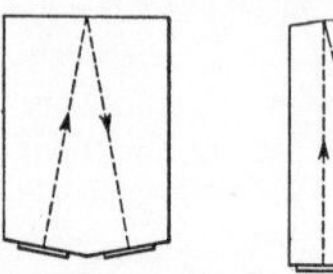

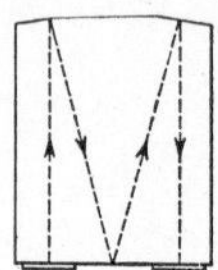

Ultrasonic delay line, and methods of placing transducers to increase path length and delay time by reflecting signal beam.

**ultrasonic detector** A mechanical, electrical, thermal, or optical device for detecting and measuring ultrasonic waves.

**ultrasonic equipment** Equipment that generates a-c energy at frequencies above about 20 kc and uses that energy to excite or drive an electromechanical transducer for the production and transmission of ultrasonic energy for industrial, scientific, medical, or other purposes.

**ultrasonic flaw detector** An ultrasonic generator and detector used together much as in radar, to determine the distance to a wave-reflecting internal crack or other flaw in a solid object.

**ultrasonic frequency** A frequency lying above the audio-frequency range. The term is commonly applied to elastic waves propagated in gases, liquids, or solids.

**ultrasonic generator** A generator consisting of an oscillator driving an electroacoustic transducer, used to produce acoustic waves above about 20 kc.

**ultrasonic grating constant** The distance between diffracting centers of an acoustic wave used to produce optical diffraction spectra.

**ultrasonic light diffraction** The formation of optical diffraction spectra when a beam of light is passed through a longitudinal ultrasonic wave field. The diffraction results from the periodic variation of the light refraction in the ultrasonic field.

**ultrasonic light valve** A light valve consisting essentially of a quartz crystal placed in the center of a container filled with a transparent liquid such as carbon tetrachloride. The crystal is excited at an ultrasonic frequency of about 10 mc to set up compressive waves that make the liquid act as a diffraction grating. Light projected through the cell produces a series of fine illuminated lines on a screen. When the crystal is modulated also with video frequencies, these lines are broken up into light and dark picture elements that can serve as a reproduced television image. In this system, all picture elements are projected simultaneously.

**ultrasonic material dispersion** The production of suspensions or emulsions of one material in another due to the action of high-intensity ultrasonic waves.

**ultrasonics** The branch of acoustics that deals with vibrations and acoustic waves above about 20 kc.

**ultrasonic sounding** Determining ocean depth with a fathometer by measuring the time interval between the transmission of an ultrasonic wave downward from the surface of the water and the arrival of the echo reflected from the bottom of the ocean.

**ultrasonic space grating** A periodic spatial variation of the index of refraction of a medium due to the presence of acoustic waves.

**ultrasonic storage cell** *Ultrasonic delay line.*

**ultrasonic stroboscope** A light interrupter whose action is based on the modulation of a light beam by an ultrasonic field.

**ultrasonic therapy** Use of ultrasonic vibrations for therapeutic purposes.

**ultrasonic trainer** A radar trainer in which ultrasonic waves are directed against a molded relief map under water, to simulate responses occurring in actual radar operation during flight over the terrain represented by the map.

**ultrasonic waves** Elastic waves having a frequency above about 20 kc.

**ultraviolet lamp** A lamp providing a high proportion of ultraviolet radiation, such as various forms of mercury-vapor lamps.

**ultraviolet radiation** The portion of the electromagnetic spectrum between visible violet light (about 0.38 micron or 3,800 angstroms) and the x-ray region at about 0.02 micron or 200 angstroms. Also called black light (deprecated).

**umbilical connector** The quick-disconnect connector that terminates an umbilical cable and mates with the umbilical connector receptacle.

**umbilical cord** A cable capable of being disconnected quickly, used to test a missile up to the instant of launching, and sometimes also to feed in last-minute target data. The cord may be severed by such means as a trigger-action spring, explosive charge, or guillotine blade.

**umbra** The region of total shadow behind an object in a beam of radiation. A straight line drawn from any point in this region to any point in the source passes through the object.

**umbrella antenna** An antenna in which the radiating wires run downward at an angle in some or all directions from a central tower or from a wire running between two towers, somewhat like the ribs of an open umbrella.

**unbalanced circuit** A circuit whose two sides are electrically unlike.

**unbalanced line** A transmission line in which the voltages on the two conductors are not equal with respect to ground. A coaxial line is an example.

**unbalanced output** An output in which one of the two input terminals is substantially at ground potential.

**uncage** To disconnect the erection circuit of a displacement gyroscope system.

**uncertainty** The estimated amount by which an observed or calculated value may depart from the true value.

**uncharged** Having a normal number of electrons, and hence having no electric charge.

**unclassified** Not having a security classification.

**unconditional jump** A digital-computer instruction that interrupts the normal process of obtaining instructions in an ordered sequence, and specifies the address from which the next instruction must be taken. Also called unconditional transfer.

**unconditional transfer** *Unconditional jump.*

**undamped oscillation** *Steady-state oscillation.*

**undamped wave** A continuous wave produced by oscillations having constant amplitude.

**underbunching** A condition representing less than optimum bunching in a velocity-modulation tube.

**undercurrent relay** A relay designed to operate when its coil current falls below a predetermined value.

**underdamping** *Periodic damping.*

**underlap** The amount by which the effective height of a facsimile scanning spot falls short of the nominal width of the scanning line.

**undermodulation** Insufficient modulation at a transmitter, due to electrical limitations or improper adjustment of the modulator.

**underthrow distortion** Distortion occurring in facsimile when the maximum signal amplitude is too low.

**undervoltage alarm** An alarm that gives an aural and/or visual indication when the voltage in an equipment falls below a specified value.

**undervoltage relay** A relay designed to operate when its coil voltage falls below a predetermined value.

**underwater ambient noise** The portion of the underwater background noise that is due to disturbances in the water other than the signal and manmade noise.

**underwater antenna** An antenna placed and used under water.

**underwater background noise** Underwater sound other than the desired signal, including underwater ambient noise and that inherent in the hydrophone and its associated equipment.

**underwater mine** A mine designed to be located under water and exploded by means of propeller vibration, magnetic attraction, contact, or remote control.

**underwater-mine coil** A coil and associated equipment used to detect changes in the magnetic field at an underwater mine due to a passing ship.

**underwater-mine depth compensator** A hydrostatically actuated device that increases the sensitivity of the firing mechanism when an underwater mine exceeds a predetermined depth.

**underwater signal** An underwater disturbance that is to be detected, such as sonar transmissions from a projector, machinery noise, and propeller noise.

**underwater sound communication set** The components and items required to provide underwater communication by utilizing sonic or ultrasonic waves in water.

**underwater sound projector** A transducer used to produce sound waves in water. Also called projector.

**undisturbed-one output** A one output of a magnetic cell to which no partial-read pulses have been applied since that cell was last selected for writing.

**undisturbed-zero output** A zero output of a magnetic cell to which no partial-write pulses have been applied since that cell was last selected for reading.

**unfired tube** The condition of a tr, atr, or pre-tr tube during which there is no r-f glow discharge at either the resonant gap or resonant window.

**ungrounded** Without an intentional connection to ground except through voltmeters or other high-impedance devices.

**unguided** Not subject to guidance or control during flight.

**uniconductor waveguide** A waveguide consisting of a cylindrical or rectangular metallic surface surrounding a homogeneous dielectric medium.

**unidentified flying object** [abbreviated ufo] A reported flying object that cannot be identified at the time of the sighting.

**unidirectional** 1. Flowing in only one direction, such as direct current. 2. Radiating in only one direction.

**unidirectional antenna** An antenna that has a single well-defined direction of maximum gain.

**unidirectional coupler** A directional coupler that samples only one direction of transmission.

**unidirectional current** A current that flows in the same direction at all times.

**unidirectional hydrophone** A unidirectional microphone that responds to waterborne sound waves.

**unidirectional microphone** A microphone that is responsive predominantly to sound incident from one hemisphere, without picking up sounds from the sides or rear.

**unidirectional pulse** A pulse in which pertinent departures from the normally constant value occur in one direction only. Also called single-polarity pulse (deprecated).

**unidirectional pulse train** A pulse train in which all pulses rise in the same direction. Also called single-polarity pulse train (deprecated). A unidirectional pulse train may contain bidirectional pulses.

**unidirectional voltage** A voltage whose polarity, but not necessarily magnitude, is constant.

**unifilar** Having or using only one fiber, wire, or thread.

**uniform plane wave** A plane wave in which the electric and magnetic field vectors have constant amplitude over the equiphase surfaces. Such a wave can only be found in free space at an infinite distance from the source.

**uniform waveguide** A waveguide in which the physical and electrical characteristics do not change with distance along the axis of the guide.

**unijunction transistor** An n-type bar of semiconductor with a p-type alloy region on one side. Connections are made to base contacts at either end of the bar and to the p region. The transistor has a thyratron-like characteristic between the terminal of

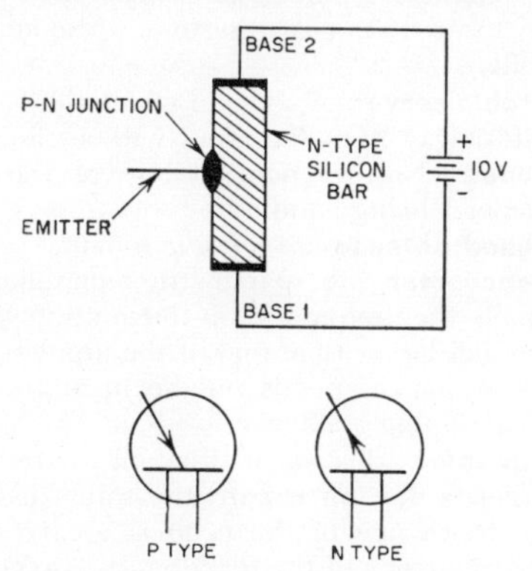


Unijunction transistor construction for n-type unit, and symbols for both types.

the p region and the negative base terminal. It can be triggered on by applying a voltage between the base terminals or by changing the bias voltage on the p region. It is turned off by resetting the bias on the p region. Also called double-base diode, double-base junction diode, and filamentary transistor.

**unilateral-area track** A sound track in which only one edge of the opaque area is modulated in accordance with the recorded signal. A second edge may be modulated by a noise-reduction device.

**unilateral bearing** A bearing obtained with a radio direction finder having unilateral response, eliminating the chance of a 180° error.

**unilateral conductivity** Conductivity in only one direction, as in a perfect rectifier.

**unilateral device** A device that transmits energy in one direction only.

**unilateralization** Use of an external feedback circuit in a high-frequency transistor amplifier to prevent undesired oscillation by canceling both the resistive and reactive changes produced in the input circuit by internal voltage feedback. With neutralization, only the reactive changes are canceled.

**unilateral transducer** A transducer in which the waves at its outputs cannot affect its inputs.

**unimpeded harmonic operation** Operation of a magnetic amplifier in such a way that the impedance of the control circuit is substantially zero, permitting essentially unrestricted flow of all harmonic currents in the control circuit.

**unipolar machine** *Homopolar generator.*

**unipolar transistor** A transistor that utilizes charge carriers of only one polarity. The input signal is used to modulate an electric field that in turn varies the effective cross-sectional area of a tiny bar of

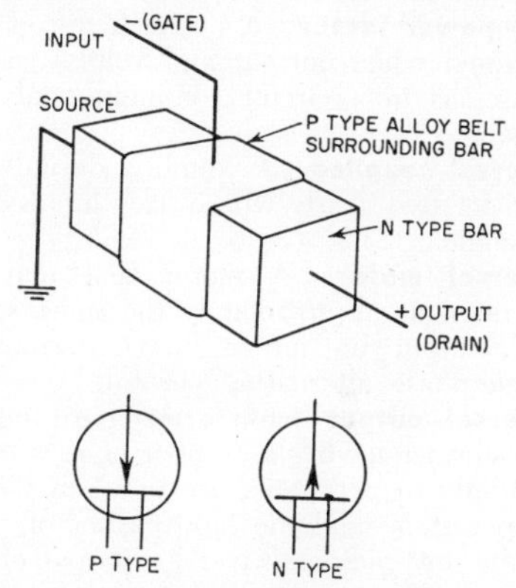


Unipolar transistor construction for n-type unit, and symbols for both types.

semiconductor material, thereby varying the resistance that controls the output current. Also called field-effect transistor, fieldistor, and transtrictor.

**unipole** A hypothetical antenna that radiates or receives signals equally well in all directions. Also called isotropic antenna.

**unipotential cathode** *Indirectly heated cathode.*

**unit** 1. An assembly or device capable of independent operation, such as a radio receiver, cathode-ray oscilloscope, or computer subassembly that performs some inclusive operation or function. 2. A quantity adopted as a standard of measurement.

**unit area acoustic impedance** *Specific acoustic impedance.*

**unit charge** The electric charge that will exert a repelling force of 1 dyne on an equal and like charge 1 centimeter away in a vacuum, assuming that each charge is concentrated at a point.

**unitized construction** Construction of equipment in subassemblies that can be manufactured and tested separately and are readily replaceable as individual units within the equipment of which they are a part.

**unit magnetic pole** A magnetic pole that will repel an equal magnetic pole of the same sign with a force of 1 dyne if the two poles are placed 1 centimeter apart in a vacuum.

**unit operator** A symbolic operator that leaves every other operator unchanged.

**unity coupling** Perfect magnetic coupling between two coils, so all the magnetic flux produced by the primary winding passes through the entire secondary winding.

**unity power factor** A power factor of 1.0, obtained when current and voltage are in phase, as in a circuit containing only resistance.

**universal coupling** A coupling designed to connect two shafts whose axes intersect at an angle.

**universal motor** A motor that may be operated at approximately the same speed and output on either direct current or single-phase alternating current.

**universal output transformer** An output transformer having a number of taps on its winding, to permit its use between the a-f output stage and the loudspeaker of practically any radio receiver by proper choice of connections.

**universal receiver** *A-c/d-c receiver.*

**universal shunt** *Ayrton shunt.*

**unloaded antenna** An antenna having no added inductance or capacitance.

**unloaded applicator impedance** The complex impedance measured at the point of application of dielectric heating, without the load material in position, at a specified frequency.

**unloaded Q** The $Q$ of a switching tube, unloaded by either the generator or the termination. Also called intrinsic $Q$.

**unloading machine** A machine for removing fuel from a nuclear reactor.

**unmodified scatter** Radiation that is scattered without a change in photon energy.

**unmodulated** Having no modulation, as during moments of silence in a radio program or a disk recording.

**unmodulated groove** A groove made in a mechanical recording medium when no signal is applied to the cutter. Also called blank groove.

**unpack** To separate packed computer information into a sequence of separate words or elements.

**unpolarized light** A beam of light in which the planes of vibration of the photons are oriented at random about the axis of the beam.

**unquenched spark gap** A spark gap having no special means for deionization.

**unstabilized antenna** An antenna mounted directly on the structure of a vessel or aircraft, with no means for offsetting roll and pitch.

**unstable** Capable of undergoing spontaneous change, as in a radioactive nuclide or an excited nuclear system. Also called labile.

**unstable servo** A servo in which the output drifts away from the input without limit.

**untuned** Not resonant at any of the frequencies being handled.

**untuned antenna** *Aperiodic antenna.*

**up-converter** A parametric amplifier in which the frequency of the output signal is much larger than that of the input signal. Power gain depends on the magnitude of the pumping voltage.

**up-doppler** The sonar situation wherein the target is moving toward the transducer, so the frequency of the echo is greater than the frequency of the reverberations received immediately after the end of the outgoing ping. Opposite of down-doppler.

**upper sideband** The higher of two frequencies or groups of frequencies produced by a modulation process.

**uranides** 1. Elements in the last row of the periodic system, in oxidation state +6. The compounds of the elements beyond uranium in this state are generally isostructural with the corresponding ones of uranium. 2. Elements beyond actinium in the periodic system, having a transition group in which uranium is a prominent member and having the same pattern of oxidation states.

**uranium** [symbol U] A radioactive element. Atomic number is 92. Naturally occurring uranium is a mixture of 99.28% $U^{238}$, 0.71% $U^{235}$, and 0.00580% $U^{234}$. The nucleus of $U^{235}$ is capable of absorbing a neutron and thereupon undergoing fission into two highly radioactive fragments that fly apart with great energy, releasing neutrons. The fact that fission is induced by one neutron but releases more than one makes possible a chain reaction.

**uranium age** The age of a mineral as calculated from the numbers of ionium atoms present originally, now, and when equilibrium is established with uranium.

**uranium-radium series** *Uranium series.*

**uranium reactor** A nuclear reactor in which the principal fuel is uranium. The uranium may be natural, with the naturally occurring ratio of one atom of $U^{235}$ to about 139 atoms of $U^{238}$, or may be enriched to have a higher proportion of fissile $U^{233}$ or $U^{235}$ uranium atoms.

**uranium series** The series of nuclides resulting from the decay of $U^{238}$, including uranium I, II, $X_1$, $X_2$, Y, and Z, and radium A, B, C, C′, C″, D, E, E″, F, and G. Also called uranium-radium series.

**urea plastic** A thermosetting plastic having good dielectric qualities and good strength. Used for radio receiver cabinets and instrument housings.

**URSI** Abbreviation for Union Radio Scientifique Internationale, the French title for International Radio Scientific Union.

**USAF** Abbreviation for United States Air Force.

**useful beam** The part of the primary radiation that passes through the aperture, cone, or other collimator used in radiology.

**useful line** *Available line.*

**uuf** Alternate abbreviation for *micromicrofarad.*

# V

**V** Symbol used on diagrams to designate a voltmeter or an electron tube.

**v** Abbreviation for *volt.*

**V-1** [abbreviation of Vergeltungswaffe eins, meaning revenge weapon one] A German missile first launched across the English Channel on June 13, 1944. Range was about 150 miles and speed about 360 miles per hour at an altitude of about 2,500 feet.

**V-2** [abbreviation of Vergeltungswaffe zwei, meaning revenge weapon two] A German ballistic missile first launched against England September 8, 1944. Range was about 200 miles. It was launched vertically and reached a speed of 3,600 miles per hour, then tilted in the direction of its target and plunged earthward without power.

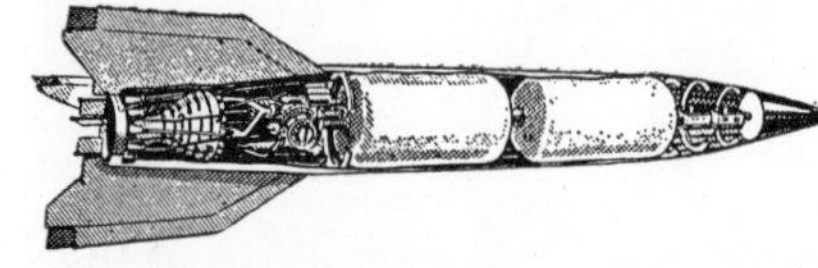

V-2 ballistic missile construction. Fuel tanks are in center, and warhead is in pointed nose, ahead of gyroscope control system.

**va** Abbreviation for *volt-ampere.*

**vacuum** An enclosed space from which practically all air has been removed.

**vacuum capacitor** A capacitor having separated metal plates or cylinders mounted in an evacuated glass envelope to obtain a high breakdown voltage rating.

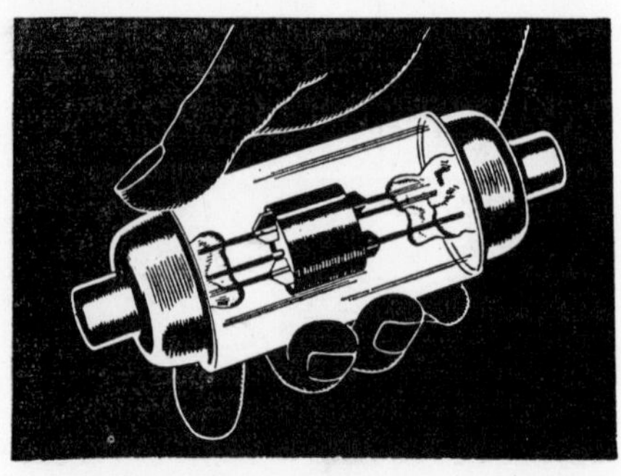

Vacuum capacitor used in radio transmitters.

**vacuum deposition** *Vacuum metallizing.*

**vacuum forepump** A vacuum pump capable of lowering the pressure down to about 0.001 millimeter of mercury, which is low enough for operation of a diffusion pump.

**vacuum gage** A device that indicates the absolute gas pressure in a vacuum system, in millimeters of mercury or microns of mercury. One micron is the pressure that will support a column of mercury one-thousandth of a millimeter high.

**vacuum-impregnated** Impregnated with an insulating compound while in a vacuum, to insure penetration of the compound into the layers of a capacitor or between the turns of a coil.

**vacuum-leak detector** An instrument used to detect and locate leaks in a high-vacuum system. A mass spectrometer is frequently used for this purpose.

**vacuum metallizing** Deposition of a metal coating on a plastic or other object by evaporating the metal in a vacuum chamber containing the object. Also called vacuum deposition.

**vacuum phototube** A phototube that is evacuated to such a degree that its electrical characteristics are essentially unaffected by gaseous ionization. In a gas phototube, some gas is intentionally introduced.

**vacuum relay** A sensitive relay having its contacts mounted in a highly evacuated glass housing, to permit handling r-f voltages as high as 20,000 volts without flashover between contacts even though contact spacing is but a few hundredths of an inch when open.

**vacuum seal** An airtight junction.

**vacuum spectrograph** A spectrograph in which the optical path is in a vacuum and a reflection grating is usually used in place of a dispersive prism. Used for measurements in the extreme infrared and ultraviolet ranges, where lenses and air would absorb the radiation.

**vacuum switch** A switch having its contacts in an evacuated envelope to minimize sparking.

**vacuum tank** The airtight metal chamber that contains the electrodes of a mercury-arc rectifier or similar tube and in which the rectifying action takes place.

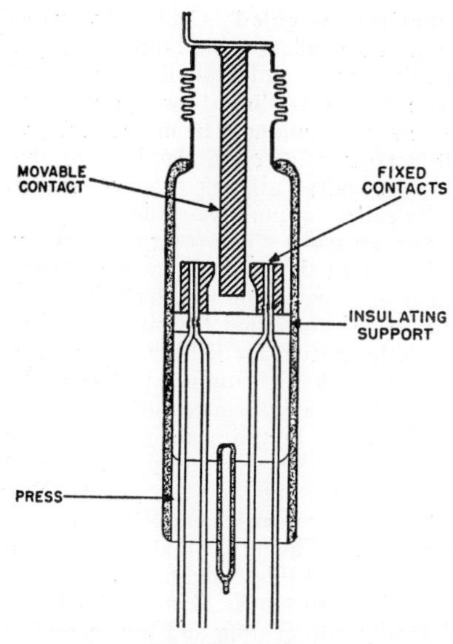


Vacuum switch.

**vacuumtight** *Hermetic.*

**vacuum tube** An electron tube evacuated to such a degree that its electrical characteristics are essentially unaffected by the presence of residual gas or vapor.

**vacuum-tube amplifier** An amplifier employing one or more vacuum tubes to control the power obtained from a local source.

**vacuum-tube electrometer** An electrometer in which the ionization current in an ionization chamber is amplified by a special vacuum triode having an input resistance above 10,000 megohms.

**vacuum-tube keying** A code-transmitter keying system in which a vacuum tube is connected in series with the anode supply lead of the final stage. The grid of the keying tube is connected to its filament through the transmitting key. When the key is open, the tube blocks, interrupting the anode supply to the output stage. Closing the key allows anode current to flow through the keying tube and the output tubes.

**vacuum-tube modulator** A modulator employing a vacuum tube as a modulating element for impressing an intelligence signal on a carrier.

**vacuum-tube oscillator** A circuit utilizing a vacuum tube to convert d-c power into a-c power at a desired frequency.

**vacuum-tube rectifier** A rectifier in which rectification is accomplished by the unidirectional passage of electrons from a heated electrode to one or more other electrodes within an evacuated space.

**vacuum-tube transmitter** A transmitter in which electron tubes are utilized to convert the applied electric power into r-f power.

**vacuum-tube voltmeter** [abbreviated vtvm] An electronic voltmeter that uses vacuum tubes along with or without semiconductor devices. Also called tube voltmeter.

**valence** A number representing the proportion in which an atom is able to combine directly with other atoms. It generally depends on the number and arrangement of electrons in the outermost shell of each type of atom.

**valence band** An energy band, occurring in the spectrum of a solid crystal, in which lie the energies of the valence electrons that bind the crystal together.

**valence bond** The bond formed between the electrons of two or more atoms.

**valence electron** *Conduction electron.*

**valence shell** The electrons that form the outermost shell of an atom.

**value** The magnitude of a quantity.

**valve** British term for *electron tube.*

**vanadium** [symbol V] A metallic element. Atomic number is 23.

**Van Allen layer** *Radiation belt.*

**Van de Graaff generator** An electrostatic generator widely used as an accelerator. An endless moving belt of insulating material collects electric charges by induc-

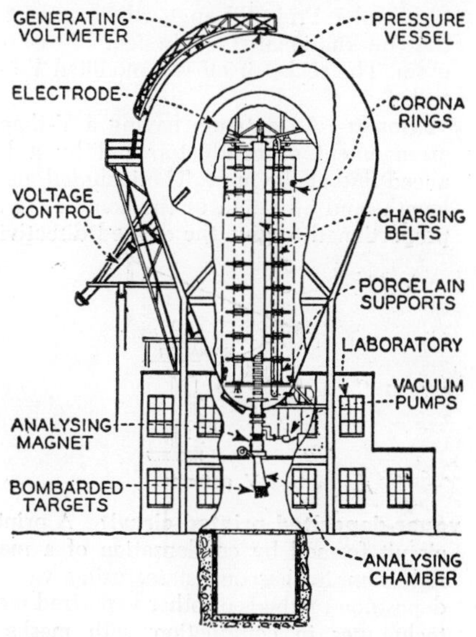


Van de Graaff generator 47 feet high, capable of generating up to 3,000,000 volts d-c.

tion and discharges them inside a large hollow spherical electrode to produce voltages as high as 9,000,000 volts. Used for accelerating electrons, protons, and other nuclear particles. Also called statitron.

**vane attenuator** *Flap attenuator.*

**vane-type anode** A magnetron anode resembling that used in rising-sun magnetrons except that all vanes are the same size and shape.

**vane-type instrument** A measuring instrument utilizing the force of repulsion between fixed and movable magnetized iron vanes, or the force existing between a coil and a pivoted vane-shaped piece of soft iron, to move the indicating pointer.

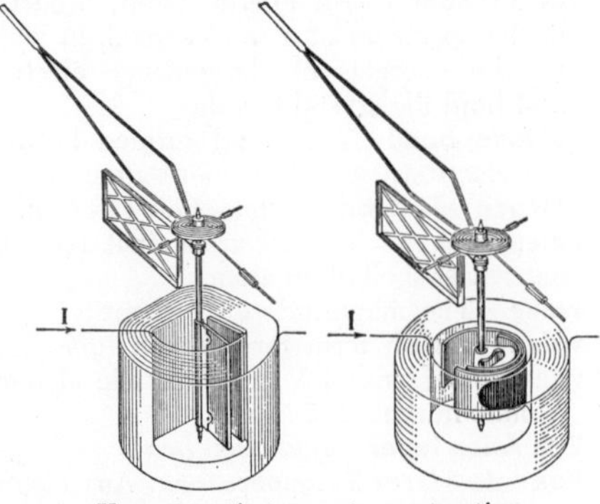

Vane-type instrument construction

**Vanguard** A satellite-carrying rocket developed by U. S. Naval Research Laboratory under Project Vanguard, as a part of the International Geophysical Year program. The rocket itself is a modified Viking rocket.

**V antenna** An antenna having a V-shaped arrangement of conductors fed by a balanced line at the apex. The included angle, length, and elevation of the conductors are proportioned to give the desired directivity.

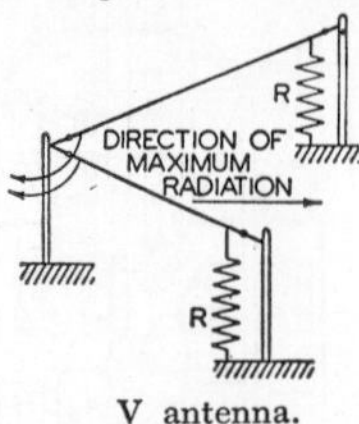

V antenna.

**vapor-deposited printed circuit** A printed circuit formed by condensation of a material from its gaseous state, using vacuum deposition methods or other vaporized metal techniques in conjunction with masks or other pattern-forming means.

**vaporization-cooled** Cooled by vaporization of a nonflammable liquid having a low boiling point and high dielectric strength. The liquid is flowed or sprayed on hot electronic equipment in an enclosure, where it vaporizes, carrying the heat to the enclosure walls, radiators, or a heat exchanger. Also called evaporative-cooled.

**vapor pressure** The pressure of the vapor of a liquid that is kept in confinement, as in a mercury-vapor rectifier tube.

**var** 1. Abbreviation for *visual-aural range.* 2. Abbreviation for *volt-ampere reactive.*

**varactor** A semiconductor device characterized by a voltage-sensitive capacitance that resides in the space-charge region at the surface of a semiconductor bounded by an insulating layer. A varactor may be used for parametric amplification.

**variable** A quantity that may assume a number of distinct values.

**variable-area track** A sound track divided laterally into opaque and transparent areas. A sharp line of demarcation between these areas corresponds to the waveform of the recorded signal.

**variable attenuator** An attenuator for reducing the strength of an a-c signal either continuously or in steps, without causing appreciable signal distortion, by maintaining a substantially constant impedance match.

**variable-capacitance pickup** A phonograph pickup in which the stylus produces a variation in capacitance that is used to frequency-modulate an oscillator. The output signal of the oscillator is converted to an audio voltage by a detector. Also called f-m pickup and frequency-modulated pickup.

**variable capacitor** A capacitor whose capacitance can be varied by moving one set of metal plates with respect to another.

**variable-carrier modulation** *Controlled-carrier modulation.*

**variable coupling** Inductive coupling that can be varied by moving one coil with respect to another.

**variable-cycle operation** A computer operation in which the cycles of action may be of different lengths, as in an asynchronous computer.

**variable-density sound track** A constant-width sound track in which the average light transmission varies along the longitudinal axis in proportion to some characteristic of the applied signal.

**variable-depth sonar** [abbreviated vds] Sonar in which the projector and receiving transducer are mounted in a watertight pod that can be lowered below a vessel to an optimum depth for minimizing thermal effects when detecting underwater targets.

**variable-focal-length lens** A television camera lens system whose focal length can be changed continuously during use while maintaining sharp focusing and a constant aperture, to give the effect of gradually moving the camera toward or away from the subject. The Zoomar lens is an example.

**variable-frequency oscillator** [abbreviated vfo] An oscillator whose frequency can be varied over a given range.

**variable inductance** A coil whose inductance value can be varied. Also called inductometer.

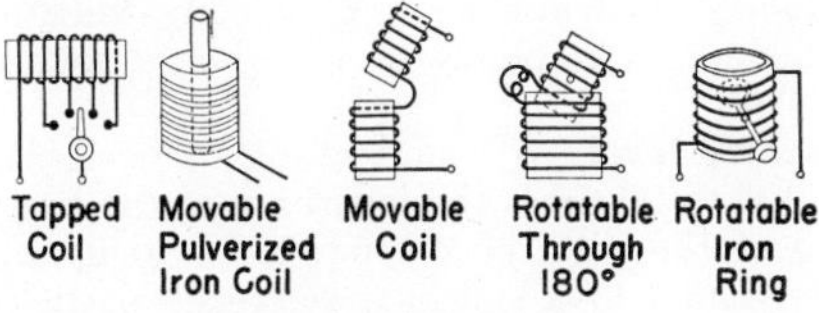

Variable inductances.

**variable-inductance pickup** A phonograph pickup that depends for its operation on the variation of its inductance.

**variable-iris waveguide coupler** A microwave component used to couple a waveguide to the external input or output cavity of a klystron. It permits simple matching adjustments over a wide tuning range without use of stub tuners or matching sections.

**variable-mu tube** An electron tube in which the amplification factor varies in a predetermined manner with control-grid voltage. This characteristic is achieved by making the spacing of the grid wires vary

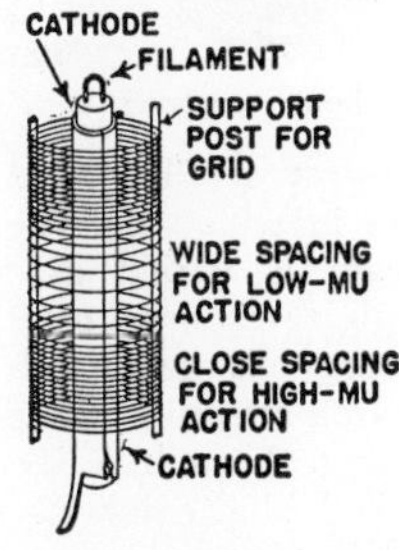

Variable-mu tube construction.

regularly along the length of the grid, in such a manner that a very large negative grid bias is required to block anode current completely. Also called remote-cutoff tube and supercontrol tube.

**variable-reluctance microphone** A microphone that depends for its operation on variations in the reluctance of a magnetic circuit. Also called magnetic microphone and reluctance microphone.

**variable-reluctance pickup** A phonograph pickup that depends for its operation on variations in the reluctance of a magnetic circuit due to the movements of an iron stylus assembly that is a part of the magnetic circuit. The reluctance variations alternately increase and decrease the flux through two series-connected coils, inducing in them the desired a-f output voltage. Also called magnetic cartridge, magnetic pickup, and reluctance pickup.

**variable-resistance pickup** A phonograph pickup that depends for its operation on the variation of a resistance.

**variable resistor** *Rheostat.*

**variable-time fuze** [abbreviated vt fuze] *Proximity fuze.*

**variable transformer** An iron-core transformer having provisions for varying its output voltage over a limited range or continuously from zero to maximum output voltage, generally by means of a contact arm moving along exposed turns of the secondary winding. It may be an autotransformer. Also called adjustable transformer and continuously adjustable transformer.

**Variac** Trademark of General Radio Co. for their line of variable transformers.

**variocoupler** An r-f transformer having provisions for varying the coupling between the two windings. Its construction is like that of a variometer, but its coils are not connected together.

**variolosser** A variable-loss circuit used to improve the signal-to-noise ratio of a communication channel. At the transmitting end the variolosser is connected so its loss increases as input signal strength increases. At the receiving terminal the variolosser is in a circuit that restores the original dynamic range of the signal.

**variometer** A variable inductance having two coils in series, one mounted inside the other, with provisions for rotating the inner coil in order to vary the total inductance of the unit over a wide range.

**varistor** A two-electrode semiconductor device having a voltage-dependent nonlinear resistance. Its resistance drops as the applied voltage is increased. The resistance also varies in a predictable manner with temperature.

**Varley loop test** A method of using a

Wheatstone bridge to determine the distance from the test point to a fault in a telephone or telegraph line or cable.

**varmeter** An instrument for measuring reactive power in vars. Also called reactive volt-ampere meter.

**varnished cambric** Linen or cotton fabric that has been impregnated with varnish or insulating oil and baked. Used for insulating purposes, especially for between-layer insulation in transformers.

**V band** A band of frequencies in the millimeter region, extending from 46,000 to 56,000 mc.

**V-beam radar** A volumetric radar system that uses two fan beams to determine the distance, bearing, and height of a target. One beam is vertical and the other inclined. The beams intersect at ground level and rotate continuously about a vertical axis. The time difference between the arrivals of the echoes of the two beams is a measure of target elevation.

**vds** Abbreviation for *variable-depth sonar.*

**vector** 1. A quantity having both magnitude and direction, represented as a line terminated by an arrowhead. Often used to portray the amplitude and phase of a sinusoidal signal. Also called vector quantity. 2. To guide a pilot, navigator, aircraft, or missile from one point to another within a given time by means of a direction communicated to the craft.

**vectorcardiogram** *Vector electrocardiogram.*

**vector diagram** An arrangement of vectors showing the magnitude and phase relations between two or more alternating quantities having the same frequency.

**vector electrocardiogram** The two-dimensional or three-dimensional presentation of cardiac electric activity that results from displaying lead pairs against each other rather than against time. Also called vectorcardiogram.

**vector quantity** *Vector.*

**vectorscope** A cathode-ray oscilloscope that displays both the phase and amplitude of an applied signal with respect to a reference signal.

**velocity** 1. A vector quantity denoting both the direction and the speed of a linear motion, or denoting the direction of rotation and the angular speed in the case of rotation. 2. The time rate of change of a position vector of a point with respect to an inertial frame, the axes of which are usually fixed with respect to the earth.

**velocity antiresonance** The condition wherein a small change in the frequency of a sinusoidal force applied to a body or system causes an increase in velocity at the driving point, or the frequency is such that the absolute value of the driving-point impedance is a maximum.

**velocity correction** A method of register control in which the velocity of the web is changed gradually.

**velocity factor** The ratio of the velocity of propagation in any medium to the velocity of propagation in free space. The velocity of an r-f current is slightly less in a conductor than it would be in free space.

**velocity-focusing mass spectrograph** *Velocity spectrograph.*

**velocity hydrophone** A velocity microphone that responds to waterborne sound waves.

**velocity level** A sound rating in decibels, equal to 20 times the logarithm to the base 10 of the ratio of the particle velocity of the sound to a specified reference particle velocity.

**velocity-limiting servo** A servomechanism in which the maximum velocity of the servo is the chief limit on performance.

**velocity microphone** A microphone whose electric output depends on the velocity of the air particles that form a sound wave. Examples are hot-wire microphone and ribbon microphone.

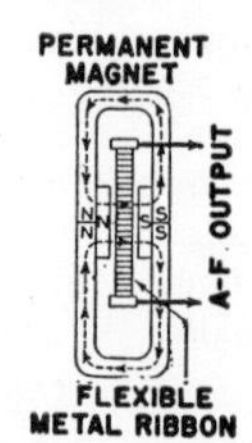


Velocity microphone construction and operation.

**velocity-modulated oscillator** An electron-tube structure in which the velocity of an electron stream is varied as the stream passes through a resonant cavity called a buncher. Energy is extracted from the bunched electron stream at a higher energy level in passing through a second cavity resonator called the catcher. Oscillations are sustained by coupling energy from the catcher cavity back to the buncher cavity.

**velocity-modulated tube** An electron-beam tube in which the velocity of the electron stream is alternately increased and decreased within a period comparable to the local transit time.

**velocity modulation** 1. Modulation in which a time variation in velocity is impressed on the electrons of a stream. 2. A television system in which the intensity of the electron beam remains constant throughout a scan and the velocity of the spot at the screen is varied to produce changes in picture brightness. Not in general use.

**velocity of light** A physical constant equal to 2.99779 × $10^{10}$ centimeters per second. This corresponds to 186,280 statute miles per second, 161,750 nautical miles per second, and 328 yards per microsecond. All electromagnetic radiation travels at this same speed in free space.

**velocity resonance** *Resonance.*

**velocity sorting** Any process of selecting electrons according to their velocities.

**velocity spectrograph** A mass spectrograph in which only positive ions having a certain velocity pass through all three slits and enter a chamber where they are deflected by a magnetic field in proportion to their charge-to-mass ratio. Also called velocity-focusing mass spectrograph.

**vented baffle** *Reflex baffle.*

**vented pressure pickup** A pressure pickup in which a vent reestablishes a reference pressure just before each use.

**Verdet's constant** A Faraday-effect constant that determines the angle of rotation of plane-polarized light when passing through certain materials in a magnetic field.

**verification** Automatic comparison of one data transcription with another transcription of the same data, to reveal errors.

**verifier** A punched-card machine or auxiliary computer device that automatically provides verification.

**vernier** Any device, control, or scale used to obtain a fine adjustment or to increase the precision of a measurement.

**vernier capacitor** A small variable capacitor placed in parallel with a larger tuning capacitor to provide a finer adjustment after a larger unit has been set approximately to the desired position.

**vernier dial** A tuning dial in which each complete rotation of the control knob causes only a fraction of a revolution of the main shaft, permitting fine and accurate adjustment.

**vertical antenna** A vertical metal tower, rod, or suspended wire used as an antenna.

**vertical blanking** Blanking of a television picture tube during the vertical retrace.

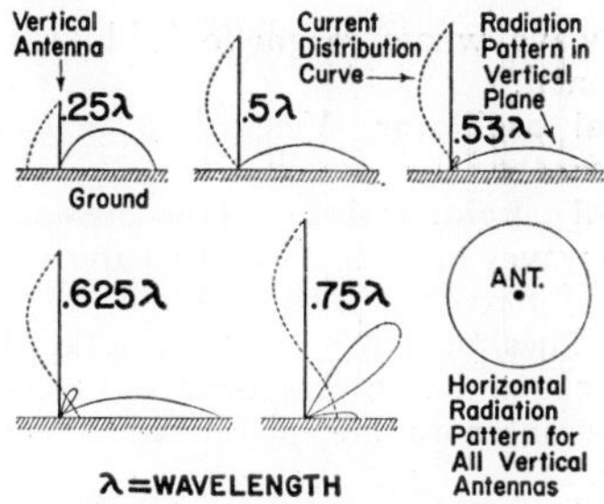


Vertical-antenna radiation patterns in vertical plane for various heights (thin solid curves) and current distribution curves (dotted).

**vertical blanking pulse** The rectangular pulse that is transmitted at the end of each field of a television signal to cut off the beam current of the picture tube while the beam is returning to the top of the screen for the start of the next field.

**vertical centering control** The centering control provided in a television receiver or cathode-ray oscilloscope to shift the position of the entire image vertically in either direction on the screen.

**vertical compliance** The ability of a stylus to move freely in a vertical direction while in the groove of a phonograph record.

**vertical convergence control** The control that adjusts the amplitude of the vertical dynamic convergence voltage in a color television receiver.

**vertical definition** *Vertical resolution.*

**vertical deflection electrodes** The pair of electrodes that moves the electron beam up and down on the fluorescent screen of a cathode-ray tube employing electrostatic deflection.

**vertical deflection oscillator** The oscillator that produces, under control of the vertical synchronizing signals, the sawtooth voltage waveform that is amplified to feed the vertical deflection coils on the picture tube of a television receiver. Also called vertical oscillator.

**vertical hold control** The hold control that changes the free-running period of the vertical deflection oscillator in a television receiver, so the picture remains steady in the vertical direction.

**vertical linearity control** A linearity control that permits narrowing or expanding the height of the image on the upper half of the screen of a television picture tube, to give linearity in the vertical direction so circular objects appear as true circles. Usually mounted at the rear of the receiver.

**vertically polarized wave** A linearly polarized wave whose magnetic field vector is horizontal.

**vertical oscillator** *Vertical deflection oscillator.*

**vertical polarization** Transmission of radio waves in such a way that the electric lines of force are vertical, while the magnetic lines of force are horizontal. With this polarization, transmitting and receiving dipole antennas are placed in a vertical plane.

**vertical radiator** A transmitting antenna that is perpendicular to the earth.

**vertical recording** A type of disk recording in which the groove modulation is perpendicular to the surface of the recording medium, so the cutting stylus moves up and down rather than from side to side during recording. Also called hill-and-dale recording.

**vertical resolution** The number of distinct horizontal lines, alternately black and white, that can be seen in the reproduced image of a television or facsimile test pattern. Vertical resolution is primarily fixed by the number of horizontal lines used in scanning. Also called vertical definition.

**vertical retrace** The return of the electron beam to the top of the screen at the end of each field in television.

**vertical stylus force** *Stylus force.*

**vertical sweep** The downward movement of the scanning beam from top to bottom of the picture being televised.

**vertical synchronizing pulse** One of the six pulses that are transmitted at the end of each field in a television system to keep the receiver in field-by-field synchronism with the transmitter. Also called picture synchronizing pulse.

**very high frequency** [abbreviated vhf] A Federal Communications Commission designation for the band from 30 to 300 mc in the radio spectrum.

**very-long-range radar** Radar whose maximum line-of-sight range is greater than 800 miles for a target having an area of 1 square meter perpendicular to the radar beam.

**very low frequency** [abbreviated vlf] A Federal Communications Commission designation for the band from 10 to 30 kc in the radio spectrum.

**very-short-range radar** Radar whose maximum line-of-sight range is less than 50 miles for a target having an area of 1 square meter perpendicular to the radar beam.

**vestigial sideband** The transmitted portion of an amplitude-modulated sideband that has been largely suppressed by a filter having a gradual cutoff in the neighborhood of the carrier frequency. The other sideband is transmitted without much suppression.

**vestigial-sideband filter** A filter that is inserted between a transmitter and its antenna to suppress part of one of the sidebands.

**vestigial-sideband transmission** A type of radio signal transmission for amplitude modulation in which the normal complete sideband on one side of the carrier is transmitted, but only a part of the other sideband is transmitted. Used for the visual transmitter in television, where the lower sideband extends only 0.75 mc below the carrier and the upper sideband extends 4 mc. Also called asymmetrical-sideband transmission.

**vestigial-sideband transmitter** A transmitter in which one sideband and a portion of the other are intentionally transmitted.

**vfo** Abbreviation for *variable-frequency oscillator.*

**vfr** Abbreviation for *visual flight rules.*

**vfr conditions** Weather conditions equal to or better than the minimum prescribed for flights under visual flight rules.

**vhf** Abbreviation for *very high frequency.*

**vhf homing adapter** A homing adapter operating in the vhf range.

**vhf omnirange** [abbreviated vor] An omnirange operating in the band from 112 to 118 mc to provide bearing information for aircraft. It emits a nondirectional reference modulation and a signal whose character varies with rotation through 360° in a horizontal plane. The receiving equipment interprets the two signals in terms of the bearing between the receiver and the signal source.

**vhf/uhf direction finder** A ground-based radio direction finder capable of being used alone or in conjunction with airport surveillance radar.

**vibrating capacitor** A capacitor whose capacitance is varied in a cyclic manner to produce an alternating voltage proportional to the charge on the capacitor. Used in a vibrating-reed electrometer.

**vibrating-reed electrometer** An instrument using a vibrating capacitor to measure a small charge, often in combination with an ionization chamber.

**vibrating-reed frequency meter** A frequency meter consisting of steel reeds having different and known natural frequencies, all excited by an electromagnet carrying the alternating current whose frequency is to be measured. Also called Frahm frequency meter, reed frequency meter, and tuned-reed frequency meter.

**vibrating-reed rectifier** An electromagnetic device for rectifying an alternating current by reversing the connections between the power line and load each time the alternating current reverses in direction. The reversing contacts are on a vibrating reed of magnetic material that is acted on by a coil carrying the alternating current, so the reed moves in synchronism with the current.

**vibration** A periodic change in the position of an object, such as a pendulum or a diaphragm of a loudspeaker energized by a sinusoidal current. Also called oscillation.

**vibration galvanometer** An a-c galvanometer in which the natural oscillation frequency of the moving element is equal to the frequency of the current being measured.

**vibration meter** An instrument used to measure the displacement, velocity, and acceleration associated with mechanical vibration. In one form it consists of a piezoelectric vibration pickup having uniform response from 2 to 1,000 cps, feeding an amplifier having an indicating meter at its output. Also called vibrometer.

**vibration pickup** A pickup designed to respond to mechanical vibrations rather than to sound waves. In one type, twisting or bending of a Rochelle-salt crystal generates a voltage that varies in accordance with the vibration being analyzed.

**vibrato** A musical embellishment that depends primarily on periodic variations of frequency which are often accompanied by variations in amplitude and waveform.

**vibrator** An electromagnetic device for converting a direct voltage into an alternating voltage. It contains a vibrating armature on which are contacts that reverse

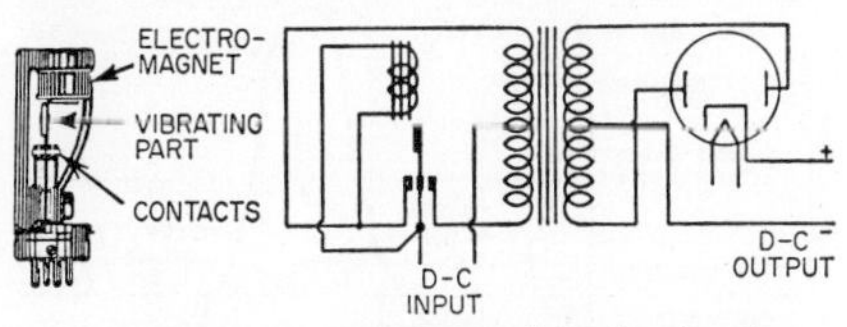


Vibrator construction and basic circuit.

the direction of current flow during each vibration. Used to convert the voltage of a storage battery into a low alternating voltage that can be stepped up by a power transformer and rectified.

**vibrator-type inverter** A device that uses a vibrator and an associated transformer or other inductive device to change d-c input power to a-c output power.

**vibratory gyroscope** An instrument that utilizes a vibrating rod or tuning fork to perform certain equilibrium or directional functions of the spinning gyroscope.

**vibrometer** *Vibration meter.*

**vibrotron** A triode electron tube having an anode that can be moved or vibrated by an externally applied force.

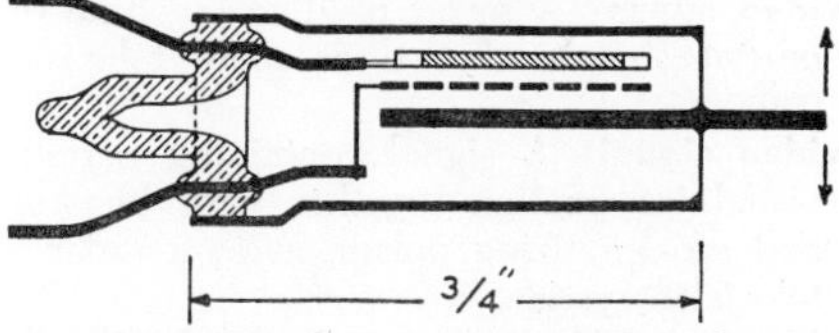


Vibrotron construction. Anode is rod that projects through flexible metal diaphragm at right end of tube. Moving rod up or down moves anode away from or toward grid and cathode, causing anode current to vary in proportion to frequency and amplitude of external motion.

**video** [Latin for "I see"] 1. Pertaining to picture signals or to the sections of a television system that carry these signals in either unmodulated or modulated form. 2. Pertaining to the demodulated radar receiver output that is applied to a radar indicator.

**video amplifier** A wideband amplifier capable of amplifying video frequencies in radar and television.

**video circuit** A broadband circuit carrying intelligence that could become visible.

**video converter** A converter consisting of a television camera tube and a cathode-ray tube between which may be inserted a transparent overlay. The image on the overlay is illuminated by the radial scan of the cathode-ray tube and converted to video signals by the camera tube, for superimposition on a radar ppi display or other radar applications.

**video detector** The detector that demodulates video i-f signals in a television receiver.

**video frequency** One of the frequencies existing in the output of a television camera when an image is scanned. It may be any value from almost zero to well over 4 mc.

**video-frequency amplifier** An amplifier capable of handling the entire range of frequencies that comprise a periodic visual presentation in television, facsimile, or radar.

**video gain control** A control that adjusts the gain of a video amplifier, as for varying the intensity of the echoes on a radar ppi screen to get maximum contrast between desired echoes and undesired clutter.

**video integration** A method of utilizing the redundancy of repetitive signals to improve the output signal-to-noise ratio, by summing the successive video signals.

**video mapping** A mapping procedure in which a chart of an area is electronically superimposed on a radar display.

**video mixer** A mixer used to combine the output signals of two or more television cameras.

**video signal** A signal containing periodic visual information together with blanking and synchronizing pulses, as in a radar or television system.

**video stretching** A procedure whereby the duration of a video pulse is increased in a navigation system.

**Videotape** Trademark of Ampex Corp. for its line of video tape recorders.

**video tape recording** [abbreviated vtr] A method of recording television video signals on magnetic tape for later rebroadcasting of television programs.

**video waveform** The portion of the television signal waveform that corresponds to visual information. Synchronizing pulses are not included.

**vidicon** A camera tube in which a charge-density pattern is formed by photoconduction and stored on a photoconductor surface that is scanned by an electron beam, usually of low-velocity electrons. Used chiefly in industrial television cameras.

**viewfinder** An auxiliary optical or electronic device attached to a television camera so the operator can see the scene as the camera sees it.

**viewing screen** *Screen.*

**Viking** A liquid-fuel research rocket first launched in May, 1949, and used in modified form as the first stage of the Vanguard satellite-carrying rocket.

**Villari effect** The change in magnetic induction that occurs when a magnetostrictive material is mechanically stressed.

**Vinylite** Trademark of Bakelite Corp. for their vinyl-resin plastics.

**vinyl resin** A soft plastic material used in making long-playing phonograph records.

**virginium** Former name for *francium.*

**virgin neutron** A neutron from any source, before it makes a collision.

**virgin neutron flux** A flux of neutrons that have had no collisions and therefore have lost none of the energy with which they were born.

**virtual cathode** The locus of a space-charge-potential minimum such that only some of the electrons approaching it are transmitted, the remainder being reflected back to the electron-emitting cathode.

**virtual height** The apparent height of an ionized layer, as determined from the time interval between the transmitted signal and the ionospheric echo at vertical incidence.

**virtual level** An energy level of a compound nucleus, characterized by a lifetime that is long compared with the transit times of the nucleons across nuclear dimensions at energies corresponding to the excitation in question. Also called virtual state.

**virtual ppi reflectoscope** *Reflection plotter.*

**virtual quantum** A quantum or photon in an intermediate state in which energy is not conserved.

**virtual state** *Virtual level.*

**viscometer** An instrument used to measure the degree to which a liquid resists a change in shape. Also called viscosimeter.

**viscosimeter** *Viscometer.*

**viscous-damped arm** A phonograph pickup arm that is mechanically damped by a highly viscous liquid in such a way that the arm floats gently down to a record when dropped.

**visibility factor** The ratio of the minimum signal input power detectable by ideal instruments connected to the output of a receiver, to the minimum signal power detectable by a human operator through a display connected to the same receiver. Also called display loss.

**visible radiation** Radiation having wavelengths ranging from about 3,800 to 7,800 angstroms, corresponding to the visible spectrum of light.

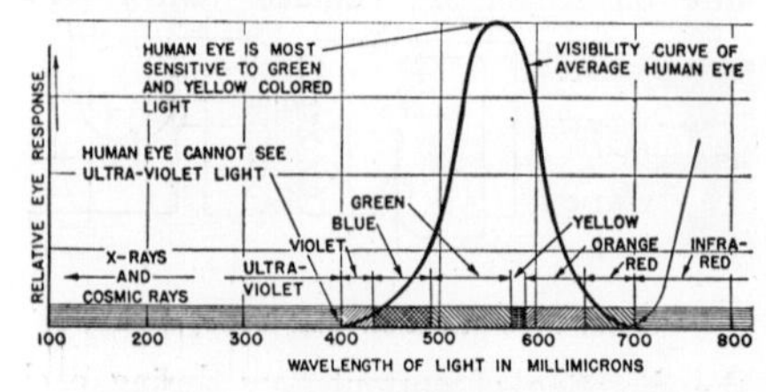


Visible radiation.

**visual-aural range** [abbreviated var] A vhf radio range that provides one course for display to the pilot on a zero-center left-right indicator and another course, at right angles to the first, in the form of aural A-N radio range signals. The A-N aural signals provide a means for differentiating between the two directions of the visual course.

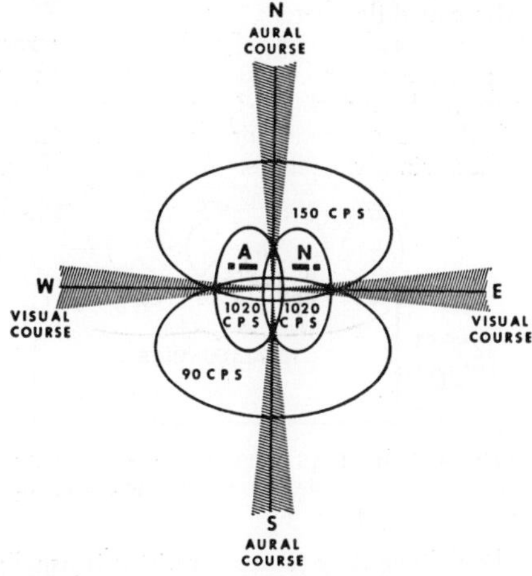


Visual-aural range radiation patterns, courses, and signals.

**visual broadcast service** A service rendered by stations broadcasting images for general public reception, such as by television broadcast stations and facsimile broadcast stations.

**visual carrier frequency** The frequency of the television carrier that is modulated by picture information.

**visual flight rules** [abbreviated vfr] Regulations governing flying of aircraft when visibility is good up to a specified altitude.

**visual radio range** Any range facility whose course is flown by visual instrumentation not associated with aural reception.

**visual range** The range of beta particles in an absorber, usually aluminum, as estimated by visual inspection of breaks in the absorption curve of the material.

**visual signal** The picture portion of a television signal.

**visual transmitter** The radio equipment used to transmit the video part of a television program. Also called picture transmitter. The visual and aural transmitters together are called a television transmitter.

**visual transmitter power** The peak power output when transmitting a standard television signal.

**vlf** Abbreviation for *very low frequency.*

**vocoder** A device for producing synthetic speech.

**vodas** [Voice-Operated Device Anti-Singing] A voice-operated switching device used in transoceanic radiotelephone circuits to suppress echoes and singing sounds automatically. It connects a subscriber's line automatically to the transmitting station as soon as he starts speaking and simultaneously disconnects it from the receiving station, thereby permitting the use of one radio channel for both transmitting and receiving without appreciable switching delay as the parties alternately talk.

**voder** [Voice Operation DEmonstratoR] An electronic system that uses electron tubes and filters, controlled through a keyboard, to produce voice sounds artificially.

**vogad** [Voice-Operated Gain-Adjusting Device] An automatic gain control circuit used to maintain a constant speech output level in long-distance radiotelephony.

**voice coil** The coil that is attached to the diaphragm of a moving-coil loudspeaker and moves through the air gap between the pole pieces due to interaction of the fixed magnetic field with that associated with the a-f current flowing through the voice coil. Also called loudspeaker voice coil and speech coil (British).

**voice frequency** An audio frequency in the range essential for transmission of speech of commercial quality, from about 300 to 3,400 cps. Also called speech frequency.

**voice-frequency carrier telegraphy** Carrier telegraphy in which the modulated currents are transmitted over a voice-frequency telephone channel.

**volatile storage** Computer storage in which information cannot be retained without continuous power dissipation. If power is turned off, the stored information vanishes. Delay-line memories and electrostatic storage tubes are examples.

**volt** [abbreviated v] The unit of voltage, potential, and electromotive force. One volt will send a current of one ampere through a resistance of one ohm. Named after the Italian physicist Alessandro Volta, 1745–1829.

**Volta effect** *Contact potential.*

**voltage** [symbol $E$] The term most often used to designate electric pressure that exists between two points and is capable of producing a flow of current when a closed circuit is connected between the two points. Voltage is measured in volts, millivolts, microvolts, and kilovolts. The terms electromotive force, potential, potential difference, and voltage drop are all often called voltage.

**voltage amplification** The ratio of the magnitude of the voltage across a specified load impedance to the magnitude of the input voltage of the amplifier or other transducer feeding that load. Often expressed in decibels by multiplying the common logarithm of the ratio by 20.

**voltage amplifier** An amplifier designed primarily to build up the voltage of a signal, without supplying appreciable power.

**voltage-amplifier tube** A tube designed primarily for use in a voltage amplifier, hence having high gain but delivering very little output power.

**voltage attenuation** The ratio of the magnitude of the voltage across the input of a transducer to the magnitude of the voltage delivered to a specified load impedance connected to the transducer. Often expressed in decibels by multiplying the common logarithm of the ratio by 20.

**voltage-controlled capacitor** A capacitor whose capacitance value can be changed by varying an externally applied bias voltage, as in a silicon capacitor. Used for automatic frequency control, remote control of tuning, and frequency modulation.

**voltage-controlled magnetic amplifier** A magnetic amplifier in which the flux change during the resetting interval is related to the input signal voltage and is essentially independent of control-circuit current.

**voltage-controlled oscillator** An oscillator whose frequency of oscillation can be varied by changing an applied voltage.

**voltage cutoff** The electrode voltage that reduces the anode current, beam current, or some other electron-tube characteristic to a specified low value.

**voltage divider** A tapped resistor, adjustable resistor, potentiometer, or a series arrangement of two or more fixed resistors connected across a voltage source. A desired fraction of the total voltage is obtained from the intermediate tap, movable contact, or resistor junction. Also called potential divider.

**voltage doubler** A transformerless rectifier circuit that gives approximately double the output voltage of a conventional half-wave vacuum-tube rectifier by charging a capacitor during the normally wasted half-cycle and discharging it in series with the output voltage during the next half-cycle. Also called doubler.

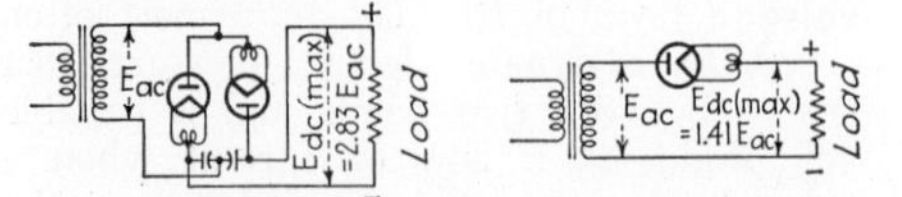


Voltage doubler, shown in comparison with conventional half-wave rectifier circuit that gives only half as much d-c output voltage for a given a-c input voltage.

**voltage drop** The voltage developed across a component or conductor by the flow of current through the resistance or impedance of that component or conductor. Often called simply voltage. Also called drop.

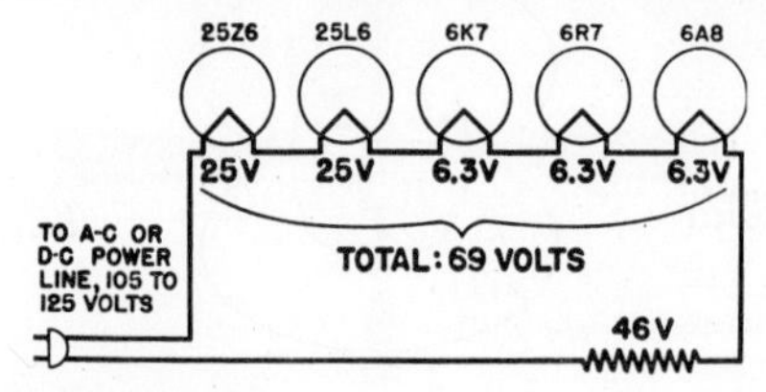


Voltage drop of 46 volts across series resistor cuts line voltage down to value required by tube filaments in series.

The voltage drop across a resistor is usually called an *IR* drop, while that in a conductor is usually called a resistance drop.

**voltage feed** Excitation of a transmitting antenna by applying voltage at a voltage loop or antinode.

**voltage feedback** Feedback in which the voltage drop across part of the load impedance acts in series with the input signal voltage.

**voltage gain** The difference between the output signal voltage level in decibels and the input signal voltage level in decibels. This value is equal to 20 times the common logarithm of the ratio of the output voltage to the input voltage. The voltage gain is equal to the amplification factor of the tube or transistor only for a matched load.

**voltage generator** A two-terminal circuit element in which the terminal voltage is independent of the current through the element.

**voltage gradient** The voltage per unit length along a resistor or other conductive path.

**voltage jump** An abrupt change or discontinuity in tube voltage drop during operation of a glow-discharge tube.

**voltage loop** An antinode at which voltage is a maximum.

**voltage multiplier** *Instrument multiplier.*

**voltage node** A point having zero voltage in a stationary wave system, as in an antenna or transmission line. A voltage node exists at the center of a half-wave antenna.

**voltage-range multiplier** *Instrument multiplier.*

**voltage rating** The maximum sustained voltage that can safely be applied to an electric device without risking the possibility of electric breakdown. Also called working voltage.

**voltage-reference tube** A gas tube in which the tube voltage drop is approximately constant over the operating range of current and is also relatively stable with time at fixed values of current and temperature.

**voltage reflection coefficient** The ratio of the complex electric field strength or voltage of a reflected wave to that of the incident wave.

**voltage-regulating transformer** A power transformer designed to deliver an essentially constant output voltage over a wide range of input voltage values.

**voltage regulation** The ratio of the difference between no-load and full-load output voltage of a device to the full-load output voltage, expressed as a percentage.

**voltage regulator** A device that maintains the terminal voltage of a generator or other voltage source within required limits despite variations in input voltage or load. Also called automatic voltage regulator.

**voltage-regulator tube** A glow-discharge tube in which the tube voltage drop is approximately constant over the operating range of current. Used to maintain an essentially constant direct voltage in a circuit despite changes in line voltage or load. Also called vr tube.

**voltage relay** A relay that functions at a predetermined value of voltage.

**voltage saturation** *Anode saturation.*

**voltage-sensitive resistor** A resistor whose value varies markedly with applied voltage over at least a portion of its voltage range. It may consist of one or more mineral crystals or two or more metallic oxide disks, but does not have rectifying properties.

**voltage-stabilizing tube** A gas-filled tube normally working with a glow discharge in the part of the characteristic where the voltage is practically independent of current.

**voltage standard** A voltage source whose value is known to a high degree of accuracy. A standard cell is an example.

**voltage standing-wave ratio** [abbreviated vswr] The ratio of the amplitude of the electric field or voltage at a voltage minimum to that at an adjacent maximum in a stationary-wave system, as in a waveguide, coaxial cable, or other transmission line.

**voltage standing-wave ratio meter** [abbreviated vswr meter] An electrically operated instrument that indicates voltage standing-wave ratios and is calibrated in voltage ratios.

**voltage transformer** An instrument transformer whose primary winding is connected in parallel with a circuit in which the voltage is to be measured or controlled. Also called potential transformer.

**voltage-tunable tube** An oscillator tube whose operating frequency can be varied by changing one or more of the electrode voltages, as in a backward-wave magnetron.

**voltage-type telemeter** A telemeter that employs the magnitude of a single voltage as the translating means.

**voltaic cell** A primary cell consisting of two dissimilar metal electrodes in a solution that acts chemically on one or both of them to produce a voltage.

**voltaic pile** An early form of primary battery, consisting of a pile of alternate pairs of dissimilar metal disks, with moistened pads between pairs.

**voltammeter** An instrument that may be used either as a voltmeter or ammeter.

**volt-ampere** [abbreviated va] The unit of apparent power in an a-c circuit containing reactance. Apparent power is equal to the voltage in volts multiplied by the current in amperes, without taking phase into consideration.

**volt-ampere meter** An instrument for measuring the apparent power in an a-c circuit.

**volt-ampere reactive** [abbreviated var] The unit of reactive power. This name was adopted by the International Electrotechnical Commission in 1930. Also called reactive volt-ampere.

**Volta's law** The contact voltage developed between two dissimilar conductors is the same whether the contact is direct or through one or more intermediate conductors.

**voltmeter** An instrument for measuring voltage. Its scale may be calibrated in volts or related smaller or larger units. A voltmeter that indicates millivolt values is often called a millivoltmeter. Similarly, a microvoltmeter indicates microvolt values and a kilovoltmeter indicates kilovolt values.

**voltmeter-ammeter** A voltmeter and an ammeter combined in a single case but having separate terminals.

**voltmeter sensitivity** The ratio of the total resistance of a voltmeter to its full-scale reading in volts, expressed in ohms per volt.

**volt-ohm-milliammeter** A test instrument having a number of different ranges for measuring voltage, current, and resistance. It consists essentially of a single meter having a number of different scales, with a selector switch that connects the meter into the correct measuring circuit for a measurement. Also called circuit analyzer, multimeter, and multiple-purpose tester.

**volume** 1. The magnitude of a complex a-f current as measured in volume units (vu) on a standard volume indicator. 2. The intensity of a sound.

**volume compressor** An a-f control circuit that limits the volume range of a radio program at the transmitter, to permit using a higher average per cent modulation without risk of overmodulation. Also used when making disk recordings, to permit a closer groove spacing without overcutting. Also called automatic volume compressor.

**volume control** A potentiometer used to vary the loudness of a reproduced sound by varying the a-f signal voltage at the input of the audio amplifier.

**volume dose** *Integral absorbed dose.*

**volume expander** An a-f control circuit sometimes used to increase the volume range of a radio program or recording by making weak sounds weaker and loud sounds louder. The expander counteracts volume compression at the transmitter or recording studio. Also called automatic volume expander.

**volume indicator** A standardized instrument for indicating the volume of a complex electric wave such as that corresponding to speech or music. The reading in volume units (vu) is equal to the number of decibels above a reference level. The sensitivity is adjusted so the reference level of 0 vu is indicated when the instrument is connected across a 600-ohm resistor that is dissipating a power of 1 milliwatt at 1,000 cps.

**volume ionization** The average ionization density in a given volume of ionizing particles.

**volume lifetime** The average time interval between the generation and recombination of minority carriers in a homogeneous semiconductor.

**volume-limiting amplifier** An amplifier containing an automatic device that functions only when the input volume exceeds a predetermined level, and then reduces the gain in such a manner that the output volume stays substantially constant despite further increases in input volume. The normal gain of the amplifier is restored when the input volume returns below the predetermined limiting level.

**volume magnetostriction** The change in the volume of a magnetostrictive material when subjected to a magnetic field.

**volume recombination** Recombination of positive and negative ions at low energies throughout the volume of an ionization chamber, or recombination of free electrons and holes in the volume of a semiconductor.

**volume recombination rate** The time rate at which free electrons and holes recombine within the volume of a semiconductor.

**volumetric radar** A radar capable of providing three-dimensional position data for a multiplicity of targets.

**volume unit** [abbreviated vu] A unit used to specify the a-f power level in decibels above a reference level of 1 milliwatt (0.001 watt), as measured with a standard volume indicator. A volume unit is equal to a decibel only when changes in power are involved or when the decibel value has this same reference level. It is unnecessary to specify the reference level when dealing in volume units because the level is a part of the definition.

**volume-unit meter** [abbreviated vu meter] A meter calibrated to read a-f power levels directly in volume units.

**volume velocity** The rate of flow of a medium through a specified area due to a sound wave.

**vor** Abbreviation for *vhf omnirange.*

**vor receiver** An aircraft radio receiver designed to receive signals from a vhf omnirange and interpret them for the aircraft crew.

**vortac** An air navigation system that combines vhf omnirange and tacan equipment.

**vowel articulation** The per cent articulation obtained for vowels, usually combined with consonants into meaningless syllables.

**vr tube** *Voltage-regulator tube.*

**vswr** Abbreviation for *voltage standing-wave ratio.*

**vswr meter** Abbreviation for *voltage standing-wave ratio meter.*

**vt fuze** Abbreviation for *variable-time fuze*

**vtr** Abbreviation for *video tape recording.*

**vtvm** Abbreviation for *vacuum-tube voltmeter.*

**vu** Abbreviation for *volume unit.*

**vulcanized fibre** A laminated plastic made by chemically treating layers of 100% rag-content paper to gelatinize the paper and fuse the layers into a solid mass. When dried under pressure, it forms a hard, tough material having good electrical properties along with mechanical strength and dimensional stability. The British spelling, "fibre," is commonly used for this product because of its British origin.

**vu meter** Abbreviation for *volume-unit meter.*

# W

**w** Abbreviation for *watt.*

**Wac Corporal** A sounding rocket used for carrying instruments to high altitudes.

**WADC** Abbreviation for *Wright Air Development Center.*

**wafer socket** An electron-tube socket consisting of one or two wafers of insulating material, having holes in which are spring metal clips that grip the terminal pins of a tube.

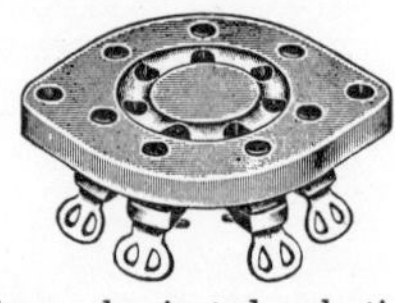

Wafer sockets made from laminated plastic (left) and from molded steatite (right).

**Wagner ground** A ground connection used with an a-c bridge to minimize stray capacitance errors when measuring high impedances. A potentiometer is connected across the bridge supply oscillator, with its movable tap grounded.

**walkie-lookie** A complete portable television camera and transmitter used for remote pickups, as at political conventions or golf matches. The signals are transmitted to a microwave receiver at a local control point, and go from there over standard coaxial cable or microwave links to the television station.

**walkie-talkie** A compact portable combination radio transmitter and receiver that can be carried by one man, usually strapped over the back, and used for communication over medium distances.

**wall absorption** The decrease in beta-ray or gamma-ray output due to absorption in the radioactive material itself.

**wall effect** The contribution to the ionization in an ionization chamber by electrons liberated from the walls.

**wall outlet** An outlet mounted on a wall, from which electric power can be obtained by inserting the plug of a line cord.

**wamoscope** [WAve-MOdulated oscilloSCOPE] A special cathode-ray oscilloscope tube that detects, amplifies, and displays a microwave radar signal, thereby serving as a complete radar receiver.

**wander** *Scintillation.*

**warble-tone generator** An a-f signal generator whose frequency is varied cyclically at a subaudio rate over a fixed range.

**warhead** The section of a bomb, guided missile, ballistic missile, torpedo, or other missile that contains the explosive, chemical, or other charge intended to damage the enemy.

**warmup time** The time interval between application of power to a system and the instant at which the system is stabilized and ready to perform its intended function.

**war surplus** A military article of supply or piece of equipment that has been declared surplus because it is obsolete, unserviceable, or excess to current and reserve military requirements.

**watch** The service performed by a qualified operator when on duty in the radio room of a vessel. Also called radio watch.

**water-activated battery** A primary battery that contains the electrolyte but requires the addition of or immersion in water before it is usable.

**water-boiler reactor** A homogeneous reactor that uses enriched uranium as fuel and ordinary water as moderator. The fuel is uranyl sulfate dissolved in water.

**water calorimeter** A calorimeter that measures r-f power in terms of the rise in temperature of water in which the r-f energy is absorbed.

**water-cooled tube** An electron tube that is cooled by circulating water through or around the anode structure.

**water load** A matched waveguide termination in which the electromagnetic energy is absorbed in water. The resulting rise in the temperature of the water is a measure of the output power.

**water monitor** A monitoring device used to detect and measure waterborne radioactivity.

**watt** [abbreviated w] The practical unit of electric power. In a d-c circuit, the power in watts is equal to volts multiplied by amperes. In an a-c circuit, the true power in watts is effective volts multiplied by effective amperes, then multiplied by the circuit power factor. There are 746 watts in 1 horsepower. The term is named after James Watt, Scottish inventor.

**wattage rating** A rating expressing the maximum power that a device can safely handle continuously.

**watthour** The practical unit of electric energy, equal to a power of 1 watt absorbed continuously for 1 hour. One kilowatt-hour is equal to 1,000 watthours.

**watthour meter** A meter that measures and registers the integral, with respect to time, of the active power of the circuit in which it is connected. The unit of measurement is usually the kilowatthour.

**wattless power** *Reactive power.*

**wattmeter** A meter that measures electric power in watts.

**watt-second** The amount of energy corresponding to 1 watt acting for 1 second. One watt-second is equal to 1 joule.

**wave** A propagated disturbance for which the intensity at any point in the medium is a function of time, and the intensity at a given instant is a function of the position of the point. A wave may be electric, electromagnetic, acoustic, mechanical, or any other form.

**wave amplitude** The magnitude of the maximum change from zero of a characteristic of a wave.

**wave analyzer** An instrument for measuring the amplitude and frequency of the various components of a complex current or voltage wave.

**wave angle** The angle in azimuth and elevation at which a radio wave arrives at a receiving antenna or leaves a transmitting antenna.

**wave antenna** A directional antenna composed of a system of parallel horizontal conductors from one half-wavelength to several wavelengths long, with the far ends terminated to ground by the characteristic impedance of the antenna. Also called Beverage antenna.

**waveband** A band of frequencies, such as that assigned to a particular radio communication service.

**waveband switch** A multiposition switch used to change the tuning range of a receiver or transmitter from one waveband to another.

**wave clutter** *Sea clutter.*

**wave converter** A converter used to change one type of wave to another in a waveguide.

**wave equation** An equation that describes a particular wave motion through a medium.

**wave filter** A transducer for separating waves on the basis of their frequency. It introduces relatively small insertion loss to waves in one or more frequency bands and relatively large insertion loss to waves of other frequencies.

**waveform** The shape of a wave, as obtained by plotting a characteristic of the wave with respect to time.

**waveform-amplitude distortion** *Frequency distortion.*

**waveform monitor** A cathode-ray oscilloscope having a time base suitable for viewing the waveform of the video signal in a television system. Also called A scope.

**waveform synthesizer** A signal generator whose output is variable as to frequency, phase, harmonic content, and harmonic amplitude.

**wavefront** The portion of a wave envelope that is between the virtual zero point and the point at which the wave reaches its crest value, as measured either in time or distance.

**wave function** A set of solutions to Maxwell's equations for wave propagation in a homogeneous isotropic region.

**wave group** The resultant of two or more wave trains of different frequency traversing the same path.

**waveguide** A rectangular or circular metal pipe having a predetermined cross-section, specifically designed to guide or conduct high-frequency electromagnetic waves through its interior, or any other equivalent system of material boundaries capable of guiding waves. Also called guide.

**waveguide attenuator** An attenuator designed for use in a waveguide to produce attenuation by any means, including absorption and reflection.

**waveguide bend** A section of waveguide in which the direction of the longitudinal axis is changed. An E-plane bend in a rectangular waveguide is bent along the narrow dimension, while an H-plane bend is bent along the wide dimension. Also called waveguide elbow.

**waveguide component** A device designed to be connected at specified ports in a waveguide system.

**waveguide connector** A mechanical device for electrically joining and locking together separable mating parts of a waveguide system. Also called waveguide coupling.

**waveguide coupling** *Waveguide connector.*

**waveguide cutoff wavelength** The wavelength corresponding to the cutoff frequency of a waveguide. Below this frequency the attenuation rises rapidly.

**waveguide directional coupler** A directional coupler consisting of two parallel waveguides having a common wall in which two slots are cut. These permit extraction of a portion of the r-f energy traveling in one direction through the main waveguide while rejecting energy traveling in the opposite direction.

**waveguide elbow** *Waveguide bend.*

**waveguide filter** A filter made up of waveguide components, used to change the amplitude-frequency response characteristic of a waveguide system.

**waveguide flange** *Flange.*

**waveguide gasket** A gasket that maintains electric continuity between two mating sections of waveguide.

**waveguide junction** *Junction.*

**waveguide lens** A microwave lens in which the required phase changes result from transmission through suitable waveguide elements.

**waveguide mode suppressor** A waveguide filter that suppresses undesired modes of propagation in a waveguide.

**waveguide phase shifter** A device for adjusting the phase of the output signal of a waveguide system with respect to the phase of the input signal.

**waveguide plunger** *Piston.*

**waveguide post** A cylindrical rod placed in a transverse plane of a waveguide to serve substantially as a shunt susceptance. Also called post.

**waveguide probe** *Probe.*

**waveguide radiator** An open-ended waveguide, with or without a flaring horn, used to radiate electromagnetic energy to a reflector or out into space.

**waveguide resonator** *Cavity resonator.*

**waveguide seal** A seal used over the end of a waveguide to prevent entrance of moisture without appreciably attenuating radio frequencies.

**waveguide shim** A thin, resilient metal sheet inserted between waveguide components to ensure electric continuity.

**waveguide shutter** A waveguide section containing an adjustable mechanical barrier that can be set to block or divert r-f energy.

**waveguide slug tuner** A quarter-wavelength dielectric slug that projects into a waveguide for tuning purposes. It is usually adjustable as to position and depth of penetration.

**waveguide stub** An auxiliary section of waveguide having an essentially nondissipative termination, joined at some angle with a main section of waveguide.

**waveguide stub tuner** A waveguide tuning or detuning device consisting of an adjustable piston mounted in a waveguide stub.

**waveguide switch** A switch designed for mechanically positioning a waveguide section so as to couple it to one of several other sections in a waveguide system.

**waveguide taper** A section of tapered waveguide.

**waveguide tee** *Tee junction.*

**waveguide transformer** A waveguide component that provides impedance transformation.

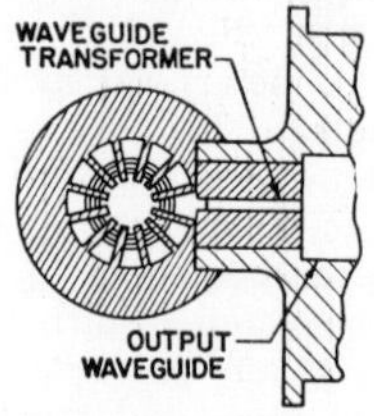


Waveguide transformer used between type 4J57 magnetron and output waveguide.

**waveguide tuner** An adjustable tuner used in a waveguide system to provide impedance transformation.

**waveguide twist** A waveguide section in which there is progressive rotation of the cross-section about the longitudinal axis.

**waveguide wavelength** The distance along a uniform waveguide, at a given frequency and for a given mode, between similar points at which a signal component differs in phase by $2\pi$ radians.

**waveguide window** A thin conducting metal window placed transversely in a waveguide for impedance matching. For an inductive window, the edges of the slit in the window are parallel to the electric field in the lowest mode in the waveguide. For a capacitive window the edges of the slit are perpendicular to the electric field.

**wave heating** Heating of a material by energy absorption from a traveling electromagnetic wave.

**wave impedance** The ratio of the transverse electric field to the transverse magnetic field in a waveguide.

**wave interference** The variation of wave amplitude with distance or time, caused by the superposition of two or more waves having the same or nearly the same frequency.

**wavelength** The distance between points having corresponding phase in two consecutive cycles of a periodic wave. The wavelength in meters is approximately equal to 300 divided by the frequency in megacycles. Wavelengths of light are specified in microns (1 micron = $10^{-6}$ meter), millimicrons (1 millimicron = $10^{-9}$ meter), and angstroms (1 angstrom = $10^{-10}$ meter).

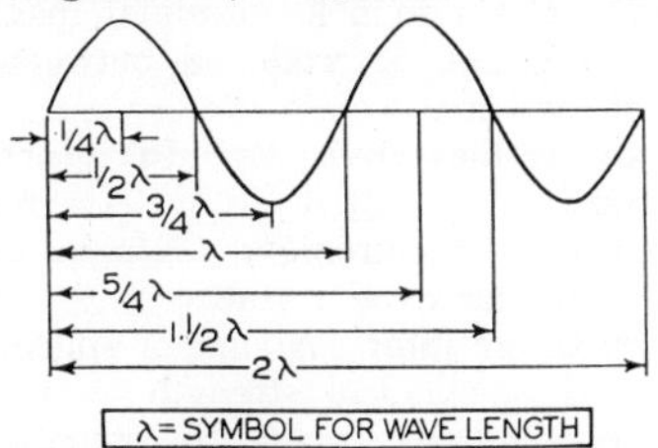

Wavelength values for sinusoidal wave.

**wavelength shifter** A photofluorescent compound used with a scintillator material to increase the wavelengths of the optical photons, thereby permitting more efficient use of the photons by the phototube or photocell.

**wave mechanics** A theory that assigns wave characteristics to the components of atomic structure and seeks to interpret physical phenomena in terms of hypothetical waveforms. Introduced by Schroedinger in 1926.

**wavemeter** An instrument for measuring the wavelength of an r-f wave. Since wavelength is related to frequency, a wavemeter also serves as a frequency meter.

**wave normal** A unit vector that is perpendicular to an equiphase surface, with its positive direction taken on the same side of the surface as the direction of propagation. In isotropic media, the wave normal is in the direction of propagation.

**wave propagation** *Propagation.*

**wave tail** The portion of a wave envelope that is between the crest and the end of the envelope.

**wave tilt** The forward inclination of the waveform of radio waves arriving along the ground. Its value depends on the electric constants of the ground.

**wave train** A series of wave cycles produced by the same disturbance.

**wave trap** A resonant circuit connected to the antenna system of a receiver to suppress signals at a particular frequency, such as that of a powerful local station that is interfering with reception of other stations. Also called trap.

**wave trough** The minimum value of the envelope of a progressive wave.

**wave-type microphone** A microphone that depends on wave interference for its directivity.

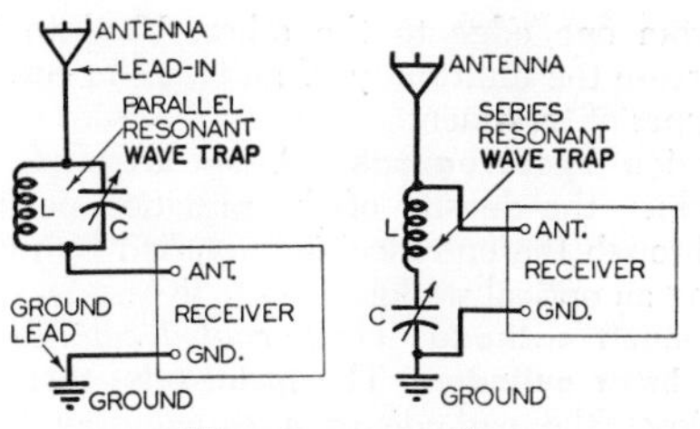

Wave-trap connections.

**wave velocity** The velocity of propagation of an electromagnetic wave, equal to $3 \times 10^{10}$ cm per second in free space. In a waveguide, the rate of energy transfer is called the group velocity and is less than the wave velocity, while the velocity of the electric wave is called the phase velocity and may be greater than the wave velocity.

**wax** A blend of waxes with metallic soaps, used in mechanical recording.

**wax master** Deprecated term for *wax original.*

**wax original** An original recording made on a wax surface and used to make a master. Also called wax master (deprecated).

**way point** A selected point on a radio navigation course line having some particular significance. Also called check point.

**WCEMA** Abbreviation for *West Coast Electronic Manufacturers Association.*

**weak coupling** *Loose coupling.*

**weapon system** The complete system required to deliver a weapon to its target, including production, storage, transport, launchers, aircraft, and guidance equipment.

**weather transmitting set** The components and items required to operate a complete electronic set for automatically measuring and transmitting weather data by radio.

**weber** The practical unit of magnetic flux. It is the amount of flux that, when linked with a single turn of wire, will induce 1 volt in the turn as it decreases uniformly to zero in 1 second. One weber is equal to $10^8$ maxwells.

**wedge** 1. A waveguide termination consisting of a tapered length of dissipative material introduced into the guide, such as carbon. 2. A convergent pattern of equally spaced black and white lines, used in a television test pattern to indicate resolution. 3. An optical filter in which the transmission decreases continuously or in steps from one end to the other.

**wedge filter** A radiation filter so constructed that its thickness or transmission characteristics vary continuously or in steps from one edge to the other. Used to increase the uniformity of radiation in certain types of treatment.

**wedge spectrograph** A spectrograph in which the density of the radiation passing through the entrance slit is varied by moving an optical wedge.

**Wehnelt cathode** *Oxide-coated cathode.*

**Wehnelt cylinder** The metal tube that encloses the cathode of a cathode-ray tube and serves to concentrate the electrons emitted in all directions from the cathode.

**weighting** The artificial adjustment of measurements to account for factors that, in the normal use of the device, would otherwise be different from conditions during the measurements. As an example, background noise measurements may be weighted by applying factors or by introducing networks to reduce measured values in inverse ratio to their interfering effects.

**weighting function** A measure of the relative effect on reactivity of localized changes in nuclear properties as a function of position and property change.

**Weissenberg method** An x-ray crystal analysis method in which the crystal is rotated in the beam of x-rays and the film is moved parallel to the axis of rotation. The crystal is surrounded by a sleeve having a slot that passes a line-shaped beam of x-rays, to give positive identification of each spot or line on the pattern.

**welding** A process of joining metals by the application of heat, pressure, or both.

**welding current** The current that is sent through a joint to produce the heat needed to make a weld.

**welding cycle** The complete series of events involved in making a weld.

**welding transformer** A high-current, low-voltage power transformer used to supply current for welding.

**weld interval** The total of all heating and cooling times when making a single multiple-impulse weld by means of resistance welding.

**weld-interval timer** A timer that controls the heating and cooling intervals when making multiple-impulse welds.

**weld time** The time that welding current flows through the work when making a weld.

**weld timer** A timer that controls only the weld time.

**well counter** A radiation counter having a heavy tubular lead shield closed at one end, in which the radiation detector and the radioactive sample are inserted to reduce the effect of background radiation. Generally used with scintillation counters having large crystals, for the purpose of counting beta particles.

**Wertheim effect** When a ferromagnetic wire is twisted in a longitudinal magnetic field, a voltage is produced between the ends of the wire.

**West Coast Electronic Manufacturers Association** [abbreviated WCEMA] An organization of electronic manufacturers located in the far western states.

**Western Union joint** A joint or splice having good mechanical strength as well as good conductivity, made by crossing the cleaned ends of two wires, then winding the end of each wire around the other wire and soldering the joint.

**Weston standard cell** A standard cell used as a highly accurate voltage source for calibrating purposes. The positive electrode is mercury, the negative electrode is cadmium, and the electrolyte is a saturated cadmium sulfate solution. The Weston standard cell has a voltage of 1.018636 volts at 20°C.

**Westrex stereo pickup** *Stereo pickup.*

**wet cell** A cell whose electrolyte is in liquid form.

**wet electrolytic capacitor** An electrolytic capacitor employing a liquid electrolyte.

**wetting agent** A substance that decreases the surface tension of a liquid, to make the liquid spread and adhere better.

**Wheatstone bridge** A four-arm bridge, all arms of which are predominantly resistive. Used for measuring resistance. Also called resistance bridge.

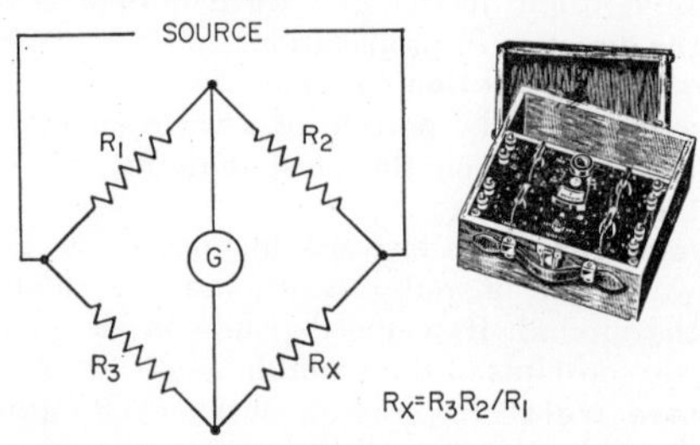


Wheatstone bridge circuit, equation, and commercial example.

**wheel static** Interference encountered in auto-radio installations due to static electricity developed by friction between the tires and the street.

**whip antenna** A flexible vertical rod antenna, used chiefly on vehicles. Also called fishpole antenna.

**whisker** *Catwhisker.*

**whistler** An effect caused when an elec-

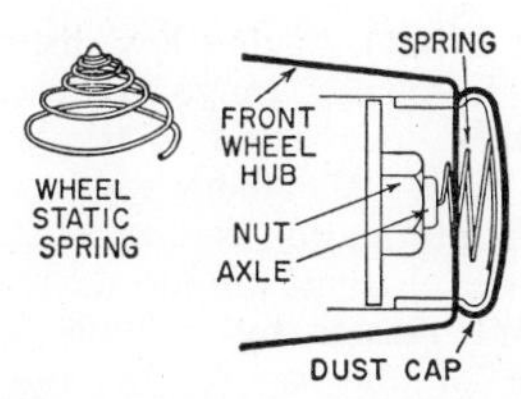


Wheel static suppressor spring and method of installation in dust cap.

trical disturbance caused by a lightning discharge travels out along lines of magnetic force of the earth's field and is reflected back to its origin from a magnetically conjugate point on the earth's surface. The characteristic drawn-out descending pitch of the whistler is a dispersion effect due to the greater velocity of the higher-frequency components of the disturbance. Radio signals can be transmitted along whistler paths from the northern hemisphere to the southern hemisphere.

**white** *Picture white.*

**White Alice** *Alice.*

**white compression** The reduction in picture-signal gain at levels corresponding to light areas, with respect to the gain at the level for midrange light values. The over-all effect of white compression is to reduce contrast in the highlights of the picture. Also called white saturation.

**white level** The carrier signal level corresponding to maximum picture brightness in television and facsimile.

**white light** Any radiation producing the same color sensation as average noon sunlight.

**white noise** Random acoustic or electric noise having equal energy per cycle over a specified total frequency band. The electric disturbance caused by random movements of free electrons in a conductor or semiconductor is one example; another is the frequency spectrum of white light.

**white-noise record** A phonograph record used to test the frequency response of audio reproduction systems. A recording of acoustic white noise extending over the entire band of audio frequencies is reduced in bandwidth progressively in steps as the top limit of the bandwidth at each step is announced.

**white object** An object that reflects all wavelengths of light with substantially equal high efficiencies and with considerable diffusion.

**white peak** A peak excursion of the picture signal in the white direction.

**white recording** A form of amplitude-modulation recording in which the maximum received power corresponds to the minimum density of the record medium. In a frequency-modulated white recording the lowest received frequency corresponds to the minimum density of the record medium.

**white saturation** *White compression.*

**white signal** The signal produced at any point in a facsimile system by scanning a minimum-density area of the subject copy.

**white-to-black amplitude range** The ratio of signal voltage for picture white to signal voltage for picture black at a given point in a facsimile system using positive amplitude modulation. Generally expressed in decibels. For negative amplitude modulation the reverse ratio, of black to white, is used.

**white-to-black frequency swing** The difference between the signal frequencies corresponding to picture white and picture black at any point in a facsimile system employing frequency modulation.

**white transmission** Amplitude-modulated transmission in which maximum transmitted power corresponds to minimum density of the subject copy, or frequency-modulated transmission in which the lowest transmitted frequency corresponds to the minimum density of the subject copy.

**whole step** *Whole tone.*

**whole tone** The interval between two sounds whose basic frequency ratio is approximately equal to the sixth root of 2. Also called whole step.

**wide-angle horizon sensor** *Horizon sensor.*

**wide-angle lens** An optical lens having a large angular field, generally greater than 80°.

**wideband amplifier** An amplifier that will pass a wide range of frequencies with substantially uniform amplification.

**wideband axis** The direction of the phasor representing the fine chrominance primary (the I signal) in color television.

**wideband dipole** A dipole having a low ratio of length to diameter, to give resonance over a relatively wide frequency band.

**wideband ratio** The ratio of the occupied frequency bandwidth to the intelligence bandwidth in a multiplex system.

**width** 1. The horizontal dimension of a television or facsimile picture. 2. The time duration of a pulse.

**width control** The control that adjusts the width of the pattern on the screen of a cathode-ray tube in a television receiver or oscilloscope.

**Wiedemann effect** The twist produced in a current-carrying wire when placed in a longitudinal magnetic field.

**Wiedemann-Franz law** The ratio of the thermal conductivity to the electric conductivity is proportional to the absolute temperature for all metals.

**Wien bridge** A four-arm a-c bridge used to measure capacitance or inductance in terms of resistance and frequency. Bridge balance depends on frequency.

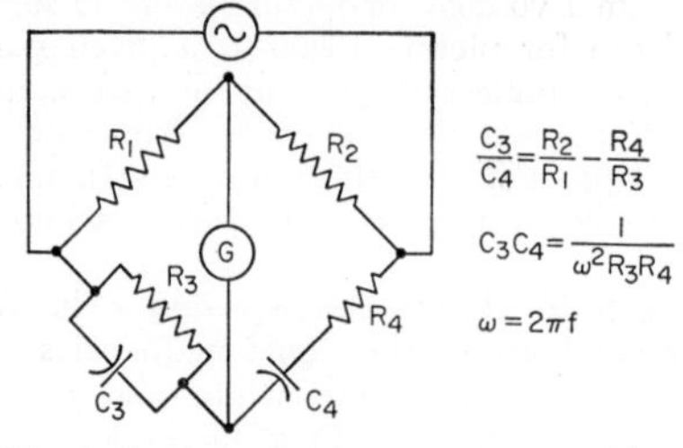


Wien bridge for measuring capacitance, with equations.

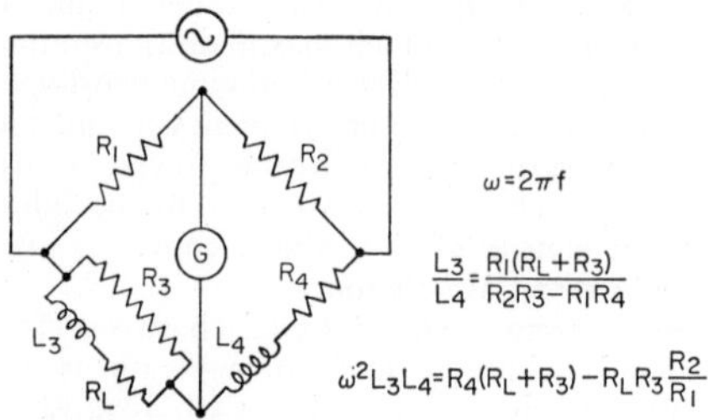


Wien bridge for measuring inductance, with equations.

**Wien-bridge oscillator** A phase-shift feedback oscillator that uses a Wien bridge as the frequency-determining element.

**Wien's displacement law** The wavelength of the peak radiation is inversely proportional to the absolute temperature of a blackbody. As temperature rises, the peak of the spectral energy distribution curve is shifted toward the short-wavelength end of the spectrum.

**Wigner nuclides** The most important class of mirror nuclides, comprising pairs of odd-mass-number isobars for which the atomic number and the neutron number differ by one.

**Wigner theorem** In a nuclear collision involving a transfer of energy, the angular momentum of electron spin is conserved.

**willemite** A natural fluorescent mineral consisting chiefly of zinc orthosilicate, used for screens of cathode-ray tubes.

**Williamson amplifier** An audio amplifier circuit developed by D. T. N. Williamson in Great Britain, having low distortion and a wide frequency range. It uses tetrode output tubes connected as triodes.

**Williams tube** A cathode-ray storage tube in which information is stored as a pattern of electric charges produced, maintained, read, and erased by suitably controlled scanning of the screen by the electron beam. Developed in Great Britain by F. C. Williams.

**Wilson chamber** A cloud chamber containing air supersaturated with water vapor by sudden expansion, in which rapidly moving nuclear particles such as alpha or beta rays produce ionization tracks by condensation of vapor on the ions produced by the rays. These tracks may be observed or photographed through a suitable window.

**Wimshurst machine** An electrostatic generator consisting of two glass disks rotating in opposite directions, having sectors of tinfoil and collecting combs so arranged that static electricity is produced for charging Leyden jars or discharging across a gap.

**wind charger** A wind-driven d-c generator used for charging storage batteries.

**wind-driven generator** A generator that derives its power from wind acting on its own propeller.

**winding** 1. One or more turns of wire forming a continuous coil for a transformer, relay, rotating machine, or other electric device. 2. A conductive path, usually of wire, that is inductively coupled to a magnetic storage core or cell.

**Windom antenna** A multiband transmitting antenna that provides satisfactory operation on even harmonics of its fundamental frequency, such as for operation on several harmonically related amateur bands. One version is a wire one half-wavelength long at the fundamental frequency, with a 300-ohm twin-line feeder connected about 35% off center.

**window** 1. A confusion reflector consisting of strips of chaff, wire, or bars cut to give resonance at expected enemy radar frequencies, and dropped in clusters from aircraft or expelled from shells or rockets as a radar countermeasure. Called window because the first pieces used were the size of small panes of glass. Later it was found that strips worked just as well and took less foil. 2. A hole in a partition between two cavities or waveguides, used for coupling. 3. An aperture for the passage of particles or radiation in a nuclear reactor. 4. An energy range of relatively high transparency in the total neutron cross section of a material. Such windows arise from interference between potential and resonance scattering in elements of intermediate atomic weight, and can be of importance in neutron shielding.

**window corridor** A region in which window has been dropped as a radar countermeasure.

**window rocket** A rocket filled with window that is to be expelled at a desired height.

**wiping contact** A switch or relay contact designed to move laterally with a wiping motion after it touches a mating contact. Also called self-cleaning contact and sliding contact.

**wire** A single bare or insulated metallic conductor having solid, stranded, or tinsel construction, designed to carry current in an electric circuit.

**wire communication** Communication over a wire, as distinguished from radio communication.

**wired radio** The transmission of modulated r-f carrier signals as currents flowing through wires, rather than as radio waves traveling through space. Telephone wires are sometimes used for this purpose, as in the British system of rediffusion.

**wire gage** A system of numerical designations of wire sizes. The American wire gage starts with 0000 as the largest size, going to 000, 00, 0, 1, 2, and up to 40 and beyond for the smallest sizes.

**wire guidance** Control of a guided missile by sending control signals over wires pulled by the missile. Used chiefly in simple short-range antitank missiles.

**wireless** British term for radio. Used in this country chiefly when the word radio might be misinterpreted, as in the term wireless record player.

**wireless record player** An electric phonograph connected to modulate an r-f oscillator that feeds a small built-in antenna, used to broadcast a phonograph recording across a room or into another room of a home for reception and reproduction by a radio receiver that is tuned to the frequency at which the oscillator is operating.

**wirephoto** 1. A photograph transmitted over wires to a facsimile receiver. 2. *Facsimile.*

**wire recorder** A magnetic recorder that utilizes a round stainless steel wire about 0.004 inch in diameter instead of magnetic tape.

**wiresonde** An apparatus for gathering meteorological data at low altitudes, in which meteorological data such as temperature and humidity are transmitted over wire to ground-recording devices by sensing and sending apparatus carried aloft by a captive balloon.

**wirewound resistor** [symbol WW] A resistor employing as the resistance element a length of high-resistance wire or ribbon, usually Nichrome, wound on an insulating form.

**wirewound rheostat** A rheostat in which a sliding or rolling contact moves over resistance wire that has been wound on an insulating core.

**wire-wrap connection** A solderless connection made by wrapping several turns of bare wire around a sharp-corner rectangular terminal under tension, using either a power tool or hand tool. Also called solderless wrapped connection and wrapped connection.

**wiring harness** An array of insulated conductors bound together by lacing cord, metal bands, or other binding, in an arrangement suitable for use only in specific equipment for which the harness was designed. It may include terminations.

**Wizard** An Air Force surface-to-air guided missile system.

**wobbulator** A signal generator in which a motor-driven variable capacitor is used to vary the output frequency periodically between two known limits, as required for displaying a frequency-response curve on the screen of a cathode-ray oscilloscope.

**wolfram** Term usually used for tungsten outside the United States.

**woofer** A large loudspeaker designed to reproduce low audio frequencies at relatively high power levels. Usually used in combination with a crossover network and a high-frequency loudspeaker called a tweeter.

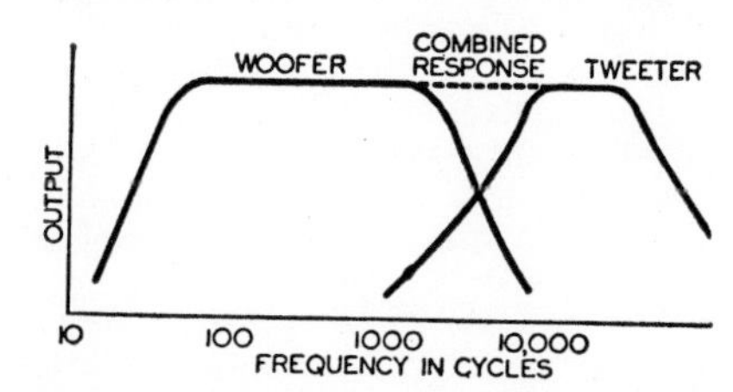


Woofer response curve, shown with tweeter curve and combined flat response of loudspeaker system.

**word** An ordered set of characters that is treated, stored, and transported by computer circuits as a unit. Word lengths may be either fixed or variable, depending on the computer.

**word time** *Minor cycle.*

**work** *Load.*

**work coil** *Load coil.*

**work function** The minimum energy needed to remove an electron from the Fermi level of a metal to infinity. Usually expressed in electron-volts.

**working** Carrying on radio communication with a station by means of telegraphy, telephony, or facsimile for a purpose other than calling.

**working storage** A portion of the internal storage of a computer that is reserved for the data upon which operations are being performed, including partial results.

**working voltage** *Voltage rating.*

**world geographic reference system** [abbreviated georef] A geographic reference system used by the U. S. Air Force for aircraft position reports, target designations, and other tactical air operations.

**wow** A low-frequency flutter. When caused by an off-center hole in a disk record or an eccentric turntable drive wheel, it occurs once per revolution of the turntable.

**wrapped connection** *Wire-wrap connection.*

**Wratten filter** A gelatin or glass filter designed to have specific light-transmission characteristics.

**Wright Air Development Center** [abbreviated WADC] An air development center at Wright-Patterson Air Force Base, Dayton, Ohio. Formerly called Wright Field.

**Wright-Patterson Air Force Base** An Air Force base at Dayton, Ohio, formerly two separate bases known as Wright Field and Patterson Field.

**wrinkle finish** A lacquer or varnish finish that may be applied with a brush or spray and that dries with an attractive wrinkled surface. Often used on panels and cabinets of electronic equipment.

**write** To introduce information into some form of storage in a computer.

**write pulse** A pulse that causes information to be introduced into a magnetic cell or cells for storage purposes in a computer.

**write time** The time interval between the instant at which information is ready for storage and the instant at which storage is completed in a computer. Also called access time.

**writing speed** The rate of writing on successive storage elements in a charge-storage tube.

**WW** Symbol used on diagrams to designate a *wirewound resistor.*

**WWV** The call letters of a radio station maintained by the National Bureau of Standards to provide standard radio and audio frequencies and other technical services such as precision time signals and radio propagation disturbance warnings. The station broadcasts on 2.5, 5, 10, 15, 20, 25, 30, and 35 mc at various times.

**WWVH** The National Bureau of Standards radio station at Maui, Hawaii, broadcasting services similar to those of WWV on 5, 10, and 15 mc.

**wye junction** A junction of waveguides such that the guide axes form a Y.

# X

**X** Symbol for *reactance.* Inductive reactance is designated as $X_L$ and capacitive reactance as $X_C$.

**x axis** 1. A reference axis in a quartz crystal. 2. The horizontal axis on a cathode-ray oscilloscope screen or on a graph. The corresponding vertical axis is the *y* axis.

**X band** A radio-frequency band extending from 5,200 to 11,000 mc, corresponding to wavelengths of 5.77 to 2.75 centimeters.

**x cut** A quartz-crystal cut made in such a manner that the *x* axis is perpendicular to the faces of the resulting slab.

**xenon** [symbol Xe] A gaseous element. It is one of the rare gases used in some thyratrons and other gas-discharge tubes. Atomic number is 54.

**xerography** A type of photography in which a selenium-coated surface takes an electrostatic image when exposed to an optical image. The image is developed by dusting with oppositely charged fine dark powder that adheres to the charged areas. The powder image is then transferred to paper and fixed by heating.

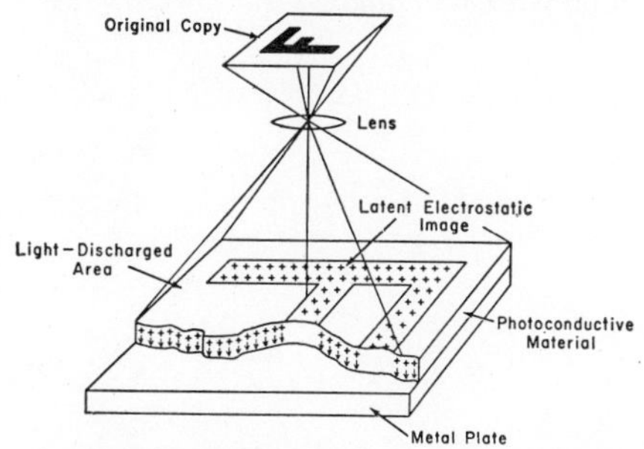


Xerography principle. Charges leak down through photoconductive selenium to grounded metal plate wherever light strikes.

**X guide** A surface-wave transmission line consisting of a dielectric structure having an X-shaped cross-section.

**x plate** One of the two deflection electrodes used to deflect the electron beam horizontally in an electrostatic cathode-ray tube.

**x-ray** 1. A penetrating electromagnetic radiation similar to light but having much shorter wavelengths (from about $10^{-7}$ to $10^{-10}$ centimeter or 0.1 to 100 angstroms), between ultraviolet rays and gamma rays. Usually generated by accelerating electrons to high velocity and suddenly stopping them by collision with a metal target. The resulting bombardment of the atoms in the target causes the atoms to lose energy, and this energy is radiated as x-rays of definite wavelength. X-rays are also produced by transitions of atoms from higher to lower energy states. Properties of x-rays include ionization of a gas through which they pass, penetration of all solids in varying degrees, production of secondary x-rays when stopped by material bodies, and action on fluorescent screens and photographic film. Photons originating in the nucleus of an atom are generally called gamma rays, and photons originating outside the nucleus are called x-rays. Also called roentgen ray. 2. To photograph with x-rays. 3. *X-ray photograph.*

**x-ray analysis** Determination of the internal structure of crystalline solids by means of the diffraction pattern produced when x-rays are passed through the material.

**x-ray diffraction** Diffraction of a beam of x-rays by the regular atomic lattice of a crystal. A characteristic diffraction pattern is obtained for each crystalline material.

**x-ray diffractometer** An instrument used in x-ray analysis to measure the intensities of the diffracted beams at different angles.

**x-ray film** A film base coated, usually on both sides, with an emulsion designed for use with x-rays.

**x-ray goniometer** An instrument that determines the positions of the electric axes of a quartz crystal by reflecting x-rays from the atomic planes of the crystal.

**x-ray hardness** The penetrating ability of x-rays. It is an inverse function of the wavelength.

**x-ray photograph** A radiograph made with x-rays. Also called roentgenogram (deprecated) and x-ray.

**x-ray spectrogram** A record of an x-ray diffraction pattern.

**x-ray spectrograph** An x-ray spectrometer

equipped with photographic or other recording apparatus.

**x-ray spectrometer** An instrument for producing the x-ray spectrum of a material and measuring the wavelengths of the various components.

**x-ray spectrum** The spectrum of x-rays, arranged according to wavelength, produced by electron bombardment of a target as in an x-ray tube. It consists of a continuous spectrum on which are superimposed certain groups of much sharper lines characteristic of the element used as target. These lines, such as the K line, L line, and M line, correspond to transitions between the inner energy levels of the atom.

**x-ray structure** The atomic structure of a substance as determined by x-ray diffraction patterns.

**x-ray therapy** Medical treatment by controlled application of x-rays. It is one type of radiotherapy.

**x-ray thickness gage** A thickness gage used for measuring and indicating the thickness of moving cold-rolled sheet steel during the rolling process without making contact with the sheet. An x-ray beam directed through the sheet is absorbed in proportion to the thickness of the material and its atomic number.

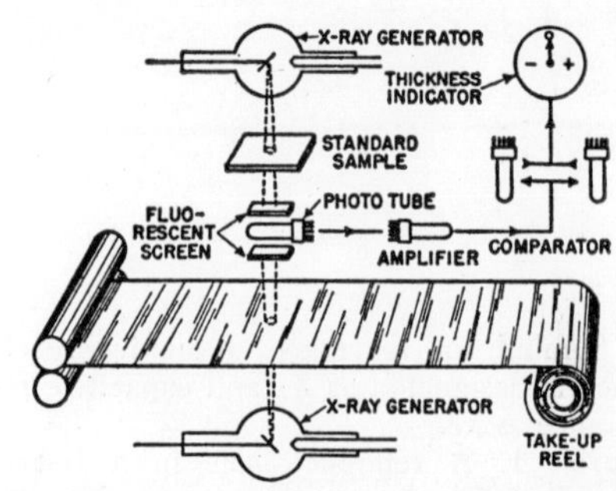


X-ray thickness gage for moving strip of metal.

**x-ray tube** A vacuum tube designed to produce x-rays by accelerating electrons to a high velocity by means of an electrostatic field, then suddenly stopping them by collision with a target.

**x unit** A unit of wavelength equal to 0.001 angstrom or $10^{-11}$ centimeter. Used for specifying wavelengths of x-rays and other highly penetrating radiations.

**X wave** *Extraordinary-wave component.*

**x-y recorder** A recorder that traces on a chart the relation of two variables, neither of which is time. Sometimes the chart is moved in proportion to time and one of the variables is so controlled that it increases in proportion to time.

# Y

**Y** Symbol for *admittance.*

**Yagi antenna** An end-fire antenna array having maximum radiation in the direction of the array line. It has one dipole connected to the transmission line and a number of equally spaced unconnected dipoles mounted parallel to the first in the same horizontal plane to serve as directors and reflectors.

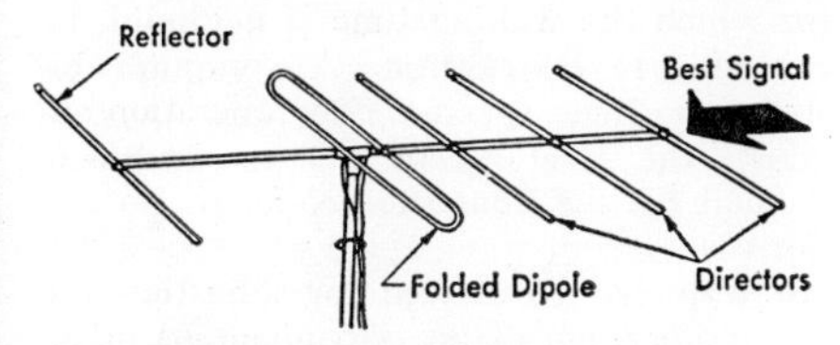

Yagi antenna for television reception.

**Y antenna** *Delta-matched antenna.*

**y axis** 1. A line perpendicular to two opposite parallel faces of a quartz crystal. 2. The vertical axis on a cathode-ray oscilloscope screen or on a graph.

**Y circulator** A circulator in which three identical rectangular waveguides are joined to form a symmetrical Y-shaped configuration, with a ferrite post or wedge at its center. Power entering any waveguide will emerge from only one adjacent waveguide.

**Y connection** *Y network.*

**Y cut** A quartz-crystal cut such that the $y$ axis is perpendicular to the faces of the resulting slab.

**yield** The number of molecules produced or converted per 100 electron-volts of energy absorbed.

**Y junction** A waveguide in which the longitudinal axes of the waveguide form a Y.

**Y network** A star network having three branches. Also called Y connection.

**yoke** *Deflection yoke.*

**y plate** One of the two deflection electrodes used to deflect the electron beam vertically in an electrostatic cathode-ray tube.

**y-ray** The electromagnetic radiation emitted by nuclei, as in a betatron.

**yrneh** [henry spelled backward] A unit of reciprocal inductance.

**Y signal** *Luminance signal.*

**ytterbium** [symbol Yb] A rare-earth metallic element. Atomic number is 70.

**yttrium** [symbol Y] A rare-earth metallic element. Atomic number is 39.

**Yukawa kernel** *Diffusion kernel.*

**Yukawa potential** A nuclear potential used in the meson theory of nuclear forces to specify the interaction between two nucleons.

# Z

**Z** 1. Symbol for *atomic number.* 2. Symbol for *impedance.*

**z axis** The optical axis of a quartz crystal. It is perpendicular to both the $x$ and $y$ axes.

**z-axis modulation** *Intensity modulation.*

**z-cut crystal** A quartz crystal cut in such a way that the $z$ axis is perpendicular to the faces of the resulting slab.

**Zebra time** Mean time at the Greenwich meridian. Used in communication and for synchronized reckonings. The hour 2400 Zebra time is 1900 EST, 1800 CST, 1700 MST, 1600 PST, 1400 Hawaiian standard time, 1000 Sydney standard time, 0900 Tokyo standard time, 0800 Manila standard time, 0300 Moscow standard time, and 0100 Berlin standard time. Also called Z time.

**Zeeman effect** The increase in the number of spectrum lines produced by a light source when in a strong magnetic field.

**zener breakdown** Nondestructive breakdown in a semiconductor, occurring when the electric field across the barrier region becomes high enough to produce a form of field emission that suddenly increases the number of carriers in this region. It has now been proved that the properties of so-called zener diodes are due to avalanche breakdown rather than zener breakdown.

**zener diode** Former name of *avalanche diode,* used before it was found that the Zener effect had no significant role in the operation of diodes of this type.

**Zener effect** The effect that is responsible for zener breakdown in a semiconductor.

**zener voltage** The reverse voltage at which zener breakdown occurs at a semiconductor junction, allowing the reverse current to increase suddenly.

**zeppelin antenna** A horizontal antenna that is some multiple of a half-wavelength long and is fed at one end by one lead of a two-wire transmission line that is also some multiple of a half-wavelength long.

**zero** Nothing. Most computers use both plus and minus zero. A positive binary zero is usually indicated by the absence of digits or pulses in a word, while negative binary zero (in a computer operating on one's complements) is indicated by a pulse in every pulse position in a word. In a coded decimal machine, decimal zero and binary zero may not have the same representation.

**zero-access storage** A computer storage for which the waiting time is negligible.

**zero-address instruction** A computer instruction that specifies an operation in which the locations of the operands are defined by the computer code, so no address is needed.

**zero adjuster** A device for adjusting the pointer position of an instrument or meter to read zero when the electrical quantity is zero.

**zero beat** The condition in which a circuit is oscillating at the exact frequency of an input signal, so no beat tone is produced or heard.

**zero-beat reception** *Homodyne reception.*

**zero bias** The condition in which the control grid and cathode of an electron tube are at the same d-c voltage.

**zero-cut crystal** A quartz crystal that has been cut in such a way that its temperature coefficient with respect to frequency is essentially zero.

**zero-frequency component** The d-c component of a complex waveform.

**zero level** The reference level used for comparing sound or signal intensities. In a-f work, a power of 0.006 watt is generally used as zero level. In sound, the threshold of hearing is generally assumed as the zero level.

**zero method** *Null method.*

**zero output** 1. The voltage response obtained from a magnetic cell in a zero state by a reading or resetting process. 2. The integrated voltage response obtained from a magnetic cell in a zero state by a reading or resetting process. A ratio of a one output to a zero output is a one-to-zero ratio.

**zero-point energy** The kinetic energy remaining in a substance at a temperature of absolute zero.

**zero potential** An expression usually applied to the potential of the earth, as a convenient reference for comparison.

**zero-power reactor** An experimental nuclear reactor operated at low neutron flux and at a power level so low that no forced cooling is required. Fission product activity in the fuel is then sufficiently low to permit handling the fuel after use.

**zero shift** The output of a balanced magnetic amplifier for zero control signal, due to drift.

**zero stability** The maximum zero shift that occurs over a given period of time during given changes in operating conditions of a balanced magnetic amplifier.

**zero state** A state of a magnetic cell wherein the magnetic flux through a specified cross-sectional area has a negative value, when determined from an arbitrarily specified direction for negative flux. A state wherein the magnetic flux has a positive value, when similarly determined, is a one state.

**zero-subcarrier chromaticity** The chromaticity that is intended to be displayed when the subcarrier amplitude is zero in a color television system.

**zero-suppression** The elimination of nonsignificant zeros to the left of the integral part of a quantity before printing the results of a computer operation.

**zero time reference** The time reference for the schedule of events in one cycle of radar operation.

**zeta** [Zero-Energy Thermonuclear Apparatus] British experimental equipment for investigating controlled fusion. The pinch effect associated with heavy current pulses in a doughnut-shaped tube containing deuterium at low pressure is used to heat the deuterium atoms to temperatures over 5,000,000°C.

**zinc** [symbol Zn] A bluish-white metallic element. Atomic number is 30.

**zinc orthosilicate** The mineral willemite, used in making fluorescent screens for cathode-ray tubes. When bombarded by an electron beam, it glows with a green tint.

**zinc telluride** [symbol ZnTe] A semiconductor having a forbidden-band gap of 2.2 electron-volts and a maximum operating temperature of 780°C when used in a transistor.

**zirconium** [symbol Zr] A metallic element. Atomic number is 40. Used as a structural material in nuclear reactors because it has good mechanical properties at high temperature and, when pure, has a low absorption cross section for thermal neutrons.

**Z marker beacon** *Zone marker.*

**Zn Te** Symbol for *zinc telluride.*

**zone leveling** The passage of one or more molten zones along a semiconductor body for the purpose of distributing impurities uniformly throughout the material.

**zone marker** A vhf radio station designed to radiate signals vertically in a cone-shaped pattern to define a zone above a radio range station. Also called Z marker beacon.

**zone of silence** *Skip zone.*

**zone purification** The passage of one or more molten zones along a semiconductor for the purpose of reducing the impurity concentration of part of the ingot. The semiconductor crystal is slowly moved through zones of intense heat. The crystal melts a portion at a time. As the molten region moves from one end of the crystal to another, the impurities move with it and congregate at one end of the crystal, where they can be sawed off.

**zoning** The displacement of various portions of the lens or surface of a microwave reflector so the resulting phase front in the near field remains unchanged. Also called stepping.

**zoom** To enlarge a portion of a television picture rapidly at the transmitter, optically or electronically.

**Zoomar** Trademark of Television Zoomar Corp. for their variable-focal-length lens.

**Z time** *Zebra time.*